ENCYCLOPEDIA OF
EARTH AND SPACE SCIENCE
VOLUME I

TIMOTHY KUSKY, Ph.D.

Katherine Cullen, Ph.D., Managing Editor

An imprint of Infobase Publishing

For Daniel and Shoshana

ENCYCLOPEDIA OF EARTH AND SPACE SCIENCE

Facts On File, Inc.
An imprint of Infobase Publishing
132 West 31st Street
New York NY 10001

Library of Congress Cataloging-in-Publication Data
Kusky, Timothy M.
Encyclopedia of Earth and space science / Timothy Kusky ; managing editor, Katherine Cullen.
p. cm.
Includes bibliographical references and index.
ISBN 978-0-8160-7005-3 (set : acid-free paper) 1. Earth sciences—Encyclopedias. 2. Space sciences—Encyclopedias. I. Cullen, Katherine E. II. Title.
QE5.K845 2010
550.3—dc22   2009015655

Facts On File books are available at special discounts when purchased in bulk quantities for businesses, associations, institutions, or sales promotions. Please call our Special Sales Department in New York at (212) 967-8800 or (800) 322-8755.

You can find Facts On File on the World Wide Web at http://www.factsonfile.com

Text design by Annie O'Donnell
Illustrations by Dale Williams
Photo research by Suzanne M. Tibor
Composition by Hermitage Publishing Services
Cover printed by Times Offset (M) Sdn Bhd, Shah Alam, Selangor
Book printed and bound by Times Offset (M) Sdn Bhd, Shah Alam, Selangor
Date printed: May 2010
Printed in Malaysia

This book is printed on acid-free paper.

CONTENTS

Acknowledgments

I would like to express appreciation to Frank K. Darmstadt, executive editor, for his critical review of this manuscript, wise advice, patience, and professionalism, and Katherine E. Cullen for her expert editing. Thank you to Richard Garratt, Dale Williams, and the graphics department, who created the illustrations that accompany the entries in this work, and to Suzie Tibor for performing the photo research. I express deep thanks to Dr. Lu Wang for help in preparing this manuscript through its many drafts and stages. Many sections of the work draw from my own experiences doing scientific research in different parts of the world, and it is not possible to thank the hundreds of colleagues whose collaborations and work I have related in this book. Their contributions to the science that allowed the writing of this volume are greatly appreciated. I have tried to reference the most relevant works or, in some cases, more recent sources that have more extensive reference lists. Any omissions are unintentional. Finally, I would especially like to thank my wife and my children, Shoshana and Daniel, for their patience during the long hours spent at my desk preparing this book. Without their understanding this work would not have been possible.

INTRODUCTION

Encyclopedia of Earth and Space Science is a two-volume reference intended to complement the material typically taught in high school Earth science and astronomy classes, and in introductory college geology, atmospheric sciences, and astrophysics courses. The substance reflects the fundamental concepts and principles that underlie the content standards for Earth and space science identified by the National Committee on Science Education Standards and Assessment of the National Research Council for grades 9–12. Within the category of Earth and space science, these include energy in the Earth system, geochemical cycles, origin and evolution of the Earth system, and origin and evolution of the universe. The National Science Education Standards (NSES) also place importance on student awareness of the nature of science and the process by which modern scientists gather information. To assist educators in achieving this goal, other subject matter discusses concepts that unify the Earth and space sciences with physical science and life science: science as inquiry, technology and other applications of scientific advances, science in personal and social perspectives including topics such as natural hazards and global challenges, and the history and nature of science. A listing of entry topics organized by the relevant NSES Content Standards and an extensive index will assist educators, students, and other readers in locating information or examples of topics that fulfill a particular aspect of their curriculum.

Encyclopedia of Earth and Space Science emphasizes physical processes involved in the formation and evolution of the Earth and universe, describes many examples of different types of geological and astrophysical phenomena, provides historical perspectives, and gives insight into the process of scientific inquiry by incorporating biographical profiles of people who have contributed significantly to the development of the sciences. The complex processes related to the expansion of the universe from the big bang are presented along with an evaluation of the physical principles and fundamental laws that describe these processes. The resulting structure of the universe, gal-axies, solar system, planets, and places on the Earth are all discussed, covering many different scales of observation from the entire universe to the smallest subatomic particles. The geological characteristics and history of all of the continents and details of a few selected important areas are presented, along with maps, photographs, and anecdotal accounts of how the natural geologic history has influenced people. Other entries summarize the major branches and subdisciplines of Earth and space science or describe selected applications of the information and technology gleaned from Earth and space science research.

The majority of this encyclopedia comprises 250 entries covering NSES concepts and topics, theories, subdisciplines, biographies of people who have made significant contributions to the earth and space sciences, common methods, and techniques relevant to modern science. Entries average more than 2,000 words each (some are shorter, some longer), and most include a cross-referencing of related entries and a selection of recommended further readings. In addition, one dozen special essays covering a variety of subjects—especially how different aspects of earth and space sciences have affected people—are placed along with related entries. More than 300 color photographs and line art illustrations, including more than two dozen tables and charts, accompany the text, depicting difficult concepts, clarifying complex processes, and summarizing information for the reader. A glossary defines relevant scientific terminology. The back matter of *Encyclopedia of Earth and Space Science* contains a geological timescale, tables of conversion between different units used in the text, and the periodic table of the elements.

I have been involved in research and teaching for more than two decades. I am honored to be a Distinguished Professor and Yangtze Scholar at China's leading geological institution, China University of Geosciences, in Wuhan. I was formerly the P. C. Reinert Endowed Professor of Natural Sciences and am the founding director of the Center for Environmental Sciences at St. Louis University. I am actively involved in research, writing, teaching, and advising students.

My research and teaching focus on the fields of plate tectonics and the early history of the Earth, as well as on natural hazards and disasters, satellite imagery, mineral and water resources, and relationships between humans and the natural environment. I have worked extensively in North America, Asia, Africa, Europe, the Middle East, and the rims of the Indian and Pacific Oceans. During this time I have authored more than 25 books, 600 research papers, and numerous public interest articles, interviews with the media (newspapers, international, national and local television, radio, and international news magazines), and I regularly give public presentations on science and society. Some specific areas of current interest include the following:

- Precambrian crustal evolution
- tectonics of convergent margins
- natural disasters: hurricanes, earthquakes, volcanoes, tsunami, floods, etc.
- drought and desertification
- Africa, Madagascar, China
- Middle East geology, water, and tectonics

I received bachelor and master of science degrees from the Department of Geological Sciences at the State University of New York at Albany in 1982 and 1985, respectively, then continued my studies in earth and planetary sciences at the Johns Hopkins University in Baltimore. There I received a master of arts in 1988 and a Ph.D. in 1990. During this time I was also a graduate student researcher at the NASA Laboratory for Terrestrial Physics, Goddard Space Flight Center. After this I moved to the University of California at Santa Barbara where I did postdoctoral research in Earth-Sun-Moon dynamics in the Department of Mechanical Engineering. I then moved to the University of Houston for a visiting faculty position in the department of geosciences and allied geophysical laboratories at the University of Houston. In 1992 I moved to a research professor position in the Center for Remote Sensing at Boston University and also took a part-time appointment as a research geologist with the U.S. Geological Survey. In 2000 I moved to St. Louis University, then was appointed to a distinguished professor position at China University of Geoscience in 2009.

I have tried to translate as much of this experience and knowledge as possible into this two-volume encyclopedia. It is my hope that you can gain an appreciation for the complexity and beauty in the earth and space sciences from different entries in this book, and that you can feel the sense of exploring, learning, and discovery that I felt during the research related here, and that you enjoy reading the different entries as much as I enjoyed writing them for you.

Entries Categorized by National Science Education Standards for Content (Grades 9–12)

When relevant an entry may be listed under more than one category. For example, Alfred Wegener, one of the founders of plate tectonic theory, is listed under both Earth and Space Science Content Standard D: Origin and Evolution of the Earth System, and Content Standard D: History and Nature of Science. Subdisciplines are listed separately under the category Subdisciplines, which is not a NSES category, but are also listed under the related content standard category.

Science as Inquiry
(Content Standard A)
astronomy
astrophysics
biosphere
climate
climate change
Coriolis effect
cosmic microwave background
    radiation
cosmology
Darwin, Charles
ecosystem
Einstein, Albert
environmental geology
evolution
Gaia hypothesis
geological hazards
global warming
greenhouse effect
hydrocarbons and fossil fuels
ice ages
life's origins and early evolution
mass extinctions
origin and evolution of the Earth
    and solar system
origin and evolution of the
    universe
ozone hole
plate tectonics
radiation
sea-level rise

Earth and Space Science
(Content Standard D):
Energy in the Earth System
asthenosphere
atmosphere
aurora, aurora borealis, aurora
    australis
black smoker chimneys
climate
climate change
clouds
convection and the Earth's
    mantle
Coriolis effect
cosmic microwave background
    radiation
cosmic rays
Earth
earthquakes
Einstein, Albert
El Niño and the Southern
    Oscillation (ENSO)
electromagnetic spectrum
energy in the Earth system
Gaia hypothesis
geodynamics
geological hazards

geomagnetism, geomagnetic
    reversal
geyser
global warming
greenhouse effect
hot spot
hurricanes
ice ages
large igneous provinces, flood
    basalt
magnetic field, magnetosphere
mantle plumes
mass wasting
meteorology
Milankovitch cycles
monsoons, trade winds
ocean currents
paleomagnetism
photosynthesis
plate tectonics
precipitation
radiation
radioactive decay
subduction, subduction zone
Sun
thermodynamics
thermohaline circulation
thunderstorms, tornadoes
tsunami, generation mechanisms
volcano

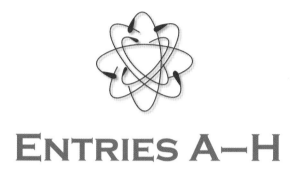

ENTRIES A–H

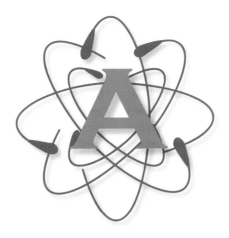

accretionary wedge   Plate tectonic theory recognizes that the surface of the Earth is broken up into a few dozen rigid plates that are all moving relative to one another by sliding along a partially molten zone deep within the mantle. These plates can have one of three types of boundaries with each other, including divergent, convergent, and transform. Divergent margins form where the plates are moving apart, convergent margins form where the plates are moving toward each other, and transform (or strike-slip) margins form where the plates are sliding past each other. Along convergent plate margins, one tectonic plate is typically pushed or subducted beneath another plate along deep oceanic trenches. In most cases a dense oceanic plate is subducted beneath a less dense, overriding continental plate, and a chain of volcanoes known as a volcanic arc forms on the overriding plate. Accretionary wedges are structurally complex parts of these subduction zone systems that form on the landward side of the trench from material scraped off from the subducting plate, as well as trench fill sediments. They typically have wedge-shaped cross sections and one of the most complex internal structures of any tectonic element known on Earth. Parts of accretionary wedges are characterized by numerous thin units of rock layers that are repeated by numerous faults, known as thrust faults, along which the same unit may be stacked upon itself many times. Other parts or other wedges are characterized by a relatively large section of rocks with relatively few faults, and still other sections are dominated by folded units, packages of rocks. They also host rocks known as tectonic mélanges that are complex mixtures of blocks and thin slivers of rocks surrounded by thrust faults. The rock types in these mélanges are quite diverse and typically include greywacke, basalt, chert, and limestone, characteristically encased in a matrix of a different rock type (such as shale or serpentinite). Some accretionary wedges contain small blocks or layers of high-pressure, low-temperature metamorphic rocks (known as blueschists) that have formed deep within the wedge where pressures are high and temperatures are low because of the insulating effect of the cold subducting plate. These high-pressure rocks were brought to the surface by structural processes.

Accretionary wedges grow by the gradual process of scraping sedimentary and volcanic rock material from the trench and subducting plate, which constantly pushes new material in front of and under the wedge as plate tectonics drives plate convergence. The type and style of material offscraped and incorporated into the wedge depends on the type of material near the surface on the subducting plate. Subducting plates with thin layers of deep-sea sediment such as chert on their basaltic surface yield packages in the accretionary wedge dominated by basalt and chert rock types, whereas subducting plates with thick sequences of greywacke sediments yield packages (thrust slices of rock from the subducting plate) in the accretionary wedge dominated by greywacke. Prisms of accreted rock at convergent plate boundaries may also grow by a process known as underplating, where packages are added to the base of the accretionary wedge, a process that typically causes folding of the overlying parts of the wedge. The fronts or toes of accretionary wedges are also characterized by material slumping off of the steep slope of the wedge into the trench. This material can then be recycled back into the accretionary wedge to form even more complex structures. The processes of offscraping and underplating work together and rotate rock layers and structures to steeper orientations. In this way rock layers rotate from an orientation that is near horizontal at the toe of the wedge, to near vertical at the back of the wedge.

Accretionary wedges are thought to behave mechanically somewhat as if they were piles of sand or snow bulldozed in front of a plow. They grow into a triangular wedge shape in cross section that increases its slope until it becomes oversteepened and mechanically unstable, which then causes the toe of the wedge to advance by thrusting, or the top of the wedge to collapse by normal faulting. Either of these two processes can reduce the slope of the wedge and lead it to become more stable. In addition to the evidence for thrust faulting in accretionary wedges, structural geologists have documented many examples of normal faults where the tops of the wedges have collapsed, supporting models of extensional collapse of oversteepened wedges.

Accretionary wedges are forming above nearly every subduction zone on the planet. However, these accretionary wedges presently border open oceans that have not yet closed by plate tectonic processes. Eventually the movements of the plates and continents will cause the accretionary wedges to become involved in plate collisions that will dramatically change the character of the accretionary wedges. They are typically overprinted by additional shortening, faulting, folding, and high-temperature metamorphism, and intruded by magmas related to arcs and collisions. These later events, coupled with the initial complexity and variety, make identification of accretionary wedges in ancient mountain belts difficult, and prone to uncertainty.

DESCRIPTION OF A TYPICAL ACCRETIONARY WEDGE: SOUTHERN ALASKA'S CHUGACH TERRANE

Southern Alaska is underlain by a complex assemblage of accreted terranes, including the Wrangellia superterrane (consisting of three separate terranes called the Peninsular, Wrangellia, and Alexander terranes), and farther outboard, the Chugach–Prince William superterrane. During much of the Meso-

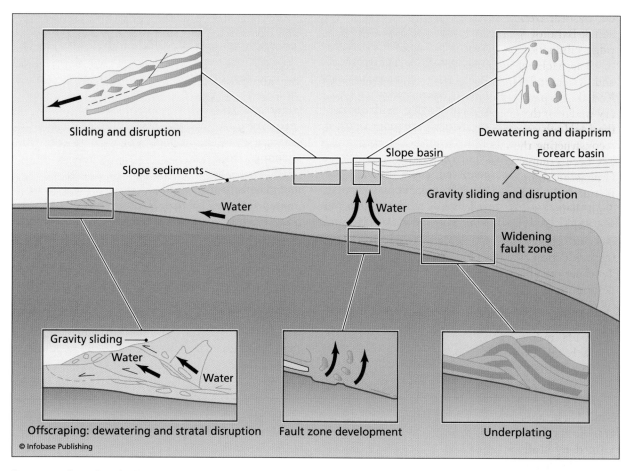

Cross section of typical accretionary wedge, showing material being offscraped at the toe of the wedge and underplated beneath the wedge. Water escapes upward through the accretionary wedge, causing the wedge material to become denser and more compacted.

zoic, the two superterranes formed a magmatic arc and accretionary wedge, respectively, above a circum-Pacific subduction zone. The Border Ranges fault forms the boundary between the Wrangellia and Chugach–Prince William superterranes; it initiated as a subduction thrust but has been reactivated in various places as a strike-slip or normal fault. On the Kenai Peninsula the Chugach terrane contains two major units. The unit located farther inland, the McHugh complex, is composed mainly of basalt, chert, argillite, and greywacke, as well as several large ultramafic massifs. Pinhead-sized marine fossils called radiolarians from McHugh cherts throughout south-central Alaska range in age from middle Triassic to middle Cretaceous. The interval during which the McHugh complex formed by subduction and accretion is not well known but probably spanned most of the Jurassic and Cretaceous. The McHugh has been thrust seaward on the Eagle River/Chugach Bay fault over a relatively coherent tract of trench turbidites assigned to the Upper Cretaceous Valdez Group. After the protracted episode of subduction-accretion that built the Chugach terrane, the accretionary wedge was cut by near-trench intrusive rocks, assigned to the Sanak-Baranof plutonic belt, probably related to ridge subduction.

The McHugh complex of south-central Alaska and its lateral equivalent, the Uyak complex of Kodiak, are part of the Mesozoic/Cenozoic accretionary wedge of the Chugach terrane. The vast extent of the McHugh complex has proven to be of value in reconstructing the tectonics of the Pacific realm and has been compared with similar tracts such as the Franciscan complex of California and the Shimanto Belt of Japan. The evolution of the McHugh and its equivalents can be broken down into three broad, somewhat overlapping phases: (1) origin of igneous and sedimentary rocks; (2) incorporation into the subduction complex ("accretion"), and attendant deformation and metamorphism; and (3) younger deformations.

Few fossil ages have been reported from the McHugh complex, but at several places on the Kenai Peninsula radiolarian chert depositionally overlies pillow basalt. Precise radiolarian age calls show that the base of the chert varies in age from middle Triassic to middle Cretaceous. Greywacke depositionally overlying chert has yielded Early Jurassic radiolarians. These ages are readily explained by a stratigraphic model in which the McHugh basalts were formed by seafloor spreading, the overlying cherts were deposited on the ocean floor as it was conveyed toward a trench, and the argillite and greywacke record deposition in the trench, just prior to subduction-accretion. The timing of subduction-accretion

is not well known but probably spanned most of the Jurassic and Cretaceous.

Limestones within the McHugh complex are of two categories. A limestone clast in McHugh conglomerate has yielded conodonts with a possible age range of Late Mississippian to Early Pennsylvanian. This clast could have been shed from the Wrangellia terrane. Most of the dated limestones, however, are tectonic blocks typically occurring as severely extended strings of boudins that have yielded Permian fusulinids or conodonts. Both the fusulinids and conodonts are of shallow-water, tropical, Tethyan affinity; the fusulinids are quite distinct from those of Wrangellia. The limestone blocks might represent the tops of seamounts that were decapitated at the subduction zone. If so, some of the ocean floor offscraped to form the McHugh complex must have formed in the Paleozoic.

The seaward part of the Chugach terrane is underlain by the Valdez group of Late Cretaceous age. On the Kenai Peninsula it includes medium- and thin-bedded greywacke turbidites, black argillite, and minor pebble to cobble conglomerate. These strata were probably deposited in a deep-sea trench and accreted shortly thereafter. Most of the Valdez group consists of relatively coherent strata, deformed into regional-scale tight- to isoclinal folds, cut by a slaty cleavage. The McHugh complex and Valdez group are juxtaposed along a thrust, which in the area of Turnagain Arm has been called the Eagle River fault, and on the Kenai Peninsula is known as the Chugach Bay thrust. Beneath this thrust is a mélange of partially to thoroughly disrupted Valdez group turbidites. This monomict mélange, which is quite distinct from the polymict mélanges of the McHugh complex, can be traced for many kilometers in the footwall of the Eagle River thrust and its along-strike equivalents.

In early Tertiary time, the Chugach accretionary wedge was cut by near-trench intrusive rocks forming the Sanak-Baranof plutonic belt. The near-trench magmatic pulse migrated 1,370 miles (2,200 km) along the continental margin, from about 63–65 million years ago at Sanak Island in the west, to about 50 million years ago at Baranof Island in the east. The Paleogene near-trench magmatism was related to subduction of the Kula-Farallon spreading center.

Mesozoic and Cenozoic rocks of the accretionary wedge of south-central Alaska are cut by abundant late brittle faults. Along Turnagain Arm near Anchorage, four sets of late faults are present: a conjugate pair of east-northeast-striking dextral and northwest-striking sinistral strike-slip faults, northnortheast-striking thrusts, and less abundant westnorthwest-striking normal faults. All four fault sets are characterized by quartz ± calcite ± chlorite fibrous slickenside surfaces and appear to be approximately coeval. The thrust- and strike-slip faults together resulted in subhorizontal shortening perpendicular to strike, consistent with an accretionary wedge setting. Motion on the normal faults resulted in extension of the wedge but is of uncertain tectonic significance. Some of the late brittle faults host gold-quartz veins that are the same age as nearby near-trench intrusive rocks. By implication, the brittle faulting and gold mineralization are probably related to ridge subduction.

Scattered fault-bounded ultramafic-mafic complexes in southern Alaska stretch 600 miles (1,000 km) from Kodiak Island in the south to the Chugach Mountains in the north. These generally consist of dunite +/- chromite, several varieties of peridotite, which grade upward into gabbronorites. These rocks are intruded by quartz diorite, tonalite, and granodiorite. Because of general field and mineralogic similarities, these bodies are generally regarded as having a similar origin and are named the Border Ranges ultramafic-mafic complex (BRUMC). The BRUMC includes six bodies on Kodiak and Afognak Islands, plus several on the Kenai Peninsula (including Red Mountain) and other smaller bodies. In the northern Chugach Mountains the BRUMC includes the Eklutna, Wolverine, Nelchina, and Tonsina complexes, and the Klanelneechena complex in the central Chugach Mountains.

Some models for the BRUMC suggest that all these bodies represent cumulates formed at the base of an intraoceanic arc sequence, and were formed at the same time as volcanic rocks now preserved on the southern edge of the Wrangellian composite terrane located in the Talkeetna Mountains. Some of the ultramafic massifs on the southern Kenai Peninsula, however, are not related to this arc, but represent deep oceanic material accreted in the trench. The ultramafic massifs on the Kenai Peninsula appear to be part of a dismembered assemblage that includes the ultramafic cumulates at the base, gabbroic-basalt rocks in the center, and basalt-chert packages in the upper structural slices. The ultramafic massifs may represent pieces of an oceanic plate subducted beneath the Chugach terrane, with fragments offscraped and accreted during the subduction process. There are several possibilities as to what the oceanic plate may have been, including normal oceanic lithosphere, an oceanic plateau, or an immature arc. Alternatively, the ultramafic/mafic massifs may represent a forearc or suprasubduction zone ophiolite, formed seaward of the incipient Talkeetna (Wrangellia) arc during a period of forearc extension.

See also ASIAN GEOLOGY; CONVERGENT PLATE MARGIN PROCESSES; DEFORMATION OF ROCKS; MÉLANGE; PLATE TECTONICS; STRUCTURAL GEOLOGY.

FURTHER READING

Bradley, Dwight C., Timothy M. Kusky, Peter Haeussler, D. C. Rowley, Richard Goldfarb, and S. Nelson. "Geologic Signature of Early Ridge Subduction in the Accretionary Wedge, Forearc Basin, and Magmatic Arc of South-Central Alaska." In *Geology of a Transpressional Orogen Developed During a Ridge–Trench Interaction Along the North Pacific Margin,* edited by Virginia B. Sisson, Sarah M. Roeske, and Terry L. Pavlis. *Geological Society of America* Special Paper 371 (2003): 19–50.

Bradley, Dwight C., Timothy M. Kusky, Peter Haeussler, S. M. Karl, and D. Thomas Donley. "Geologic Map of the Seldovia Quadrangle, United States Geological Survey Open File Report 99-18, scale 1:250,000, with marginal notes, 1999." Available online. URL: http://wrgis.wr.usgs.gov/open-file/of99-18. Accessed October 25, 2008.

Burns, L. E. "The Border Ranges Ultramafic and Mafic Complex, South-Central Alaska: Cumulate Fractionates of Island Arc Volcanics." *Canadian Journal of Earth Science* 22 (1985): 1,020–1,038.

Connelly, W. "Uyak Complex, Kodiak Islands, Alaska—A Cretaceous Subduction Complex." *Geological Society of America Bulletin* 89 (1978): 755–769.

Cowan, Darrel S. "Structural Styles in Mesozoic and Cenozoic Mélanges in the Western Cordillera of North America." *Geological Society of America Bulletin* 96 (1985): 451–462.

Hatcher, Robert D. *Structural Geology, Principles, Concepts, and Problems.* 2nd ed. Englewood Cliffs, N.J.: Prentice Hall, 1995.

Hudson, Travis. "Calc-Alkaline Plutonism along the Pacific Rim of Southern Alaska: Circum-Pacific Terranes." *Geological Society of America Memoir* 159 (1983): 159–169.

Kusky, Timothy M., and Dwight C. Bradley. "Kinematics of Mélange Fabrics: Examples and Applications from the McHugh Complex, Kenai Peninsula, Alaska." *Journal of Structural Geology* 21, no. 12 (1999): 1,773–1,796.

Kusky, Timothy M., Dwight C. Bradley, D. Thomas Donley, D. C. Rowley, and Peter Haeussler. "Controls on Intrusion of Near-Trench Magmas of the Sanak-Baranof Belt, Alaska, during Paleogene Ridge Subduction, and Consequences for Forearc Evolution." In "Geology of a Transpressional Orogen Developed During a Ridge—Trench Interaction Along the North Pacific Margin," edited by Virginia B. Sisson, Sarah M. Roeske, and Terry L. Pavlis. *Geological Society of America Special Paper* 371 (2003): 269–292.

Kusky, Timothy M., Dwight C. Bradley, Peter Haeussler, and S. Karl. "Controls on Accretion of Flysch and Mélange Belts at Convergent Margins: Evidence from the Chugach Bay Thrust and Iceworm Mélange,

Chugach Terrane, Alaska." *Tectonics* 16, no. 6 (1997): 855–878.

Kusky, Timothy M., Dwight C. Bradley, and Peter Haeussler. "Progressive Deformation of the Chugach Accretionary Complex, Alaska, during a Paleogene Ridge-Trench Encounter." *Journal of Structural Geology* 19, no. 2 (1997): 139–157.

Plafker, George, and H. C. Berg. "Overview of the Geology and Tectonic Evolution of Alaska." In *The Geology of Alaska, Decade of North American Geology, G-1*, edited by G. Plafker and H. C. Berg. Boulder, Colo.: Geological Society of America, 1994, 389–449.

Plafker, George, James C. Moore, and G. R. Winkler. "Geology of the Southern Alaska Margin." In *The Geology of Alaska, Decade of North American Geology, G-1*, edited by G. Plafker and H. C. Berg. *Geological Society of America* (1994): 989–1,022.

van der Pluijm, Ben A., and Stephen Marshak. *Earth Structure: An Introduction to Structural Geology and Tectonics.* Boston: WCB-McGraw Hill, 1997.

African geology   The continent of Africa consists of several old nuclei of very old (Archean) rocks called cratons that were welded together along younger (Proterozoic) mountain belts called orogenic belts that formed during collision of the cratons in the Late Precambrian. The cratons include the intensely studied Kalahari craton, comprising two Archean cratons known as the Kaapvaal and Zimbabwe cratons, plus the less well-known Congo and West African cratons. The Madagascar craton, which used to be attached to the African continent, lies off the coast of East Africa. These cratons are sutured along orogenic belts colloquially known as *Pan African orogens,* a term that is sometimes used to refer to the belts of rocks affected by complex igneous, metamorphic, and structural events that cut across Africa and many other continental masses between about 1,000 and 500 million years ago. The northern and southern margins of the African continent are affected by Paleozoic-Mesozoic deformation and mountain building, and the eastern side of the continent is experiencing active rifting and breakup into microplates, one of which extends through Madagascar and links with the Indian-Australian ridge.

KALAHARI CRATON

Southern Africa's Kalahari craton is composed of two older cratons, the Kaapvaal and Zimbabwe, that collided and were sutured 2.5 billion years ago along the Limpopo belt and have acted as a single craton since that time. For times before 2.5 billion years ago, therefore, the two parts (Kaapvaal and Zimbabwe cratons) are discussed separately, but from

True-color composite satellite image of Africa from data collected by the Thematic Mapper instrument on an American Landsat satellite. The Sahara is the brown (dry) area across the northern part of the continent, the Congo basin is lush green in the center, and the steppes of southern Africa are in the south. Madagascar lies off the southeastern coast. *(Earth Satellite Corporation/Photo Researchers Inc.)*

the Proterozoic onward, most geologists refer to the amalgamated cratons as the Kalahari craton.

Kaapvaal Craton, South Africa

The Archean Kaapvaal craton of southern Africa contains some of the world's oldest and most intensely studied Archean rocks, yet nearly 86 percent of the craton is covered by younger rocks. The craton covers approximately 363,000 square miles (585,000 km²) near the southern tip of the African continent. The craton is bordered on the north by the high-grade Limpopo mobile belt, initially formed when the Kaapvaal and Zimbabwe cratons collided at 2.6 billion years ago. On its southern and western margins the craton is bordered by the Namaqua-Natal Proterozoic orogens, and it is overlapped on the east by the Lebombo sequence of Jurassic rocks recording the breakup of Gondwana.

Most of the rocks composing the Archean basement of the Kaapvaal craton are granitoids and gneisses, along with less than 10 percent greenstone belts known locally as the Swaziland Supergroup. The oldest rocks are found in the Ancient Gneiss complex of Swaziland, where a 3.65–3.5 billion-year-old bimodal gneiss suite consisting of interlayered tonalite-trondhjemite-granite and amphibolite

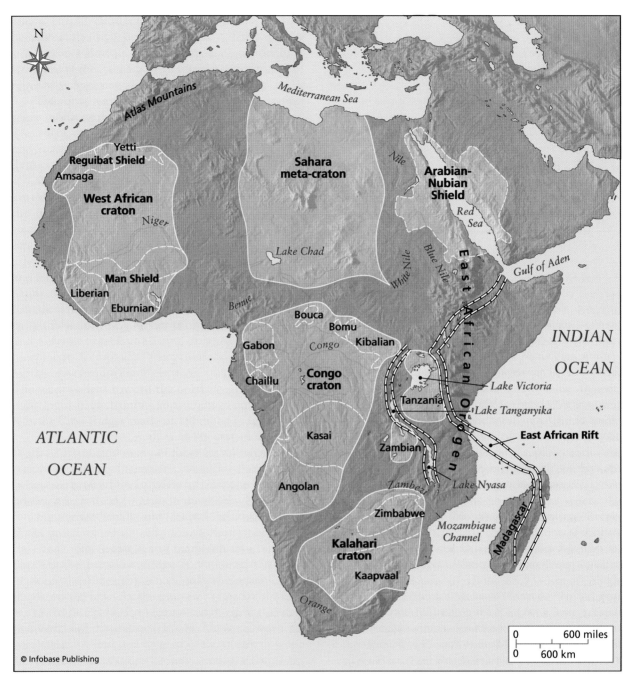

Map of Africa showing cratons, orogens, rifts, and main geographic elements. The Kalaharai craton in the south comprises the Zimbabwe and Kaapvaal cratons, the Congo craton occupies much of central Africa, and the West African craton is located in northwest Africa. The East African Orogen includes the Arabian-Nubian shield in the north and the Mozambique Belt in the south, whereas the active East African Rift cuts from the Gulf of Aden past Lake Victoria to the Mozambique Channel, and offshore to Madagascar. The Atlas Mountains are located in northwest Africa. *(modeled from Alan Goodwin, 1991)*

are complexly folded together with migmatitic gneiss, biotite-hornblende tonalitic gneiss, and lenses of 3.3–3.0 billion-year-old quartz monzonite. Several folding and deformation events are recognized from the Ancient Gneiss complex, whose history spans a longer interval of 700 million years, longer than the entire Phanerozoic.

There are six main greenstone belts in the Kaapvaal craton, the most famous of which is the Barberton greenstone belt. Although many studies have attempted to group all of the greenstone sequences of the Kaapvaal craton into the term *Swaziland Supergroup*, there is little solid geochronologic or other evidence that any of these complexly deformed belts are contemporaneous or related to each other, so this usage is not recommended. Other greenstone belts include the Murchison, Sutherland, Amalia, Muldersdrif, and Pietersburg belts. U-Pb (Uranium-Lead) isotopic ages from these belts span the interval from 3.5 to 3.0 billion years ago, a period of 500 million years. The greenstone belts include structurally repeated and complexly folded and metamorphosed sequences of tholeiitic basalts, komatiites, picrites, cherts (or metamorphosed felsic mylonite), felsic lava, clastic sediments, pelites, and carbonates. Possible partial ophiolite sequences have been recognized in some of these greenstone belts, particularly in the Jamestown section of the Barberton belt.

One of the long-held myths about the structure of greenstone belts in the Kaapvaal craton is that they represent steep synclinal keels of supracrustal rocks squeezed between diapiric granitoids. Detailed structural studies of the Murchison greenstone belt have established, however, that there is a complete lack of continuity of strata from either side of the supposed syncline of the Murchison belt, and that the structure is much more complex than the pinched-synform model predicts. Downward-facing structures and fault-bounded panels of rocks with opposing directions of younging (the direction toward the younger beds) and indicators of isoclinal folding have been documented, emphasizing that the "stratigraphy" of this and other belts cannot be reconstructed until the geometry of deformation is better understood; early assumptions of a simple synclinal succession are invalid.

Detailed mapping in a number of greenstone belts in the Kaapvaal craton has revealed early thrust faults and associated recumbent nappe-style folds. Most do not have any associated regional metamorphic fabric or axial planar cleavage, making their identification difficult without very detailed structural mapping. In some cases late intrusive rocks have utilized the zone of structural weakness provided by the early thrusts for their intrusion.

A complex series of tectonic events is responsible for the present structural geometry of the greenstone belts of the Kaapvaal craton. Early regional recumbent folds, thrust faults, inverted stratigraphy, juxtaposition of deep and shallow water facies, nappes, and precursory olistostromes related to the northward tectonic emplacement of the circa 3.5 Ga Barberton greenstone collage on gneissic basement have been documented. The thrusts may have been zones of high fluid pressure resulting from hydrothermal circulation systems surrounding igneous intrusions, and are locally intruded by syn-tectonic 3.43–3.44 billion-year-old felsic igneous rocks. Confirmation of thrust-style age relationships comes from recent

U-Pb zircon work, which has shown that older (circa 3.482 ± 5 Ga) Komatii Formation rocks lie on top of younger (circa 3.453 ± 6 Ga) Theespruit Formation.

The Pietersburg greenstone belt is located north of the Barberton and Murchison belts, near the high-grade Limpopo belt. Greenschist to amphibolite facies oceanic-affinity basaltic pillow lavas, gabbros, peridotites, tuffs, metasedimentary rocks, and banded iron formation are overlain unconformably by a terrestrial clastic sequence deposited during a second deformation event marked by northward-directed thrusting between 2.98 and 2.69 billion years ago. Coarse clastic rocks deposited in intermontaine basins are imbricated with the oceanic affinity rocks and were carried piggyback on the moving allochthon. Syn-thrusting depositional troughs became tightened into synclinal structures during the evolution of the thrust belt, and within the coarse-clastic section it is possible to find thrusts that cut local unconformities, and unconformities that cut thrusts.

The granite-greenstone terrane is overlain unconformably by the 3.1 billion-year-old Pongola Supergroup that has been proposed to be the oldest well-preserved continental rift sequence in the world. Deposition of these shallow-water tidally influenced sediments was followed by a widespread granite intrusion episode at 3.0 billion years ago. The next major events recorded include the formation of the West Rand Group of the Witwatersrand basin on the cratonward side of an Andean arc around 2.8 billion years ago, then further deposition of the extremely auriferous sands of the Central Rand Group in a collisional foreland basin formed when the Zimbabwe and Kaapvaal cratons collided. This collision led to the formation of a continental extensional rift province in which the Ventersdorp Supergroup was deposited at 2.64 billion years ago, with the extension occurring at a high angle to the collision. The latest Archean through Early Proterozoic history of the Kaapvaal craton is marked by deposition of the 2.6–2.1 billion-year-old Transvaal Supergroup in a shallow sea, perhaps related to slow thermal subsidence following Ventersdorp rifting. The center of the Witwatersrand basin is marked by a large circular structure called the Vredefort dome. This structure, several tens of kilometers wide, is associated with shock metamorphic structures, melts, and extremely high-pressure phases of silica, suggesting that it represents a meteorite impact structure.

The Bushveld complex is the world's largest layered mafic-ultramafic intrusion, located near the northern margin of the Kaapvaal craton. The complex occupies an area of 40,000 square miles (65,000 km^2) and intrudes Late Archean-Early Proterozoic rocks of the Transvaal Supergroup. Isotopic studies using a variety of methods have yielded age estimates of 2.0–2.1 billion years, with some nearby intrusions yielding ages as young as 1.6 billion years. The complex consists of several lobes with a conelike form, and contains numerous repeating cycles of mafic, ultramafic, and lesser felsic rocks. Several types of ores are mined from the complex, including chromite, platinum-group metals, cobalt, nickel, copper, and vanadiferous iron ores. Nearly 70 percent of the world's chrome reserves are located in the Bushveld complex. The mafic phases of the complex include dunite, pyroxenite, harzburgite, norite, anorthosite, gabbro, and diorite. The center of the complex includes felsic rocks, including granophyres and granite.

Much of the Kaapvaal craton is covered by rocks of the Karoo basin, including fluvial-deltaic deposits and carbonaceous deposits including coal. The top of the Karoo Sequence includes mafic and felsic lavas that were erupted soon before the breakup of Gondwana 200 million years ago.

Witwatersrand Basin

The Witwatersrand basin on South Africa's Kaapvaal craton is one of the best known of Archean sedimentary basins, and it contains some of the largest gold reserves in the world, accounting for more than 55 percent of all the gold ever mined. Sediments in the basin include a lower flysch-type sequence, and an upper molassic facies, both containing abundant silicic volcanic detritus. The strata are thicker and more proximal on the northwestern side of the basin that is at least locally fault-bounded. The Witwatersrand basin is a composite foreland basin that developed initially on the cratonward side of an Andean arc, similar to retroarc basins forming presently behind the Andes. A continental collision between the Kaapvaal and Zimbabwe cratons 2.7

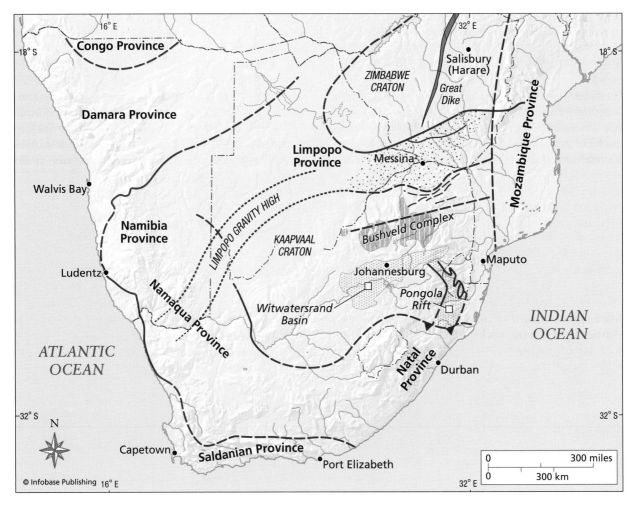

Map of the Kaapvaal and Zimbabwe cratons, showing the Great Dike on the Zimbabwe craton, the Witwatersrand and Pongola basins on the Kaapvaal craton, the Bushveld complex, and Limpopo Province separating the two cratons. Other major tectonic elements of southern Africa are also shown.

Mineralized conglomerate forming gold ore from the Archean Witwatersrand basin on the Kaapvaal craton
(*Brooks Kraft/Corbis*)

billion years ago caused further subsidence and deposition in the Witwatersrand basin. Regional uplift during this later phase of development placed the basin on the cratonward edge of a collision-related plateau, now represented by Limpopo Province. There are many similarities between this phase of development of the Witwatersrand basin and basins such as the Tarim and Tsaidam, north of the Tibetan Plateau.

The Witwatersrand basin is an elongate trough filled predominantly by 2.8–2.6 billion-year-old clastic sedimentary rocks of the West Rand and Central Rand Groups, together constituting the Witwatersrand Supergroup. These are locally, in the northwestern parts of the basin, underlain by the volcano sedimentary Dominion Group. The structure strikes in a northeasterly direction parallel with, but some distance south of, the high-grade gneissic terrane of Limpopo Province. The high-grade metamorphism, calc-alkaline plutonism, uplift, and cooling in the Limpopo are of the same age as and closely related to the evolution of the Witwatersrand basin. Strata dip inward with dips greater on the northwestern margin of the basin than on the southeastern margin. The northwestern margin of the basin is

a steep fault that locally brings gneissic basement rocks into contact with Witwatersrand strata to the south. Dips are vertical to overturned at depth near the fault, but only 20° near the surface, demonstrating that this is a thrust fault. A number of folds and thrust faults are oriented parallel to the northwestern margin of the basin.

The predominantly clastic sedimentary fill of the Witwatersrand basin has been divided into the West Rand and the overlying Central Rand Groups, which rest conformably on the largely volcanic Dominion Group. The Dominion Group was deposited over approximately 9,000 square miles (15,000 km^2), but it is correlated with many similar volcanic groups along the northern margin of the Kaapvaal craton. The Dominion Group and its correlatives, and a group of related plutons, have been interpreted as the products of Andean arc magmatism, formed above a 2.8 billion-year-old subduction zone that dipped beneath the Kaapvaal craton. The overlying West Rand and Central Rand Groups were deposited in a basin at least 50,000 square miles (80,000 km^2). Stratigraphic thicknesses of the West Rand Group generally increase toward the fault-bounded northwestern margin of the basin, whereas thicknesses of the Central Rand Group increase toward the center of the basin. Strata of both groups thin considerably toward the southeastern basin margin. The northeastern and southwestern margins are poorly defined, but some correlations with other strata (such as the Godwan Formation) indicate that the basin was originally larger than the present basin. Strata originally deposited north of Johannesburg are buried, removed by later uplift, omitted by igneous intrusion, and cut out by faulting.

The West Rand Group consists of southeastward tapering sedimentary wedges that overlie the Dominion Group, and were deposited directly on top of granitic basement in many places. The maximum thickness of the West Rand Group, 25,000 feet (7,500 m), occurs along the northern margin of the basin, and the group thins southeast to a preserved thickness of 2,700 feet (830 m) near the southern margin. Shale and sandstone in approximately equal proportions characterize the West Rand Group, and a thin horizon of mafic volcanics is locally present. This volcanic horizon thickens to 800 feet (250 m) near the northern margin of the basin, but is absent in the south. The West Rand Group contains mature quartzites, minor chert, and sedimentation patterns indicating both tidal and aeolian reworking. Much of the West Rand Group is an ebb-dominated tidal deposit later influenced by beach-swash deposition. More shales are preserved near the top of the group. Overall, the West Rand Group preserves a transition from tidal flat to beach, then deeper water deposition indicates a deepening of the Witwatersrand basin during deposition. Upper formations in the West Rand Group contain magnetic shales and other fine-grained sediments suggestive of a distal shelf or epicontinental sea environment of deposition.

The lower West Rand Group records subsidence of the Witwatersrand basin since the sediments grade vertically from beach deposits to a distal shallow marine facies. This transition means that the water was becoming deeper during deposition, showing that there was active subsidence of the basin. Since there is a lack of coarse, immature, angular conglomerate and breccia-type sediments, the subsidence was probably accommodated by gentle warping and flexure, not by faulting. A decreasing rate of subsidence and/or a higher rate of clastic sediment supply is indicated by the progressively shallower-water facies deposits in the upper West Rand Group. Numerous silicic volcanic clasts in the West Rand Group indicate that a volcanic arc terrane to the north was contributing volcanic detritus to the Witwatersrand basin. Additionally, the presence of detrital ilmenite, fuchsite, and chromite indicate that an ultramafic source such as an elevated greenstone belt was also contributing detritus to the basin.

The Central Rand Group was deposited conformably on top of the West Rand Group and attains a maximum preserved thickness of 9,500 feet (2,880 m) northwest of the center of the basin, and north of the younger Vredefort impact structure. Sediments of the Central Rand Group consist of coarse-grained graywackes and conglomerates along with subordinate quartz sandstone interbedded with local lacustrine or shallow marine shales and siltstones. The conglomerates are typically poorly sorted and have large clasts with well-rounded shapes, while the smaller pebbles have angular to subangular shapes. Paleocurrent indicators show that the sediments prograded into the basin from the northwestern margin in the form of a fan-delta complex. This is economically important because numerous goldfields in the Central Rand Group are closely associated with major entry points into the basin. Some transport of sediments along the axis of the basin is indicated by paleocurrent directions in a few locations. A few volcanic ash (tuffaceous) horizons and a thin mafic lava unit are found in the Central Rand Group in the northeast part of the basin. The great dispersion of unimodal paleocurrent directions derived from most of the Central Rand Group indicates that these sediments were deposited in shallow-braided streams on coalescing alluvial fans. The paleorelief is estimated at 20 feet (6 m) in areas proximal to the source, and 1–2 feet (0.5 m) in more distal areas. Some of the placers in the Central Rand Group have planar upper surfaces, commonly associated with pebbles and heavy placer mineral concentrations, which may be attributed to reworking by tidal currents. Clasts in the conglomerates include vein quartz, quartz arenite, chert, jasper, silicic volcanics, shales and schists, and other rare rocks.

The Central Rand Group contains a large amount of molassic-type sediments disposed as sand and gravel bars in coalesced alluvial fans and fluvial systems. The West Rand/Central Rand division of the Witwatersrand basin into a lower flysch-type sequence and an upper molasse facies is typical of foreland basins. Extensive mining of paleoplacers for gold and uranium has enabled mapping of the dendritic paleodrainage patterns and the points of entry into the basin to be determined. The source of the Central Rand sediments was a mountain range located to the northwest of the basin, and this range contained a large amount of silicic volcanic material.

The growth of folds parallel to the basin margin during sedimentation and the preferential filling of synclines by some of the mafic lava flows in the basin indicate that folding was in progress during Central Rand Group sedimentation. Deformation of this kind is diagnostic of flexural foreland basins, and studies show that the depositional axis of the basin migrated southeastward during sedimentation, with many local unconformities related to tilting during flexural migration of the depositional centers.

The Witwatersrand basin exhibits many features characteristic of foreland basins, including an asymmetric profile with thicker strata and steeper dips toward the mountainous flank, a basal flysch sequence overlain by molassic-type sediments, and thrust faults bounding one side of the basin. Compressional deformation was in part syn-sedimentary and associated folds and faults strike parallel to the basin margins, and the depositional axis migrated away from the thrust front with time, as in younger foreland basins. Stratigraphic relationships within the underlying Dominion Group, the presence of silicic volcanic clasts throughout the stratigraphy, and minor lava flows within the basin suggest that the foreland basin was developed behind a volcanic arc, partly preserved as the Dominion Group. Sediments of the West Rand Group are interpreted as deposited in an actively subsiding foreland basin developed adjacent to an Andean margin and fold thrust belt.

Deformation in Limpopo Province and the northern margin of the Kaapvaal craton are related to a collision between the northern Andean margin of the Kaapvaal craton with a passive margin developed on the southern margin of the Zimbabwe craton that began before 2.64 billion years ago, when Ventersdorp rifting, related to the collision, commenced. It is possible that some of the rocks in the Witwatersrand basin, particularly the molasse of the Central Rand Group, may represent erosion of a collisional plateau developed as a consequence of this collision. The plateau would have been formed in the region between the Witwatersrand basin and Limpopo Province, a region characterized by a deeply eroded gneiss terrane. A major change in the depositional style occurs in the Witwatersrand basin between the Central Rand and West Rand Groups, and this break may represent the change from Andean arc retroarc foreland basin sedimentation to collisional plateau erosion–related phases of foreland basin evolution.

Paleoplacers in the Witwatersrand basin have yielded more than 850 million tons of gold, dwarfing all the world's other gold placer deposits put together. Many of the placer deposits (called *reefs* in local terminology) preserved detrital gold grains on erosion surfaces, along foreset beds in cross-laminated sandstone and conglomerate, in trough cross-beds, in gravel bars, and as detrital grains in sheet sands. Most of the gold is located close to the northern margin of the basin in the fluvial channel systems. Some of the gold flakes in more distal areas were trapped by stromatolite-like filamentous algae, and some appear to have even been precipitated by types of algae, although it is more likely that these are fine, recrystallized grains trapped by algal filaments. Besides gold more than 70 other ore minerals are recognized in the Witwatersrand basin; most are detrital grains, and others are from metamorphic fluids. The most abundant detrital grains include pyrite, uraninite, brannerite, arsenopyrite, cobaltite, chromite, and zircon. Gold-mining operations in the Witwatersrand employ more than 300,000 people and have led to the economic success of South Africa.

ZIMBABWE CRATON

The Zimbabwe craton is a classic granite greenstone terrane. In 1971 Clive W. Stowe, in a Ph.D. dissertation and publication from the University of London, proposed a division of the Zimbabwean (then Rhodesian) craton into four main tectonic units. His first unit includes remnants of older gneissic basement in the central part of the craton, including the Rhodesdale, Shangani, and Chilimanzi gneissic complexes. Stowe's second (northern) unit includes mafic and ultramafic volcanics overlain by a mafic/felsic volcanic sequence, iron formation, phyllites, and conglomerates of the Bulawayan Group all overlain by sandstones of the Shamvaian Group. The third, or southern, unit consists of mafic and ultramafic lavas of the Bulawayan Group, overlain by sediments of the Shamvaian Group. The southern unit is folded about east-northeast axes. Stowe defined a fourth unit in the east, including remnants of schist and gneissic rocks, enclosed in a sea of younger granitic rocks. Stowe was a leader in recognizing complex structures in greenstone belts, stating in his dissertation and 1974 paper that the Selukwe greenstone belt "appears to be part of an imbricated and overturned lower limb of a large recumbent fold, resting allochthonously on a gneissic basement." In 1979 John F. Wilson, a geologist from the Geological Survey of Rhodesia (now Zimbabwe), proposed a regional correlation between the greenstone belts in the craton. His general comparison of the compositions of the upper volcanics in the greenstone belts resulted in a distinction between the greenstone belts located in the western part of the craton from those in the eastern section. The greenstone belts to the west of his division are composed of dominantly calc-alkaline rock suites including basalt, andesite, and dacite flows and pyroclastic rocks. This western section includes bimodal volcanic rocks consisting of tholeiite and magnesium-rich pillow basalt and massive flows, with some peridotitic rocks alternating with dacite flows, tuffs, and agglomerates. The eastern section of the Zimbabwe craton is characterized by pillowed and massive tholeiitic basalt flows and less abundant magnesium-rich basalts and their metamorphic equivalents. The eastern section contains a number of phyllites, banded iron formations, local conglomerate, and grit and rare limestone. Wilson identified an area of well-preserved 3.5 billion-year-old gneissic rocks and greenstones in the southern part of the province, and named this the Tokwe segment. He suggested that this may be a "mini-craton," and that the rest of the Zimbabwe craton stabilized around this ancient nucleus.

Despite these early hints that the Zimbabwe craton may be composed of a number of distinct terranes, much of the work on rocks of the Zimbabwe craton has been geared toward making lithostratigraphic correlations between these different belts, and attempting to link them all to a single supergroup-style nomenclature. Many workers attempted to pin the presumably correlatable 2.7 billion-year-old stratigraphy of the entire Zimbabwe craton to an unconformable relationship between older gneissic

rocks and overlying sedimentary rocks exposed in the Belingwe greenstone belt. More recently, Timothy Kusky, Axel Hoffman, and others have emphasized that the sedimentary sequence unconformably overlying the gneissic basement may be separated from the mafic/ultramafic magmatic sequences by a regional structural break, and that the presence of a structural break in the type of stratigraphic section for the Zimbabwe craton casts doubt on the significance of any lithostratigraphic correlations across Stowe's divisions of the craton.

Central Gneissic Unit (Tokwe Terrane)

Three and a half billion-year-old gneissic and greenstone rocks are well exposed in the area between Masvingo (Fort Victoria), Zvishavane (Shabani), and Shurugwi (Selukwe), in the Tokwe segment. The circa 3.5–3.6 billion-year-old Mashaba tonalite forms a relatively central part of this early gneissic terrane,

The Great Dike, a 2.5 billion-year-old magmatic intrusion in Zimbabwe, seen from the space shuttle *Endeavour* during mission STS-54, January 13–19, 1993 *(NASA/Photo Researchers, Inc.)*

and other rocks include mainly tonalitic to granodioritic, locally migmatitic gneissic units such as the circa 3.475 billion-year-old Tokwe River gneiss, Mushandike granitodiorite (2.95 billion years old), 3.0 billion-year-old Shabani gneiss, and 3.5 billion-year-old Mount d'Or tonalite. Similar rocks extend in both the northeast and southwest directions, but they are less well exposed and intruded by younger rocks in these directions. The Tokwe terrane probably extends to the northeast to include the area of circa 3.5 Ga greenstones and older gneissic rocks southeast of Harare. The Tokwe segment represents the oldest known portion of the Tokwe terrane, which was acting as a coherent terrane made up of 3.5–2.95 billion-year-old tectonic elements by circa 2.9 Ga.

The 3.500–2.950 billion-year-old Tokwe terrane also contains numerous narrow greenstone belt remnants, which are typically strongly deformed and multiply folded along with interlayered gneiss. The area in the northeasternmost part of the central gneissic terrane southeast of Harare best exhibits this style of deformation, although it continues southwest through Shurugwi. In the Mashava area west of Masvingo, ultramafic rocks, iron formations, quartzites, and mica schist are interpreted as 3.5 billion-year-old greenstone remnants tightly infolded with the ancient gneissic rocks. The 3.5 billion-year-old Shurugwi (Selukwe) greenstone belt was the focus of Clive W. Stowe's classic studies in the late 1960s, in which he identified Alpine-type inverted nappe structures and proposed that the greenstone belt was thrust over older gneissic basement rocks, forming an imbricated and inverted mafic/ultramafic allochthon. This was subsequently folded and intruded by granitoids during younger tectonic events.

The Tokwe terrane is in many places unconformably overlain by a heterogeneous assemblage of volcanic and sedimentary rocks known as the Lower Greenstones. In the Belingwe greenstone belt this Lower Greenstone assemblage is called the Mtshingwe Group, composed of mafic, ultramafic, intermediate and felsic volcanic rocks, pyroclastic deposits, and a wide variety of sedimentary rocks. Isotopic ages on these rocks range from 2.9 to 2.83 billion years, and the rocks are intruded by the 2.83 billion-year-old Chingezi tonalite. The Lower Greenstones are also well developed in the Midlands (Silobela), Filabusi, Antelope–Lower Gwanda, Shangani, Bubi, and Gweru-Mvuma greenstone belts. The upper part of the Lower Greenstones has yielded U-Pb ages of 2.8 billion years in the Gweru greenstone belt, and 2.79 billion years in the Filabusi belt.

The Buhwa and Mweza greenstone belts contain the thickest section of 3 billion-year-old shallow-water sedimentary rocks in the Zimbabwe craton. The Buhwa belt contains a western shelf succession and an eastern deeper water basinal facies association. The shelf sequence is up to 2.5 miles (4 km) thick and includes units of quartzite and quartz sandstone, shale, and iron formation, whereas the eastern deep-water association consists of strongly deformed shales, mafic-ultramafic lavas, chert, iron formation, and possible carbonate rocks. The Buhwa greenstone belt is intruded by the Chipinda batholith, which has an estimated age of 2.9 billion years. Shelf-facies rocks may have originally extended along the southeastern margin of the Tokwe terrane into Botswana, where a similar assemblage is preserved in the Matsitama greenstone belt. Rocks of the Matsitama belt include interlayered quartzites, iron formations, marbles and metacarbonates, and quartzofeldspathic gneisses in a 6–12 mile (10–20 km) thick structurally imbricated succession. The strong penetrative fabric in this belt may be related to deformation associated with the formation of the Limpopo belt to the south, but early nappes and structural imbrication that predate the regional cleavage-forming event are also recognized. The Matsitama belt (Mosetse Complex) is separated from the Tati belt to the east by an accretionary gneiss terrane (Motloutse Complex) formed during convergence of the two crustal fragments. The Tati and Vumba greenstone belts (Francistown granite-greenstone complex) were overturned before penetrative deformation, possibly indicating that they represent lower limbs of large regional nappe structures. The mafic, oceanic-affinity basalts of the Tati belt are overlain by andesites and other silicic igneous rocks, and intruded by syn-tectonic granitoids, typical of magmatic arc deposits. Similar arc-type rocks occur in the lower Gwanda greenstone belts to the east. The Lower Gwanda and Antelope greenstone belts are allochthonously overlain by basement gneisses thrust over the greenstones prior to granite emplacement.

A second sequence of sedimentary rocks lies unconformably over the Lower Greenstone assemblage and overlaps onto basement gneisses in several greenstone belts, most notably in the Belingwe belt, where the younger sequence is known as the Manjeri Formation. The Manjeri Formation contains conglomerates and shallow-water sandstones and locally carbonates at the base, and ranges stratigraphically up into cherts, argillaceous beds, graywacke, and iron formation. The top of the Manjeri Formation is marked by a regional fault. The Manjeri Formation is between 800 and 2,000 feet (250–600 m) thick along most of the eastern side of the Belingwe belt, except where it is cut out by faulting, and it thins northward to zero meters north of Zvishavane. It is considerably thinner on the western edge of the belt. On the scale of the Belingwe belt, the Manjeri Formation thickens toward the southeast, with some variation in structural thickness attributed to either sedimentary or tectonic ramping. The age of the Manjeri Formation is poorly constrained and may be diachronous across strike. However, the Manjeri Formation must be younger than the unconformably underlying circa 2.8 billion-year-old Ga Lower Greenstones, and it must be older than or in part contemporaneous with the thrusting event that emplaced circa 2.7 billion-year-old magmatic rocks of the Upper Greenstones over the Manjeri Formation. The Manjeri Formation overlaps onto the gneissic basement of the Tokwe terrane on the eastern side of the Belingwe belt, and at Masvingo, and rests on older (3.5 billion-year-old Sebakwian Group) greenstones at Shurugwi. Regional stratigraphic relationships suggest that the Manjeri Formation forms a southeast-thickening sedimentary wedge that prograded onto the Tokwe terrane.

Northern Belt (Zwankendaba Arc)

The northern volcanic terrane includes the Harare (Salisbury), Mount Darwin, Chipuriro (Sipolilo), Midlands (Silobela and Que Que), Chegutu (Gatoma), Bubi, Bulawayo, and parts of the Filabusi and Gwanda greenstone belts. These contain a lower volcanic series overlain by a calc-alkaline suite of basalts, andesites, dacites, and rhyolites. Pyroclastic, tuffaceous, and volcaniclastic horizons are common. Also common are iron formations, and other sedimentary rocks including slates, phyllites, and conglomerate. In the Bulawayo-Silobela area (Mulangwane Range), the top of the upper volcanics include a series of porphyritic and amgdaloidal andesitic and dacitic agglomerate and other pyroclastic rocks.

U-Pb ages from felsic volcanics of the northern volcanic belt include 2.696, 2.698, 2.683, 2.702, and 2.697 billion years. Isotopic data from the Harare-Shamva greenstone belt and surrounding granitoids suggest that the greenstones evolved on older continental crust between 2.715 and 2.672 billion years ago. The age of deformation is constrained by a 2.667 billion-year-old syn-tectonic gneiss, a 2.664 billion-year-old late syn-tectonic intrusion, and 2.659 billion-year-old shear zone-related gold mineralization. Other posttectonic granitoids yielded U-Pb zircon ages of 2.649, 2.618, and 26.01 billion years. Isotopic data for the felsic volcanics suggest that the felsic magmas were derived from a melt extracted from the mantle 200 million years before volcanism and saw considerable interaction between these melts and older crustal material.

Southern Belt

The southern belt of tholeiitic mafic-ultramafic–dominated greenstones structurally overlies shallow-water sedimentary sequences and gneissic rocks in parts of the Belingwe, Mutare (Umtali), Masvingo (Fort Victoria), Buhwa, Mweza, Antelope, and Lower Gwanda belts. The most extensively studied of these is the Belingwe belt, which many workers have used as a stratigraphic archetype for the entire Zimbabwe craton. The allochthonous Upper greenstones are here discussed separately from the structurally underlying rocks of the Manjeri Formation that rest unconformably on Tokwe terrane gneissic rocks.

The Archean Belingwe greenstone belt in southern Zimbabwe has proven to be one of the most important Archean terranes for testing models for the early evolution of the Earth and the formation of continents. It has been variously interpreted to contain a continental rift, arc, flood basalt, and structurally emplaced ophiolitic or oceanic plateau rocks. It is a typical Archean greenstone belt, being an elongate belt with abundant metamorphosed mafic rocks and metasediments, deformed and metamorphosed at greenschist to amphibolite grade. The basic structure of the belt is a refolded syncline, although debate has focused on the significance of early folded thrust faults.

The 3.5 billion-year-old Shabani-Tokwe gneiss Complex forms most of the terrain east of the belt and underlies part of the greenstone belt. The 2.8–2.9 billion-year-old Mashaba tonalite and Chingezi gneiss are located west of the belt. These gneissic rocks are overlain unconformably by a 2.8 billion-year-old group of volcanic and sedimentary rocks known as the Lower Greenstones or Mtshingwe Group, including the Hokonui, Bend, Brooklands, and Koodoovale Formations. These rocks, and the eastern Shabani-Tokwe gneiss, are overlain unconformably by a shallow water sedimentary sequence known as the Manjeri Formation, consisting of quartzites, banded iron formation, graywacke, and shale. A major fault is located at the top of the Manjeri Formation, and the Upper Greenstones structurally overlie the lower rocks being everywhere separated from them by this fault. The significance of this fault, whether a major tectonic contact or a fold accommodation–related structure, has been the focus of considerable scientific debate. The 2.7 billion-year-old Upper Greenstones, or the Ngezi Group, includes the ultramafic-komatiitic Reliance Formation, the four-mile (6-km) thick tholeiitic pillow lava–dominated Zeederbergs Formation, and the sedimentary Cheshire Formation. All of the units are intruded by the 2.6 billion-year-old Chibi granitic suite.

The Lower Greenstones have been almost universally interpreted to be deposits of a continental rift or rifted arc sequence. However, the tectonic significance of the Manjeri Formation and Upper Greenstones has been debated. The Manjeri Formation is certainly a shallow-water sedimentary sequence that rests unconformably over older greenstones and gneisses. Correlated with other shallow-water sedimentary rocks across the southern craton, it may represent the remnants of a passive-margin type of sedimentary sequence.

Geochemical studies have suggested that the komatiites of the Reliance Formation in Belingwe could not have been erupted through continental crust, but rather that they are similar to intraplate basalts and distinct from midocean ridge and convergent margin basalts. The geochemistry of the Ngezi Group in the Belingwe greenstone belt suggests that it could be a preserved oceanic plateau and that there was no evidence for them to have been derived from a convergent margin.

The top of the Manjeri Formation is marked by a fault, the significance of which has been disputed. Some scientists have suggested that it may be a fault related to the formation of the regional syncline, formed in response to the rocks in the center of the belt being compressed and moving up and out of the syncline. Work on the sense of movement on the fault zone, however, shows that the movement sense is incompatible with such an interpretation, and that the fault is a folded thrust fault that placed the Upper Greenstones over the Manjeri Formation. Therefore, the tectonic setting of the Upper Greenstones is unrelated to the rocks under the thrust fault, and the Upper Greenstones likely were emplaced from a distant location. The overall sequence of rocks in the Upper Greenstones, including several kilometers of mafic and ultramafic lavas, is very much like rock sequences found in contemporary oceanic plateaus or thick oceanic crust, and such an environment seems most likely for the Upper Greenstones in Belingwe and other nearby greenstone belts of the Zimbabwe craton.

Cratonwide Overlap Assemblage (Shamvaian Group)

The Shamvaian Group consists of a sequence of coarse clastic rocks that overlie the Upper Greenstones in several locations. These conglomerates, arkoses, and graywackes are well known from the Harare, Midlands, Masvingo, and Belingwe greenstone belts. The Cheshire Formation, the top unit of the Belingwe greenstone belt, consists of a heterogeneous succession of sedimentary rocks including conglomerate, sandstone, siltstone, argillite, limestone, cherty limestone, stromatolitic limestone, and minor banded iron formation. The Shamvaian Group is intruded by the circa 2.6 billion-year-old Chilimanzi Suite granites, providing an upper age limit on deposition. In the Bindura-Shamva greenstone belt the Shamvaian Group is 1.2 miles (2 km) thick, beginning with basal conglomerates and grading up into a

thick sandstone sequence. Tonalitic clasts in the basal conglomerate have yielded igneous ages of 3.2, 2.9, 2.8, and 2.68 billion years. Felsic volcanics associated with the Shamvaian Group in several greenstone belts have ages of 2.66 to 2.64 billion years.

Chilimanzi Suite

The Chilimanzi suite of K-rich granitoids is one of the last magmatic events in the Zimbabwe craton, with reported ages of 2.57 to 2.6 billion years. These granites appear to be associated with a system of large intracontinental shear zones that probably controlled their position and style of intrusion. These relatively late structures are related to north-northwest to south-southeast shortening and associated southwestward extrusion of crust during the continental accretion and collision as recorded in the Limpopo belt.

Accretion of the Archean Zimbabwe Craton

The oldest part of the Zimbabwe craton, the Tokwe terrane, preserves evidence for a complex series of tectonomagmatic events ranging in age from 3.6 to 2.95 billion years ago. These events resulted in complex deformation of the Sebakwian greenstones and intervening gneissic rocks. This may have involved convergent margin accretionary processes that led to the development of the Tokwe terrane as a stable continental nuclei by 2.95 billion years ago.

A widespread unit of mixed volcanic and sedimentary rocks was deposited on the Tokwe terrane at circa 2.9 billion years ago. These lower greenstones include mafic and felsic volcanic rocks, coarse conglomerates, sandstones, and shales. The large variation in volcanic and sedimentary rock types, along with the rapid and significant lateral variations in stratigraphic thicknesses that typify the Lower Greenstones, are characteristic of rocks deposited in continental rift or rifted arc settings. The Tokwe terrane was subjected to rifting at 2.9 Ga, leading to the formation of widespread graben in which the Lower Greenstones were deposited. The southeastern margin of the Tokwe terrane may have been rifted from another, perhaps larger fragment at this time, along a line extending from the Buhwa-Mweza greenstone belts to the Mutare belt, allowing a thick sequence of passive margin–type sediments (preserved in the Buwha greenstone belt) to develop on this rifted margin. Age constraints on the timing of the passive margin development are not good, but appear to fall within the range of 3.09 to 2.86 billion years ago. By 2.7 billion years ago, a major marine transgression covered much of the southern half of the Tokwe terrane, as recorded in shallow-water sandstones, carbonates, and iron formations of the Manjeri-type units preserved in several greenstone belts. The Manjeri-type units overlap the basement of the Tokwe terrane in several places (e.g., Belingwe, Masvingo), and lie unconformably over the circa 3.5 and 2.9 billion-year-old greenstones. Regional stratigraphic relationships suggest that the Manjeri Formation forms a southeast thickening sedimentary wedge that prograded onto the Tokwe terrane, in a manner analogous to the Ocoee-Chilhowee and correlative Sauk Sequence shallow-water progradational sequence of the Appalachians, and similar sequences in other mountain ranges. The progradation could have been driven by sedimentary or tectonic flexural loading of the margin of the Tokwe terrane, but most evidence points to the latter cause. The top of the Manjeri-type units represents a regional detachment surface, on which allochthonous units of the southern greenstones were emplaced. Loading of the passive margin by these thrust sheets would have induced flexural subsidence and produced a foreland basin that migrated onto the Tokwe terrane.

The 2.7 billion-year-old greenstones are divided into a northwestern arc-like succession, and a southeastern allochthonous succession. The northwestern arc succession contains lavas with strong signatures of eruption through older continental crust, and the arc appears to be a continental margin type of magmatic province. In contrast, the southern greenstones are allochthonous and were thrust in place along a shear zone that is well exposed in several places, including the Belingwe belt. These southern greenstones have a stratigraphy reminiscent of thick oceanic crust, suggesting that they may represent an oceanic plateau that was obducted onto the Tokwe terrane 2.7 billion years ago. All of the southern greenstones are distributed in a zone confined to about 100 miles (150 km) from the line of passive margin-type sediments extending from Mweza-Buhwa to Mutare. This "Umtali line" may represent the place where an ocean or back arc basin opened between 2.9 and 2.8 billion years ago, then closed at 2.7 billion years, and forms the root zone from which the southern greenstones were obducted. This zone contains numerous northeast-striking mylonitic shear zones in the quartzofeldspathic gneisses. Closure of the Sea of Umtali at circa 2.7 billion years ago deposited a flysch sequence of graywacke-argillite turbidites that forms the upper part of the Manjeri Formation, and formed a series of northeast-striking folds.

The latest Archean tectonic events to affect the Zimbabwe craton is associated with deposition of the Shamvaian group, and intrusion of the Chilimanzi suite granitoids at circa 2.6–2.57 billion years ago. These events appear to be related to a collision of the now-amalgamated Zimbabwe craton with northern Limpopo Province, as the Zimbabwe and Kaapvaal cratons collided. Interpretations of the Limpopo orogeny suggest that the Central Zone of the Limpopo Province collided with the Kaapvaal craton at circa 2.68 billion years ago, and that this orogenic collage collided with the southern part of the Zimbabwe craton at 2.58 billion years ago. Deposition of the Shamvaian Group clastic sediments occurred in a foreland basin related to this collision, and the intrusion of the Chilimanzi suite occurred when this foreland became thickened by collisional processes, and was cut by sinistral intracontinental strike-slip faults. Late folds in the Zimbabwe craton are oriented roughly parallel to the collision zone, and appear contemporaneous with this collision. The map pattern of the southern Zimbabwe craton shows some interference between folds of the early generation (related to the closure of the Sea of Umtali) and these late folds related to the Limpopo orogeny.

CONGO CRATON

The Congo craton includes a generally poorly known and underexplored region of Archean rocks that are exposed around the Congo basin in central Africa. The craton extends from the Kasai region of the Democratic Republic of the Congo into Sudan, Angola, Zambia, Gabon, and Cameroon, and is known for containing many greenstone belts, albeit deeply weathered under thick profiles of laterite soil. Most parts of the Congo craton are known from separate regions that outcrop around the Congo basin, and these areas are generally known as cratons or blocks, even though they are likely continuous at depth.

The Kasai (and northeast Angolan) block is exposed over an area 270 miles (450 km) across by 210 miles (350 km) north-south in Kasai, Democratic Republic of the Congo, Lunda, and Angola. The area is overlain by a thick Phanerozoic cover, so most of the Archean rocks are restricted to river valleys. The Archean rocks are divided into three main divisions, including the circa 3.4 billion-year-old Luanyi tonalitic-granodioritic gneiss; two belts of younger, strongly metamorphosed (granulite facies) rocks; and a still younger granitoid and migmatite complex known as the Dibaya Complex. The age of the granulite facies events are constrained to be between 2.77 and 2.84 billion years, with a lower pressure and temperature (retrogressive) metamorphic event at 2.68 billion years ago. The late-stage Dibaya Complex includes calc-alkaline granites and gneisses that are strongly deformed and mylonitized locally, with deformation and migmatization at 2.68 billion years ago. These different assemblages are cut by the undeformed circa 2.59 Ga Malafundi granites.

The Gabon-Chaillu block consists of two roughly elliptical areas each a couple of hundred miles (several hundred km) across, in Cameroon,

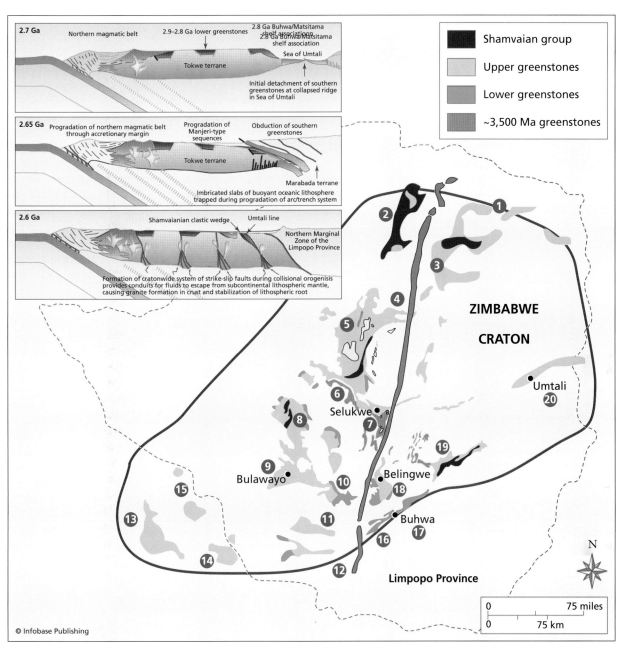

Map of the Zimbabwe craton showing the distribution of the old gneissic Tokwe terrane, the northern and southern magmatic belts, and shelf-type associations. Numbers correspond to individual greenstone belts: 1 Mount Darwin, 2 Chipuriro, 3 Harare, 4 Chegutu, 5 Midlands, 6 Gweru-Mvuma, 7 Shurugwe, 8 Bubi, 9 Bulawayo, 10 Filabusi, 11 Gwanda, 12 Antelope, 13 Lower Gwanda, 14 Tati, 15 Vumba, 16 Mweza, 17 Buhwa, 18 Belingwe, 19 Masvingo, 20 Mutare. Inset is a tectonic cross section showing evolution of the Zimbabwe craton, with the northern magmatic belt evolving as an Andean-style arc above a major subduction zone from 2.7 to 2.6 billion years ago, while the southern magmatic belts represent where a small ocean basin closed in the same interval.

Equatorial Guinea, Gabon, and Congo. Rocks in this block include an assemblage of 2.8–3.2 billion-year-old charnockite, migmatite, gneiss, greenstone belts, and late-stage 2.7 billion-year-old granitoid plutons. Greenstone belts in the block include complexly folded pillow basalts, rhyolites, quartzite, and banded iron formation, in structural contact with granitoid gneiss and granodiorite. High-grade metamorphism occurred at 2.9 Ga, corresponding to other high-grade metamorphic events known across central Africa. Much of the northern part of the block was reworked by strong northeast-trending folds, faults, and regional metamorphism at 500 million years ago, in a part of the Central African belt, related to the late Proterozoic–early Paleozoic amalgamation of the supercontinent of Gondwana.

The Kibalian block covers an area about 500 miles (800 km) long by 300 miles (500 km) wide in northern Zaire and the southern Central African Republic, to Lake Mobutu in Uganda. The Kibale and Uele Rivers flow across the block, providing good exposures of the basement rocks. The block is bordered in the north by folded Late Proterozoic rocks, by the West Nile gneiss Complex on the west, and on the south by strata of the Congo basin. The Kibalian block contains a granite-greenstone assemblage, with granitoids falling into three groups. An older tonalite-trondhjemite-granodiorite group (TTG) has an age of 2.8 billion years, and younger granites have been dated to be 2.5 billion years old.

The greenstone and schist belts form structurally complex assemblages of mafic schists, intruded by 2.9 billion-year-old tonalites. Of the 11 major greenstone belts, most have a similar rock assemblage, including the lower Kibalian sequence, consisting of mafic to intermediate volcanic rocks and banded iron formation, overlain by the upper Kibalian sequence, consisting of andesite, quartzite, and banded iron formation. At Mambasa the lower Kibalian sequence is thought to overlie unconformably the 3.35 Ga Ituri metasedimentary-rich basement gneiss, and is in turn intruded by the 2.5 Ga old Mambasa granite.

A 600 × 300–mile (1,000 × 500–km) area on the central plateau of Tanzania and east of Lake Victoria is known as the Tanzania block. The southeastern part of the block near Dodoma contains mainly granitoids and migmatitic gneisses, with remnants of schist or greenstone belts. These schist belts contain assemblages of quartzite, banded iron formation, schists that locally bear corundum, amphibolite, and mafic and ultramafic gneiss. The granitoid gneisses are about 2.6 billion years old, and the craton is intruded by circa 1.8 Ga late-stage granites. Kimberlite pipes locally bring up fragments of older, circa 3.1 Ga gneiss, the oldest rocks recognized in the Tanzania block. Schist belts in the central plateau region of Tanzania are very similar to the schist belts of southeast Uganda, and this forms the basis of correlating the Tanzania and Kibalian blocks as part of the larger Congo craton.

WEST AFRICAN CRATON

Archean rocks form large sections of the basement rocks between the Gulf of Guinea and the Atlas Mountains. These rocks form parts of the Man and Reguibat shields of the West African craton, and parts of the Tuareg shield.

The Man shield forms the southern part of the West African craton along the Gulf of Guinea and includes the Archean Liberian (Kenema-Man) domain in the west and the Eburnean (Baoule-Mossi) domain in the east. The Liberian domain occupies most of Sierra Leone, Liberia, Guinea, Ivory Coast, and part of Guinea Bissau. Rocks in this domain include many structurally complex greenstone remnants consisting of ultramafic rocks, mafic volcanic rocks and banded iron formations, granitoid gneiss, and granites, forming a classical granite-greenstone terrain. Some of the gneisses have been dated to be 3.2–3.0 billion years old, whereas the greenstones are intruded by granites that are 2.7 billion years old. Greenstone belts in the west are generally metamorphosed to amphibolite facies, comparatively large, being up to 80 miles (130 km) long, and up to 4 miles (6.5 km) in structural thickness. Greenstone belts in the southeast are smaller, being up to 25 miles (40 km) long, thinner, and show a wide range of metamorphic grades from greenschist to granulite facies. Greenstone belts in the east have more quartzite and pelite than those in the west, and are structurally discordant with basement gneiss and granitoids. Structures across the Man shield generally strike northerly and formed in a strong tectonic event at 2.75 billion years ago, known as the Liberian orogeny, that is superimposed on structures from an older event, known as the Leonean orogeny.

The westernmost part of the Man shield consists of three narrow belts of Proterozoic-Paleozoic tectonic activity, known as the Rokelides. Conglomerates, sandstones, arkose, and volcanic rocks of the Rokell River Group unconformably overlie the older Kenema basement and are increasingly deformed and metamorphosed to the west. The western margin of the Rokell River Group is marked by large thrusts where recumbently folded klippen of intensely deformed and metamorphosed sedimentary and volcanic rocks of the late Archean-Proterozoic Marampa Group were thrust to the east over the Man shield. Farther west, a 180-mile (300-km) long belt of granulite facies Archean metasedimentary rocks of the Kasila Group represent the core of a deeply eroded orogen. The eastern boundary of the Kasila Group is a 3-mile (5-km) wide mylonite belt, interpreted as an Archean suture that formed when the West African craton collided with the Guiana shield of South America.

PAN-AFRICAN BELTS AND THE EAST AFRICAN OROGEN

The East African Orogen encompasses the Arabian-Nubian shield in the north and the Mozambique belt in the south. These and several other orogenic

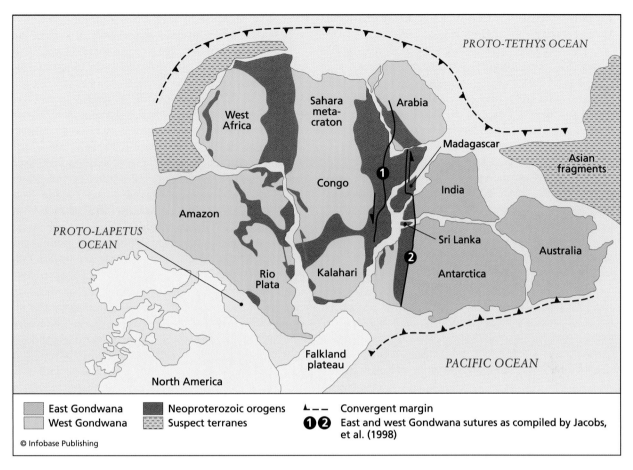

Map of the East African orogen and other Pan-African belts that formed as ocean basins closed to form the supercontinent of Gondwana (modeled from T. Kusky, M. Abdelsalam, R. Tucker, and R. Stern, 2003)

belts are commonly referred to as Pan-African belts, recognizing that many distinct belts in Africa and other continents experienced deformation, metamorphism, and magmatic activity in the general period of 800–450 Ma. Other definitions of the Pan-African orogens are more restrictive, and consider them to be confined to a complex collisional system between the Congo and Kalahari cratons in this time interval, thus including in Africa the Gareip belt, the Kaoko belt, Damara orogen, Lufilian arc, Zambezi belt, Malawi orogen, Mozambique belt, and Luria arc. Pan-African tectonothermal activity in the Mozambique belt was broadly contemporaneous with magmatism, metamorphism, and deformation in the Arabian-Nubian shield, and the two are broadly equivalent. The difference in lithology and metamorphic grade between the two belts has been attributed to the difference in the level of exposure, with the Mozambican rocks interpreted as lower crustal equivalents of the rocks in the Arabian-Nubian shield. Neoproterozoic closure of the Mozambique Ocean collapsed an accretionary collage of arc and microcontinental

terranes and sutured east and west Gondwana along the length of the East African orogen.

The formation of Gondwana at the end of the Precambrian and the dawn of the Phanerozoic by the collision of cratons including the Congo, Kalahari, India, Antarctica, and South American blocks represents one of the most fundamental problems being studied in earth sciences today. Studies of Gondwana link many different fields, and there are currently numerous and rapid changes in understanding of events related to the assembly of Gondwana. One of the most fundamental and most poorly understood aspects of the formation of Gondwana is the timing and geometry of closure of the oceanic basins that separated the continental fragments that amassed to form the Late Proterozoic supercontinent. Final collision between East and West Gondwana most likely occurred during closure of the Mozambique Ocean, forming the East African orogen.

Recent geochronologic data indicate the presence of two major "Pan-African" tectonic events within East Africa. The East African Orogeny (800–

650 Ma) represents a distinct series of events within the Pan-African of central Gondwana, responsible for the assembly of greater Gondwana. Collectively, paleomagnetic and age data indicate that another later event at 550 Ma (Kuunga Orogeny) may represent the final suturing of the Australian and Antarctic segments of the Gondwana continent.

ATLAS MOUNTAINS

The Atlas Mountains are a series of mountains and plateaus in northwest Africa extending about 1,500 miles (2,500 km) in southwest Morocco, northern Algeria, and northern Tunisia. The highest peak in the Atlas is Jabel Toubkal, at 13,665 feet (4,168 m) in southwest Morocco. The Atlas Mountains are dominantly folded sedimentary rocks uplifted in the Jurassic, and related to the Alpine system of Europe. The Atlas consists of several ranges separated by fertile lowlands in Morocco, from north to south including the Rif Atlas, Middle Atlas, High Atlas (Grand Atlas), and Anti Atlas. The Algerian Atlas consists of a series of plateaus including the Tell and Saharan Atlas rimming the Chotts Plateau, then converging in Tunisia. The Atlas form a climatic barrier between the Atlantic and Mediterranean basins and the Sahara, with rainfall falling on north-facing slopes but arid conditions dominating on the rain-shadow, south-facing slopes. The Atlas are rich in mineral deposits including coal, iron, oil, and phosphates. The area is also used extensively for sheep grazing, with farming in the more fertile intermountain basins.

EAST AFRICAN RIFT SYSTEM

Extensional plate tectonic forces are presently breaking Africa apart, with parts of eastern Africa rifting away from the main continent. The rift valley that separates these two sections is known as the East African rift system, or the Great Rift Valley, extending from the Ethiopian Afar region, through two segments known as the eastern and western rifts that bend around Lake Victoria, then extend to the Mozambique Channel.

The Main Ethiopian and North-Central Afar rifts are part of the continental East African rift system. These two kinematically distinct rift systems, typical of intracontinental rifting, are at different stages of evolution. In the north and east the continental rifts meet the oceanic rifts of Red Sea and Gulf of Aden, respectively, both of which have propagated into the continent. Seismic refraction and gravity studies indicate that the thickness of the crust in the Main Ethiopian rift is less than or equal to 18.5 miles (30 km). In Afar the thickness varies from 14 to 16 miles (23–26 km) in the south and to 8.5 miles (14 km) in the north. The plateau on both sides of the rift

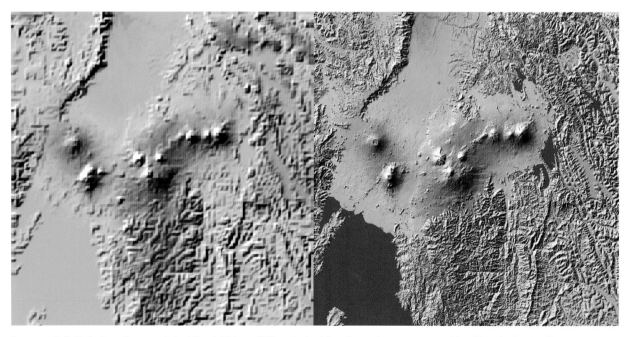

Image of digital elevation model of East African Rift at Lake Kivu from data generated by Shuttle Radar Topography Mission. Area shown covers parts of the Democratic Republic of the Congo, Rwanda, and Uganda. Elevation is color coded, progressing from green at lower elevations through yellow to brown at higher elevations. A false sun in the NW (upper left, pixelated area) causes topographic shading. Lake Kivu lies in the East African Rift, which forms a smooth lava and sediment–filled trough in the area. Two volcanic complexes are shown in the rift, including the Nyiragongo volcano (the one closer to the lake), which erupted in 2002. Virunga volcanic chain extends east of the rift.

has a crustal thickness of 21.5–27 miles (35–44 km). Geologic and geodetic studies indicate separation rates of 0.1–0.2 inches (3–6 mm) per year across the northern sector of the Main Ethiopian rift between the African and Somali plates. The rate of spreading between Africa and Arabia across the North-Central Afar rift is relatively faster, about 0.8 inches (20 mm) per year. Paleomagnetic directions from Cenozoic basalts on the Arabian side of the Gulf of Aden indicate seven degrees of counterclockwise rotation of the Arabian plate relative to Africa, and clockwise rotations of up to 11 degrees for blocks in eastern Afar. The initiation of extension on both sides of the southernmost Red Sea rift, Ethiopia, and Yemen appear coeval, with extension starting between 22 and 29 million years ago.

The Ethiopian Afar region is one of the world's largest, deepest regions below sea level that is subaerially exposed on the continent, home to some of the earliest known hominid fossils. The Afar is a hot, arid region where the Awash River drains northward out of the East African rift system, and is evaporated in Lake Abhe before it reaches the sea. This unique and spectacular region is located in eastern Africa in Ethiopia and Eritrea, between Sudan, Somalia, and across the Red Sea and Gulf of Aden from Yemen. The region is so topographically low because it is located at a tectonic triple junction, where three main plates are spreading apart, causing regional subsidence. The Arabian plate is moving northeastward away from the African plate, and the Somali plate is moving, at a much slower rate, to the southeast away from Africa. The southern Red Sea and north-central Afar Depression form two parallel north-northwest-trending rift basins, separated by the Danakil Horst, related to the separation of Arabia from Africa. Of the two rifts, the Afar depression is exposed at the surface, whereas the Red Sea rift floor is submerged below the sea. The north-central Afar rift is complex, consisting of many grabens and horsts. The Afar Depression merges southward with the northeast-striking Main Ethiopian rift, and eastward with the east-northeast-striking Gulf of Aden. The Ethiopian Plateau bounds it on the west. Pliocene volcanic rocks of the Afar stratoid series and the Pleistocene to recent volcanics of the Axial Ranges occupy the floor of the Afar Depression. Miocene to recent detrital and chemical sediments are intercalated with the volcanics in the basins.

South of the Ethiopian rifts, the eastern branch of the main East African rift strikes southward through Kenya, forming Lake Turkana across the Kenya highlands, and forming the famous Ngorongoro crater, where millions of wildlife gather for scarce water in the deep rift valley. The western branch of the main East Africa rift strikes southward from Ethiopia and

Sudan into Uganda, forming a series of deep, steep-sided lakes including Lakes Albert, Edward, Kivu, Tanganyika, and Malawi. Some of these lakes are more than a mile (1.6 km) deep, and are fed by drainage systems that remain on the floor of the rift, while drainage on the rift shoulders carries water away from the central rift.

See also ARCHEAN; BASIN, SEDIMENTARY BASIN; CONVERGENT PLATE MARGIN PROCESSES; CRATON; DEFORMATION OF ROCKS; DESERTS; DIVERGENT PLATE MARGIN PROCESSES; GREENSTONE BELTS; MADAGASCAR; OROGENY; PRECAMBRIAN.

FURTHER READING

Antrobus, E. S. A., ed. *Witwatersrand Gold 100 Years.* Johannesburg: Geological Society of South Africa, 1986.

Bickle, Mike J., and Euan G. Nisbet, eds. *The Geology of the Belingwe Greenstone Belt, Geological Society of Zimbabwe Special Publication* 2. Rotterdam, Netherlands: A.A. Balkema, 1993(a).

Burke, Kevin, William S. F. Kidd, and Timothy M. Kusky. "Archean Foreland Basin Tectonics in the Witwatersrand, South Africa." *Tectonics* 5 (1986): 534–535.

———. "Is the Ventersdorp Rift System of Southern Africa Related to a Continental Collision between the Kaapvaal and Zimbabwe Cratons at 2.64 Ga ago?" *Tectonophysics* 115 (1985): 1–24.

———. "The Pongola Structure of Southeastern Africa: The World's Oldest Recognized Well-Preserved Rift?" *Journal of Geodynamics* 2 (1985): 35–50.

Geology of the World. "An Overview of the Best Regional Geology Resources—Region by Region. Textbooks, Papers, Excursions, Videos, Websites. Africa." Available online. URL: http://www.geology-of-the-world.com/africa.htm. Last modified August 22, 2008.

Goodwill, Alan M. *Precambrian Geology, The Dynamic Evolution of the Continental Crust.* New York: Academic Press, 1991

Hoffman, A., and Timothy M. Kusky. "The Belingwe Greenstone Belt: Ensialic or Oceanic?" In *Precambrian Ophiolites and Related Rocks,* edited by Timothy M. Kusky. Amsterdam, Netherlands: Elsevier, 2004.

Kusky, Timothy M., and Julian Vearncombe. "Structure of Archean Greenstone Belts." In *Tectonic Evolution of Greenstone Belts,* edited by Maarten J. de Wit and Lewis D. Ashwal. Oxford: Oxford Monograph on Geology and Geophysics, 1997.

Kusky, Timothy M., and William S. F. Kidd. "Remnants of an Archean Oceanic Plateau, Belingwe Greenstone Belt, Zimbabwe." *Geology* 20, no. 1 (1992): 43–46.

Kusky, Timothy M., and Pamela A. Winsky. "Structural Relationships along a Greenstone/Shallow Water Shelf Contact, Belingwe Greenstone Belt, Zimbabwe." *Tectonics* 14, no. 2 (1995): 448–471.

Kusky, Timothy M. "Tectonic Setting and Terrane Accretion of the Archean Zimbabwe Craton." *Geology* 26 (1998): 163–166.

Stowe, Clive W. "The Structure of a Portion of the Rhodesian Basement South and West of Selukwe." Ph.D. diss., University of London, 1968.

———. "Alpine Type Structures in the Rhodesian Basement Complex at Selukwe." *Journal of the Geological Society of London* (1974): 411–425.

Tankard, Anthony J., M. P. A. Jackson, Ken A. Eriksson, David K. Hobday, D. R. Hunter, and W. E. L. Minter. *Crustal Evolution of Southern Africa: 3.8 Billion Years of Earth History.* New York: Springer-Verlag, 1982.

Tesfaye, Sansom, David Harding, and Timothy M. Kusky. "Early Continental Breakup Boundary and Migration of the Afar Triple Junction, Ethiopia." *Geological Society of America Bulletin* 115 (2003): 1,053–1,067.

de Wit, Maarten J., Chris Roering, Rojer J. Hart, Richard A. Armstrong, Charles E. J. de Ronde, R. W. E. Green, Marian Tredoux, E. Pederdy, and R. A. Hart. "Formation of an Archean Continent." *Nature* 357 (1992): 553–562.

Alpamayo Peak in the Cordilleras Mountains, Peruvian Andes *(Galyna Andrushko, Shutterstock, Inc.)*

Andes Mountains The Andes are a 5,000-mile (8,000-km) long mountain range in western South America, running generally parallel to the coast, between the Caribbean coast of Venezuela in the north and Tierra del Fuego in the south. The mountains merge with ranges in Central America and the West Indies in the north, and with ranges in the Falklands and Antarctica in the south. Many snow-covered peaks rise more than 22,000 feet (6,000 m), making the Andes the second-tallest mountain belt in the world, after the Himalayan chain. The highest range in the Andes is the Aconcauga on the central and northern Argentine-Chilean border. The high, cold Atacama desert is located in the northern Chile sub-Andean range, and the high Altiplano Plateau is situated along the great bend in the Andes in Bolivia and Peru.

The southern part of South America consists of a series of different terranes (belts of distinctive rocks) added to the margin of the supercontinent of Gondwana in late Proterozoic and early Proterozoic times. Subduction and the accretion of oceanic terranes continued through the Paleozoic, forming a 155-mile (250-km) wide accretionary wedge. The Andes developed as a continental margin volcanic arc system on the older accreted terranes, formed above a complex system of subducting plates from the Pacific Ocean. They are geologically young, having been uplifted mainly in the Cretaceous and Tertiary (roughly the past 100 million years), with active volcanism, uplift, and earthquakes. The specific nature of volcanism, plutonism, earthquakes, and uplift are found to be strongly segmented in the Andes, and related to the nature of the subducting part of the plate, including its dip and age. Regions above places where the subducting plate dips more than 30 degrees have active volcanism, whereas regions above places where the subduction zone is subhorizontal do not have active volcanoes.

The Altiplano is a large, uplifted plateau in the Bolivian and Peruvian Andes of South America. The plateau has an area of about 65,536 square miles (170,000 km²), and an average elevation of 12,000 feet (3,660 m) above sea level. The Altiplano is a sedimentary basin caught between the mountain ranges of the Cordillera Oriental on the east and the Cordillera Occidental on the west. The Altiplano is a dry region with sparse vegetation, and scattered salt flats. Villagers grow potatoes and grains, and a variety of minerals are extracted from the plateau and surrounding mountain ranges.

Lake Titicaca, the largest high-altitude lake navigable to large vessels in the world, is located at the northern end of the Altiplano. Sitting at 12,500 feet (3,815 m) above sea level, the lake straddles the border between Peru and Boliva. The lake basin is situated between Andean ranges on the Altiplano plateau, and is bordered to the northeast by some of the highest peaks in the Andes in the Cordillera Real, where several mountains rise to over 21,000 feet (6,400 m). Covering 3,200 square miles, Lake Titicaca is the largest freshwater lake in South America, although it is divided into two parts by the Strait of Tiquina. The body of water north of the strait is called Chucuito in Bolivia and Lake Grande in Peru, and south of the strait the smaller body of water is called Lake Huinaymarca in Bolivia and Lake Pequeno in Peru. Most of the lake is 460–600 feet (140–180 m) deep, but reaches 920 feet (280 m) deep near the northeast corner of the lake. The lake is fed by many short tributaries from surrounding mountains, and is drained by the Desaguadero River, which flows into Lake Poopo. However, only 5 percent of water loss is through this single outlet—the remainder is lost by evaporation in the hot, dry air of the Altiplano. Lake levels fluctuate on seasonal and several longer-time cycles, and the water retains a relatively constant temperature of 56°F (14°C) at the surface, but cools to 52°F (11°C) below a thermocline at 66 feet (20 m). Salinity ranges from 5.2–5.5 parts per thousand.

Lake Titicaca, translated variously as Rock of the Puma or Craig of Lead, has been the center of culture since pre-Inca times (600 years before present [b.p.]), and its shoreline is presently covered by Indian villages and terraced rice fields. Some of the oldest civilizations are preserved in ruins around Lake Titicaca, including those at Tiahuanaco, on the southern end of the lake, and others on the many islands in the lake. Ruins of a temple on Titicaca Island mark the spot where Inca legends claim that Manco Capac and Mama Ocllo, the founders of the Inca dynasty, were sent to Earth by the Sun.

The northern Andes are drained to the east by the world's second-longest river, the Amazon, stretching 3,900 miles (6,275 km) from the foothills of the Andes to the Atlantic Ocean. The southern Andes are drained to the east by the Paraná River, and a number of smaller rivers run down the steep western slope of the Andes to the Pacific Ocean. The Amazon begins where the Ucayili and Maranon tributaries merge, and drains into the Atlantic near the city of Belém. The Amazon carries the most water and has the largest discharge of any river in the world, averaging 150 feet (45 m) deep. Its drainage basin amounts to about 35 percent of South America, covering 2,500,000 square miles (6,475,000 km²). The Amazon lowlands in Brazil include the largest tropical rain forest in the world. In this region the Amazon is a muddy, silt-rich river with many channels that wind around numerous islands in a complex maze. The delta region of the Amazon is marked by numerous fluvial islands and distributaries, as the muddy waters of the river get dispersed by strong currents and waves into the Atlantic. A strong tidal bore, up to 12 feet (3.7 m) high, runs up to 500 miles (800 km) upstream.

The Amazon River basin occupies a sediment-filled rift basin, between the Precambrian crystalline basement of the Brazil and Guiana shields. The area hosts economic deposits of gold, manganese, and other metals in the highlands, and detrital gold in lower elevations. Much of the region's economy relies on the lumber industry, with timber, rubber, vegetable oils, Brazil nuts, and medicinal plants sold worldwide.

Spanish commander Vincent Pinzon was probably the first European in 1500 to explore the lower part of the river basin, followed by the Spanish explorer Francisco de Orellana in 1540–41. Orellana's tales of tall, strong female warriors gave the river its name, borrowing from Greek mythology. Further exploration by Pedro Teixeira, Charles Darwin, and Louis Agassiz led to greater understanding of the river's course, peoples, and environment, and settlements did not appear until steamship service began in the middle 1800s.

See also CONVERGENT PLATE MARGIN PROCESSES; PLATE TECTONICS; SOUTH AMERICAN GEOLOGY.

FURTHER READING

Moores, Eldridge, and Robert Twiss. *Tectonics*. New York: W.H. Freeman, 1995.

Antarctica The southern continent, Antarctica, is located nearly entirely below the Antarctic Circle (66° 33′ 39″) and is distributed asymmetrically around the south pole. The continent covers approximately 5.46 million square miles (14 million km²), is nearly completely covered in ice, and has several large ice shelves extending off the mainland into surrounding oceans. Antarctica is surrounded by relatively isolated waters of the Southern Ocean, comprised of southern reaches of the Atlantic, Pacific, and Indian Oceans. Antarctica is the fifth-largest continent, covering an area equal to 57 percent of North America, or nearly 1.5 times the size of the United States including Alaska. The Russian explorers Mikhail Lazarev and Fabian Gottlieb von Bellingshausen first discovered the continent in 1820, and the Scottish cartographer John Bartholomew named it in 1890. In 1959 12 countries (later joined by more, to bring the total to 46) signed the Antarctic Treaty, prohibiting military activity and mining in Antarctica, and promoting cooperative scientific and environmental work.

The ice sheet covering Antarctica is the world's largest reservoir of fresh water (although frozen) and averages more than a mile (1.6 km) thick. The weight of this ice causes the underlying continent to be depressed by more than 1.5 miles (2.5 km). A small amount of rock is exposed in the Transantarctic Mountains and in the Dry Valleys area.

The Transantarctic Mountains divide the continent in two, stretching from the Ross Sea to the Wedell Sea. Western Antarctica (using the Greenwich meridian that runs nearly along the Transantarctic Mountains) is covered by the West Antarctic Ice Sheet, which some climatologists warn could collapse, raising sea levels by 10 feet (3 m) or more in a short time. The Antarctic Peninsula, Marie Byrd Land, and the area east and north of the Transantarctic Mountains are part of West Antarctica. The eastern part of Antarctica is a large Precambrian craton known as East Antarctica, with ages extending to at least 3 billion years. The rocks of the ancient craton, however, were reworked in younger mountain-building events, including an Early Paleozoic event during which East Antarctica was incorporated into Gondwana.

Most of western Antarctica was built up through the accretion of microplates that include the Ellsworth Mountains terrane, the Antarctic Peninsula, Marie Byrd Land, and an unnamed block of igneous and metamorphic rocks. Compared with the subdued (subglacial) topography of East Antarctica, western Antarctica has relatively rugged, mountainous topography.

The Transantarctic Mountains are up to 15,000 feet (4,570 m) high, and were formed during the Ross orogeny 500 million years ago. In contrast, the Ellsworth Mountains reach 16,000 feet (4,880 m) and were formed about 190 million years ago in the Early Mesozoic. The Antarctic Peninsula is the youngest addition to Antarctica, formed mostly in the Late Mesozoic to Early Cenozoic Andean orogeny (80–60 million years ago). Most activity in the Transantarctic Mountains was in the period of the breakup of Gondwana, as this region became a convergent margin continuous with the Andes of South America.

GEOLOGY, PALEONTOLOGY, AND PALEOCLIMATE

The Precambrian basement rocks of East Antarctica comprise the East Antarctic craton, with Archean cores surrounded by Proterozoic orogenic belts with younger deformation and metamorphism than the Archean blocks. These Archean cores include coastal areas of western Dronning Maud Land, Enderby Land, the Prince Charles Mountains, and the Vestfold Hills. The rocks in the Vestfold Hills include 2.5 billion-year-old gneisses derived from igneous rocks, and some indications of 2.8 billion-year-old gneisses in the area. The best known part of the East Antarctic craton is the Napier Complex in Enderby Land. This Archean granulite-gneiss belt includes 2.8 to 3.0 billion-year-old metamorphosed igneous gneisses, with some indication that initial igneous activity in the area may extend back to 3.8 billion years. Several deformation and metamorphic events are recorded at about 2.8 billion years ago, and a very high temperature metamorphic event recorded at 2.51–2.47 billion years ago. Additional deformation is recorded

Transantarctic Mountains (North Victoria Land) in Antarctica *(Hinrich Baesemann/dpa/Landov)*

as East Antarctica became part of Gondwana, near the end of the Precambrian.

East Antarctica was part of Gondwana in the Early Paleozoic, resting in equatorial latitudes and accepting marine deposits of fossiliferous limestones from the tropical seas. These limestones are rich in trilobite and invertebrate fossils. West Antarctic was in the northern hemisphere in the Paleozoic, and not yet sutured with East Antarctica. By the Devonian, Gondwana, with East Antarctica, had drifted into southern latitudes and experienced a cooler climate, but still has a good fossil record of land plants in sandstone and siltstone beds exposed in the Ellsworth and Pensacola Mountains. The end of the Devonian (360 million years ago) witnessed a major glaciation as Gondwana became centered on the South Pole. Despite the glaciation, East Antarctica remained vegetated, and by the Permian many swamps across Antarctica were flourishing with the fernlike *Glossopteris* fauna, known throughout much of Gondwana. By the end of the Permian the climate over much of Gondwana had turned hot and dry.

The end-Permian warming caused the polar ice caps on Eastern Gondwana, including Antarctica, to melt, and the continent became a vast desert. Still, seed ferns and giant reptiles, including *Lystrosaurus,* inhabited the land, and thick beds of sandstone and shale were deposited on the East Antarctic platform. The Antarctic Peninsula was forming during the Jurassic (206–146 million years ago), and beech trees began to take over the floral assemblage. West Antarctica had accreted to the East Antarctic craton and was covered in conifer forests through the Cretaceous, gradually replaced by the beech trees toward the end of the period. The seas around Antarctica were inhabited by ammonites.

Modern-day Antarctica began to take shape in the Cenozoic. The Antarctic Peninsula and Western Antarctica are an extension of the Andes of South America.

CLIMATE AND ICE CAP

The climate of Antarctica is the coldest, driest, and windiest on Earth, with the lowest recorded temperature being -129°F (-89°C) from the Vostok weather station, located at two miles (three km) elevation in Antarctica. Despite being covered in ice, the climate of Antarctica is best described as a dry or polar desert, since the amount of precipitation is so low, with the South Pole receiving fewer than four inches (10 cm) of rainfall equivalent each year. Although it is covered in ice, interior Antarctica is technically the largest desert on Earth. Temperatures show a considerable range, from -112°F to -130°F (-90°C to -90°C) in interior winters, to 41°F to 59°F (5°C to 15°C) along the coastline in summer. In general, the eastern

Ice calving from an ice front off Adelaide Island, Antarctica *(British Antarctic Survey/Photo Researchers, Inc.)*

part of the continent is colder than the western part, because it has a higher elevation.

Although 98 percent of Antarctica is covered in ice, a few places are ice-free. The Dry Valleys are the largest area on Antarctica not covered by ice. The Dry Valleys, located near McMurdo Sound on the side of the continent closest to New Zealand, have a cold desert climate and receive only four inches (10 cm) of precipitation per year, overwhelmingly in the form of snow. The Dry Valleys are one of the coldest, driest places on Earth and are used by researchers from the National Aeronautics and Space Administration (NASA) as an analog for conditions on Mars. No vegetation exists in the Dry Valleys, but a number of unusual microbes live in the frozen soils and form cyanobacterial mats in places. In the Southern Hemisphere summer, glaciers in the surrounding Transantarctic Mountains release significant quantities of meltwater so that streams and lakes form over the thick permafrost in the valleys.

The edge of the continent is often hit by strong katabatic winds, formed when high-density air forms over the ice cap, and then moves rapidly downhill, typically along glaciated valleys, at times reaching hurricane strength in force. High-density air often forms over the ice cap because the ice cools the air through radiative cooling effects, making it denser. This dense air then finds the lowest points to flow downhill, and because of the high elevation of central Antarctica the winds pick up enormous speed through gravitational energy, until they roar out of the coastal valleys as exceptionally cold channels of hurricane-force winds.

ANTARCTIC ICE CAP AND GLOBAL WARMING

The Antarctic ice cap is huge, containing more than 70 percent of the fresh water on the planet. It is about the same size as the Laurentide ice sheet that covered the northern part of North America in the last ice age. If the ice in the Antarctic ice cap all melted, sea levels would rise by 230 feet (70 m), yet there is no evidence that the south polar ice cap is melting, and it has been stable for about the past 5 million years. Many models predict that global warming may increase precipitation in Antarctica and actually cause the ice cap to increase in volume and lower sea levels by 0.04 inches (0.09 cm) per year.

The ice cap consists of a vast area of ice more than one mile (1.6 km) thick, covering nearly all of East Antarctica in Queen Maud Land and Wilkes Land. The geology of this region is understood through nunataks, isolated peaks piercing through the ice cap mostly near the coast, and in the Transantarctic Mountains. Likewise, most of West Antarctica is covered by ice, including Marie Byrd Land, Ellsworth Land, Palmer Land, and the Antarctic Peninsula. The large Ross Ice Shelf is located between Marie Byrd Land and the Transantarctic Mountains, while on the other side of the continent, the Ronne Ice Shelf fills the space between the Antarctic Peninsula and the Transantarctic Mountains.

Global warming is not significantly affecting most of Antarctica, since the interior of the continent is isolated from the global climate system. The ice cap in central Antarctica is presently growing in volume, whereas some of the peripheral ice shelves, such as along the northern parts of the Antarctic Peninsula, are losing volume. For instance, in 2003 parts of the Larsen ice shelf on the northern Antarctic Peninsula began collapsing from a combination of global warming and other cyclical processes. Farther south on the Peninsula, the Wilkins ice shelf lost 220 square miles (570 km²) of ice in 2008, but it is still not well established whether these giant collapses result from global warming or whether similar processes have existed for many thousands of years. In support of the latter idea is the observation that the overall amount of sea ice around Antarctica has remained stable over the past 30 years, although there is considerable variation month to month and year to year.

See also CONVERGENT PLATE MARGIN PROCESSES; CRATON; GLACIER, GLACIAL SYSTEMS; GLOBAL WARMING; GONDWANA, GONDWANALAND.

FURTHER READING

Craddock, Campbell. *Antarctic Geoscience*. Madison: University of Wisconsin Press, 1982.
McKnight, T. L., and Darrel Hess. "Katabatic Winds." In *Physical Geography: A Landscape Appreciation*, 131–32. Upper Saddle River, N.J.: Prentice Hall, 2000.
Stonehouse, B., ed. *Encyclopedia of Antarctica and the Southern Oceans*. New York: John Wiley & Sons, 2002.

Arabian geology   The Arabian Peninsula can be classified into two major geological provinces, including the Precambrian Arabian shield and the Phanerozoic cover. The Arabian shield comprises the core and deep-lying rocks of the Arabian Peninsula, a landmass of near trapezoidal shape bounded by three water bodies. The Red Sea bounds it from the west, the Arabian Sea and the Gulf of Aden from the south, and the Arabian Gulf and Gulf of Oman bound it on the east.

The Precambrian Shield is located along the western and central parts of the peninsula. It narrows in the north and the south but widens in the central part of the peninsula. The shield lies between latitudes 12° and 30° north and between longitudes 34° and 47° east. The Arabian shield is considered part of the Arabian-Nubian shield formed in the upper Proterozoic Era and stabilized in the Late Proterozoic around 600 million years ago. The shield has since subsided and been covered by thick deposits of Phanerozoic continental shelf sediments along the margins of the Tethys Ocean. Later in the Tertiary the Red Sea rift system rifted the Arabian-Nubian shield into two fragments.

Phanerozoic cover rocks unconformably overlie the eastern side of the Arabian shield, forming the Tuwaiq Mountains, and these rocks dip gently toward the east. Parts of the Phanerozoic cover are found overlying parts of the Precambrian shield, such as the Quaternary lava flows of Harrat Rahat in the middle and northern parts of the shield, as well as some sandstones, including the Saq, Siq, and Wajeed sandstones in different parts of the shield. The Phanerozoic rocks are well exposed again in tectonic uplifts in the Oman (Hajar) Mountains in the east, where the geology is well known.

TECTONIC MODELS OF THE ARABIAN SHIELD

The Arabian shield includes an assemblage of Middle to Late Proterozoic rocks exposed in the western and central parts of the Arabian Peninsula and overlapped to the north, east, and south by Phanerozoic sedimentary cover rocks. Several parts of the shield are covered by Tertiary and Quaternary lava flows that were extruded along with rifting of the Red Sea starting about 30 million years ago. Rocks of the Arabian shield may be divided into assemblages of Middle to Late Proterozoic stratotectonic units, volcanosedimentary, and associated mafic to intermediate intrusive rocks. These rocks are divided into two major categories, the layered rocks and the intrusive rocks. Researchers variously interpret these assemblages as a result of volcanism and magmatism in continental basins or above subduction zones. More recently workers suggested that many of these assemblages belong to Late Proterozoic volcanic-arc

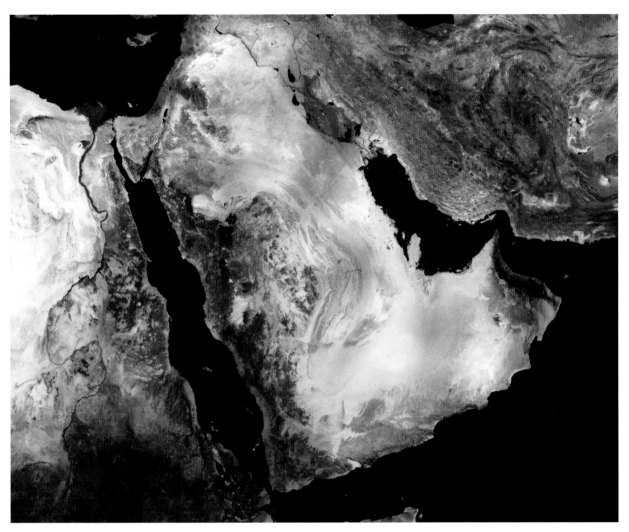

American Landsat image of Arabia. The Rub'a Khali (Empty Quarter) desert forms the great yellow sand sheet in the southern part of the peninsula, the Arabian shield forms the dark-colored terrane in the west, and the Semail ophiolite (oceanic crust and lithosphere) forms the dark area in the southeast. The fertile Mesopotamia area (in dark green, between the Tigris and Euphrates Rivers) separates Arabia from the Zagros Mountains of Iran. *(Earth Satellite Corporation/Photo Researchers, Inc.)*

systems that comprise distinct tectonic units or terranes, recognized following definitions established in the North America cordillera.

Efforts in suggesting models for the evolution of the Arabian shield started in the 1960s. Early workers suggested that the Arabian shield experienced three major orogenies in the Late Proterozoic Era. They also delineated four classes of plutonic rocks that evolved in chemistry from calc-alkaline to peralkaline through time. In the 1970s a great deal of research emerged concerning models of the tectonic evolution of the Arabian shield. Two major models emerged from this work, including mobilistic plate tectonic models and a nonmobilistic basement-tectonic model.

The main tenet of the plate tectonic model is that the evolution of the Arabian shield started and took place in an oceanic environment, with the formation of island arcs over subduction zones in a huge oceanic basin. On the contrary, the basement-tectonic model considers that the evolution of the Arabian shield started by the rifting of an older craton or continent to form intraoceanic basins that became the sites of island arc systems. In both models, late stages of the formation of the Arabian-Nubian shield are marked by the sweeping together and collision of the island arcs systems, thrusting of the ophiolites onto continents, and cratonization of the entire orogen, forming one craton attached to the African craton. Most subsequent investigators in the 1970s supported one of these two models and tried to gather evidence to support that model.

As more investigations, mapping, and research were carried out in the 1980s and 1990s, a third model invoking microplates and terrane accretion

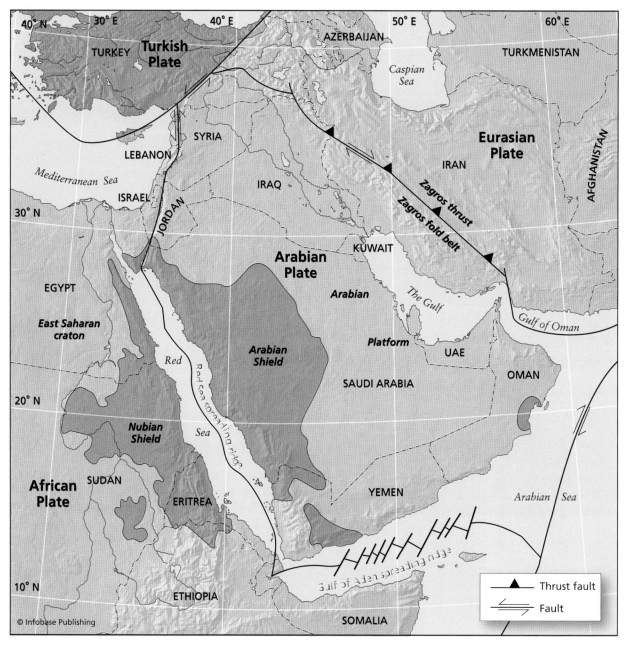

Simple map showing the geology of the Arabian Peninsula

was suggested. This model suggests the existence of an Early to mid-Proterozoic (2.0–1.63 billion-year-old) craton that was extended, rifted, then dispersed, causing the development of basement fragments that were incorporated as allochthonous microplates into younger tectonostratigraphic units. The tectonostratigraphic units include volcanic complexes, ophiolite complexes, and marginal-basin and fore-arc stratotectonic units that accumulated in the intraoceanic to continental-marginal environments that resulted from rifting of the preexisting craton. These rocks, including the older continental fragments, constitute five large and five small tectonostratigraphic terranes

accreted and swept together between 770 and 620 million years ago to form a neocraton on which younger volcanosedimentary and sedimentary rocks were deposited. Most models developed in the period since the early 1990s represent varieties of these three main classical models, along with a greater appreciation of the formation of the supercontinent of Gondwana in the formation of the Arabian-Nubian shield.

GEOLOGY OF THE ARABIAN SHIELD
Peter Johnson of the U.S. Geological Survey and his coworkers have synthesized the geology of the

Arabian shield and proposed a general classification of the geology of the Arabian shield that attempts to integrate and resolve the differences between the previous classifications. According to this classification, the layered rocks of the Arabian shield are divided into three main units separated by periods of regional tectonic activity (orogenies). This gives an overall view that the shield was created through three tectonic cycles. These tectonic cycles include early, middle, and late Upper Proterozoic tectonic cycles.

The early Upper Proterozoic tectonic cycle covers the period older than 800 million years and includes the oldest rock groups that formed before and up to the Aqiq orogeny in the south and up to the Tuluhah orogeny in the north. In this general classification the Aqiq and Tuluhah orogenies are considered part of one regional tectonic event, or orogeny, that is given a combined name of Aqiq-Tuluhah orogeny.

The middle Upper Proterozoic tectonic cycle is considered to have taken place between 700 and 800 Ma. It includes the Yafikh orogeny in the south and the Ragbah orogeny in the north. These two orogenies were combined into one regional orogeny, the Yafikh-Ragbah orogeny.

The late Upper Proterozoic tectonic cycle took place in the period between 700 and 650 Ma. It includes the Bishah orogeny in the south and the Rimmah orogeny in the north. These two orogenies are combined into one regional orogeny, the Bishah-Rimmah orogeny.

CLASSIFICATION OF ROCK UNITS

The layered rocks in the Arabian shield are classified into three major rock units, each of them belonging to one of the three tectonic cycles mentioned above. These major layered rock units are the lower, middle, and upper layered rock units.

The lower layered rock unit covers those rock groups that formed in the early upper Proterozoic tectonic cycle (older than 800 Ma) and includes rocks with continental affinity. The volcanic rocks that belong to this unit are characterized by tholeiitic basalt compositions and by the domination of basaltic rocks older than 800 Ma. The rock groups of this unit are located mostly in the southwestern and eastern parts of the shield.

The rock groups of the lower layered unit include rocks formed in an island arc environment and characterized by basic tholeiitic volcanic rocks (Baish and Bahah Groups) and calc-alkaline rocks (Jeddah Group). In some places these rocks overlie highly metamorphosed rocks of continental origin (Sabia Formation and Hali schists) considered to have been brought into the system either from a nearby craton such as the African craton, or from

microplates rifted from the African plate such as the Afif microplate.

The middle layered rock unit includes the layered rock groups that formed during the middle upper Proterozoic tectonic cycle between 700 and 800 Ma ago. The volcanic rocks are predominately intermediate igneous rocks characterized by a calc-alkaline nature. These rocks are found in many parts of the shield with a greater concentration in the north and northwest, and scattered outcrops in the southern and central parts of the shield.

The upper-layered rock unit includes layered rock groups that formed in the late upper Proterozoic tectonic cycle in the period between 700 and 560 Ma ago and are predominately calc-alkaline, alkaline intermediate, and acidic rocks. These rock groups are found in the northeastern, central, and eastern parts of the shield.

INTRUSIVE ROCKS

Intrusive rocks that cut the Arabian shield are divided into three main groups, called (from oldest to youngest) Pre-orogenic, Syn-orogenic, and Post-orogenic.

The preorogenic intrusions cut through the lower-layered rocks unit only and not the other layered rock units. They are considered older than the middle-layered rock unit but younger than the lower-layered rock unit. These intrusions are characterized by their calcic to calc-alkaline composition. They are dominated by gabbro, diorite, quartz-diorite, trondhjemite, and tonalite. These intrusions are found in the southern, southeastern, and western parts of the shield and coincide with the areas of the lower-layered rocks unit. These intrusions are assigned ages between 700 and 1,000 million years. Geochemical signatures including strontium isotope ratios show that these intrusions were derived from magma that came from the upper mantle.

The synorogenic intrusions cut the lower and the layered rock units, as well as the preorogenic intrusions but do not cut or intrude the upper-layered rock units. These intrusions are considered older than the upper-layered rocks unit and younger than the preorogenic intrusions, as well as the lower- and the middle-layered rocks units, and they are assigned ages between 620 and 700 million years. Their chemical composition is closer to the granitic calc-alkaline to alkaline field than the preorogenic intrusions. These intrusions include granodiorite, adamalite, monzonite, granite, and alkali granite, with lesser amounts of gabbro and diorite in comparison with the preorogenic intrusions. The general form of these intrusions is batholithic bodies that cover wide areas. They are found mostly in the eastern, northern, and northeastern parts of the Arabian shield. The initial strontium ratio of these intrusions is higher than that

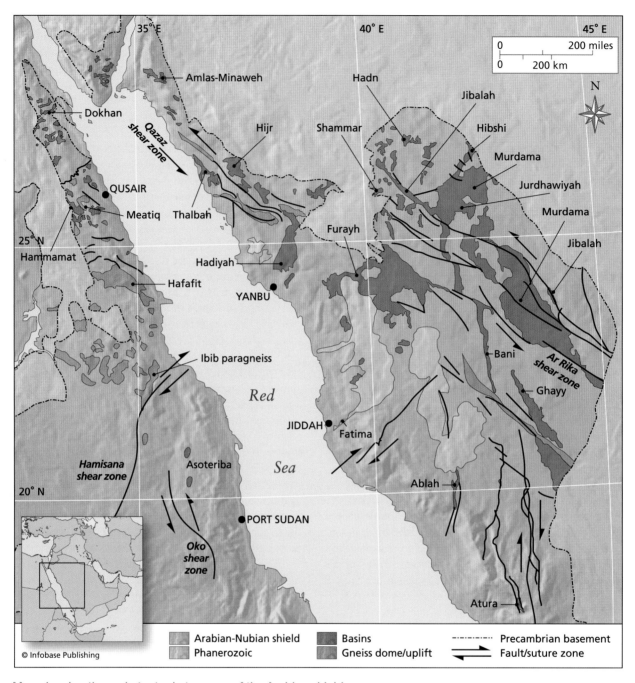

Map showing the main tectonic terranes of the Arabian shield

of the preorogenic intrusions and indicates that these intrusions were derived from a magma generated in the lower crust.

Postorogenic intrusions cut through the three upper Proterozoic–layered rocks units as well as the pre- and synorogenic intrusions. These are assigned ages between 620 and 550 million years. They form circular, elliptical, and ringlike bodies that range in chemical composition from alkaline to peralkaline. These intrusions include peralkaline granites such as riebeckite granite, alkaline syenite, pink granite, biotite granite, monzogranite, and perthite-biotite granite.

Ringlike bodies and masses of gabbro are also common, and the postorogenic magmatic suite is bimodal in silica content. These intrusions are scattered in the Arabian shield, but they are more concentrated in the eastern, northern, and central parts of the shield. The initial strontium ratio of the postorogenic intrusions ranges between 0.704 and 0.7211, indicating that these intrusions were derived from a magma generated in the lower crust.

OPHIOLITE BELTS

Mafic and ultramafic rocks that comply with the definition of the ophiolite sequence are grouped into six major ophiolitic belts. Four of these belts strike north, while the other two strike east to northeast. These ophiolite belts include

- Amar-Idsas ophiolite belt
- Jabal Humayyan-Jabal Sabhah ophiolite belt
- Bijadiah-Halaban ophiolite belt
- Hulayfah-Hamdah "Nabitah" ophiolite belt
- Bir Umq-Jabal Thurwah ophiolite belt
- Jabal Wasq-Jabal Ess ophiolite belt

These rocks were among other mafic and ultramafic rocks considered as parts of ophiolite sequences, but later only these six belts were considered to comply with the definition of ophiolite sequences. However, the sheeted dike complex of the typical ophiolite sequence is not clear or absent in some of these belts, suggesting that the dikes may have been obscured by metamorphism, regional deformation, and alteration. These belts are considered to represent suture zones where convergence between plates or island arc systems took place, and are considered as the boundaries between different tectonic terranes in the shield.

NAJD FAULT SYSTEM

One of the most striking structural features of the Arabian shield is the existence of a fault system in a zone 185 miles (300 km) wide with a length of nearly 750 miles (1,200 km), extending from the southeastern to the northwestern parts of the shield. This system was generated just after the end of the Hijaz tectonic cycle, and it was active from 630 to 530 Ma, making it the last major event of the Precambrian in the Arabian shield. These faults are left-lateral strike-slip faults with a 150-mile (250 km) cumulative displacement on all faults in the system.

The main rock group formed during and after the existence of the Najd fault system is the Jibalah Group. This group formed in the grabens that were formed by the Najd fault system and are the youngest rock group of the Precambrian Arabian shield. The Jibalah Group formed between 600 and 570 Ma ago. The Jibalah Group is composed of coarse-grained clastic rocks and volcanic rocks in the lower parts, stromatolitic and cherty limestone and argillites in the middle parts, and fine-grained clastic rocks in the upper parts. These rocks were probably deposited in pull-apart basins that developed in extensional bends along the Najd fault system.

TECTONIC EVOLUTION OF THE ARABIAN SHIELD

The Arabian shield is divided into five major terranes and tectonostratigraphic units separated by four major suture zones, many with ophiolites along them. The five tectonic terranes include the Asir, Al-Hijaz, Midyan, Afif, and Ar-Rayn. The first three terranes are interpreted as interoceanic island arc terranes, while the Afif terrain is considered continental, and the Ar-Rayn terrain is considered to be probably continental. The four suture zones include the Bir Omq, Yanbu, Nabitah, and Al-Amar-Idsas belts. These suture zones represent the collision and suturing that took place between different tectonic terranes in the Arabian shield. For example, the Bir Omq belt represents the collision and suturing between two island arc terranes of Al-Hijaz and Asir, while the Yanbu suture zone represents the collision zone between the Midyan and Al-Hijaz island arc terranes. The Nabitah zone represents collision and suturing between a continental microplate (Afif) in the east and island arc terranes (Asir and Al-Hijaz) in the west; Al-Amar Idsas suture represents a collision and suturing zone between two continental microplates, Afif and Ar-Rayn.

Five main stages are recognized in the evolution of the Arabian shield, including rifting of the African craton (1,200–950 million years ago), formation of island arcs over oceanic crust (950–715 million years ago), formation of the Arabian shield craton from the convergence and collision of microplates with adjacent continents (715–640 million years ago), continental magmatic activity and tectonic deformation (640–550 million years ago), and epicontinental subsidence (550 million years ago).

Information about the rifting stage (1,200–950 million years ago) is limited, but the Mozambique belt in the African craton underwent rifting in the interval between 1,200 and 950 million years ago. This rifting formed an oceanic basin along the present northeastern side of the African craton. This was a part of the Mozambique Ocean that separated the facing margins of East and West Gondwana. Alternatively there may have been more than one ocean basin, separated by rifted microcontinental plates such as the Afif microcontinental plate.

The island arc formation stage (950–715 million years ago) is characterized by the formation of oceanic island arcs in the oceanic basins formed in the first stage. The stratigraphic records of volcanic and sedimentary rocks in the Asir, Al-Hijaz, and some parts of the Midyan terranes present rocks with ages between 900 and 800 million years. These rocks are of mafic or bimodal composition, and are considered products of early island arcs, particularly in the Asir terrain. These rocks show mixing or the involvement of rocks and fragments formed in the previous stage of rifting of the African craton.

The formation of island arc systems did not take place at the same time, but rather different arc systems evolved at different times. The Hijaz terrain is considered the oldest island arc, formed between 900 and 800 million years ago. This terrane may have encountered continental fragments now represented by the Khamis Mushayt Gneiss and Hali Schist, which are considered parts of, or derived from, the old continental crust from the previous stage of rifting.

Later on in this stage (760–715 million years ago) three island arc systems apparently formed simultaneously. These are the Hijaz, Tarib, and Taif island arc systems. These island arc systems evolved and formed three crustal plates, including the Asir, Hijaz, and Midyan plates. Later in this stage the Amar Andean arc formed between the Afif plate and Ar-Rayn plate, and it is considered part of the Ar-Rayn plate. Oceanic crustal plateaus may have been involved in the formation of the oceanic crustal plates in this stage.

In the collision stage (715–640 million years ago) the five major terranes that formed in the previous stages were swept together and collisions took place along the four suture zones mentioned above. The collision along these suture zones did not take place at the same time. For example, the collision along the Hijaz and Taif arcs occurred around 715 million years ago, and the collision along Bir Omq suture zone took place between 700 and 680 million years ago, while the island arc magmatic activity in the Midyan terrain continued until 600 million years ago. The collision along Nabitah suture zone was diachronous along strike. The collision started in the northern part of the Nabitah suture between Afif and the Hijaz terranes at about 680 to 670 million years ago, and at the same time the southern part of the suture zone was still experiencing subduction. Further collision along the Nabitah suture zone shut off the arc in the south, and the Afif terrain collided with the Asir terrain. As a result, the eastern Afif plate and the western island arc plates of the Hijaz and Asir were completely sutured along the Nabitah orogenic belt by 640 Ma. In this stage three major magmatic arcs developed, and later on in this stage they were shut off by further collision. These arcs include the Furaih magmatic arc that developed on the northern part of the Nabitah suture zone and on the southeastern part of the Hijaz plate, the Sodah arc that developed on the eastern part of the Afif plate, and an Andean-type arc on the eastern part of the Asir plate.

The Ar-Rayn collisional orogeny along the Amar suture was between the two continental plates of Afif and Ar-Rayn, and took longer than any other collisions in the shield (from 700 to 630 million years ago). Many investigators suggest that the Ar-Rayn terrain is part of a bigger continent, which extends under the eastern Phanerozoic cover and is exposed in Oman. This terrane may have collided with or into the Arabian shield from the east and was responsible for the development of the Najd left-lateral fault system.

By 640 million years ago the five major terranes had collided with each other, forming the four mentioned suture zones, and the Arabian shield was stabilized. Since then, the shield behaved as one lithospheric plate until the rifting of the Red Sea. Orogenic activity inside the Arabian shield, however, continued for a period of about 80 million years after collision, during which time the Najd fault system developed as the last tectonic event in the Arabian shield in the late Proterozoic Era.

After development of the Najd fault system, tectonic activity in the Arabian shield ended, and the Arabian-Nubian shield subsided and was peneplained, as evidenced by the existence of epicontinental Cambro-Ordovician sandstone covering many parts of the shield in the north and the south. The stratigraphic records of the Phanerozoic cover show that the Arabian shield has been tectonically stable with the exception of ophiolite obduction in Oman and collision along the margins of the plate during the closure of the Tethys Sea until rifting of the Red Sea in the Tertiary.

PHANEROZOIC COVER OF THE ARABIAN SHIELD: OMAN MOUNTAINS

The northern and eastern parts of the Arabian Peninsula are composed of a series of sandstone, limestone, siltstone, evaporates, and rare volcanic rocks deposited in the Paleozoic, Mesozoic, and Cenozoic. These rocks are known as the Arabian platform, the youngest rocks of which consist of unconsolidated sands, silts, gravels, and sabkha deposits such as those that cover much of Kuwait, the Gulf States, and upper layers on the Arabian platform.

The Phanerozoic rocks of the Arabian platform dip to the east very gently, and gradually increase in thickness from a few feet where they overlie the Precambrian Arabian shield in the east, to more than six miles (10 km) in thickness in Oman, eastern Saudi Arabia, and beneath Kuwait. Since some of the thickest sections of the Arabian platform are known from Oman, these rocks are described in detail using the exposures in the northern Oman (Hajar) Mountains as examples.

The Oman, or Hajar, Mountains in northern Oman and the United Arab Emirates are located on the northeastern margin of the Arabian plate,

60–120 miles (100–200 km) from the active deformation front in the Gulf of Oman between Arabia and the Makran accretionary wedge of Asia. They are made up of five major structural units ranging in age from Precambrian to Miocene. These include the pre-Permian basement, Hajar Unit, Hawasina nappes, Semail ophiolite and metamorphic sole, and postnappe structural units.

The Hajar Mountains reach up to 1.8 miles (three km) high, displaying many juvenile topographic features such as straight mountain fronts and deep, steep-walled canyons that may reflect active tectonism causing uplift of these mountains. The present height and ruggedness of the Hajar mountainous area is a product of Cretaceous ophiolite obduction, Tertiary extension, and rejuvenated uplift and erosion that was initiated at the end of the Oligocene and continues to the present. The Sayq Plateau southwest of Muscat is 1.2–1.8 miles (2–3 km) in elevation. Jabal Shams on the margin of the Sayq Plateau is the highest point in Arabia, rising more than 1.8 miles (3 km) in the central Hajar Mountains. The heights decrease gradually northward, reaching 1.2 miles (2 km) on the Musandam peninsula. There the mountain slopes drop directly into the sea.

Pre-Permian rocks are exposed mainly in the Jabal Akhdar, Saih Hatat, and Jabal J'Alain areas. The oldest structural unit includes a Late Proterozoic basement gneiss correlative with the Arabian-Nubian shield, overlain by a Late Proterozoic/Ordovician volcano-sedimentary sequence. The latter is divided into the Late Proterozoic/Cambrian Huqf Group and the Ordovician Haima Group. The Huqf Group is composed mainly of diamictites, siltstone, graywacke, dolostone, and intercalated mafic volcanics. The Ordovician Haima Group consists of a series of sandstones, siltstones, quartzites, skolithos-bearing sandstones, and shales, interpreted as subtidal to intertidal deposits.

The Hajar Unit represents the main part of the Permian/Cretaceous Arabian platform sequence that formed on the southern margin of the Neo-Tethys Ocean. These carbonates form most of the rugged peaks of Jabal Akhdar, form a rim around the southwestern parts of Saih Hatat, and continue in several thrust sheets in the Western Hajar region. They are well exposed on the Musandam peninsula. The Hajar Unit contains the Akhdar, Sahtan, Kahmmah, and Waisa Groups of mainly carbonate lithologies, overlain by the Muti Formation in the eastern Hajar, and the equivalent Ruus al Jibal, Elphinstone, Musandam, and Thamama Groups on the Musandam peninsula.

The Hawasina nappes consist of a series of Late Permian/Cretaceous sedimentary and volcanic rocks deposited in the Hawasina basin, between the Arabian continental margin and the open Neo-Tethys Ocean. The Hawasina nappes include the Hamrat Duru, Al Aridh, Kawr, and Umar Groups. Chaotic deposits of the Baid Formation are interpreted as a foundered carbonate platform. The Hamrat Duru Group includes radiolarian chert, gabbro, basaltic and andesitic pillow lava, carbonate breccia, shale, limestone, and sandstone turbidites. The Al Aridh Group contains an assemblage of basaltic andesite, hyaloclastite and pillow lavas, micrites, pelagic carbonates, carbonate breccias, chert, and turbidites. This is overlain by the Kawr Group, which includes basalts, andesites, and shallow marine carbonates. The Umar Group contains basaltic and andesitic pillow lavas, cherts, carbonate breccias, and micrites.

SEMAIL OPHIOLITE

The Semail nappe forms the largest ophiolitic sheet in the world, and it is divided into numerous blocks in the northern Oman Mountains. The Semail ophiolite contains a complete classic ophiolite stratigraphy, although parts of it are unusual in that it contains two magmatic sequences, including upper and lower units. The upper magmatic unit grades downward from radiolarian cherts and umber of the Suhaylah Formation, to basaltic and andesitic pillow lavas locally intruded by trondhjemites, through a sheeted diabase dike unit, and into massive and layered gabbros, and finally into cumulate gabbro, wehrlite, dunite, and clinopyroxenite. This upper magmatic sequence grades down from basaltic pillow lavas, into a sheeted dike complex, through isotropic then layered gabbros, then into cumulate gabbro and dunite. The Mohorovicic discontinuity is well exposed throughout the northern Oman Mountains, separating the crustal and the mantle sequences. The mantle sequence consists of tectonized harzburgite, dunite, and lherzolite, cut by pyroxenite dikes, and local chromite pods.

The metamorphic sole, or dynamothermal aureole, of the Semail ophiolite formed through metamorphism of rocks immediately under the basal thrust, heated and deformed during emplacement of the hot allochthonous sheets. In most places it consists of two units including a lower metasedimentary horizon, and an upper unit of banded amphibolites. The metamorphic grade increases upward through the unit to upper amphibolite facies near the contact with the Semail nappe.

Postnappe units consist of Late Cretaceous and Tertiary rocks. The Cretaceous Aruma Group consists of a lower unit of Turonian-Santonian polymict conglomerate, sandstone and shale of the Qahlah Formation, and an upper unit of Campanian-Maastrictian marly limestone and polymict breccia (Thaqab Formation). The Tertiary Hadhramaut

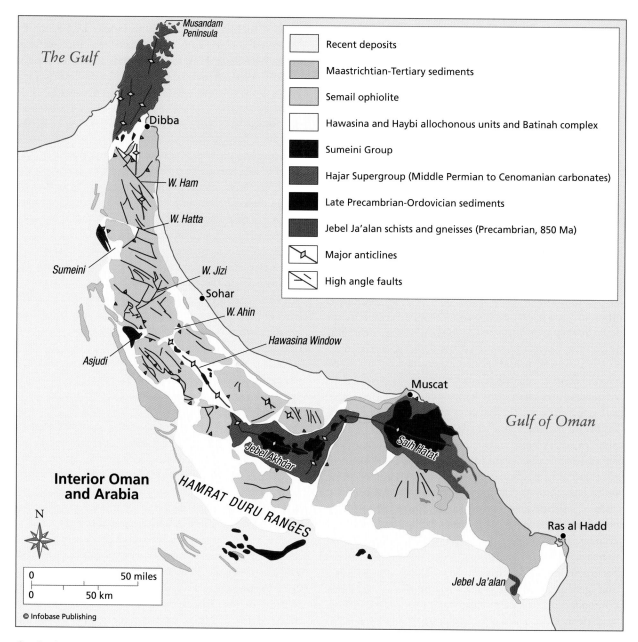

Geologic map of Oman. The world's largest ophiolite, the semail, is shown in green colors.

Group comprises Paleocene to Eocene limestones, marly limestone, dolostone, conglomerate, and sandstones that outcrop along the southern edge of the Batinah coastal plain at the border with the northeast flank of the Hajar Mountains.

QUATERNARY GEOLOGY OF NORTHERN ARABIA

Several levels of Quaternary fluvial terraces are preserved along the flanks of the Hajar Mountains in Oman. These can be divided in most places into an older lower-cemented terrace and an upper younger-uncemented terrace group. The lower-cemented terrace is one of the youngest geological units and has been used as a time marker to place constraints on the ages of structures. The terraces are younger than and unconformably overlie most faults and folds, but in several places faults and fracture intensification zones cut through the Quaternary terraces, providing some of the best evidence for the young age of some of the faults along the northeastern edge of the Arabian plate. These terraces grade both northward and southward into coalesced alluvial fans, forming bajada flanking the margins of the mountains. The northern alluvial plains grade into a narrow coastal plain along the Gulf of Oman.

The Oman (Hajar) Mountains are situated at the northeastern margin of the Arabian plate. This plate is bounded to the south and southwest by the active spreading axes of the Gulf of Aden and the Red Sea. On the east and west its border is marked by transcurrent fault zones of the Owen Fracture Zone and the Dead Sea Transform. The northern margin of the plate is marked by a complex continent-continent to continent-oceanic collision boundary along the Zagros and Makran fold and thrust belts.

Rocks of the Hajar Supergroup preserve a history of Permian through Cretaceous subsidence of the Arabian platform on the margin of the Neo-Tethys Ocean. Formations that now comprise the Hawasina nappes have biostratigraphic ages of 260–95 Ma, interpreted to have been deposited on the continental slope and in abyssal environments of the Neo-Tethys Ocean. By about 100 Ma ago, spreading in the Neo-Tethys generated the oceanic crust of the Semail ophiolite, which was detached in the oceanic realm and thrust over adjacent oceanic crust soon after its formation. Metamorphic ages for the initiation of thrusting range from 105 Ma to 89 Ma. The ophiolitic nappes moved toward the Arabian margin, forming the high-grade metamorphic sole during transport, and progressively scraping off layers of the Hawasina sediments and incorporating them as thrust nappes to the base of the ophiolite. The ophiolite reached the Arabian continental margin and was thrust over it before 85–75 Ma, as indicated by greenschist facies metamorphism in the metamorphic sole and by deformation of the Arabian margin sediments. Initial uplift of the dome-shaped basement cored antiforms of Jabel Akhdar and Saih Hatat may have been initiated during the late stages of the collision of the ophiolite with the Arabian passive margin, and may have been localized by preexisting basement horst and graben structures. The location and geometry of these massive uplifts is probably controlled by basement ramps. Uplift of these domes was pronounced during the Oligocene/Miocene, as shown by tilting of Late Cretaceous/Tertiary formations on the flanks of the domes. Uplift of the domes may have begun in the Oligocene, resulting from the propagation of a fault beneath the southern limbs of the folds. The uplift of the domes includes a complex history, involving several different events. Some uplift of the domes continues at present, whereas much of the Batinah coastal plain is subsiding.

In most of the Hajar Mountains, the Hawasina nappes structurally overlie the Hajar Supergroup, and form a belt of north or northeastward dipping thrust slices. On the southern margins of Jabal Akhdar, Saih Hatat, and other domes, however, the Hawasina form south-dipping thrust slices. Major valleys typically occupy the contact between the Hajar Supergroup and the Hawasina nappes, because of the many, easily erodable shale units within the Hawasina nappes. Several very large [~6 mile (10 km) scale] allochthonous limestone blocks known as the "Oman Exotics" are also incorporated into melange zones within the Hawasina nappes. These form light-colored, erosionally resistant cuestas, including Jabal Kawr and several smaller mountains south of Al Hamra.

South and southwest of the belt of ophiolite blocks, sediments of the Hamrat Duru Group are complexly folded and faulted in a regional foreland-fold-thrust belt and then grade into the Suneinah foreland basin. The Hamrat Duru rocks include radiolarian cherts, micritic limestones, turbiditic sandstones, shales, and calcarenite, all complexly folded and thrust faulted in an 18.5-mile (30-km)-wide fold/thrust belt.

A belt of regional anticlinal uplifts brings up carbonates of the Hajar Supergroup in the central part of the basin, as exposed at Jabal Salakh. These elongate anticlinal domes have gentle to moderate dips on their flanks, and are cut by several thrust faults that may be linked to a deeper system. This could be a blind thrust, or the folds could be flower structures developed over deep strike-slip faults. South of the Jabal Salakh fold belt, the surface is generally flat and covered by Miocene/Pliocene conglomerates of the Barzaman Formation, and cut by an extensive network of Quaternary channels of the active alluvial plain.

Tertiary/Quaternary uplift of the northern Oman Mountains may account for the juvenile topography of the area. One of the best pieces of evidence for young uplift of the Northern Oman Mountains comes from a series of uplifted Quaternary marine terraces, best exposed in the Tiwi area 31–62 miles (50–100 km) southeast of Muscat. The uplift is related to the contemporaneous collision between the northeastern margin of the Arabian plate and the Zagros fold belt and the Makran accretionary prism. The Hajar Mountains lie on the active forebulge of this collision, and the fault systems are similar to those found in other active and ancient forebulge environments. The amount of Quaternary uplift, estimated between 300 and 1,600 feet (100–500 m), is also similar to uplift in other forebulge environments developed on continental margins. This Quaternary uplift is superimposed on an older, Cretaceous/Tertiary (Oligocene) topography.

CENOZOIC GEOLOGY OF NORTHERN ARABIAN PLATE: KUWAIT

Kuwait is located in the northwest corner of the Arabian Gulf between 28°30′ and 30° north latitude, and 46°30′ to 48°30′ east longitude. It is approximately 10,700 square miles (17,818 km²); the extreme north-south distance is 120 miles (200 km), the east-west distance is 100 miles (170 km).

To the south it shares a border with Saudi Arabia; to the west and north, it shares a border with Iraq. The semiarid climate of Kuwait is characterized by two seasons: a long, hot, humid summer, and a relatively cold, short winter. Summer temperatures range from 84.2 to 113°F (29–45°C), with relatively high humidity. The prevailing shamal winds from the northwest bring severe dust and sand storms from June to early August, with gusts up to 60 miles per hour (100 km/hr). Winter temperatures range from 46.4 to 64.4°F (8–18°C). Occasionally samum winds (meaning poison wind, describing the extremely hot and dry winds from the Sahara that can reach 130°F [55°C] bring more heat to people's bodies than can be removed by transpiration, and they lead to many cases of heatstroke. These winds come from the southwest during November. Annual precipitation averages 4.5 inches (11.4 mm) and rapidly infiltrates the sandy soil, leaving no surface water except in a few depressions. Most of the limited rainfall occurs in sudden squalls during the winter season.

Most of Kuwait is a flat, sandy desert. There is a gradual decrease in elevation from an extreme of 980 feet (300 m) in the southwest near Shigaya to sea level. The southeast is generally lower than the northwest. There are no mountains or rivers. The country can be divided into roughly two parts, including a hard, flat stone desert in the north with shallow depressions and low hills running northeast to southwest. The principal hills in the north are Jal al-Zor (475 feet or 145 m) and the Liyah ridge. Jal al-Zor runs parallel to the northern coast of Kuwait Bay for a distance of 35 miles (60 km). The southern region is a treeless plain covered by sand. The Ahmadi Hills (400 feet or 125 m) are the sole exception to the flat terrain. Along the western border with Iraq lies Wadi Al-Batin, one of the few valleys in Kuwait. The only other valley of note is Ash Shaqq, a portion of which lies within the southern reaches of the country. Small playas, or enclosed basins, are covered intermittently with water. During the rainy season they may be covered with dense vegetation; during the dry season they are often devoid of all vegetation. Most playas range between 650 and 985 feet (200–300 m) in length, with depths from 16 to 50 feet (5–15 m).

There are few sand dunes in Kuwait, occurring mainly near Umm Al-Neqqa and Al-Huwaimiliyah. The dunes at Umm Al-Neqqa are crescent-shaped barchan dunes with an average width of 550 feet (170 m) and average height of 25 feet (8 m). Those near Al-Huwaimiliyah are smaller, averaging 65 feet (20 m) wide and 7 feet (2 m) height, and are clustered into longitudinal dune belts. Both mobile and stable sand sheets occur in Kuwait. A major mobile sand belt crosses Kuwait in a northwest to southeast direction, following the prevailing wind pattern. Smaller sheets occur in the Al-Huwaimiliyah area, in the Al-Qashaniyah in the northeast, and in much of the southern region.

During the past few years overgrazing and an increase of motor vehicles in the desert have caused great destruction to the desert vegetation. Stabilized vegetated sheets have changed to mobile sheets as the protective vegetation is destroyed. The largest stabile sheet occurs at Shugat Al-Huwaimiliyah. Recently smaller sheets have begun to develop at Umm Al-Neqqa and Burgan oil field owing to an increase in desert vegetation resulting from a prohibition of traffic.

Kuwait Bay is a 25-mile (40-km) long indentation of marshes and lagoons. The coast is mostly sand interspersed with sabkhas and gravel. Sabkhas are flat, coastal areas of clay, silt, and sand often encrusted with salt. The northern portion of the bay is very shallow, averaging fewer than 15 feet (5 m). This part of the shore consists mostly of mud flats and sandy beaches. The more southern portion is relatively deep, with a bed of sand and silicic deposits. Most of the ports are situated in the southern area.

Kuwaiti territory includes 10 islands. Most are covered by scrub and a few serve as breeding grounds for birds. Bubiyan, the largest island, measuring approximately 600 square miles (1,000 km²), is a low, level bare piece of land with mud flats along much of its north and west coasts that are covered during high tides. It is connected to the mainland by a concrete causeway. To its north lies Warba, another low-lying island covered with rough grass and reeds. East of Kuwait Bay and on the mud flats extending from Bubiyan lie three islands: Failaka, Miskan, and Doha. Failaka is the only inhabited island belonging to Kuwait. A small village, located near an ancient shrine, is set on a 30-foot (9-m) hill at the northwest point of the island. The rest of the island is flat with little vegetation. There are a few trees in the center of the island, and date trees are grown in the village. West of Kuwait City in the bay are two islets: Al-Qurain and Umm Al-Naml. On the south side of the gulf lie three more small islands: Qaruh, Kubbar, and Umm Al-Maradim. The last two are surrounded by reefs on three sides.

Kuwait occupies one of the most petroleum-rich areas in world, situated in a structurally simple region on the Arabian platform in the actively subsiding foreland of the Zagros Mountains to the north and east. Principal structural features of Kuwait include two subsurface arches (Kuwait arch and Dibdibba arch) and the fault-bounded Wadi Al-Batin. Faults defining Wadi Al-Batin are related to Tertiary extension in the region. The Kuwait and Dibdibba arches have no geomorphic expression, whereas the younger Bahra anticline and Ahmadi ridge have a surface expression and are structurally superimposed on the

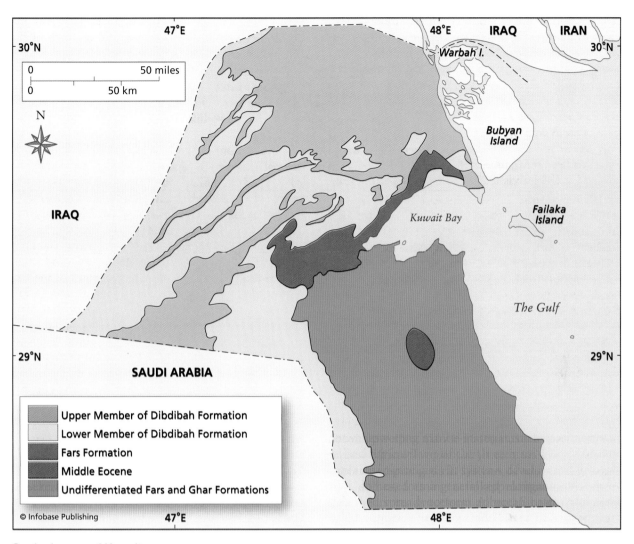

Geologic map of Kuwait

Kuwait arch. Major hydrocarbon accumulations are associated with the Kuwait arch. The subsurface stratigraphy of Kuwait includes a nearly continuous section of Arabian platform sediments ranging in age from Cambrian through Holocene, although the pre-Permian rocks are poorly known. The Permian through Miocene section is 3.5–4 miles (6–7 km) thick in Kuwait but thickens toward the northeast. These rocks include continental and shallow marine carbonates, evaporites, sandstones, siltstones, and shales, with less common gravels and cherts. Plio-Pleistocene sand and gravel deposits of the Dibdibba Formation outcrop in northwestern Kuwait, and Miocene sands, clay, and nodular limestone of the Fars and Ghar Formation outcrop in the southeast. A small area of Eocene limestone and chert (Dammam Formation) outcrops south of Kuwait City on the Ahmadi Ridge.

The structural arches in Kuwait are part of a regional set of north-trending arches known as the Arabian folds, along which many of the most important oil fields in the Arabian Gulf are located. These arches are at least mid-Cretaceous. The orientation of the Arabian folds has been interpreted to be inherited from older structures in the Precambrian basement, with possible amplification from salt diapirism. The north-south trends may continue northward beneath the Mesopotamian basin and the Zagros fold belt.

The northwest trending anticlinal structures of the Ahmadi ridge and Bahra anticline are younger than the Arabian folds, and related to the Zagros collision, initiated in post-Eocene times. These younger folds seem to have a second-order control on the distribution of hydrocarbon reservoirs in Kuwait, as oil wells (and after the First Gulf War in 1990, oil lakes) are concentrated in northwest trending belts across the north-striking Kuwait arch. The Kuwait arch has a maximum structural relief in the region between Burgan and Bahra, with closed structural contours around the Wafra, Burgan, Magwa, and Bahra areas, and a partial closure indicating a domal structure beneath Kuwait City and Kuwait Bay. The superposition of the Kuwait arch and the shallow anticlinal structure of the Ahmadi ridge forms a total structural relief of at least one mile (1.6 km).

The northwest-trending Dibdibba arch represents another subsurface anticline in western Kuwait. The ridge is approximately 45 miles (75 km) long, and is an isolated domal structure, but has not to date yielded any significant hydrocarbon reservoirs.

Wadi Al-Batin is a large valley, 4–6 miles (7–10 km) wide, with relief of up to 185 feet (57 m). In the upper valley of the wadi, the valley sides are steep, but in southwestern Kuwait few ravines have steep walls taller than 15 feet (5 m). The wadi has a length of more than 45 miles (75 km) in Kuwait, and extends 420 miles (700 km) southwestward into Saudi Arabia, where it is referred to as Wadi Ar-Rimah. The ephemeral drainage in the wadi drains from the southwest, and has transported Quaternary and Tertiary gravels consisting of igneous and metamorphic rock fragments from the Saudi Arabian and Syrian deserts during Pleistocene pluvial episodes. The wadi widens toward the northeast and becomes indistinguishable from its surroundings northwest of Kuwait City. Ridges made of Dibdibba gravel define paleodrainage patterns of a delta system draining Wadi Al-Batin, and many of these gravel ridges stand out as prominent lineaments. Some of these gravel ridges are marked by faults on at least one side, suggesting a structural control on the drainage pattern.

Numerous small and several relatively large faults are revealed on seismic reflection lines across the wadi, and hydrological pumping tests show a break in the drawdown slope at the faults. The steep Miocene–late Eocene faults parallel to the wadi have displaced the block in the center of the wadi upward by 15–20 feet (25–35 m) relative to the strata outside the wadi, and those displacements die out toward the northeast.

See also CONVERGENT PLATE MARGIN PROCESSES; DESERTS; GONDWANA, GONDWANALAND; OPHIOLITES; PALEOZOIC; PETROLEUM GEOLOGY; PHANEROZOIC; PROTEROZOIC.

FURTHER READING

Abdelsalam, Mohamed G., and Robert J. Stern. "Sutures and Shear Zones in the Arabian-Nubian Shield." *Journal of African Earth Sciences* 23, no.3 (1996): 289–310.

Al-Lazki, A. I., Don Seber, and Eric Sandvol. "A Crustal Transect Across the Oman Mountains on the Eastern Margin of Arabia." *GeoArabia* 7, no. 1 (2002): 47–78.

Al-Shanti, A. M. S. *The Geology of the Arabian Shield.* Jeddah, Saudi Arabia: Center for Scientific Publishing, King AbdlAziz University, 1993.

Boote, D. R. D., D. Mou, and R. I. Waite. "Structural Evolution of the Suneinah Foreland, Central Oman Mountains." In *The Geology and Tectonics of the Oman Region,* edited by A. H. F. Robertson, M. P. Searle, and A. C. Reis. Geological Society of London Special Publication 49 (1990): 397–418.

Brown, Glen F., Dwight L. Schmidt, and Curtis A. Huffman Jr. *Geology of the Arabian Peninsula, Shield Area of Western Saudi Arabia.* United States Geological Survey Professional Paper 560-A, 1989.

Directorate of General Geologic Surveys and Mineral Investigation. Tectonic map of Iraq, scale 1: 1,000,000, 1984.

Geological Society of Oman. "Information on the Geology of Oman." Available online. URL: http://www.gso.org.om/. Accessed October 9, 2008.

Glennie, Ken. *The Geology of the Oman Mountains: An Outline of Their Origin.* 2nd ed. Buchs, U.K.: Scientific Press. 2005.

Glennie, Ken W., M. G. A. Boeuf, M. W. Hughes-Clarke, Stuart M. Moody, W. F. H. Pilar, and B. M. Reinhardt. *The Geology of the Oman Mountains.* Amsterdam, Netherlands: Verhandelingen van het Koninklijk Nederlands Geologisch Mijnbouwkundig Genootschap, deel 31 (NE ISSM 0075-6741) (1974).

Johnson, Peter R., Erwin Scheibner, and Alan E. Smith. "Basement Fragments, Accreted Tectonostratigraphic Terranes, and Overlap Sequences: Elements in the Tectonic Evolution of the Arabian Shield, Geodynamics Series." *American Geophysical Union* 17 (1987): 324–343.

Kusky, Timothy M., Mohamed Abdelsalam, Robert Tucker, and Robert Stern, eds. "Evolution of the East African and Related Orogens, and the Assembly of Gondwana." *Special Issue of Precambrian Research* (2003): 81–85.

Kusky, Timothy M., and Mohamed Matsah. "Neoproterozoic Dextral Faulting on the Najd Fault System, Saudi Arabia, Preceded Sinistral Faulting and Escape Tectonics Related to Closure of the Mozambique Ocean." In *Proterozoic East Gondwana: Supercontinent Assembly and Break-up,* edited by M. Yoshida, Brian F. Windley, S. Dasgupta, and C. Powell, 327–361. London: Geological Society of London, Special Publication, 2003.

Kusky, Timothy M., Cordula Robinson, and Farouk El-Baz. "Tertiary and Quaternary Faulting and Uplift of the Hajar Mountains of Northern Oman and the U.A.E." *GeoArabia* 162 (2005): 1–18.

Kuwait Oil Company. *Geological Map of the State of Kuwait, Scale 1:250,000.* 1981.

Milton, D. I. "Geology of the Arabian Peninsula, Kuwait." Washington, D.C.: United States Government Printing Office, Reston, Va.: Geological Survey Professional Paper 560-D (1967).

Saudi Geological Survey. "The National Geologic Survey of the Kingdom of Saudi Arabia home page." Available online. URL: http://www.sgs.org.sa/. Updated August 28, 2008.

Searle, Mike, and J. Cox. "Tectonic Setting, Origin, and Obduction of the Oman Ophiolite." *Geological Society of America Bulletin* 111 (1999): 104–122.

Stern, Robert J. "Arc Assembly and Continental Collision in the Neoproterozoic East African Orogen: Implications for Consolidation of Gondwanaland." *Annual Review of Earth and Planetary Sciences* 22 (1994): 319–351.

Stocklin, Jovan, and M. H. Nabavi. Tectonic Map of Iran, Geologic Survey of Iran, scale 1:2,500,000, 1973.

Stoeser, Douglas B., and Camp, Victor E. "Pan-African Microplate Accretion of the Arabian Shield." *Geological Society of America Bulletin* 96 (1985): 817–826.

Stoeser, Douglas B., and John S. Stacey. "Evolution, U-Pb Geochronology, and Isotope Geology of the Pan-African Nabitah Orogenic Belt of the Saudi Arabian Shield." In *The Pan-African Belts of Northeast Africa and Adjacent Areas,* edited by S. El Gaby and R. O. Greiling, 227–288. Braunschweig, Germany: Friedr Vieweg and Sohn, 1988.

U.S. Geological Survey. Central Region Energy Resources Team. Maps Showing Geology, Oil and Gas Fields and Geological Provinces of the Arabian Peninsula, by Richard M. Pollastro, Amy S. Karshbaum, and Roland J. Viger, U.S. Geological Survey Open-File Report 97–470B, version 2. Available online. URL: http://pubs.usgs.gov/of/1997/ofr-97-470/OF97-470B/arabGmap.html. Updated February 5, 2008.

Warsi, Waris E. K. "Gravity Field of Kuwait and Its Relevance to Major Geological Structures." *American Association of Petroleum Geologists Bulletin* 74 (1990): 1610–1622.

Archean (Archaean) Earth's first geological eon for which there is an extensive rock record, the Archean also preserves evidence for early primitive life-forms. The Archean is second of the four major eons of geological time: the Hadean, Archean, Proterozoic, and Phanerozoic. Some time classification schemes use an alternative division of early time, in which the Hadean, Earth's earliest eon, is considered the earliest part of the Archean. The Archean encompasses the one and one-half billion-year-long (Ga = giga année, or 10^9 years) time interval from the end of the Hadean eon to the beginning of the Proterozoic eon. In most classification schemes it is divided into three parts called eras, including the Early Archean (4.0–3.5 Ga), the Middle Archean (3.5–3.1 Ga), and the Late Archean, ranging up to 2.5 billion years ago.

Gneisses are strongly deformed rocks with a strong layering formed by the parallel alignment of deformed and flat minerals. The oldest known rocks on Earth are the 4.0 billion-year-old Acasta gneisses from northern Canada that span the Hadean-Archean boundary. Single zircon crystals from the Jack Hills and Mount Narryer in western Australia have been dated to be as old as 4.1–4.3 billion years. The oldest well-documented and extensive sequence of rocks on Earth is the Isua belt, located in western Greenland, estimated to be 3.8 billion years old. Life on Earth originated during the Archean, with the oldest known fossils coming from the 3.5 billion-year-old Apex chert in western Australia, and possible older traces of life found in the 3.8 billion-year-old rocks from Greenland.

Archean and reworked Archean rocks form more than 50 percent of the continental crust and are present on every continent. Most Archean rocks are found in cratons or as tectonic blocks in younger orogenic belts. Cratons are low-relief, tectonically stable parts of the continental crust that form the nuclei of many continents. Shields are the exposed parts of cratons, other parts of which may be covered by younger platformal sedimentary sequences. Archean rocks in cratons and shields are generally divisible into a few basic types. Relatively low-metamorphic-grade greenstone belts consist of deformed metavolcanic and metasedimentary rocks. Most Archean plutonic rocks are quartz and feldspar–dominated tonalites, trondhjemites, granodiorites, and granites that intrude or are in structural contact with strongly deformed and metamorphosed sedimentary and volcanic rocks in greenstone belt associations. Together these rocks form the granitoid-greenstone association that characterizes many Archean cratons. Granite-greenstone terranes are common in parts of the Canadian shield, South America, South Africa, and Australia. Low-grade cratonic basins are preserved in some places, including southern Africa and parts of Canada. High-grade metamorphic belts are also common in Archean cratons, and these generally include granitic, metasedimentary, and metavolcanic gneisses that were deformed and metamorphosed at middle to deeper crustal levels. Some well-studied Archean high-grade gneiss terranes include the Lewisian and North Atlantic Province, the Limpopo belt of southern Africa, the Hengshan of north China, and parts of southern India.

The Archean witnessed some of the most dramatic changes in Earth in the history of the planet. During the Hadean, the planet experienced frequent impacts of asteroids, some of which were large enough to melt parts of the outer layers of Earth and vaporize the atmosphere and oceans. Any attempts by life to get a foothold on the planet in the Hadean would have been difficult, and if any organisms were to survive this early bombardment, they would have to have been sheltered in some way from these dra-

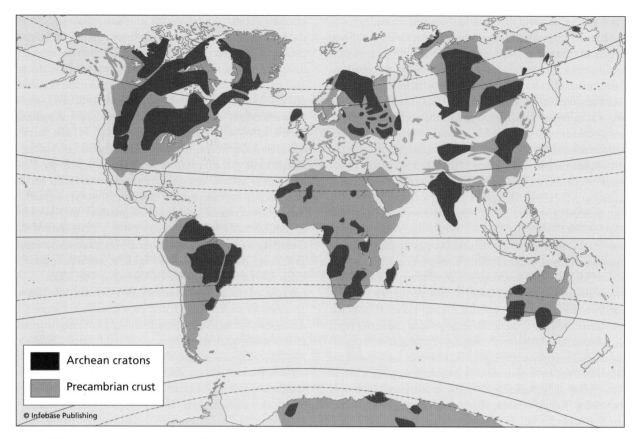

Map of the world showing the distribution of Archean cratons, Precambrian crust, and Phanerozoic rocks (beige) *(modified from Timothy Kusky and Ali Polat)*

matic changes. Early atmospheres of Earth were blown away by asteroid and comet impacts, and by strong solar winds from an early T-Tauri phase of the Sun's evolution. Free oxygen was either not present or present in much lower concentrations, and the atmosphere evolved slowly to a more oxygenic condition.

Earth was also producing and losing more heat during the Archean than in more recent times, and the patterns, styles, and rates of mantle convection and the surface style of plate tectonics must have reflected these early conditions. Heat was still left over from early accretion, core formation, late impacts, and decay of some short-lived radioactive isotopes such as iodine 129. In addition, the main heat-producing radioactive decay series were generating more heat then than now, since more of these elements were present in older half-lives. In particular uranium 235, uranium 238, thorium 232, and potassium 40 were cumulatively producing two to three times as much heat in the Archean as at present. Since scientists know from the presence of rocks that formed in the Archean that the planet was not molten then, this heat must have been lost by convection of the mantle. It is possible that the temperatures and geothermal

gradients were 10–25 percent hotter in the mantle during the Archean, but most of the extra heat was likely lost by more rapid convection and by the formation and cooling of oceanic lithosphere in greater volumes. The formation and cooling of oceanic lithosphere is presently the most efficient mechanism of global heat loss through the crust, and this mechanism was even more efficient in times of higher heat production. A highly probable scenario for removing the additional heat is that more ridges were present, producing thicker piles of lava, and moving at faster rates in the Archean compared to the present. There is currently much debate and uncertainty about the partitioning of heat loss among these mechanisms, and it is also possible that changes in mantle viscosity and plate buoyancy would have led to slower plate movements in the Archean as compared with the present.

GRANITE-GREENSTONE TERRANES

Archean granitoid-greenstone terrains are one of the most distinctive components of Archean cratons. About 70–80 percent of the Archean crust consists of granitoid material, most of which are compositionally tonalites and granodiorites (comprised of the minerals quartz, feldspar, and biotite). Many of these are intrusive into metamorphosed and deformed volcanic and sedimentary rocks in greenstone belts. Greenstone belts are generally strongly deformed and metamorphosed, linear to irregularly shaped assemblages of volcanic and sedimentary rocks. They derive their name from the green-colored metamorphic minerals chlorite and amphibole, reflecting the typical greenschist to amphibolite facies metamorphism of these belts. Early 19th century South African workers preferred to use the name schist belt for this assemblage of rocks, in reference to the generally highly deformed nature of the rocks. Volcanic rocks in greenstone belts most typically include basalts flows, many of which show pillow structures where they are not too intensely deformed, and lesser amounts of ultramafic, intermediate, and felsic rocks. Ultramafic volcanic rocks with quench-textures and high magnesium oxide (MgO) contents, known as komatiites, are much more abundant in

Satellite image of the Pilbara craton in northwestern Australia, showing prominent light-colored igneous batholiths that intruded darker colored greenstone belts consisting of metamorphosed and strongly deformed volcanic and sedimentary rocks *(Nick Short)*

Archean greenstone belts than in younger orogenic belts, but they are generally only a minor component of greenstone belts. Some literature leads readers to believe that Archean greenstone belts are dominated by abundant komatiites; however, this is not true. There have been an inordinate number of studies of komatiites in greenstone belts since they are such an unusual and important rock type, but the number of studies does not relate to the abundance of the rock type. Sedimentary rocks in greenstone belts are predominantly greywacke-shale sequences (or their metamorphic equivalents), although conglomerates, carbonates, cherts, sandstones, and other sedimentary rocks are found in these belts as well.

Suites of granitoid rock now deformed and metamorphosed to granitic gneisses typically intrude the volcanic and sedimentary rocks of the greenstone belts. The deformation of the belts has in many cases obscured the original relationships between many greenstone belts and gneiss terrains. Most of the granitoid rocks appear to intrude the greenstones, but in some belts older groups of granitic gneisses have been identified. In these cases it has been important to determine the original contact relationships between granitic gneisses and greenstone belts, as this relates to the very uncertain tectonic setting of the Archean greenstones. If contact relationships show that the greenstone belts were deposited unconformably over the granitoid gneisses, then it can be supposed that greenstone belts represent a kind of continental tectonic environment unique to the Archean. In contrast, if contact relationships show that the greenstone belts were faulted against or thrust over the granitoid gneisses, then the greenstone belts may be allochthonous (far-traveled) and represent closed ocean basins, island arcs, and other exotic terrains similar to orogenic belts of younger ages.

Before the mid 1980s and 1990s, many geologists believed that many if not most greenstone belts were deposited unconformably over the granitoid gneisses, based on a few well-preserved examples at places including Belingwe, Zimbabwe; Point Lake, Yellowknife, Cameron River, and Steep Rock Lake, Canada; and in the Yilgarn of western Australia. However, more recent mapping and structural work on these contact relationships have revealed that all of them have large-scale thrust fault contacts between the main greenstone belt assemblages and the granitoid gneisses, and these belts have since been reinterpreted as allochthonous oceanic and island arc deposits similar to those of younger mountain belts.

The style of Archean greenstone belts varies in an age-dependent manner. Belts older than 3.5 billion years have sediments including chert, banded iron formation, evaporites, and stromatolitic carbonates, indicating shallow-water deposition, and contain only very rare conglomerates. They also have more abundant komatiites than younger greenstone belts. Younger greenstone belts seem to contain more intermediate volcanic rocks such as andesites, and have more deep-water sediments and conglomerates. They also contain banded iron formations, stromatolitic carbonates, and chert. Since so few early Archean greenstone belts are preserved, it is difficult to know whether these apparent temporal variations represent real-time differences in the style of global tectonics or are a preservational artifact.

GRANULITE-GNEISS BELTS

High-grade granitoid gneiss terrains form the second main type of Archean terrain. Examples include the Limpopo belt of southern Africa, the Lewisian of the North Atlantic Province, the Hengshan of north China, and some less-well-documented belts in Siberia and Antarctica. The high-grade gneiss assemblage seems similar in many ways to the lower-grade greenstone belts, but more strongly deformed and metamorphosed, reflecting burial to 12.5–25 miles (20–40 km) depth. Strongly deformed mylonitic gneisses and partially melted rocks known as migmatites are common, reflecting the high degrees of deformation and metamorphism. Most of the rocks in the high-grade gneiss terranes are metamorphosed sedimentary rocks, including sandstones, greywackes, carbonates, as well as layers of volcanic rocks. Many are thought to be strongly deformed continental margin sequences with greenstone-type assemblages thrust over them, deformed during continent-continent collisions. Most high-grade gneiss terrains have been intruded by several generations of mafic dikes, reflecting crustal extension. These are typically deformed into boudins (thin layers in the gneiss), making them difficult to recognize. Some high-grade gneiss terrains also have large layered mafic/ultramafic intrusions, some of which are related to the mafic dike swarms.

The strong deformation and metamorphism in the Archean high-grade gneiss terranes indicates that they have been in continental crust that has been thickened to double crustal thicknesses of about 50 miles (80 km), and some even more. This scale of crustal thickening is typically associated with continental collisions and/or thickened plateaus related to Andean-style magmatism. High-grade gneiss terrains are therefore typically thought to represent continent-continent collision zones.

CRATONIC BASIN ASSOCIATION

A third style of rock association also typifies the Archean but is less common than the previous two associations. The cratonic basin association is characterized by little-deformed and -metamorphosed sequences of clastic and carbonate sedimentary rocks, with a few intercalated volcanic horizons. This category of Archean sequence is best developed on South Africa's Kaapvaal craton, and includes the Pongola, Witwatersrand, Ventersdorp, and Transvaal Supergroups. These groups include sequences of quartzites, sandstones, arkoses, carbonates, and volcanic rocks, deposited in shallow marine to lacustrine basins. They are interpreted to represent rift, foreland basin, and shallow marine cratonic eperic sea-type deposits, fortuitously preserved although only slightly metamorphosed and deformed. Several shallow-water carbonate shelf associations are also preserved, including the 3.0-Ga Steep Rock platform in Canada's Superior Province, and possibly also the 2.5-Ga Hamersley Group in western Australia.

EARLY LIFE

Life clearly had already established itself on Earth by the Early Archean. The geologic setting and origin of life are topics of intense current interest, research, and thought by scientists and theologians. Any models for the origin of life need to explain some observations about early life from Archean rock sequences.

Evidence for early life comes from two separate lines. The first includes remains of organic compounds and chemical signatures of early life, and the other line consists of fossils, microfossils, and microstructures. The best organic evidence for early life comes from kerogens, which are nonsoluble organic compounds or the nonextractable remains of early life that formed at the same time as the sediments in which they are found. Other extractable organic compounds such as amino acids and sugars may also represent remains of early life, but they are soluble in water and may have entered the rocks after the deposition of the sediments. Therefore most work on the biochemistry of early life has focused on the nonextractable kerogens. Biological activity changes the ratio of some isotopes, most notably carbon 13/carbon 12, producing a distinctive biomarker that is similar in Archean through present-day life. Such chemical evidence of early life has been documented in Earth's oldest sedimentary rocks, the 3.8-Ga Isua belt in Greenland.

The earliest known fossils come from the 3.5–3.6-Ga Apex chert of the Pilbara craton in western Australia. Three distinctive types of microfossils have been documented from the Apex chert. These include spheroidal bodies, 5–20 microns (one micron = 10^{-6} m) in diameter, some of which have been preserved in the apparent act of cell division. These microfossils are similar to some modern cyanobacteria, and show most clearly that unicellular life existed on Earth by 3.5 Ga. Simple rod-shaped microfossils up to one micron long are also present in the Apex chert, and their shapes and characteristics are also remarkably similar to modern bacteria. Less distinctive filamentous structures up to several microns long may also be microfossils, but they are less convincing than the spheroidal and rod-shaped bodies. All of these show, however, that simple, single-celled, probably prokaryotic life-forms were present on Earth by 3.5 billion years ago, 1 billion years after Earth formed.

Stromatolites are a group of generally dome-shaped or conical mounds, or sheets of finely laminated sediments produced by organic activity. They were most likely produced by cyanobacteria (formerly called blue-green algae) that alternately trapped sediment with filaments that protruded above the sediment/water interface and secreted a carbonate layer during times when little sediment was passing to be trapped. Stromatolites produced a distinctive layering by preserving this alternation between sediment trapping and secretion of carbonate layers. Common in the Archean and Proterozoic record, stromatolites show that life was thriving in many places in shallow water and was not restricted to a few isolated locations. The oldest stromatolites known are in 3.6 billion-year-old sediments from the Pilbara craton of western Australia, with many examples in early, middle, and late Archean rock sequences. Stromatolites seem to have peaked in abundance in the Middle Proterozoic, and largely disappeared in the Late Proterozoic with the appearance of grazing metazoans.

See also ATMOSPHERE; CRATON; ORIGIN AND EVOLUTION OF THE EARTH AND SOLAR SYSTEM; PHANEROZOIC; PRECAMBRIAN; PROTEROZOIC.

FURTHER READING

Burke, Kevin, William S. F. Kidd, and Timothy M. Kusky. "Archean Foreland Basin Tectonics in the Witwatersrand, South Africa." *Tectonics* 5, no. 3 (1986): 439–456.

———. "The Pongola Structure of Southeastern Africa: The World's Oldest Recognized Well-Preserved Rift." *Journal of Geodynamics* 2, no. 1 (1985): 35–50.

———. "Is the Ventersdorp Rift System of Southern Africa Related to a Continental Collision Between the Kaapvaal and Zimbabwe Cratons at 2.64 Ga ago?" *Tectonophysics* 11 (1985): 1–24.

Kusky, Timothy M. "Structural Development of an Archean Orogen, Western Point Lake, Northwest Territories." *Tectonics* 10, no. 4 (1991): 820–841.

———. "Evidence for Archean Ocean Opening and Closing in the Southern Slave Province." *Tectonics* 9, no. 6 (1990): 1,533–1,563.

———. "Accretion of the Archean Slave Province." *Geology* 17 (1989): 63–67.

Kusky, Timothy M., and Ali Polat. "Growth of Granite-Greenstone Terranes at Convergent Margins and

Stabilization of Archean Cratons." In *Tectonics of Continental Interiors,* edited by Stephen Marshak and Ben van der Pluijm, 43–73. *Tectonophysics* 305 (1999).

Kusky, Timothy M., and Peter J. Hudleston. "Growth and Demise of an Archean Carbonate Platform, Steep Rock Lake, Ontario Canada." *Canadian Journal of Earth Sciences* 36 (1999): 1–20.

Kusky, Timothy M., and Julian Vearncombe. "Structure of Archean Greenstone Belts." Chap. 3 in *Tectonic Evolution of Greenstone Belts,* edited by Maarten J. de Wit and Lewis D. Ashwal, 95–128. Oxford: Oxford Monograph on Geology and Geophysics, 1997.

Kusky, Timothy M., and Pamela A. Winsky. "Structural Relationships along a Greenstone/Shallow Water Shelf Contact, Belingwe Greenstone Belt, Zimbabwe." *Tectonics* 14, no. 2 (1995): 448–471.

Kusky, Timothy M., and William S. F. Kidd. "Remnants of an Archean Oceanic Plateau, Belingwe Greenstone Belt, Zimbabwe." *Geology* 20, no. 1 (1992): 43–46.

Kusky, Timothy M., ed. *Precambrian Ophiolites and Related Rocks, Developments in Precambrian Geology* 13. Amsterdam: Elsevier, 2004.

McClendon, John H. "The Origin of Life." *Earth Science Reviews* 47 (1999): 71–93.

Schopf, William J. *Cradle of Life: The Discovery of Earth's Earliest Fossils.* Princeton, N.J.: Princeton University Press, 1999.

Asian geology   Asia is one of the most geologically and geomorphologically diverse continents on Earth, stretching from the Arctic to tropical regions, and from the world's highest peaks to broad plains, deserts, steeps, and deep basins. The range in age of features in Asia spans most of geological time, from 3.8 to 3.5 billion-year-old gneiss in China, to active volcanoes and sedimentary deposits across Asia. The India-Asia collision has resulted in Asia's being the most tectonically active region of continental crust in the world.

CHINA

China contains some of the most complex geology in the world, ranging from a number of ancient Archean cratons, to active tectonic belts, and offshore marine basins. The geomorphology changes from deep marine basins, to coastal plains, flat steppes, deserts, mountains, and the highest plateau of uplifted crust in the world. With such diversity it is fortunate that China has a long history of geological exploration and records, although much of this is not easily accessible to the Western world. Metallurgical exploration and workings go back to prehistoric times in China, and the oldest natural gas well in the world was dug in Sichuan Province in the 12th cen-

tury. Paleontology started in China, with the studies of the scholar Yen Cheng-ching, and ideas about mountain building processes may have first originated in China as shown by the works of Chu Hsi, who described mountains as features uplifted from oceans, which then became eroded by streams, forming sedimentary basin deposits. Geology became a formal discipline in China with the establishment of a geology department at Peking University in 1909, through the efforts of Dr. F. Solger from Germany and Dr. Amadeus Grabau from the United States, who studied the stratigraphy of China for many years, and became known as the father of Chinese geology. Grabau also authored numerous books and proposed ideas on the origin of mountains, continental crust, and cycles in Earth history, based mainly on his studies in China.

GEOMORPHOLOGY

China is geomorphologically diverse, consisting of about 33 percent mountains, 25 percent plateaus (including Tibet), 20 percent basins, and 10 percent hilly terrain. In general, the land surface slopes from the high regions including Tibet in the west, to the 1,100-mile (1,800-km) coastline in the east.

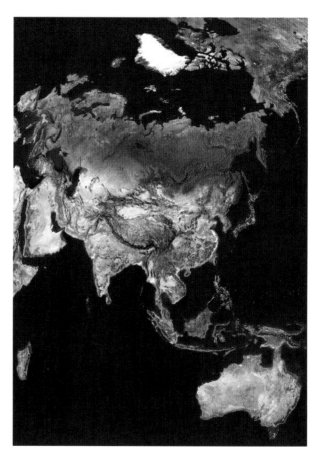

Satellite image of Asia *(M-Sat Ltd. Photo Researchers, Inc.)*

There are three main physiographic provinces of China based on elevation. The Tibetan, or Qinghai-Xizang Plateau, in the south rises generally to more than 2.5 miles (4 km) above sea level, including the Himalaya Mountains, which rise above 3.7 miles (6 km), with many peaks surpassing 5 miles (8 km) above sea level. Mount Everest (Jolmo Lungma) is the highest peak in the world, reaching 29,133 feet (8,882 m) above sea level. A series of basins and plateaus are located north and east of the Tibetan, or Qinghai-Xizang, Plateau, with elevations between 0.5 and 1.3 miles (1–2 km). The most important plateaus in this region include the Inner Mongolia Plateau or steppe, the loess plateau, and the Yunnan-Guizhou Plateau, whereas the large basins include the Sichuan, Junggar, and Tarim. Eastern China consists mostly of hills lower than a half-mile (1 km) and broad plains, including the Northeast, Lower, Northern, and Upper Chang Jiang (Yangtze River) plains, whereas the low mountains include the Shandong and Southern Hills.

Mountains (the word *shan* means "mountain" in Chinese) in China are divided into different groups according to their orientations, or trend (measured relative to geographic north). These include the E-W, or Altaid, trend, with major belts including the Tienshan, Kunlun, Tanglha, Kangkar Tesi, Qinling, Himalaya, Yinshan, and Nanling. Interestingly, the age of the activity becomes younger from north to south, reflecting processes related to the India-Asia collision. The NE-SW (Cathysian) trending mountains include the Greater Kingham, Taihang, Changbai, and Wui, mostly in eastern China. The N-S trending mountains include the Helan, Luban, and Hengduan. Other mountains trend NW-SE, such as the Karakoram and Altai Mountains in the west, and the Lesser Kinghan in the northeast.

The Tibet, or Qinghai-Xizhang, Plateau is the world's largest region of thickened, uplifted crust; it formed in response to the collision of India with Asia after the Cretaceous. Many rivers in Asia have their source on this plateau, and changing climate conditions with loss of glacial ice imperils many of these rivers. Many tectonically controlled lakes on the Tibet Plateau have formed in fault-controlled basins largely since the Eocene. In southwestern China the Yunnan-Guizhou Plateau has an average elevation of about 1.2 miles (2 km), but decreases to the east. This uplifted plateau is comprised largely of limestones and other soluble rocks, so as the uplift proceeded and groundwater levels dropped, a spectacular karst landscape developed across much of this plateau, forming some of the world's most impressive karst landscapes, such as the stone forest of Kunning and the karst towers of Guilin. Many Mesozoic red beds lie in the western part of this plateau. The Great Wall of China was built along the north edge of the loess plateau, which has an average elevation of 0.6 miles (1 km) over an area of 156,000 square miles (400,000 km^2), encompassing the Yinshan, Qinling, Qilian, and Taishan Mountains. Wind-blown dust and silt on the plateau and extending eastward to the Yellow Sea form the thickest and most extensive loess deposits in the world. In most places the loess is 150–250 feet (50–80 m) thick, but in places is up to 3,200 (975 m) feet thick. The dust originates in the Gobi and Ordos desert basins to the west, with much being deposited during the Quaternary, formed during the Pleistocene ice ages when strong winds blew from the NW to the SE. The plateau of Inner Mongolia has a similar height to the loess plateau, but this region is being actively extended and eroded, after a period of uplift in the Cenozoic.

China has many large basins of diverse origin. The Tarim, Junggar, and Qiadam basins in the west are relic back-arc basins developed on the margins of the Paleotethys Ocean, although the basins have complex older histories. The Tarim basin, one of the largest interior basins in the world, has a Precambrian basement, overlain by thick Mesozoic and Cenozoic rocks, whereas the Junggar basin has Paleozoic basement overlain by Mesozoic and Cenozoic basinal sequences. The Sichuan basin has a complex history, including Precambrian basement overlain by Paleozoic and Mesozoic red beds, including foreland basin deposits from the Longmen and Qinling orogens. The Ordos basin has Precambrian platformal sediments overlain by Paleozoic through Tertiary sedimentary layers. In the east the Bohai, or North China, basin is a complex rift-pull-apart basin of Paleogene age that has huge hydrocarbon reserves, including the Shenli, Dagong, and Renqin oil fields. In the north the Songliao basin is a Late Mesozoic rift basin that developed behind an active margin. On the eastern coast of China the East and South China Seas form passive margin-type basins formed in back-arc environments above the Pacific seduction system.

One of the world's great deserts, the Gobi, located in central Asia, encompasses more than 500,000 square miles (1,295,000 km^2) in Mongolia and northern China. The desert covers the region from the Great Khingan Mountains northwest of Beijing to the Tien Shan north of Tibet, but the desert is expanding at an alarming rate, threatening the livelihood of tens of thousands of farmers and nomadic sheepherders every year. Every spring dust from the Gobi covers eastern China, Korea, and Japan, and may extend at times around the globe. Northwesterly winds have removed almost all the soil from land in the Gobi, depositing it as thick loess

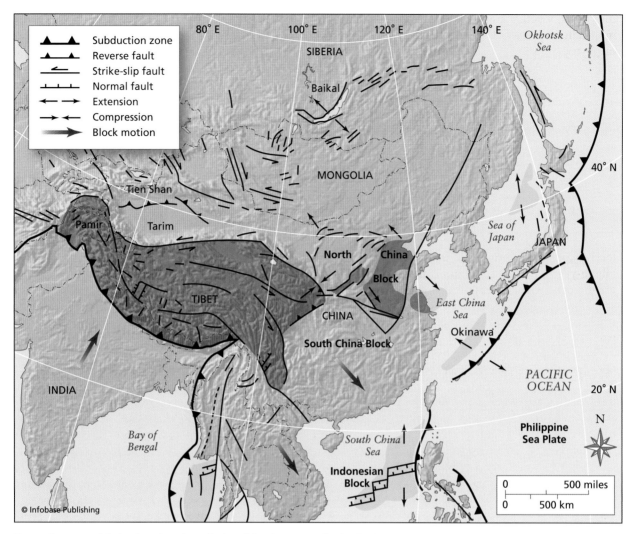

Tectonic map of Asia showing the relationships between the India-Asia collision, escape of the Indonesian and South China blocks seaward, and extension from Siberia to the Pacific margin, including opening of the Sea of Japan and the South China Sea *(modeled after T. Kusky, M. Zhai, and W. J. Xiao)*

in eastern China. Most of the Gobi is situated on a high plateau resting 3,000–5,000 feet (900–1,500 m) above sea level, and it contains numerous alkaline sabkhas and sandy plains in the west. Regions in the Gobi include abundant steppes, high mountains, forests, and sandy plains. The Gobi has yielded many archaeological, paleontological, and geological finds, including early stone implements, dinosaur eggs, and mineral deposits and precious stones including turquoise and jasper.

ARCHEAN CRATONS

China includes several blocks of ancient continental crust known as cratons. These include the North China craton, South China craton (including the Cathaysia and Yangtze blocks), and the Tarim block. Parts of northeastern and southern China in these cratons are well known for Late Proterozoic–age sed-

imentary deposits that host some of the world's most spectacular early animal fossils. One of the best studied sequences is the Doushantuo Formation of phosphatic sedimentary rocks exposed in south China's Guizhou Province, dated to be 580–600 million years old. The Sinian fauna is therefore older than the well-known Ediacaran metazoan fauna, and is currently the oldest-known assemblage of multicellular animal fossils on the Earth. The macrofossil assemblages are associated with prokaryotic and eukaryotic microfossils, and display remarkably well-preserved cellular and tissue structures and even remnants of organic material known as kerogen. Many of the fossils are unusual acritarchs, organic walled fossils with peripheral processes such as spines, hairs, and flagellum, which cannot be confidently placed into any living plant or animal group classification. Scientists are currently debating the origin of some of the fossils, whether they may be metazoan embryos, multicellular algae, filamentous bacteria, acritarchs, or phytoplankton.

North China Craton and Tarim Block

The North China craton occupies about 1 million square miles (1.7 million km^2) in northeastern China, Inner Mongolia, the Yellow Sea, and North Korea, and apparently shares an early geological history with the poorly known Tarim block to the west. It is bounded by the Qinling-Dabie Shan orogen to the south, the Yinshan-Yanshan orogen to the north, the Longshoushan belt to the west, and the Qinglong-Luznxian and Jiao-Liao belts to the east. The North China craton includes a large area of intermittently exposed Archean crust, including circa 3.8–2.5 billion-year-old gneiss, tonalite, trondhjemite, and granodiorite. Other areas include granite, migmatite, amphibolite, ultramafite, mica schist and dolomitic marble, graphitic and other metasedimentary gneiss, banded iron formation (BIF), and metaarkose. The Archean rocks are overlain by the 1.85–1.40 billion-year-old Mesoproterozoic Changcheng (Great Wall) system. In some areas in the central part of the North China craton 2.40–1.90 billion-year-old Paleoproterozoic sequences deposited in cratonic rifts are preserved.

The North China craton is divided into two major blocks separated by the Neoarchean Central orogenic belt, in which virtually all isotopic ages on the rocks fall between 2.55 and 2.50 billion years. The Western block, also known as the Ordos block, is a stable craton with a thick mantle root, no earthquakes, low heat flow, and lack of internal deformation since the Precambrian. In contrast, the Eastern block is atypical for a craton in that it has numerous earthquakes, high heat flow, and a thin lithosphere reflecting the lack of a thick mantle root. The North China craton is one of the world's most unusual cratons in that it had a thick tectosphere (subcontinental lithospheric mantle) developed in the Archean, which was present through the Ordovician, as shown by deep xenoliths preserved in Ordovician kimberlites. The eastern half of the root, however, appears to have delaminated or otherwise disappeared during Paleozoic, Mesozoic, or Cenozoic tectonism. This is demonstrated by Tertiary basalts that bring up mantle xenoliths of normal "Tertiary mantle" with no evidence of a thick root. The processes responsible for the loss of this root are enigmatic but are probably related to the present-day high-heat flow, Phanerozoic basin dynamics, and orogenic evolution.

The Central orogenic belt includes belts of tonalite-trondhjemite-granodiorite, granite, and supracrustal sequences metamorphosed from granulite to greenschist facies. It can be traced for about 1,000 miles (1,600 km) from west Liaoning to west Henan. Widespread high-grade regional metamorphism including migmatization occurred throughout the Central orogenic belt between 2.6 and 2.5 billion years ago, with final uplift of the metamorphic terrain at 1.9–1.8 billion years ago associated with extensional tectonism or a collision on the northern margin of the craton. Amphibolite to greenschist-grade metamorphism predominates in the southeastern part of the Central orogenic belt, but the northwestern part of the orogen is dominated by granulite-facies to amphibolite-facies rocks, including some high-pressure assemblages (10–13 kilobars at 850 ± 50°C). The high-pressure assemblages can be traced for more than 400 miles (700 km) along a linear belt trending east-northeast. Internal (western) parts of the orogen are characterized by thrust-related horizontal foliations, flat-dipping shear zones, recumbent folds, and tectonically interleaved high-pressure granulite migmatite and metasediments. It is widely overlain by sediments deposited in rifts and continental shelf environments, and intruded by several dike swarms (2.4–2.5, and 1.8–1.9 billion years ago). Several large anorogenic granites with ages of 2.2–2.0 billion years are identified within the belt. Recently two linear units have been documented within the belt, including a high-pressure granulite belt in the west and a foreland-thrust fold belt in the east. The high-pressure granulite belt is separated by normal-sense shear zones from the Western block, which is overlain by thick metasedimentary sequences younger than 2.4 billion years that metamorphosed 1.86 billion years ago.

The Hengshan high-pressure granulite belt is about 400 miles (700 km) long, consisting of several metamorphic terrains, including the Hengshan, Huaian, Chengde, and west Liaoning complexes. The high-pressure assemblages commonly occur as inclusions within intensely sheared tonalite-trondhjemite-granodiorite (2.6–2.5 billion years) and granitic gneiss (2.5 billion years) and are widely intruded by K-granite (2.2–1.9 billion years) and mafic dike swarms (2.40–2.45 Ga, 1.77 billion years). Locally, khondalite and turbiditic slices are interleaved with the high-pressure granulite rocks, suggesting thrusting. The main rock type is garnet-bearing mafic granulite with characteristic plagioclase-orthopyroxene corona around the garnet, which shows rapid exhumation-related decompression. A constant-temperature decompressive pressure-temperature-time path can be documented within the rocks, and the peak pressures and temperatures are in the range of 1.2–1.0 GPa, at 1,290–1,470°F (700–800°C). At least three types of geochemical patterns are shown by mafic rocks of the high-pressure granulites, indicating a tectonic setting of active continental margin or island arc. The high-pressure granulites were formed through subduction-collision, followed by rapid rebound-extension, recorded by 2.5–2.4 billion-year old mafic dike swarms and rift-related sedimentary sequences in the Wutai Mountai-Taihang Mountain areas.

The Qinglong foreland basin and fold-thrust belt is north- to northeast-trending and is now preserved as several relict-folded sequences (Qinglong, Fuping, Hutuo, and Dengfeng). Its general sequence from bottom to top can be further divided into three subgroups of quartzite-mudstone-marble, turbidite, and molasse, respectively. The lower subgroup of quartzite-mudstone-marble is well preserved in central sections of the Qinglong foreland basin (Taihang Mountain), with flat-dipping structures, interpreted as a passive margin developed before 2.5 Ga on the Eastern block. It is overlain by lower-grade turbidite and molasse-type sediments. The western margin of Qinglong foreland basin is intensely reworked by thrusting and folding and is overthrust by the overlying orogenic complex (including the tonalitic-trondhjemitic-granodiorotic gneiss, ophiolites, accretionary prism sediments). To the east its deformation becomes weaker in intensity. The Qinglong foreland basin is intruded by a gabbroic dike complex consisting of 2.4 billion-year-old diorite and is overlain by graben-related sediments and flood basalts. In the Wutai and North Taihang basins, many ophiolitic blocks are recognized along the western margin of the foreland thrust-fold belt. These consists of pillow lava, gabbroic cumulates, and harzburgite. The largest ophiolitic thrust complex imbricated with foreland basin sedimentary rocks is up to 5 miles (10 km) long, preserved in the Wutai-Taihang Mountains.

Several dismembered Archean ophiolites have been identified in the Central orogenic belt, including some in Lioning Province, at Dongwanzi, north of Zunhua, and at Wutai Mountain. The best studied of these are the Dongwanzi and Zunhua ophiolitic terranes. The Zunhua structural belt of eastern Hebei Province preserves a cross section through most of the northeastern part of the Central orogenic belt. This belt is characterized by highly strained gneiss, banded iron formation, 2.6–2.5 billion-year-old greenstone belts, and mafic to ultramafic complexes in a high-grade ophiolitic mélange. The belt is intruded by widespread 2.6–2.5 billion-year-old tonalite-trondhjemite gneiss and 2.5 billion-year-old granites, and is cut by ductile shear zones. The Neoarchean high-pressure granulite belt (Chengde-Hengshan HPG) strikes through the northwest part of the belt. The Zunhua structural belt is thrust over the Neoarchean Qianxi-Taipingzhai granulite-facies terrane, consisting of high-grade metasedimentary to charnockitic gneiss forming several small dome-like structures southeast of the Zunhua belt. The Zunhua structural belt clearly cuts across the dome-like Qian'an-Qianxi structural patterns to the east. The Qian'an granulite-gneiss dome (3.8–2.5 billion years old) forms a large circular dome in the southern part of the area and is composed of tonalitic-trondhjemitic gneiss and biotite granite. Meso-archean (2.8–3.0 billion years old) and Paleoarchean (3.50–3.85 billion years old) supracrustal sequences outcrop in the eastern part of the region. The Qinglong Neoarchean amphibolite to greenschist-facies supracrustal sequence strikes through the center of the area and is interpreted to be a foreland fold-thrust belt, intruded by large volumes of 2.4 billion-year-old diorite in the east. The entire North China craton is widely cut by at least two Paleoproterozoic mafic dike swarms (2.5–2.4, 1.8–1.7 billion years old), associated with regional extension. Mesozoic-Cenozoic granite, diorite, gabbro, and ultramafic plugs occur throughout the NCC and form small intrusions in some of the belts.

The largest well-preserved sections of the Dongwanzi ophiolite are located approximately 120 miles (200 km) northeast of Beijing in the northeastern part of the Zunhua structural belt, near the villages of Shangyin and Dongwanzi. The belt consists of prominent amphibolite-facies mafic-ultramafic complexes in the northeast sector of the Zunhua structural belt. The southern end of the Dongwanzi ophiolite belt near Shangyin is complexly faulted against granulite-facies gneiss, with both thrust faults and younger normal faults present. The main section of the ophiolite dips steeply northwest, is approximately 30 miles (50 km) long, and is 3–6 miles (5–10 km) wide. A U/Pb-zircon age of 2.505 billion years for two gabbro samples from the Dongwanzi ophiolite shows that this is the oldest, relatively complete ophiolite known in the world. Parts of the central belt, however, are intruded by a mafic/ultramafic Mesozoic pluton with related dikes.

A high-temperature shear zone intruded by the 2.4 Ga old diorite and tonalite marks the base of the ophiolite. Exposed ultramafic rocks along the base of the ophiolite in the ophiolite include strongly foliated and lineated dunite and layered harzburgite. Aligned pyroxene crystals and generally strong deformation of serpentinized harzburgite resulted in strongly foliated rock. Harzburgite shows evidence for early high-temperature deformation. This unit is interpreted to be part of the lower residual mantle, from which the overlying units were extracted.

The cumulate layer represents the transition zone between the lower ultramafic cumulates and upper mafic assemblages. The lower part of the sequence consists of layers of pyroxenite, dunite, wehrlite, lherzolite and websterite, and olivine gabbro-layered cumulates, all formed by heavy crystals sinking through the magma and settling on the bottom of the magma chamber. Many layers grade from dunite at the base, through wehrlite, and are capped by clinopyroxene. Basaltic dikes cut through the cumulates and are similar mineralogically and texturally to dikes in the upper layers.

The gabbro complex of the ophiolite is up to three miles (5 km) thick and grades up from a zone of mixed layered gabbro and ultramafic rocks to one of strongly layered gabbro that is topped by a zone of isotropic gabbro. Thicknesses of individual layers vary from centimeter to meter scale and include clinopyroxene and plagioclase-rich layers. Layered gabbros from the lower central belt alternate between fine-grained layers of pyroxene and metamorphic biotite that are separated by layers of metamorphic biotite intergrown with quartz. Biotite and pyroxene layers show a random orientation of grains. Coarse-grained veins of feldspar and quartz are concentrated along faults and fractures. Plagioclase feldspar shows core replacement and typically has irregular grain boundaries. The gabbro complex of the ophiolite

has been dated by the U-Pb method on zircons to be 2,504 ± 2 million years old.

The sheeted dike complex is discontinuous over several kilometers. More than 70 percent of the dikes are unusual in that only one half of each dike is preserved. In most tectonic settings dikes have fine-grained chill margins on both margins where they intruded and cooled against country rocks, crystallizing quickly and forming two parallel chill margins with finer-grained crystals than in the slower-cooling interior of the dikes. Dikes in ophiolites often show dikes with only one chill margin, where new dikes successively intrude the center of the previously intruded dike, in a setting where the crust is being extended and filled by new dikes. When the next dike intruded, it intruded along the center of the last dike, and this happened several times in a row, leaving a dike complex with chill margins preserved preferentially on one side. One-way chill margins are preferentially preserved on their northeast side of the dikes in the Dongwanzi ophiolite. Gabbro screens are common throughout the complex and increase in number and thickness downward, marking the

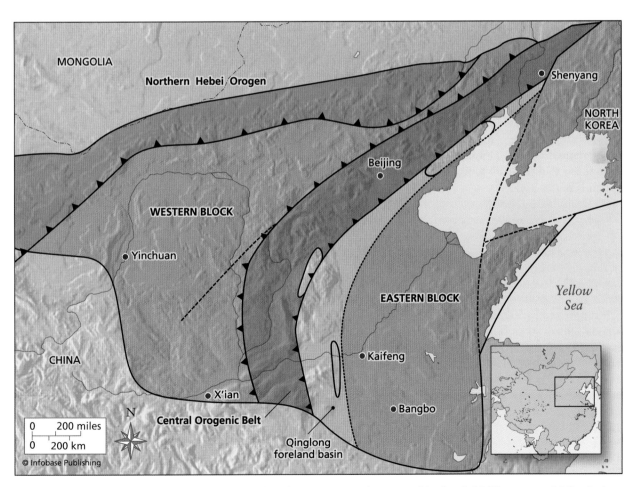

Tectonic map of the North China craton showing the eastern and western blocks, 2.5 billion-year-old Central Orogenic Belt, and the 1.9 billion-year-old North Hebei Orogen *(modeled after T. Kusky and J. H. Li)*

transition from the dike complex to the fossil magma chamber. In some areas the gabbro is cut by basaltic-diabase dikes, but in others it cuts through xenoliths of diabase, suggesting comagmatic formation.

The upper part of ophiolite consists of altered and deformed pillow basalts, pillow breccias, and interpillow sediments (chert and banded iron formations). Many of the pillows are interbedded with more massive flows and cut by sills; however, some well-preserved pillows show typical lower cuspate and upper lobate boundaries that define stratigraphic younging. Pyroxenes from pillow lavas from the ophiolite have been dated by the Lu-Hf method to be 2.5 billion years old, the same age as estimated for the gabbro and mantle sections.

The base of the ophiolite is strongly deformed, and intruded by the 2.391 billion-year-old Cuizhangzi diorite-tonalite complex. The Dongwanzi ophiolite is associated with a number of other amphibolite-facies belts of mafic plutonic and extrusive igneous rocks in the Zunhua structural belt. These mafic-to-ultramafic slices and blocks can be traced regionally over a large area from Zunhua to West Liaoning (about 120 miles or 200 km). Much of the Zunhua structural belt is interpreted as a high-grade ophiolitic mélange, with numerous tectonic blocks of pillow lava, BIF, dike complex, gabbro, dunite, serpentinized harzburgite, and podiform chromitite in a biotite-gneiss matrix, intruded extensively by tonalite and granodiorite. Cross-cutting granite has yielded an age of 2.4 billion years. Blocks in the mélange correlate with the Dongwanzi and other ophiolitic fragments in the Zunhua structural belt. This correlation is supported by the isotopic system of Rhenium (Re)-Osmium (Os), since Re-Os age determinations on several of these blocks reveal that they are 2.54 billion years old.

The Eastern and Western blocks of the North China craton collided at 2.5 billion years ago during an arc/continent collision, forming a foreland basin on the Eastern block, a granulite facies belt on the Western block, and a wide orogen between the two blocks. This collision was followed rapidly by post-orogenic extension and rifting that formed mafic dike swarms and extensional basins along the Central orogenic belt, and led to the development of a major ocean along the north margin of the craton. An arc terrane developed in this ocean and collided with the north margin of the craton by 2.3 Ga, forming an 850-mile (1,400-km) long orogen known as the Inner Mongolia–Northern Hebei orogen. A 1,000-mile (1,600-km) long granulite-facies terrain formed on the southern margin of this orogen, representing a 120-mile (200-km) wide uplifted plateau formed by crustal thickening. The orogen was converted to an Andean-style convergent margin between 2.20 and 1.85 billion years ago, recorded by belts of plutonic

rocks, accreted metasedimentary rocks, and a possible back arc basin. A pulse of convergent deformation is recorded at 1.9–1.85 billion years across the northern margin of the craton, perhaps related to a collision outboard of the Inner Mongolia–Northern Hebei orogen, and closure of the back arc basin. This event caused widespread deposition of conglomerate and sandstone of the basal Changcheng Series in a foreland basin along the north margin of the craton. At 1.85 billion years the tectonics of the North China craton became extensional, and a series of aulacogens and rifts propagated across the craton, along with the intrusion of mafic dike swarms. The northern granulite facies belt underwent retrograde metamorphism, and was uplifted during extensional faulting. High-pressure granulites are now found in the areas where rocks were metamorphosed to granulite facies and exhumed two times, at 2.5 and 1.8 billion years ago, respectively, exposing rocks that were once at lower crustal levels. Rifting led to the development of a major ocean along the southwest margin of the craton, where oceanic records continue until 1.5 billion years ago.

South China Craton

The South China craton is divided into the Yangtze craton and the Cathaysia block, joined together in a plate collision in the late Mesoproterozoic. Radiometric dating of two ophiolite suites in the eastern part of the suture between the Yangtze and Cathaysia blocks gives ages of 1.03 to 1.02 billion years for the age of this collision. Late Mesoproterozoic suture zones have been also reported from the northwestern and northern margins of the Yangtze block. An ophiolitic mélange in western Sichuan (the northwestern margin) that contains gabbro and diabase has been dated isotopically to be 1.01 billion years old. Zircon ages of 1.3 and 1.0 billion years have been determined from igneous intrusions at the northern margin of the Yangtze block. The late Mesoproterozoic continental collision belts in and around the South China block corresponded to one of the central Grenvillian sutures of the Rodinian supercontinent that brought together Australia, Yangtze, and Cathaysia-Laurentia at about 1 billion years ago.

The stratigraphy, tectonic evolution, and paleomagnetic features of the southern North China craton differ from those of the South China craton and suggest that north China became a part of the Rodinia supercontinent about 1 billion years ago, when the southern border of the North China craton lay adjacent to Siberia. The south and north China blocks became parts of the Rodinia supercontinent independently about 1 billion years ago.

The breakup of Rodinia was probably initiated by a mantle plume that rose beneath the supercontinent 820 million years ago. The Kangdian rift in the southwestern border of the South China block and the Nanhua rift along the boundary between the Yangtze and Cathaysia blocks formed as a result of the breakup of Rodinia. The 800–700 million-year-old intrusion ages of granitic gneisses in the northern border of the South China block are also associated with rifting of the South China block from Rodinia. The separation of the South China block from Rodinia must have occurred after 750 million years ago. Likewise, on the southern borders of the North China block, rather weak magmatic activity and metamorphism at about 800–600 million years ago has been identified, which may relate to the rifting of the North China block from Rodinia.

Following the breakup of Rodinia during the late Neoproterozoic, several microcontinents and immature island arcs amalgamated through collision and accretion to form Gondwana during the Pan-African orogeny. The South and North China cratons probably converged with each other during the early Cambrian and lay close to Australia by the mid and late Cambrian. The Paleo-Tethys Ocean opened in the Devonian, moving the North and South China blocks northward, and the Paleo-Tethys Ocean between two blocks disappeared in the Middle Triassic by subduction of all oceanic crust in the ocean. The presence of 450–400 million-year-old ophiolites, arc-related 470–435 million-year-old metamorphism, and 210–330 million-year-old eclogites in the Qinling-Dabie-Sulu collision belt indicate that opening of the Paleo-Tethys had started in the mid-Ordovician instead of the Devonian.

CENTRAL ASIAN OROGENIC BELT

The Paleoasian, or Turkestan, Ocean was present on the northern side of the North China craton and Tarim block throughout the Paleozoic, with Paleotethys to the south. Several subduction zones were active during this interval, leading to continental growth through accretion of terranes along the northern margin of the cratons and the generation of arc-magmas. These terranes north of the Archean cratons host more than 900 late Paleozoic to Early Triassic plutons, formed during closure of the Paleo-Asian Ocean at the end of the Permian. Closure is marked by the Solonker suture and 300–250 million-year-old south-directed subduction beneath the accreted terranes along the northern side and the northern margin of the cratonic blocks. Continued convergence from the north during the Triassic and Jurassic caused postcollisional thrusting and considerable crustal thickening on the northwest side of the cratons. The northeastern margin of the North China craton with a Permian shelf sequence collided with the Khanka block in the Late Permian to Early Trias-

sic, as indicated by syncollisional granites. Many of the subsequent later Mesozoic granitoids, metamorphic core complexes, and extensional basins, located south of the Solonker suture in the northern part of the craton and the adjacent Paleozoic accretionary orogen, may be related to postcollisional Jura-Cretaceous collapse of the massive Himalayan-style Solonker orogen and plateau.

DABIE SHAN–SULU TONGBAI CENTRAL CHINA OROGEN

The Qinling-Dabie-Sulu, or Dabie Shan, is the world's largest ultrahigh pressure metamorphic belt, containing Triassic (220–240 Ma) high-pressure, low-temperature (eclogite) facies metamorphic rocks formed during the collision of the North and South China cratons. Most remarkably, the orogen contains coesite and diamond-bearing eclogite rocks, indicating metamorphic burial to depths exceeding 60 miles (100 km). The Dabie Shan metamorphic belt stretches from the Tanlu fault zone between Shanghai and Wuhan, approximately 1,250 miles (2,000 km) to the west northwest to the Qaidam basin north of the Tibetan Plateau. The orogen is only 30–60 miles (50–100 km) wide in most places, and it separated the North China craton on the north from the Yangtze craton (also called the South China block) on the south. A small tectonic block or terrane known as the South Qinling is wedged between the North and South China cratons in the orogen and is thought to have collided with the North China craton in the Triassic, before the main collision.

The Qinling-Dabie-Sulu orogen is marked by numerous terranes forming the irregular suture between the North China and South China cratons. It is a major part of the E-W-trending Central China orogen that extends for 900 miles (1,500 km) eastward from the Kunlun Range, to the Qinling Range, and then 370 miles (600 km) farther east through the Tongbai-Dabie Range. Its easternmost extent, offset by movement along the Tan-Lu fault system, continues northeastward through the Sulu area of the Shandong Peninsula then into South Korea. The Sulu belt may extend through the southern part of South Korea. The intermittent presence of ultrahigh-pressure diamonds, eclogites, and felsic gneisses indicates very deep subduction along a cumulative 2,500-mile (> 4,000-km) long zone of collisional orogenesis.

The rifting and collisional history throughout the Paleozoic of the North China and Tarim cratons with blocks and orogens to the south, such as the North Qinling terrane, the South Qinling terrane, and eventually (in the Triassic) the South China craton, have been complicated and controversial. In the early Paleozoic northward subduction of the Qaidam-South Tarim plate (possibly connected with the South China plate) took place beneath the active southern margin of the NCC. The North China craton, probably together with the Tarim block, collided with the South Tarim-Qaidam block in the Devonian, then with the South China block in the Permo-Triassic. This latter collision resulted in exposure of ultrahigh-pressure rocks from approximately 60–120 miles (100–200 km) depth in the Dabie Shan, and westward sliding or escape of the South Tarim-Qaidam block, and caused uplift of a large plateau (Huabei Plateau) in the eastern North China craton. Younger extrusion tectonics related to Himalayan collisions farther west resulted in approximately 300 miles (500 km) of left-lateral motion along the Altyn-Tagh fault, separating the North China craton from the South Tarim-Qaidam block, slicing and sliding to the west the arc that formed on the southern margin of the craton during early Paleozoic subduction.

The terrane accretion and eventual "continent-continent" collision along the southern margin of the North China and Tarim cratons are defined by a geometrically irregular suture, defining a diachronous convergence with a complex spatial and temporal pattern. Many models of extrusion tectonics, such as eastward, vertical (upward), and lateral, have been proposed for the Qinling-Dabie orogen in the last decade. Vertical movement along a paleosubduction zone was important to Triassic uplift of the ultrahigh-pressure rocks in the eastern part of the orogen. An orogen-parallel, eastward extrusion occurred diachronously between 240 and 225–210 Ma. Cretaceous to Cenozoic unroofing was initially dominated by eastward tectonic escape in the early Cretaceous and then by Pacific subduction in the mid-Cretaceous. The Triassic Dabie high-pressure (HP)-UHP metamorphic rocks were originally located beneath the Foping dome, located in the narrowest part of the Qinling belt, and these rocks were extruded eastward to their present-day location.

TIBETAN PLATEAU

The Tibetan Plateau is the largest high area of thickened continental crust on Earth, with an average height of 16,000 feet (4,880 m) over 470,000 square miles (1,220,000 km²). Bordered on the south by the Himalayan Mountains, the Kunlun Mountains in the north, the Karakoram on the west, and the Hengduan Shan on the east, Tibet is the source of many of the largest rivers in Asia. The Yangtze, Mekong, Indus, Salween, and Brahmaputra Rivers all rise in Tibet, and flow through Asia, forming the most important source of water and navigation for huge regions.

Southern Tibet merges into the foothills of the northern side of the main ranges of the Himalaya, but are separated from the mountains by the deeply incised river gorges of the Indus, Sutlej, and Yarlung Zangbo (Brahmaputra) Rivers. Central and northern Tibet consists of plains and steppes that are about 3,000 feet (1,000 m) higher in the south than the north. Eastern Tibet includes the Transverse ranges (the Hengduan Shan) that are dissected by major faults in the river valleys of the northwest-southeast–flowing Mekong, Salween, and Yangtze Rivers.

Tibet has a high plateau climate, with large diurnal and monthly temperature variations. The center of the plateau has an average January temperature of 32°F (0°C), and an average June temperature of 62°F (17°C). The southeastern part of the plateau is affected by the Bay of Bengal summer monsoons, whereas other parts of the plateau experience severe storms in fall and winter months.

Geologically, the Tibetan Plateau is divided into four terranes, including the Himalayan terrane in the south, and the Lahasa terrane, the Qiangtang terrane, and Songban-Ganzi composite terrane in the north. The Songban-Ganzi terrane includes Triassic flysch and Carboniferous-Permian sedimentary rocks, and a peridotite-gabbro-diabase sill complex that may be an ophiolite, overlain by Triassic flysch. Another fault-bounded section includes Paleozoic limestone and marine clastics, probably deposited in an extensional basin. South of the Jinsha suture, the Qiangtang terrane contains Precambrian basement overlain by Early Paleozoic sediments that are up to 12 miles (20 km) thick. Western parts of the Qiangtang terrane contain Gondwanan tillites, and Triassic-Jurassic coastal swamp and shallow marine sedimentary rocks. Late Jurassic–Early Cretaceous deformation uplifted these rocks, before they were unconformably overlain by Cretaceous strata.

The Lhasa terrane collided with the Qiangtang terrane in the Late Jurassic and formed the Bangong suture, containing flysch and ophiolitic slices, that now separates the two terranes. It is a composite terrane containing various pieces that rifted from Gondwana in the Late Permian. Southern parts of the Lhasa terrane contain abundant Upper Cretaceous to Paleocene granitic plutons and volcanics, as well as Paleozoic carbonates, and Triassic-Jurassic shallow marine deposits. The center of the Lhasa terrane is similar to the south but with fewer magmatic rocks, whereas the north contains Upper Cretaceous shallow marine rocks that onlap the Upper Jurassic–Cretaceous suture.

The Himalayan terrane collided with the Lhasa terrane in the Middle Eocene, forming the ophiolite-decorated Yarlungzangbo suture. Precambrian metamorphic basement is thrust over Sinian through Tertiary strata, including Lower Paleozoic carbonates and Devonian clastics, overlain unconformably by Permo-Carboniferous carbonates. The Himalayan terrane contains Lower Permian Gondwanan flora,

An 80-mile (130-km) wide view of the Himalayas from *International Space Station,* January 28, 2004—Mount Everest is shown in the upper center, Makalu is at the top left. The view looks south: Tibet is at the bottom, and Nepal is the land beyond Everest. *(NASA/Photo Researchers, Inc.)*

and probably represents the northern passive margin of Mesozoic India, with carbonates and clastics in the south, thickening to an all clastic continental rise sequence in the north.

The Indian plate rifted from Gondwana and started its rapid (3.2–3.5 inches per year, 80–90 mm/yr) northward movement about 120 million years ago. Subduction of the Indian plate beneath Eurasia until about 70 million years ago formed the Cretaceous Kangdese batholith belt, containing diorite, granodiorite, and granite. Collision of India with Eurasia at 50–30 million years ago formed the Lhagoi-Khangari belt of biotite and alkali granite, and the 20–10 million-year-old Himalayan belt of tourmaline-muscovite granites.

Tertiary faulting in Tibet is accompanied by volcanism, and the plateau is presently undergoing east-west extension with the formation of north-south graben associated with hot springs, and probably deep magmatism. Seismic reflection profiling has detected some regions with unusual characteristics beneath some of these grabens, interpreted by some seismologists as regions of melt or partially molten crust.

Much research has focused on the timing of the uplift of the Tibetan Plateau and modeling the role this uplift has had on global climate. The plateau strongly affects atmospheric circulation, and many models suggest that the uplift may contribute to global cooling and the growth of large continental ice sheets in latest Tertiary and Quaternary times. In addition to immediate changes to air-flow patterns around the high plateau, the uplift of large amounts of carbonate platform and silicate rocks exposes them to erosion. The weathering of these rocks causes the exposed rocks to react with atmospheric carbon dioxide, which combines these ions to produce bicarbonate ions such as are used to form calcium carbonate ($CaCO_3$), drawing down the atmospheric carbon dioxide levels and contributing to global cooling.

The best estimates of the time of collision between India and Asia is between 54 and 49 million years ago. Since then convergence between India and Asia has continued, but at a slower rate of 1.6–2.0 inches per year (40–50 mm/yr), and this convergence has resulted in intense folding, thrusting, shortening, and uplift of the Tibetan Plateau. Timing the uplift to specific altitudes is difficult, and considerable debate has centered on how much earlier than 50 million years ago the plateau reached its current height of 16,404 feet (5 km). Most geologists would now agree that this height was attained by 13.5 million years ago, and that any additional height increase is unlikely since the strength of the rocks at depth has been exceeded, and the currently active east-west extensional faults are accommodating any additional height increase by allowing the crust to flow laterally.

When the plateau reached significant heights, it began to deflect regional air-flow currents that in turn deflect the jet streams, causing them to meander and change course. Global weather patterns were strongly changed. In particular, the cold polar jet stream is now at times deflected southward over North America, northwest Europe, and other places where ice sheets have developed. The uplift increased aridity in Central Asia by blocking moist airflow across the plateau, leading to higher summer and cooler winter temperatures. The uplift also intensified the Indian Ocean monsoon over what it was before the uplift, because the height of the plateau intensifies temperature-driven atmospheric flow as higher and lower pressure systems develop over the plateau during winter and summer. This has increased the amount of rainfall along the front of the Himalayan Mountains, where some of the world's heaviest rainfalls have been reported, as the Indian monsoons are forced over the high plateau. The cooler temperatures on the plateau led to the growth of glaciers, which in turn reflect back more sunlight, further adding to the cooling effect.

Paleoclimate records show that the Indian Ocean monsoon underwent strong intensification 7–8 million years ago, in agreement with some estimates of the time of uplift, but younger than other estimates. The effects of the uplift would be different if the uplift had occurred rapidly in the Late Pliocene-Pleistocene (as suggested by analysis of geomorphology, paleokarst, and mammal fauna), or if the uplift had occurred gradually since the Eocene (based on lake sediment analysis). Most geologists accept analysis of data that suggest that uplift began about 25 million years ago, with the plateau reaching its current height by 14 or 15 million years ago. These estimates are based on the timing of the start of extensional deformation that accommodated the exceptional height of the plateau, sedimentological records, and uplift histories based on geothermometry and fission track data.

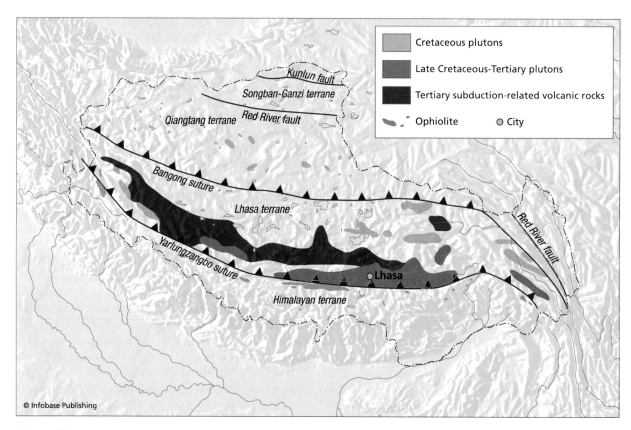

© Infobase Publishing

Map of Tibet showing different terranes that make up the high plateau

HIMALAYA MOUNTAINS

The world's tallest mountains, as well as those exhibiting the greatest vertical relief over short distances, form the Himalaya range that developed in the continent-continent collision zone between India and Asia. The range extends for more than 1,800 miles (3,000 km) from the Karakorum near Kabul (Afghanistan), past Lhasa in Tibet, to Arunachal Pradesh in the remote Assam province of India. Ten of the world's 14 peaks that rise to more than 26,000 feet (8,000 m) are located in the Himalaya, including Mount Everest, 29,035 feet (8,850 m), Nanga Parbat, 26,650 feet (8,123 m), and Namche Barwa, 25,440 feet (7,754 m). The rivers that drain the Himalaya include some with the highest sediment outputs in the world, including the Indus, Ganges, and Brahmaputra. The Indo-Gangetic plain on the southern side of the Himalaya represents a foreland basin filled by sediments eroded from the mountains and deposited on Precambrian and Gondwanan rocks of peninsular India. The northern margin of the Himalaya is marked by the world's highest and largest uplifted plateau, the Tibetan Plateau.

The Himalaya is one of the youngest mountain ranges in the world but has a long and complicated history. This history is best understood in the context of five main structural and tectonic units within the ranges. The Subhimalaya includes the Neogene Siwalik molasse, bounded on the south by the Main Frontal Thrust, which places the Siwalik molasse over the Indo-Gangetic plain. The Lower or Subhimalaya is thrust over the Subhimalaya along the Main Boundary Thrust and consists mainly of deformed thrust sheets derived from the northern margin of the Indian shield. The High Himalaya is a large area of crystalline basement rocks, thrust over the Subhimalaya along the Main Central Thrust. Further north, the High Himalaya sedimentary series, or Tibetan Himalaya, consists of sedimentary rocks deposited on the crystalline basement of the High Himalaya. Finally, the Indus-Tsangpo suture represents the suture between the Himalaya and the Tibetan Plateau to the north.

Sedimentary rocks in the Himalaya record events on the Indian subcontinent, including a thick Cambrian-Ordovician through Late Carboniferous/Early Permian Gondwanan sequence, followed by rocks deposited during rifting and subsidence events on the margins of the Tethys and Neotethys Oceans. The collision of India with Asia was in progress by the Early Eocene. This collision exposed the diverse rocks in the Himalaya, revealing a rich geologic history that extends back to the Precambrian, where shield rocks of the Aravalli Delhi cratons are intruded by 500 million-year-old granites. Subduction of Tethyan oceanic crust along the southern margin of Tibet formed an Andean-style arc represented by the Transhimalaya batholith that extends west into the Kohistan island arc sequence, in a manner similar to the Alaskan range—Aleutians of western North America. The obduction of ophiolites and high-pressure (blueschist facies) metamorphism dated to have occurred around 100 million years ago is believed to be related to this subduction. Thrust stacks began stacking up on the Indian subcontinent, and by the Miocene, deep attempted intracrustal subduction of the Indian plate beneath Tibet along the Main Central Thrust formed high-grade metamorphism and generated a suite of granitic rocks in the Himalaya. After 15–10 million years ago, movements were transferred to the south to the Main Frontal Thrust, which is still active.

CENOZOIC TECTONICS OF ASIA

Many large Mesozoic and Cenozoic basins cover the eastern North China craton and extend northward into Mongolia. The development of these large basins was concentrated in two time periods, the Jurassic to Cretaceous and the Cretaceous to present. An overall NW-SE–trending extensional stress field during formation of these basins was related to changes in convergence rates of India-Eurasia and Pacific-Eurasia combined with mantle upwelling. Two stages of basin formation may have been related to lithosphere erosion that began in the Early Jurassic or to subduction of the Kula plate beneath eastern China in Jurassic-Cretaceous time and later subduction of the Pacific plate. Geophysical and geochemical data show that the areas of thinner lithosphere correspond to the deepest Cenozoic basins. Kimberlites found in these basins provide the only direct source of information about the underlying mantle.

The Cretaceous-Tertiary Tieling basin in northern Liaoning Province hosts Mesozoic-Tertiary kimberlites. Phanerozoic lithosphere beneath the Tan-Lu fault was replaced by hotter, more fertile material that may be related to the Tertiary rifting of the Shanxi highlands. Furthermore, the Eocene Luliang kimberlites imply Phanerozoic-type mantle was in place by the end of the Cretaceous. Another kimberlite within a narrow Cenozoic basin lying along the Tan-Lu fault in Tieling County shows similar Phanerozoic-type mantle related to rifting. Garnet temperatures at shallow depths indicate that significant cooling occurred after the Phanerozoic mantle was emplaced beneath this area.

Cenozoic Extension in the Shanxi Graben and Bohai Sea Basins

Cenozoic extensional deformation in the central North China craton is localized in two elongate graben systems surrounding the Ordos block: the S-shaped Weihe-Shanxi graben system (Shanxi grabens for short) to the east and southeast and the arc-shaped Yinchuan-Hetao graben system to the northwest. The southwestern margin of this block corresponds to a zone of compression, through which the North China craton is in direct contact with the Tibetan Plateau. The subsidence in these grabens began during the Eocene and extended to the whole graben system during the Pliocene. The Shanxi graben system was the last to be initiated in northern China at about 6 million years ago. These two extensional domains show differences in the thickness of the crust and lithosphere, which change sharply across the eastern edge of the Taihangshan Massif on the eastern side of the Shanxi graben system. The Shanxi graben system consists of a series of en echelon depressions bounded by normal faults. The S-shaped geometry of the Shanxi graben system has two broad extensional domains in the north and south, and a narrow transtensional zone in the middle. Satellite image interpretation and field analyses of active fault morphology show predominantly active normal faulting. Right-lateral strike-slip motion along faults that strike more northerly suggest that the Shanxi graben system is a right lateral transtensional shear zone, or alternatively, an oblique divergent boundary between blocks within northern China.

North-northeast–oriented initial extension along the footwall of range-frontal fault zones in northern Shanxi predates the Pliocene opening of the Shanxi graben and may be coincident with the Miocene Hannuoba basalt flow. The direction of extension that prevailed during the initiation and evolution of the Shanxi graben system shows a northward clockwise rotation, from N300–330°E along its southern and middle portion to N330–350°E across the northern part. Late Quaternary active fault morphology implies that the opening of the Shanxi graben system proceeded by northward propagation. This opening mode reflects a counterclockwise rotation of the Taihangshan Massif with respect to the Ordos block around a pole located outside the block.

During the Miocene, the regions of rifting in northern China were subjected to regional subsidence and the eruption of widespread basalt flows. Basalt volcanism, dated to have occurred between 25 and 10 million years ago, was extensive in Mongolia and eastern China, including in the areas of the above rifts. This volcanism was related to extension in response to rollback of the subducted Pacific plate beneath eastern Asia. Miocene normal faulting occurred particularly in the offshore part of the Bohai Sea basin, where this normal fault set strikes more easterly.

Miocene extension in north China may have shared a common mechanism with that of the opening of the Japan Sea. First, the opening of the Japan Sea began at the end of the Oligocene, around 28 million years ago, or earlier to the Middle Miocene, about 18 million years ago; the youngest dredged basaltic volcanic rocks were dated as 11 million years old. Second, the spreading direction of the Japan Sea is roughly N-S to NNE-SSW, consistent with the Miocene stretching direction in northern China. Finally, the same extensional stress regime trending ENE to NE has been documented in northeastern Japan (east of the Japan Sea) based on the direction of dike swarms and dated at 20–15 million years old.

PACIFIC PLATE SUBDUCTION

Subduction along the Pacific margin of China was active from 200 to 100 Ma, soon after closure of the ocean basins on the northern side of the North China craton. Westward-directed oblique subduction was responsible for the generation of arc magmas, deformation, and possibly mantle hydration during this interval. Although the duration and history of Mesozoic subduction beneath the eastern margin of China is not well known, the active margin stepped outward by the Cenozoic, from which a better record is preserved. Numerous plate reconstructions for the Cenozoic of Asia and the eastern Pacific basin show that a wide scenario of different plates, convergence rates, and angles of subduction definitely relate to some of the processes of basin formation, magmatism, and deformation in easternmost China.

The long-lived subduction beneath eastern China may have led to the formation of the many Mesozoic basins in this region. When oceanic lithosphere subducts, it adds water to and thereby weakens the upper mantle. It lowers the melting temperature, and decreases its strength. This may be the principal cause of the fragmentation of the oceanic lithosphere in the western Pacific.

ZAGROS AND MAKRAN MOUNTAINS

The Zagros are a system of folded mountains in western and southern Iran, extending about 1,100 miles (177 km) from the Turkish-Russian-Iranian border, to Zendam fault north of the Strait of Hormuz. The Makran Mountains extend east from the Zagros, through the Baluchistan region of Iran, Pakistan, and Afghanistan. The mountains form the southern and western borders of the Iranian Plateau and Dasgt-e Kavir and Lut Deserts. The northwestern Zagros are forested and snow-capped, and include many volcanic cones, whereas the central Zagros are characterized by many cylindrical folded ridges and interridge basins. The southwest Zagros and Makran ranges are characterized by more subdued topography with bare rock, sand dunes, and lowland salt marshes. Many major oil fields are located in the western foothills of the central Zagros, where many salt domes have punctured through overlying strata creating many oil traps.

Southwestern central Iran has been an active continental margin since the Mesozoic, with at least three main phases of magmatic activity related to

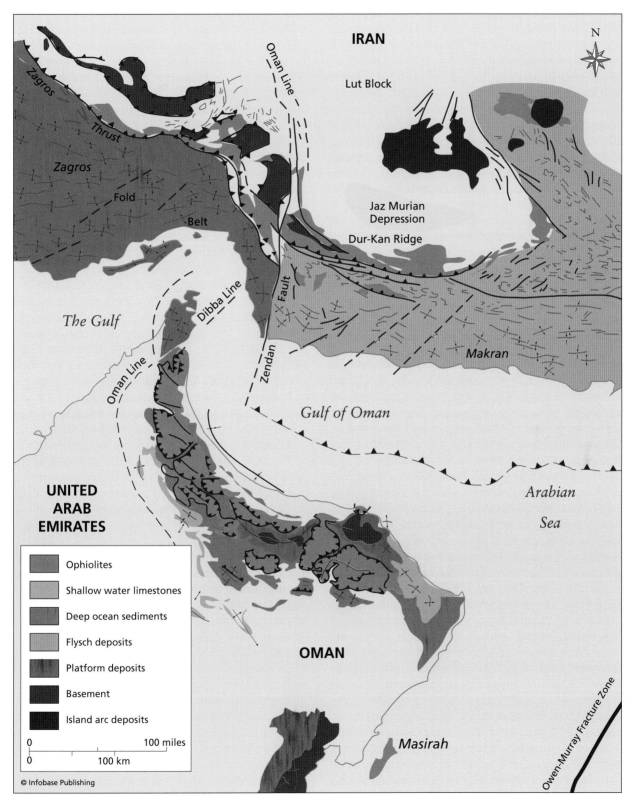

Tectonic map of parts of Iran, Afghanistan, and Arabia showing the Zagros, Makran, and Oman Mountains

subduction of Tethyan oceanic crust beneath the mountain ranges. Late Cretaceous magmatism in the Makran formed above subducting oceanic crust related to the Oman ophiolite preserved on the Arabian continental margin. In the late Eocene the axis of active magmatism shifted inland away from the Mesozoic magmatic belt, but then shifted back during the Oligocene-Miocene. The Oligocene-Miocene magmas are also related to subduction of oceanic crust, suggesting that the Arabian-Iranian collision did not begin until the Miocene. Most of the southern Zagros consists of folded continental margin sediments of the Arabian platform, deformed since the Miocene and mostly since the Pliocene. In contrast, the Makran is an oceanic accretionary wedge consisting of folded Cretaceous to Eocene flysch- and ribbon-chert–bearing mélange resting above the subducting oceanic crust of the Gulf of Oman. A large ophiolitic sheet is thrust over the ophiolitic mélange and flysch and is part of a large ophiolitic belt that stretches the length of the Makran-Zagros ranges, falling between the Cenozoic volcanics and accretionary wedge/folded platform rocks of the Makran and Zagros. The main differences between the Zagros and the Makran exist because continent/continent collision has begun in the Zagros, but has not yet begun in the Makran.

Iran is seismically active, as shown by the devastating magnitude 6.7 earthquake that destroyed the ancient walled fortress city of Bam on December 26, 2003, killing an estimated 50,000 people. The Zagros belt is extremely active, where thrust-style earthquakes occur beneath a relatively ductile layer of folded sedimentary rocks on the surface. The Makran accretionary wedge is also seismically active, especially along the boundary where the subduction zone and upper-plate accretionary wedge meet. The boundary between the Makran and Zagros is a structurally complex region where many strike-slip faults, including the Zendan fault and related structures, rupture to the surface. The Bam earthquake was a strike-slip earthquake, related to this system of structures. The central Iranian plateau is also seismically active, and experiences large-magnitude earthquakes that rupture to the surface.

ENVIRONMENTAL DISASTER OF THE ARAL SEA

The Aral Sea is a large inland sea in southwestern Kazahkstan and northwest Uzbekistan, east of the Caspian Sea. The Aral Sea is fed by the Syr Darya and Amu Darya Rivers, which flow from the Hindu Kush and Tien Shan Mountains to the south, and is very shallow, attaining a maximum depth of only 220 feet (70 m). In the latter half of the 20th century the Soviet government diverted much of the water from the Syr Darya and Amu Darya for irrigation, and

this has had dramatic effects on the inland sea. In the 1970s the Aral was the world's fourth-largest lake, covering 26,569 square miles (68,000 km^2). It has an average depth of 52.5 feet (16 m), and was the source of about 45,000 tons of carp, perch, and pike fish caught each year. Since the diversion of the rivers, the Aral has shrunk dramatically, retreating more than 31 miles (50 km) from its previous shore, lowering the average depth to fewer than 30 feet (9 m), reducing its area to fewer than 15,376 square miles (40,000 km^2), and destroying the fishing industry in the entire region. Furthermore, since the lake bottom has been exposed, winds have been blowing the salts from the evaporated water around the region, destroying local farming. The loss of evaporation from the sea has even changed the local climate, reducing rainfall and increasing temperatures, all of which exacerbate the problems in the region. Disease and famine have followed, devastating the entire Central Asian region.

See also ARABIAN GEOLOGY; ATMOSPHERE; CONVERGENT PLATE MARGIN PROCESSES; CRATON; FLOOD; INDIAN GEOLOGY; PRECAMBRIAN; RUSSIAN GEOLOGY; SUPERCONTINENT CYCLES.

FURTHER READING

Berberian, F., and Manuel Berberian. "Tectono-plutonic episodes in Iran." In *Zagros-Hindu Kush-Himalaya Geodynamic Evolution,* edited by Harsh K. Gupta and Frances M. Delany. Washington, D.C.: American Geophysical Union Geodynamics Series 3, 1981.

Berberian, Manuel. "Active Faulting and Tectonics of Iran." In *Zagros-Hindu Kush-Himalaya Geodynamic Evolution,* edited by Harsh K. Gupta and Frances M. Delany. Washington, D.C.: American Geophysical Union Geodynamics Series 3, 1981.

Dewey, John F., Robert M. Shakleton, Chen Fa Chang, and Yi Ying Sun. "The Tectonic Evolution of the Tibet plateau." In *Tectonic Evolution of the Himalayas and Tibet,* edited by Robert Shakleton, John F. Dewey, and Brian F. Windley. London: Philosophical Transactions of the Royal Society of London Series A., 1988.

Glennie, Ken W., M. W. Hughes-Clarke, M. G. A. Boeuf, W. F. H. Pilaar, and B. M. Reinhardt. "Inter-relationship of Makran-Oman Mountains belts of convergence." In *The Geology and Tectonics of the Oman Region,* edited by A. H. F. Robertson, Mike P. Searle, and Alison C. Ries. London: Geological Society Special Publication 49, 1990.

Kusky, Timothy M., Jianghai Li, and Robert T. Tucker. "The Archean Dongwanzi Ophiolite Complex, North China Craton: 2.505 Billion Year Old Oceanic Crust and Mantle." *Science* 292 (2001): 1,142–1,145.

Kusky, Timothy M., and Jianghai Li. "Paleoproterozoic Tectonic Evolution of the North China Craton." *Journal of Asian Earth Sciences* 22 (2003): 383–397.

Li, Jianghai, Timothy M. Kusky, and Xiongnan Huang. "Archean Podiform Chromitites and Mantle Tectonites in Ophiolitic Mélange, North China Craton: A Record of Early Oceanic Mantle Processes." *GSA Today* 12 (2002): 4–11.

Molnar, Peter. "The Geologic History and Structure of the Himalaya." *American Scientist* 74 (1986): 144–154.

Molnar, Peter, and P. Tapponier. "Active Tectonics of Tibet." *Journal of Geophysical Research* 83 (1978): 5,361–5,375.

Okay, Aral I., and A. M. Celal Sengor. "Evidence for Intracontinental Thrust-Related Exhumation of the Ultra-High Pressure Rocks in China." *Geology* 20 (1992): 411–414.

Raymo, Maureen E., and William F. Ruddiman. "Tectonic Forcing of Late Cenozoic Climate." *Nature* 359 (1992): 117–122.

Royden, L. H., B. C. Burchfiel, and R. D. van der Hilst. "The Geological Evolution of the Tibetan Plateau." *Science* 321 (2008): 1,054–1,058.

Ruddiman, William F. *Tectonic Uplift and Climate Change.* New York: Plenum Press, 1997.

asteroid Asteroids and comets are space objects that orbit the Sun. When these objects enter the Earth's atmosphere, they may make a streak of light known as a meteor, and if they are not burned up in the atmosphere, the remaining rocky or metallic body is known as a meteorite. By definition, asteroids are a class of minor planets that have a diameter fewer than 620 miles (1,000 km), whereas planets are roughly spherical objects that orbit the Sun and have a diameter greater than 620 miles (1,000 km). Comets are partly icy bodies that may have rocky cores and typically orbit the Sun in highly elliptical paths. Most streaks of light known as meteors are produced by microscopic or dust-sized particles entering the atmosphere. Larger objects falling to Earth from space make a larger streak of light, known as a fireball, as they burn up on entry through the atmosphere.

Earth formed through the accretion of many asteroids about 4.5 billion years ago. As the solar system was condensing from a spinning disk of gas and dust particles, these particles began to collide, sometimes sticking to other particles. Gradually the particles became bigger and bigger until they were large rocky and metallic asteroids, which collided with each other until some became so large that they began collecting other asteroids through their gravitational attraction. These protoplanets swept their orbits clear of other asteroids and gradually grew larger in the process. Many of the protoplanets grew large enough to start melting internally, producing layered planetary bodies with dense cores and lighter crusts. At this late stage many protoplanets, including Earth, experienced collisions with other large protoplanets, causing catastrophic melting and fragmentation of the earliest planetary crusts. As this late-stage accretion period ended, Earth went through a period called the late bombardment, when many smaller meteorites were still falling to Earth, causing local disruption of the crust. Since then the number of meteorites that hit Earth has been gradually decreasing, but in the rare events when medium or large objects from space hit Earth, the results can be devastating.

Meteorites are rocky objects from space that strike Earth. When meteorites pass through Earth's atmosphere, they get heated and their surfaces become ionized, causing them to glow brightly, forming a streak moving across the atmosphere known as a shooting star or fireball. If the meteorite is large enough, it may not burn up in the atmosphere and will then strike Earth. Small meteorites may just crash on the surface, but the rare, large object can excavate a large impact crater, or do worse damage. At certain times of the year Earth passes through parts of the solar system that are rich in meteorites, and the night skies become filled with shooting stars and fireballs, sometimes as frequent as several per minute. Examples of these high-frequency meteorite encounters, known as meteor showers, include the Perseid showers that appear around August 11 and the Leonid showers that appear about November 14.

A large body of evidence now suggests that an impact with a space object, probably a meteorite, caused the extinction of the dinosaurs and 65 percent of all the other species on the planet at the end of the Cretaceous Period 66 million years ago. The meteor impact crater is apparently preserved at Chicxulub on Mexico's Yucatán Peninsula, and the impact occurred at a time when the world's biosphere was already stressed, probably by massive amounts of volcanism, global atmospheric change, and sea-level fall. The volcanic fields that were being laid down for a few million years before the mass extinction and death of the dinosaurs are preserved as vast lava plains in western India known as the Deccan traps. Impacts and massive volcanism can both dramatically change the global climate on scales that far exceed the changes witnessed in the past few thousands of years, or in the past hundred years as a result of human activities such as burning fossil fuels. These changes have dramatic influences on evolution and the extinction of species, and current research suggests that impacts and volcanism have been responsible for most of the great extinctions of geological time.

Impacts cause earthquakes of unimaginable magnitude, thousands of times stronger than any ever observed on Earth by humans. If a meteorite lands in the ocean, it can form giant tsunamis hundreds if not thousands of feet (m) tall that sweep across ocean basins in minutes, and run up hundreds of miles (km) onto the continents. Impacts kick up tremendous amounts of dust and hot flaming gases that scorch the atmosphere and fill it with Sun-blocking dust clouds for years. Global fires burn most organic matter in a global fireball, and these fires are followed by a period of dark deep freeze, caused by the atmospheric dust blocking out warming sunlight. This may be followed rapidly by a warm period after the dust settles, caused by the extra CO_2 released in the atmosphere by the impact. These severe and rapid changes in atmospheric and oceanic temperature and chemistry kill off many of the remaining life-forms in the oceans.

Collisions of asteroids with each other in the early history of the solar system led to the formation of planets but left many asteroids and other debris scattered at different places in the solar system. After the initial periods of accretion of the planets, the bombardment of the planets by asteroids decreased, but some bodies still fall to the planets as meteorites. Most are small and burn up before they hit the surface, but others have inflicted tremendous disruption on the planet. However, collisions of comets with Earth are largely responsible for bringing the lighter, volatile elements, including those that make up the planet's air and oceans, to Earth. It is also probable that asteroids or comets brought primitive organic molecules or even life to the planet.

ORBITS OF ASTEROIDS AND COMPOSITIONAL CLASSIFICATION OF METEORITES

Understanding of the composition of meteorites and asteroids has evolved with time. Early studies relied on meteorites, the bodies that had fallen to Earth, since no space travel or observations in space were possible during the previous centuries. Recently space missions to asteroids and remote sensing have enabled observations of asteroids in space to be integrated with the data from samples taken from meteorites collected on Earth.

The ice content in asteroids generally increases as the distance of their orbits from the Sun increases. Geologists classify meteorites collected on Earth according to the composition (reflected in the types of minerals present) and how much they have been metamorphosed, or changed by events that have subjected the meteorites to higher temperatures and pressures. Many meteorites were part of planetesimals (small planets) that had formed iron-rich cores, and then were probably destroyed in a catastrophic collision with another large asteroid early in the history of the solar system, spreading the asteroid debris

of both planetesimals across the solar system. Other asteroids, comets, and meteorites appear not to have ever been part of larger planets, and may represent some of the primordial matter from the solar nebula from which the solar system formed.

Meteorites are classified on the basis of their composition and structure. The aim of such a classification scheme is to group together all the known bodies that may share a common parent body, whether it was a large asteroid, comet, planet, or moon. This is achieved by placing known asteroids into groups and subgroups based on their important mineralogical, physical, chemical, and isotopic properties.

Meteorites are made from material similar to that which makes up the Earth, including common silicate minerals, plus iron and nickel metals. Some meteorites also contain small lumps of material called chondrules, which represent melt droplets that formed before the meteorite fragments were accreted to asteroids, and thus represent some of the oldest material in the solar system. Some meteorites also contain presolar grains, often comprising tiny carbon crystals in the form of diamonds.

Traditional classification schemes for meteorites broke them into three main groups based on composition. These groups include stony meteorites composed mosty of rocky material, iron meteorites composed mostly of metallic material, and mixtures called stony-iron meteorites. These groups were then divided into subgroups; for instance, the stony meteorites were divided into chondrites and achrondrites according to whether or not they contained chondrules. The iron meteorites were divided into textural groups including structures such as octahedrites, hexahedrites, and ataxites. More recently these textures have not been used for classification but only for descriptive purposes, and the iron meteorites are further divided according to their chemistry. Stony-iron meteorites were divided into pallasites and mesosiderites.

More modern schemes use a simpler classification, in which meteorites are classified as either chondrites or nonchondrites. The nonchrondrites are divided into primitive and differentiated types. The differentiated nonchondrites have three groups, including the achondrites, stony-irons, and irons.

Recognizing meteorites on Earth is difficult since they have similar minerals to many Earth rocks. Many meteorites develop a fused crust on their surface from the heat during entry through Earth's atmosphere. Most of the meteorites from Earth used for classifications have been collected from Antarctica, where the only place for rocks on the ice fields to come from is space. Most, about 85 percent of all meteorites falling on Earth, are chondrites, thought to represent largely primitive solar material. For the nonchrondrites, many classifications and descriptions are aimed at determining whether they come from larger parent bodies that broke up during early impacts, and if so, what the characteristics of this parent body may have been.

Chondrites

Chondrites are meteorites that have chemical compositions similar to that of the Sun. Since the Sun makes up about 99 percent of the mass of the solar system, it is assumed that the composition of the Sun represents the average composition for the entire solar system, and that this average composition resembles the original composition of the solar system when it was formed. Therefore since chondrites and the Sun have similar compositions, chondrites are thought to have very primitive compositions that are close to the average starting material that formed the solar system.

Chondritic meteorites contain small round nodules called chondrules that consist of a mixture of crystals and glass. Most interpretations for the origin of chondrules suggest that they represent small droplets of liquid that condensed during the earliest stages of the formation of the solar system, before they were incorporated into the meteorites. Chondrules that have been dated yield isotopic ages of 4.568 billion years, the time that the solar system began to condense from the solar nebula. Thus, chondrules represent small remnants of the earliest solar system material. Other chondritic meteorites, such as the famous Allende meteorite, that have been dated also give ages of 4.566 billion years, similar to that obtained from the chondrites.

Many experiments have been done on chondrules to determine their exact components and the conditions in which they formed since they are interpreted to have formed during the early stages of the formation of the solar system. Knowing the conditions of their formation yields information about the conditions in the early solar system and solar nebula. Most experiments show that the chondrules formed at temperatures of at least 2,700°F (1,500°C) and that the chondrules cooled rapidly. Some chondrules contain unusual minerals. One group of these contains a suite of very high-temperature minerals and is called calcium-aluminum inclusions (CAIs), typically exhibiting textures like concentric skins of an onion. Experiments on the temperature of formation of these CAIs indicate that they formed at temperatures of at least 3,100°F (1,700°C) and underwent slow cooling. From these extraordinarily high temperatures scientists infer that these CAIs represent the oldest parts of the oldest fragments of the early solar system.

Most chondritic meteorites consist of mixtures of chondrules and the minerals olivine and pyroxene, and show little evidence of being heated or metamorphosed since they formed. Some, however, show textures like partial melting that indicate they were heated to temperatures of up to 1,800°F (1,000°C) after they formed. Still others are cut by veins that have minerals with water in their structures such as carbonates, sulfates, and magnetite. Thus water existed in the asteroid belt in the early solar system. The range in the amount of heating of chondrites likely reflects that they were incorporated into a larger body, with the higher temperature heating happening deeper inside this now-destroyed asteroid or protoplanetary body.

Isotopic dating techniques have shown that most chondritic meteorites cooled within 60 million years of the time of the formation of the solar system. Some time after that, impacts in the asteroid belt between the orbits of Mars and Jupiter broke the larger protoplanets or asteroid parent bodies into smaller pieces now preserved as the asteroid belt. Some chondrites are composed of strongly fragmented rock called breccia, produced by these early collisions in the asteroid belt. Calculations of the pressures needed to produce these breccias indicate that the pressures reached 75 giga Pascals, or the equivalent of 750,000 times the atmospheric pressure on Earth.

Chondritic meteorites are divided into a number of classes with similar compositions and textures, thought to represent formation in similar parts of the solar system. These in turn are divided into groups thought to represent fragments of the same parent body.

The main classes of chondrites include the ordinary chondrites, carbonaceous chondrites, and enstatite chondrites. Some classifications add further subdivisions based on the type of alteration or metamorphism, such as alteration by water, or metamorphism by late heating. Ordinary chondrites are thought to have formed in parent bodies that were 100–120 miles (165–200 km) in diameter. Carbonaceous chondrites contain organic material such as hydrocarbons in rings and chains, and amino acids. Even though these organic molecules can serve as building blocks of life, no life has been found on any meteorites unless they were contaminated on Earth. Enstatite chondrites contain small sulfide minerals that indicate very rapid cooling, suggesting that the original material came from deep in a larger body that was broken apart by strong impacts, and the deep material cooled quickly in space after the violent collisions. The enstatite chondrites typically have impact breccias in their structures.

Achondrites

Achondrite meteorites resemble typical igneous rocks found on Earth. They formed by crystallizing from a

Computer artwork of main asteroid belt of the solar system (not to scale), between orbits of Mars and Jupiter *(Mark Garlick/Photo Researchers, Inc.)*

silicate magma and are remnants of larger bodies in the solar system that were large enough to undergo differentiation and internal melting. These meteorites do not contain chondrules or remnant pieces of the early solar system since they underwent melting and recrystallization.

Some achondrites have been shown to have origins on the Earth's moon and on Mars. They formed by crystallization from magma on these bodies, and were ejected from the gravitational fields of these bodies during large-impact events. The debris from these impacts then floated in space until being captured by the gravitational field of Earth, where they fell as meteorites. These meteorites are named for the places they have fallen on Earth, and include shergottites (Shergotty [Shergahti], India), nakhlites (El Nakhla, Egypt), and chassignites (Chassigny, France) (collectively named SNC meteorites after these three falls). SNC achondrites have ages between 150 million and 1.3 billion years, billions of years younger than other meteorites and the age of the solar system. These ages mean that the SNC meteorites must have come from a large planet that was able to remain hot and sustain

magma for a considerable time after formation at 4.56 billion years ago. Chemical analysis of the SNC meteorites revealed that most match the bulk chemistry of Mars, confirming the link. Analysis of the damage done to the surface of these meteorites by cosmic rays as they were in space has yielded estimates for the time of transit from the ejection during impact on Mars to the landing on Earth at fewer than 2 million years. Thus most of the SNC meteorites originated from meteorite impacts on Mars in the past 1–20 million years. An estimated billion tons of material has landed on Earth that was originally ejected by meteorite impact on Mars. A smaller number of SNC meteorites have been shown to come from the Earth's moon.

Other achondrites formed on other bodies that have been destroyed by giant impacts. For example, the howardites, eucrites, and diogenites are thought to have formed in one body, the largest remnant of which is the asteroid 4 Vesta, currently orbiting the Sun in the asteroid belt. The eucrites and diogenites represent basaltic magma produced on this early protoplanet and destroyed by a large impact at 4.4 billion years ago, within a hundred million years of the formation of the solar system. Eucrites are basalts that contain the minerals clinopyroxene and plagioclase, diogenites contain orthopyroxene that formed layers of dense crystals called cumulates, and howardites are breccias of these rocks that formed during the giant impact that destroyed the parent achondrite body.

A number of unusual achondrites have no known parent bodies. These include the acapulcoites, angrites, brachinites, lodranites, and urelites. Some of these are relatively primitive—for instance, the urelites formed early in the solar system evolution, were heated inside a large parent body and crystallized at 2,300°F (1,250°C), were destroyed in a massive impact, then cooled at 50°F (10°C) per hour in the cold vacuum of space. One arubite shows a remarkably fast cooling rate of 1,800°F (980°C) per hour, probably coming from deep within the parent body then being suddenly frozen in space. Brachinites show some of the earliest igneous activity known from any asteroid body, showing the earliest time at which planets may have begun accreting in the early solar system. The crystallized magmas from these bodies have given ages of 4.564 billion years, meaning that the accretion of planetesimals to a size that could partially melt from the solar nebula happened within 5 million years.

Iron Meteorites

Iron meteorites consist of iron, 5–20 percent nickel, and some minor metals, and they contain almost no silicate minerals. They are thought to represent the differentiated cores of large planetesimals or protoplanets that formed in the early solar system in the asteroid belt, grew large enough to melt partially and differentiate into core and mantle and crust, then were broken apart by large impacts exposing the core material to outer space. Iron meteorites therefore represent valuable samples of the cores of planetary bodies.

Iron meteorites are famous for exhibiting a criss-cross texture known as the Widmannstatten texture best shown in polished metallic surfaces. This is produced by intergrown blades of iron and nickel minerals; the size of the blades is related to the cooling rate of the minerals. The slower the cooling, the larger the crystals grow, so it is possible to calculate the cooling rate of the early planetesimals using the Widmannstatten texture, and from that infer the size of the early planetesimals. A range of different cooling rates and sizes has been determined, from 210°F (100°C) per hour in a body of fewer than 48 miles (80 km) in diameter, to cooling at 9°F (5°C) per year in a body up to 210 miles (340 km) in diameter. Just like the chondrites, the iron meteorites are estimated to have formed into the parent bodies or planetesimals within 5 million years of the formation of the solar system. Additional analysis of the cosmic ray interaction of the surfaces of the iron meteorites can tell the amount of time that they have been exposed and traveling through space. In this case the exposure age, dating from the time of breakup of the parent body to the time the meteorites fell to Earth, is remarkably different from other meteorites. The iron meteorite parent bodies apparently broke up by impacts only between 200 million and 1 billion years ago, and thus had survived as large bodies in the asteroid belt for 3.5 to 4.5 billion years.

Iron meteorites are divided into groups on the basis of texture, how the iron and nickel minerals are intergrown, and the sizes of the crystals. The main groups include octahedrites, hexahedrites, and ataxites. They are further divided into a large number of classes based on their chemical characteristics, which indicate that they formed in at least several different parent bodies.

Stony-Iron Meteorites

As the name implies, stony-iron meteorites consist of mixtures of metal and silicate (rocky) components, resembling a cross between achondrites and iron meteorites. They are thought to come from the part of a planetesimal or parent body near the boundary of the core and mantle, incorporating parts of each in the meteorite.

Stony-iron meteorites are classified into pallasites and mesosiderites. Pallasites contain a mixture of Widmannstatten-textured iron phases and large yellow to green olivine crystals, formed along the core-mantle boundary of the parent body. Mesosiderites consist of silicate minerals with inclusions of iron that melted in the core of the parent body. Mesosiderites are puzzling, because the silicate phases are magmatic rocks that formed near the surface of the parent body, and the iron phases are melts from the core of the body. Ages for mesosiderites range from 4.4 to 4.56 billion years, so they formed early in the history of the solar system. Some models suggest that they are breccias from bodies that only partially differentiated in the planetsimal forming stages of the solar system. These bodies broke up by impact by 4.2 billion years ago, then apparently were reassembled by gravity within 10–170 million years. The parent body for the mesosiderites was between 120 and 240 miles (200–400 km) in diameter.

ORBITS AND LOCATION OF ASTEROIDS IN THE SOLAR SYSTEM

Asteroids have been named and numbered in the order they were discovered, since the first discovery of asteroid 1 Ceres in 1801. Most of the large asteroids are located in the main asteroid belt between Mars and Jupiter and are thought to have originated by numerous collisions between about 50 large planetesimals in the early history of the solar system. In this belt the vast number of relatively large bodies meant that there were many mutual collisions, and they never coalesced into a single large body like the other planets. There are 33 known asteroids with diameters larger than 120 miles (200 km), and 1,200 known with diameters larger than 19 miles (30 km). The largest is 1 Ceres (590 miles, or 960 km), and the other large asteroids in this belt include 2 Pallas (350 miles, 570 km), 4 Vesta (326 miles, 525 km), 10 Hygiea (279 miles, 450 km), 15 Eunomia (169 miles, 272 km), and Juno (150 miles, 240 km).

Like planets, asteroids orbit the Sun and rotate on their axes, although many asteroids have more of a tumbling motion than the spinning typical of planets. Many asteroids have unusual shapes, and it is not unusual for asteroids to resemble familiar objects, like giant dog bones (216 Kleopatra, 135 miles, 217 km long) tumbling through space. Typical periods of rotation (corresponding to the length of a day) range from 3 hours to several Earth days, with most falling around periods of 9 hours. Many asteroids are really collections of blocks of rubble all rotating in the same place. These are thought to be asteroids that were once single solid masses, but were strongly fractured and broke apart while in orbit.

Inner Solar System Asteroids

Relatively few asteroids orbit inside the orbit of Jupiter, since the numerous planets and proximity to the Sun in this region exert many forces that cause the orbits of bodies to become unstable. Most asteroids that end up orbiting in the inner solar system (between Jupiter and the Sun) are deflected there by collisions in the other asteroid belts, and have short lifetimes before they collide with a planet, are pulled into the Sun, or are deflected back to outer space.

Despite these forces, a couple of places in the inner solar system exist where gravitational physics lets asteroids have stable orbits. Called resonances, these stable orbits are an effect of the gravity and different orbital periods of the larger planets in the inner solar system. Some locations (the resonances) are at stable locations for bodies that orbit at specific rates relative to the larger planets. The largest planet, Jupiter, forms several resonances within which many asteroids orbits, generally known as Jovian asteroids. These include the Trojans, which have an orbital period equal to Jupiter's; another group known as the Hildas, which orbit three times for every two orbits of Jupiter; and the Thule asteroids, which orbit four times for every three orbits of Jupiter. Just as the resonances represent stable places for asteroids to orbit, the gravitational forces of large planets such as Jupiter cause some places to be particularly unstable for asteroids. These places devoid of asteroids are known as Kirkwood Gaps (after their discoverer, Daniel Kirkwood, who first described them in 1886).

Gravitational physics results in additional locations that represent stable orbits for asteroids in the inner solar system. These stable orbits were first described by the French mathematician Joseph-Louis Lagrange in the late 18th century. Lagrange showed that for a two-body system (e.g., the Sun and a large planet like Jupiter) there are five orbital positions inside the orbit of the planet where the gravitational pull of the two large bodies is equal to the centripetal force acting on other bodies between them, and that the two forces cancel each other and make these stable orbits. Three of the five Lagrange orbits turn out to be unstable over long geological periods, but two yield long-term stable orbits for asteroids. These two so-called Lagrange points turn out to be in the orbit of the planet, located 60 degrees ahead and 60 degrees behind the position of the planet. Asteroids in the solar system located in these points are called Trojans, and many asteroids in the inner solar system fall into this category. Trojans exist for the Sun-Jupiter gravitational system, for the Sun-Mars system, and for many other planets in the solar system. The most abundant Trojans orbit in synchroneity with Jupiter, then with Mars, on either side of the main asteroid belt. The Sun-Earth system has dense concentrations of dust at the Trojan points but no know asteroids.

Gravitational physics predicts that there are two main stable regions in the inner solar system where many asteroids may exist with stable orbits for long periods of time. The first location is inside the orbit of Mercury, but no asteroids have yet been identified in this region. Asteroids that may be in this region are named Vulcan objects, but have proven difficult to observe because of their proximity to the Sun. The second large stable region for asteroids in the inner solar system is known as the main belt, divided into several subbelts, between the orbits of Mars and Jupiter.

Main Asteroid Belt

By far the largest number of asteroids in the inner solar system are located in the main asteroid belt between the orbits of Mars and Jupiter. There are estimated to be between 1.1 and 2 million asteroids in this belt with diameters greater than half a mile (1 km), comprising about 95 percent of the known asteroids. There are probably billions of objects in this belt with diameters of less than half a mile (1 km). Despite the large number of objects in the main asteroid belt, the total mass of all asteroids in this belt is less than one-tenth of 1 percent of the mass of the Earth, and represents a mass less than the Earth's moon.

In 1993 the *Galileo* spacecraft flew about 1,500 miles (2,400 km) from asteroid Ida at a relative velocity of 28,000 mph (12.4 km/sec), in the second encounter of an asteroid by a spacecraft. At the time of nearest pass, the asteroid and spacecraft were 274 million miles (441 million km) from the Sun. Ida is about 32 miles (52 km) in length, more than twice as large as Gaspra, the first asteroid observed by *Galileo* in October 1991. Ida is an irregularly shaped asteroid placed by scientists in the S class (believed to be like stony or stony-iron meteorites). It is a member of the Koronis family, presumed fragments left from the breakup of a precursor asteroid in a catastrophic collision. It has numerous craters, including many degraded craters larger than any seen on Gaspra. The extensive cratering dispels theories about Ida's surface being geologically youthful.

The main asteroid belt is located between 1.7 to 4 astronomical units (A.U., or the distance from the Sun to Earth) from the Sun, in an unusually large gap, between the planets of Mars and Jupiter. Early models for this gap and the asteroids suggested that perhaps there was once a planet in this gap that was destroyed by a tremendous impact that resulted in the formation of the asteroids. Some of the asteroids obviously come from once-larger objects, since their metallic material comes from a differentiated core of a planetlike body, and these differentiated cores need a planet of at least a few hundred miles (few hundred km) to form. It is now clear, however, that the huge gravitational forces from Jupiter are so large that they would have prevented a large planet from ever forming in that gap. Models for most of these asteroids suggest that some formed a number of several-hundred-mile (km) diameter planetesimals, but that none ever reached true planet size. These planetesimals then repeatedly crashed into each other, forming the numerous odd-shaped fragments in the belts. Other asteroids in the main asteroid belt have compositions that show that they were never part of larger bodies, and that they represent material that is "left over" from the solar nebula, still floating freely in space. One of the classes of asteroids, known as carbonaceous chondrites, shows unmetamorphosed (never heated or put under high pressure) minerals, demonstrating that they were never deep in a planetary interior and never located next to the Sun. The compositions of the main-belt asteroids show that they formed in about the same region in which they are located with respect to distance from the Sun.

The main asteroid belt contains several belts of asteroids, and these show variations with distance from the Sun, thought to represent variations in the original solar nebula and in the different parent bodies that broke up during formation of the solar system. One of the main differences is that asteroids inside 2.5 A.U. have no water, whereas beyond 2.5 A.U. the asteroids have water, the amount increasing with distance from the Sun. Not coincidentally, the planets inside the asteroid belt are rocky, whereas those outside the belt are gaseous and icy.

The innermost group of asteroids in the main belt, just beyond the orbit of Mars at 1.52 A.U., are the Hungaria objects, orbiting between 1.78 and 2.0 A.U. These are followed outward by the Flora family between 2.1 and 2.3 A.U., then the main part of the main asteroid belt between 2.3 and 3.25 A.U. The Koronis asteroids are located in this main section of the main belt, including more than 200 identified large bodies orbiting around 3 A.U. A gap lies outward of the Koronis asteroids, then the Cybele family asteroids orbit the Sun at a distance of 3.5 A.U., with properties suggesting they originated from a single large planetsimal destroyed early in the history of the solar system. The outermost part of the main asteroid belt is occupied by the Hilda asteroids, orbiting in a resonance from Jupiter at 3.9 to 4.2 A.U. Beyond the Hilda family is a gap, followed by the orbit of Jupiter and its Trojans.

Aten, Apollo, and Amor-Class Asteroids

In addition to the stable belts numerous asteroids in the inner solar system have unstable orbits, and these bodies present the greatest threat to Earth and its inhabitants. The orbits of some of these asteroids cross the paths of other planets, including Earth, and pose an even greater risk of impact. These asteroids are classified according to their distance from the Sun.

Aten asteroids orbit less than one A.U. and an orbital period of less than one year, and some of these cross Earth's orbit. Many of the more than 220 known Aten objects are being tracked, as they pose a high risk of impact with Earth. Apollo asteroids are similar to Atens in that they cross Earth's orbit, except they have periods longer than one year. Of the more than 1,300 known Apollo asteroids, the largest is 1685 Toto, which is 7.5 miles (12 km) across. Collision of this object with Earth would be cataclysmic. More than 13 Apollo asteroids have diameters greater than 3 miles (5 km), all of which are larger than the asteroid that hit Earth at the end of the Cretaceous, killing the dinosaurs and causing a mass extinction event. Another class of inner planet–crossing asteroids are the Amors, which orbit between Earth and Mars but do not cross Earth's orbit. There are more than 1,200 known Amor objects, and the moons of Mars (Deimos and Phobos) may be Amors that were gravitationally captured by Mars.

The asteroid deemed the most threatening to Earth (among known objects) is 4179 Toutatis. This asteroid measures a mile (1.6 km) across and orbits in a plane only one-half a degree different from Earth's. Because the asteroid that hit Chicxulub and killed the dinosaurs was only 6 miles (10 km) across, the devastation potential of a collision of Earth with Toutatis is colossal. The possibility of a collision is not that remote—on September 29, 2004, Toutatis passed by Earth at only four times the distance to the Moon. The next-largest object in nearly coplanar orbits with Earth are less than 0.6 miles (1 km) in diameter, but a collision with these would also produce an impact crater greater than 15 miles (25 km) across. Impact statistics predict that objects this size should still be hitting Earth three times per million years.

Outer Solar System Asteroids

The outer solar system, beyond the orbit of Jupiter, is awash in asteroids, most of which are icy compared with the rocky and metallic bodies of the inner solar system. In addition to the Trojans around Jupiter, a group of about 13,000 asteroids with highly eccentric and inclined orbits cross the path of Jupiter, in positions that cause relatively frequent collisions and deflections of the asteroids into the inner solar system. Asteroids whose orbits are inside the orbit of a planet generally do not hit that planet, but only hit planets closer to the Sun than its orbit. This is an artifact of the great gravitational attraction of the Sun, constantly pulling these objects closer in toward the center of the solar system.

Centaurs, a group of asteroids with highly eccentric orbits that extend beyond yet cross the orbits of Jupiter and Saturn, can thus potentially collide with these planets. Many of these are large bodies thought to have been deflected inward from the Kuiper belt, into unstable orbits that have them on an eventual collision course with the giant planets, or to be flung into the inner solar system. Coming from so far out in the solar system, Centaurs are icy bodies. One Centaur, Chiron, is about 50 miles (85 km) in diameter, is classified as a minor planet, and exhibits a cometary tail when at its perihelion but not along other parts of its orbit. Chiron therefore is classified as both an asteroid and a comet.

Trans-Neptunian objects are a class of asteroid that orbit beyond the orbit of Neptune at 30 A.U., and beyond into the Kuiper belt, extending from 30 to 49 A.U. Beyond the Kuiper belt is a gap of about 11 A.U. containing relatively few asteroids before the beginning of the Oort Cloud. The total number of objects in this belt is unknown but undoubtedly large, because many are being discovered as the ability to detect objects at this distance increases. More than a thousand Trans-Neptunian objects are currently documented.

Kuiper Belt

The Kuiper belt contains many rocky bodies and is thought to be the origin of many short-period comets. Few of the known Kuiper belt objects have been shown to have frozen water, however, but may have ices of other compositions. The total mass of asteroids in the Kuiper belt is estimated at about 20 percent of Earth's mass, about 100 times as much as the mass of the main asteroid belt.

Formerly classified as a planet, Pluto, and its moon, Charon, are Kuiper belt objects. Pluto is one of the larger Kuiper belt objects, but its size is not anomalous for the belt—there are thought to be thousands of Kuiper belt objects with diameters greater than 600 miles (1,000 km), 70,000 asteroids or comets with diameters greater than 60 miles (1,000 km) and half a million objects with diameters greater than 30 miles (50 km). Saturn's moon Phoebe is an icy captured asteroid that was deflected inward from the Kuiper belt and gravitationally captured by Saturn. Its surface shows a mixture of dusty rock debris and ice.

Oort Cloud

The Oort Cloud is a roughly spherical region containing many comets and other objects, extending from about 60 A.U. to beyond 50,000 A.U., or about 1,000 times the distance from the Sun to Pluto, or about one light year. This distance of the outer edge of the solar system is also about one quarter of the way to the closest star neighbor, Proxima Centauri. The Oort Cloud is thought to be the source for long-period comets and Halley-type comets that enter the inner solar system. It contains rocky as well as icy bodies.

The Oort Cloud can be divided into two main segments including the inner, doughnut-shaped segment from 50 to 20,000 A.U., and the outer spherical shell extending from 20,000 to at least 50,000 A.U. Some estimates place the outer limit of the Oort Cloud at 125,000 A.U. The inner part of the Oort Cloud is also known as the Hills Cloud. It is thought to be the source of Halley-type comets, whereas the outer Oort Cloud is the source of the long-period comets that visit the inner solar system. The Hills, or inner Oort Cloud, contains much more material than the outer Oort Cloud, yet the outer cloud contains trillions of comets and bodies larger than 0.8 mile (1.3 km) across, spaced tens of millions of miles (km) apart. The total mass of the outer Oort Cloud is estimated to be several Earth masses.

COMPOSITION AND ORIGIN OF ASTEROIDS, METEORITES, AND COMETS

The asteroid belt is thought to have originated as a group of larger planetesimals that began to be formed within 5 million years of the formation of the solar system, and that soon afterward the planetesimals began to be broken apart by mutual collisions among them. Some collisions happened within a few tens to hundreds of millions of years after initial formation of the bodies; others have happened within the past 200 million years. Most of these collisions are induced by the gravitational forces between Jupiter and the Sun.

The composition of asteroids is determined through remote sensing methods, typically using reflection spectra from the surfaces of asteroids. Presently no samples have been returned from exploration missions to asteroids, so it is difficult to correlate directly their composition with meteorites. The remote sensing studies of asteroids reveal that they have a diverse range of compositions and closely match the range of meteorite compositions found on Earth. In this way some meteorites have been matched to remnants of their parent bodies in the asteroid belt. For instance, asteroid 4 Vesta has the same composition as and is thought to be the largest remnant of the parent body for the howardite, eucrite, and diogenite classes of achondrites.

The composition of asteroids changes gradually with distance from the Sun. The asteroids closest to Mars are classified as S-type silicate bodies and resemble ordinary chondrites. These are followed outward by more abundant B- and C-types, containing some water-rich minerals, and appear to be carbonaceous chondrites. D- and P-types rise in abundance outward, but these do not have any known correlatives in meteorites that have fallen to Earth. These dark objects appear rich in organic material.

The outer solar system asteroids, including those in the Oort Cloud, are thought to be the remnants of the original protoplanetary disc that the solar system formed from 4.6 to 4.5 billion years ago. Many of the objects in the Oort Cloud may have initially been closer to the Sun, but moved outward from gravitational perturbations by the outer planets. The current mass of the Oort Cloud, three to four Earth masses, is much less than the 50–100 Earth masses estimated to have been ejected from the solar system during its formation. It is possible that the outer edges of the Oort Cloud interact gravitationally with the outer edges of other Oort Clouds from nearby star systems, and that these gravitational interactions cause comets to be deflected from the cloud into orbits that send them into the inner solar system.

Bombardment of Earth by comets early in its history may have brought large quantities of water and organic molecules to the planet. In some models for the evolution of the early Earth, most of the volatiles initially on the planet were blown away by a strong solar wind associated with a T-Tauri phase of solar evolution, and the present-day atmosphere and oceans were brought to Earth by comets. Small microcomets continue to bombard Earth constantly, bringing a constant stream of water molecules to Earth from space.

SUMMARY

Modern classification schemes for the composition of asteroids and meteorites divide them into either chondrites or nonchondrites. The nonchondrites are divided into primitive and differentiated types. The differentiated nonchondrites have three groups, including the achondrites, stony-irons, and irons, based on the chemistry and texture of the meteorites and reflecting their origin. Chondrites have compositions similar to the Sun, and represent the average composition of the solar system, thought to be close to the original composition of the solar nebula. Many chondrites contain chondrules, which are small, originally liquid melt drops of the original material that condensed to form the solar system 4.6 billion years ago. Some chondrites contain calcium-aluminum inclusions, which may represent presolar system material. One class of chondrites, carbonaceous condrites, contain complex organic molecules. The variation in chondritic meteorites is thought to represent formation at different depth in a large, 100–120-mile (165–200-km) wide asteroid destroyed in a catastrophic collision early in the history of the solar system, dispersing the fragments across the solar system.

The nonchondrites are divided into primitive and differentiated types. The differentiated nonchondrites have three groups, including the achondrites, stony-irons, and irons. Achondrites are rocky silicate igneous rocks, whereas the irons consist of mixtures of iron and nickle. Stony-irons represent a transitional group. These meteorites are also thought to have formed in several parent bodies destroyed by collisions in what is now the asteroid belt, but these bodies were initially large enough (120–240 miles [200–400 km]) across that they were able to differentiate into crust, mantle, and core. The irons are from the core of these bodies, the achondrites from the mantle and crust, and the stony-irons from the transition zone. Some unusual achondrites have been shown to have been ejected from the Moon and Mars during impacts, eventually landing on Earth.

Most meteorites are thought to come from the asteroid belt, where 1–2 million asteroids with diameters greater than 0.6 miles (1 km) are orbiting the Sun between Mars and Jupiter. They may get pushed into Earth-crossing orbits after being deflected by collisions in the asteroid belt or by gravitational perturbations during complex orbital dynamics. Spectral measurements of some of the asteroids show that their compositions correlate with the meteorites sampled on Earth, and a crude gradation of compositions in the asteroid belt is thought to represent both the original distribution of different parent bodies that broke up during collisions and the initial compositional trends across the solar nebula. Asteroids closer to the Sun are rockier, with more silicates and metals, whereas those farther out have more ices of nitrogen, methane, and water.

Several different groups of asteroids have unstable orbits that cross the paths of the planets in the inner solar system. These objects represent grave dangers to life on Earth, as any impacts with large objects are likely to be catastrophic. These Earth and Mars orbit-crossing asteroids are classified according to their increasing distance from the Sun into Aten-, Apollo-, and Amor-class asteroids. Some of these asteroids are being tracked, to monitor the risk to life on Earth, since collisions of asteroids of this size are known to cause mass extinction events, such as the Cretaceous-Tertiary extinction that killed the dinosaurs. Major impacts occur on Earth about every 300,000 years.

The outer solar system also hosts belts of asteroids, and the number and mass of these objects pales in comparison with the amount of material in the inner solar system. There are many names for asteroids and other bodies orbiting in specific regions, but the bodies of most significance include the Trans-Neptunian objects that orbit beyond the orbit of Neptune at 30 A.U. and into the Kuiper belt, that extends to about 49 A.U. Beyond this there is a relatively empty gap before the beginning of the Oort Cloud at 60 A.U. Most objects in the Kuiper belt and the Oort Cloud consist of mixtures of rock and ice, and are the source region for comets. There are thought to be thousands of Kuiper belt objects with diameters greater than 600 miles (1,000 km), 70,000 asteroids or comets with diameters greater than 60 miles (1,000 km), and half a million objects with diameters greater than 30 miles (50 km).

The Oort Cloud represents the outer reaches of the solar system, and may actually extend into the Oort Cloud of the nearby star system, Proxima Centauri. There are thought to be trillions of comets in the Oort Cloud over 0.8 miles (1.3 km) in diameter, totally several Earth masses. The Oort Cloud is the source of long-period comets, with orbits longer than 200 years. Comets typically have a rocky core and emit jets of ices consisting of methane, water, and ammonia, and other ice compounds. Many comets are coated by a dark surface consisting of complex organic molecules, and these may be the source for much of the carbon and volatile elements on the Earth that presently make up much of the atmosphere and the oceans. Some scientists speculate that comets may be responsible for bringing the complex organic molecules to Earth that served as the building blocks for life.

See also ASTRONOMY; ASTROPHYSICS; COMET; METEOR, METEORITE; ORIGIN AND EVOLUTION OF THE EARTH AND SOLAR SYSTEM; SOLAR SYSTEM.

FURTHER READING

Albritton, C. C. Jr. *Catastrophic Episodes in Earth History*. London: Chapman and Hale, 1989.

Alvarez, Walter. *T Rex and the Crater of Doom*. Princeton, N.J.: Princeton University Press, 1997.

Angelo, Joseph A. *Encyclopedia of Space and Astronomy*. New York: Facts On File, 2006.

Chaisson, Eric, and Steve McMillan. *Astronomy Today*. 2nd ed. Upper Saddle River, N.J.: 2007.

Chapman, C. R., and D. Morrison. "Impacts on the Earth by Asteroids and Comets: Assessing the Hazard." *Nature* 367 (1994): 33–39.

Cox, Donald, and James Chestek. *Doomsday Asteroid: Can We Survive?* New York: Prometheus Books, 1996.

Dressler, B. O., R. A. F. Grieve, and V. L. Sharpton, eds. *Large Meteorite Impacts and Planetary Evolution*. (1994): 348. Boulder, Colo.: Geological Society of America Special Paper 293.

Elkens-Tanton, Linda T. *Asteroids, Meteorites, and Comets*. New York: Facts On File, 2006.

Erickson, J. *Asteroids, Comets, and Meteorites: Cosmic Invaders of the Earth*. New York: Facts On File, 2003.

Hodge, Paul. *Meteorite Craters and Impact Structures of the Earth*. Cambridge: Cambridge University Press 1994.

Krinov, E. L. *Giant Meteorites*. Oxford: Pergamon Press, 1966.

Lunar and Planetary Laboratory, University of Arizona. "Students for the Exploration and Development of Space (SEDS)." Available online. URL: http://seds.lpl.arizona.edu/nineplanets/nineplanets/meteorites.html. Accessed October 26, 2008.

National Aeronautic and Space Administration (NASA). NASA's Web site on Lunar and Planetary Science, including information about all the planets, major asteroids, near Earth asteroid tracking systems, and current and past missions to asteroids. Available online. URL: http://nssdc.gsfc.nasa.gov/planetary/planets/asteroidpage.html Accessed October 26, 2008.

Wasson, John T. *Meteorites: Their Record of Early Solar-System History*. New York: W.H. Freeman, (1985): 267.

asthenosphere   The asthenosphere is the layer of the Earth's mantle between the lithosphere and the mesosphere. Its depth in the Earth ranges from about 155 miles (250 km) to zero miles below the midocean ridges, and 31 to 62 miles (50–100 km) below different parts of the continents and oceans. Some old continental cratons have deep roots that extend deeper into the asthenosphere. The asthenosphere is characterized by small amounts (1–10 percent) of partial melt that greatly reduces the strength of the layer and is thought to accommodate much of the movement of the plates and vertical isostatic motions. The name derives from the Greek for "weak sphere." S-wave seismic velocities clearly demarcate the asthenosphere and show a dramatic drop through the asthenosphere because of the partial melt present in this zone. Because of this the asthenosphere is also known as the low-velocity zone, and it shows the greatest attenuation, or weakening, of seismic waves anywhere in the Earth.

The asthenosphere is composed of the rock type peridotite, consisting primarily of the mineral olivine, with smaller amounts of the minerals orthopyroxene, clinopyroxene, and other accessory minerals including spinels such as chromite. The term *peridotite* is a general term for many narrowly defined ultramafic rock compositions including harzburgite, lherzolite, websterite, wehrlite, dunite, and pyroxenite. Peridotites are not common in the continental crust but are common in the lower cumulate section of ophiolites, in the mantle, and in continental layered intrusions and ultramafic dikes. Peridotites have unstable compositions under shallow crustal metamorphic conditions, and in the presence of shallow-surface hydrating weathering conditions, they commonly become altered to serpentinites through the addition of water to the mineral structures.

The asthenosphere is flowing in response to heat loss in the deep Earth, and geologists are currently debating the relative coupling between the flowing asthenosphere and the overlying lithosphere. In some models the convection in the asthenosphere exerts a considerable mantle drag force on the base of the lithosphere, and significantly influences plate motions. In other models the lithosphere and asthenosphere are thought to be largely uncoupled, with the driving forces for plate tectonics being more related to the balance between the gravitational ridge push force, slab pull force, slab drag force, transform resistance force, and subduction resistance force. There is also a current debate on the relationship between upper-mantle (asthenosphere) convection and convection in the mesosphere. Some models propose double or several layers of convection, whereas other models purport that the entire mantle is convecting as a single layer.

See also CONVECTION AND THE EARTH'S MANTLE; ENERGY IN THE EARTH SYSTEM; MANTLE; PLATE TECTONICS.

astronomy Astronomy is the study of celestial objects and phenomena that originate outside the Earth's atmosphere. The name is derived from the Greek words *astron* for star and *nomos* for law and includes the study of stars, planets, galaxies, comets, interstellar medium, the large-scale structure of the universe, and the natural laws that describe these features. Astronomy is also concerned with the chemistry and meteorology of stellar objects, the physics of motion, and the evolution of the universe through time.

A BRIEF HISTORY OF ASTRONOMY

Ancient cultures were fascinated with the heavens, and astronomy developed into one of the earliest sciences as these cultures formalized their studies of the night skies. Much of the work of these early astronomers focused on observations and predictions of the motions of objects visible to the naked eye, and some cultures erected large monuments that likely have astronomical significance. Early Jewish, Chinese, and other cultures established calendars based on observations and calculations of the Moon cycles, and these calendars became essential for determining seasons and knowing when to plant crops. By the year 1000 B.C.E. Chinese astronomers had calculated Earth's obliquity to the ecliptic, or the tilt of the planet's axis relative to the orbital plane about the Sun. Early astronomers included the study of astrology, celestial navigation, and time calculations such as making calendars in their field, but modern professional astronomy is equivalent to astrophysics, with branches of observational and theoretical astronomy.

A revolution in astronomy and science was marked by Polish astronomer Nicolaus Copernicus's (1473–1543) proposal in his book *De revolutionibus orbium coelestium* (On the revolutions of the heavenly spheres), where he proposed that the Earth is not the center of the universe as most previous scientists believed, but that the Earth and other planets orbit around the Sun. Astronomy changed from its classical period to its modern period with the invention of the telescope in the late 16th century. Some of the early Islamic scholars described the optics of lenses required for telescopes, but the first surviving instruments are from the Netherlands, invented by eye spectacle makers Hans Lippershey and Zacharias Janssen of Middleberg, and Jacob Matius of Alkaamar. In 1602 Galileo Galilei, a physicist from Tuscany, improved on these designs and produced a telescope that earned him the nickname father of modern observational astronomy. These designs were further improved by the English physicist Isaac Newton in 1668.

The German astronomer and natural philosopher Johannes Kepler (1571–1630) further described and refined the laws of planetary motion. Newton further explained these laws in his law of universal gravitation, still used as a general approximation for most gravity-driven processes. Albert Einstein's general theory of relativity more accurately describes gravity, but Newton's laws work for most applications. Further significant advances in astronomy came with the inventions of new technologies, including photography and the spectroscope. Spectroscopic observations of the Sun by Bavarian optician Joseph von Fraunhofer (1787–1826) showed about 600 spectral bands present, which were correlated with different elements by the German physicist Gustav Kirchhoff in 1859. Spectral observations of other stars revealed similar spectral bands, and hence similar compositions to the Sun.

SUBDISCIPLINES OF MODERN ASTRONOMY

One relatively newer goal of modern astronomy is to describe and characterize objects in the distant universe, with the Milky Way galaxy being recognized as a distinct and related group of stars only in the 20th century. This realization was followed by recognition of the expansion of the universe as described by Hubble's law, as well as distant objects such as quasars, pulsars, radio galaxies, black holes, and neutron stars.

The field of observational astronomy is based on data received from electromagnetic radiation from celestial objects and is divided into different sub-fields based on the wavelengths being studied. Radio astronomy deals with interpreting radiation received from celestial objects where the radiation has a wavelength greater than one millimeter, and is commonly used to study supernovae, interstellar gas, pulsars, and galactic nuclei. Radio astronomy uses wave theory to interpret these signals, since these long wavelengths are more easily assigned wavelengths and amplitudes than shorter wavelength forms of radiation. Most radio emissions from space received on Earth are a form of synchrotron radiation, produced when electrons oscillate in a magnetic field, although some is also associated with thermal emission from celestial objects, and interstellar gas is typically associated with 21-cm radio waves.

Infrared astronomy works with infrared wavelengths (longer than the wavelength of red light) and is used primarily to study areas such as planets and circumstellar disks that are too cold to radiate in the visible wavelengths of the electromagnetic spectrum. The longer infrared wavelengths are able to penetrate dust clouds, so infrared astronomy is also useful for observing processes such as star formation in molecular clouds and galactic cores blocked from observations in the visible wavelengths. Infrared astronomy observatories must be located in outer space or in high dry locations since the Earth's atmosphere is associated with significant infrared emissions.

Optical astronomy, the oldest form of observational astronomy, uses light recorded from the visible wavelengths. Most optical astronomy is now completed by using digital recording apparatus, speeding analysis. Ultraviolet astronomy (observations in the ultraviolet wavelengths) is used to study thermal radiation and the emission of spectral lines from hot blue stars, planetary nebula, supernova, and active galactic nuclei. Like infrared observatories, ultraviolet observation stations must be located in the upper atmosphere or in space, since ultraviolet rays are strongly absorbed by Earth's atmosphere.

The study and analysis of celestial objects at X-ray wavelengths is known as X-ray astronomy. X-ray emitters include some binary star systems, pulsars, supernova remnants, elliptical galaxies, galaxy clusters, and active galactic nuclei. X-rays are produced by celestial objects by thermal and synchrotron emission (generated by the oscillation of electrons around magnetic fields), but are absorbed by the Earth's atmosphere, so they must be observed from high-altitude balloons, rockets, or space. The study of the shortest wavelengths of the electromagnetic spectrum, known as gamma-ray astronomy, can so far be observed only by indirect observations of gamma ray bursts from objects including pulsars, neutron stars, and black holes near galactic nuclei.

See also ASTROPHYSICS; BLACK HOLES; CONSTELLATION; COSMOLOGY; GALAXIES; GALAXY CLUSTERS; ORIGIN AND EVOLUTION OF THE UNIVERSE; UNIVERSE.

FURTHER READING

Chaisson, Eric, and Steve McMillan. *Astronomy Today.* 6th ed. Upper Saddle River, N.J.: Addison-Wesley, 2007.

Comins, Neil F. *Discovering the Universe.* 8th ed. New York: W. H. Freeman, 2008.

Snow, Theodore P. *Essentials of the Dynamic Universe: An Introduction to Astronomy.* 4th ed. St. Paul, Minn.: West Publishing Company, 1991.

astrophysics Astrophysics is the branch of astronomy that examines the behavior, physical properties, and dynamic processes of celestial objects and phenomena. Astrophysics includes study of the luminosity, temperature, density, chemical composition, and other characteristics of celestial objects and aims at understanding the physical laws that explain these characteristics and behavior of celestial systems. Astrophysics is related to observational astronomy, as well as cosmology, the study of the theories related to the very large-scale structure and evolution of the universe. Astrophysicists study these systems using principles from different subfields in physics and astronomy, including thermodynamics, mechanics, electromagnetism, quantum mechanics, relativity, nuclear and particle physics, and atomic and molecular physics.

Much of astrophysics is founded on formulating theories based on observational astronomy using principles of quantum mechanics and relativity. Theoretical astrophysicists use analytical models and complex computational and numerical models of the behavior of celestial systems to understand better the origin and evolution of the universe and to test for unpredicted phenomena. In general theoretical models of celestial behavior are tested with the observations and constraints from astronomical studies, and the agreement (or lack thereof) between the model and the observed behavior is used to refine the models of celestial evolution.

Current topics of research in astrophysics include celestial and stellar dynamics and evolution, the large-scale structure of the universe, cosmology and the origin and evolution of the universe, models for galaxy formation, the physics of black holes, quasars, and phenomena such as gravity waves, and implications and tests of models of general relativity.

See also ASTRONOMY; BLACK HOLES; CONSTELLATION; COSMOLOGY; GALAXIES; GALAXY CLUSTERS; ORIGIN AND EVOLUTION OF THE UNIVERSE; UNIVERSE.

FURTHER READING

Chaisson, Eric, and Steve McMillan. *Astronomy Today.* 6th ed. Upper Saddle River, N.J.: Addison-Wesley, 2007.

Comins, Neil F. *Discovering the Universe.* 8th ed. New York: W. H. Freeman, 2008.

Encyclopedia of Astronomy and Astrophysics. CRC Press, Taylor and Francis Group. Available online. URL: http://eaa.crcpress.com/. Accessed October 24, 2008.

ScienceDaily. "Astrophysics News." ScienceDaily LLC. Available online. URL: http://www.sciencedaily.com/news/space_time/astrophysics/. Accessed October 24, 2008.

Snow, Theodore P. *Essentials of the Dynamic Universe: An Introduction to Astronomy.* 4th ed. St. Paul, Minn.: West Publishing Company, 1991.

atmosphere   Thin sphere around the Earth consisting of the mixture of gases we call air, held in place by gravity. The most abundant gas is nitrogen (78 percent), followed by oxygen (21 percent), argon (0.9 percent), carbon dioxide (0.036 percent), and minor amounts of helium, krypton, neon, and xenon. Atmospheric (or air) pressure is the force per unit area (similar to weight) that the air above a certain point exerts on any object below it. Atmospheric pressure causes most of the volume of the atmosphere to be compressed to 3.4 miles (5.5 km) above the Earth's surface, even though the entire atmosphere is hundreds of kilometers thick.

The atmosphere is always moving, because the equator receives more of the Sun's heat per unit area than the poles. The heated air expands and rises to where it spreads out, then it cools and sinks, and gradually returns to the equator. This pattern of global air circulation forms Hadley cells that mix air between the equator and midlatitudes. Similar circulation cells mix air in the middle to high latitudes, and between the poles and high latitudes. The effects of the Earth's rotation modify this simple picture of the atmosphere's circulation. The Coriolis effect causes any freely moving body in the Northern Hemisphere to veer to the right, and toward the left in the Southern Hemisphere. The combination of these effects forms the familiar trade winds, easterlies and westerlies, and doldrums.

The atmosphere is divided into several layers, based mainly on the vertical temperature gradients that vary significantly with height. Atmospheric pressure and air density both decrease more uniformly with height, and therefore are not a useful way to differentiate between different atmospheric layers.

The lower 36,000 feet (11,000 m) of the atmosphere, the troposphere, is where the temperature generally decreases gradually, at about 70°F per mile

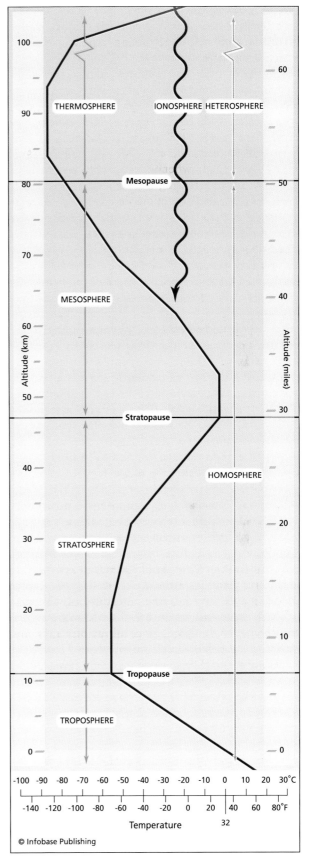

Structure of the atmosphere showing various layers and temperature profile with height

(6.4°C per km), with increasing height above the surface. This is because the Sun heats the surface, which in turn warms the lower part of the troposphere. Most of the familiar atmospheric and weather phenomena occur in the troposphere.

Above the troposphere is a boundary region known as the tropopause, marking the transition into the stratosphere. The stratosphere in turn continues to a height of about 31 miles (50 km). The base of the stratosphere contains a region known as an isothermal, where the temperature remains the same with increasing height. The tropopause is generally at higher elevations in summer than winter and is also the region where the jet streams are located. Jet streams are narrow, streamlike channels of air that flow at high velocities, often exceeding 115 miles per hour (100 knots). Above about 12.5 miles (20 km), the isothermal region gives way to the upper stratosphere, where temperatures increase with height, back to near surface temperatures at 31 miles (50 km). The heating of the stratosphere is due to ozone at this level absorbing ultraviolet radiation from the Sun.

The mesosphere lies above the stratosphere, extending between 31 and 53 miles (50–85 km). An isothermal region known as the stratopause separates the stratosphere and mesosphere. The air temperature in the mesosphere decreases dramatically above the stratopause, reaching a low of -130°F (-90°C) at the top of the mesosphere. The mesopause separates the mesosphere from the thermosphere, a hot layer where temperatures rise to more than 150°F (80°C). The relatively few oxygen molecules at this level absorb solar energy and heat quickly, and temperatures may change dramatically in this region in response to changing solar activity. The thermosphere continues to thin upward, extending to about 311 miles (500 km) above the surface. Above this level, atoms dissociate from molecules and are able to shoot outward and escape the gravitational pull of Earth. This far region of the atmosphere is sometimes referred to as the exosphere.

In addition to the temperature-based division of the atmosphere, it is possible to divide the atmosphere into different regions based on their chemical and other properties. Using such a scheme, the lower 46.5–62 miles (75–100 km) of the atmosphere may be referred to as the homosphere, which contains a well-mixed atmosphere with a fairly uniform ratio of gases from base to top. In the overlying heterosphere, the denser gases (oxygen and nitrogen) have settled to the base, whereas lighter gases (hydrogen and helium) have risen to greater heights, resulting in chemical differences with height.

The upper parts of the homosphere and the heterosphere contain a large number of electrically charged particles known as ions. Also called the ionosphere, this region strongly influences radio transmissions and the formation of the aurora borealis and aurora australis.

The production and destruction or removal of gases from the atmospheric system occur at approximately equal rates, although some gases are gradually increasing or decreasing in abundance, as described below. Soil bacteria and other biologic agents remove nitrogen from the atmosphere, whereas decay of organic material releases nitrogen back to the atmosphere. However, decaying organic material removes oxygen from the atmosphere by combining with other substances to produce oxides. Animals also remove oxygen from the atmosphere by breathing, whereas photosynthesis returns oxygen to the atmosphere.

Water vapor is an extremely important gas in the atmosphere, but it varies greatly in concentration (0–4 percent) from place to place and from time to time. Though water vapor is normally invisible, it becomes visible as clouds, fog, ice, and rain when the water molecules coalesce into larger groups. In the liquid or solid state, water constitutes the precipitation that falls to Earth and is the basis for the hydrologic cycle. Water vapor is also a major factor in heat transfer in the atmosphere. A kind of heat known as latent heat is released when water vapor turns into solid ice or liquid water. This heat, a major source of atmospheric energy, is a major contributor to the formation of thunderstorms, hurricanes, and other weather phenomena. Water vapor may also play a longer-term role in atmospheric regulation, as it is a greenhouse gas that absorbs a significant portion of the outgoing radiation from the Earth, causing the atmosphere to warm.

Carbon dioxide (CO_2), although small in concentration, is another very important gas in the Earth's atmosphere. Carbon dioxide is produced during decay of organic material, from volcanic outgassing, deforestation, burning of fossil fuels, and cow, termite, and other animal emissions. Plants take up carbon dioxide during photosynthesis, and many marine organisms use it for their shells, made of $CaCO_3$ (calcium carbonate). When these organisms (for instance, phytoplankton) die, their shells can sink to the bottom of the ocean and be buried, removing carbon dioxide from the atmospheric system. Like water vapor, carbon dioxide is a greenhouse gas that traps some of the outgoing solar radiation reflected from the earth, causing the atmosphere to warm up. Because carbon dioxide is released by the burning of fossil fuels, its concentration is increasing in the atmosphere as humans consume more fuel. The concentration of CO_2 in the atmosphere has increased by 15 percent since 1958, enough to cause considerable global warming. Estimates predict that the concentration of CO_2 will increase by another 35 percent by the end of the 21st century, further enhancing global warming.

Other gases also contribute to the greenhouse effect, notably methane (CH_4), nitrous oxide (NO_2) and chlorofluorocarbons (CFCs). Methane concentration is increasing in the atmosphere and is produced by the breakdown of organic material by bacteria in rice paddies and other environments, termites, and the stomachs of cows. Produced by microbes in the soil, NO_2 is also increasing in concentration by 1 percent every few years, even though it is destroyed by ultraviolet radiation in the atmosphere. Chlorofluorocarbons have received much attention since they are long-lived greenhouse gases increasing in atmospheric concentration as a result of human activity. Chlorofluorocarbons trap heat like other greenhouse gases, and also destroy ozone (O_3), a protective blanket that shields the Earth from harmful ultraviolet radiation. Chlorofluorocarbons were used widely as refrigerants and as propellants in spray cans. Their use has been largely curtailed, but since they have such a long residence time in the atmosphere, they are still destroying ozone and contributing to global warming, and will continue to do so for many years.

Ozone is found primarily in the upper atmosphere where free oxygen atoms combine with oxygen molecules (O_2) in the stratosphere. The loss of ozone has been dramatic in recent years, even leading to the formation of "ozone holes" with virtually no ozone present above the Arctic and Antarctic in the fall. There is currently debate about how much of the ozone loss is human-induced by chlorofluorocarbon production, and how much may be related to natural fluctuations in ozone concentration.

Many other gases and particulate matter play important roles in atmospheric phenomena. For instance, small amounts of sulfur dioxide (SO_2) produced by the burning of fossil fuels mixes with water to form sulfuric acid, the main harmful component of acid rain. Acid rain is killing the biota of many natural lake systems, particularly in the northeastern United States in areas underlain by granitic-type rocks, and it is causing a wide range of other environmental problems across the world. Other pollutants are major causes of respiratory problems and environmental degradation, and the major increase in particulate matter in the atmosphere in the past century has increased the hazards and health effects from these atmospheric particles.

AIR PRESSURE

The weight of the air above a given level is known as air pressure. This weight produces a force in all directions caused by constantly moving air molecules bumping into one another and other objects in the atmosphere. The air molecules in the atmosphere are constantly moving with each air molecule averaging a remarkable 10 billion collisions per second with other air molecules near the Earth's surface. The density of air molecules is highest near the surface, decreases rapidly upward in the lower 62 miles (100 km) of the atmosphere, then decreases slowly upward to above 310 miles (500 km). Gravity pulls air molecules toward the Earth, and they are therefore more abundant closer to the surface. Pressure, including air pressure, is measured as the force divided by the area over which it acts. The air pressure is greatest near the Earth's surface and decreases with height because there is a greater number of air molecules near the Earth's surface (the air pressure represents the sum of the total mass of air above a certain point). A one-square-inch column of air extending from sea level to the top of the atmosphere weighs about 14.7 pounds (6.67 kg). The typical air pressure at sea level is therefore 14.7 pounds per square inch (2.62 kg per square cm). It is commonly measured in units of millibars (mb) or hectopascals (hPa), and also in inches of mercury. Standard air pressure in these units equals 1,013.25 mb, 1,013.25 hPa, and 29.92 in of mercury. Air pressure is equal in all directions, unlike some pressures (such as a weight on one's head) that act in one direction. This explains why objects and people are not crushed or deformed by the pressure of the overlying atmosphere.

Air pressure also changes in response to temperature and density, as expressed by the following gas law:

$$pressure = temperature \times density \times constant$$

where the gas constant is equal to 2.87×10^6 erg/g K.

From this gas law it is apparent that at the same temperature, air at a higher pressure is denser than air at a lower pressure. Therefore high-pressure regions of the atmosphere are characterized by denser air, with more molecules of air than areas of low pressure. These pressure changes are caused by wind that moves air molecules into and out of a region. When more air molecules move into an area than move out, the area is called an area of net convergence. Conversely, in areas of low pressure, more air molecules are moving out than in, and the area is one of divergence. If the air density is constant and the temperature changes, the gas law states that at a given atmospheric level, as the temperature increases, the air pressure decreases. With these relationships, if either the temperature or the pressure is known, the other can be calculated.

If the air above a location is heated, it will expand and rise; if air is cooled, it will contract, become denser, and sink closer to the surface. Therefore the air pressure decreases rapidly with height in the cold column of air because the molecules are packed closely to the surface. In the warm column of air the air pressure will be higher at any height than in the cold column of air, because the air has expanded and more of the original air molecules are above the specific height than in the cold column. Therefore warm air masses at high height are generally associated with high-pressure systems, whereas cold air aloft is generally associated with low pressure. Heating and cooling of air above a location induces the air pressure to change in that location, causing lateral variation in air pressure across a region. Air will flow from high-pressure areas to low-pressure areas, forming winds.

The daily heating and cooling of air masses by the Sun can in some situations cause the opposite effect, if it is not overwhelmed by effects of the heating and cooling of the upper atmosphere. Over large continental areas, such as the southwestern United States, the daily heating and cooling cycle is associated with air pressure fall and rise, as expected from the gas law. As the temperature rises in these locations the pressure decreases, then increases again in the night when the temperature falls. Air must flow in and out of a given vertical column on a diurnal basis for these pressure changes to occur, as opposed to having the column rise and fall in response to the temperature changes.

ROLE OF THE ATMOSPHERE IN GLOBAL CLIMATE

Interactions among the atmosphere, hydrosphere, biosphere, and lithosphere control global climate. Global climate represents a balance between the amount of solar radiation received and the amount of this energy retained in a given area. The planet receives about 2.4 times as much heat in the equatorial regions compared to the polar regions. The atmosphere and oceans respond to this unequal heating by setting up currents and circulation systems that redistribute the heat more equally. These circulation patterns are in turn affected by the ever-changing pattern of the distribution of continents, oceans, and mountain ranges.

The amounts and types of gases in the atmosphere can modify the amount of incoming solar radiation, and hence global temperature. For instance, cloud cover can cause much of the incoming solar radiation to be reflected back to space before being trapped by the lower atmosphere. On the other hand, greenhouse gases allow incoming short wavelength solar radiation to enter the atmosphere, but trap this radiation when it tries to escape in its longer wavelength reflected form. This causes a buildup of heat in the atmosphere, and can lead to a global warming known as the greenhouse effect.

The amount of heat trapped in the atmosphere by greenhouse gases has varied greatly over Earth's history. One of the most important greenhouse gases is carbon dioxide (CO_2). Plants, which release O_2 to the atmosphere, now take up CO_2 by photosynthesis. In the early part of Earth's history (in the Precambrian before plants covered the land surface), photosynthesis did not remove CO_2 from the atmosphere, with the result that CO_2 levels were much higher than at present. Marine organisms remove atmospheric CO_2 from ocean surface water (which is in equilibrium with the atmosphere) and use the CO_2 along with calcium to form their shells and mineralized tissue. These organisms make $CaCO_3$ (calcite), which is the main component of limestone, a rock composed largely of the dead remains of marine organisms. Approximately 99 percent of the planet's CO_2 is presently removed from the atmosphere/ocean system because it is locked up in rock deposits of limestone on the continents and on the seafloor. If this amount of CO_2 were released back into the atmosphere, the global temperature would increase dramatically. In the early Precambrian, when this CO_2 was free in the atmosphere, global temperatures averaged about 550°F (290°C).

The atmosphere redistributes heat quickly by forming and redistributing clouds and uncondensed water vapor around the planet along atmospheric circulation cells. Oceans are able to hold and redistribute more heat because of their greater amount of water, but they redistribute this heat more slowly than the atmosphere. Surface currents form in response to wind patterns, but deep ocean currents that move more of the planet's heat follow courses more related to the bathymetry (topography of the seafloor) and the spinning of the Earth than they are related to surface winds.

The balance of incoming and outgoing heat from the Earth has determined the overall temperature of the planet through time. Examination of the geological record has enabled paleoclimatologists to reconstruct intervals when the Earth had glacial periods, hot dry episodes, hot wet, or cold dry cycles. In most cases the Earth has responded to these changes by expanding and contracting its climate belts. Warm periods see an expansion of the warm subtropical belts to high latitudes, and cold periods see an expansion of the cold climates of the poles to low latitudes.

HADLEY CELL

Hadley cells are the globe-encircling belts of air that rise along the equator and drop moisture as they rise in the Tropics. As the air moves away from the equator at high elevations, it cools, becomes drier, then descends at 15–30°N and S latitude, where it either returns to the equator or moves toward the poles.

The locations of the Hadley cells move north and south annually in response to the changing apparent seasonal movement of the Sun. High-pressure systems form where the air descends, characterized by stable clear skies and intense evaporation because the air is so dry. Another pair of major global circulation belts is formed as air cools at the poles and spreads toward the equator. Cold polar fronts form where the polar air mass meets the warmer air that has circulated around the Hadley cell from the Tropics. In the belts between the polar front and the Hadley cells, strong westerly winds develop. The position of the polar jet stream (formed in the upper troposphere), which is partly fixed in place in the Northern Hemisphere by the high Tibetan Plateau and the Rocky Mountains, controls the position of the polar front and extent of the west-moving wind. Dips and bends in the jet stream path are known as Rossby waves, and these partly determine the location of high- and low-pressure systems. These Rossby waves tend to be semistable in different seasons and have predictable patterns for summer and winter. If the pattern of Rossby waves in the jet stream changes significantly for a season or longer, storm systems may track to different locations than normal, causing local droughts or floods. Changes in this global circulation can also change the locations of regional downwelling, cold dry air. This can cause long-term drought and desertification. Such changes can persist for periods of several weeks, months, or years, and may explain several of the severe droughts that have affected Asia, Africa, North America, and elsewhere.

JET STREAMS

Jet streams are high-level, narrow, fast-moving currents of air typically thousands of kilometers long, hundreds of kilometers wide, and a couple of miles (several kilometers) deep. Jet streams typically form near the tropopause, six to nine miles (10–15 km) above the surface, and can reach speeds of 115–230 miles per hour (100–200 knots). Rapidly moving cirrus clouds often reveal the westerly jet streams moving air from west to east. Several jet streams are common—the subtropical jet stream forms about eight miles (13 km) above the surface, at the poleward limit of the tropical Hadley cell, where a tropospheric gap develops between the circulating Hadley cells. The polar jet stream forms at about a six-mile (10-km) height, at the tropospheric gap between the cold polar cell and the midlatitude Ferrel cell. The polar jet stream is often associated with polar front depressions. The jet streams, especially the subtropical jet, are fairly stable and drive many of the planet's weather systems. The polar jet stream tends to meander and develop loops more than the subtropical jet. A third common jet stream often develops as an easterly flow, especially over the Indian subcontinent during the summer monsoon.

ATMOSPHERIC EVOLUTION

Considerable uncertainty exists about the origin and composition of the Earth's earliest atmosphere. Many models assume that methane and ammonia dominated the planet's early atmosphere, instead of nitrogen and carbon dioxide, as it is presently. The gases that formed the early atmosphere could have come from outgassing by volcanoes, from extraterrestrial sources (principally cometary impacts), or, most likely, both. Alternatively, comets may have brought organic molecules to Earth. A very large late impact is thought to have melted outer parts of the Earth, formed the Moon, and blown away the earliest atmosphere. The present atmosphere must therefore represent a later, secondary atmosphere formed after this late impact.

The earliest atmosphere and oceans of the Earth probably formed from early degassing of the interior by volcanism within the first 50 million years of Earth history. It is likely that our present atmosphere is secondary, in that the first, or primary, atmosphere would have been vaporized by the late great impact that formed the Moon, if it survived being blown away by an intense solar wind when the Sun was in a T-Tauri stage of evolution. The primary atmosphere would have been composed of gases left over from accretion, including primarily hydrogen, helium, methane, and ammonia, along with nitrogen, argon, and neon. Since the atmosphere has much less than the expected amount of these elements, however, and is quite depleted in these volatile elements relative to the Sun, it is thought the primary atmosphere has been lost to space.

Gases are presently escaping from the Earth during volcanic eruptions, and also being released by weathering of surface rocks. The secondary atmosphere was most likely produced from degassing of the mantle by volcanic eruptions, and perhaps also by cometary impact. Gases released from volcanic eruptions include N, S, CO_2, and H_2O, closely matching the suite of volatiles that the present atmosphere and oceans comprise. But there was little or no free oxygen in the early atmosphere, as oxygen was not produced until later, by photosynthetic life.

The early atmosphere was dense, with H_2O, CO_2, S, N, HCl. The mixture of gases in the early atmosphere would have made greenhouse conditions similar to that presently existing on Venus. But, since the early Sun during the Hadean Era was approximately 25 percent less luminous than today, the atmospheric greenhouse kept temperatures close to their present range, where water is stable

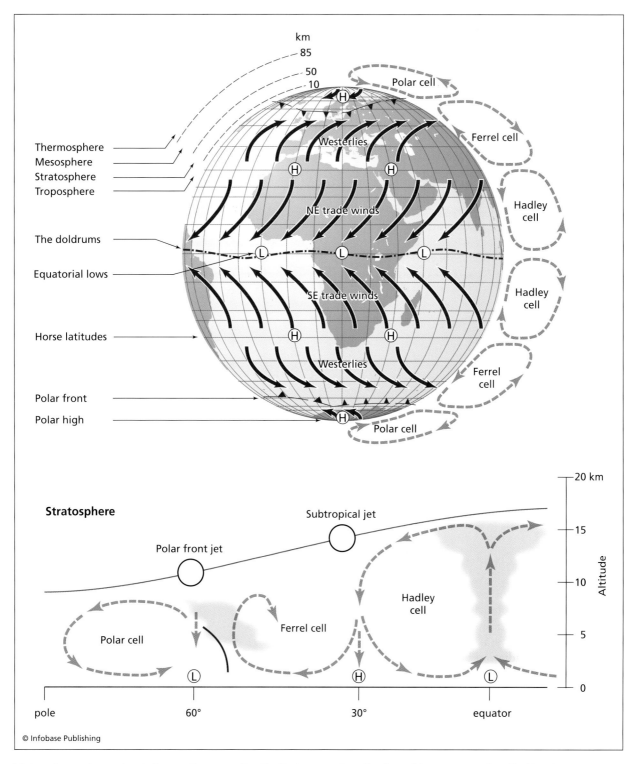

Major atmosphere circulation patterns on the Earth, in map view (top), and in cross section (bottom)

and life can form and exist. As the Earth cooled, water vapor condensed to make rain that chemically weathered igneous crust, making sediments. Gases dissolved in the rain made acids, including carbonic acid (H_2CO_3), nitric acid (HNO_3), sulfuric acid (H_2SO_4), and hydrochloric acid (HCl). These acids were neutralized by minerals (which are bases) that became sediments, and chemical cycling began. These waters plus dissolved components became the early hydrosphere, and chemical reactions gradually began changing the composition of the atmosphere, getting close to the dawn of life.

During the early Archean, the Sun was only about 70 percent as luminous as it is presently, so the Earth must have experienced a greenhouse warming effect to keep temperatures above the freezing point of water, but below the boiling point. Increased levels of carbon dioxide and ammonia in the early atmosphere could have acted as greenhouse gases, accounting for the remarkable maintenance of global temperatures within the stability field of liquid water, allowing the development of life. Much of the carbon dioxide that was in the early atmosphere is now locked up in deposits of sedimentary limestone, and in the planet's biomass. The carbon dioxide that shielded the early Earth and kept temperatures in the range suitable for life to evolve now forms the bodies and remains of those very life-forms.

See also AURORA, AURORA BOREALIS, AURORA AUSTRALIS; CLIMATE; CLIMATE CHANGE; GREENHOUSE EFFECT; WEATHERING.

FURTHER READING

Ahrens, C. Donald. *Meteorology Today.* 7th ed. Pacific Grove, Calif.: Brooks/Cole, 2002.

Ashworth, William, and Charles E. Little. *Encyclopedia of Environmental Studies, New Edition.* New York: Facts On File, 2001.

Bekker, Andrey, H. Dick Holland, P. L. Wang, D. Rumble III, H. J. Stein, J. L. Hannah, L. L. Coetzee, and Nick Beukes, "Dating the Rise of Atmospheric Oxygen." *Nature* 427 (2004): 117–120.

Kasting, James F. "Earth's Early Atmosphere." *Science* 259 (1993): 920–925.

aurora, aurora borealis, aurora australis

Auroras Borealis and Aurora Australis are glows in the sky sometimes visible in the Northern and Southern Hemispheres, respectively. They are informally known as the northern lights and the southern lights. The glows are strongest near the poles, and originate in the Van Allen radiation belts, regions where high-energy charged particles of the solar wind that travel outward from the Sun are captured by the Earth's magnetic field. The outer Van Allen radiation belt consists mainly of protons, whereas the inner Van Allen belt consists mainly of electrons. At times electrons spiral down toward Earth near the poles along magnetic field lines and collide with ions in the thermosphere, emitting light in the process. Light in the aurora is emitted between a base level of about 50–65 miles (80–105 km), and an upper level of about 125 miles (200 km) above the Earth's surface.

The solar wind originates when violent collisions between gases in the Sun emit electrons and protons that escape the gravitational pull of the Sun and travel through space at about 250 miles per second

(more than 1 million km/hr) as a plasma known as the solar wind. When these charged particles move close to Earth, they interact with the magnetic field, changing its shape in the process. The natural undisturbed state of the Earth's magnetic field is broadly similar to a bar magnet, with magnetic flux lines (of equal magnetic intensity and direction) coming out of the south polar region, and returning into the north magnetic pole. The solar wind deforms or distorts this ideal state into an elongate teardrop-shaped configuration known as the magnetosphere. The magnetosphere has a rounded compressed side facing the Sun, and a long tail (magnetotail) on the opposite side that stretches past the orbit of the moon. The magnetosphere shields the Earth from many of the charged particles from the Sun by deflecting them around the edge of the magnetosphere, causing them to flow harmlessly into the outer solar system.

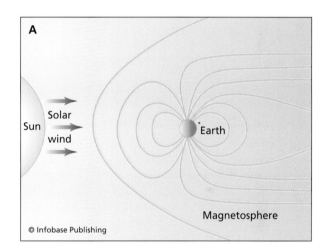

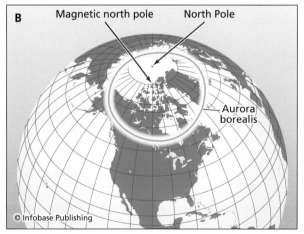

(A) Drawing of magnetosphere, showing asymmetric shape created by distortion of the Earth's magnetic field by the solar wind (B) Earth, showing typical auroral ring with the greatest intensity of auroral activity about 20–30° from the magnetic pole, where magnetic field lines are most intense

Active aurora borealis arc in Alaska *(Roman Krochuk, Shutterstock, Inc.)*

The Sun periodically experiences periods of high activity when many solar flares and sunspots form. During these periods the solar wind is emitted with increased intensity, and the plasma is emitted with greater velocity, in greater density, and with more energy than in its normal state. During these periods of high solar activity the extra energy of the solar wind distorts the magnetosphere and causes more electrons to enter the Van Allen belts, causing increased auroral activity.

When the electrons from the magnetosphere are injected into the upper atmosphere, they collide with atoms and molecules of gases there. The process involves the transfer of energy from the high-energy particles of the magnetosphere to the gas molecules from the atmosphere, which become excited and temporarily jump to a higher energy level. When the gas molecules return to their normal, regular energy level, they release radiation energy in the process. Some of this radiation is in the visible spectrum, forming the aurora borealis in the Northern Hemisphere and the aurora australis in the Southern Hemisphere.

Auroras typically form waving sheets, streaks, and glows of different colors in polar latitudes. The colors originate because different gases in the atmosphere emit different characteristic colors when excited by charged particles from the magnetosphere, and the flickering and draperies are caused by variations in the magnetic field and incoming charged particles. The auroras often form rings around the magnetic poles, being most intense where the magnetic field lines enter and exit the Earth at 60–70° latitude.

See also MAGNETIC FIELD, MAGNETOSPHERE; SUN.

Australian geology The geologic history of Australia spans almost all of the history of the Earth, hosting the oldest known terrestrial mineral grains dated to be 4.4 billion years old. The region contains active deposition of lake sediments in the desert interior and some of the world's most diverse carbonate reefs located offshore its northeast coastline. The geology of Australia can be divided into provinces of several ages, including the Archean Pilbara, Yilgarn, Kimberly, and Gawler cratons, which are encased in Proterozoic orogenic belts including the Musgrave orogen and Arunta Inlier. Paleozoic orogens include the Lachlan and Tasman orogens in the east, whereas the northern and northeastern edges of the Australia plate are involved in active convergent tectonic activity. Mesozoic to recent sedimentary basins in Australia include the Perth and Bowen basins; Sydney, Gunnedah, and Ipswich basins; and the large active desert drainage system of Lake Eyre in the Australian midcontinent.

ARCHEAN CRATONS

Archean rocks form the core of the Australian continent and include the cratonic nuclei of the Yilgarn, Plibara, Gawler, and Kimberly cratons. Archean rocks may also underlie portions of some of the Proterozoic basins and orogens, but less is known about the rocks at great depths in Australia.

The Pilbara craton, located in northwestern Australia, contains mainly low- to medium-grade Archean rocks with the metavolcanic and metasedimentary rocks confined to relatively narrow belts between broad domal granitoid-gneiss domes typically 50–60 miles (~100 km) in width. Early ideas that the greenstones and metasedimentary rocks were simply deposited on top of older granitoids then later deformed by folding as their density caused them to sink into rheologically soft granitoids have proven to be myths. Detailed structural analysis has shown that the greenstones were emplaced structurally upon the gneissic and granitoid rocks, then deformed several times before the late open folding caused by the doming of the granitoids.

The granitoid rocks of the domal structures are of four basic types: older migmatitic, gneissic and foliated granodiorites, tonalites, and trondhjemites, in turn intruded by coarse-grained porphyritic granodiorite and then unfoliated post-tectonic granites. The older gneissic rocks range in age from 3.5 to 3.3 billion years, whereas the younger intrusives are between 3.05 and 2.85 billion years old. The domes in the Pilbara are not formed by intrusion-related processes, but rather reflect complex, large-scale folding events. Much of the doming occurred at 3.0–2.95 billion years ago in a cratonwide event.

The metavolcanic and sedimentary rocks between the domal granitoids, known as the Pilbara Supergroup, comprise three groups: the Warrawoona, George Creek, and Whim Creek sequences. Nowhere can it be shown that these rocks were deposited on the older gneissic rocks, though some groups argue that the entire sequence was deposited on continental basement, and others argue the supergroup is an allochthonous (exotic and far traveled) assemblage emplaced on the gneisses by thrusting and tectonic processes. Some of the rock sequences within the Pilbara Supergroup are calc-alkaline volcanic assemblages that resemble younger island arc sequences, and others are tholeiitic mafic volcanic and plutonic sequences that resemble younger ocean floor assemblages. These rocks are significantly disrupted and repeated along many thin shear zones and interca-

lated with sedimentary rocks in a manner like many younger accretionary prisms found at convergent margins, so some interpretations of the Pilbara suggest that it may represent an ancient accretionary orogen, formed at 3.5 billion years ago and disrupted by collisional tectonic events between 3.5 and 2.8 billion years ago.

The southern margin of the Pilbara craton and northern margin of the Yilgarn is covered by thick sedimentary deposits of the Hamersley and Nabberu basins, including spectacular banded-iron formations (BIF). The Mount Bruce Supergroup of the Hamersley basin includes thick deposits of clastic and chemical sediments divided into the lower Fortescue Group containing mafic volcanics, the 1.3-mile (2.5-km) thick Middle Hamersley Group consisting of banded-iron formation, shale, dolostone, and fewer diabasic intrusions and felsic volcanics. The uppermost rocks in the basin include shales, sandstones, and glacial deposits of the Turee Creek Group. The Hamersley Group is marked by thin layers of BIF that are remarkably continuous over thousands of square miles (km^2) and contain roughly 30 percent iron, making these rocks a significant economic resource. The Hamersley basin is deformed into a regional synclinorium structure, and only weakly metamorphosed, with dips of strata typically fewer than 10 degrees. The ages of these rocks include an estimate of 2.49 billion years for the Hamersley Group, and all groups in the basin are cut by intrusives dated between 2.4 and 2.3 billion years old.

The Yilgarn craton occupies the southwestern part of the Australian continent, covering an area

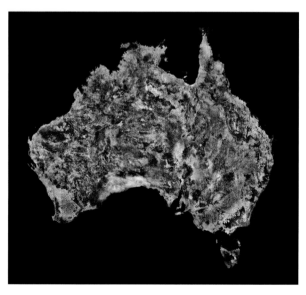

Satellite image mosaic of Australia, consisting of more than 1,000 merged images *(Earth Satellite Corporation/Photo Researchers, Inc.)*

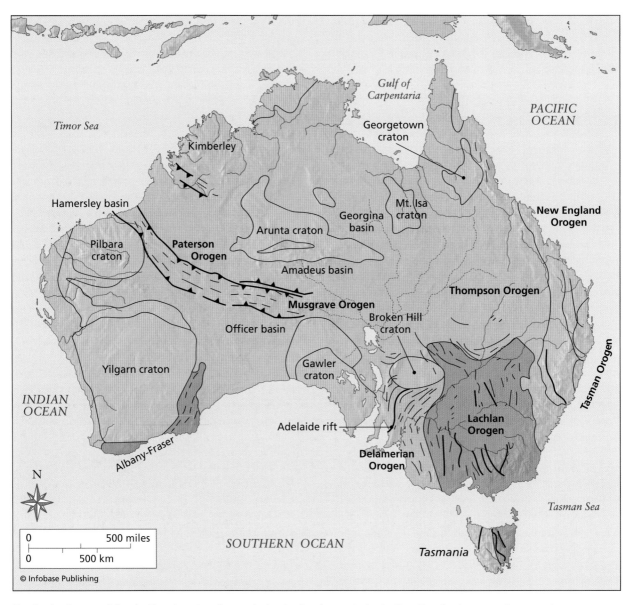

Geological map of Australia, showing the main tectonic elements including the Archean cratons and younger orogens

600 miles (1,000 km) long by 420 miles (700 km) across. The craton is subdivided into four major provinces including a dominantly gneissic terrane, the Western Gneiss Terrane in the southwest, and then low-to-medium–grade granite-greenstone terranes of the Murchison Province in the northwest, the Southern Cross Province in the center of the craton, and the Eastern Goldfields in the east. All of the belts in the Yilgarn craton were affected by a regional metamorphic and plutonic intrusion event at 2.7–2.6 billion years ago.

The Western Gneiss Terrane consists mainly of quartzofeldspathic gneiss derived from sedimentary protoliths, then intruded by migmatitic to porphyritic granitoids. The Narryer Gneiss Complex in the northern part of the terrane includes layers of metamorphosed conglomerates, sandstones, pelites, and carbonate rocks and is interpreted as a strongly metamorphosed shallow water sedimentary sequence. Clastic zircons from the Narrier Complex have yielded many ancient zircons with ages from 3.6 to 3.5 billion years ago, and one sample dated by Australian geologist Simon Wilde has yielded an age of 4.4 billion years, a mere 100 million years after the formation of the Earth. Geochemical analysis of this zircon grain by Simon Wilde and his colleagues has shown that the zircon was derived from a rock that interacted with the early hydrosphere of the Earth, showing that oceans existed on Earth by 4.4 billion years ago.

The Murchison, Southern Cross, and Eastern Goldfields Provinces are all dominated by different types of granitoid and gneissic rocks, with about 30 percent of the outcrop area consisting of greenstone belts and metasedimentary terrains. Most of these strike roughly north-northwest and have broad synclinal structures disrupted by numerous faults, reflecting their complex history. These rocks include tholeiitic basalts and ultramafic rocks near the base of most successions, with felsic volcanic rocks and clastic sedimentary rocks at the tops of the successions. Ages on the volcanics range from 3.05 to 2.69 billion years. Some models for the volcanic groups suggest that the repetitive sequences from mafic to felsic volcanics are depositional, whereas others have suggested that thrust faults repeat the stratigraphic sequence.

Late-stage major ductile transcurrent shear zones cut the Yilgarn craton and form many of the boundaries between different belts and terranes. The 186-mile (300-km) long Koolyanobbing shear zone in the Southern Cross Province is a four- to nine-mile (6- to 15-km) wide zone with a gradation from foliated granitoid, through protomylonite, mylonite, to ultramylonite, from the edge to the center of the shear zone. Shallowly plunging lineations and a variety of kinematic indicators show that the shear zone is a major sinistral fault, but regional relationships suggest that it does not represent a major crustal boundary or suture. Fault fabrics both overprint and appear coeval with late stages in the development of the regional metamorphic pattern, suggesting that the shear zone was active around 2.7 to 2.65 Ga.

The granitoid intrusives in the Yilgarn craton account for about 70 percent of the outcrop area. These include 2.9–2.6 billion-year-old tonalitic to granodioritic phases, and 2.7–2.6 billion-year-old granodiorite to granite. The older granitoids have geochemical affinities to convergent margin arc magmas, whereas the younger granitoids may be related to post-collisional melting such as characterizes many younger convergent and collisional mountain belts in Phanerozoic orogens.

The Gawler craton is a relatively small block located in areas surrounding the Eyre Peninsula in south-central Australia, and is younger than the Pilbara and Yilgarn. It contains rocks that formed in convergent margins between 2.5 and 1.5 billion years ago, then became relatively stable after an orogenic event between 1.9 and 1.84 billion years ago, probably reflecting the incorporation of the block into the Rodinian supercontinent.

Hamersley Gorge on the margin of the Pilbara craton, Australia, showing red banded-iron formation *(imagebroker/Alamy)*

The Kimberly block of northern Australia is thought to be an Archean craton, but it is covered by thick deposits of the Kimberly basin. This block is bounded on the southeast by the Halls Creek belt and on the southwest by the King Leopold belt, both of which are Proterozoic orogenic belts that experienced strong deformation in the Barramundi orogeny at 1.85 billion years ago. After this major deformation event the Kimberly block was covered by up to three miles (five km) of uniformly bedded quartz sandstones, shales, limestones, and flood basalts.

PROTEROZOIC GNEISS BELTS AND BASINS

The Archean cratons of Australia are welded together by several Proterozoic orogenic belts, the most important of which include the Musgrave orogen and its continuation to the west as the Paterson orogen that together link north and south Australia. The Capricorn orogen is located between the Pilbara and Yilgarn cratons; convergent tectonism across this belt joined those cratons and their flanking sedimentary basin sequences in the Paleoproterozoic at around 2.2 billion years ago, with remnants preserved in the Bangemall basin, Gascoyne Complex, and the Glengarry, Yerrida and Padbury basins. Other Paleoproterozoic orogenic segments in central Australia are deeply buried by younger Proterozoic-Palaeozoic rocks of the Officer and Amadeus basins.

In eastern Australia rocks in the Mount Isa Complex were complexly deformed into fold-thrust belt structures in the Paleoproterozoic, while rocks farther south in the Broken Hill Inlier were experiencing high-grade metamorphism and polyphase deformation.

PHANEROZOIC OROGENS AND BASINS

The main area of Phanerozoic deformation and activity in Australia is along the east coast, in the Lachlan fold belt and Tasman orogen. The Lachlan fold belt contains Cambrian ophiolitic sequences that were thrust on top of the Australian continent in the Ordovician in the Lachlan Orogeny. This orogeny was associated with many classical Alpine-type events including the formation of flysch and molasse belts, strongly deformed zones with serpentinitic and ophiolitic mélange, and affected a large part of the New South Wales region of Australia. Tectonic activity continued in this belt through the Silurian with the formation of volcanic arcs in the New England orogen and the intrusion of belts of granitic batholiths. The high topography formed in the east during the Early Paleozoic was significantly eroded in the Devonian, with thick clastic sequences reaching into the continental interior.

In the Carboniferous eastern Australia collided with parts of South America and New Zealand as part of the amalgamation of the Gondwanan supercontinent; this collision formed high, Tibetan-style mountain ranges on the east coast. These ranges have since been nearly completely eroded, and just their deeper-level roots remain as testimony to this event.

The Permian-Triassic saw the establishment of major subduction zones along the east coast in the Hunter-Bowen Orogeny, which was initiated as an arc colliding with Australia and then conversion of this margin to a convergent tectonic setting, with related deformation continuing until the Middle Triassic at 230–225 million years ago. A major glaciation event in the Permian caused accelerated erosion of these mountain ranges, particularly in central and western Australia. Glacial tillite deposits from this event cover large parts of central Australia.

The environment of the Jurassic changed such that most of western Australia experienced tropical weathering in a savanna to jungle setting, and several offshore oil basins formed including the Gippsland, Bass, and Otway basins in Victoria. Coal-bearing strata were laid down across northern Australia, while passive margin sedimentation continued in the Perth basin in the west.

Antarctica rifted from Australia in the Jurassic. Rift-sedimentation and subsidence continued in the Cretaceous and developed into seafloor spreading and the separation of Tasmania from the Australian mainland. These rifted to passive margins, then developed extensive coral reefs in the northeast, and rare intraplate volcanic centers formed through the Tertiary.

GREAT BARRIER REEF

As the largest coral reef in the world, the Great Barrier Reef forms a 1,250-mile (2,010-km) long breakwater in the Coral Sea along the northeast coast of Queensland, Australia. The reef has been designated a World Heritage area, the world's largest such site. The reef comprises several individual reef complexes including 2,800 individual reefs stretching from the Swain reefs in the south to the Warrier reefs along the southern coast of Papua New Guinea. Many reef types are recognized including fringing reefs, flat platform reefs, and elongate ribbon reefs. The reef complexes are separated from the mainland of Queensland by a shallow lagoon ranging from 10 to 100 miles (16–160 km) wide.

There are more than 400 types of coral known on the Great Barrier Reef, as well as 1,500 species of fish, 400 species of sponges, and 4,000 types of mollusk, making it one of the world's richest sites in terms of faunal diversity. Additionally, the reefs are home to animals including numerous sea anemones, worms, crustaceans, echinoderms, and an endangered mammal known as the dugong. Sea turtles feed on abundant algae and sea grass, and the reef is frequented by humpback whales that migrate from Antarctic waters to have babies in warm waters. Hundreds of bird species have breeding colonies in the islands and cays among the reefs, and these birds include beautiful herons, pelicans, osprey, eagles, and shearwaters.

The reefs also hide dozens of shipwrecks and have numerous archaeological sites of significance to the Aboriginal and Torres Strait Islander peoples.

LAKE EYRE

The center of Australia is covered by a shallow, frequently dry salt lake that occupies the lowest point on the continent, at 39 feet (12 m) below sea level. The lake occupies 3,430 square miles (8,884 km²), but the drainage basin is one of the world's largest internally draining river systems covering 1.2 million square miles (1.93 million km²), with no outlet to the sea. All water that enters the Lake Eyre basin flows into the lake and eventually evaporates, leaving salts behind. Lake Eyre is located in the driest part of Australia, where the evaporation potential is 8.175 feet (2.5 m), but the annual precipitation is only half an inch (1.25 centimeters). However, flows in the river system are highly variable and unpredictable, since rare rainfall events may cause flash flooding. All rivers in the system are ephemeral, typically with no water in the system. Aridity increases downstream toward the lake,

and the basin is characterized by huge braided stream networks, floodplains, and waterholes. The stream systems leading into Lake Eyre are one of the largest unregulated river systems in the world.

See also ARCHEAN; BASIN, SEDIMENTARY BASIN; CONVERGENT PLATE MARGIN PROCESSES; CRATON; DESERTS; DIVERGENT PLATE MARGIN PROCESSES; GONDWANA, GONDWANALAND; GREENSTONE BELTS; HISTORICAL GEOLOGY; OPHIOLITES; PASSIVE MARGIN; PLATE TECTONICS.

FURTHER READING

Condie, Kent C., and Robert Sloan. *Origin and Evolution of Earth: Principles of Historical Geology.* Upper Saddle River, N.J.: Prentice Hall, 1997.

Goodwin, Alan M. *Precambrian Geology.* London: Academic Press, 1991.

Johnson, David. *Geology of Australia.* Cambridge: Cambridge University Press, 2004.

Kusky, Timothy M. *Precambrian Ophiolites and Related Rocks.* Amsterdam, Netherlands: Elsevier, 2004.

Wilde, Simon, John W. Valley, William M. Peck, and Colin M. Graham. "Evidence from Detrital Zircons for the Existence of Continental Crust and Oceans on the Earth 4.4 Gyr Ago." *Nature* 409 (2001): 175–178.

Windley, Brian F. *The Evolving Continents.* 3rd ed. Chinchester, U.K.: John Wiley & Sons, 1995.

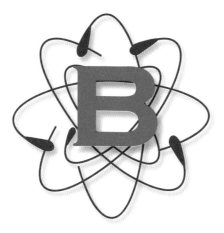

basin, sedimentary basin A depression in the surface of the Earth or other celestial body is known as a basin. When this depression becomes filled with sediments, it is known as a sedimentary basin. There are many types of basins, including depressed areas with no outlet or with no outlet for deep levels (such as lakes, oceans, seas, and tidal basins), and areas of extreme land subsidence (such as volcanic calderas or sinkholes). In contrast, drainage basins include the total land area that contributes water to a stream. Drainage (river or stream) basins are geographic areas defined by surface slopes and stream networks where all the surface water that falls in the drainage basin flows into that stream system or its tributaries. Groundwater basins are areas where all the groundwater is contained in one system, or flows toward the same surface water basin outlet. Impact basins are circular depressions excavated instantaneously during the impact of a comet or asteroid with the Earth or other planetary surface.

Areas of prolonged subsidence and sediment accumulation are known as sedimentary basins, even though they may not presently be topographically depressed. Several types of sedimentary basins exist and are classified by their shape and relationships to bordering mountain belts or uplifted areas. Foreland basins are elongate areas on the stable continent sides of orogenic belts, characterized by a gradually deepening, generally wedge-shaped basin, filled by clastic and lesser amounts of carbonate and marine sedimentary deposits. The sediments are coarser-grained and of more proximal varieties toward the mountain front, from where they were derived. Foreland basins may be several hundred feet to about 12 miles (100 meters to 20 km) deep and filled entirely by sedimentary rocks, and are therefore good sites for hydro-

carbon exploration. Many foreland basins have been overridden by the orogenic belts from where they were derived, producing a foreland fold-thrust belt, and parts of the basin incorporated into the orogen. Many foreland basins show a vertical profile from a basal continental shelf type of assemblage, made up dominantly of limestone, upward to a greywacke/shale flysch sequence, into an upper conglomerate/sandstone sequence known as molasse.

Rift basins are elongate depressions in the Earth's surface in which the entire thickness of the lithosphere has ruptured in extension. They are typically bounded by normal faults along their long sides, and display rapid lateral variation in sedimentary facies and thicknesses. Rock types deposited in the rift basins include marginal conglomerate, fanglomerate, and alluvial fans, grading basinward into sandstone, shale, and lake evaporite deposits. Volcanic rocks may be intercalated with the sedimentary deposits of rifts, and in many cases include a bimodal suite of basalts and rhyolites, some with alkaline chemical characteristics.

Several other less common types of sedimentary basins form in different tectonic settings. For instance, pull-apart rift basins and small foreland basins may form along bends in strike-slip fault systems, and many varieties of rift and foreland basins form in different convergent margin and divergent margin tectonic settings.

FORELAND BASINS

Foreland basins are wedge-shaped sedimentary basins that form on the continentward side of fold-thrust belts, filling the topographic depression created by the weight of the mountain belt. Most foreland basins have asymmetric, broadly wedge-shaped profiles with the deeper side located toward the mountain range, and a flexural bulge developed about 90 miles (150 km) from the foothills of the mountains where the deformation front is located. The Indo-Gangetic plain on the south side of the Himalaya Mountains is an example of an active foreland basin, whereas some ancient examples include the Cretaceous Canadian Rockies Alberta foreland basin, the Cenozoic flysch basins of the Alps, and the Ordovician and Devonian clastic wedges in the Appalachian foreland basins. Foreland basins are characterized by asymmetric subsidence, with greater amounts near the thrust front. Typical amounts of sudsidence fall in the range of about 0.6 miles (1 km) every 2 to 5 million years.

Deformation such as folding, thrust faulting, and repetition of stratigraphic units may affect foreland basins near the transition to the mountain front. These types of foreland basins appear to have formed largely by the flexure of the lithosphere by the weight of the mountain range, with the space created by the flexure filled in by sediments eroded from the uplifted mountains. Sedimentary facies typically grade from fluvial/alluvial systems near the mountains to shallow marine clastic environments farther away from the mountains, with typical deposition of flysch sequences by turbidity currents. These deposits may be succeeded laterally by distal black shales, then shallow water carbonates over a cross-strike distance of several hundred miles (kilometers). There is also often a progressive zonation of structural features across the foreland basin, with contractional deformation (folds and faults) affecting the region near the mountain front, and normal faulting affecting the area on the flexural bulge a few tens to hundreds of kilometers from the deformation front. Sedimentary facies and structural zones all may migrate toward the continent in collisional foreland basins.

A second variety of foreland basin is found on the continentward side of noncollisional mountain belts such as the Andes, and these are sometimes referred to as retroarc foreland basins. They differ from the collisional foreland basins described above in that the mountain ranges are not advancing on the foreland, and the basin subsidence is a response to the weight of the mountains, added primarily by magmatism.

Another variety of foreland basins, known as extensional foreland basins, include features such as impactogens and aulacogens, which are extensional basins that form at wide angles to the mountain front. Impactogens form during the convergence, whereas aulacogens are reactivated rifts that formed during earlier ocean opening. Many of these basins have earlier structural histories, including formation as a rift at a high angle to an ocean margin.

These rifts are naturally oriented at wide angles to the mountain ranges when the oceans close, and become sites of enhanced subsidence, sedimentation, and locally additional extension. The Rhine graben in front of the Alpine collision of Europe is a well-known example of an aulacogen.

RIFTS

Active rift systems may exhibit very steep escarpments that drop from the rift shoulders to the base of the rift valley floor, typically forming an elongate depression that may extend for hundreds or even thousands of miles. The world's best known example of a continental rift is the East African rift, extending from the Ethiopian Afar to Mozambique. Other spectacular examples include Lake Baikal in Siberia, the Rio Grande in the desert southwest of Arizona and New Mexico, and the Alaotra rift in Madagascar. Most of these rifts have coarse-grained sediments deposited along their margins, and fine-grained and even lake sediments in their centers. Volcanic centers are sporadically developed.

Many rifts in continents are associated with incipient breaking apart of the continent to form an oceanic basin. These types of rift system typically form three arms that develop over domed areas above upwelling mantle material, such as is observed in east Africa. Two of the three arms may link with other three-pronged rift systems developed over adjacent domes, forming a linked elongate rift system that then spreads to form an ocean basin. This type of development leaves behind some failed rift arms that will come to reside on the margins of young oceans when the successful rift arms begin to spread. These failed rift arms then become sites of increased sedimentation and subsidence, and also tend to be low-lying areas, and form the tectonic setting where many of the world's major rivers flow (for example, the Nile, Amazon, and Mississippi). Other rifts form at high angles to collisional mountain belts, and still others form in regions of widespread continental extension such as the basin and range province of the southwestern United States.

PULL-APART BASIN

Pull-apart basins are elongate depressions that develop along extensional steps on strike-slip faults. Pull-apart basins are features that develop in transtensional regions, in which the principal stresses are compressional, but some areas within the region are under extension due to the obliquity of the major stress direction with respect to the plane of failure. This results in extension of the crust along releasing bends, leading to a break in the crust and the formation of basins. Some pull-apart basins show several progressive stages in their formation. Others initiate along a fracture, and progress into lazy Z or S shapes, and finally progress into a basin that ranges in length-to-width ratio from 2:1 to 10:1. These types of basins are characterized by steep sides on major fault boundaries with normal faults developing on their shorter sides. Continuous movement along the major faults tends to offset deposits from their source inlet to the basin. These basins are characterized by rapid deposition and rapid facies changes along or across the width of the basin and gradual facies change along the longest axis of the basin. Pull-apart basin deposits are typically made mostly of coarse fanglomerate, conglomerate, sandstone, shales, and shallow water limestones and evaporites. Bimodal volcanics and volcanic sediments are also found interbedded within the basin deposits. These bimodal volcanics are typical of those found in rift settings, but here they are in a transtensional regime. Transcurrent faults can penetrate down deep into the crust, reaching the upper mantle and providing a conduit for magma.

See also CONVERGENT PLATE MARGIN PROCESSES; DIVERGENT PLATE MARGIN PROCESSES; DRAINAGE BASIN (DRAINAGE SYSTEM); OCEAN BASIN; PLATE TECTONICS; TRANSFORM PLATE MARGIN PROCESSES.

FURTHER READING

Allen, Philip, A., and John R. Allen. *Basin Analysis, Principles and Applications.* Oxford: Blackwell Scientific Publications, 1990.

Bradley, Dwight C., and Timothy M. Kusky. "Geologic Methods of Estimating Convergence Rates During Arc-Continent Collision." *Journal of Geology* 94 (1986): 667–681.

Mann, Paul, Mark R. Hempton, Dwight C. Bradley, and Kevin Burke. "Development of Pull-Apart Basins." *Journal of Geology* 91 (1983): 529–554.

Reading, Harold G. "Characteristics and Recognition of Strike-Slip Fault Systems." In *Sedimentation in Oblique-Slip Mobile Zones*, edited by Peter F. Balance and Harold G. Reading, 7–26. International Association of Sedimentology Special Publication 4, 1980.

beaches and shorelines   A beach is an accumulation of sediment exposed to wave action along a coastline, whereas the shoreline environment is a more encompassing area including beaches, islands, and near-shore areas that are in some way affected by coastal processes. The beach extends from the limit of the low-tide line to the point inland where the vegetation and landforms change to that typical of the surrounding region. This may be a forest, a cliff, dune, or lagoon. Many beaches merge imperceptibly with grasslands, or forests, whereas others end abruptly at cliffs or other permanent features, including artificial seawalls that have been built in many places in the past century. A beach may occupy bays between headlands, it may form elongate strips attached (or detached, in the cases of barrier islands) to the mainland, or it may form spits that project out into the water. To understand a beach it is necessary also to consider the nearshore environment, the area extending from the low-tide line out across the surf zone. The nearshore environment may include sandbars, typically separated by troughs. The width of nearshore environments is variable, depending on the slope of the seafloor, wave dynamics, and availability of sediment. Most nearshore environments include an inner sandbar located about 100–165 feet (30–50 m) offshore, and another bar about twice as far offshore. The inner bar is often cut by rip channels that allow water that piles up between the bar and beach to escape back to sea, often generating dangerous rip currents that can drag unsuspecting swimmers rapidly out to sea.

In the eastern United States, Florida is known for wide sandy beaches, the Outer Banks of the Carolinas are famous for barrier island beaches, and Maine is well known for its beautiful rocky shorelines. The western coast has many rocky shorelines in Washington, Oregon, and California, whereas the Gulf of Mexico has low relief beaches, barrier islands, and some mangrove-dominated shorelines.

Most sandy beaches develop typical profiles that change through the seasons and include several zones. These are the ridge and runnel, foreshore, backshore, and storm ridge. The ridge and runnel is the most seaward part of the beach, characterized by a small sandbar called a ridge, and a flat-bottom trough called the runnel, and is typically fewer than 30 feet (10 m) wide. The runnel is covered by water at high tide and has many small sand ripples that get extensively burrowed into by worms, crabs, and other beach life.

The foreshore, or beach face, is a flat, seaward-sloping surface that grades seaward into the ridge and runnel, or the intertidal zone if the ridge and runnel are not present. A narrow zone of gravel or broken shells may be present at the small slope-break between the foreshore and the ridge and runnel. The foreshore contains the swash and backwash zone, where waves move sand diagonally up the beach face parallel to the wave incidence direction, and gravity pulls the water and sand directly down the beach face parallel to the slope. This diagonal, then beach-perpendicular motion produces a net transport of sand and water along the beach, known as longshore drift and longshore currents.

The backshore extends from a small ridge and change in slope at the top of the foreshore known as a berm, to the next feature (dune, seawall, forest, lagoon) toward the land. This area is generally flat

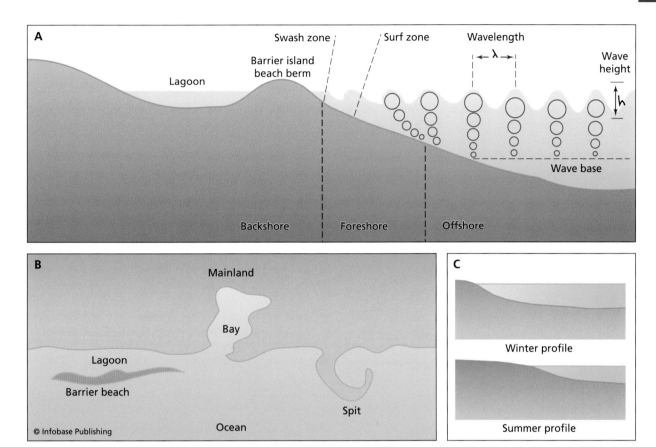

(A) Diagram of beach profile showing the major elements from the backshore to offshore; (B) different types of coastal environments, including barrier beaches, bays, and spits; (C) typical beach profiles in summer and winter showing how large winter storms erode the beach and smaller summer waves rebuild the beach

or gently landward sloping. The backshore area is usually dry and above the high-water mark except during large storms, so the backshore area is mainly affected and shaped by wind. Some backshore areas are characterized by multiple berms, and others have none. On gravel beaches, found in high-energy environments, the backshore area may be replaced by a storm ridge marked by a ridge of gravel that may be several to 10 feet (3 m) high. These ridges form because incoming waves have the velocity to move gravels up the beach face, but since these gravels are porous, the water sinks into the gravel before it can drag the gravel back down the beach face, causing its accumulation in a large ridge.

Beaches are highly variable in the width and heights of these various zones. Some beaches are steep, whereas others are flat. Beaches that have flat slopes are said to be dissipative in that they take the energy from waves and gradually dissipate it across the intertidal zone. These types of beaches often have multiple sand bars in the nearshore environment. Reflective beaches are those with steep gradients, and these tend to take much of the wave energy and reflect it back to sea. Reflective beaches do not gener-

ally have nearshore bars and are erosive. Dissipative beaches tend to be depositional, as they are actively accreting sediment.

The shape of a beach is largely controlled by the nature of the waves, tides, currents, and, to a lesser extent, wind. Waves move the sediment onshore, and are then transported along the beach face by the longshore currents, and perhaps blown to the backshore by wind. Tides change the areas to which waves direct their energy vertically up and down, bringing the sediment alternatively to different sections of the beach. Out of all these processes, the currents produced by the waves on the beach are the most important. These currents include longshore currents, rip currents, onshore-offshore currents produced in the swash zone, and combined currents.

Beaches are very dynamic environments and are always changing, being eroded and redeposited constantly from day to day and from season to season. They are typically eroded to thin strips, known as storm beaches, by strong winter storms and built up considerably during summer, when storms tend to be less intense. The wide summer beaches are known as accretionary beaches. The processes controlling this

BEAUTY AND THE BEACH: RETHINKING COASTAL LIVING

Civilized societies have built villages, cities, and industrial sites near the sea for thousands of years. Coastal settings offer beauty and commercial convenience but also invite disaster with coastal storms, tsunami, and rising sea levels. In 2004 and 2005 the world witnessed two furious incursions of the sea into heavily populated coastal regions, killing hundreds of thousands of people and causing trillions of dollars in damage. Coastal communities are experiencing early stages of a new incursion, as global sea levels slowly and inexorably rise, increasing the likelihood of additional, even more devastating disasters. These events demand serious reconsideration of priorities about further developing fragile and changing coastlines. Most pressing is a scientific reevaluation of the wisdom of rebuilding areas such as New Orleans, Louisiana, where sinking areas presently far below sea level doom residents to further, more serious disasters and tremendous loss of life. Allowing large, generally poor segments of the population to exist at great risk of death and property loss is socially irresponsible. Reconstruction funds may be better used to relocate large parts of the nation's population that have been displaced by coastal disasters to safer regions.

The year 2005 began with cleanup and recovery efforts from the tragic December 26, 2004, earthquake and tsunami that devastated coastal regions of the Indian Ocean. One of the worst natural disasters of the 21st century unfolded following a magnitude 9.0 earthquake off the northern Sumatra coast. Within minutes of the earthquake a mountain of water 100 feet (30 m) tall was ravaging northern Sumatra, sweeping into coastal villages and resort communities with a fury that crushed all in its path, removing buildings and vegetation, and in many cases eroding shoreline areas down to bedrock. Scenes of destruction and devastation rapidly moved up the coast of nearby Indonesia, then across the Indian Ocean to India and Africa. Buildings, vehicles, trees, boats, and other debris in the water formed projectiles that smashed into other structures at 30 miles (50 km) per hour, leveling all in their path, and killing nearly a quarter million people.

Areas in the United States at greatest risk for tsunamis are along the Pacific coast, including Hawaii, Alaska, Washington, Oregon, and California. Although most tsunamis are generated by earthquakes, others are generated by landslides, volcanic eruptions, meteorite impacts, and possibly gas releases from the deep ocean. Any of these events may happen at any time, in any of the world's oceans, including the Gulf of Mexico, which is prone to tsunamogenic submarine landslides.

Hurricanes Katrina (2005) and Rita (2005) devastated the Gulf Coast, inundating New Orleans with up to 23 feet (7 m) of water. Large sections of the city are uninhabitable, having been destroyed by floods and subsequent decay by contaminated water and toxic mold. The natural human inclination to respond to the disaster is to rebuild the city grander and greater than before, yet years after the disaster fewer than half of the residents of the city have been able to return to their former homes. This is not the most scientifically sound response, and could lead to even greater human catastrophes and financial loss in the future. New Orleans is located on a coastal delta in a basin that is up to 12 feet (3–4 m) below sea level and is sinking at rates of up to an inch (several mm to 2 cm) per year, so that much of the city could be 3–7 feet (1–2 m) farther below sea level by the end of the century. As New Orleans continues to sink, tall levees built to keep the Gulf, Mississippi River, and Lake Pontchartrain out of the city have to be repeatedly raised, and the higher they are built the greater the likelihood of failure and catastrophe.

Flood protection levees that reach 20 feet (6 m) tall built along the Mississippi keep the river level about 25–30 feet (4–5 m) above sea level at New Orleans. If these levees were to be breached, water from the river would quickly fill in

seasonal change are related to the relative amounts of energy in summer and winter storms—summer storms (except for hurricanes) tend to have less energy than winter storms, so they produce waves with relatively short wavelengths and heights. These waves gradually push the offshore and nearshore sands up to the beach face, building the beach throughout the summer. In contrast, winter storms have more energy with longer wavelength, higher amplitude waves. These large waves break on the beach, erode the beach face, and carry the sand seaward, depositing it in the nearshore and offshore environments. In some cases, especially along the rocky Pacific coasts, storms may remove all the sand from beaches, leaving only a rocky bedrock bench behind until the small summer waves can restore the beach. Storm beaches, however, tend to be temporary conditions as the wave energy decreases right after the storms. Even between winter storms the beach may tend to rebuild itself to a wider configuration.

BARRIER ISLANDS
Barrier islands are narrow linear mobile strips of sand up to about 30–50 feet (10–15 m) above sea level, and typically form chains located a few to tens of miles offshore along many passive margins. They

the 6–12 foot (2–4 m) deep depression with up to 25 feet (8 m) of water and leave a path of destruction where the torrents of water raged through the city. These levees also channel the sediments that would naturally get deposited on the flood plain and delta far out into the Gulf of Mexico, with the result being that the land surface of the delta south of New Orleans has been sinking below sea level at an alarming rate. A total land area the size of Manhattan is disappearing every year, meaning that New Orleans will be directly on the Gulf by the end of the century. Alarming poststorm assessments of damage from Hurricanes Katrina and Rita push that estimate forward by years.

The projected setting of the city in 2100 is in a bowl up to 30 feet (5 m) below sea level, directly on the hurricane-prone coast, and south of Lake Ponchartrain (by then part of the Gulf). The city will need to be surrounded by 50–100 foot (15–30 m) tall levees that will make the city look like a fish tank submerged off the coast. The levee system will not be able to protect the city from hurricanes any stronger than Katrina. Hurricane storm surges and tsunami could easily initiate catastrophic collapse of any levee system, initiating a major disaster. Advocates of rebuilding are suggesting elevating buildings on stilts or platforms, but forget that the city will be 3–6 meters below sea level by 2090, and that storm surges may reach 30–35 feet (10 m) above sea level. A levee failure in this situation would be catastrophic, with a debris-laden wall of water 45–50 feet (15 m) tall sweeping through the city at 30

miles (50 km) per hour, hitting these buildings-on-stilts with the force of Niagara Falls, and causing a scene of devastation like the Indian Ocean tsunami.

Sea-level rise is rapidly becoming one of the major global hazards that humans must deal with, since most of the world's population lives near the coast in the reach of the rising waters. The current rate of rise of an inch (a couple cm) every 10 years will have enormous consequences. Many of the world's large cities, including New York, Houston, New Orleans, and Washington, D.C., have large areas located within 10–20 feet (a few meters) of sea level. If sea levels rise even a few feet (1 m), many of the city streets will be underwater, not to mention basements, subway lines, and other underground facilities. New Orleans will be the first under, lying a remarkable 10–15 feet (3–5 m) below the projected sea level on the coast at the turn of the next century. At this point governments should not be rebuilding major coastal cities in deep holes along the sinking, hurricane-prone coast. Governments, planners, and scientists must begin to make more sophisticated plans for action during times of rising sea levels. The first step would be to use the reconstruction money for rebuilding New Orleans as a bigger, better, stronger city in a location where it is above sea level, and will last for more than a couple of decades, saving the lives and livelihoods of hundreds of thousands of people.

New Orleans is sinking farther below sea level every year and getting closer

to the approaching shoreline. Sea level is rising, and more catastrophic hurricanes and floods are certain to occur in the next 100 years. Americans must decide whether to spend hundreds of billions of tax dollars to rebuild a city with historic and emotional roots where it will be destroyed again, or to move the bulk of the city to a safer location before subsidence increases and another disaster strikes. The costs of either decision will be enormous. The latter makes more sense and will eventually be inevitable. The city could be moved in the slump following the destruction by Hurricane Katrina, saving lives, or residents could wait until an unexpected category five superhurricane makes a direct hit and kills hundreds of thousands of people. Katrina was a warning, New Orleans is sinking below sea level, and it is time to move to high and dry ground.

FURTHER READING

Beatley, Timothy, David J. Brower, and Anna K. A. Schwab. *Introduction to Coastal Management.* Washington, D.C.: Island Press, 1994.

Davis, R., and D. Fitzgerald. *Beaches and Coasts.* Malden, Mass.: Blackwell, 2004.

Kusky, T. M. *The Coast: Hazardous Interactions within the Coastal Environment.* New York: Facts On File, 2008.

Williams, S. J., K. Dodd, and K. K. Gohn. *Coasts in Crisis.* Reston, Va.: U.S. Geological Survey Circular 1075, 1990.

are separated from the mainland by the back-barrier region, which is typically occupied by lagoons, shallow bays, estuaries, or marshes. Barriers are built by vertical accumulation of sand from waves and wind action. Barrier islands are so named because they form a natural protection of the shoreline from the forces of waves, tsunami, tides, and currents from the main ocean. Many barrier islands have become heavily developed, however, as they offer beautiful beaches and resort-style living. The development of barrier islands is one of the most hazardous trends in coastal zones, since barriers are simply mobile strips of sand that move in response to changing sea

levels, storms, coastal currents, and tides. Storms are capable of moving the entire sandy substrate out from underneath tall buildings.

The size of barrier islands ranges from narrow and discontinuous strips of sand that may be only a few hundred feet wide, to large islands that extend many miles in width and length. The width and length is determined by the amount of sediment available, as well as a balance between wave and tidal energy. Most barriers are built of sand, either left over from glaciations, as in New England, eroded from coastal cliffs, or deposited by rivers along deltas such as at the end of the Mississippi River in the Gulf

Photo of waves crashing on beach *(Stephanie Coffman, Shutterstock, Inc.)*

of Mexico. Barrier island systems need to be discontinuous, to allow water from tidal changes to escape back to sea along systems of tidal inlets.

Subenvironments of barriers are broadly similar to those of beaches; they include the beach, barrier interior, and landward interior. The beach face of a barrier is the most dynamic part of the island, absorbing energy from waves and tides, and responding much as beaches on the mainland do. The backside of the beach on many barrier islands is marked by a long frontal or foredune ridge, followed landward by secondary dunes. Barrier islands that have grown landward with time may be marked by a series of linear ridges that mark the former positions of the shoreline and foredune ridges, separated by low areas called swales. The landward margins of many barriers merge gradually into mud flats, or salt marshes, or may open into lagoons, bays, or tidal creeks.

About 15 percent of the world's coastlines have barrier islands offshore, with most located along passive-margin continental shelves, which have shallow slopes and a large supply of sediment available to build the barriers. In the United States the eastern seaboard and Gulf of Mexico exhibit the greatest development of barrier island systems. It seems that areas with low tidal ranges in low to middle climate zones have the most extensively developed barrier systems.

Barrier systems are of several types. Barrier spits are attached to the mainland at one end and terminate in a bay or the open ocean on the other end. They are most common along active tectonic coasts, although Cape Cod in Massachusetts is one of the better-known examples of a spit formed along a passive continental margin. Some spits have ridges of sand that curve around the end of the spit that terminates in the sea, reflecting its growth. These are known as recurved spits. Sandy Hook, at the northern end of the New Jersey coast, is a recurved spit. Spits form as longshore currents carry sediment along a coastline, and the coastline makes a bend into a bay. In many cases the currents that carry the sand continue straight and carry the sediment offshore, depositing it in a spit that juts out from the mouth of the bay. Many other subcategories of spits are known and classified according to specific shape. Some, known as tombolos, may connect offshore islands with the mainland, whereas others have cuspate forms or jut outward into the open water.

In some cases barriers grow completely across a bay and seal off the water inside it from the ocean. These are known as welded barriers and are most common along rocky coasts such as in New England and Alaska. Welded barriers seem also to form preferentially where tidal energy is low, as this prevents the tides from creating tidal channels that allow salty water to circulate into the bay. Some also form during onshore migration of barriers during times of sea-level rise, when the barrier sands get moved into progressively narrowing bays as they are forced to move inland. Since they are cut off from the ocean, bays that form behind welded barriers tend to be brackish or even filled with freshwater.

Barriers form by a variety of different mechanisms in different settings, but the most common mechanisms include the growth and accretion of spits that become breached during storms, growth as offshore sandbars, and as submergence of former islands during times of sea-level rise. Barriers are constantly moving and respond to storms, currents, waves, and sea-level rise by changing their position and shape. Barriers moving onshore are known as retrograding barriers; they move by a process of rolling over, where sand on the outer beach face is moved to the backshore, then overrun by the next sand from the beach face. A continuation of this process leads the barrier to roll over itself as it migrates onshore. Prograding barriers are building themselves seaward, generally through a large sediment supply, whereas aggrading barriers are simply growing upward in place as sea levels rise.

COASTAL DUNES

Many coastal areas have well-developed sand dunes in the backshore area, some of which reach heights of several tens or even hundreds of feet (tens of meters). The presence or absence of dunes, and their shape and height, is mostly controlled by the amount of sediment supply available, although wind strength and type and distribution of vegetation also play significant roles. Dunes are fragile ecosystems that can easily be changed by disturbing the vegetation or beach dynamics, yet their importance is paramount to protecting inland areas from storm waves and surges, tsunami, and other hazards from the ocean.

Most coastal dunes are of the linear type, known as foredunes, which form elongate ridges parallel to the beach just landward of the foreshore. In some cases numerous foredunes are present, with the ones closest to land being the oldest and the younger ones forming progressively seaward of these older dunes.

Sand dunes in the backbeach area are built by the windblown accumulation of sand derived from the foreshore area. The sands may grow far into the backshore environment, in some cases extending miles inland if not obstructed by vegetation, cliffs, or constructions such as buildings or seawalls. Vegetation is extremely effective at stabilizing mobile sand, and many examples of sand being trapped by plants are visible on beaches of the world.

Dunes are built by the slow accumulation of sand moved by wind, but may be rapidly eroded by storm surges and wave attack when the sea surface is elevated on storm surges. A single storm can remove years of dune growth in a few hours, transporting the dune sand offshore or along shore. Examples of this process were all too clear from Hurricanes Katrina in Louisiana and Mississippi in 2005 and Ike along the Texas coast in 2008. Likewise, tsunamis can remove entire dune fields in a single devastating event, as seen in many places during the Indian Ocean tsunami of December 2004. Rising sea levels pose a huge threat to many existing coastal dune fields, since a rise in sea level of one foot (0.3 m) on flat terrain can be equated with a 100-foot (305 m) landward migration of the shoreline, and the removal of the dune field from one location to an area farther inland, or to its complete elimination.

COASTAL LAGOONS

Lagoons are a special, rather rare class of restricted coastal bays that are separated from the ocean by an efficient barrier that blocks any tidal influx, and they do not have significant freshwater influx from the mainland. Water enters lagoons mainly from rainfall and occasional storm wash-over. Evaporation from the lagoon causes their waters to have elevated salinity and distinctive environments and biota.

Most lagoons are elongate parallel to the coast and separated from the ocean by a barrier island or in some cases by a reef. They are most common in dry or near-desert climates, since freshwater runoff needs to be very limited to maintain lagoon conditions. Lagoons are therefore common along coasts including the Persian Gulf, North Africa, southeast Africa, Australia, Texas, Mexico, and southern Brazil.

Many lagoons show large seasonal changes in salinity, with nearly fresh conditions during rainy seasons and extremely salty conditions as the waters evaporate and even dry up in the dry seasons. Normal marine and estuarine organisms cannot tolerate such wide variations in salinity, so typically large numbers of a relatively few specialized species of organisms are found in lagoons. Some species of fish, such as the killifish, can regulate the salinity in their bodies to match that of the outside waters, so they are well suited for the lagoon environment. Certain species of gastropods (snails) are also very tolerant to variations in salinity, and are found in large quantities in some lagoons.

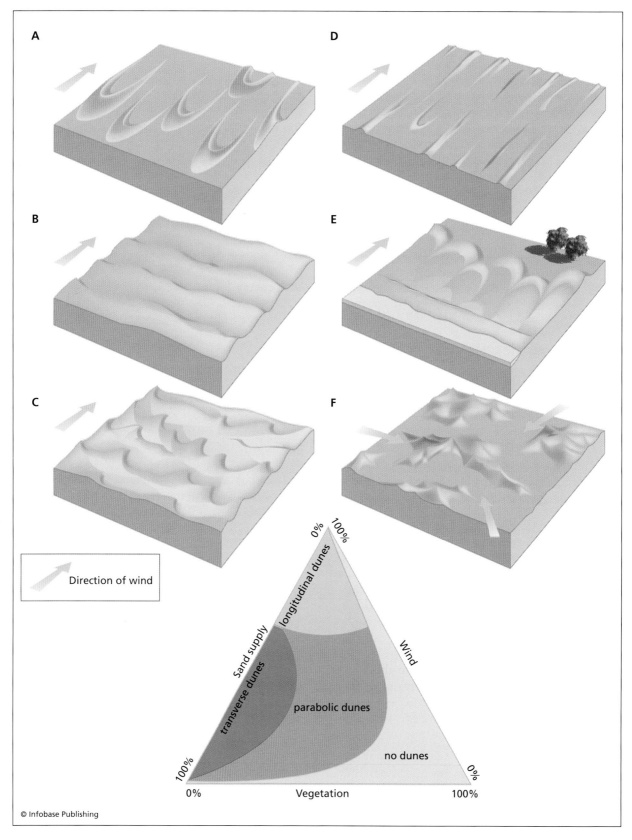

Block diagram of types of dunes including (A) barchan, (B) transverse, (C) barchanoid, (D) linear, (E) parabolic, and (F) star. Graph (triangular) illustrates which types of dunes form under different conditions of wind, sand supply, and vegetative cover.

As the water in lagoons evaporates in summer months, it deposits chemical sediments known as evaporates and carbonates. These typically include a sequence of minerals from aragonite to calcite to gypsum to halite. Many ancient lagoon environments are recognized by the presence of this repeating sequence of evaporate and carbonate minerals in the rock record.

Lagoons are not significantly influenced by waves or tides, and are dominated by effects of the wind. Winds can induce circulation in lagoons or even waves during windstorms. Small wind tides in lagoons may transport more water to one side of the lagoon, and deposit fine-grained sediments on this one side as the waters retreat when the wind dies out. During large ocean storms tidal surges may overtop the barrier to the lagoon, bringing a surge of seawater and sediment into the lagoon. During storms, and during the daily sea breeze cycle, sand from the beach and coastal dunes can be transported into lagoons. This can be a major contributor to sediment accumulation in some lagoons, and in some examples sand dunes from the beach are moving landward into lagoons, migrating over lagoonal sediments and vegetation.

Sediments deposited in lagoons include layers of chemical sediments that precipitated from the water as it evaporated, leaving behind the elements initially dissolved in the water as sedimentary layers. These sediments are most commonly fine-grained, clay-sized calcite and aragonite, and micrite, a form of carbonate mud. Many lagoons are covered by mucky micrite layers that have green slimy microorganisms known as cyanobacteria, or blue-green algae, growing along the edges of the lagoon in the mud and forming matlike pads surrounding the central, water-filled part of the lagoon. Many times these mudflats and algal pads are dried out and cracked by the Sun, forming thin flakes that can be blown around by the wind. Lagoons also have sediments such as sand grains carried by the wind, and the skeletal and other remains of the organisms that lived in the lagoon. Sand washed into lagoons from storms often forms small, fan-shaped bodies known as wash-over fans that cover parts of the lagoon on its seaward side.

TIDAL INLETS

Tidal inlets are breaks in barrier island systems that allow water, nutrients, organisms, ships, and people easy access and exchange between the high-energy open ocean and the low-energy back-barrier environment consisting of bays, lagoons, tidal marshes, and creeks. Most tidal inlets are within barrier island systems, but others may separate barrier islands from rocky or glacial headlands. Tidal inlets are extremely important for navigation between sheltered ports on the back-barrier bays and the open ocean; thus they are the sites of many coastal modifications such as jetties, breakwaters, and dredged channels to keep the channels stable and open.

Strong tidal currents move water into and out of tidal inlets as the tides wax and wane, and also carry out of the channel large amounts of sediment brought in by waves and long shore transport. Never try to swim in a tidal inlet. As tides rise on the ocean side of tidal inlets the water rises faster than on the inside of the inlet, since the inlet is narrow and it takes a long time for the water to move into the restricted environment behind the barrier. The difference in elevation causes the water to flow into the inlet with a strong current, called a flood-tidal current. As the tide falls outside the inlet, the reverse happens—as the tide falls quickly on the outside of the barrier, the sea surface is higher inside the inlet and a strong current known as an ebb-tidal current then flows out of the inlet, returning the water to the ocean. Considering the amount of time that it takes for water to flow into and out of tidal inlets, it is apparent that times of high and low tide may be considerably different on the two sides of barrier systems connected by tidal inlets.

The sides of tidal inlets are often marked by curved sand ridges of recurved spits, formed as waves are refracted into the barrier and push the sand into ridges. The strongest currents in tidal inlets are found where the inlet is the narrowest, a place with the deepest water called the inlet throat. Water rushes at high velocity into and out of this throat, carrying sand into and out of the back-barrier environment. Since the velocity of the water decreases after it passes through the throat, large lobes and sheets of sand are typically deposited as tidal deltas on both the inside and the outside of tidal inlets. The delta deposited by the incoming (flood) tide on the inside or landward side of the inlet is known as a flood-tidal delta, whereas the delta deposited on the outside of the inlet by the ebb tide is known as an ebb-tidal delta.

Tidal inlets form by a variety of mechanisms. The most common is during the formation and evolution of barrier systems along coastal platforms on passive margins, where barrier islands emerged as glaciers retreated and sea levels rose onto the continental shelves in Holocene times. Sea levels rose more slowly about 5,000 years ago, and enhanced coastal erosion provided abundant sand to create the barrier island systems. Continued rising sea levels plus diminished sediment supplies and the many modifications of the shoreline by humans has led to increased erosion along much of the world's coastlines. With

Sketch showing features of a barrier island, tidal inlet, and lagoon coastal system. Note the positions of the small deltas on either side of the tidal inlet, the coastal march, and beach-dune ridge. *(modeled after R. Davis and D. Fitzgerald).*

this trend many barriers have been breached or cut through during storms. Typically this happens when an incoming storm erodes the foredune ridge, and waves top the barrier island, washing sand into the back barrier region, often making a shallow channel through the barrier. As the storm and elevated tides recede, the water in the back barrier bay, lagoon, or tidal marsh is left high, then begins to escape quickly through the new shallow opening, deepening it rapidly. If the tides can continue to keep this channel open, a new tidal inlet is established. Many tidal inlets along the Outer Banks barrier islands of North Carolina have formed in this way. Any homes or roads that were in the way are gone.

Tidal inlets may also form by longshore currents, building a spit across a bay or drowned river valley. As the spit grows across an open bay, the area open to the sea gradually becomes narrower until it begins to host strong tidal currents, when it becomes deep and reaches an equilibrium between the amount of sediment transported to the inlet by longshore drift and the amount of sand moved out of the inlet by tidal currents and waves.

INTERTIDAL FLATS

Many coastlines have flat areas within the tidal range sheltered from waves, dominated by mud, and devoid of vegetation that are accumulating sediment, known as intertidal (or just tidal) flats. The width of tidal flats depends on the tidal range and the shape and morphology of the coastline or bay where they are located. Some large bays with large tidal ranges, such as the Bay of Fundy in eastern Canada, are dominated by tidal flats. Tidal flats are typically flat areas cut by many channels and dominated by mud and sandy sediment. Many may have layers of shell debris and ecosystems of organisms specially adapted to this environment. They are alternately covered at high water and exposed to the atmosphere at low tide.

The sand, mud, and shell fragment layers that most tidal flats comprise are distributed in ways that reflect the distribution of energy in this environment. Sands are typically located near the base of the intertidal zone where energy from tides is the highest, and these sands gradually merge with and then give way to muds toward the upper parts of the zone, and away from the ocean. At any given location within the tidal flats the sediments tend to be rather uniform, because of the similar conditions that persist and repeat at any given location.

Modern tidal flat environments are inhabited by a variety of organisms specially adapted to this harsh environment, including specialized snails, worms, amphipods, oysters, mussels, and other bivalves. Many of these organisms survive by burrowing into the mud for protection; in doing so they destroy the fine-scale layering in the mud by a process called bioturbation. The mud in many tidal flats is also inhabited by microscopic filamentous cyanobacteria (commonly called blue-green algae) that produce slimy mats that cover many mud surfaces and hold the mud particles together during the ebb and flow of the tides. These mucaceous mats also trap mud and other sediments, helping to build up the sedimentary accumulation in these environments.

Tidal flats often expose sediments in which the sedimentary particles are arranged in specific, peculiar, and repeating forms known as sedimentary structures. Familiar types of sedimentary structures include sedimentary layers; ripples, produced by currents moving the sedimentary particles as sets of small waves; mega-ripples, large ripples formed by unusually strong currents; mudcracks, produced by muddy sediments being dried by the Sun and shrinking and cracking; and other structures produced by organisms. These latter structures include burrows from worms, bivalves, and other organisms, trails, and footprints.

Most tidal flats are cut by a network of tidal channels that may contain water even at low tides. These form a network of small to large channels but differ from normal streams in that they carry water in different directions with the ebb and flood of the tides. As the tide rises into tidal flats strong currents that range up to several feet (1 m) per second bring the tidal wave through these channels, moving sediments throughout the flats. Tidal flats tend to build themselves slowly seaward, out from the bays or estuaries that they initially grow around. They may eventually fill in the bays and estuaries up to the sand dunes or berm in the back-beach area or coastal barrier. Most tidal flats are not significantly affected by waves. Exceptionally large flats, however, such as the Wadden Sea area of the North Sea on Germany's coast are significantly affected by waves for several hours of each high-tide stand.

COASTAL WETLANDS AND MARSHES
Many bays, estuaries, and coastal tidal flats are bordered inland by a vegetated intertidal area containing grasses or shrublike mangrove swamps. Mangroves do not tolerate freezing conditions so are found only at low latitudes, whereas salt marshes are found at all latitudes. These coastal wetlands or salt marshes host a range of water salinities, from salty and brackish to nearly fresh. As estuaries age or mature, they tend to become progressively filled in first by tidal flats, then by salt marshes or coastal wetlands. Thus the degree to which estuaries are filled in can indicate the state of their maturity.

Salt marshes form on the upper part of the intertidal zone where organic rich sediments are rarely disturbed by tides, providing a stable environment for grasses to take root. The low marsh area is defined as the part of the marsh that ranges from the beginning of vegetation to the least mean high tide. The high marsh extends from the mean high tide up to the limit of tidal influence. Different genera and species of grass form at different latitudes and on different continents, but in North America high parts of salt marshes are dominated by Juncus grasses, known also as needle- or black-rush, which can be 5 to 6 feet (2 m) tall, with sharp, pointed ends. Low parts of salt marshes tend to be dominated by dense growths of knee-high Spartina grasses.

Salt marshes must grow upward to keep up with rising sea levels. To do this they accumulate sediments derived from storm floods moving sediment inland from the beach environment, from river floods bringing in sediment from the mainland, and from the accumulation of organic material that grew and lived in the salt marshes. When plants in salt marshes are suddenly covered by sediment from storms or floods, they quickly recover by growing up through the new sediment, thereby allowing the marsh to survive and grow upward. Some salt marshes grow upward so efficiently that they raise themselves above tidal influence and eventually become a freshwater woodland environment. With the increasing rate of sea-level rise predicted for the next century, however, many scientists are concerned that sea level will start to rise faster than marsh sedimentation can keep pace with it. This problem is particularly exacerbated in places where the normal supply of river and flood sediments is cut off, for instance, by levees along rivers. If this happens, many of the fragile and environmentally unique coastal marsh settings will disappear. Marshes are among the most productive of all environments on Earth; they serve as nurseries for many organisms and are large producers of oxygen through photosynthesis. The disappearance of coastal marshes is already happening at an alarming rate in places such as the Mississippi River delta, where coastal subsidence, loss of delta replenishment, together with sea-level rise leads to more than 0.39 inches (1 cm) of relative sea-level rise each year. Salt marshes are disappearing at an alarming rate along the Mississippi River delta, as discussed in a later chapter.

Many coastal marshes in low latitudes are covered with dense mangrove tidal forest ecosystems, known also as mangals. These have fresh to brackish water and are under tidal influence. Mangrove stands have proven to be extremely effective protective barriers against invaders from the sea, including hostile armies, storm surges, and tsunamis. The destruction of many coastal mangrove forests in recent years has proven catastrophic to some regions, such as areas inundated by the 2004 Indian Ocean tsunami that were once protected by mangroves. Many local governments removed the mangroves to facilitate development and shrimp farming, but when the tsunami hit, it swept far inland in areas without mangroves, and was effectively stopped in places where the mangroves were still undisturbed. There are many examples of places where mangrove-dominated coasts have withstood direct hits from hurricanes and storm surges, yet protected the coastline to the extent that there was little detectable change after the storm.

Several dozen or more types of mangroves are known, occurring on many coasts of North America, Africa, South America, India, Southeast Asia, and elsewhere around the Pacific. Mangroves prefer protected, low-energy coasts such as estuaries, lagoons, and back-barrier areas. Mangroves develop extensive root systems and propagate by dropping seeds into the water, where they take root and spread. Mangrove stands have also been known to be uprooted by storms, float to another location, and take root in the new setting.

The extensive root network of mangrove stands slows many tidal currents and reduces wave energy by a factor of 10, forming lower-energy conditions inside the mangrove forest. These lower-energy conditions favor the deposition of sediment, enhancing seaward growth of the mangrove forest.

DELTAS

Deltas are low flat deposits of alluvium at the mouths of streams and rivers that form broad triangular- or irregular-shaped areas that extend into bays, oceans, or lakes. They are typically crossed by many distributaries from the main river and may extend for a considerable distance underwater. When a stream enters the relatively still water of a lake or the ocean, its velocity and its capacity to hold sediment drop suddenly. Thus the stream dumps its sediment load there, and the resulting deposit is known as a delta. The term *delta* was first used for these deposits by Herodotus in the fifth century B.C.E., for the triangular-shaped alluvial deposits at the mouth of the Nile River. The stream first drops the coarsest material, then progressively finer material further out, forming a distinctive sedimentary deposit. In a study of several small deltas in ancient Lake Bonneville in Utah, Nevada, and Idaho, Grover Karl Gilbert in 1890 recognized that the deposition of finer-grained material farther away from the shoreline also created a distinctive vertical sequence in delta deposits. The resulting foreset layer is thus graded from coarse nearshore to fine offshore. The bottomset layer consists of the finest material, deposited far out. As this material continues to build outward, the

stream must extend its length and forms new deposits, known as topset layers, on top of all this. Topset beds may include a variety of subenvironments, both subaqueous and subaerial, formed as the delta progrades seaward.

Most of the world's large rivers such as the Mississippi, the Nile, and the Ganges, have built enormous deltas at their mouths, yet all of these are different in detail. Deltas may have various shapes and sizes or may even be completely removed, depending on the relative amounts of sediment deposited by the stream, the erosive power of waves and tides, the climate, and the tectonic stability of the coastal region. The distributaries and main channel of the rivers forming deltas typically move to find the shortest route to the sea, and this causes the shifting of the active locus of deposition on deltas. Inactive areas, which may form lobes or just parts of the delta, typically subside and are reworked by tidal currents and waves. High-constructive deltas form where the fluvial transport dominates the energy balance on the delta. These deltas are typically elongate, such as the modern delta at the mouth of the Mississippi, shaped like a bird's foot, or they may be lobate, such as the older Holocene lobes of the Mississippi that have now largely subsided below sea level.

High-destructive deltas form where the tidal and wave energy is high and much of the fluvial sediment gets reworked before it is finally deposited. In wave-dominated high-destructive deltas sediment typically accumulates as arcuate barriers near the mouth of the river. Examples of wave-dominated deltas include the Nile and the Rhône deltas. In tide-dominated high-destructive deltas, tides rework the sediment into linear bars that radiate from the mouth of the river, with sands on the outer part of the delta sheltering a lower-energy area of mud and silt deposition inland from the segmented bars. Examples of tide-dominated deltas include the Ganges and the Kikari and Fly River deltas in the Gulf of Papua, New Guinea. Other rivers drain into the sea in places where the tidal and wave current is so strong that these systems completely overwhelm the fluvial deposition, removing most of the delta. The Orinoco River in South America has had its sediment deposits transported southward along the South American coast, with no real delta formed at the mouth of the river.

Where a coarse sediment load of an alluvial fan dumps its load in a delta, the deposit is known as a fan-delta. Braid-deltas are formed when braided streams meet local base level and deposit their coarse-grained load.

Deltas create unique, diverse environments where fresh and saltwater ecosystems meet, and swamps, beaches, and shallow marine settings are highly varied. Deltas also form some of the world's greatest hydrocarbon fields, as the muds and carbonates make good source rocks and the sands make excellent trap rocks.

GLACIATED COASTS

Glaciated and recently deglaciated coastlines offer a variety of environments that are significantly different from other coastal features so far discussed. Some coastlines, such as many in Antarctica, Greenland, and Alaska, have active glaciers that reach the sea, whereas other coasts, such as from New England northward into Canada, Scandinavia, and parts of the Far East have recently been deglaciated (within the past 18,000 years).

The primary effects of glaciers on coastlines include the carving out of wide U-shaped glacial valleys and erosion of loose material overlying bedrock, the deposition of huge quantities of sediment especially near the termini of glaciers, and lowering of global sea levels during periods of widespread glaciation. In addition, many coastal areas that had thick ice sheets on them were depressed by the weight of the glaciers, and have been slowly rebounding upward since the weight of the glaciers was removed. This glacial rebound causes coastal features to move seaward and former beaches and coastlines to be uplifted.

When glaciers move across the land surface, they can erode bedrock by a combination of grinding and abrasion, plucking material away from the bedrock, and ice wedging where water penetrates cracks, expands as it freezes, and pushes pieces of bedrock away from its base. The material removed from the bedrock and overburden is then transported with the glacier to its end point, often at the coast, where it may be deposited as a pile of gravel, sand, and boulders known as a glacial moraine. Some glacial moraines are relatively small and outline places where individual glaciers flowed out of valleys and ended at the sea. These form where the glaciers were relatively small and were confined to valleys. Other glacial moraines are huge, and mark places where continental ice sheets made their farthest movement southward, depositing vast piles of sand and gravel at their terminus. On the eastern seaboard of the United States, New York's Long Island and Massachusetts's Cape Cod, Martha's Vineyard, and Nantucket Island represent the complex terminal moraine from the Pleistocene ice sheets. In places like New England that were covered by large continental ice sheets, the glaciers tended to scour the surface to the bedrock, leaving behind irregular and rocky coasts characterized by promontories and embayments, islands, but only rare sandy beaches.

Depositional features on deglaciated coasts are varied. *Glacial drift* is a general term for all sediment deposited directly by glaciers, or by glacial meltwater in streams, lakes, and the sea. Till is glacial drift that was deposited directly by the ice. It is a nonsorted random mixture of rock fragments. Glacial marine drift is sediment deposited on the seafloor from floating ice shelves or bergs, and may include many isolated pebbles or boulders that were initially trapped in glaciers on land, then floated in icebergs that calved off from tidewater glaciers. These rocks melted out while over open water, and fell into the sediment on the sea bottom. These isolated dropstones are often one of the hallmarks of ancient glaciation in rock layers that geologists find in the rock record. Stratified drift is deposited by meltwater and may include a range of sizes, deposited in different fluvial or lacustrine environments.

Terminal or end moraines are ridgelike accumulations of drift deposited at the farthest point of travel of a glacier's terminus. Terminal moraines may be found as depositional landforms at the bases of mountain or valley glaciers marking the locations of the farthest advance of that particular glacier, or may be more regional in extent, marking the farthest advance of a continental ice sheet. There are several different categories of terminal moraines, some related to the farthest advance during a particular glacial stage, and others referring to the farthest advance of a group of or all glacial stages in a region. Continental terminal moraines are typically succeeded poleward by a series of recessional moraines marking temporary stops in the glacial retreat or even short advances during the retreat. They may also mark the boundary between a glacial outwash terrain and a knob and kettle or hummocky terrain toward more poleward latitudes from the moraine. The knob and kettle terrain is characterized by knobs of outwash gravels and sand separated by depressions filled with finer material. Many of these kettle holes were formed when large blocks of ice were left by the retreating glacier, and the ice blocks melted later, leaving large pits where the ice once was. Kettle holes are typically filled with lakes; many regions characterized by many small lakes have a recessional kettle hole origin.

Glacial erratics are glacially deposited rock fragments with compositions different from underlying rocks. In many cases the erratics are composed of rock types that do not occur in the area they are resting in, but are found only hundreds or even thousands of miles away. Many glacial erratics in the northern part of the United States can be shown to have come from parts of Canada. Sediment deposited by streams washing out of glacial moraines, known as outwash, is typically deposited by braided streams. Many of these glacial outwash braided streams form on broad plains known as outwash plains. When glaciers retreat, the load is diminished, and a series of outwash terraces may form.

Drumlins are teardrop-shaped accumulations of till that are up to about 150 feet (50 m) in height, and tend to occur in groups of many drumlins. These have a steep side that faces in the direction that the glacier advanced from and a back side with a more gentle slope. Drumlins are thought to form beneath ice sheets and record the direction of movement of the glacier. Drumlin coasts are found on the eastern side of Nova Scotia and in Massachusetts Bay, including many in Boston Harbor. A final common depositional landform of glaciers found on many coasts are eskers, elongate ridges of sands and gravel that may extend many miles but be only a few tens of feet (several m) wide. These represent the paths of meltwater streams that flowed inside and underneath the glaciers, depositing the sand and gravel in the stream bed, which got left behind as the glacier retreated.

Coastlines that were mountainous when the glaciers advanced had their valleys deepened by the glaciers carving out their floors and sides, creating fjords. Fjords are steep-sided glacial valleys that open to the sea. Southern Alaska has numerous fjords that have active tidewater glaciers in them, which are now experiencing a phase of rapid retreat. The Hudson River valley and Palisades just north of New York City comprise a fjord formed in the Pleistocene, and many fjords are found in Scandinavia, New Zealand, Greenland, Chile, and Antarctica.

ROCKY COASTS

Rocky coastlines are most common along many convergent tectonic plate boundaries and on volcanic islands, but may also be found on recently deglaciated coasts and along other uplifted coasts such as southern Africa and recently uplifted coasts such as along the Red Sea. Rocky coasts are the most common type of coastline in the world, forming on the order of 75 percent of the world's coasts. The morphology of rocky coastlines is determined mainly by the type of rock, its internal structure and tectonic setting, as well as the physical, chemical, and biological processes operating on the coastline. Coastlines with mountains and steep slopes under the sea tend to have large waves, since little of the wave energy is dissipated by shallow water as the waves approach. These large waves erode the coast and also transport any sand that tends to accumulate offshore, so it is rare to find sandy beaches on steep, rocky coastlines. Some tropical islands have exposures of jagged limestone along their coastlines. Much of this limestone formed as shallow water mud and reefs, and was exposed above sea level when sea levels fell during the Pleistocene.

The rates of geological processes and change along rocky coastlines are much slower than along sandy beaches, so it is often difficult to notice change over individual lifetimes. Rocky coastlines are experiencing erosion over geological time periods, however, through a combination of waves, rain, ice wedging during the freeze-thaw cycle, and chemical and biological processes. Waves that continuously pound on rocky coastlines are the most effective erosive agents, slowly wearing down the rock and, in some cases, quarrying away large boulders. When waves carry sand and smaller rocks, these particles are thrown against the coastal rocks, causing significant abrasion and erosion. Abraded rock surfaces tend to be smooth, whereas those eroded by wave quarrying are irregular.

In higher latitudes subject to the freeze-thaw cycle, water often gets trapped in cracks and joints in the rock and then freezes. Since water expands by 9 percent when it freezes, this creates large stresses on the rock around the crack, often enough to expand the crack and eventually contribute to causing blocks of rock to separate from the main rocky coast and become a boulder.

Biological processes also contribute to the erosion of rocks along rocky coasts. Microscopic blue-green algae burrow into limestone, using the calcium carbonate ($CaCO_3$) as a nutrition source and causing the limestone to be more easily weathered away, a fraction of an inch (mm) at a time. Other organisms, such as sea urchins, abalone, chitons, and other invertebrates, bore into rocky substrate, slowly eroding the coast. Rocks along the coast are also subject to chemical weathering, like other rocks in other environments. Limestones may be dissolved by acid rain, and feldspars in granites and other rocks may be converted by hydrolysis into soft, easily eroded clay.

The relative strength of these processes is determined by several factors, including rock type, climate, wave energy, rock structure, tidal range, and sea level. Soft rocks such as sandstones are easily eroded, whereas granites weather much more slowly and typically form headlands along rocky coasts. Highly fractured rocks tend to break and erode faster, especially in climates with a significant freeze-thaw cycle. In humid wet climates chemical erosion may be more important than the freeze-thaw cycle. Wave height and energy is important, as waves exert their greatest erosive power at just above the mean high-water level and are very effective at slowly removing rocks, seawalls, and other structures, particularly in places where sandy beaches are absent. Sandy beaches absorb wave energy, so when beaches are absent the waves are much more erosive. Areas with small tidal ranges tend to focus the wave energy on a small area, whereas areas with large tidal ranges tend to change the area being attacked by waves. Thus tides and relative sea level also influence the effectiveness of waves in eroding the coast.

Many rocky coasts are bordered by steep cliffs, many of which are experiencing active erosion. The erosion is a function of waves' undercutting the base of the cliffs and oversteepening the slopes, which then collapse to form a pile of boulders that are then broken down by wave action. On volcanic islands, such as Hawaii and Cape Verde, some large, amphitheater-shaped cliffs were formed by giant landslides when large sections of these islands slumped into the adjacent ocean, creating the cliffs and generating tsunamis. In contrast, cliffs of unconsolidated gravels and sand attempt to recover to the angle of repose by rain water erosion or slumping from the top of the cliff. This erosion can be dramatic, with many tens of feet removed during single storms. The material eroded from the cliffs replenished the beaches, and without the erosion the beaches would not exist. Coarser material is left behind as it cannot be transported by the waves or tidal currents while the finer-grained material is carried out to sea. The remaining coarse-grained deposits typically form a rocky beach with a relatively flat platform known as a wave-cut terrace.

Some rocky shorelines are marked by relatively flat bedrock platforms known as benches, ranging from a few tens of feet (m) to thousands of feet (1 km) wide, typically followed inland by cliffs. These benches may be horizontal, gently seaward dipping, or inclined as many as 30 degrees toward the sea. Benches are formed by wave abrasion and quarrying of material away from cliffs, and develop along with cliff retreat from the shoreline. The waves must have enough energy to remove the material that falls from the cliffs, and the waves abrade the surface during high tides. Benches developed on flat-lying sedimentary rocks tend to be flat, whereas those developed on other types of rocks may be more rugged. An unusual type of chemical weathering may also play a role in the formation of wave-cut benches. The alternate wetting and drying leaves sea salts behind that promote weathering of the rock, making it easier for the waves to remove the material.

Along many coastlines, wave-cut benches or platforms may be found at several different levels high above sea level. These marine terraces generally form where tectonic forces are uplifting the coastline, such as along some convergent margins, and can be used to estimate the rates of uplift of the land if the ages of the various marine terraces can be determined.

A variety of other unusual erosional landforms are found along rocky shorelines, particularly where cliffs are retreating. Sea stacks are isolated columns of rock left by retreating cliffs, with the most famous

being the Twelve Apostles, along the southern coast of Australia. Arches are sometimes preserved in areas of seastacks that have developed in horizontally layered sedimentary rocks, where waves erode tunnels through headlands.

REEF SYSTEMS

Reefs are wave-resistant, framework-supported carbonate or organic mounds generally built by carbonate-secreting organisms, or in some usages the term may be used for any shallow ridge of rock lying near the surface of the water. Reefs contain a plethora of organisms that together build a wave-resistant structure to just below the low-tide level in the ocean waters and provide shelter for fish and other organisms. The spaces between the framework are typically filled by skeletal debris, which together with the framework become cemented together to form a wave-resistant feature that shelters the shelf from high-energy waves. Reef organisms (presently consisting mainly of zooxanthellae) can survive only in the photic zone, so reef growth is restricted to the upper 328 feet (100 m) of the seawater.

Reefs are built by a wide variety of organisms, today including red algae, mollusks, sponges, and cnidarians (including corals). The colonial Scleractinia corals are presently the principal reef builders, producing a calcareous external skeleton characterized by radial partitions known as septa. Inside the skeleton are soft-bodied animals called polyps, containing symbiotic algae essential for the life cycle of the coral and the building of the reef structure. The polyps contain calcium bicarbonate that is broken down into calcium carbonate, carbon dioxide, and water. The calcium carbonate is secreted to the reef building its structure, whereas the algae photosynthesize the carbon dioxide, producing food for the polyps.

There are several different types of reefs, classified by their morphology and relationship to nearby landmasses. Fringing reefs grow along and fringe the coast of a landmass and are often discontinuous. They typically have a steep outer slope, an algal ridge crest, and a flat, sand-filled channel between the reef and the main shoreline. Barrier reefs form at greater distances from the shore than fringing reefs, and are generally broader and more continuous than fringing reefs. They are among the largest biological structures on the planet—for instance, the Great Barrier Reef of Australia is 1,430 miles (2,300 km) long. A deep, wide lagoon typically separates barrier reefs from the mainland. All of these reefs show a zonation from a high-energy side on the outside or windward side of the reef, grow fast, and have a smooth outer boundary. In contrast, the opposite side of the reef receives little wave energy and may be irregular and poorly developed, or grade into a lagoon. Many reefs also show a vertical zonation in the types of organisms present, from deepwater to shallow levels near the sea surface.

Atolls or atoll reefs form circular-, elliptical-, or semicircular-shaped islands made of coral that rise from deep water; atolls surround central lagoons, typically with no internal landmass. Some atolls do have small central islands, and these, as well as parts of the outer circular reef, are in some cases covered by forests. Most atolls range in diameter from half a mile to more than 80 miles (1–130 km), and are most common in the western and central Pacific Ocean basin and in the Indian Ocean. The outer margin of the semicircular reef on atolls is the most active site of coral growth, since it receives the most nutrients from upwelling waters on the margin of the atoll. On many atolls coral growth on the outer margin is so intense that the corals form an overhanging ledge from which many blocks of coral break off during storms, forming a huge pile of broken reef debris at the base of the atoll called talus slope. Volcanic rocks, some of which lie more than half a mile (1 km) below current sea level, underlay atolls. Since corals can grow only in very shallow water fewer than 65 feet (20 m) deep, the volcanic islands must have formed near sea level, grown coral, and subsided over time, with the corals growing at the rate that the volcanic islands were sinking.

Charles Darwin proposed such an origin for atolls in 1842 based on his expeditions on the HMS *Beagle* from 1831 to 1836. He suggested that volcanic islands were first formed with their peaks exposed above sea level. At this stage coral reefs were established as fringing reef complexes around the volcanic island. He suggested that with time the volcanic islands subsided and were eroded, but that the growth of the coral reefs was able to keep up with the subsidence. In this way, as the volcanic islands sank below sea level, the coral reefs continued to grow and eventually formed a ring circling the location of the former volcanic island. When Darwin proposed this theory in 1842, he did not know that ancient, eroded volcanic mountains underlay the atolls he studied. More than 100 years later, drilling confirmed his prediction that volcanic rocks would be found beneath the coralline rocks on several atolls.

With the advent of plate tectonics in the 1970s the cause of the subsidence of the volcanoes became apparent. When oceanic crust is created at midocean ridges, it is typically about 1.7 miles (2.7 km) below sea level. With time, as the oceanic crust moves away from the midocean ridges, it cools and contracts, sinking to about 2.5 miles (4 km) below sea level. In many places on the seafloor small volcanoes form on the oceanic crust a short time after the main part of the crust formed at the midocean ridge. These volcanoes may stick above sea level a few hundred meters. As the oceanic crust moves away from the midocean ridges, these volcanoes subside below sea level. If the volcanoes happen to be in the tropics where corals can grow, and if the rate of subsidence is slow enough for the growth of coral to keep up with subsidence, then atolls may form where the volcanic island used to be. If corals do not grow or cannot keep up with subsidence, then the island subsides below sea level and the top of the island gets scoured by wave erosion, forming a flat-topped mountain that continues to subside below sea level. These flat-topped mountains are known as guyots, many of which were mapped during exploration of the seafloor associated with military operations of World War II.

Reefs are extremely sensitive and diverse environments and cannot tolerate large changes in temperature, pollution, turbidity, or water depth. Reefs have also been subject to mining, destruction for navigation and even sites of testing nuclear bombs in the Pacific. Thus human-induced and natural changes in the shoreline environment pose a significant threat to the reef environment.

See also CORAL; DELTAS; ESTUARY; HURRICANES; OCEAN BASIN; SEA-LEVEL RISE.

FURTHER READING

Beatley, Timothy, David J. Brower, and Anna K. A. Schwab. *Introduction to Coastal Management.* Washington, D.C.: Island Press, 1994.

Davis, R., and D. Fitzgerald. *Beaches and Coasts.* Malden, Mass.: Blackwell, 2004.

Dean, C. *Against the Tide: The Battle for America's Beaches.* New York: Columbia University Press, 1999.

Dolan, Robert, Paul J. Godfrey, and William E. Odum. "Man's Impact on the Barrier Islands of North Carolina." *American Scientist* 61 (1973): 152–162.

Kaufman, W., and Orrin H. Pilkey Jr. *The Beaches are Moving.* Durham, N.C.: Duke University Press, 1983.

King, C. A. M. *Beaches and Coasts.* London: Edward Arnold, 1961.

Komar, Paul D., ed. *CRC Handbook of Coastal Processes and Erosion.* Boca Raton, Fla.: CRC Press, 1983.

Longshore, David. *Encyclopedia of Hurricanes, Typhoons, and Cyclones, New Edition.* New York: Facts On File, 2008.

Nordstrom, K. F., N. P. Psuty, and R. W. G. Carter. *Coastal Dunes: Form and Process.* New York: John Wiley & Sons, 1990.

Pilkey, O. H., and W. J. Neal. *Coastal Geologic Hazards.* In *The Geology of North America*, Volume 1–2, *The Atlantic Continental Margin*, edited by R. E. Sheridan and J. A. Grow. Boulder, Colo.: Geological Society of America, 1988.

U.S. Army Corps of Engineers Engineer Research and Development Center home page. Available online. URL: http://www.erdc.usace.army.mil/. Updated August 22, 2008.

U.S. Army Corps of Engineers home page. Available online. URL: http://www.usace.army.mil/. Updated September 17, 2008.

Williams, Jeffress, Kurt A. Dodd, and Kathleen K. Gohn. *Coasts in Crisis.* Reston, Va. United States Geological Survey, Circular 1075, 1990.

benthic, benthos The benthic environment includes the ocean floor and the benthos are those organisms that dwell on or near the seafloor. Bottom-dwelling benthos organisms include large plants that grow in shallow water, as well as animals that dwell on the seafloor at all depths.

Many of the sediments on the deep seafloor are derived from erosion of the continents and carried to the deep sea by turbidity currents, carried by wind (e.g., volcanic ash), or released from floating ice. Other sediments, known as deep-sea oozes, include pelagic sediments derived from marine organic activity. When small organisms such as diatoms die in the ocean, their shells sink to the bottom and over time can make significant accumulations. Calcareous ooze occurs at low to middle latitudes where warm water favors the growth of carbonate-secreting organisms. Calcareous oozes are not found in water more than 2.5–3 miles (4–5 km) deep because this water is under such high pressure that it contains a lot of dissolved CO_2, which dissolves carbonate shells. The depth below which all calcium-bearing shells and tests dissolve is known as the calcium carbonate compensation depth. Siliceous ooze is produced by organisms that use silicon to make their shell structure.

The benthic world is amazingly diverse, yet parts of the deep seafloor are less explored than the surface of the Moon. Organisms that live in the benthic community generally use one or more of three main strategies for living. Some attach themselves to anchored surfaces and get food by filtering it from the seawater. Other organisms move freely about on the ocean

Benthic organisms including crabs, anemones, clams, and mussels on ocean floor *(TheSupe87, 2008, Shutterstock, Inc.)*

bottom and get their food by predation. Still others burrow or bury themselves in the ocean bottom sediments and obtain nourishment by digesting and extracting nutrients from the benthic sediments. All the benthic organisms must compete for living space and food, with other factors including light levels, temperature, salinity, and the nature of the bottom controlling the distribution and diversity of some organisms. Species diversification is related to the stability of the benthic environment. Areas that experience large variations in temperature, salinity, and water agitation tend to have low species diversification, but may have large numbers of a few different types of organisms. In contrast, stable environments tend to show much greater diversity, with a larger number of species present.

There are a large number of different benthic environments. Rocky shore environments in the intertidal zone have a wide range of conditions from alternately wet and dry to always submerged, with wave agitation and predation being important factors. These rocky shore environments tend to show a distinct zonation in benthos, with some organisms inhabiting one narrow niche and other organisms in others. Barnacles and other organisms that can firmly attach themselves to the bottom do well in wave-agitated environments, whereas certain types of algae prefer areas from slightly above the low-tide line to about 33 feet (10 m) depth. The area around the low-tide mark tends to be inhabited by abundant organisms, including snails, starfish, crabs, mussels, sea anemones, urchins, and hydroids. Tide pools are highly variable environments that host specialized plants and animals including crustaceans, worms, starfish, snails, and seaweed. The subtidal environment may host lobster, worms, mollusks, and even octopus. Kelp, brown benthic algae, inhabit the subtidal zone in subtropical to subpolar waters and can grow down to a depth of about 130 feet (40 m), often forming thick underwater forests that may extend along a coast for many kilometers.

Sandy and muddy bottom benthic environments often form at the edges of deltas, sandy beaches, marshes, and estuaries. Many of the world's temperate to tropical coastlines have salt marshes in the intertidal zone and beds of sea grasses growing just below the low-tide line. Surface-dwelling organisms in these environments are known as epifauna, whereas organisms that bury themselves in the bottom sand and mud are called infauna. Many of these organisms obtain nourishment either by filtering seawater that they pump through their digestive system or by selecting edible particles from the seafloor. Deposit-feeding bivalves such as clams inhabit the area below the low-tide mark, whereas other deposit feeders may inhabit the intertidal zone. Other

organisms that inhabit these environments include shrimp, snails, oysters, tube-building crustaceans, and hydroids.

Coral reefs are special benthic environments that require warm water greater than 64.4°F (18°C) to survive. Colonial animals secrete calcareous skeletons, placing new active layers on top of the skeletons of dead organisms, and thus build the reef structure. Encrusting red algae, as well as green and red algae, produce the calcareous cement of the coral reefs. The reef hosts a huge variety and number of other organisms, some growing in symbiotic relationships with the reef builders, others seeking shelter or food among the complex reef. Upwelling waters and currents bring nutrients to the reef. The currents release more nutrients produced by the reef organisms. Some of the world's most spectacular coral reefs include the Great Barrier Reef, off the northeast coast of Australia, reefs along the Red Sea and Indonesia, and reefs in the Caribbean and south Florida.

Unique forms of life were recently discovered deep in the ocean near hot vents located along the midocean ridge system. The organisms that live in these benthic environments are unusual in that they get their energy from chemosynthesis of sulfides exhaled by hot hydrothermal vents, and not from photosynthesis and sunlight. The organisms that live around these vents include tube worms, sulfate-reducing chemosynthetic bacteria, crabs, giant clams, mussels, and fish. The tube worms grow to enormous size, some being 10 feet (3 m) long and 0.8–1.2 inches (2–3 cm) wide. Some of the bacteria that live near these vents include the most heat-tolerant (thermophyllic) organisms recognized on the planet, living at temperatures of up to 235°F (113°C). They are thought to be some of the most primitive organisms known, being both chemosynthetic and thermophilic, and may be related to some of the oldest life-forms that inhabited the Earth.

The deep seafloor away from the midocean ridges and hot vents is also inhabited by many of the main groups of animals that inhabit the shallower continental shelves. The number of organisms on the deep seafloor is few, however, and the animals tend to be much smaller than those at shallower levels. Some deepwater benthos similar to the hot-vent communities have recently been discovered living near cold vents above accretionary prisms at subduction zones, near hydrocarbon vents on continental shelves, and around decaying whale carcasses.

See also BEACHES AND SHORELINES; BLACK SMOKER CHIMNEYS; CONTINENTAL MARGIN.

binary star systems   Most stars are parts of systems that include two or more stars that rotate in

NASA image captured by *Chandra X-Ray Observatory* of two white dwarf stars in binary star system J0806 *(UPI Photo/NASA/Landov)*

orbit around each other. When the system consists of two stars, it is known as a binary star system. Larger groups of stars are known as multiple star systems, or star clusters. Optical doubles are stars that appear to be binaries but are actually not related and just appear to be close in their visible configuration. In binary systems the two stars rotate about their common center of mass (the center of mass of both stars combined) and are held in place by the mutual gravitation attraction between them.

Binary star systems are classified according to how they appear to astronomers on Earth. Simple visual binaries are systems in which the two stars are far enough apart to be visibly distinct when viewed through a telescope from Earth, and each star is bright enough to be monitored separately from the other. In other cases the binary system may be too far, or the stars too close or small, to be visibly distinct from Earth, but the rotation of the stars around each other can be detected spectroscopically by observing shifts in the frequency and wavelength of a wave for an observer moving relative to the source of the waves, known as Doppler shifts, as each star alternately moves toward and away from the observer on Earth as the stars rotate around

each other. The Doppler shift is recorded as a shift toward the blue end of the spectrum as the star moves toward the observer, and a redshift as the star moves away. Binary systems that can be detected only by using these spectroscopic Doppler shifts are known as spectroscopic binaries, and they are of two main types. Double-line spectroscopic binaries contain two distinct sets of spectral lines, one for each star, that shift back and forth from blue to redshifts as the star moves alternately toward and away from the observer. In these systems both stars are large and bright enough to be distinguished spectroscopically. In other systems one star may be too small or faint to be distinguished from the other, and the result is a single-line system in which one set of spectroscopic lines is observed to shift back and forth, caused by the stars rotating around each other even though they are too close to be resolved individually.

A rare class of binary star systems is known as eclipsing binaries. In these systems the orbital plane of the binary system is aligned nearly head-on with the line of sight from Earth, so as each star passes in front (in the line of sight) of the other, it blocks the light coming from the blocked star, and the amount of light observed from Earth alternately changes as each star passes periodically in front of the other. Observations of eclipsing binaries can yield information about each star's mass, orbits, orbital periods, radii, and luminosity or brightness.

The range in the orbital periods of binary star systems is very large, spanning from hours to centuries. Knowledge of the orbital periods, plus the distance to the binary system, can be used to determine additional physical properties of the binary system, such as the combined mass of the stars. If the distance of each star from the center of mass of the system can be measured, then the individual masses of each star can also be determined. Calculations based on observations of binary star systems have formed the basis for most of what is known about the masses of stars in the solar system.

See also ASTRONOMY; ASTROPHYSICS; EINSTEIN, ALBERT; ELECTROMAGNETIC SPECTRUM; UNIVERSE.

FURTHER READING

Chaisson, Eric, and Steve McMillan. *Astronomy Today.* 6th ed. Upper Saddle River, N.J.: Addison-Wesley, 2007.

Comins, Neil F. *Discovering the Universe.* 8th ed. New York: W. H. Freeman, 2008.

Snow, Theodore P. *Essentials of the Dynamic Universe: An Introduction to Astronomy.* 4th ed. St. Paul, Minn.: West, 1991.

biosphere The biosphere encompasses the part of the Earth that is inhabited by life, and includes parts of the lithosphere, hydrosphere, and atmosphere. Life evolved more than 3.8 billion years ago, and has played an important role in determining the planet's climate and insuring that it does not venture out of the narrow window of parameters that allow life to continue. In this way the biosphere functions as a self-regulating system that interacts with chemical, erosional, depositional, tectonic, atmospheric, and oceanic processes on the Earth.

Most of the Earth's biosphere depends on photosynthesis for its primary source of energy, driven ultimately by energy from the Sun. Plants and many bacteria use photosynthesis as their primary metabolic strategy, whereas other microorganisms and animals rely on photosynthetic organisms as food for their energy, and thus use solar energy indirectly. Most of the organisms that rely on solar energy live, by necessity, in the upper parts of the oceans (hydrosphere), lithosphere, and lower atmosphere. Bacteria are the dominant form of life on Earth (comprising about 5×10^{30} cells), and also live in the greatest range of environmental conditions. Some of the important environmental parameters for bacteria include temperature, between -41 to 235°F (-5 to 113°C), pH levels from 0 to 11, pressures between a near vacuum and 1,000 times atmospheric pressure, and supersaturated salt solutions to distilled water.

Bacteria and other life-forms exist with diminished abundance to several miles (kilometers) or more beneath the Earth's surface, deep in the oceans, and some bacterial cells and fungal spores are found in the upper atmosphere. The lack of nutrients and the lethal levels of solar radiation above the shielding effects of atmospheric ozone limit life in the upper atmosphere.

Soils and sediments in the lithosphere contain abundant microorganisms and invertebrates at shallow levels. Bacteria exist at much deeper levels and are being found in deeper and deeper environments as exploration continues. Bacteria are known to exist to about two miles (3.5 km) in pore spaces and cracks in rocks, and deeper in aquifers, oil reservoirs, and salt and mineral mines. Deep microorganisms do not rely on photosynthesis, but rather use other geochemical or geothermal energy to drive their metabolic activity.

The hydrosphere and especially the oceans teem with life, particularly in the near-surface photic zone environment where sunlight penetrates. At greater depths below the photic zone most life is still driven by energy from the Sun, as organisms rely primarily on food provided by dead organisms that filter down from above. In the benthic environment of the seafloor there may be as many as 10 billion (10^{10}) bacteria per milliliter of sediment. Bacteria also exist beneath the level that oxygen can penetrate, but the bacteria at these depths are anaerobic, primarily sulfate-reducing varieties. Bacteria are known to exist to greater than 2,789 feet (850 m) beneath the seafloor.

In 1977 a new environment for a remarkable group of organisms was discovered on East Pacific Rise and observed directly in 1979 by geologist Peter Lonsdale and his team from Woods Hole Oceanographic Institute in Massachusetts using the deep-sea submarine ALVIN. The organisms survive on the seafloor along the midocean ridge system, where hot hydrothermal vents spew heated nutrient-rich waters into the benthic realm. In these environments seawater circulates into the ocean crust where it is heated near oceanic magma chambers. This seawater reacts with the crust and leaches chemical components from the lithosphere, then rises along cracks or conduits to form hot black and white smoker chimneys that spew the nutrient-rich waters at temperatures of up to 662°F (350°C). Life has been detected in these vents at temperatures of up to 235°F (113°C). The vents are rich in methane, hydrogen sulfide, and dissolved reduced metals such as iron that provide a chemical energy source for primitive bacteria. Some of the bacteria around these vents are sulfate-reducing chemosynthetic thermophyllic organisms, living at high temperatures using only chemical energy and therefore exist independently of photosynthesis. These and other bacteria are locally so great in abundance that they provide the basic food source for other organisms, including spectacular worm communities, crabs, giant clams, and even fish.

See also ATMOSPHERE; BENTHIC, BENTHOS; BLACK SMOKER CHIMNEYS; SUPERCONTINENT CYCLES.

FURTHER READING

Raven, Peter, and Linda Berg. *Environment.* New York: John Wiley & Sons, 2008.

black holes   The final stage of stellar evolution for stars with a large mass may be a black hole, a superdense collection of matter that has collapsed from a giant star or stars, and has such a strong gravity field that nothing can escape from it, not even light. Black holes are known by physicists as a singularity, a point with zero radius and infinite density. These dense but invisible objects form when a star has at least three solar masses left in its core after it has completed burning its nuclear fuel. This stage of stellar evolution is typically marked by the star experiencing a supernova explosion, after which, if enough mass is left over, the star's nucleus collapses to a small point and warps space-time, forming a black hole. Black holes have such a strong gravitational field that they apparently draw material into them that is never to be seen again.

Black holes are one of the possible end states of old stars. Stars that have a low total mass (less than 1.4 solar masses) end their evolution as a white dwarf, whereas stars with masses between 1.4 and three solar masses may end their life cycle as a small dense mass known as a neutron star. When the mass of the dying star is greater than three solar masses, the star collapses as the nuclear fuel runs out. The gravitational attraction of the mass is so great that electrons and even neutrons cannot support the core against its own gravity, so it continues to collapse into what is called a singularity. There is no force known in nature that is strong enough to resist the gravitational attraction of a collapsing star once the pressure is so great that the neutrons degenerate and collapse. The force is so strong that not even light can escape from inside a black hole, hence the name.

The concept of a black hole as an infinitesimally small singularity with infinite mass is difficult to comprehend, in part because it is not adequately explained by the classical Newtonian laws of physics, or gravitational theory. To understand fully the workings of black holes it is necessary to move into the realm of quantum mechanics and Albert Einstein's theory of relativity. A quantitative treatment of relativity is beyond the scope of this book, but many aspects of Einstein's theories can be understood qualitatively. To understand how black holes work, it is necessary to know that nothing can travel faster than the speed of light, and that the gravitational force acts on everything, including electromagnetic radiation, or light.

To understand black holes, it is necessary to understand the concept of escape velocity. Objects on Earth must move 6.8 miles (11 km) per second to escape the pull of the Earth's gravitational attraction and move into open space. Escape velocity v_e for any planetary or stellar object is proportional to the square root of the mass—of the body being escaped from divided by the square root of its radius r, which is the distance from the center of the body being escaped from and the location where the moving object is escaping. This can be written as

$$v_e \sqrt{\frac{2GM}{r}}$$

where G is the gravitational constant, $6.67428 \times 10^{-11} m^3 kgs^{-1}$. This relationship means that for objects denser and smaller than the Earth but with the same mass, the escape velocity would need to be faster for any object to escape its gravitational field. As a massive object such as a huge star begins to collapse, therefore, the escape velocity required for anything to leave its gravitational field rises rapidly as the star shrinks from a large radius to a small object a fraction of its original size. If the star shrinks to a quarter of its original size, the escape velocity doubles—and as a star experiences a rapid collapse after a supernova, the escape velocity rises to such extremely high values that it is virtually impossible for any object to escape the star's gravitational pull. If an object the size of the Earth were to collapse to about 1/3 inch (1 cm), the escape velocity would be 186,000 miles per second (300,000 km/sec), the speed of light. From Einstein's theory of relativity, which states that even light is attracted by gravity, it becomes clear that at some point during the collapse of massive stars even light will no longer be able to escape the gravitational pull of the body, and the collapsed star will become dark forever. Since some large stars collapse to a size smaller than an elementary particle, the escape velocity becomes infinitely high, and the gravitational attraction becomes stronger and stronger. At this point the black hole can pull objects in, but nothing can ever escape. That is the meaning of the term *black hole.* The only way to detect a black hole is by its immense gravitational field, which can deflect light as it bends toward the huge gravitational pull. Astronomers use sophisticated measurements to tell when a star moves optically behind a black hole, and they can measure the deflection of the light. This allows for the determination of some of the physical properties (like mass, charge, and angular momentum) of the black hole.

Every object with a specific mass has a critical radius at which the escape velocity equals the speed of light. When the object is compressed to that radius, nothing can escape its gravitational pull—not even light—and the object becomes invisible. This critical radius is known as the Schwarzschild radius, named after the German physicist Karl Schwarzschild, who first described this phenomenon. The Sun has a Schwarzschild radius of about 9.8 feet (3 m), but stars with the mass of the Sun do not usually collapse to become black holes since they are too small. The smallest stellar objects that form black holes have about three solar masses, and the Schwarzschild radius for these stars is about 5.6 miles (9 km).

Another concept useful for understanding black holes is that of the event horizon, which is the surface of an imaginary sphere with a radius equal to the Schwarzschild radius, centered on a collapsing star. The event horizon is an imaginary surface, but can be thought of as the surface of the black hole, since beyond the event horizon, no event that happens can ever be heard, seen, or detected by any known means. The event horizon does not represent the size of the material that collapsed to form the black hole. Since this material should theoretically collapse to a tiny singularity, it merely represents the radius past which the gravitational pull of the dense black hole at the center of the sphere is so strong that nothing can escape once inside that radius.

Black holes are said to warp the space-time continuum in the way we understand it from a classical Newtonian mechanical way. According to Einstein's theory of relativity, all matter tends to warp space in its vicinity, and objects respond to this warp by changing their direction of movement as they approach other objects. Newtonian physics would describe this as a gravitational pull, whereas relativity theory suggests that the objects are just following the curved space that was distorted by the nearby massive object. The more massive the objects, the more they curve the space. In the case of black holes the warping of space is extreme because of the huge mass in the black hole, and at the event horizon, space is actually folded over upon itself, such that objects that cross the event horizon disappear from space forever.

As material falls into a black hole, the gravitational stresses are so great that they distort and tear apart objects as they plunge toward the event horizon. These objects become heated and emit radiation, so the regions surrounding black holes are

sometimes emitters of strong radiation. Once the material crosses the event horizon, however, nothing can escape, not light, not radiation, and the mass is never seen or heard from again.

The gravity fields of black holes are so strong that it is virtually impossible to get close to one without being physically torn apart by the strong gravity, and the difference in the strength of the gravity from one end of any approaching object (or person) and the other end. Nonetheless it is interesting and informative to discuss what it might be like to approach, and even enter, a black hole. The first thing an outside observer of an object approaching a black hole would notice is that light, and other electromagnetic radiation coming from the object, shows a redshift (toward longer wavelengths) that increases as the object gets closer to the event horizon. This is not a Doppler shift caused by the motion of the object, as the object near the black hole would exhibit the redshift even if it were motionless with respect to the observer. This redshift is a quantum mechanical effect known as a gravitational redshift. Einstein's general theory of relativity shows that as photons (light) try to escape a strong gravitational field, they have to use up some of their energy. Photons are light, and they always move at the speed of light, so this loss of energy is equated with a decrease in frequency, or a lengthening of the wavelength of the light (or other electromagnetic radiation) coming from the object approaching the black hole. The distant observer measures this as a redshift. Interestingly, an observer on the object emitting the radiation would see no redshift, and the radiation (light) would have the same energy and wavelength as when it was emitted. These gravitational redshifts have been measured on light coming from many dense objects in the universe, and objects even the size of Earth and the Sun have detectable gravitational redshifts. The largest, by far, are from black holes.

Black holes distort the space-time continuum. Another strange quantum mechanical effect explained in Einstein's theory of general relativity is time dilation near massive objects such as black holes. A distant observer looking at a clock on the object approaching the black hole would notice that the clock (and time itself) moves progressively slower and slower as the object approaches the black hole's event horizon. When the object is at the event horizon, the clock (and time) would appear to stand still, and from the perspective of the outside observer, the object would be frozen at the event horizon forever, never entering past it. However, from the perspective of anyone on the object approaching the event horizon, there is no difference in the way time passes; each second seems like one second. The object and observer would simply pass through the event horizon and notice nothing different (assuming they could withstand the strong gravitational forces). Time dilation is difficult to understand but can be thought of in the same way that the redshift of electromagnetic radiation occurs. If time is considered to be measured, for instance, as the passage of a wavelength of light, each second corresponding to the passage of one wave crest, then as the wavelengths are increased by the gravitational redshift, the time is also gradually expanded until it appears to stop by the outside observer.

No one really knows what happens inside the event horizon of a black hole. The laws of physics do not adequately explain such dense small objects as singularities, and some new concepts are being investigated by physicists, such as a merging of the laws of quantum mechanics and general relativity into the field of quantum gravity—but these investigations are incomplete. There are many ideas, some approaching science fiction, that have been proposed for what may happen near the singularity at the center of the black hole's event horizon. Some models suggest that new states of matter are created; others have suggested that black holes may be gateways for matter and energy to enter other universes or to travel in time.

Black holes are difficult to detect, since they are invisible. Their huge gravitational field, however, and the energy released by matter outside the event horizon as it falls into the black hole may be detectable. There are several good candidates for possible black holes in the Milky Way Galaxy. The best may be a massive but invisible body in a binary star system known as Cygnus X-1. This possible black hole is orbiting with a supergiant star companion, and is known as a powerful X-ray source (presumably from the material approaching the event horizon). This binary star system has an orbital diameter of 12.4 million miles (20 million km) and an orbital period of 5.6 days, and the mass of the system is 30 times that of the Sun. Calculations show that the invisible component of this binary system has a mass of 5–10 times that of the Sun, enough to have formed a black hole. In this system it appears that hot gases are flowing from the supergiant star into the black hole companion, and this is the source of the X-ray radiation. Other calculations show that the invisible part of this binary star is small, less than 186,000 miles (300,000 km) across, and other calculations show that it is likely less than 186 miles (300 km) across. Thus Cygnus X-1 is one of the most likely candidates for a black hole in the Milky Way Galaxy. There are nearly a dozen other black hole candidates in the Milky Way Galaxy, and as the observational powers of physicists increase with new space-borne telescopes, more and more are being discovered. What

is needed is a breakthrough in the field of quantum gravity to understand what may really happen underneath the event horizon.

See also ASTRONOMY; ASTROPHYSICS; BINARY STAR SYSTEMS; DWARFS (STARS); EINSTEIN, ALBERT; STELLAR EVOLUTION.

FURTHER READING

"Black Holes, Gravity's Relentless Pull." Support provided by the National Aeronautics and Space Administration (NASA). Available online. URL: http://hubblesite.org/explore_astronomy/black_holes/home.html. Accessed October 9, 2008.

Chaisson, Eric, and Steve McMillan. *Astronomy Today*. 6th ed. Upper Saddle River, N.J.: Addison-Wesley, 2007.

Comins, Neil F. *Discovering the Universe*. 8th ed. New York: W. H. Freeman, 2008.

Snow, Theodore P. *Essentials of the Dynamic Universe: An Introduction to Astronomy*. 4th ed. St. Paul, Minn.: West, 1991.

black smoker chimneys   Black smoker chimneys are hydrothermal vent systems that typically form near active magmatic systems along the mid-ocean ridge system, approximately 2 miles (3 km) below sea level. They were first discovered by deep submersibles exploring the oceanic ridge system near the Galápagos Islands in 1979, and many other examples have been documented since then, including a number along the mid-Atlantic ridge.

Black smokers are hydrothermal vent systems that form by seawater percolating into fractures in the seafloor rocks near the active spreading ridge, where the water gets heated to several hundred degrees Celsius. This hot pressurized water leaches minerals from the oceanic crust and extracts other elements from the nearby magma. The superheated water and brines then rise above the magma chamber in a hydrothermal circulation system and escape at vents on the seafloor, forming the black smoker hydrothermal vents. The vent fluids are typically rich in hydrogen sulfides (H_2S), methane, and dissolved reduced metals, such as iron. The brines may escape at temperatures greater than 680°F (360°C), and when these hot brines come into contact with cold seawater, many of the metals and minerals in solution rise in plumes, since the hot fluids are more buoyant than the colder seawater. The plumes are typically about 0.6 miles (1 km) high and 25 miles (40 km) wide and can be detected by temperature and chemical anomalies, including the presence of primitive helium 3 isotopes derived from the mantle. These plumes may be rich in dissolved iron, manganese, copper, lead, zinc, cobalt, and cadmium, which

Black smoker chimney from the East Pacific Rise showing tube worms feeding at base of the chimney *(Science Source/Photo Researchers, Inc.)*

rain out of the plumes, concentrating these elements on the seafloor. Manganese remains suspended in the plumes for several weeks, whereas most of the other metals are precipitated as sulfides (e.g., pyrite, FeS_2; chalcopyrite, $CuFeS_2$; sphalerite, ZnS), oxides (e.g., hematite, Fe_2O_3), orthohydroxides (e.g., goethite, FeOOH), or hydroxides (e.g., limonite, $Fe(OH)_3$). A group of related hydrothermal vents that form slightly farther from central black smoker vents, known as white smokers, typically have vent temperatures between 500 and 572°F (260–300°C).

On the seafloor along active spreading ridges the hydrothermal vent systems form mounds that are typically 164–656 feet (50–200 m) in diameter, and some are more than 66 feet (20 m) high. Clusters of black smoker chimneys several meters high may occupy the central area of mounds and deposit iron-copper sulfides. White smoker chimneys typically form in a zone around the central mound, depositing iron-zinc sulfides and iron oxides. Some mounds on the seafloor have been drilled to determine their

internal structure. The Trans-Atlantic Geotraverse (TAG) hydrothermal mound on the mid-Atlantic ridge is capped by central chimneys made of pyrite, chalcopyrite, and anhydrite, overlying massive pyrite breccia, with anhydrite-pyrite and silica-pyrite–rich zones found a few to tens of meters below the surface. Below this the host basalts are highly silicified, then at greater depths form a network of chloritized breccia. White smoker chimneys made of pyrite (FeS_2) and sphalerite (ZnS) rim the central mound. In addition to the sulfides, oxides, hydroxides, and orthohydroxides, including several percent copper and zinc, the TAG mound contains minor amounts of gold.

Seafloor hydrothermal mounds and particularly the black smoker chimneys host a spectacular community of unique life-forms, found only in these environments. Life-forms include primitive sulfate-reducing thermophilic bacteria, giant worms, giant clams, crabs, and fish, all living off the chemosynthetic metabolism made possible by the hydrothermal vent systems. Life at the black smokers draws energy from the internal energy of the Earth (not the Sun), via oxidation in a reducing environment. Some of the bacteria living at these vents are the most primitive organisms known on Earth, suggesting that early life may have resembled these chemosynthetic thermophilic organisms.

Black smoker chimneys and the entire hydrothermal mounds bear striking similarities to volcanogenic massive sulfide (VMS) deposits found in Paleozoic and older ophiolite and arc complexes including the Bay of Islands ophiolite in Newfoundland, the Troodos ophiolite in Cyprus, and the Semail ophiolite in Oman. Even older VMS deposits are common in Archean greenstone belts, and these are typically basalt or rhyolite-hosted chalcopyrite, pyrite, sphalerite, copper-zinc-gold deposits that many workers have suggested may be ancient seafloor hydrothermal vents. Interestingly, complete hydrothermal mounds with preserved black and white smoker chimneys have been reported recently from the 2.5 billion-year-old North China craton, in the same belt that the world's oldest well-preserved ophiolite is located.

The tectonic setting for the origin of life on the early Earth is quite controversial. Some favor environments in shallow pools, some favor deep ocean environments where the organisms could get energy from the chemicals coming out of seafloor hydrothermal vents. The discovery of black smoker types of hydrothermal vents in Archean ophiolite sequences is significant because the physical conditions at these midocean ridges more than 2.5 billion years ago would have permitted the inorganic synthesis of amino acids and other prebiotic organic molecules. Some scientists think that the locus of precipitation and synthesis for life might have been in small iron-sulfide globules, such as those that form around black smokers. Black smoker chimneys may provide a window into the past and the origin of life on Earth.

See also ASIAN GEOLOGY; BENTHIC, BENTHOS; BIOSPHERE; GREENSTONE BELTS; OPHIOLITES.

FURTHER READING

Scott, Steven. "Minerals on Land, Minerals in the Sea." *Geotimes* 47, no. 12 (2002): 19–23.

Bowen, Norman Levi (1887–1956) *Canadian Petrologist, Geologist* Dr. Norman Levi Bowen was one of the most brilliant igneous petrologists of the 20th century. Although he was born in Ontario, Canada, he spent most of his productive research career at the Geophysical Laboratories of the Carnegie Institute in Washington, D.C. Bowen studied the relationships between plagioclase feldspars and iron-magnesium silicates in crystallizing and melting experiments. From these experiments he derived the continuous and discontinuous reaction series explaining the sequence of crystallization and melting of these minerals in magmas. He also showed how magmatic differentiation by fractional crystallization can result from a granitic melt from an originally basaltic magma through the gradual crystallization of mafic minerals, leaving the felsic melt behind. Similarly he showed how partial melting of one rock type can result in a melt with a different composition than the original rock, typically forming a more felsic melt than the original rock and leaving a more mafic residue (or restite) behind. Bowen also worked on reactions between rocks at high temperatures and pressures, and the role of water in magmas. In 1928 Bowen published his pioneering book, *The Evolution of Igneous Rocks*.

N. L. Bowen is most famous for his works on the origin of igneous rocks, through the processes of magmatic differentiation by partial melting and magmatic differentiation by fractional crystallization. The phrase "magmatic differentiation by partial melting" refers to the process of forming magmas with differing compositions through the incomplete melting of rocks. For magmas formed in this way the composition of the magma depends on both the composition of the parent rock and the percentage of melt. If a rock melts completely, the magma has the same composition as the rock. However, rocks contain many different minerals, all of which melt at different temperatures. So if a rock is slowly heated, the resulting melt or magma will initially have the composition of the first mineral that melts, and then the first plus the second minerals that melt, and so

on. If the rock melts completely, the magma will eventually end up with the same composition as the starting rock, but this does not always happen. Oftentimes the rock only partially melts, so that the minerals with low melting temperatures contribute to the magma, whereas the minerals with high melting temperatures did not melt and are left as a residue (or restite). In this way the end magma can have a different composition than the rock from which it was derived.

Just as rocks partially melt to form different liquid compositions, magmas may solidify to different minerals at different times to form different solids (rocks). This process also results in the continuous change in the composition of the magma—if one mineral is removed, the resulting composition is different. If some process removes these solidified crystals that have been removed from the system of melts, a new magma composition results.

The removal of crystals from the melt system may occur by several processes, including the squeezing of melt away from the crystals or by sinking of dense crystals to the bottom of a magma chamber. These processes lead to magmatic differentiation by fractional crystallization, as first described by Bowen, who systematically documented how crystallization

Norman Levi Bowen *(Queens University Archives)*

of the first minerals changes the composition of the magma and leads to the formation of progressively more silicic rocks with decreasing temperature.

See also IGNEOUS ROCKS; PETROLOGY AND PETROGRAPHY.

FURTHER READING

Bowen, Norman Levi. "Progressive Metamorphism of Siliceous Limestone and Dolomite." *Journal of Geology* 48, no. 3 (1940): 225–274.

———. "Recent High-Temperature Research on Silicates and Its Significance in Igneous Geology." *American Journal of Science* 33 (1937): 1–21.

Bowen, Norman Levi, and John Frank Schairer. "The Problem of the Intrusion of Dunite in the Light of the Olivine Diagram." *International Geological Congress* 1 (1936): 391–396.

Brahe, Tycho (1546–1601) Danish *Nobleman, Astronomer* Tycho Brahe was born as Tyge Ottesen Brahe on December 14, 1546, in Scania, a region of Denmark now part of Sweden. He was born to nobility, the son of Otte Brahe and Beate Bille, at his family's ancestral home, Knutstorp Castle. His father was a nobleman in the court of the Danish king, and he had an older and a younger sister, and a twin brother who died soon after birth. Tycho's uncle, Danish nobleman Jorgen Brahe, took him from his parents when he was two years old, and from then he lived at his uncle's home at Tosterup Castle. Brahe was educated at a Latin school from age six until he was 12. He enrolled at the University of Copenhagen in 1559 at age 13 and studied law but gradually became more interested in astronomy. One of the defining moments in his career was at Copenhagen, when he witnessed an eclipse on August 21, 1560, at the precise time that was predicted by his professors and astronomers. Tycho purchased a book, called an ephemeris, that gives tables and positions of astronomical objects in the sky at different times. He studied this and many other texts for several years until he came to realize in 1563 that most astronomical texts of the time disagreed with one another. He wrote that astronomy could not progress by the types of haphazard observations that he was reading and suggested that long-term systematic study of the heavens was needed. With the naked eye and the help of his sister Sophia he made many measurements of the stars and planets and improved many astronomical instruments. Brahe's work preceded the invention of the telescope, however, and later observations by German astronomer Johannes Kepler using the telescope proved to be more accurate than Brahe's.

In 1565 Brahe's uncle Jorgen Brahe died after a night of heavy drinking with Frederick II, king of Denmark, when both men fell off a bridge into a river. Jorgen saved Frederick but caught pneumonia and later died. The next year Brahe and his friends were attending a dance at their professor's house and, after drinking considerably, he fought a duel with his fellow nobleman Manderup Parsbjerg in which Brahe lost part of his nose. This caused him to spend the rest of his life wearing prosthetic noses to hide his disfigurement. In 1571 Brahe's father died following a long illness, after which the son started building an astronomical observatory and alchemistry laboratory at Herrevad Abbey near Ljungbyhed, Scania (present-day southwest Sweden).

At the age of 26 Tycho Brahe fell in love with a commoner, Kirsten Jorgensdatter, and they moved to Copenhagen and were married three years later. They had eight children, six of whom lived to adulthood, and the couple stayed together until Brahe died in 1601 at the age of 54. Brahe became quite wealthy and hosted many parties in his castle, at which his tame pet elk and dwarf Jepp, who acted as court jester, were usually present. Brahe thought the dwarf was clairvoyant and had him spend meals under the table. At one of Brahe's parties the elk consumed too much beer and fell down a flight of stairs and died. On October 10, 1601, Brahe became ill at a banquet and died 11 days later. His death was a mystery for years, many having thought he died of bacteria from drinking too much at the banquet. Since manners at the time did not allow one to get up and leave during a meal, his bladder was thought to have stretched and caused infection. But exhumation of his remains has shown that he more likely died of mercury poisoning, either accidentally by swallowing mercury-tainted medicine, or possibly by being murdered. The book *Heavenly Intrigue: Johannes Kepler, Tycho Brahe, and the Murder Behind One of History's Greatest Scientific Discoveries* by Joshua Gilder and Anne-Lee Gilder (2005) speculates that Johannes Kepler is likely to have poisoned Brahe, having the means, motive, and opportunity.

SCIENTIFIC DISCOVERIES

Tycho Brahe grew up in a scientific era in which Aristotelian ideas that the universe was unchanging followed the principle of celestial immutability. While based at his observatory at Herrevad, Brahe made many observations on the planets and stars. On November 11, 1572, he made a discovery that changed the thinking about the universe. He observed a very bright star that appeared in the constellation Cassiopeia and showed that no star had been visible in that location before. At first other astronomers suggested that the bright object must be located below the orbit of the Moon, since Aristotle showed that the heavens were unchanging. Tycho

Brahe's systematic observations showed clearly that the object was far away and thus outside the Moon's orbit—it was a new object not previously observed. His continued observations over the next several months showed that the new star did not move relative to the planets. The world was convinced that the heavens now were changeable, and he published his observations in *De Stella Nova* in 1573, naming his new star a "nova." This nova, or new star, is now known to have been the supernova SN1572.

In addition to discovering the Cassiopeia supernova, Brahe became known as the best observational astronomer of the era before telescopes. He published in a star catalog many accurate observations on the positions of planets and stars. He proposed a model for the motion of the stars and planets in which the Sun orbited Earth, and the other planets orbited the Sun, a model that was a compromise between the Ptolemaic (named after the Greek astronomer Claudius Ptolemaeus, 83–168) view of the universe and Copernicus's heliocentric model, in which the planets orbited the Sun. This became known as the Tychonic System, even though it was originally proposed by the Greek philosopher Ponticus Heraclides in the fourth century B.C.E. The Tychonic system explained most of the observations of the motion of the planets, while still satisfying the powerful Catholic Church, which held that the Earth was the center of the universe. Brahe's observations were considered valid for some time after his death, as Kepler used Brahe's measurements of the motion of Mars to calculate the laws of planetary motion and to argue that the Copernican model with the Sun at the center of the universe was correct and that the other planets and Sun and stars were not orbiting Earth.

See also ASTRONOMY; SUPERNOVA.

FURTHER READING

Brahe, Tycho. "Astronomiae Instauratae Mechanica." 1598. European Digital Library Treasure.

Gilder, Joshua, and Anne-Lee Gilder. *Heavenly Intrigue: Johannes Kepler, Tycho Brahe, and the Murder Behind One of History's Greatest Scientific Discoveries.* New York: Doubleday, 2004.

Hawking, Stephen. *The Illustrated On the Shoulders of Giants: The Great Works of Physics and Astronomy.* Philadelphia: Running Press, 2004.

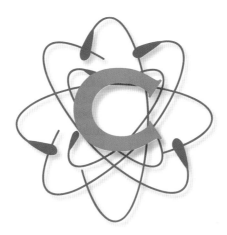

Cambrian The first geologic period of the Paleozoic Era and the Phanerozoic Eon, the Cambrian began 544 million years (Ma) ago and ended 505 million years ago. It is preceded by the Late Proterozoic Eon and succeeded by the Ordovician Period. The Cambrian System refers to the rocks deposited during this period. The Cambrian is named after Cambria, the Roman name for Wales, where the first detailed studies of rocks of this age were completed.

The Cambrian is sometimes called the age of invertebrates. Until this century scientists thought the Cambrian marked the first appearance of life on Earth. As the oldest period of the Paleozoic Era, meaning "ancient life," scientists now recognize the Cambrian as the short period in which a relatively simple pre-Paleozoic fauna suddenly diversified in one of the most remarkable events in the history of life. For the 4 billion years before the Cambrian explosion, life consisted mainly of single-celled organisms, with the exception of the remarkable Late Proterozoic soft-bodied Ediacaran (Vendian) fauna, sporting giant sea creatures that all went extinct by or in the Cambrian and have no counterpart on Earth today. The brief 40-million-year Cambrian saw the development of multicelled organisms, as well as species with exoskeletons, including trilobites, brachiopods, arthropods, echinoderms, and crinoids.

At the dawn of the Cambrian most of the world's continents were distributed within 60°N/S of the equator, and many of the continents that now form Asia, Africa, Australia, Antarctica, and South America were joined together in the supercontinent of Gondwana. These continental fragments had broken off an older supercontinent (Rodinia) between 700 and 600 million years ago, then joined together in the new configuration of Gondwana, with the final ocean closure of the Mozambique Ocean between East and West Gondwana occurring along the East African orogen at the Precambrian-Cambrian boundary. Even though the Gondwana supercontinent had formed only at the end of the Proterozoic, it was already breaking up and dispersing different continental fragments by the Cambrian.

Closure of the Mozambique Ocean in Neoproterozoic times sutured East and West Gondwana and intervening arc and continental terranes along the length of the East African orogen. Much active research in the earth sciences is aimed at providing a better understanding of this ancient mountain belt and its relationships to the evolution of crust, climate, and life at the end of the Precambrian and the opening of the Phanerozoic. There have been numerous rapid changes in our understanding of events related to the assembly of Gondwana. The East African orogen encompasses the Arabian-Nubian shield in the north and the Mozambique belt in the south. These and several other orogenic belts are commonly referred to as Pan-African belts, as many distinct belts in Africa and other continents experienced deformation, metamorphism, and magmatic activity in the general period 800–450 Ma. Pan-African tectonic activity in the Mozambique belt was broadly contemporaneous with magmatism, metamorphism, and deformation in the Arabian-Nubian shield. Geologists attribute the difference in lithology and metamorphic grade between the two belts to the difference in the level of exposure, with the Mozambican rocks interpreted as lower crustal equivalents of the rocks in the Arabian-Nubian shield.

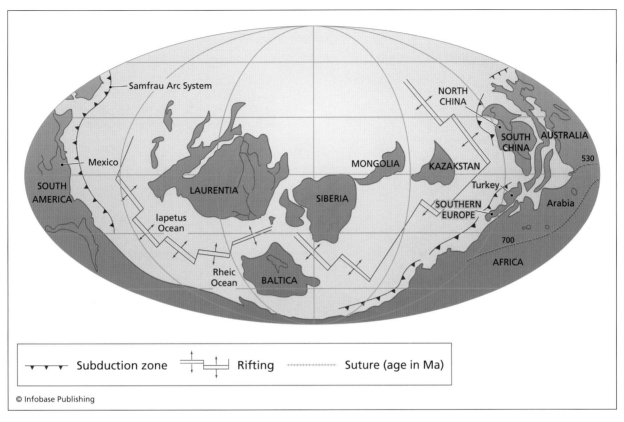

Paleogeographic plate reconstruction of the world during Late Precambrian times *(modeled after Kent Condie and Robert Sloan)*

The timing of Gondwana's amalgamation coincides remarkably with the Cambrian explosion of life, which has focused the research of many scientists on relating global-scale tectonics to biologic and climatic change. The dramatic biologic, climatic, and geologic events that mark Earth's transition into the Cambrian are likely linked to the distribution of continents and the breakup and reassembly of a supercontinent. The formation and dispersal of supercontinents causes dramatic changes in the Earth's climate, and changes the distribution of environmental settings for life to develop within. Plate tectonics and the formation and breakup of the supercontinents of Rodinia and Gondwana set the stage for life to diversify during the Cambrian explosion, bringing life from the primitive forms that dominated the Precambrian to the diverse fauna of the Paleozoic. The breaking apart of supercontinents creates abundant shallow and warm-water inland seas, as well as shallow passive margins along the edges of the rifted fragments. As rifting separated continental fragments from Rodinia, they moved across warm oceans, and new life-forms developed on these shallow passive margins and inland seas. When these "continental icebergs" carrying new life collided with the supercontinent Gondwana, the new life-forms could rapidly expand and diversify, then compete with the next organism brought in by the next continent. This process happened over and over again, with the formation and breakup of the two Late Proterozoic–Cambrian supercontinents of Rodinia and Gondwana.

North America began rifting away from Gondwana as it was forming, with the rifting becoming successful enough to generate rift-type volcanism by 570 million years ago and an ocean named Iapetus by 500 million years ago. The Iapetus Ocean saw some convergent activity in the Cambrian but experienced major contractional events during the Middle Ordovician. North and south China had begun rifting off of Gondwana in the Cambrian, as did Kazahkstan, Siberia, and Baltica (Scandinavia). The margins of these continental fragments subsided and accommodated the deposition of thick, carbonate passive-margin sequences that heralded the rapid development of life, and some of which are now petroleum provinces.

A Middle Cambrian sequence of fine-grained turbidites near Calgary, Alberta, Canada has yielded a remarkable group of extremely well-preserved fauna. In 1909 Charles D. Walcott discovered the

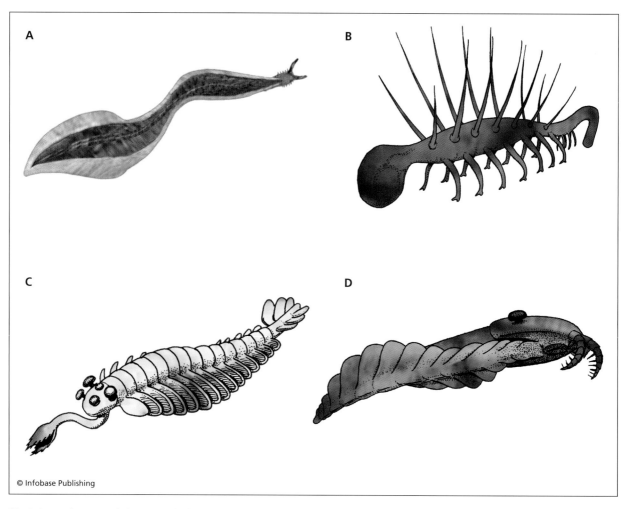

© Infobase Publishing

Sketches of some of the remarkable fauna from the Burgess shale, including (a) Pikaia, (b) Hallucigenia, (c) Opabinia, and (d) Anomocaris

Burgess shale, which preserved organisms deposited in a lagoon that was buried suddenly in anaerobic muds, which so well preserved the organisms that even the soft parts show fine detail. Fossils from the Burgess shale and related rocks have revealed much of what we know about the early life-forms in the Cambrian and represent the earliest-known fauna. The Burgess shale has yielded some of the best-preserved jellyfish, worms, sponges, brachiopods, trilobites, arthropods, mollusks, and first invertebrate chordates.

One of the important steps for the rapid expansion of life-forms in the Cambrian was the rapid radiation of acritarchs, small spores of planktonic algae such as green algae or dinoflagellates. The acritarchs were the primary source of food and the base of the food chain for the higher animals that later developed. Acritarchs appeared first during the Late Proterozoic, but 75 percent of all their taxa became extinct in the Late Proterozoic glaciation. The Late Proterozoic–Cambrian transition also saw the first

appearance of trace fossils of soft-bodied organisms, showing that they developed and rapidly diversified in this period. Traces of worm paths are most common where they searched for food or burrowed into soft sediments.

Shelly fossils first appeared slightly later in the Cambrian, during the Tommotian, a 15-million-year stage added to the base of the Cambrian timescale in the 1970s. Most of the early shelly fossils were small (1–2 mm) conical shells, tubes, plates, or spicules made of calcite or calcium phosphate. They represent different phyla, including mollusks, brachiopods, armored worms, sponges, and archeocyathid reefs. The next major phase of the Cambrian radiation saw calcite shells added to trilobites, enabling their widespread preservation. Trilobites rapidly became abundant, forming about 95 percent of all preserved Cambrian fossils. But since trilobites form only 10 percent of the Burgess shale, this number could be biased in favor of the hard-shelled trilobites over other soft-bodied organisms. Trilobites experienced

Mid-Cambrian sea life reconstruction *(Tom McHugh/Photo Researchers, Inc.)*

five major extinctions in the Cambrian, each one followed by an adaptive radiation and expansion of new species into vacant ecological niches. Other arthropods that appeared in the Cambrian include crustaceans (lobsters, crabs, shrimp, ostracods, and barnacles) and chelicerates (scorpions, spiders, mites, ticks, and horseshoe crabs).

Mollusks also appeared for the first time at the beginning of the Cambrian, with the first clam (pelecypod) by the end of the Cambrian. Snails (gastropods) also emerged, including those with multiple gas-filled chambers. The cephalopods are other mollusks that had gas-filled chambers that appeared in the Cambrian. Echinoderms with hard skeletons first appeared in the Early Cambrian; these include starfish (asteroids), brittle stars (ophiuroids), sea urchins (echinoids), and sea cucumbers (holothuroids).

See also GONDWANA, GONDWANALAND; PALEOZOIC; PRECAMBRIAN; SUPERCONTINENT CYCLES.

FURTHER READING

Cowrie, J. W., and M. D. Braiser, eds. *The Precambrian-Cambrian Boundary.* Oxford: Clarendon Press, 1989.

Gould, Steven J. *Wonderful Life: Burgess Shale and the Nature of History.* New York: W.W. Norton, 1989.

Kusky, Timothy M., Mohamed Abdelsalam, Robert Tucker, and Robert Stern, eds. "Evolution of the East African and Related Orogens, and the Assembly of Gondwana." *Special Issue of Precambrian Research* (2003): 81–85.

Lipps, Jere H., and Philip W. Signor, eds. *Origin and Early Evolution of the Metazoa.* New York: Plenum, 1992.

McMenamin, Mark A. S., and Diana L. S. McMenamin. *The Emergence of Animals: The Cambrian Breakthrough.* New York: Columbia University Press, 1989.

Prothero, Donald R., and Robert H. Dott. *Evolution of the Earth.* 6th ed. Boston: McGraw Hill, 2002.

carbon cycle The carbon cycle is a complex series of processes in which the element carbon makes a continuous and complex exchange between the atmosphere, hydrosphere, lithosphere and solid Earth, and biosphere. Carbon is one of the fundamental building blocks of Earth, with most life-forms consisting of organic carbon, and inorganic carbon dominating the physical environment. The carbon cycle is driven by energy flux from the Sun and plays a major role in regulating the planet's climate.

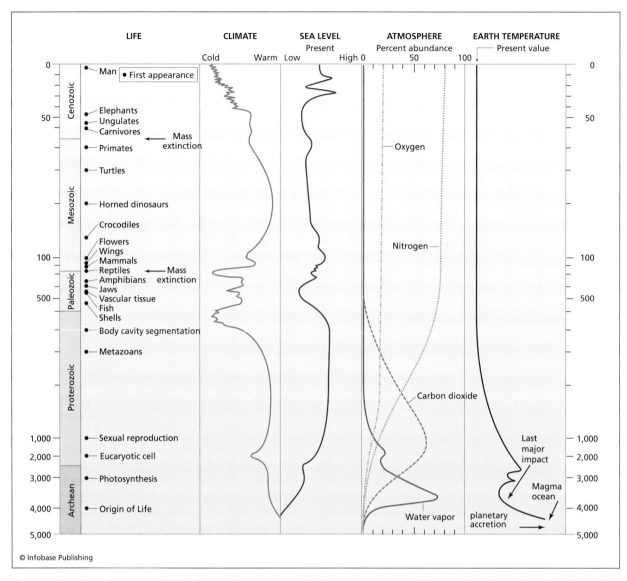

Graphs showing changes of CO₂, O₂, and temperature in the atmosphere with time *(modeled after Kent Condie and Robert Sloan)*

Several main processes control the flux of carbon on the Earth, and these processes are presently approximately balanced. Assimilation and dissimilation of carbon, by photosynthesis and respiration by life, cycles about 10^{11} metric tons of carbon each year. Some carbon is simply exchanged between systems as carbon dioxide (CO_2), and other carbon undergoes dissolution or precipitation as carbonate compounds in sedimentary rocks.

Atmospheric carbon forms the long-lived compounds carbon dioxide and methane, and the compound carbon monoxide, which has a short atmospheric residence time. Global temperatures and the amount of carbon (chiefly as CO_2) in the atmosphere are closely correlated, with more CO_2 in the atmosphere causing higher temperatures. Whether increased carbon flux in the atmosphere from the carbon cycle forces global warming or whether global warming causes an increase in the carbon flux remains undetermined. Since the industrial revolution, humans have increased CO_2 emissions in the atmosphere, and this has also caused measurable global warming, showing that increased carbon flux can control global temperatures.

The oceans are the largest carbon reservoir on the planet, containing more than 60 times as much carbon as the atmosphere. Dissolved inorganic carbon forms the largest component, followed by the more mobile dissolved organic carbon. The oceans are stratified into three main layers. The well-mixed surface layer is about 246 feet (75 m) thick and overlies the thermocline, a stagnant zone characterized by decreasing temperature and increasing density to its base at about 0.6-mile (1-km) depth. Below this lie the deep cold-bottom waters where dissolved CO_2 transferred by descending cold saline waters in polar regions may remain trapped for thousands of years. Cold polar waters contain more CO_2 because gases are more soluble in colder water. Some, perhaps large amounts, of this carbon becomes incorporated in gas hydrates, which are solid, icelike substances made of cases of ice molecules enclosing gas molecules like methane, ethane, butane, propane, carbon dioxide, and hydrogen sulfide. Gas hydrates have recently been recognized as a huge global energy resource, with reserves estimated to be at least twice that of known fossil fuel deposits. However, gas hydrates form at high pressures and cold temperatures, and extracting them from the deep ocean without releasing huge amounts of CO_2 into the atmosphere may be difficult.

Carbon is transferred to the deep ocean by its solubility in seawater, whereas organic activity (photosynthesis) in the oceanic surface layer accounts for 30–40 percent of the global vegetation flux of carbon. About 10 percent of the carbon used in respiration in the upper oceanic layer is precipitated out and sinks to the lower oceanic reservoir.

The majority of Earth's carbon is locked up in sedimentary rocks, primarily limestone and dolostone. This stored carbon reacts with the other reservoirs at a greatly reduced rate (millions and even billions of years) compared with the other mechanisms discussed here. Some cycles of this carbon reservoir are related to the supercontinent cycle and the weathering of carbonate platforms when they are exposed by continental collisions.

Earth's living biomass, the decaying remains of this biomass (litter), and soil all contain significant carbon reserves that interact in the global carbon cycle. Huge amounts of carbon are locked in forests, as well as in arctic tundra. Living vegetation contains about the same amount of carbon as is in the atmosphere, whereas the litter or dead biomass contains about twice the amount in the living biomass. Plants absorb an estimated 100 gigatons of carbon a year and return about half of this to the atmosphere by respiration. The remainder is transformed to organic carbon and incorporated into plant tissue and soil organic carbon.

Understanding the global carbon cycle is of great importance for predicting and mitigating climate change. Climatologists, geologists, and biologists are just beginning to understand and model the consequences of changes to parts of the system induced by changes in other parts of the system. For instance, a current debate centers on how plants respond to greater atmospheric CO_2. Some models indicate that they may grow faster under enhanced CO_2, tending to pull more carbon out of the atmosphere in a planetary self-regulating effect known as the fertilization effect. Many observations and computer models are being performed to investigate the effects of natural and human-induced changes (anthropogenic) in the global carbon cycle, and to understand better what the future may hold for global climates.

See also GEOCHEMICAL CYCLES; GLOBAL WARMING; HYDROCARBONS AND FOSSIL FUELS.

FURTHER READING

Berner, Elizabeth Kay, and Robert Berner. *Global Environment: Water, Air, and Geochemical Cycles.* Upper Saddle River, N.J.: Prentice Hall, 1994.

Brantley, Susan, James D. Kubicki, and Art White. *Kinetics of Global Geochemical Cycles.* New York: Springer, 2008.

Carboniferous The Carboniferous is a Late Paleozoic geologic period in which the Carboniferous System of rocks was deposited between 355 and 285 million years (Ma) ago. The system was named after coal-bearing strata in Wales and has the distinction of being the first formally established stratigraphic system. In the United States it is customary to use the divisions Mississippian Period (355–320 Ma) and Pennsylvania Period (320–285 Ma), whereas Europeans and the rest of the world refer to the entire interval of time as the Carboniferous Period and divide the rocks deposited in the period into two subsystems, the Upper and Lower, and five series.

The Carboniferous is known as the age of amphibians and the age of coal. The supercontinent Pangaea straddled the equator in the early Carboniferous, with warm climates dominating the southern (Gondwana) and northern (Laurasia) landmasses. In the Lower Carboniferous giant seed ferns and great coal forests spread across much of Gondwana and Laurasia, and most marine fauna that developed in the Lower Paleozoic flourished. Brachiopods, however, declined in number and species. Fusulinid foraminifera appeared for the first time. Primitive amphibians roamed the Lower Carboniferous swamps, along with swarms of insects including giant dragonflies and cockroaches.

In the Early Carboniferous (Mississippian), Gondwana was rotating northward toward the northern Laurentian continent, closing the Rheic Ocean. Continental fragments that now make up much of Asia were rifting from Gondwana, and the west coasts of North and South America were subduction-type convergent margins open to the Panthallassic Ocean. Several arc and other collisions with North America were underway, including the Antler Orogeny in the

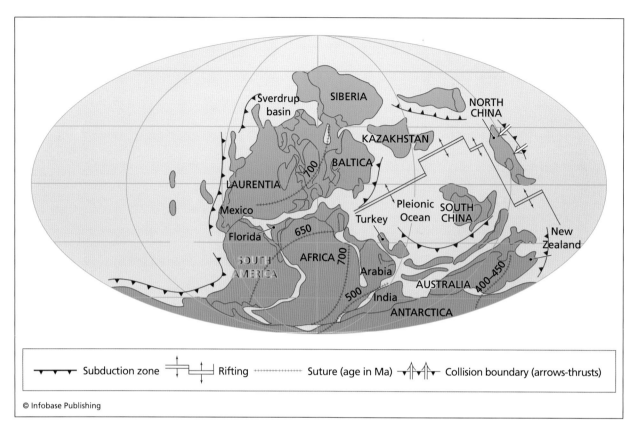

Paleogeographic map of Gondwana in the Carboniferous *(modeled after Kent Condie and Robert Sloan)*

western United States. The Hercynian Orogeny in Europe marked the collision between Baltica, southern Europe, and Africa. In the Late Carboniferous (Pennsylvanian), Laurentia and Gondwana finally collided, forming the single large landmass of Pangaea. This collision produced the Alleghenian Orogeny in the Appalachians of the eastern United States and the Ouachita Orogeny in the southern United States and South America, and formed the ancestral Rocky Mountains. In Asia Kazakhstan collided with Siberia, forming the Altai Mountains. Several microcontinents were rifted off the Gondwana continents to be accreted to form much of present-day Asia.

Global climates in the Carboniferous ranged from tropical around much of Laurentia and northern Gondwana, to polar on southern Gondwana, which experienced glaciation in the Pennsylvanian. This widespread glaciation formed in response to Gondwana migrating across the South Pole and is characterized by several advances and retreats and glacial deposits on Africa, Australia, South America, and India. Coal formed at both high and low latitudes in the Pennsylvanian, reflecting the warm climates from easterly trade winds around the closing Rheic Ocean and future opening of the Tethys Ocean. Most of the coal deposits formed in foreland basins associated with continental collisions.

Many sedimentary deposits of the Carboniferous age worldwide show development in a repetitive cycle, including accumulation of organic material (vegetation), deposition of carbonates, deposition of clastic sands, and erosion to sea level and soil development. These types of sedimentary deposits have become known as cyclothems; they reflect a uniform fluctuation of sea level by 500–650 feet (150–200 m). Analysis of the ages of each cyclothem has led to the recognition that each cycle represents 300,000 years, but the cause of the repetitive cycles remains a mystery. They may be related to cyclical variations in orbital parameters (Milankovitch cycles) or to variations in the intensity of the southern glaciation.

Extinctions in the Late Devonian paved the way for rapid expansion of new marine invertebrate forms in many ecological niches. Radiations in the brachiopods, ammonoids, bryozoans, crinoids, foraminifera, gastropods, pelcypods, and calcareous algae became widespread. Crinoids were particularly abundant in the Mississippian, forming dense submarine gardens, along with reefs made of bryozoans and calcareous algae. Fusulinid foraminifera with distinctive coiled forms evolved at the beginning of the Pennsylvanian and serve as a useful index fossil since they evolved so quickly and are abundant in many environments.

Artwork of Carboniferous landscape, including scale trees, ferns, seed ferns, and giant dragonflies *(Publiphoto/ Photo Researchers, Inc.)*

Land plants originated in the Devonian and saw additional diversification in the Carboniferous. Chordates, a prominent gymnosperm with long, thin leaves, flourished in the Mississippian, whereas conifers appeared in the Late Pennsylvanian. The tropical coal forests of the Pennsylvanian had trees that were more than 100 feet (30 m) tall, including the prominent Lepidodendron and Calamites trees and the seed-bearing Glossopteris shrub, which covered much of the cooler parts of Gondwana. Warm climates in the low-latitude coal swamps led to a flourishing fungi flora. The dense vegetation of the Carboniferous led to high levels of atmospheric oxygen, estimated to have made up about 35 percent of the gases in the atmosphere, compared with present-day levels of 21 percent.

The insects radiated in the Early Pennsylvanian and included the wingless hexapods and the primitive Paleoptera, ancestors of the modern dragonfly and mayfly. A giant Pennsylvanian dragonfly had a wingspan of 24 inches (60 cm) and preyed largely on other insects. Exopterygota, primitive crickets and cockroaches, appeared in the Pennsylvanian. Endopterygota, the folding-wing insects including flies and beetles, did not appear until the Permian.

The Carboniferous is famous for the radiation of amphibians. By the end of the Mississippian 10 different amphibian families had appeared, living mostly in water and feeding on fish. Eryops and other amphibians of this time resembled crocodiles, and include relatives of modern frogs and salamanders. Embolomeres evolved into large (up to 13 feet, or 4 m) eel-like forms with small legs, some living on land and eating insects. Leopospondyls remained in the water, eating mollusks and insects. The earliest known reptile, westlothiana, evolved from the amphibians in the Late Mississippian by 338 Ma ago. The transition from amphibians to reptiles occurred quickly, within a few tens of millions of years after the origin of amphibians. Amniotes are four-legged animals (tetrapods) that produced eggs similar to the modern bird egg, and include reptiles with scales. The rise of amniotes is a major evolutionary step, since the older amphibians underwent an early tadpole stage in which the young are vulnerable to prey. In contrast, the eggs of the amniotes and later reptiles provided enough food for the growth of the embryo in a safer environment and lessened the dependence on water, allowing them to move further inland. Descendants of the amniotes include mammals and birds.

The evolutionary transition between reptiles and mammals is gradual, with more intermediate evolutionary steps known than for any other high-order taxa. Like many other major evolutionary periods in Earth history, this evolutionary step occurred during a supercontinental amalgamation, enabling many species to compete. Many species intermediate between reptiles and mammals (the so-called mammal-like reptiles), and these dominated the land fauna for about 100 million years until the period of the dinosaurs began in the Permian. Mammal-like reptiles include two orders: Pelycosaurs and Therasids. The mammal-like reptiles had evolved into true

mammals by this time but did not become dominant until the dinosaurs were killed off at the end of the Cretaceous.

See also GONDWANA, GONDWANALAND; MILANKOVITCH CYCLES; PALEOCLIMATOLOGY; PALEOZOIC; PANGAEA.

FURTHER READING

University of California, Berkeley. "The Carboniferous." Available online. URL: http://www.ucmp.berkeley.edu/carboniferous/carboniferous.html. Accessed October 30, 2008.

cave systems, caves   Caves are underground openings or passageways in rock that are larger than individual spaces between the constituent grains of the rock. The term *cave* is often reserved for spaces that are large enough for people to enter. Some scientists use the term to describe any rock shelter, including overhanging cliffs. Many caves are small pockets along enlarged or widened cavities, whereas others are huge open underground spaces. The largest cave in the world is the Sarawak Chamber in Borneo, with a volume of 65 million cubic feet (20 million cubic meters). The Majlis Al Jinn (Khoshilat Maqandeli) Cave in Oman is the second-largest known cave; it is big enough to hold several of the sultan of Oman's royal palaces, with a 747 flying overhead. Its main chamber exceeds 13 million cubic feet (4 million cubic meters), larger than the biggest pyramid at Giza. Other large caves include the world's third-, fourth-, and fifth-largest caves, the Belize Chamber, Salle de la Verna, and the largest "Big Room" of Carlsbad Cavern, a chamber 4,000 feet long (1,200 m), 625 feet wide (190 m), and 325 feet high (100 m). Each of these has a volume of at least 3 million cubic feet (1 million cubic meters). Some caves form networks of linked passages that extend for many miles. Mammoth Cave in Kentucky, for instance, has at least 300 miles (485 km) of interconnected passageways. While the caves are forming, water flows through these passageways in underground stream networks.

The formation of caves and sinkholes in karst regions begins with a process of dissolution. Rainwater that filters through soil and rock may work its way into natural fractures or breaks in the rock, and chemical reactions that remove ions from the limestone slowly dissolve and carry away in solution parts of the limestone. Fractures are gradually enlarged, and new passageways are created by groundwater flowing in underground stream networks through the rock. Dissolution of rocks is the most effective if the rocks are limestone and the water is slightly acidic (acid rain greatly helps cave formation). Carbonic

Limestone cave stalagmites and stalactites in Israel *(Joshua Haviv, Shutterstock, Inc.)*

acid (H_2CO_3) in rainwater reacts rapidly with the limestone (at typical rates of a few millimeters per thousand years), creating open spaces, cave and tunnel systems, and interconnected underground stream networks.

Many caves are decorated with natural deposits of minerals that coat the cave walls, hang from the cave ceiling, or protrude upward from the floor. Speleothems are any secondary mineral deposit formed in a cave by the action of groundwater. Most speleothems are made of carbonate minerals such as calcite, aragonite, or dolomite, but some are made of silicates and evaporites. Dripstone and flowstone are the most common carbonate speleothems. Yellow, brown, orange, tan, green, and red colors in dripstone and flowstone are formed through staining by organic compounds, oxides derived from overlying clays and soils, and rarely by ionic substitution in the carbonate minerals. Dripstone forms where water enters the cave through joints, bedding planes, or other structures and degasses carbon dioxide (CO_2) from water droplets, forming a small ring of calcite before each drop breaks free and falls into the cave. Each succeeding drop deposits another small ring of calcite, eventually forming a hollow tube called a straw stalactite. Additional growth can occur on the outside of the straw stalactite, forming a wedge-shaped, hanging calcite deposit. Drops that fall to the cave floor below deposit additional calcite, forming a mound-shaped stalagmite. These have no central canal but consist of a series of layers deposited over each other and typically are symmetric about a vertical axis. Flowstone is a massive secondary carbonate deposit formed by water that moves as sheet flows over cave walls and floors. The water deposits layered and terraced carbonate with complex and bizarre shapes, with shapes and patterns determined by the flow rates of the water and the shape of the cave walls, shelves, and floor. Draperies are layered deposits with furled forms intermediate between dripstone and flowstone.

Less common types of speleothems include shields, massive platelike forms that protrude from cave walls. They are fed by water that flows through a medial crack separating two similar sides of the shield, with the crack typically parallel to regional joints in the cave.

Some speleothems have erratic forms not controlled by joints, walls, or other structures. Helictites are curved stalactitelike forms with a central canal; anthodites are clusters of radiating crystals such as aragonite and a variety of botryoidal forms that resemble beads or corals. Moonmilk is a wet powder or wet pasty mass of calcite, aragonite, or magnesium carbonate minerals. Travertine forms speleothems in cave systems in which the waters are saturated in carbon dioxide.

Evaporite minerals form deposits in some dry, dusty caves where the relative humidity drops to below 90 percent and the waters have dissolved anions. Gypsum is the most common evaporite mineral found as a speleothem, with magnesium, sodium, and strontium sulfates being less common. Phosphates, nitrates, iron minerals, and even ice form speleothems in other less common settings.

See also KARST.

FURTHER READING

Drew, D. *Karst Processes and Landforms.* New York: Macmillan Education Press, 1985.

Ford, D., and P. Williams. *Karst Geomorphology and Hydrology.* London: Unwin-Hyman, 1989.

Jennings, J. N. *Karst Geomorphology.* Oxford: Basil Blackwell, 1985.

The Karst Waters Institute. Available online. URL: http://www.karstwaters.org/. Accessed December 10, 2007.

White, William B. *Geomorphology and Hydrology of Karst Terrains.* Oxford: Oxford University Press, 1988.

Cenozoic The Cenozoic Era marks the emergence of the modern Earth, starting at 66 million years ago and continuing until the present. Also spelled Cainozoic and Kainozoic, the term *Cenozoic* is taken from the Greek meaning recent life and is commonly referred to as the age of the mammals. Cenozoic divisions include the Tertiary (Paleogene and Neogene) and Quaternary Periods, and the Paleocene, Eocene, Oligocene, Miocene, Pliocene, Pleistocene, and Holocene Epochs.

Modern ecosystems developed in the Cenozoic, with the appearance of mammals, advanced mollusks, birds, modern snakes, frogs, and angiosperms, such as grasses and flowering weeds. Mammals developed rapidly and expanded to inhabit many different environments. Unlike the terrestrial fauna and flora, the marine biota underwent only minor changes, with the exception of the origin and diversification of whales.

CRETACEOUS-TERTIARY BOUNDARY

The Cenozoic began after a major extinction at the Cretaceous-Tertiary boundary, marking the boundary between the Mesozoic and Cenozoic Eras. This extinction event was probably caused by a large asteroid impact that hit Mexico's Yucatán Peninsula near Chicxulub at 66 million years ago. Dinosaurs, ammonites, many marine reptile species, and a large number of marine invertebrates suddenly died off,

and the planet lost about 26 percent of all biological families and numerous species. Some organisms were dying off slowly before the dramatic events at

Cenozoic timescale

the close of the Cretaceous, but a clear, sharp event occurred at the end of this time of environmental stress and gradual extinction. Iridium anomalies have been found along most of the clay layers that mark this boundary, considered by many to be the "smoking gun," indicating an impact origin for the cause of the extinctions. An estimated one-half million tons of iridium are present in the Cretaceous-Tertiary boundary clay, equivalent to the amount that would be contained in a meteorite with a 6-mile (9.5-km) diameter. Some scientists have argued that volcanic processes within the Earth can produce iridium, and an impact is not necessary to explain the iridium anomaly. But, the presence of other rare elements and geochemical anomalies along the Cretaceous-Tertiary boundary supports the idea that a huge meteorite hit the Earth at this time.

Many features found around and associated with an impact crater on the Yucatán Peninsula suggest that this site is the crater associated with the death of the dinosaurs. The Chicxulub crater is about 66 million years old and lies half-buried beneath the waters of the Gulf of Mexico and half on land. Tsunami deposits of the same age are found in inland Texas, much of the Gulf of Mexico, and the Caribbean, recording a huge tsunami perhaps several hundred feet (a hundred meters) high that was generated by the impact. The crater is at the center of a huge field of scattered spherules that extends across Central America and through the southern United States. The large structure is the right age to be the crater that resulted from the impact at the Cretaceous-Tertiary boundary, recording the extinction of the dinosaurs and other families.

The 66 million-year-old Deccan flood basalts, also known as traps, cover a large part of western India and the Seychelle Islands in the Indian Ocean. They are associated with the breakup of India from the Seychelles during the opening of the Indian Ocean. Slightly older flood basalts (90–83 million years old) are associated with the breaking away of Madagascar from India. The volume of the Deccan traps is estimated at 5,000,000 cubic miles (20,841,000 km³), and the volcanics are thought to have erupted within about 1 million years, starting slightly before the great Cretaceous-Tertiary extinction. Most scientists now agree that the gases released during eruption of the flood basalts of the Deccan traps stressed the global biosphere to such an extent that many marine organisms had gone extinct, and many others were stressed. Then the massive Chicxulub impactor hit the planet, causing the massive extinction including the end of the dinosaurs. Faunal extinctions have been correlated with the eruption of the Deccan flood basalts at the Cretaceous-Tertiary (K-T) boundary. There is still considerable debate about the relative significance of flood basalt volcanism and impacts

of meteorites for the K-T boundary. Most scientists would now agree, however, that the global environment was stressed shortly before the K-T boundary by volcanic-induced climate change, and then a huge meteorite hit the Yucatán Peninsula, forming the Chicxulub impact crater, causing the massive K-T boundary extinction and the death of the dinosaurs.

CENOZOIC TECTONICS AND CLIMATE

Cenozoic global tectonic patterns are dominated by the opening of the Atlantic Ocean, closure of the Tethys Ocean and formation of the Alpine-Himalayan Mountain System, and mountain building in western North America. Uplift of mountains and plateaus and the movement of continents severely changed oceanic and atmospheric circulation patterns, altering global climate patterns.

As the North and South Atlantic Oceans opened in the Cretaceous, western North America was experiencing contractional orogenesis. In the Paleocene (66–58 Ma) and Eocene (58–37 Ma), shallow dipping subduction beneath western North America caused uplift and basin formation in the Rocky Mountains, with arc-type volcanism resuming from later Eocene through late Oligocene (about 40–25 Ma). In the Miocene (starting at 24 Ma), the Basin and Range Province formed through crustal extension, and the formerly convergent margin in California was converted into a strike-slip or transform margin, causing the initial formation of the San Andreas fault.

The Cenozoic saw the final breakup of Pangaea and closure of the tropical Tethys Ocean between Eurasia and Africa, Asia, and India and a number of smaller fragments that moved northward from the southern continents. Many fragments of Tethyan Ocean floor (ophiolites) were thrust upon the continents during the closure of Tethys, including the Semail ophiolite (Oman), Troodos (Cyprus), and many Alpine bodies. Relative convergence between Europe and Africa, and Asia and Arabia plus India continues to this day, and is responsible for the uplift of the Alpine-Himalayan chain of mountains. The uplift of these mountains and the Tibetan Plateau has had important influences on global climate, including changes in the India Ocean monsoon and the cutting off of moisture that previously flowed across southern Asia. Vast deserts such as the Gobi were thus born.

The Tertiary began with generally warm climates, and nearly half of the world's oil deposits formed at this time. By the mid-Tertiary (35 Ma) the Earth began cooling again, and this culminated in the ice house climate of the Pleistocene, with many glacial advances and retreats. The Atlantic Ocean continued to open during the Tertiary, which helped lower global temperatures. The Pleistocene experienced many fluctuations between warm and cold climates, called glacial and interglacial stages (the Earth is currently in the midst of an interglacial stage). These fluctuations are rapid—for instance, in the past 1.5 million years the Earth has experienced 10 major and 40 minor periods of glaciation and interglaciation. The most recent glacial period peaked 18,000 years ago, when huge ice sheets covered most of Canada and the northern United States, and much of Europe.

The human species developed during the Holocene Epoch (since 10,000 years ago). The Holocene is just part of an extended interglacial period in the planet's current ice house event, raising important questions about how humans will survive if climate suddenly changes back to a glacial period. Since 18,000 years ago the climate has warmed by several degrees, sea level has risen 500 feet (150 m), and atmospheric CO_2 has increased. Some of the global warming is human induced. One scenario of climate evolution is that global temperatures will rise, causing some of the planet's ice caps to melt, raising the global sea level. This higher sea level may increase the Earth's reflectance of solar energy, suddenly plunging the planet into an ice house event and a new glacial advance.

See also CLIMATE; CLIMATE CHANGE; MASS EXTINCTIONS; PLATE TECTONICS.

FURTHER READING

Pomerol, Charles. *The Cenozoic Era: Tertiary and Quaternary.* Chichester, U.K.: Ellis Horwood, 1982.

Proterero, Donald, and Robert Dott. *Evolution of the Earth*, 6th ed. New York: McGraw Hill, 2002.

Stanley, Steven M. *Earth and Life through Time.* New York: Freeman, 1986.

climate Climate refers to the average weather of a place or area and its variability over a period of years. The term *climate* is derived from the Greek work *klima*, meaning inclination and referring specifically to the angle of inclination of the Sun's rays, a function of latitude. The average temperature, precipitation, cloudiness, and windiness of an area determine its climate. Factors that influence climate include latitude; proximity to oceans or other large bodies of water that could moderate the climate; topography, which influences prevailing winds and may block precipitation; and altitude. All of these factors are linked together in the climate system of any region on the Earth. The global climate is influenced by many additional factors. The rotation of the Earth and latitudinal position determine where a place is located with respect to global atmospheric and oceanic circulation currents. Chemical interactions between seawater and magma significantly change the amount of carbon dioxide in the oceans and atmosphere and may change global temperatures. Pollution from humans also changes the amount of greenhouse gases in the atmosphere, which may be contributing to global warming. Climatology is the field of science concerned with climate, including both present-day and ancient climates. Climatologists study a variety of problems, ranging from the classification and effects of present-day climates through to the study of ancient rocks to determine ancient climates and their relationship to plate tectonics. An especially important field actively studied by climatologists is global climate change, with many studies focused on the effects that human activities have had and will continue to have on global climate. Many of these models require powerful supercomputers and computer models known as global circulation models. These models input various parameters at thousands or millions of grid points on a model Earth, and demonstrate how changing one or more variables (e.g., carbon dioxide, or CO_2, emissions) will affect the others.

Classifications of climate must account for the average, extremes, and frequencies of the different meteorological elements. Of the many different ways to classify climate, the most modern classifications are based on the early work of the German climatologist Wladimir Koppen. His classification (initially published in 1900) was based on the types of vegetation in an area, assuming that vegetation tended to reflect the average and extreme meteorological changes in an area. He divided the planet into different zones such as deserts, tropics, rain forest, tundra, and the like. In 1928 Norwegian meteorologist Tor Bergeron modified Koppen's classification to include the types of air masses that move through an area and how they influence vegetation patterns. The British meteorologist George Hadley made another fundamental understanding of the factors that influence global climate in the 18th century. Hadley proposed a simple, convective type of circulation in the atmosphere where heating by the Sun causes the air to rise near the equator and move poleward, where the air sinks back to the near surface, then returns to the equatorial regions. We now recognize a slightly more complex situation, in that there are three main convecting atmospheric cells in each hemisphere, named Hadley, Ferrel, and Polar cells. These play important roles in the distribution of different climate zones, as moist or rainy regions are located, in the tropics and at temperate latitudes, where the atmospheric cells are upwelling and release water. Deserts and dry areas are located around zones where the convecting cells downwell, bringing descending dry air into these regions.

The rotation of the Earth sets up systems of prevailing winds that modify the global convective atmospheric (and oceanic) circulation patterns. The spinning of the Earth sets up latitude-dependent airflow patterns, including the trade winds and westerlies. In addition, uneven heating of the Earth over land and ocean regions causes regional airflow patterns such as rising air over hot continents that must be replenished by air flowing in from the sides. The Coriolis force results from the rotation of the Earth and causes any moving air mass in the Northern Hemisphere to be deflected to the right and masses in the Southern Hemisphere to be deflected to the left. These types of patterns tend to persist for long periods of time and move large masses of air around the planet, redistributing heat and moisture and regulating the climate of any region.

Temperature, largely determined by latitude, is a major factor in the climate of any area. Polar regions experience huge changes in temperature between winter and summer months, largely a function of the wide variations in amount of incoming solar radiation and length of days. The proximity to large bodies of water such as oceans influences temperature, as water heats up and cools down much more slowly than land surfaces. Proximity to water therefore moderates temperature fluctuations. Altitude also influences temperature, with temperature decreasing with height.

Climate may change in cyclical or long-term trends, as influenced by changes in solar radiation, orbital variations of the Earth, amount of greenhouse gases in the atmosphere, or other phenomena such as El Niño or La Niña.

See also ATMOSPHERE; CLIMATE CHANGE; EL NIÑO AND THE SOUTHERN OSCILLATION (ENSO); GLACIER, GLACIAL SYSTEMS; GREENHOUSE EFFECT; ICE AGES; MILANKOVITCH CYCLES; PLATE TECTONICS; SEA-LEVEL RISE.

FURTHER READING

Ahrens, C. D. *Meteorology Today: An Introduction to Weather, Climate, and the Environment.* 6th ed. Pacific Grove, Calif.: Brooks/Cole, 2000.

Douglas, B., M. Kearney, and M. and S. Leatherman. *Sea Level Rise: History and Consequence.* San Diego, Calif.: Academic Press, International Geophysics Series 75, 2000.

Intergovernmental Panel on Climate Change 2007. *Climate Change 2007: The Physical Science Basis. Contributions of Working Group I to the Fourth Assessment Report of the Intergovernmental Panel on Climate Change* (Solomon, S., D. Qin, M. Manning, Z. Chen, M. Marquis, K. B. Averyt, M. Tignor, and H. L. Miller, eds.) Cambridge: Cambridge University Press, 2007. Also available online. URL: http://www.ipcc.ch/index.htm. Accessed October 10, 2008.

National Aeronautic and Space Administration (NASA). "Earth Observatory." Available online. URL: http://earthobservatory.nasa.gov/. Accessed October 9, 2008, updated daily.

climate change The Earth has experienced many episodes of dramatic climate change, with different periods in Earth history seeing the planet much hotter or much colder than the present. There have been periods when the entire planet was covered in ice in a frozen, seemingly perpetual winter, then other times the Earth's surface was scorchingly hot and dry, and others when much of the planet felt like a hot, wet sauna. Scientists, including those from the Intergovernmental Panel on Climate Change, warn that the planet is currently experiencing global warming at a rapid pace, and there will be significant consequences for the people and ecosystems on the planet, as explained in sections below.

Many different variables control climate and can change the planet rapidly from one condition to another. Most of these are related to variations in the amount of incoming solar radiation caused by astronomical variations in the Earth's orbit. Other variables that can strongly influence long-term climate change include the amount of heat retained by the atmosphere and oceans, and on timescales of tens to hundreds of millions of years, the distribution of landmasses as they move about the planet from plate tectonics. Each of these changes operates with different time cycles, alternately causing the climate to become warmer and colder.

Significant long-term climate changes include the gradual alteration of the Earth's atmosphere from a global hothouse dominated by carbon dioxide (CO_2) and other greenhouse gases when the Earth was young to an atmosphere rich in nitrogen and oxygen over the next couple billion years. Fortunately, during the early history of the Earth the Sun was less luminous, and the planet was not exceedingly hot. The motion of the continents has over time alternately placed them over the poles, which causes the continent to be covered in snow, reflecting more heat back to space and causing global cooling. Plate tectonics also has a complex interaction with concentrations of CO_2 in the atmosphere, for instance, by uplifting carbonate rocks to be exposed to the atmosphere during continental collisions. The calcium carbonate ($CaCO_3$) then combines with atmospheric CO_2, depositing it in the oceans. Thus continental collisions and times of supercontinent formation are associated with drawdown and reduction of CO_2 from the atmosphere, global cooling, and sea-level changes.

Orbital variations are the main cause of climate variations on more observable geological timescales.

The main time periods of these variations induce alternations of hotter and colder times, varying with frequencies of 100,000, 41,000, 23,000, and 19,000 years. To understand the complexity of natural climate variations, the contributions from each of these factors must be added together, forming a complex curve of climate warming and cooling trends. Built on top of these long-term climate variations that can change rapidly are shorter-term variations caused by changes in ocean circulation, sunspot cycles, and, finally, the contribution in the last couple of hundred years from the industry of humans, called anthropogenic changes. Deciphering which of these variables causes a particular percentage of the present global warming is no simple matter, and many political debates focus on who is to blame. Perhaps it is just as appropriate to focus on how we humans need to respond to global warming. Coastal cities may need to be moved, crop belts are migrating, climate zones are changing, river conditions will change, and many aspects of life that we are used to will be different. Scientists are expending considerable effort to understand the climate history of the past million years in order to predict the future.

NATURAL LONG-TERM CLIMATE CHANGE

Many controls operate to change the Earth's climate on different timescales. Some cause the global temperature to rise and fall within a time interval between warming and cooling influences of billions to hundreds of millions of years; others operate on time frames of millions to tens of millions of years. These slowly operating forces include the sluggish evolution of the composition of the planet's atmosphere from an early greenhouse atmosphere when the Earth had recently formed to its present-day composition. During the earliest history of the solar system, the Sun was about 30 percent less luminous, so the temperatures on Earth's surface were not as high as they could have been, given the early greenhouse conditions. Changes in solar luminosity have been significant in Earth history, and will be significant again in the future.

Plate tectonics exhibits different types of controls and with different timescales of influence on changing the atmospheric composition and climate. One type of influence of plate tectonics is on a planetary scale—plate tectonics goes through intervals of time in which seafloor spreading and volcanism is very active and periods when it is less active. During the active times the volcanism releases a lot of carbon dioxide and other greenhouse gases into the atmosphere, causing global warming. During inactive times global cooling can result. These changes operate on timescales of tens to hundreds of millions of years. Periods of very active seafloor spreading are often associated with periods of breakup of large continental landmasses known as supercontinents, and thus breakup of continents is often associated with global warming. Periods of less active seafloor spreading are often associated with continental amalgamations, formation of supercontinents, and global cooling.

When continents collide this process uplifts large sections of carbonate rocks from passive margins and exposes them to atmospheric weathering. When the calcium carbonate ($CaCO_3$) in these rocks is broken down by chemical weathering the carbonate ion (CO_3^{2-}) is dissolved by rainwater, and the free calcium ion (Ca^{2+}) then combines with atmospheric CO_2 to form new layers of limestone in the ocean, while drawing down CO_2 from the atmosphere and causing global cooling.

Scientists have shown that the interaction between these different long-term drivers of global climate is largely responsible for the long-term fluctuations in global climate on the billions to tens of millions of years timescales. Many aspects of these changes are not yet understood by geologists and paleoclimatologists, but the mechanisms described above seem fairly well understood and represent the most likely explanation for the causes of the changes.

Role of the Atmosphere in Climate Change

Interactions between the atmosphere, hydrosphere, biosphere, and lithosphere control global climate. Global climate represents a balance between the amount of solar radiation received and the amount of this energy retained in a given area. The planet receives about 2.4 times as much heat in the equatorial regions as in the polar regions. The atmosphere and oceans respond to this unequal heating by setting up currents and circulation systems that redistribute the heat more equally. These circulation patterns are in turn affected by the ever-changing pattern of the distribution of continents, oceans, and mountain ranges.

The amounts and types of gases in the atmosphere can modify the amount of incoming solar radiation, and hence global temperature. For instance, cloud cover can cause much of the incoming solar radiation to be reflected to space before being trapped by the lower atmosphere. On the other hand, greenhouse gases allow incoming short-wavelength solar radiation to enter the atmosphere, but trap this radiation when it tries to escape in its longer-wavelength reflected form. This causes a buildup of heat in the atmosphere and can lead to a phenomenon known as the greenhouse effect.

The amount of heat trapped in the atmosphere by greenhouse gases has varied greatly over Earth's history. One of the most important greenhouse gases

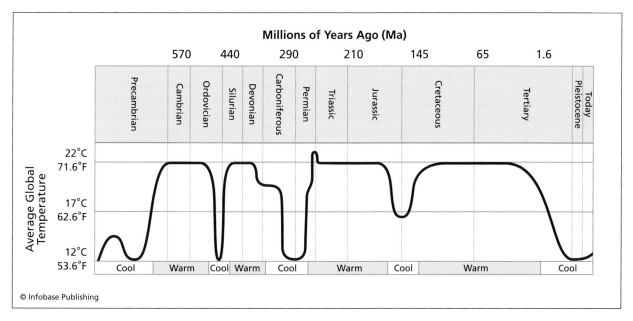

Plot showing how the average temperature on the surface of the Earth has changed with time over the past several hundreds of millions of years. These represent slow, long-term changes in global temperature.

is carbon dioxide (CO_2). Plants, which release oxygen gas (O_2) to the atmosphere, take up CO_2 during photosynthesis. In the early part of Earth's history (in the Precambrian, before plants covered the land surface), photosynthesis did not remove CO_2 from the atmosphere, with the result that CO_2 levels were much higher than at present. Marine organisms also take up atmospheric CO_2 by removing it from the ocean surface water (which is in equilibrium with the atmosphere) and use the CO_2 along with calcium to form their shells and mineralized tissue. These organisms make $CaCO_3$ (calcite is the most common mineral form of calcium carbonate), which is the main component of limestone, a rock composed largely of the dead remains of marine organisms. The atmosphere-ocean system presently has approximately 99 percent of the planet's CO_2 locked up in rock deposits of limestone on the continents and on the seafloor. If this amount of CO_2 were released back into the atmosphere, the global temperature would increase dramatically. In the early Precambrian, when this CO_2 was free in the atmosphere global temperatures averaged about 550°F (290°C).

The atmosphere redistributes heat quickly by forming and redistributing clouds and uncondensed water vapor around the planet along atmospheric circulation cells. Oceans are able to hold and redistribute more heat because of the greater amount of water in the oceans, but they redistribute this heat more slowly than the atmosphere. Surface currents are formed in response to wind patterns, but deep ocean currents that move more of the planet's heat follow courses that are more related to the bathym-etry (topography of the seafloor) and the spinning of the Earth than they are related to surface winds.

The balance of incoming and outgoing heat from the Earth has determined the overall temperature of the planet through time. Examination of the geological record has enabled paleoclimatologists to reconstruct periods when the Earth had glacial periods, hot dry periods, hot wet periods, or cold dry periods. In most cases the Earth has responded to these changes by expanding and contracting its climate belts. Warm periods see an expansion of the warm subtropical belts to high latitudes, and cold periods see an expansion of the cold climates of the poles to low latitudes.

Plate Tectonics and Climate

The outer layers of the Earth are broken into about a dozen large tectonic plates, extending to about 60–100 miles (100–160 km) beneath the surface. Each of these plates may be made of oceanic crust and lithosphere, continental crust and lithosphere, or an oceanic plate with a continent occupying part of the area of the plate. Plate tectonics describes processes associated with the movement of these plates along three different types of boundaries: divergent, convergent, and transform. At divergent boundaries the plates move apart from one another, and molten rock (magma) rises from the mantle to fill the space between the diverging plates. This magma makes long ridges of volcanoes along a midocean ridge system that accounts for most of the volcanism on the planet. These volcanoes emit huge quantities of carbon dioxide (CO_2) and other gases when they erupt.

There have been times in the history of Earth that midocean ridge volcanism was very active, producing huge quantities of magma and CO_2 gas, and other times when the volcanism is relatively inactive. The large quantities of magma and volcanism involved in this process have ensured that variations in midocean ridge magma production have exerted strong controls on the amount of CO_2 in the atmosphere and ocean, and thus, are closely linked with climate. Periods of voluminous magma production are correlated with times of high atmospheric CO_2, and globally warm periods. These times are also associated with times of high sea levels, since the extra volcanic and hot oceanic material on the seafloor takes up extra volume and displaces the seawater to rise higher over the continents. This rise in sea levels in turn buried many rocks that are then taken out of the chemical weathering system, slowing down reactions between the atmosphere and the weathering of rocks. Those reactions are responsible for removing large quantities of CO_2 from the atmosphere, so the rise in sea level further promotes global warming during periods of active seafloor volcanism.

Convergent boundaries are places where two plates are moving toward each other or colliding. Most plate convergence happens where an oceanic plate is pushed or subducted beneath another plate, either oceanic or continental, forming a line of volcanoes on the overriding plate. This line of volcanoes is known as a magmatic arc, and specifically as an island arc if built on oceanic crust or an Andean arc if built on continental crust. When continents on these plates collide, the rocks that were deposited along their margins, typically underwater, are uplifted in the collision zone and exposed to weathering processes. The weathering of these rocks, particularly the limestone and carbonate rocks, causes chemical reactions where the CO_2 in the atmosphere reacts with the products of weathering, and forms new carbonate ($CaCO_3$) that gets deposited in the oceans. Continental collisions are thus associated with the overall removal of CO_2 from the atmosphere and help promote global cooling.

Transform margins do not significantly influence global climate since they are not associated with large amounts of volcanism, nor do they uplift large quantities of rock from the ocean.

The timescale of variations in global CO_2 related to changes in plate tectonics are slow, and they fall under the realm of causing very long-term climate changes, in cycles ranging from millions to tens of millions of years. Plate tectonics and movement of continents has been associated with glaciations for the past few billion years, but the exact link between tectonics and climate is not clearly established. Some doubt remains as to why global temperatures dropped,

inducing the various glacial ages. The answer may be related to changes in the natural (nonbiogenic) production rate of carbon dioxide—the number one greenhouse gas. CO_2 is produced in volcanoes and in the midocean ridges, and it is lost by being slowly absorbed into the oceans. Both of these processes are very slow—about the right timescales to explain the Great Ice Ages. One theory is that more carbon dioxide is produced during times of faster-than-average rates of seafloor spreading and the subsequent increase in volcanism. During times of rapid spreading, the higher volcanic activity, coupled with higher sea levels and reduced chemical weathering of rocks, may promote global warming by enriching the CO_2 content of the atmosphere. Similarly, global cooling may result from stalled or slowed spreading.

Supercontinents and Climate

The motion of the tectonic plates periodically causes most of the continental landmasses of the planet to collide with each other, forming giant continents known as supercontinents. For much of the past several billion years, these supercontinents have alternately formed and broken up in a process called the supercontinent cycle. The last supercontinent was known as Pangaea, which broke up about 160 million years ago to form the present-day plates on the planet. Before that the previous supercontinent was Gondwana, which formed about 600–500 million years ago, and the one before that was Rodinia, formed around a billion years ago. The distribution of landmasses and formation and breakup of supercontinents has dramatically influenced global and local climate on timescales of 100 million years, with cycles repeating for the past few billion years of Earth's history. The supercontinent cycle predicts that the planet should have periods of global warming associated with supercontinent breakup, and global cooling associated with supercontinent formation. The supercontinent cycle affects sea-level changes, initiates periods of global glaciation, changes the global climate from hothouse to icehouse conditions, and influences seawater salinity and nutrient supply. All of these consequences of plate tectonics have profound influences on life on Earth.

Sea level has changed by thousands of feet (hundreds of meters) above and below current levels many times during Earth's history. In fact, sea level is constantly changing in response to a number of different variables, many of them related to plate tectonics, the supercontinent cycle, and climate. Sea level was 1,970 feet (600 m) higher than now during the Ordovician and reached a low stand at the end of the Permian. Sea levels were high again in the Cretaceous during the breakup of the supercontinent of Pangaea.

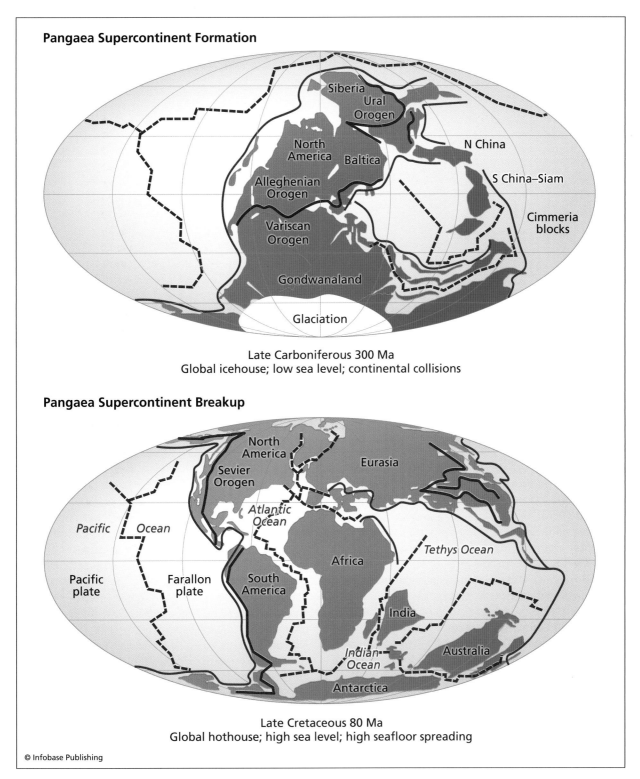

Pangaea Supercontinent Formation

Siberia
Ural
Orogen
North
America Baltica
N China
Alleghenian
Orogen
S China–Siam
Variscan
Orogen
Cimmeria
blocks
Gondwanaland
Glaciation

Late Carboniferous 300 Ma
Global icehouse; low sea level; continental collisions

Pangaea Supercontinent Breakup

North
America
Eurasia
Sevier
Orogen
Atlantic
Ocean
Pacific   Ocean
Tethys Ocean
Pacific   Farallon   Africa
plate   plate
South
America
India
Indian
Ocean   Australia
Antarctica

Late Cretaceous 80 Ma
Global hothouse; high sea level; high seafloor spreading

Maps of continental positions during cold and warm climates showing the relationship between climate and tectonics

Sea levels may change at different rates and amounts in response to different phases of the supercontinent cycle, and the sea level changes are closely related to climate. The global volume of the midocean ridges can change dramatically, either by increasing the total length of ridges or changing the rate of seafloor spreading. Either process produces more volcanism; increases the volume of volcanoes on the seafloor, raising sea levels; and puts a lot of extra CO_2 into the atmosphere, raising global temperatures. The total length of ridges typically increases during continental breakup, since continents are being rifted apart and some continental rifts can evolve into midocean ridges. Additionally, if seafloor spreading rates are increased, the amount of young, topographically elevated ridges is increased relative to the slower, older topographically lower ridges that occupy a smaller volume. If the volume of the ridges increases by either mechanism, then a volume of water equal to the increased ridge volume is displaced and sea level rises, inundating the continents. Changes in ridge volume are able to change sea levels positively or negatively by about 985 feet (300 m) from present values, at rates of about 0.4 inch (1 cm) every 1,000 years.

Continent-continent collisions, such as those associated with supercontinent formation, can lower sea levels by reducing the area of the continents. When continents collide, mountains and plateaus are uplifted, and the amount of material taken from below sea level to higher elevations no longer displaces seawater, causing sea levels to drop. The contemporaneous India-Asia collision has caused sea levels to drop by 33 feet (10 m). Times when supercontinents amalgamate are associated with times when seas drop to low levels.

Other factors, such as midplate volcanism, can also change sea levels. The Hawaiian Islands are hotspot-style midplate volcanoes that have been erupted onto the seafloor, displacing an amount of water equal to their volume. Although this effect is not large at present, at some periods in Earth's history there were many more hot spots (such as in the Cretaceous), and the effect may have been larger.

The effects of the supercontinent cycle on sea level may be summarized as follows. Continent assembly favors regression, whereas continental fragmentation and dispersal favors transgression. Regressions followed formation of the supercontinents of Rodinia and Pangaea, whereas transgressions followed the fragmentation of Rodinia, and the Jurassic-Cretaceous breakup of Pangaea.

Climate and Seasonality

Variations in the average weather at different times of the year are known as seasons, controlled by the average amount of solar radiation received at the surface in a specific place for a certain time. Several factors determine the amount of radiation received at a particular point on the surface, including the angle at which the Sun's rays hit the surface, the length of time the rays warm the surface, and the distance to the Sun. As the Earth orbits the Sun approximately once every 365 days, it follows an elliptical orbit that brings it closest to the Sun in January (91 million miles, or 147 million kilometers) and farthest from the Sun in July (94.5 million miles, or 152 million kilometers). Therefore the Sun's rays are slightly more intense in January than in July but, as any Northern Hemisphere resident can testify, this must not be the main controlling factor determining seasonal warmth since winters in the Northern Hemisphere are colder than summers. Where the Sun's rays hit a surface directly, at right angles to the surface, they are most effective at warming the surface since they are not being spread out over a larger area on an inclined surface. Also, where the Sun's rays enter the atmosphere directly, they travel through the least amount of atmosphere, so are weakened much less than rays that must travel obliquely through the atmosphere, which absorbs some of their energy. The Earth's rotational axis is presently inclined at 23.5° from perpendicular to the plane on which it rotates around the Sun (the ecliptic plane), causing different hemispheres of the planet to be tilted toward or away from the Sun during different seasons. During the Northern Hemisphere summer the Northern Hemisphere is tilted toward the Sun, so it receives more direct sunlight rays than the Southern Hemisphere, causing more heating in the north than in the south. Also, since the Northern Hemisphere is tilted toward the Sun in summer, it receives direct sunlight for longer periods of time than the Southern Hemisphere, enhancing this effect. On the summer solstice on June 21, the Sun's rays are directly hitting 23.5°N latitude (called the tropic of Cancer) at noon. Because of the tilt of the planet, the Sun does not set below the horizon for all points north of the Arctic Circle (66.5°N). Points farther south have progressively shorter days, and points farther north have progressively longer days. At the North Pole the Sun rises above the horizon on March 20 and does not set again until six months later, on September 22. Since the Sun's rays are so oblique in these northern latitudes, however, they receive less solar radiation than areas farther south where the rays hit more directly but for shorter times. As the Earth rotates around the Sun, it finds the Southern Hemisphere tilted at its maximum amount toward the Sun on December 21 (summer solstice in the southern hemisphere), and the situation is reversed from the Northern Hemisphere summer, so that the same effects occur in the southern latitudes.

Seasonal variations in temperature and rainfall at specific places are complicated by global atmospheric circulation cells, proximity to large bodies of water and warm or cold ocean currents, and monsoon-type effects in some parts of the world. Seasons in some places are hot and wet, others are hot and dry, cold and wet, or cold and dry.

Supercontinents affect the supply of nutrients to the oceans and thus seasonality. Large supercontinents that contain most of the planet's landmass cause increased seasonality, and thus lead to an increase in the nutrient supply through overturning of the ocean waters. During breakup and dispersal, smaller continents have less seasonality, yielding decreased vertical mixing, leaving fewer nutrients in shelf waters. Seafloor spreading also increases the nutrient supply to the ocean; the more active the seafloor spreading system, the more interaction there is between ocean waters and crustal minerals that dissolve to form nutrients in the seawater.

NATURAL MEDIUM AND SHORT-TERM CLIMATE CHANGE

Plate tectonics, supercontinents, and massive volcanism can cause climate variations on timescales of millions to billions of years. Many other variables contribute to climate variations that operate on shorter-term timescales, many of which are more observable. Variations in Earth's orbit around the Sun exhibit cyclic variations that alternately make Earth's climate warmer and colder at timescales ranging from 100,000 years down to 11,000 years. These cycles, known as Milankovitch cycles, have been convincingly shown to correlate with advances and retreats of the glaciers in the past few million years, and have operated throughout Earth history.

Changes in ocean circulation patterns caused by changes in seawater salinity and many other factors can dramatically change the pattern of heat distribution on the planet and global climate. Many ocean currents are driven by differences in temperature and salinity of ocean waters; these currents form a pattern of global circulation known as thermohaline circulation. Changes in patterns of thermohaline circulation can occur quite rapidly, perhaps even over 5–10 years, suddenly plunging warm continents into long, icy winters or warming frozen, ice-covered landscapes. Other changes in the ocean-atmosphere system also cause the local climate to change on 5–10-year timescales. The most dramatic of these is the El Niño-Southern Oscillation, which strongly affects the Pacific and the Americas but has influences worldwide.

Astronomical Forcing of the Climate

Medium-term climate changes include those that alternate between warm and cold on timescales of 100,000 years or fewer. These medium-term climate changes include the semiregular advance and retreat of the glaciers during the many individual ice ages in the past few million years. Large global climate oscillations that have been recurring at approximately a 100,000-year periodicity at least for the

past 800,000 years have marked the last 2.8 Ma. The warm periods, called interglacial periods, appear to last approximately 15,000 to 20,000 years before regressing to a cold ice age climate. The last of these major glacial intervals began ending about 18,000 years ago, as the large continental ice sheets covering North America, Europe, and Asia began retreating. The main climate events related to the retreat of the glaciers can be summarized as follows:

- 18,000 years ago: the climate begins to warm
- 15,000 years ago: advance of glaciers halts and sea levels begin to rise
- 10,000 years ago: Ice Age megafauna goes extinct
- 8,000 years ago: Bering Strait land bridge becomes drowned, cutting off migration of people and animals
- 6,000 years ago: the Holocene Maximum warm period
- So far in the past 18,000 years Earth's temperature has risen approximately 16°F (10°C), and sea level has risen 300 feet (91 m)

This past glacial retreat is but one of many in the past several million years, with an alternation of warm and cold periods apparently related to a 100,000-year periodicity in the amount of incoming solar radiation, causing the alternating warm and cold intervals. Systematic changes in the amount of incoming solar radiation, caused by variations in Earth's orbital parameters around the Sun, are known as Milankovitch cycles, after Milutin Milankovitch (1879–1958), a Serbian scientist who first clearly elucidated the relationships between the astronomical variations of the Earth orbiting the Sun and the climate cycles on Earth. These changes can affect many Earth systems, causing glaciations, global warming, and changes in the patterns of climate and sedimentation. Milankovitch's main scientific work was published by the Royal Academy of Serbia in 1941, during World War II. He calculated that the effects of orbital eccentricity, wobble, and tilt combine every 40,000 years to change the amount of incoming solar radiation, lowering temperatures and causing increased snowfall at high latitudes. His results have been widely used to interpret the climatic variations, especially in the Pleistocene record of ice ages, and also in the older rock record.

Astronomical effects influence the amount of incoming solar radiation; minor variations in the path of the Earth in its orbit around the Sun and the inclination or tilt of its axis cause variations in the amount of solar energy reaching the top of the atmosphere. These variations are thought to be responsible for the

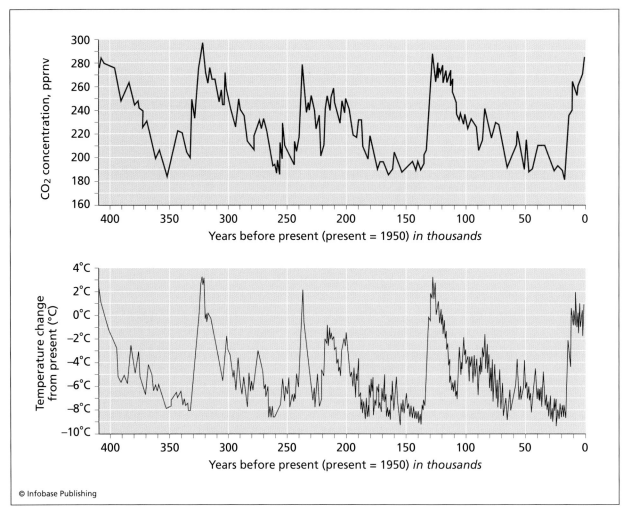

Temperature and CO_2 changes in past 400,000 years based on Antarctic ice cores

© Infobase Publishing

advance and retreat of the Northern and Southern Hemisphere ice sheets in the past few million years. In the past two million years alone the Earth has seen the ice sheets advance and retreat approximately 20 times. The climate record as deduced from ice-core records from Greenland and isotopic tracer studies from deep ocean, lake, and cave sediments suggest that the ice builds up gradually over periods of about 100,000 years, then retreats rapidly over a period of decades to a few thousand years. These patterns result from the cumulative effects of different astronomical phenomena.

Several movements are involved in changing the amount of incoming solar radiation. The Earth rotates around the Sun following an elliptical orbit, and the shape of this elliptical orbit is known as its eccentricity. The eccentricity changes cyclically with time with a period of 100,000 years, alternately bringing the Earth closer to and farther from the Sun in summer and in winter. This 100,000-year cycle is about the

same as the general pattern of glaciers advancing and retreating every 100,000 years in the past 2 million years, suggesting that this is the main cause of variations within the present-day ice age. Presently the Earth's orbit is in a period of low eccentricity (~3 percent), and this yields a seasonal change in solar energy of ~7 percent. When the eccentricity is at its peak (~9 percent), "seasonality" reaches ~20 percent. In addition a more eccentric orbit changes the length of seasons in each hemisphere by changing the length of time between the vernal and autumnal equinoxes.

The Earth's axis is presently tilting by 23.5°N/S away from the orbital plane, and the tilt varies between 21.5°N/S and 24.5°N/S. The tilt, also known as obliquity, changes by plus or minus 1.5°N/S from a tilt of 23°N/S every 41,000 years. When the tilt is greater, there is greater seasonal variation in temperature. For small tilts winters would tend to be milder and summers cooler. This would lead to more glaciation.

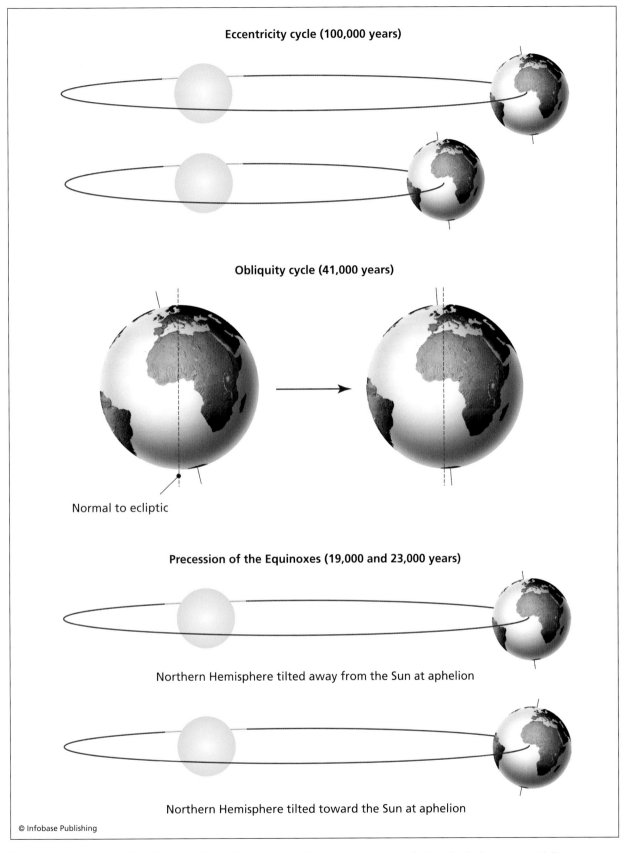

Eccentricity cycle (100,000 years)

Obliquity cycle (41,000 years)

Normal to ecliptic

Precession of the Equinoxes (19,000 and 23,000 years)

Northern Hemisphere tilted away from the Sun at aphelion

Northern Hemisphere tilted toward the Sun at aphelion

© Infobase Publishing

Orbital variations that lead to variation in the amount of incoming solar radiation, including eccentricity, obliquity (tilt), and precession of the equinoxes

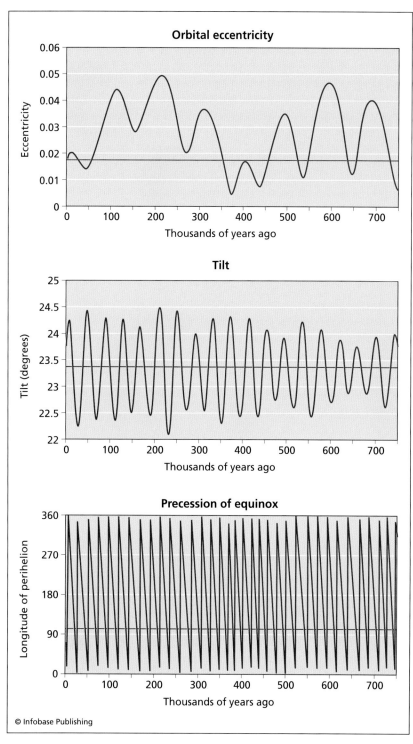

© Infobase Publishing

Milankovitch cycles related to changes in eccentricity, obliquity (tilt), and precession of the equinoxes. All of these effects act together, and the curves need to be added to each other to obtain a true accurate curve of the climate variations because all these effects act at the same time.

Wobble of the rotation axis describes a motion much like a top rapidly spinning and rotating with a wobbling motion, such that the direction of tilt toward or away from the Sun changes, even though the tilt amount stays the same. This wobbling phe-

nomenon is known as precession of the equinoxes, and it places different hemispheres closest to the Sun in different seasons. This precession changes with a double cycle, with periodicities of 23,000 years and 19,000 years. Presently the precession of the equinoxes is such that the Earth is closest to the Sun during the Northern Hemisphere winter. Due to precession the reverse will be true in ~11,000 years. This will give the Northern Hemisphere more severe winters.

Because each of these astronomical factors acts on different timescales, they interact in a complicated way (Milankovitch cycles, as described previously). Understanding these cycles, climatologists can make predictions of where the Earth's climate is heading, whether the planet is heading into a warming or cooling period, and whether populations need to plan for sea-level rise, desertification, glaciation, sea-level drop, floods, or droughts. When all the Milankovitch cycles (alone) are taken into account, the present trend should be toward a cooler climate in the Northern Hemisphere, with extensive glaciation. The Milankovitch cycles may help explain the advance and retreat of ice over periods of 10,000 to 100,000 years. They do not explain what caused the Ice Age in the first place.

The pattern of climate cycles predicted by Milankovitch cycles is further complicated by other factors that change the climate of the Earth. These include changes in thermohaline circulation, changes in the amount of dust in the atmosphere, changes caused by reflectivity of ice sheets, changes in concentration of greenhouse gases, changing characteristics of clouds, and even the glacial rebound of land that was depressed below sea level by the weight of glaciers.

Milankovitch cycles have been invoked to explain the rhythmic repetitions of layers in some sedimentary rock sequences. The cyclical orbital variations cause cyclical climate variations, which in turn are reflected in the cyclical deposition of specific types of sedimentary layers in sensitive environments. There are numerous examples of sedimentary sequences where stratigraphic and age control are sufficient to detect cyclical variation on the timescales of Milankovitch cycles; studies of these layers have proven consistent with a control of sedimentation by the planet's orbital variations. Some examples of Milankovitch-forced sedimentation have been documented from the Dolomite Mountains of Italy, the Proterozoic Rocknest Formation of northern Canada, and from numerous coral reef environments.

Predicting the future climate on Earth involves very complex calculations, including inputs from the long- and medium-term effects described in this entry, and some short-term effects such as sudden changes caused by human inputs of greenhouse gases into the atmosphere, and effects such as unpredicted volcanic eruptions. Nonetheless, most climate experts expect that the planet will continue to warm on the hundreds-of-years timescale. But judging by the recent geological past, we can reasonably expect that the planet could be suddenly plunged into another ice age, perhaps initiated by sudden changes in ocean circulation, following a period of warming. Climate is one of the major drivers of mass extinction, so the question remains whether the planet will be able to cope with rapidly fluctuating temperatures, dramatic changes in sea level, and enormous shifts in climate and agriculture belts.

Thermohaline Circulation and Climate

Variations in formation and circulation of ocean water may cause some of the thousands of years to decadal scale variations in climate. Cold water forms in the Arctic and Weddell Seas. This cold, salty water is denser than other water in the ocean, so it sinks to the bottom and gets ponded behind seafloor topographic ridges, periodically spilling over into other parts of the oceans. The formation and redistribution of North Atlantic cold bottom water accounts for about 30 percent of the solar energy budget input to the Arctic Ocean every year. Eventually this cold bottom water works its way to the Indian and Pacific Oceans, where it upwells, gets heated, and returns to the North Atlantic. Thermohaline circulation is the vertical mixing of seawater driven by density differences caused by variations in temperature and salinity. Variations in temperature and salinity are found in waters that occupy different ocean basins and those found at different levels in the water column. When the density of water at one level is greater than or equal to that below that level, the water column becomes unstable and the denser water sinks, displacing the deeper, less-dense waters below. When the dense water reaches the level at which it is stable, it tends to spread out laterally and form a thin sheet, causing intricately stratified ocean waters. Thermohaline circulation is the main mechanism responsible for the movement of water out of cold polar regions and exerts a strong influence on global climate. The upward movement of water in other regions balances the sinking of dense cold water, and these upwelling regions typically bring deep water, rich in nutrients, to the surface. Thus regions of intense biological activity are often associated with upwelling regions.

The coldest water on the planet is formed in the polar regions, with large quantities of cold water originating off the coast of Greenland and in the Weddell Sea of Antarctica. The planet's saltiest ocean water is found in the Atlantic Ocean, and this is moved northward by the Gulf Stream. As this water moves near Greenland it is cooled, then sinks to flow as a deep cold current along the bottom of the western North Atlantic. The cold water of the Weddell Sea is the densest on the planet, where surface waters are cooled to -35.4°F (-1.9°C), then sink to form a cold current that moves around Antarctica. Some of this deep cold water moves northward into all three major ocean basins, mixing with other waters and warming slightly. Most of these deep ocean currents move at a few to ten centimeters per second.

Presently, the age of bottom water in the equatorial Pacific is 1,600 years, and in the Atlantic it is 350 years. Glacial stages in the North Atlantic correlate with the presence of older cold bottom waters, approximately twice the age of the water today. This suggests that the thermohaline circulation system was only half as effective at recycling water during recent glacial stages, with less cold bottom water being produced during the glacial periods. These changes in production of cold bottom water may in turn be driven by changes in the North American ice sheet, perhaps itself driven by 23,000-year orbital (Milankovitch) cycles. Such a growth in the ice sheet would cause the polar front to shift southward, decreasing the inflow of cold saline surface water into the system required for efficient thermohaline circulation. Several periods of glaciation in the past 14,500 years (known as the Dryas) are thought to have been caused by sudden, even catastrophic injections of glacial meltwater into the North Atlantic, which would decrease the salinity and hence density of the surface water. This in turn would prohibit the surface water from sinking to the deep ocean, inducing another glacial interval.

Shorter-term decadal variations in climate in the past million years is indicated by so-called Heinrich Events, defined as specific intervals in the sedimentary record showing ice-rafted debris in the North Atlantic. These periods of exceptionally large iceberg discharges reflect that decadal-scale sea surface and atmospheric cooling are related to thickening of the North American ice sheet followed by ice-stream surges associated with the discharge of the icebergs. These events flood the surface waters with low-salinity freshwater, leading to a decrease in flux in the cold-bottom waters, and hence a short-period global cooling.

Changes in the thermohaline circulation rigor have also been related to other global climate changes. Droughts in the Sahel and elsewhere are correlated with periods of ineffective or reduced thermohaline circulation, because this reduces the amount of water drawn into the North Atlantic, in turn cooling surface waters and reducing the amount of evaporation. Reduced thermohaline circulation also reduces the amount of water that upwells in the equatorial regions, in turn decreasing the amount of moisture transferred to the atmosphere and reducing precipitation at high latitudes.

Atmospheric levels of greenhouse gases such as CO_2 and atmospheric temperatures show a correlation to variations in the thermohaline circulation patterns and production of cold-bottom waters. CO_2 is dissolved in warm surface water and transported to cold-surface water, which acts as a sink for the CO_2. During times of decreased flow from cold, high-latitude surface water to the deep ocean reservoir, CO_2 can build up in the cold polar waters, removing itself from the atmosphere and decreasing global temperatures. In contrast, when the thermohaline circulation is vigorous, cold oxygen-rich surface waters downwell, dissolving buried CO_2 and even carbonates, releasing this CO_2 into the atmosphere and increasing global temperatures.

The present-day ice sheet in Antarctica grew in the Middle Miocene, related to active thermohaline circulation that caused prolific upwelling of warm water that put more moisture in the atmosphere, falling as snow on the cold southern continent. The growth of the southern ice sheet increased the global atmospheric temperature gradients, which in turn increased the desertification of midlatitude continental regions. The increased temperature gradient also induced stronger oceanic circulation, including upwelling, removal of CO_2 from the atmosphere, lowering global temperatures, and bringing on late Neogene glaciations.

Ocean-bottom topography exerts a strong influence on dense bottom currents. Ridges deflect currents from one part of a basin to another and may restrict access to other regions, whereas trenches and deeps may focus flow from one region to another.

El Niño and the Southern Oscillation (ENSO)

El Niño–Southern Oscillation is the name given to one of the better-known variations in global atmo-

spheric circulation patterns. Global oceanic and atmospheric circulation patterns undergo frequent shifts that affect large parts of the globe, particularly those arid and semiarid parts affected by Hadley Cell circulation. It is now understood that fluctuations in global circulation can account for natural disasters including the dust bowl days of the 1930s in the midwestern United States. Similar global climate fluctuations may explain the drought, famine, and desertification of parts of the Sahel, and the great famines of Ethiopia and Sudan in the 1970s and 1980s.

The secondary air circulation phenomenon known as the El Niño-Southern Oscillation can also profoundly influence the development of drought conditions and desertification of stressed lands. Hadley cells migrate north and south with summer and winter, shifting the locations of the most intense heating. Several zonal oceanic-atmospheric feedback systems influence global climate, but the most influential is the Austro-Asian system. In normal Northern Hemisphere summers the location of the most intense heating in Austral-Asia shifts from equatorial regions to the Indian subcontinent along with the start of the Indian monsoon. Air is drawn onto the subcontinent, where it rises and moves outward to Africa and the central Pacific. In Northern Hemisphere winters the location of this intense heating shifts to Indonesia and Australia, where an intense low-pressure system develops over this mainly maritime region. Air is sucked in, moves upward, and flows back out at tropospheric levels to the east Pacific. High pressure develops off the coast of Peru in both situations, because cold, upwelling water off the coast there causes the air to cool, inducing atmospheric downwelling. The pressure gradient set up causes easterly trade winds to blow from the coast of Peru across the Pacific to the region of heating, causing warm water to pile up in the Coral Sea off the northeast coast of Australia. This also causes the sea level to be slightly depressed off the coast of Peru, and more cold water upwells from below to replace the lost water. This positive-feedback mechanism is rather stable—it enhances the global circulation, as more cold water upwelling off Peru induces more atmospheric downwelling, and more warm water piling up in Indonesia and off the coast of Australia causes atmospheric upwelling in that region.

This stable linked atmospheric and oceanic circulation breaks down and becomes unstable every two to seven years, probably from some inherent chaotic behavior in the system. At these times, the Indonesian-Australian heating center migrates eastward, and the buildup of warm water in the western Pacific is no longer held back by winds blowing westward across the Pacific. This causes the elevated warm water mass to collapse and move eastward across the Pacific, where it typically appears off the coast of Peru by the end of December. The El Niño-Southern Oscillation (ENSO) events occur when this warming is particularly strong, with temperatures increasing by 40–43°F (22–24°C) and remaining high for several months. This phenomenon is also associated with a reversal of the atmospheric circulation around the Pacific such that the dry downwelling air is located over Australia and Indonesia, and the warm upwelling air is located over the eastern Pacific and western South America.

The arrival of El Niño is not good news in Peru, since it causes the normally cold upwelling and nutrient-rich water to sink to great depths, and the fish either must migrate to better feeding locations or die. The fishing industry collapses at these times, as does the fertilizer industry that relies on the bird guano normally produced by birds (that eat fish and anchovies) that also die during El Niño events. Warm moist air replaces the normally cold dry air, and the normally dry or desert regions of coastal Peru receive torrential rains with associated floods, landslides, death, and destruction. Shoreline erosion is accelerated in El Niño events, because the warm water mass that moved in from across the Pacific raises sea levels by 4–25 inches (10–60 cm), enough to cause significant damage.

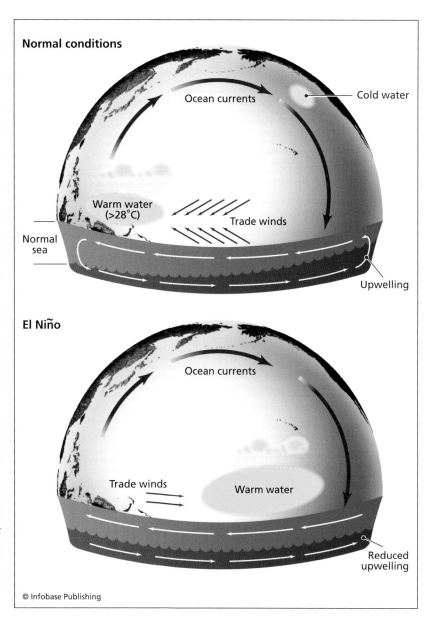

Schematic diagrams of the different patterns of ocean and air circulation over the Pacific associated with El Niño and normal conditions

The end of ENSO events also leads to abnormal conditions, in that they seem to turn on the "normal" type of circulation in a much stronger way than is normal. The cold upwelling water returns off Peru with such a ferocity that it may move northward, flooding a 1–2° band around the equator in the central Pacific ocean with water that is as cold as 68°F (20°C). This phenomenon is known as *La Niña* ("the girl" in Spanish).

The alternation between ENSO, La Niña, and normal ocean-atmospheric circulation has profound effects on global climate and the migration of different climate belts on yearly to decadal timescales, and is thought to account for about a third of all the variability in global rainfall. ENSO events may cause flooding in the western Andes and southern California, and a lack of rainfall in other parts of South America, including Venezuela, northeastern Brazil, and southern Peru. It may change the climate, causing droughts in Africa, Indonesia, India, and Australia, and is thought to have caused the failure of the Indian monsoon in 1899 that resulted in regional famine with the deaths of millions of people. Recently, the seven-year cycle of floods on the Nile has been linked to ENSO events, and famine and desertification in the Sahel, Ethiopia, and Sudan can be attributed to these changes in global circulation as well.

Major Volcanic Eruptions and Climate Change

Some of the larger, more explosive volcanic eruptions that the planet has witnessed in the past few hundred years have ejected large amounts of ash and finer particles called aerosols into the atmosphere and stratosphere, and it may take years for these particles to settle down to Earth. They get distributed about the planet by high-level winds, and they block some of the Sun's rays, which lowers global temperatures. This happens because particles and aerosol gases in the upper atmosphere tend to scatter sunlight back to space, lowering the amount of incoming solar energy. In contrast, particles that get injected only into the lower atmosphere absorb sunlight and contribute to greenhouse warming. A side effect is that the extra particles in the atmosphere also produce more spectacular sunsets and rises, as does extra pollution in the atmosphere. These effects were readily observed after the 1991 eruption of Mount Pinatubo in the Philippines, which spewed more than 172 billion cubic feet (5 billion m^3) of ash and aerosols into the atmosphere, causing global cooling for two years after the eruption. Even more spectacularly, the 1815 eruption of Tambora in Indonesia caused three days of total darkness for approximately 300 miles (500 km) from the volcano, and it initiated the famous "year without a summer" in Europe, because the ash from this eruption lowered global temperatures by more than a degree.

The amounts of gases and small airborne particles released by large volcanic eruptions such as Pinatubo, and even Tambora, are dwarfed by the amount of material placed into the atmosphere during some of Earth's most massive eruptions, known as flood basalt events. No flood basalts have been formed on Earth for several tens of millions of years, which is a good thing, since their eruption may be associated with severe changes in climate.

Scattered around the globe are numerous anomalously thick accumulations of dark lava, variously known as flood basalts, traps, or large igneous provinces. These vast outpourings of lava have different ages and represent the largest known volcanic episodes on the planet in the past several hundred million years. These deposits include continental flood basalt provinces, anomalously thick and topographically high seafloor known as oceanic plateaus, and some volcanic rifted passive margins. During eruption of these vast piles of volcanic rock, the Earth moved more material and energy from its interior in extremely short time periods than during the entire intervals between the massive volcanic events. Such large amounts of volcanism also released large amounts of volcanic gases into the atmosphere, with serious implications for global temperatures and climate, and may have contributed to some global mass extinctions. Many are associated with periods of global cooling where volcanic gases reduce the amount of incoming solar radiation and thereby bring on volcanic winters.

The largest continental flood basalt province in the United States is the Columbia River flood basalt in Washington, Oregon, and Idaho. The Columbia River flood basalt province is 6–17 million years old and contains an estimated 1,250 cubic miles (5,210 km^3) of basalt. Individual lava flows erupted through fissures or cracks in the crust, then flowed laterally across the plain for up to 400 miles (644 km).

The 66 million-year-old Deccan flood basalts, also known as traps, cover a large part of western India and the Seychelles. They are associated with the breakup of India from the Seychelles during the opening of the Indian Ocean. Slightly older flood basalts (90–83 million years) are associated with the break away of Madagascar from India. The volume of the Deccan traps is estimated at 5,000,000 cubic miles (20,840,000 km^3), and the volcanics are thought to have been erupted in about 1 million years, starting slightly before the great Cretaceous-Tertiary extinction. Most experts now agree that the gases released during this period of flood basalt volcanism aggravated the global biosphere to such an extent that many marine organisms were forced into extinction, and many others were stressed. Then the planet was hit by the meteorite that formed the massive Chicxulub impact crater on the Yucatán Peninsula (Mexico), causing the mass extinction including the end of the dinosaurs.

The breakup of east Africa along the East African rift system and the Red Sea is associated with large amounts of Cenozoic (fewer than 30 million years old) continental flood basalts. Some of the older volcanic fields are located in east Africa in the Afar region of Ethiopia, south into Kenya and Uganda, and north across the Red Sea and Gulf of Aden into Yemen and Saudi Arabia. These volcanic piles are overlain by younger (fewer than 15 million year old) flood basalts that extend both farther south into Tanzania and farther north through central Arabia, where they are known as Harrats, and into Syria, Israel, Lebanon, and Jordan.

An older volcanic province also associated with the breakup of a continent is known as the North Atlantic Igneous Province. It formed along with the breakup of the North Atlantic Ocean at 62–55 million years ago, and includes both onshore and offshore volcanic flows and intrusions in Greenland, Iceland, and the northern British Isles, including most of the Rockall Plateau and Faeroes Islands. In the south Atlantic a similar 129–134 million-year-old flood basalt was split by the opening of the ocean, and now has two parts. In Brazil the flood lavas are known as the Paraná basalts, and in Namibia and Angola of west Africa as the Etendeka basalts.

The Caribbean Ocean floor is one of the best examples of an oceanic plateau, with other major examples including the Ontong-Java Plateau, Manihiki Plateau, Hess Rise, Shatsky Rise, and Mid Pacific Mountains. All of these oceanic plateaus contain between six- and 25-mile thick piles of volcanic and subvolcanic rocks representing huge outpourings of lava. The Caribbean seafloor preserves 5–13 mile (8–21 km) thick oceanic crust formed before about 85 million years ago in the eastern Pacific Ocean. This unusually thick ocean floor was transported eastward by plate tectonics, where pieces of the seafloor collided with South America as it passed into the Atlantic Ocean. Pieces of the Caribbean oceanic crust are now preserved in Colombia, Ecuador, Panama, Hispaniola, and Cuba, and some scientists estimate that the Caribbean oceanic plateau may have once been twice its present size. In either case it represents a vast outpouring of lava that would have been associated with significant outgassing, with possible consequences for global climate and evolution.

The western Pacific Ocean basin contains several large oceanic plateaus, including the 20-mile (32-km) thick crust of the Alaskan-sized Ontong-Java Plateau, the largest outpouring of volcanic rocks on the planet. It apparently formed in two intervals, at 122 and 90 million years ago, respectively, entirely within the ocean, and represents magma that rose in a plume from deep in the mantle and erupted on the seafloor. It is estimated that the volume of magma erupted in the first event was equivalent to that of all the magma being erupted at midocean ridges at the present time. Sea levels rose by more than 30 feet (9 m) in response to this volcanic outpouring. The gases released during these eruptions are estimated to have raised average global temperatures by 23°F (13°C).

Examples of Climate Changes Caused by Flood Basalt Volcanism

The environmental impact of the eruption of large volumes of basalt can be severe. Huge volumes of sulfur dioxide, carbon dioxide, chlorine, and fluorine are released during large basaltic eruptions. Much of this gas may get injected into the upper troposphere and lower stratosphere during the eruption process, being released from eruption columns that reach 2–8 miles (3–13 km) in height. Carbon dioxide, a greenhouse gas, can cause global warming, whereas sulfur dioxide (and hydrogen sulfate) has the opposite effect and can cause short-term cooling. Many of the episodes of volcanism preserved in these large igneous provinces were rapid, repeatedly releasing enormous quantities of gases over periods of fewer than 1 million years, and released enough gas to change significantly the climate more rapidly than organisms could adapt. For instance, one eruption of the Columbia River basalts is estimated to have released 9,000 million tons of sulfur dioxide, and thousands of millions of tons of other gases, compared with the eruption of Mount Pinatubo in 1991, which released about 20 million tons (18 million tonnes) of sulfur dioxide.

The Columbia River basalts of the Pacific Northwest are instructive about how flood basalts can influence climate. These lavas continued erupting for years at a time, for approximately a million years. During this time the gases released would be equivalent to that of Mt. Pinatubo, every week, over periods maintained for decades to thousands of years at a time. The atmospheric consequences are sobering. Sulfuric acid aerosols and acid from the fluorine and chlorine would form extensive poisonous acid rain, destroying habitats and making waters uninhabitable for some organisms. At the very least the environmental consequences would be such that organisms were stressed to the point that they would not be able to handle an additional environmental stress, such as a global volcanic winter and subsequent warming caused by a giant impact.

Mass extinctions have been correlated with the eruption of the Deccan flood basalts at the Cretaceous-Tertiary (K/T) boundary, and with the Siberian flood basalts at the Permian-Triassic boundary. There is still considerable debate about the relative significance of flood basalt volcanism and impacts of meteorites for extinction events, particularly at the Cretaceous-Tertiary boundary. However, most scientists would now agree that global environment was stressed shortly before the K/T boundary by volcanic-induced climate change, and then a huge meteorite hit the Yucatán Peninsula, forming the Chicxulub impact crater, causing the massive K/T boundary extinction and the death of the dinosaurs.

The Siberian flood basalts cover a large area of the Central Siberian Plateau northwest of Lake Baikal. They are more than half a mile (1 km) thick over an area of 210,000 square miles (543,900 km²) but have been significantly eroded from an estimated volume of 1,240,000 cubic miles (3,211,600 km³). They were erupted over an extraordinarily short period of fewer than 1 million years 250 million years ago, at the Permian-Triassic boundary. They are remarkably coincident in time with the major Permian/Triassic extinction, implying a causal link. The Permian/Triassic boundary at 250 million years ago marks the greatest extinction in Earth history, where 90 percent of marine species and 70 percent of terrestrial vertebrates became extinct. It has been postulated that the rapid volcanism and degassing released enough sulfur dioxide to cause a rapid global cooling, inducing a short ice age with associated rapid fall of sea level. Soon after the ice age took hold, the effects of the carbon dioxide took over and the atmosphere heated to cause global warming. The rapidly fluctuating climate postulated to have been caused by the volcanic gases is thought to have killed off many organisms that were simply unable to cope with the wildly fluctuating climate extremes.

The close relationship between massive volcanism and changes in climate that have led to mass extinctions shows how quickly life on Earth can change. The effects of massive global volcanism are much greater than any changes so far caused by humans, and operate faster than other plate tectonic and supercontinent-related changes.

How Fast Can Climate Change

Understanding how fast climate can shift from a warm period to a cold, or cold to a warm, is difficult. The record of climate indicators is incomplete and difficult to interpret. Only 18,000 years ago the planet was in the midst of a major glacial interval, and since then global average temperatures have risen 16°F (10°C) and are still rising, perhaps at a recently accelerated rate from human contributions to the atmosphere. Still, recent climate work is revealing that there are some abrupt transitions in the slow warming, in which there are major shifts in some component of the climate, where the shift may happen on scales of 10 years or fewer.

One of these abrupt transitions seems to affect the circulation pattern in the North Atlantic Ocean, where the ocean currents formed one of two different stable patterns or modes, with abrupt transitions occurring when one mode switches to the other. In the present pattern the warm waters of the Gulf Stream come out of the Gulf of Mexico and flow along the eastern seaboard of the United States, part of the British Isles, to the Norwegian Sea. This warm current is largely responsible for the mild climate of the British Isles and northern Europe. In the second mode the northern extension of the Gulf Stream is weakened by a reduction in salinity of surface waters from sources at high latitudes in the North Atlantic. The fresher water has a source in increased melting from the polar ice shelf, Greenland, and northern glaciers. With less salt, seawater is less dense and less able to sink during normal wintertime cooling.

Studies of past switches in the circulation modes of the North Atlantic reveal that the transition from one mode of circulation to the other can occur over a period of only five to 10 years. These abrupt transitions are apparently linked to increase in the release of icebergs and freshwater from continental glaciers, which upon melting contribute large volumes of freshwater into the North Atlantic, systematically reducing the salinity. The Gulf Stream presently seems on the verge of failure, or of switching modes from mode 1 to 2, and historical records show that this switch can be very rapid. If this predicted switch occurs, northern Europe and the United Kingdom may experience a significant and dramatic cooling of their climate, instead of the warming many fear.

See also ATMOSPHERE; CARBON CYCLE; CLIMATE; GLOBAL WARMING; GREENHOUSE EFFECT; ICE AGES; METEOROLOGY; MILANKOVITCH CYCLES; PALEO-CLIMATOLOGY; SEA-LEVEL RISE; THERMOHALINE CIRCULATION.

FURTHER READING

Ahrens, C. D. *Meteorology Today: An Introduction to Weather, Climate, and the Environment.* 6th ed. Pacific Grove, Calif.: Brooks/Cole, 2000.

Dawson, A. G. *Ice Age Earth,* London: Routledge, 1992.

Douglas, B., M. Kearney, and S. Leatherman. *Sea-Level Rise: History and Consequence.* San Diego, Calif.: Academic Press, International Geophysics Series 75, 2000.

Intergovernmental Panel on Climate Change 2007. *Climate Change 2007: The Physical Science Basis. Contributions of Working Group I to the Fourth Assessment Report of the Intergovernmental Panel on Climate Change.* Edited by S. Solomon, D. Qin, M. Manning, Z. Chen, M. Marquis, K. B. Averyt, M. Tignor, and H. L. Miller. Cambridge: Cambridge University Press, 2007. Also available online. URL: http://www.ipcc.ch/index.htm. Accessed October 10, 2008.

National Aeronautic and Space Administration (NASA). "Earth Observatory." Available online. URL: http://earthobservatory.nasa.gov/. Accessed October 9, 2008, updated daily.

U.S. Environmental Protection Agency. Climate Change homepage. Available online. URL: http://www.epa.gov/climatechange/. Updated September 9, 2008.

Cloud, Preston (1912–1991) *American Historical Geologist, Geobiologist* Preston Ercelle Cloud Jr. was an eminent geobiologist and paleontologist who contributed important observations and interpretations that led to greater understanding of the evolution of the atmosphere, oceans, and crust of the Earth, and most important, to understanding the evolution of life on the planet.

Born in West Upton, Massachusetts, on September 26, 1912, as a child Cloud moved to Waynesboro, Pennsylvania, where he developed a keen sense of the outdoors and the rolling hills of the Appalachians. He joined the U.S. Navy from 1930 to 1933, then enrolled in George Washington University in Washington, D.C., where he cultivated contacts at the

National Museum of Natural History. As an undergraduate student Cloud developed a solid knowledge of paleontology, learning much especially about brachiopods from the collections at the National Museum in Washington, D.C.

Preston Cloud continued his education at Yale University and received a Ph.D. in 1940 for a study of Paleozoic brachiopods. From there he moved to Missouri School of Mines in Rolla, but then returned to Yale as a Research Fellow from 1941 to 1942. During World War II Cloud was called to duty with the U.S. Geological Survey, where he worked with the wartime strategic minerals program, first mapping manganese deposits in Maine, then investigating bauxite in Alabama. After this Cloud studied the Ellenburger Limestone—an important oil reservoir—from the Lower Paleozoic section of Texas, and made accurate descriptions of the stratigraphic and paleontologic relationships in this unit.

In 1946 Cloud took a position as an assistant professor of paleontology at Harvard University but in 1948 returned to the U.S. Geological Survey to map parts of Saipan Island in the Mariana Islands in the Pacific. This work led him to publish many papers on modern carbonate and coral reef systems, including his landmark works on evolution, in which he proposed that complex, multicellular organisms evolved from many different ancestors about 700 million years ago. Through his studies of geochemical processes Cloud linked the rapid evolution of these species to a change in atmospheric chemistry in which the oxygen levels in the atmosphere climbed rapidly, helping the organisms expand into available ecological niches. Cloud was promoted to chief paleontologist with the U.S. Geological Survey from 1949 to 1959, and the department grew dramatically under his guidance.

After he resigned as head paleontologist at the survey, Cloud studied the continental shelves and coastal zone, expanding the knowledge of these regions dramatically and leading to oil exploration on the continental shelves. He then accepted a job as chairman of the Department of Geology and Geophysics at the University of Minnesota, and organized a new multidisciplinary approach to earth sciences by forming the School of Earth Sciences. While at Minnesota Cloud concentrated on the Precambrian and the first 86 percent of Earth history, the origin and development of life, and studied Precambrian outcrops from around the world in this context. Cloud became world-famous for his studies of Precambrian carbonate rocks and their fossil assemblages, which consist mostly of stromatolites, and ideas about the origin and evolution of life.

In 1965 Cloud moved to the University of California, Los Angeles, then in 1968 he moved to the Santa Barbara campus. In 1979 he retired but remained active in publishing books on life on the planet, and was also active on campus. Preston Cloud emphasized complex interrelationships among biological, chemical, and physical processes throughout Earth history. His work expanded beyond the realm of rocks and fossils, and he wrote about the limits of the planet for sustaining the exploding human population. He recognized that limited material, food, and energy resources with the expanding human activities could lead the planet into disaster. One of his most famous works in this field was his *Oasis in Space*. Cloud was elected a member of the Academy of Sciences and was active for 30 years. In 1976 Preston Cloud was awarded the Penrose Medal by the Geological Society of America, and in 1977 he was awarded the Charles Doolittle Walcott Medal by the National Academy of Sciences. The Preston Cloud Laboratory at the University of California, Santa Barbara, is dedicated to the study of pre-Phanerozoic life on Earth.

See also HISTORICAL GEOLOGY; PALEONTOLOGY, SEDIMENTARY ROCK, SEDIMENTATION.

FURTHER READING
Cloud, Preston. "Life, Time, History and Earth Resources." *Terra Cognita* 8 (1988): 211.
———. *Oasis in Space: Earth History from the Beginning.* New York: W. W. Norton, 1988.
———. "Aspects of Proterozoic Biogeology." *Geological Society of America Memoir* 161 (1983): 245–251.
———. "A Working Model of the Primitive Earth." *American Journal of Science* 272 (1972): 537–548.

clouds Clouds are visible masses of water droplets or ice crystals suspended in the lower atmosphere, generally confined to the troposphere. The water droplets and ice crystals condense from water vapor around small dust, pollen, salt, ice, or pollution particles that aggregate into cloud formations, classified according to their shape and height in the atmosphere. Luke Howard, an English naturalist, suggested the classification system still widely used today in 1803. He suggested Latin names based on 10 genera, then further divided into species. In 1887 the British naturalist Ralph Abercromby and H. Hildebrand Hildebrandsson of Sweden further divided the clouds into high, middle, and low-level types, as well as clouds that form over significant vertical distances. The basic types of clouds include the heaped cumulus, layered stratus, and wispy cirrus. If rain is falling from a cloud, the term *nimbus* is added, as in cumulonimbus, the common thunderhead cloud.

High clouds form above 19,685 feet (6,000 m) and are generally found at mid to low latitudes.

The air at this elevation is cold and dry, so the clouds consist of ice crystals, and appear white to the observer on the ground except at sunrise and sunset. The most common high clouds are the cirrus—thin, wispy clouds typically blown into thin, horsehairlike streamers by high winds. Prevailing high-level winds blow most cirrus clouds from west to east, a sign of generally good weather. Cirrocumulus clouds are small, white puffy clouds that sometimes line up in ripplelike rows and at other times form individually. Their appearance over large parts of the sky is often described as a Mackeral sky, because of the resemblance to fish scales. Cirrostratus are thin, sheetlike clouds that typically cover the entire sky. They are so thin that the Sun, Moon, and some stars can be seen through them. They are composed of ice crystals, and light that refracts through these clouds often forms a halo or sun dogs. These high clouds often form in front of an advancing storm and typically foretell of rain or snow in 12–24 hours.

Middle clouds form between 6,560 and 22,965 feet (2,000 and 7,000 m), generally in middle latitudes. They are composed mostly of water droplets, with ice crystals in some cases. Altocumulus clouds are gray, puffy masses that often roll out in waves,

Cirrus clouds over coast range at Purisima Creek Redwoods, Bay Area, California *(NOAA/Department of Commerce)*

with some parts appearing darker than others. Altocumulus are usually less than 0.62 miles (1 km) thick. They form with rising air currents at cloud level, and a morning appearance often predicts thunderstorms by late afternoon. Altostratus are thin, blue-gray clouds that often cover the entire sky, and the sun may shine dimly through, appearing as a faint, irregular disk. Altostratus clouds often form in front of storms that bring regional steady rain.

Low clouds have bases that may form below 6,650 feet (2,000 m) and are usually composed entirely of water droplets. In cold weather they may contain ice and snow. Nimbostratus are the dark gray, rather uniform-looking clouds associated with steady light to moderate rainfall. Rain from the nimbostratus clouds often causes the air to become saturated with water, and a group of thin, ragged clouds that move rapidly with the wind may form. These are known as stratus fractus, or scud clouds. Stratocumulus clouds are low, lumpy-looking clouds that form rows or other patterns, with clear sky visible between the cloud rows. The Sun may form brilliant streaming rays known as crepuscular rays through these clouds. Stratus clouds have a uniform gray appearance and may cover the sky, resembling fog but not touching the ground. They commonly appear near the seashore, especially in summer months.

Some clouds form over a significant range of atmospheric levels. Cumulus are flat-bottomed, puffy clouds with irregular, domal, or towering tops. Their bases may be lower than 3,280 feet (1,000 m). On warm summer days small cumulus clouds may form in the morning and develop significant vertical growth by the afternoon, creating a towering cumulus or cumulus congestus cloud. These may continue to develop further into the giant cumulonimbus, giant thunderheads with bases that may be as low as a few hundred meters, and tops extending to more than 39,370 feet (12,000 m) in the tropopause. Cumulonimbus clouds release tremendous amounts of energy in the atmosphere and may be associated with high winds, vertical updrafts and downdrafts, lightning, and tornadoes. The lower parts of these giant clouds are made of water droplets, the middle parts may contain both water and ice, whereas the tops may consist entirely of ice crystals.

Many types of unusual clouds form in different situations. Plieus clouds may form over rising cumulus tops, looking like a halo or fog around the cloud peak. Banner clouds form over and downwind of high mountain tops, sometimes resembling steam coming out of a volcano. Lenticular clouds form wavelike figures from high winds moving over mountains, and may form elongate, pancakelike shapes. Unusual and even scary-looking mammatus clouds form bulging, baglike sacks underneath some

Cumulus cloud over Arizona desert *(Aleksander Bochenek, Shutterstock, Inc.)*

cumulonimbus clouds, forming when the sinking air is cooler than the surrounding air. Mammatus-like clouds may also form underneath clouds of volcanic ash. Finally, jet airplanes produce condensation trails when water vapor from the jet's exhaust mixes with the cold air, which becomes suddenly saturated with water and forms ice crystals. Pollution particles from the exhaust may provide the nuclei for the ice. In dry conditions condensation, or con trails, will evaporate quickly, but in more humid conditions the con trails may persist as cirruslike clouds. With the growing numbers of jet flights in the past few decades, con trails have rapidly become a significant source of cloudiness, contributing to the global weather and perhaps climate.

Clouds greatly influence the Earth's climate. They efficiently reflect short-wavelength radiation from the Sun back into space, cooling the planet. But since they are composed of water, they also stop the longer-wavelength radiation from escaping, causing a greenhouse effect. Together these two apparently opposing effects of clouds strongly influence the climate of the Earth. In general the low- and middle-level clouds cool the Earth, whereas abundant high clouds tend to warm the Earth with the greenhouse effect.

See also ATMOSPHERE; CLIMATE; CLIMATE CHANGE; ENERGY IN THE EARTH SYSTEM.

FURTHER READING

Schaefer, Vincent J., and John A. Day. *A Field Guide to the Atmosphere: A Peterson Field Guide.* Boston: Houghton Mifflin, 1981.

comet   Comets are bodies of ice, dust, and rock that orbit the Sun and exhibit a coma (or atmosphere) extending away from the Sun as a tail when they are close to the Sun. They have orbital periods that range from a few years to a few hundred or even thousands of years. Short-period comets have orbital periods of fewer than 200 years, and most of these orbit in the plane of the ecliptic in the same direction as the planets. Their orbits take them past the orbit of Jupiter at aphelion, and near the Sun at perihelion. Long-period comets have highly elongated or eccentric orbits, with periods longer than 200 years and extending to thousands or perhaps even millions of years. These comets range far beyond the orbits of the outer planets, although they remain gravitationally bound to the Sun. Another class of comets, called single-apparition comets, have a hyperbolic trajectory that sends them past the inner solar system only once, then they are ejected from the solar system.

Before late 20th-century space probes collected data on comets, comets were thought to be composed primarily of ices and to be lone wanderers of the solar system. Now, with detailed observations, it is clear that comets and asteroids are transitional in nature, both in composition and in orbital character. Comets are now known to consist of rocky cores with ices around them or in pockets, and many have an organic-rich dark surface. Many asteroids are also made of similar mixtures of rocky material with pockets of ice. There are so many rocky/icy bodies in the outer solar system in the Kuiper belt and Oort Cloud that comets are now regarded as the most abundant type of bodies in the universe. There may be one trillion comets in the solar system, of which only about 3,350 have been cataloged. Most are long-period comets, but several hundred short-period comets are known as well.

The heads of comets can be divided into several parts, including the nucleus; the coma, or gaseous rim from which the tail extends; and a diffuse cloud of hydrogen. The heads of comets can be quite large, some larger than moons or other objects including Pluto. Most cometary nuclei range between 0.3 and 30 miles (0.5–50 km) in diameter and consist of a mixture of silicate rock, dust, water ice, and other frozen gases such as carbon monoxide, carbon dioxide, ammonia, and methane. Some comets contain a variety of organic compounds including methanol, hydrogen cyanide, formaldehyde, ethanol, and ethane, as well as complex hydrocarbons and amino acids. Although some comets have many organic molecules, no life is known to exist on or be derived from comets. These organic molecules make cometary nuclei some of the darkest objects in the universe, reflecting only 2–4 percent of the light that falls on their surfaces. This dark color may actually help comets absorb heat, promoting the release of gases to form the tail. Cometary tails can change in length, and can be 80 times larger than the head when the comet passes near the Sun.

As a comet approaches the Sun, it begins to emit jets of ices consisting of methane, water, and ammonia, and other ices. Modeling of the comet surface by astronomers suggests that the tails form when the radiation from the Sun cracks the crust of the comet and begins to vaporize the volatiles like carbon, nitrogen, oxygen, and hydrogen, carrying away dust from the comet in the process. The mixture of dust and gases emitted by the comet then forms a large but weak atmosphere around the comet, called the coma. The radiation and solar wind from the Sun causes this coma to extend outward away from the Sun, forming a huge tail. The tail is complex and consists of two parts. The first part contains the gases released from the comet forming an ion tail that gets elongated in a direction pointing directly away from the Sun and may extend along magnetic field lines

for more than 1 astronomical unit (9,321,000 miles; 150,000,000 km). The second part is the coma, or thin atmosphere from which the tail extends, which may become larger than the Sun. Dust released by the comet forms a tail with a slightly different orientation, forming a curved trail that follows the orbital path of the comet around the Sun.

Short-period comets originate in the Kuiper belt, whereas long-period comets originate in the Oort Cloud. Many comets are pulled out of their orbits by gravitational interactions with the Sun and planets or by collisions with other bodies. When these events place comets in orbital paths that cross the inner solar system, these comets may make close orbits to the Sun, and may also collide with planets, including the Earth.

Several space missions have recently investigated the properties of comets. These include Deep Space 1, which flew by Comet Borrelly in 2001. Comet Borrelly is a relative small comet, about 5 miles (8 km) at its longest point, and the mission showed that the comet consists of asteroid-like rocky material, along with icy plains from which the dust jets that form the coma were being emitted. In 1999 the National Aeronautics and Space Administration (NASA) launched the Stardust Comet Sample Return Mission, which flew through the tail of comet Wild 2 and collected samples of the tail in a silica gel and returned them to Earth in 2006. Scientists were expecting to find many particles of interstellar dust, or the extrasolar material that composes the solar nebula, but instead found little of this material; instead they found predominantly silicate mineral grains of Earthlike solar system composition. The samples collected revealed that comet Wild 2 is made of a bizarre mixture of material that includes some particles that formed at the highest temperatures in the early solar system, and some particles that formed at the coldest temperatures. To explain this, scientists have suggested that the rocky material that makes up the comet formed in the inner solar system during its early history, then was ejected to the outer bounds of the solar system beyond the orbit of Neptune, where the icy material was accreted to the comet. Calcium-aluminum inclusions, which represent some of the oldest, highest temperature parts of the early solar system, were also collected from the comet. One of the biggest surprises was the capture of a new class of organic material from the comet tail. These organic molecules are more primitive than any on Earth and than those found in any meteorites; they are known as polycyclic aromatic hydrocarbons. Some samples even contain alcohol. These types of hydrocarbons, thought to exist in interstellar space, may yield clues about the origin of water, oxygen, carbon, and even life on Earth.

Hale-Bopp comet over Billings, Montana, 1997 *(AP images)*

COMETS AND THE ORIGINS OF LIFE

Comets are rich in water, carbon, nitrogen, and complex organic molecules that originate deep in space from radiation-induced chemical processes. Many of the organic molecules in the coma of comets originated in the dust of the solar nebula at the time and location where the comets initially formed in the early history of the solar system. Comets are relatively small bodies that have preserved these early organic molecules in a cold, relatively pristine state. This has led many scientists to speculate that life may have come to Earth on a comet, early in the history of the planet. Clearly, comets both delivered organic material to the early Earth and also destroyed and altered organic material with the heat and shock from impacts. Numerical models of the impact of organic-rich comets with Earth show that some of the organic molecules could have survived the force of impact. The organic molecules in comets may be the source of the prebiotic molecules that led to the origins of life on Earth.

Studies of the chemistry and origin of the atmosphere and oceans suggest that the entire atmosphere, ocean, and much of the carbon on Earth, including that caught up in carbonate rocks like limestone, originated from cometary impact. The period of late impacts of comets and meteorites on Earth lasted about a billion years after the formation of Earth, before greatly diminishing in intensity. Life on Earth began during this time, hinting at a possible link between the transport of organic molecules to Earth by comets, and the development of these molecules into life. The early atmosphere of Earth was also carbon dioxide–rich (much of which came from comets), however, and organic synthesis was also occurring on Earth.

In addition to bringing organic molecules to Earth, the energy from impacts certainly destroyed much of any biosphere that attempted to establish itself on the early Earth. Even the late, very minor K-T impact at Chicxulub had major repercussions for life on Earth. Certainly the early bombardment characterized by many very large impacts would have had a more profound effect on life. Any life that had established itself on Earth would need to be sheltered from the harsh surface environment, perhaps finding refuge along the deep sea volcanic systems known as black smokers, where temperatures remained hot but stable, and nutrients in the form of sulfide compounds were used by early organisms for energy.

EXAMPLE OF A COMETARY IMPACT WITH EARTH

On June 30, 1908, a huge explosion rocked a remote area of central Siberia centered near the Podkamennaya (Lower Stony) Tunguska River, in an area now known as Krasnoyarsk Krai in Russia. After years of study and debate many geologists and other scientists think that this huge explosion was produced by a fragment of Comet Encke that broke off the main body and exploded in the air about five miles (eight km) above the Siberian plains.

The early morning of June 30, 1908, witnessed a huge, pipelike fireball moving across the skies of Siberia, until at 7:17 A.M. a tremendous explosion rocked the Tunguska area and devastated more than 1,160 square miles (3,000 km²) of forest. The force of the blast is estimated to have been equal to 10–30 megatons (0.91–27 megatonnes), and is thought to have been produced by the explosion, six miles (10 km) above the surface of Earth, of an asteroid or comet with a diameter of 200 feet (60 m). The energy equivalence of this explosion was close to 2,000 times the energy released during the explosion of the Hiroshima atomic bomb. More energy was released in the air blast than the impact and solid earthquakes, demonstrating that the Tunguska impacting body exploded in the air. The pattern of destruction reflects the dominance of atmospheric shock waves rather than solid earthquakes that are estimated to have been about a magnitude 5 earthquake. Atmospheric shock waves were felt thousands of miles away, and people located closer than 60 miles (100 km) from the site of the explosion were knocked unconscious; some were thrown into the air by the force of the explosion. Fiery clouds and deafening explosions were heard more than 600 miles (965 km) from Tunguska.

For a long time one of the biggest puzzles at Tunguska was the absence of an impact crater, despite all other evidence that points to an impact origin for this event. Many scientists now think that a piece of a comet, Comet Encke, broke off the main body as it was orbiting nearby Earth, and this fragment entered Earth's atmosphere and exploded about 5–6 miles (8–10 km) above the Siberian plains at Tunguska. This model was pioneered by Slovak astronomer Lubor Kresak, following earlier suggestions by the British astronomer F. J. W. Whipple in the 1930s that the bolide (a name for any unidentified object entering the planet's atmosphere) at Tunguska may have been a comet. Other scientists suggest the bolide may have been a meteorite, since comets are weaker than metallic or stony meteorites, and more easily break up and explode in the atmosphere before they hit Earth's surface. If the Tunguska bolide was a comet, it would likely have broken up higher in the atmosphere. In either case calculations show that, because of Earth's rotation, if the impact explosion happened only four hours and 47 minutes later, the city of Saint Petersburg would have been completely destroyed by the air blast. Air blasts from disintegrating meteorites or comets the size of the Tunguska explosion occur about once every 300 years on Earth, whereas smaller explosions, about the size of the nuclear bombs dropped on Japan, occur in the upper atmosphere about once per year.

All the trees in the Siberian forest in an area the size of a large city were leveled by the explosion of Tunguska, which fortunately was unpopulated at the time. But a thousand reindeer belonging to the Evenki people of the area were reportedly killed by the blast. The pattern of downed trees indicates that the projectile traveled from the southeast to the northwest as it exploded. The height of the explosion over Tunguska is about the optimal height for an explosion-induced air burst to cause maximum damage to urban areas, and calculations suggest that if the area was heavily populated at the time of the impact, at least 500,000 people would have died. Despite the magnitude and significance of this event, the Tunguska region was very remote, and no scientific expeditions to the area to investigate the explosion were mounted until 1921, 13 years after the impact, and even then the first expedition reached only the fringes of the affected area. The first scientific expedition was led by geologist Leonid Kulik, who was looking for meteorites along the Podkamennaya Tunguska River basin, and he heard stories from the local people of the giant explosion that happened in 1908, and that the explosion had knocked down trees, blown roofs off huts, and knocked people over and even caused some to become deaf from the noise. Kulik then convinced the Russian government that an expedition needed to be mounted into the remote core of the Tunguska blast area, and this expedition reached the core of the blast zone in 1927. Kulik and his team found huge tracts of flattened and burned trees, but they were unable to locate an impact crater.

In June 2007 a team of scientists from the University of Bologna suggested that the small Lake Cheko, located about 5 miles (8 km) from the epicenter of the blast, may be the impact site. Other scientists challenge this interpretation, noting that the lake has thick sediments, implying an older age than the age of the impact.

The atmospheric blast from the Tunguska explosion raced around the planet two times before diminishing. Residents of Siberia who lived within about 50 miles (80 km) of the blast site reported unusual glowing light from the sky for several weeks after the explosion. It is possible that this light was being reflected by a stream of dust particles that were ripped off a comet as it entered the atmosphere before colliding with Earth's surface. The unusual nighttime illumination was reported from across Europe and western Russia, showing the extent of the dust stream in the atmosphere.

As the fireball from the Tunguska airburst moved through the atmosphere, the temperatures at the center of the fireball were exceedingly hot, estimated to be 30 million degrees Fahrenheit (16.6 million Celsius). On the ground trees were burned and scorched, and silverware utensils in storage huts near the center of the blast zone were melted by the heat. After the impact leveled the trees for a distance of about 25 miles (40 km) around the center of the impact, forest fires ravaged the area, but typically burned only the outer surface of many trees, as if the fires were a short-lived flash of searing heat.

The type of body that exploded above Tunguska has been the focus of much speculation and investigation. One of the leading ideas is that the impact was caused by a comet that exploded in the atmosphere above Tunguska, a theory pioneered by F. J. W. Whipple in a series of papers from 1930 to 1934. In the 1960s small silica and magnetite spherules that represent melts from an extraterrestrial source were found in soil samples from Tunguska, confirming that a comet or meteorite had exploded above the site. Further analysis of the records of the airblast indicated that several pressure waves were recorded by the event. The first was the type associated with the rapid penetration of an object into the atmosphere, and at least three succeeding bursts recorded the explosions of a probably fragmented comet about five miles (eight km) above the surface.

There have been other reported explosions, or possible explosions, of meteorites above the surface of Earth, creating air blasts since the Tunguska event, although none has been as spectacular. On August 13, 1930, a body estimated to be about 10 percent the size of the Tunguska bolide exploded above the Curuca River in the Amazonas area in Brazil, but documentation of this event is poor. On May 31, 1965, an explosion with the force equivalent of 600 tons (544 tonnes) of TNT was released eight miles (13 km) above southeastern Canada, and approximately 0.4 ounce (1 g) of meteorite material was recovered from this event. Similar sized events, also thought to be from meteorite explosions at about eight miles (13 km) above the surface, were reported from southeast Canada on May 31, 1965, over Lake Huron (Michigan) on September 17, 1966, and over Alberta, Canada, on February 5, 1967. No meteorite material was recovered from any of these events. Two mysterious explosions, probably meteorites exploding, with an equivalent of about 25 tons (23 tonnes) of TNT, were reported, strangely, over the same area of Sassowo, Russia, on April 12, 1992, and July 8, 1992. A larger explosion and airburst, esti-mated to be equivalent to 10,000 tons of TNT, was reported over Lugo, Italy, on January 19, 1993, and another 25-ton (23-tonne) event over Cando, Spain, on January 18, 1994. Russia was struck again, this time in the Bodaybo region, by a 500–5,000 ton (450–4,500 tonne) equivalent blast on September 25, 2002, after a 26,000-ton (23,600-tonne) airburst from a meteorite explosion was recorded over the Mediterranean between Greece and Libya. The last reported airburst was from a high-altitude explosion, 27 miles (43 km) over Snohomish, Washington, on June 3, 2004. Clearly airbursts associated with the explosion of meteorites or comets are fairly common events, just as events as strong as the Tunguska explosion happen only about once every 300 years.

See also ASTEROID; ASTRONOMY; ASTROPHYSICS; ORIGIN AND EVOLUTION OF THE EARTH AND SOLAR SYSTEM; SOLAR SYSTEM.

FURTHER READING

Alvarez, Walter. *T Rex and the Crater of Doom.* Princeton, N.J.: Princeton University Press, 1997.

Angelo, Joseph A. *Encyclopedia of Space and Astronomy.* New York: Facts On File, 2006.

Chaisson, Eric, and Steve McMillan. *Astronomy Today.* 6th ed. Upper Saddle River, N.J.: 2007.

Chapman, C. R., and D. Morrison. "Impacts on the Earth by Asteroids and Comets: Assessing the Hazard." *Nature* 367 (1994): 33–39.

Elkens-Tanton, Linda T. *Asteroids, Meteorites, and Comets.* New York: Facts On File, 2006.

Spencer, John R., and Jacqueline Mitton. *The Great Comet Crash: The Impact of Comet Shoemaker-Levy 9 on Jupiter.* Cambridge: Cambridge University Press, 1995.

Thomas, Paul J., Christopher F. Chyba, and Christopher P. McKay, eds. *Comets and the Origin and Evolution of Life.* New York: Springer-Verlag. 1997.

constellation   Human groupings of stars in the sky into patterns, even though they may be far apart and lined up only visibly, are known as constellations. About 6,000 stars (and other points of light such as distant galaxies, planetary nebulas, quasars, etc.) are visible to the naked eye from Earth, and an irresistible tendency to see patterns and figures in these points of light has persisted for generations going back thousands of years. Peoples of many cultures have grouped these apparent configurations of points of light in the sky into patterns called constellations, the most famous of which are named after ancient Greek mythological beings. Typically most stars that make up a constellation physically exist far apart in space, and appear grouped near to one another only when viewed from Earth. Some of the earliest records of constellations date back to about 2500 B.C.E., from the Mesopotamian region, where early peoples used the patterns in the stars to help tell stories, mixing mythology, religion, cultural values, and tradition from generation to generation. There are some references to patterns of stars being grouped into constellations by the ancient Chinese and Jewish cultures that may go back to 4000 B.C.E. The peoples and stories changed with each generation, but the constellations were always there to remind the new generations that their elders were watching from above. Some cultures also used constellations for navigation and for marking the seasons. For instance, the star Polaris, which is part of the Little Dipper, indicates the north direction,

THE CONSTELLATIONS

Name/Meaning (Latin [English])	Genitive Form of Latin Name	Abbreviation	Approximate Position (Equatorial Coordinates)	
			RA(h)	δ(°)
Andromeda (name: princess)	Andromedae	And	1	+40
Aquarius (water bearer)	Aquarii	Aqr	23	-15
Aquila (eagle)	Aquilae	Aql	20	+5
Ara (altar)	Arae	Ara	17	-55
Argo Navis (ship of Argonauts), now split into the modern constellations: Carina, puppies, Pyxis, and Vela				
Aries (ram)	Arietis	Ari	3	+20
Auriga (charioteer)	Aurigae	Aur	6	+40
Boötes (berdsman)	Boötis	Boo	15	+30
Cancer (crab)	Cancri	Cnc	9	+20
Canis Major (great dog)	Canis Majoris	CMa	7	-20
Canis Minor (little dog)	Canis Minoris	CMi	8	+5
Capricornus (sea goat)	Capricorni	Cap	21	-20
Cassiopeia (name: queen)	Cassiopeiae	Cas	1	+60
Centaurus (centaur)	Centauri	Cen	13	-50
Cepheus (name: king)	Cephei	Cep	22	+70
Cetus (whale)	Ceti	Cet	2	-10
Corona Austrina (southern crown)	Coronae Australis	CrA	19	-40
Corona Borealis (northern crown)	Coronae Borealis	CrB	16	+30
Corvus (crow)	Corvi	Crv	12	-20
Crater (cup)	Crateris	Crt	11	-15
Cygnus (swan)	Cygni	Cyg	21	+40
Delphinus (dolphin)	Delphini	Del	21	+10
Draco (dragon)	Draconis	Dra	17	+65

and, with its fairly constant position in the sky, has served as a navigational aid for ages. Some cultures have used the first appearance of certain stars or constellations just above the horizon at daybreak to mark the start of different seasons, such as the harvest, spring, and end of winter. In other cases the relative positions of different constellations were used by some mythological cultures to form predic-

tions of a person's destiny, thus creating the field of astrology.

Greek astronomers later adopted many of the ancient Mesopotamian constellations, to which they added their own culture and stories, establishing a now commonly used set of 48 constellations (see the table the Constellations). These were first codified by Eudoxus of Cnidus, then Hipparchus, and finally

Name/Meaning (Latin [English])	Genitive Form of Latin Name	Abbreviation	Approximate Position (Equatorial Coordinates)	
			RA(h)	δ(°)
Equuleus (little horse)	Equulei	Equ	21	+10
Eridanus (name: river)	Eridani	Eri	3	-20
Gemini (twins)	Geminorum	Gem	7	+20
Hercules (name: hero)	Herculis	Her	17	+30
Hydra (sea serpent; monster)	Hydrae	Hya	10	-20
Leo (lion)	Leonis	Leo	11	+15
Lepus (hare)	Leporis	Lep	6	-20
Libra (scale; balance beam)	Librae	Lib	15	-15
Lupus (wolf)	Lupi	Lup	15	-45
Lyra (lyre)	Lyrae	Lyr	19	+40
Ophiuchus (serpent bearer)	Ophiuchii	Oph	17	0
Orion (name: great hunter)	Orionis	Ori	5	0
Pegasus (name : winged horse)	Pegasi	Peg	22	+20
Perseus (name: hero)	Persei	Per	3	+45
Pisces (fishes)	Piscium	Psc	1	+15
Piscis Austrinus (southern fish)	Piscis Austrini	PsA	22	-30
Sagitta (arrow)	Sagittae	Sge	20	+10
Sagittarius (archer)	Sagittarii	Sgr	19	-25
Scorpius (scorpion)	Scorpii	Sco	17	-40
Serpens (serpent)	Serpentis	Ser	17	0
Taurus (bull)	Tauri	Tau	4	+15
Triangulum (triangle)	Trianguli	Tri	2	+30
Ursa major (great bear)	Ursae Majoris	UMa	11	+50
Ursa Minor (little bear)	Ursae Minoria	UMi	15	+70
Virgo (virgin; maiden)	Virginis	Vir	13	0

THE MODERN CONSTELLATIONS

Name/Meaning (Latin [English])	Genitive Form of Latin Name	Abbreviation	Approximate Position (Equatorial Coordinates)	
			RA(h)	δ(°)
Antlia (air pump)	Antiae	Ant	10	-35
Apus (bird of paradise)	Apodis	Aps	16	-75
Caelum (sculptor's chisel)	Caeli	Cae	5	-40
Camelopardalis (giraffe)	Camelopardalis	Cam	6	+70
Canes Venatici (hunting dogs)	Canum Venaticorum	CVn	13	+40
Carina (keel)*	Carinae	Car	9	-60
Chamaeleon (chameleon)	Chamaeleontis	Cha	11	-80
Circinus (compasses)	Circini	Cir	15	-60
Columba (dove)	Columbae	Col	6	-35
Coma Berenices (Berenice's hair)	Comae Berenices	Com	13	+20
Crux (southern cross)	Crucis	Cru	12	-60
Dorado (swordfish)	Doradus	Dor	5	-65
Fornax (furnace)	Fornacis	For	3	-30
Grus (crane)	Gruis	Gru	22	-45
Horologium (clock)	Horologii	Hor	3	-60
Hydrus (water snake)	Hydri	Hyi	2	-75
Indus (Indian)	Indi	Ind	21	-55
Lacerta (Lizard)	Lacertae	Lac	22	+45
Leo Minor (little lion)	Leonis Minoris	LMi	10	+35
Lynx (lynx)	Lyncis	Lyn	8	+45
Mensa (table mountain)	Mensae	Men	5	-80
Microscopium (microscope)	Microscopii	Mic	21	-35
Monoceros (unicorn)	Monocerotis	Mon	7	-5
Musca (fly)	Muscae	Mus	12	-70
Norma (carpenter's square)	Normae	Nor	16	-50
Octans (octant; navigation device)	Octantis	Oct	22	-85
Pavo (peacock)	Pavonis	Pav	20	-65
Phoenix (Phoenix; mythical bird)	Phoenicis	Phe	1	-50
Pictor (painter's easel)	Pictoris	Pic	6	-55
Puppis (stern)*	Puppis	Pup	8	-40
Pyxis (nautical compass)*	Pyxidis	Pyx	9	-30

Name/Meaning (Latin [English])	Genitive Form of Latin Name	Abbreviation	Approximate Position (Equatorial Coordinates)	
			RA(h)	δ(°)
Reticulum (net)	Reticuli	Ret	4	-60
Sculptor (sculptor's workshop	Sculptoris	Scl	0	-30
Scutum (shield)	Scuti	Sct	19	-10
Sextans (sextant)	Sextantis	Sex	10	0
Telescopium (telescope)	Telescopii	Tel	19	-50
Triangulum Australe (southern triangle)	Trianguli Australe	TrA	16	-65
Tucana (toucan)	Tucanae	Tuc	0	-65
Vela (sail)*	Velorum	Vel	9	-50
Volans (flying fish)	Volantis	Vol	8	-70
Vulpecula (fox)	Vulpeculae	Vul	20	+25

*Originally part of ancient constellation Argo Navis (ship of Argonauts)

compiled in the work Syntaxis by Ptolemy about 150 B.C.E.

As the Roman Empire expounded, the Romans adopted the Greek constellations and spread their usage throughout the Western world. But, as the Roman Empire declined and the Dark Ages ensued, the light of the constellations was largely preserved only in the Arabic world, where the works were translated into *The Almagest,* in which many of the older observations were embellished and more detailed observations added. At the end of the Dark Ages the traditional Greek constellations experienced a revival in Europe during the Renaissance, initiating a period of rapid scientific inquiry. In 1603 the German astronomer Johann Bayer published *Uranometria,* the first major star catalog covering the entire celestial sphere visible from Earth. Bayer introduced the nomenclature of using Greek letters for the main stars in each constellation, assigning α (alpha) to the brightest, β (beta) to the second brightest, and so on, as well as named a dozen new constellations in the Southern Hemisphere. Since then other astronomers have named additional constellations, including several named by the Polish-German astronomer Johannes Hevelius, and the 18th-century French astronomer Nicolas-Louis de Lacaille. Astonomers today recognize 88 different constellations, including 47 of the 48 original Greek constellations. These are listed in the table the Modern Constellations.

See also ASTRONOMY; GALAXIES; GALAXY CLUSTERS; UNIVERSE.

FURTHER READING

Chaisson, Eric, and Steve McMillan. *Astronomy Today.* 6th ed. Upper Saddle River, N.J.: Addison-Wesley, 2007.

Comins, Neil F. *Discovering the Universe.* 8th ed. New York: W. H. Freeman, 2008.

Dibon-Smith, Richard. The Constellations Web Page. Available online. URL: http://www.dibonsmith.com/index.htm. Updated November 8, 2007.

Snow, Theodore P. *Essentials of the Dynamic Universe: An Introduction to Astronomy.* 4th ed. St. Paul, Minn.: West, 1991.

continental crust Continental crust covers about 34.7 percent of the Earth's surface, whereas exposed continents cover only 29.22 percent of the surface, with the discrepancy accounted for in the portions of continents that lie underwater on the continental shelves. Its lateral boundaries are defined by the slope break between continental shelves and slopes, and its vertical extent is defined by a jump in seismic velocities to 4.7–5 miles per second (7.6–8.0 km/s) at the Mohorovicic discontinuity. The continental crust ranges in thickness from about 12.5 to about 37 miles (20–60 km), with an average thickness of 24 miles (39 km). The continents are divided into orogens, made of linear belts of concentrated deformation, and cratons, marking the stable, typically older interiors of the continents. The distribution of elevation of continents and oceans can be portrayed on a curve showing the percentage of land at a specific elevation, versus elevation, known as the hypsometric curve, or the hypsographic curve. The curve is a cumulative frequency profile representing the statistical distribution of areas of the Earth's solid surface above or below mean sea level. The hypsometric curve is strongly bimodal, reflecting the two-tier distribution of land in continents close to sea level, and on ocean floor abyssal plains 1.9–2.5 miles (3–4 km) below sea level. Relatively little land surface is found in high mountains or in deep-sea trenches.

Most of the continental crust is now preserved in Archean cratons that form the cores of many continents. They are composed of ancient rocks that have been stable for billions of years, since the Archean. Cratons generally have low heat flow, few if any earthquakes, and no volcanism, and many are overlain by flat-lying shallow water sedimentary sequences. Continental shields are places where the cratonic crust is exposed at the surface, whereas continental platforms are places where the cratonic

rocks are overlain by shallow-water sedimentary rocks, presently exposed at the surface.

Most cratons have a thick mantle root or tectosphere, characterized seismologically and from xenolith studies to be cold and refractory, having had basaltic melt extracted from it during the Archean. Seismological studies have shown that many parts of the tectosphere are strongly deformed, with most of the minerals oriented in planar or linear fabrics. Current understanding about the origin of stable continental cratons and their roots hinges on recognizing which processes change the volume and composition of continental lithosphere with time, and how and when juvenile crust evolved into stable continental crust. Despite decades of study, several major unresolved questions remain concerning Archean tectosphere: How is it formed? Large quantities of melt extraction (ultramafic in composition, if melting occurred in a single event) are required from petrological observations, yet little of this melt is preserved in Archean cratons, which are character-

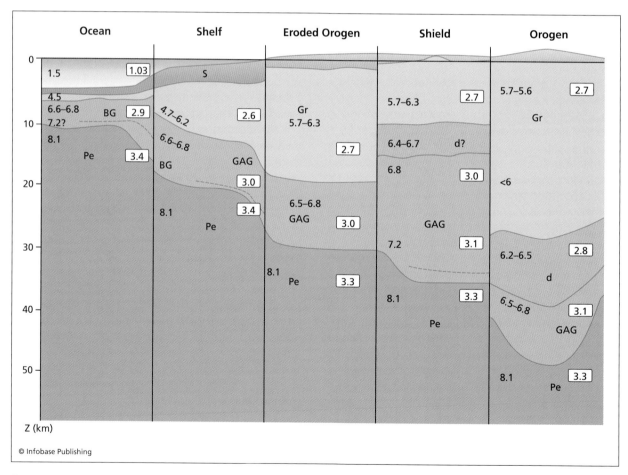

General crustal structure of different provinces as determined by seismology. Numbers in boxes represent densities in grams/cm³, and other numbers represent the seismic velocity of P waves in kilometers per second: BG = basalt-gabbro, GAG = amphibolite and granulite, Pe = peridotite, d = diorite, An Ga = anorthositic gabbro, S = sediments, Gr = granitic-gneiss upper crust, M = Mohorovicic discontinuity.

ized by highly evolved crust compositions. In what tectonic settings is it formed? Hypotheses range from intraplate, plume-generated settings to convergent margin environments. Finally, once formed, does the chemical buoyancy and inferred rheological strength of the tectosphere preserve it from disruption? Until recently most scientists would argue that cratonic roots last forever—isotopic investigations of mantle xenoliths from the Kaapvaal, Siberian, Tanzanian, and Slave cratons document the longevity of the tectosphere in these regions. However, the roots of some cratons are now known to have been lost, including the North China craton, and the processes of the loss of the tectosphere are as enigmatic as the processes that form the roots.

Orogens and orogenic belts are elongate regions that are eroded mountain ranges, and typically have abundant folds and faults. Young orogens are mountainous and include such familiar mountain ranges as the Rockies, Alps, and the slightly older Appalachians. Many Archean cratons are welded together by Proterozoic and younger orogens. In fact many Archean cratons can be divided into smaller belts that represent fragments of the planet's oldest orogenic belts.

Orogens have been added to the edges of the continental shield and cratons through processes of mountain building related to plate tectonics. Mountain belts are of three basic types, including fold and thrust belts, volcanic mountain ranges, and fault-block ranges. Fold and thrust belts are contractional mountain belts, formed where two tectonic plates collided, forming great thrust faults, folds, metamorphic rocks, and volcanic rocks. Detailed mapping of the structure in the belts can enable geologists to reconstruct their history and essentially pull them back apart. Investigations have revealed that many of the rocks in fold- and thrust-belt types of mountain ranges were deposited on the bottom of the ocean, continental rises, slope, shelves, or on ocean margin deltas. When the two plates collide, many of the sediments get scraped off and deformed, forming the mountain belts. Thus fold and thrust mountain belts mark places where oceans have closed.

Volcanic mountain ranges include Japan's Fuji and Mount St. Helens in the Cascades of the western United States. These mountain ranges are not formed primarily by deformation, but by volcanism associated with subduction and plate tectonics.

Fault-block mountains, such as the Basin and Range Province of the western United States, are formed by the extension or pulling apart of the continental crust, forming elongate valleys separated by tilted fault-bounded mountain ranges.

Every rock type known on Earth is found on the continents, so averaging techniques must be used to determine the overall composition of the crust. Estimates suggest that continental crust has a composition equivalent to andesite (or granodiorite) and is enriched in incompatible trace elements, the elements that do not easily fit into lattices of most minerals and tend to get concentrated in magmas.

The continents exhibit a broadly layered seismic structure that is different from place to place, and different in orogens, cratons, and parts of the crust with different ages. In shields the upper layer may typically be made of a few hundred meters of sedimentary rocks underlain by generally granitic types of material with seismic velocities of 3.5 to 3.9 miles per second (5.7–6.3 km/s) to depths of a couple to 6 miles (a few to 10 km), then a layer with seismic velocities of 3.9–4.2 miles per second (6.4–6.7 km/s). The lower crust is thought to be made of layered amphibolite and granulite with velocities of 4.2–4.5 miles per second (6.8–7.2 km/s). Orogens tend to have thicker, low-velocity upper layers and a lower-velocity lower crust.

Considerable debate and uncertainty surrounds the timing and processes responsible for the growth of the continental crust from the mantle. Most scientists agree that most of the growth occurred early in Earth history, since more than half of the continental crust is Archean in age, and about 80 percent is Precambrian. Some debate centers on whether early tectonic processes resembled those currently operating, or whether they differed considerably. The amount of current growth and how much crust is being recycled back into the mantle are currently poorly constrained. Most petrological models for the origin of the crust require that it be derived by a process including partial melting from the mantle, but simple mantle melting produces melts that are not as chemically evolved as the crust. Therefore the crust is probably derived through a multistage process, most likely including early melts derived from seafloor spreading and island arc magmatism, with later melts derived during collision of the arcs with other arcs and continents. Other models seek to explain the difference by calling on early higher temperatures leading to more evolved melts.

See also CRATON; GREENSTONE BELTS; OROGENY.

FURTHER READING

Taylor, Stuart Ross, and Scott M. McLennan. *Planetary Crusts*. Cambridge: Cambridge University Press, 2008.

continental drift The theory of continental drift was a precursor to plate tectonics. Proposed most clearly by Alfred Wegener in 1912, continental drift states that the continents are relatively light objects that are floating and moving freely across a substratum of oceanic crust. The theory was largely discredited because it lacked a driving mechanism, and seemed implausible if not physically impossible to most geologists and geophysicists at the time. But many of the ideas of continental drift were later incorporated into the paradigm of plate tectonics.

Early geologists recognized many of the major tectonic features of the continents and oceans. Cratons are very old, stable portions of the continents that have been inactive since the Precambrian. They typically exhibit subdued topography, including gentle arches and basins. Orogenic belts are long, narrow belts of structurally disrupted and metamorphosed rocks, typified (when active) by volcanoes, earthquakes, and folding of strata. Abyssal plains are stable, flat parts of the deep oceanic floor, whereas oceanic ridges are mountain ranges beneath the sea with active volcanoes, earthquakes, and high heat flow. To explain the large-scale tectonic features of the Earth, early geologists proposed many hypotheses, including popular ideas that the Earth was either expanding or shrinking, forming ocean basins and mountain ranges. In 1910–25 Wegener published a series of works including his 1912 treatise on *The Origin of Continents and Oceans*. Wegener proposed that the continents were drifting about the surface of the planet, and that they once fit back together to form one great supercontinent, Pangaea. To fit the coastlines of the different continental masses together to form his reconstruction of Pangaea, Wegener defined the continent/ocean transition as the outer edge of the continental shelves. The continental reconstruction proposed by Wegener showed remarkably good fits between coastlines on opposing sides of ocean basins, such as the Brazilian Highlands of South America fitting into the Niger delta region of Africa. Wegener was a meteorologist, and since he was not formally trained as a geologist, few scientists at the time believed his findings, although we now know that he was largely correct.

Most continental areas lie approximately 985 feet (300 m) above sea level, and if we extrapolate present erosion rates back in time, we find that continents would be eroded to sea level in 10–15 million years. This observation led to the application of the principle of isostasy to explain the elevation of the continents. Isostasy, which is essentially Archimedes' principle, states that continents and high topography are buoyed up by thick continental roots floating in a denser mantle, much like icebergs floating in water. The principle of isostasy states that the elevation of any large segment of crust is directly proportional to the thickness of the crust. It is significant that geologists working in Scandinavia noticed that areas that had recently been glaciated were rising quickly relative to sea level, and they equated this observation with the principle of isostatic rebound. The flow of mantle material within the zone of low viscosity beneath the continental crust accommodates isostatic rebound to compensate for the rising topography. These observations revealed that mantle material can flow at rates of a couple of inches (several centimeters) per year.

In *The Origin of Continents and Oceans* Wegener fit all the continents back together to form a Permian supercontinent, Pangaea (or "all land"). Wegener also used indicators of past climates, such as locations of ancient deserts and glacial ice sheets, and distributions of certain plant and animal species to support his ideas. Wegener's ideas found support from a famous South African geologist, Alex L. Du Toit, who in 1921 matched the stratigraphy and structure across the Pangaea landmass. Du Toit found the same plants, such as the Glossopterous fauna, across Africa and South America. He also documented similar reptiles and even earthworms across narrow belts of Wegener's Pangaea, supporting the concept of continental drift.

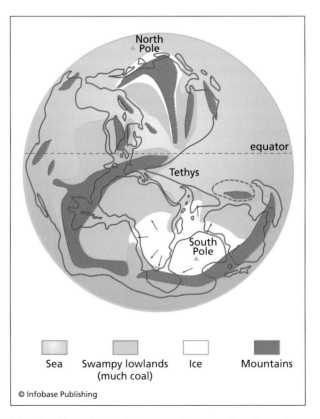

Modification of Alfred Wegener's reconstruction of Pangaea, originally from *Origin of Continents and Oceans*

Even with evidence such as the matching of geological belts across Pangaea, most geologists and geophysicists doubted the idea due to the lack of a conceivable driving mechanism, thinking it mechanically impossible for relatively soft continental crust to plow through the much stronger oceans. Early attempts at finding a mechanism were implausible and included such ideas as tides pushing the continents. Because of the lack of credible driving mechanisms, continental drift encountered stiff resistance from the geologic community, as few could understand how continents could plow through the mantle.

In 1928 British geologist Arthur Holmes suggested a driving mechanism for moving the continents. He proposed that heat produced by radioactive decay caused thermal convection in the mantle, and that the laterally flowing mantle dragged the continents with the convection cells. He reasoned that if the mantle can flow to allow isostatic rebound following glaciation, then maybe it can flow laterally as well. The acceptance of thermal convection as a driving mechanism for continental drift represented the foundation of modern plate tectonics. In the 1950s and 1960s the paleomagnetic data were collected from many continents and argued strongly that the continents had indeed been shifting, both with respect to the magnetic pole and also with respect to one another. When seafloor spreading and subduction of oceanic crust beneath island arcs was recognized in the 1960s, the model of continental drift was modified to become the new plate tectonic paradigm that revolutionized and unified many previously diverse fields of the earth sciences.

See also DU TOIT, ALEXANDER; HOLMES, ARTHUR; PLATE TECTONICS; SUPERCONTINENT CYCLES; WEGENER, ALFRED.

FURTHER READING
Moores, Eldridge M., and Robert Twiss. *Tectonics.* New York: W. H. Freeman, 1995.

continental margin Continental margins are the transition zone between thick, buoyant continental crust and the thin, dense submerged oceanic crust. There are several different types, depending on the tectonic setting. Passive, trailing, or Atlantic-type margins form where an extensional boundary evolves into an ocean basin, and new oceanic crust is added to the center of the basin between continental margins that originally faced one another. These margins were heated and thermally elevated during rifting and gradually cooled and thermally subsided for several tens of millions of years, slowly accumulating thick sequences of relatively flat sedi-

ments, forming continental shelves. Continental slopes and rises succeed these shelves seaward. The ocean/continent boundary typically occurs at the shelf/slope break on these Atlantic-type margins, where water depths average fewer than a thousand feet (a couple of hundred meters). Passive margins do not mark plate boundaries but rim most parts of many oceans, including the Atlantic and Indian, and form around most of Antarctica and Australia. Young, immature passive margins are beginning to form along the Red Sea.

Convergent, leading, or Pacific-type margins form at convergent plate boundaries. They are characterized by active deformation, seismicity, and volcanism, and some have thick belts of rocks known as accretionary prisms scraped off of a subducting plate and added to the overriding continental plate. Convergent margins may have a deep sea trench up to seven miles (11 km) deep marking the boundary between the continental and oceanic plates. These trenches form where the oceanic plate is bending and plunging deep into the mantle. Abundant folds and faults in the rocks characterize convergent margins. Other convergent margins are characterized by old eroded bedrock near the margin, exposed by a process of sediment erosion where the edge of the continent is eroded and drawn down into the trench.

A third type of continental margin forms along transform or transcurrent plate boundaries. These are characterized by abundant seismicity and deformation, and volcanism is limited to certain restricted areas. Deformation along transform margins tends to be divided into different types, depending on the orientation of bends in the main plate boundary fault. Constraining bends form where the shape of the boundary restricts motion on the fault, and are characterized by strong folding, faulting, and uplift. The Transverse Ranges of southern California form a good example of a restraining bend. Sedimentary basins and subsidence characterize bends in the opposite direction, where the shape of the fault causes extension in areas where parts of the fault diverge during movement. Volcanic rocks form in some of these basins. The Gulf of California and Salton trough have formed in areas of extension along a transform margin in southern California.

See also CONVERGENT PLATE MARGIN PROCESSES; DIVERGENT PLATE MARGIN PROCESSES; PLATE TECTONICS; TRANSFORM PLATE MARGIN PROCESSES.

FURTHER READING
Davis, R., and D. Fitzgerald. *Beaches and Coasts.* Malden, Mass.: Blackwell, 2004.
Moores, Eldridge M., and Robert Twiss. *Tectonics.* New York: W. H. Freeman, 1995.

convection and the Earth's mantle The main heat transfer mechanism in the Earth's mantle is convection, a thermally driven process where heating at depth causes material to expand and become less dense, causing it to rise while being replaced by complimentary cool material that sinks. This moves heat from depth to the surface in a very efficient cycle, since the material that rises gives off heat as it rises and cools, and the material that sinks gets heated only to rise again eventually. Convection is the most important mechanism by which the Earth is loosing heat, with other mechanisms including conduction, radiation, and advection. However, many of these mechanisms work together in the plate tectonic cycle. Mantle convection brings heat from deep in the mantle to the surface, where the heat released forms magmas that generate the oceanic crust. The midocean ridge axis is the site of active hydrothermal circulation and heat loss, forming black smoker chimneys and other vents. As the crust and lithosphere move away from the midocean ridges, it cools by conduction, gradually subsiding (according the square root of its age) from about 1.5–2.5 miles (2.5–4.0 km) below sea level. Heat loss by mantle convection is therefore the main driving mechanism of plate tectonics, and the moving plates can be thought of as the conductively cooling boundary layer for large-scale mantle convection systems.

The heat transferred to the surface by convection is produced by decay of radioactive elements, producing isotopes such as uranium 235, thorium, 232, and potassium 40, remnant heat from early heat-producing isotopes such as iodine 129, remnant heat from accretion of the Earth, heat released during core formation, and heat released during impacts of meteorites and asteroids. Early in the history of the planet at least part of the mantle was molten, and the Earth has been cooling by convection ever since. Estimating how much the mantle has cooled with time is difficult, but reasonable estimates suggest that the mantle may have been up to a couple of hundred degrees hotter in the earliest Archean.

The rate of mantle convection depends on the ability of the material to flow. The resistance to flow is a quantity measured as viscosity, defined as the ratio of shear stress to strain rate. Fluids with high viscosity are more resistant to flow than materials with low viscosity. The present viscosity of the mantle is estimated to be 10^{20}–10^{21} Pascal seconds (Pa/s) in the upper mantle and 10^{21}–10^{23} Pa/s in the lower mantle, values sufficient for allowing the mantle to convect, and complete an overturn cycle once every 100 million years. The viscosity of the mantle is temperature dependent, so the mantle may have been able to flow and convectively overturn much more quickly in early Earth history, making convection an even more efficient process and speeding the rate of plate tectonic processes.

Current ongoing debate and research concerns the style of mantle convection in the Earth. The upper mantle is relatively heterogeneous and extends to a depth of 416 miles (670 km), where there is a pronounced increase in seismic velocities. The lower mantle is more homogeneous, and extends to a region known as D″ (pronounced dee-double-prime) at 1,678 miles (2,700 km), marking the transition into the liquid outer core. One school of mantle convection thought suggests that the entire mantle, including both the upper and lower parts, convects as one unit. Another school of thought posits that the mantle convection consists of two layers, with the lower mantle convecting separately from the upper mantle. A variety of these models, presently held by the majority of geophysicists, holds that there is two-layer convection, but that subducting slabs are able to penetrate the 670-kilometer discontinuity from above, and that mantle plumes that rise from the D″ region can penetrate the 670-kilometer discontinuity from below.

The shapes that mantle convection cells take include many possible forms that are reflected to a first order by the distribution of subduction zones and midocean ridge systems. The subduction zones mark regions of downwelling, whereas the ridge system marks broad regions of upwelling. Material is upwelling in a broad planiform cell beneath the Atlantic and Indian Oceans, and downwelling in the circum-Pacific subduction zones. There is thought to be a large plumelike "superswell" beneath part

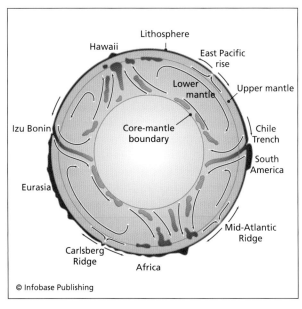

Cross section of Earth showing possible modes of mantle convection

Real data on a cutaway of Earth showing movement of deep slabs of rock in mantle. Sinking slabs are blue, mantle is yellow, and rising molten rock is red. The sinking slabs, including one (at upper left) descending from the Caribbean, are up to 930 miles (1,500 km) across and penetrate up to 1,800 miles (2,900 km) to the D″ region at the core-mantle boundary. The deep slabs can be detected by measuring the arrival times at points around the world of seismic shear waves produced by earthquakes. These waves travel faster through dense, cool rock than warm rock. *(Steve Grand, Texas University/Photo Researchers, Inc.)*

of the Pacific that feeds the planiform East Pacific rise. Mantle plumes that come from the deep mantle punctuate this broad pattern of upper-mantle convection, and their plume tails must be distorted by flow in the convecting upper mantle.

The pattern of mantle convection deep in geological time is uncertain. Some periods such as the Cretaceous seem to have had much more rigorous mantle convection and surface volcanism. More or different types or rates of mantle convection may have helped to allow the early Earth to lose heat more efficiently. Some computer models allow periods of convection dominated by plumes, and others dominated by overturning planar cells similar to the present Earth. Some models suggest cyclic relationships, with slabs pooling at the 670-kilometer discontinuity, then suddenly all sinking into the lower mantle, causing a huge mantle overturn event. Further research is needed on linking the preserved record of mantle convection in the deformed continents to help interpret the past history of convection.

See also BLACK SMOKER CHIMNEYS; DIVERGENT PLATE MARGIN PROCESSES; PLATE TECTONICS.

FURTHER READING

Schubert, Gerald, Donald L. Turcotte, and Peter Olson. *Mantle Convection in the Earth and Planets.* Cambridge: Cambridge University Press, 2001.
Turcotte, Donald L., and Gerald Schubert. *Geodynamics.* 2nd ed. Cambridge: Cambridge University Press, 2002.

convergent plate margin processes   Structural, igneous, metamorphic, and sedimentological processes that occur in the region affected by forces associated with the convergence of two or more plates are grouped under the heading of convergent plate margin processes. Convergent plate boundaries are of two fundamental types, subduction zones and collision zones. Subduction zones are in turn of two basic types, the first of which is found where oceanic lithosphere of one plate descends beneath another oceanic plate, such as in the Philippines and Marianas of the southwest Pacific. The second type of subduction zone forms where an oceanic plate descends beneath a continental upper plate, such as in the Andes of South America. The southern Alaska convergent margin is particularly interesting, as it records a transition from an ocean/continent convergent boundary to an ocean/ocean convergent boundary in the Aleutians.

Arcs have several different geomorphic zones defined largely on their topographic and structural expressions. The active arc is the topographic high with volcanoes, and the backarc region stretches from the active arc away from the trench, and it may end in an older rifted arc or continent. The forearc basin is a generally flat topographic basin with shallow to deep-water sediments, typically deposited over older accreted sediments and ophiolitic or continental basement. The accretionary prism includes uplifted, strongly deformed rocks scraped off the downgoing oceanic plate on a series of faults. The trench may be several to six miles (up to 10 or more kilometers) deep below the average level of the seafloor in the region and marks the boundary between the overriding and underthrusting plate. The outer trench slope is the region from the trench to the top of the flexed oceanic crust that forms a several hundred to one-thousand-foot (few hundred-meter) high topographic rise known as the forebulge on the downgoing plate.

Trench floors are triangular shaped in profile and typically are partly to completely filled with greywacke-shale turbidite sediments derived from erosion of the accretionary wedge. They may also be transported by currents along the trench axis for large distances, up to hundreds or even thousands of miles (thousands of kilometers) from their ultimate source in uplifted mountains in the convergent orogen.

Flysch is a term that applies to rapidly deposited deep marine syn-orogenic clastic rocks that are generally turbidites. Trenches are also characterized by chaotic deposits known as olistostromes that typically have clasts or blocks of one rock type, such as limestone or sandstone, mixed with a muddy or shaly matrix. These are interpreted as slump or giant submarine landslide deposits. They are common in trenches because of the oversteepening of slopes in the wedge. Sediments that get accreted may also include pelagic sediments initially deposited on the subducting plate, such as red clay, siliceous ooze, chert, manganiferous chert, calcareous ooze, and windblown dust.

The sediments are deposited as flat-lying turbidite packages, then gradually incorporated into the accretionary wedge complex through folding and the propagation of faults through the trench sediments. Subduction accretion is a process that accretes sediments deposited on the underriding plate onto the base of the overriding plate. It causes the rotation and uplift of the accretionary prism, which is a broadly steady-state process that continues as long as sediment-laden trench deposits are thrust deeper into the trench. Typically new faults will form and propagate beneath older ones, rotating the old faults and structures to steeper attitudes as new material is added to the toe and base of the accretionary wedge. This process increases the size of the overriding accretionary wedge and causes a seaward-younging in the age of deformation.

Parts of the oceanic basement to the subducting slab are sometimes scraped off and incorporated into the accretionary prisms. These tectonic slivers typically consist of fault-bounded slices of basalt, gabbro, and ultramafic rocks, and rarely, partial or even complete ophiolite sequences can be recognized. These ophiolitic slivers are often parts of highly deformed belts of rock known as mélanges. Mélanges are mixtures of many different rock types typically including blocks of oceanic basement or limestone in muddy, shaly, serpentinitic, or even a cherty matrix. Formed by tectonic mixing of the many different types of rocks found in the forearc, mélanges are one of the hallmark rock units of convergent boundaries.

Major differences in processes occur at Andean-style compared to Marianas-style arc systems. Andean-type arcs have shallow trenches, fewer than 3.7 miles (6 km) deep, whereas Marianas-type arcs typically have deep trenches reaching 6.8 miles (11 km) in depth. Most Andean-type arcs subduct young oceanic crust and have very shallow-dipping subduction zones, whereas Marianas-type arcs subduct old oceanic crust and have steeply dipping Benioff zones. Andean arcs have back-arc regions dominated by foreland (retroarc) fold thrust belts and sedimentary basins, whereas Marianas-type arcs typically have back-arc basins, often with active seafloor spreading. Andean arcs have thick crust, up to 43.5 miles (70 km), and big earthquakes in the overriding plate, while Marianas-type arcs have thin crust, typically only 12.5 miles (20 km), and have big earthquakes in the underriding plate. Andean arcs have only rare volcanoes, and these have magmas rich in SiO_2 such as rhyolites and andesites. Plutonic rocks are more common, and the basement is continental crust. Marianas-type arcs have many volcanoes that erupt lava low in silica content, typically basalt, and are built on oceanic crust.

Many arcs are transitional between the Andean or continental-margin types and the oceanic or Marianas types, and some arcs have large amounts of strike-slip motion. The causes of these variations have been investigated and it has been determined that the rate of convergence has little effect, but the relative motion directions and the age of the subducted oceanic crust seem to have the biggest effects. In particular old oceanic crust tends to sink to the point where it has a near-vertical dip, rolling back through the viscous mantle and dragging the arc and forearc regions of overlying Marianas-type arcs with it. This process contributes to the formation of back arc basins.

Much of the variation in the processes that occur in convergent margin arcs can be attributed to the relative convergence vectors between the overriding and underriding plates. In this kinematic approach to modeling convergent margin processes, the underriding plate may converge at any angle with the overriding plate, which itself moves toward or away from the trench. Since the active arc is a surface expression of the 68-mile (110-km) isobath on the subducted slab, the arc will always stay 68 miles (110 km) above this zone. The arc therefore separates two parts of the overriding plate that may move independently, including the frontal arc sliver between the arc and trench and the main part of the overriding plate. The frontal arc sliver is in most cases kinematically linked to the downgoing plate and moves parallel to the plate margin in the direction that contains the oblique component of motion between the downgoing and overriding plate. Different relative angles of convergence between the overriding and underriding plate determine whether or not an arc will have strike-slip motions, and the amount that the subducting slab rolls back (which is age-dependent) determines whether the frontal arc sliver rifts from the arc and causes a back arc basin to open or not. This model helps to explain why some arcs are extensional with big back arc basins, others have strike-slip dominated systems, and others are purely compressional arcs. Convergent margins also show changes in these vectors and consequent geologic

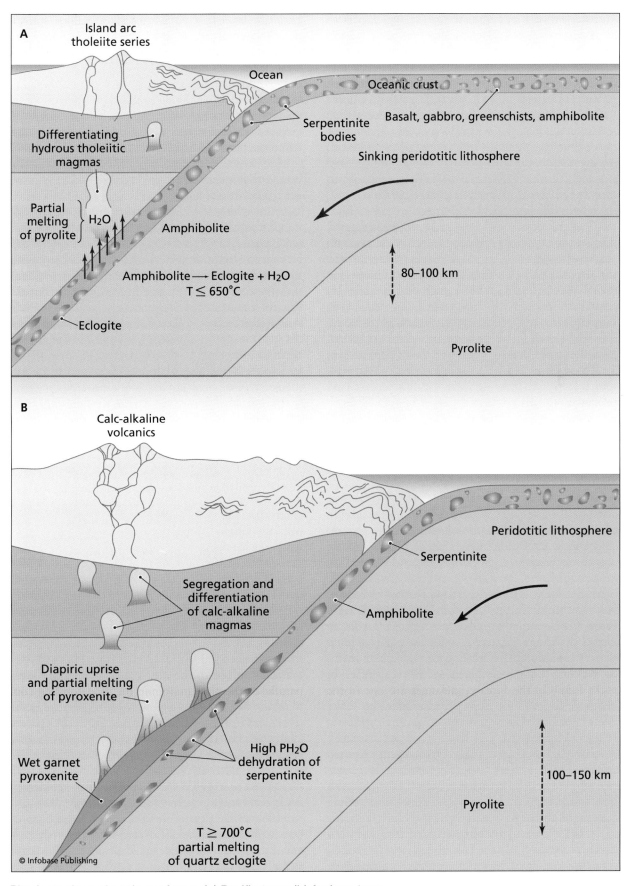

Physiography and geology of arcs: (a) Pacific-type; (b) Andean-type

processes with time, often switching quickly from one regime to the other with changes in the parameters of the subducting plate.

The thermal and fluid structure of arcs is dominated by effects of the downgoing slab, which is much cooler than the surrounding mantle and cools the forearc. Fluids released from the slab as it descends past 68 miles (110 km) aid partial melting in the overlying mantle and generate the magmas that form the arc on the overriding plate. This broad thermal structure of arcs results in the formation of paired metamorphic belts, where the metamorphism in the trench environment grades from cold and low-pressure at the surface to cold and high-pressure at depth, whereas the arc records low- and high-pressure high-temperature metamorphic facies series. One of the distinctive rock types found in trench environments is the unusual high-pressure, low-temperature blueschist facies rocks in paleosubduction zones. The presence of index minerals glaucophane (a sodic amphibole), jadeite (a sodic pyroxene), and lawsonite (Ca-zeolite) indicate low temperatures extended to depths of 12–20 miles (20–30 kilometers) (7–10 kilobars [kb]). Since these minerals are unstable at high temperatures, their presence indicates they formed in a low-temperature environment, and the cooling effects of the subducting plate offer the only known environment to maintain such cool temperatures at depth in the Earth.

Forearc basins may include several-kilometer-thick accumulations of sediments deposited in response to subsidence induced by tectonic loading or thermal cooling of forearcs built on oceanic lithosphere. The Great Valley of California is a forearc basin that formed on oceanic forearc crust preserved in ophiolitic fragments found in central California, and Cook Inlet in Alaska is an active forearc basin formed in front of the Aleutian and Alaska range volcanic arc.

The rocks in the active arcs typically include several different facies. Volcanic rocks may include subaerial flows, tuffs, welded tuffs, volcaniclastic conglomerate, sandstone, and pelagic rocks. Debris flows from volcanic flanks are common, and abundant and thick accumulations of ash deposited by winds and dropped by Plinian and other eruption columns may be present. Volcanic rocks in arcs include mainly calc-alkaline series, showing an early iron

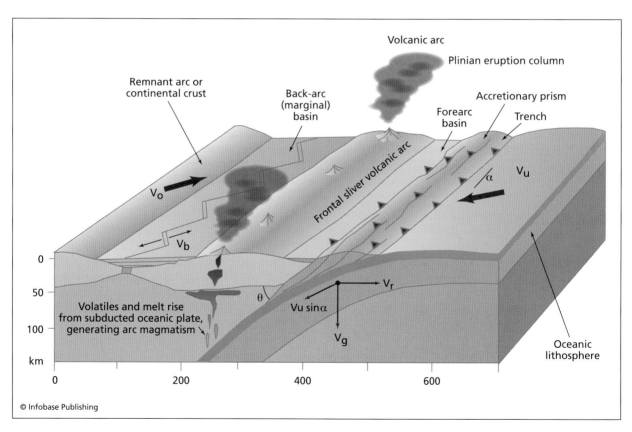

Relative motion vectors in arcs. Changes in relative motions can produce drastically different arc geology. Vu = velocity of underriding plate; Vo = velocity of overriding plate; Vb = slip vector between overriding and underriding plates; Vg = velocity of sinking; Vr = velocity of rollback. Note that Vu sin α = velocity of downdip component of subduction, and Vr = Vg cot θ.

Snow-covered Mount Fuji in Japan—a classical, active convergent margin volcano *(AP images)*

enrichment in the melt, typically including basalts, andesites, dacites, and rhyolites. Immature island arcs are strongly biased toward eruption at the mafic end of the spectrum, and may also include tholeiitic basalts, picrites, and other volcanic and intrusive series. More mature continental arcs erupt more felsic rocks and may include large caldera complexes.

Back arc and marginal basins form behind extensional arcs, or may include pieces of oceanic crust trapped by the formation of a new arc on the edge of an oceanic plate. Many extensional back arcs are found in the southwest Pacific, whereas the Bering Sea, between Alaska and the Kamchatka peninsula, is thought to be a piece of oceanic crust trapped during the formation of the Aleutian chain. Extensional back arc basins may have oceanic crust generated by seafloor spreading, and these systems closely resemble the spreading centers found at divergent plate boundaries. The geochemical signature of some of the lavas show some subtle and some not-so-subtle differences, however, with water and volatiles being more important in the generation of magmas in back arc suprasubduction zone environments.

Compressional arcs such as the Andes have tall mountains, reaching heights of more than 24,000 feet (7,315 m) over broad areas. They have little or no volcanism but much plutonism, and typically have shallow dipping slabs beneath them. Andean-type compressional arcs are characterized by thick continental crust with large compressional earthquakes, and show a foreland-style retroarc basin in the back arc region. Some compressional arc segments do not have accretionary forearcs but exhibit subduction erosion during which material is eroded and scraped off the overriding plate, and dragged down into the subduction zone. The Andes show remarkable along-strike variations in processes and tectonic style, with sharp boundaries between different segments. These variations seem to be related to what is being subducted and plate motion vectors. In areas where the downgoing slab has steep dips the overriding plate has volcanic rocks; in areas of shallow subduction there is no volcanism.

COLLISIONS

Collisions are the final products of subduction. There are several general varieties of collisions. They may be between island arcs and continents, such as the Ordovician Taconic Orogeny in eastern North America, or they may juxtapose a passive margin on one continent and an Andean margin on another. More rarely, collisions between two convergent margins occur above two oppositely dipping subduction zones, with a contemporary example extant in the Molucca Sea of Indonesia. Finally, collisions may be between two continents, such as the ongoing India/Asia collision that is affecting much of Asia.

Arc/continent collisions are the simplest of collisional events. As an arc approaches a continent, the continental margin is flexed downward by the weight of the arc, much like a ruler pushed down over the edge of a desk. The flexure induces a bulge a few hundred kilometers wide in front of the active collision zone, and this bulge migrates in front of the collision as a few hundred-meter-high broad topographic high. As the arc terrane rides up onto the continent, the thick sediments in the continental rise are typically scraped off and progressively added to the accretionary prism, with the oldest thrust faults being the ones closest to the arc, and progressively younger thrust faults along the base of the prism. Many forearc regions have ophiolitic basement, and these ophiolites get thrust upon the continents during collision events and are preserved in many arc/continent collisional orogens. The accretionary wedge grows and begins to shed olistostromes into the foredeep basin between the arc and continent, along with flysch and distal black shales. These three main facies migrate in front of the moving arc/accretionary complex at a rate equal to the convergence rate and drown any shallow-water carbonate deposition. After the arc terrane rides up the continental rise, slope, and shelf, it grinds to a halt when isostatic (buoyancy) effects do not allow continued convergence. At this stage a new subduction zone may be initiated behind the collided arc, allowing convergence to continue between the two plates.

Continent/continent collisions are the most dramatic of collisional events, with the current example of the convergence and collision of Africa, Arabia, and India with Europe and Asia affecting much of the continental landmass of the world. Continental collisions are associated with thickening of continental crust and the formation of high mountains, and deformation distributed over wide areas. The convergence between India and Asia dramatically slowed about 38 million years ago, probably associated with initial stages of that collision between 25–40 million years ago. The collision has resulted in the uplift of the Himalayan mountain chain and the Tibetan Plateau, and formed a wide zone of deformation that extends well into Siberia and includes much of Southeast Asia. Since the collision, there has been 2–2.4 inches per year (5–6 cm/yr) of convergence between India and Asia, meaning that a minimum of 775 miles (1,250 km) has had to be accommodated

in the collision zone. This convergence has been accommodated in several ways. Two large faults between India and Asia, the Main Central thrust and the Main Boundary thrust, are estimated to have 250 and 120 miles (400 and 200 km) of displacement on them, respectively, so they account for less than half of the displacement. Folds of the crust and general shortening and thickening of the lithosphere may account for some but not a large amount of the convergence. Evidence suggests that underthrusting of the Indian plate beneath Tibet and strike-slip faulting moving or extruding parts of Asia out of the collision zone toward the southwest Pacific accommodated much of the convergence.

The Tibetan Plateau and Himalayan mountain chain are about 375 miles (600 km) wide, with the crust beneath the region being about 45 miles (70 km) thick, twice that of normal continental crust. This has led to years of scientific investigation about the mechanism of thickening. Some models and data suggest that India is thrust under Asia for 600 kilometers, whereas other models and data suggest that the region has thickened by thrusting at the edges and plane strain in the center. In either case the base of the Tibetan crust has been heated to the extent that partial melts are beginning to form, and high heat flow in some rifts on the plateau is associated with the intrusions at depth. The intrusions are weakening the base of the crust, which is starting to collapse under the weight of the overlying mountains, and the entire plateau is on the verge of undergoing extension.

The collisional process is resulting in the formation of a layered differentiated lower continental crust in Tibet, with granitic melts forming a layer that has been extracted from a granulitic residue, along with strong deformation. These processes are not readily observable 30 miles (50 km) beneath Tibet, but are preserved in many old (generally Precambrian) high-grade gneiss terranes around the world thought to have formed in continental collision zones.

Continent/continent collision zones tend to have major effects on global plate motions. Convergence that used to be accommodated between the two continents must be transferred elsewhere on the planet, since all plate motions must sum to zero on Earth. Therefore continental collisions typically cause events elsewhere, such as the formation of new subduction zones and a global reorganization of plate motions.

DESCRIPTIONS OF EARTHQUAKES FROM CONVERGENT MARGINS

The world's largest earthquakes, often called "great" earthquakes, with magnitudes larger than 9, occur along convergent plate margins, especially where one oceanic plate is being subducted beneath a continental plate. This type of configuration leads to huge regions being stressed, or "bent" into a position where regions measuring hundreds of miles (hundreds of km) in length and many tens of miles (km) in depth may suddenly slip in one earthquake event, typically releasing more energy than all the other earthquakes on the planet for many years. For example, the 2004 Sumatra earthquake released more energy than all of the other earthquakes on the planet in the past 30 years. In this section some of these huge convergent margin earthquakes are described to give an understanding of the power of these events in shaping the Earth's surface. In the United States Alaska and the Pacific Northwest are the regions most at-risk for experiencing future convergent margin earthquakes.

Sumatra, 2004 (magnitude 9.0), and Indian Ocean Tsunami

One of the worst natural disasters of the 21st century unfolded on December 26, 2004, following a magnitude 9.0 (some estimates are as high as 9.3, a threefold difference in energy released) earthquake off the coast of northern Sumatra in the Indian Ocean. The earthquake was the largest since the 1964 magnitude 9.2 earthquake event in southern Alaska, and released more energy than all the earthquakes on the planet in the last 25–30 years combined. During this catastrophic earthquake a segment of the seafloor the size of California, lying above the Sumatra subduction zone trench, suddenly moved upward and seaward by several tens of feet. The slip event continued for nearly 10 minutes as the central section of the faulted area moved 65 feet (20 m) and the rupture propagated laterally more than 600 miles (1,000 km). The sudden displacement of this volume of undersea floor displaced a huge amount of water and generated the most destructive tsunami known in recorded history.

Within minutes of the initial earthquake a mountain of water more than 100 feet (30 m) high was ravaging northern Sumatra, sweeping into coastal villages and resort communities with a fury that crushed all in its path, removing buildings and vegetation, and in many cases eroding shoreline areas down to bedrock, leaving no traces of the previous inhabitants or structures. Similar scenes of destruction and devastation rapidly moved up the coast of Indonesia, where residents and tourists were enjoying a holiday weekend. Tsunami waves moved up to 12 miles (20 km) inland in northern Sumatra and elsewhere in Indonesia. Firsthand accounts and numerous videos made of the catastrophe reveal similar scenes of horror, where unsuspecting tourists and residents were enjoying beachfront playgrounds, resorts, and villages, and watched as large breaking waves appeared off the coast. Many moved toward the shore to watch with interest the high surf, then ran in panic as the sea rapidly rose beyond expectations, and walls of water engulfed entire beachfronts, rising tens of feet above hotel lobbies and washing through towns with the force of Niagara Falls. In some cases the sea retreated to unprecedented low levels before the waves struck, inducing many people to move to the shore to investigate the phenomenon; in other cases the sea waves simply came crashing inland without warning. Buildings, vehicles, trees, boats, and other debris were washed along with the ocean waters, forming projectiles that smashed at speeds of up to 30 miles per hour (50 km/hr) into other structures, leveling all in their path and killing nearly a quarter million people.

The displaced water formed a deep-water tsunami that moved at speeds of 500 miles per hour (800 km/hr) across the Indian Ocean, smashing within an hour into Sri Lanka and southern India, wiping away entire fishing communities and causing additional widespread destruction of the shore environment. South of India are many small islands including the Maldives, Chagos, and Seychelles, many of which have maximum elevations of only a few to a few tens of feet above normal sea level. As the tsunami approached these islands many wildlife species and primitive tribal residents fled to the deep forest, perhaps sensing the danger as the sea retreated and the ground trembled with the approaching wall of water. As the tsunami heights were higher than many of the maximum elevations of some of these islands, the forest protected and saved many lives in places where the tsunami caused sea levels to rise with less force than in places where the shoreline geometry caused large breaking waves to crash ashore. The tsunami traveled around the world, being measured as minor (inches) changes in sea level more than 24 hours later in the north Atlantic and Pacific.

Valdez, Alaska, 1964 (magnitude 9.2)

One of the largest earthquakes ever recorded struck southern Alaska at 5:36 P.M. on March 27, 1964, second in the amount of energy released only to the 1960 Chile earthquake. The energy released during the Valdez earthquake was more than the world's largest nuclear explosion and greater than the Earth's total average annual release of seismic energy, yet, remarkably, only 131 people died during this event. Damage is estimated at $240 million (1964 dollars), a remarkably small figure for an earthquake this size. During the initial shock and several other shocks that followed in the next one to two minutes, a 600-mile (1,000 km) long by 250-mile (400 km) wide slab of subducting oceanic crust slipped farther beneath the North American crust of southern Alaska. Ground displacements above the area that slipped were remarkable—much of the Prince William Sound and Kenai Peninsula area moved horizontally almost 65 feet (20 m), and moved upward by more than 35 feet (11.5 m). Other areas more landward of the uplifted zone were down dropped by several to ten feet. Overall almost 125,000 square miles (200,000 km²) of land saw significant movements upward, downward, and laterally during this huge earthquake.

The ground shook in most places for three to four minutes during the March 27, 1964, earthquake, but lasted for as much as seven minutes in a few places such as Anchorage and Valdez, where unconsolidated sediment and fill amplified and prolonged the shaking. In contrast, ground shaking during the 1906 San Francisco earthquake lasted less than one minute. In the 24 hours after the main earthquake rupture, 28 large aftershocks (10 larger than magnitude 6) hit the region, with epicenters distributed in an area about 50–60 miles (80–100 km) across. The aftershocks continued for months, gradually diminishing in strength but with more than 12,000 aftershocks stronger than 3.5 measured over the next three months.

The shaking caused widespread destruction in southern Alaska, damage as far away as southern California, and induced noticeable effects across the planet. Entire neighborhoods and towns slipped into the sea during this earthquake, and ground breaks, landslides, and slumps were reported across the entire region. A fault near the epicenter broke through the surface, forming a spectacular fault scarp with a displacement of more than 15 feet (3 m), uplifting beach terraces and mussel beds above the high-water mark, many parts of which rapidly eroded to a more stable configuration. Urban areas such as Anchorage suffered numerous landslides and slumps, with tremendous damage done by translational slumps where huge blocks of soils and rocks slid on curved faults down slope, in many cases toward the sea. Houses ended up in neighbors' back yards, and some homes were split in two by ground breaks. The Anchorage neighborhood of Turnagain Heights suffered extensive damage when huge sections of the underlying ground slid toward the sea on a weak layer in the bedrock known as the Bootlegger Shale, which lost cohesion during the earthquake shaking. Tsunamis swept across many towns that had just seen widespread damage by building collapses, washing buildings, vehicles, trains, petroleum tank farms, and anything in their path to higher ground. Near Valdez the tsunami broke large trees, leaving only stumps more than 100 feet (30 meters) above high-tide mark. Other tsunamis swept across the Pacific, destroying marinas as far away as southern California.

The transportation system in Alaska was severely disrupted by the earthquake. All major highways and most secondary roads suffered damage to varying degrees—186 of 830 miles (300 of 1340 km) of roads were damaged, and 83 miles (125 km) of roadway needed replacement. Seventy five percent of all bridges collapsed, became unusable, or suffered severe damage. Many railroad tracks were severed or bent by movement on faults, sliding and slumping into streams, and other ground motions. In Seward, Valdez, Kodiak, and other coastal communities, a series of 3–10 tsunami waves tore trains from their tracks, throwing them onto higher ground. The devastated shipping industry was especially difficult for Alaskans, who had used shipping for more than 90 percent of their transportation needs, and the main industry in the state is fishing. All port facilities in southern Alaska except those in Anchorage were totally destroyed by submarine slides, tsunami, tectonic uplift and subsidence, and earthquake-induced fires. Huge portions of the waterfront facilities at Seward and Valdez slid under the sea during a series of submarine landslides, with the loss of the harbor facilities and necessitating the eventual moving of the cities to higher, more stable ground. Being thrown to higher ground destroyed hundreds of boats, although no large vessels were lost. Uplift in many shipping channels formed new hazards and obstacles that had to be mapped to avoid grounding and puncturing hulls. Downed lines disrupted communication systems, and initial communications with remote communities were taken over by small, independently powered radio operators (if a similar event were to happen today, communications would likely be by cell phone). Water, sewer, and petroleum storage tanks and gas lines were broken, exploded, and generally disrupted by slumping, landslides, and ground movements. Residents were forced to obtain water and fuel that was trucked in to areas for many months while supply lines were restored. Groundwater levels generally dropped, in some cases below well levels, further compounding the problems of access to fresh water.

After the quake many agencies had to coordinate efforts to demolish unrepairable structures, move facilities and even entire communities out of danger zones, and rebuild lost buildings, roads, and railways. Municipal governments, along with state and federal authorities, helped the U.S. Army Corps of Engineers with the reconstruction effort for urban renewal, with the aim of providing affected communities with better land utilization. Some towns, such as Seward, had to be moved completely to higher, more stable ground where entirely new towns were built. Other towns, cities, and rural areas had to reconstruct the infrastructure, including gas pipelines, roads, railroad tracks, and private homes of thousands of people. Soils were tested for liquefaction potential, and homes were moved away from affected locations. Areas with high landslide risks were avoided, as were coastal areas prone to tsunami. All of these efforts led to reconstruction of the communities of southern Alaska, so that now they are safer places to live and work. Since the area is now much more densely populated than it was in 1964, however, future earthquakes that even approach the strength of the 1964 earthquake are likely to do more damage and kill more people than the 1964 catastrophe.

Southern Chile, 1960 (magnitude 9.5)

The largest earthquake ever recorded struck the Concepción area of southern Chile on May 22, 1960. This was a subduction zone earthquake, and a huge section of the downgoing oceanic slab moved during this and related precursors and aftershocks spanning a few days. The main shock was preceded by a large foreshock at 2:45 P.M. on Sunday, May 22, which was fortunate because this foreshock scared most people into the streets and away from buildings soon to collapse. Thirty minutes later at 3:15 P.M., the magnitude 9.5 event struck and affected a huge area of southern Chile, killing an estimated 3,000 to 5,700 people. Another 3,000 were injured and 2 million were left homeless in the huge area devastated by this quake and aftershocks. Massive landslides, slumps, and collapse of buildings occurred throughout the region. The Chilean government estimated property damage to be approximately $300 million. An approximately 600-mile-long by 190-mile-wide (1,000-km-long by 300-km-wide) section of the fault separating the downgoing oceanic slab from the overriding plate slipped, allowing the oceanic plate to sink farther into the mantle. The area that slipped during this event is roughly the size of California.

The main shock from this earthquake generated a series of tsunamis that ravaged the coast of Chile with 80-foot (24-m) tall waves soon after the earthquake, and these waves raced across the Pacific at 200 miles (320 km) per hour, hitting Hawaii, where 61 people were killed and damage was estimated at $75 million. The tsunami then struck Japan, killing 138 and causing another $50 million in damage. Smaller waves hit the west coast of the United States, causing about $500,000 in damage.

One of the more unusual and unexplained events to follow this earthquake was a massive eruption of the Mount Puyehue (Chile) volcano 47 hours after the main shock, presumably triggered in some way by the earthquake. Although the relationships between earthquakes and volcanic eruptions at convergent margins is not understood, recent observations of many volcanoes around the world have noted

a correlation between the passage of seismic waves, even from very distant earthquakes, and an increased amount of volcanic activity. Further research is needed to understand these phenomena.

Pakistan, October 8, 2005 (magnitude 7.6)

At 8:50 A.M. on Saturday, October 8, 2005, remote areas of northern Pakistan, north of Islamabad and neighboring Afghanistan, were hit by a major earthquake that caused catastrophic damage to a wide area, largely because of the inferior construction of buildings throughout the region. This earthquake killed more than 86,000 people and injured more than 69,000, leaving about 4 million homeless as the freezing cold of the Kashmir winter set in to the mountainous region. Worst hit was the Muzaffarabad area in Kashmir, where 80 percent of the town was destroyed and more than 32,000 buildings collapsed. Numerous landslides and rock falls blocked mountain roads, so it took many days and even weeks for rescue workers to reach remote areas.

The earthquake was initiated by motion on a thrust fault with the epicenter at 16.2 miles (26 km) depth. The thrust fault is part of a system of faults that formed in response to the collision of India with Asia, forming the Himalayan, Karakoram, Pamir, and Hindu Kush ranges. The Indian plate is moving northward at 1.6 inches (4 cm) per year, and is being pushed beneath the Asian plate, forming the high mountains and Tibetan plateau. Slip on a number of faults accommodates this plate motion and has formed a series of northwest-southeast striking-thrust faults in the Muzaffarabad area. These faults deform young Pleistocene alluvial fans into anticlinal ridges, showing that deformation in the region is active and intense, and the region is likely to suffer additional strong earthquakes.

Chi-Chi, Taiwan, 1999 (magnitude 7.3)

On September 21, 1999, a magnitude 7.3 earthquake struck the area near Chi Chi in western Taiwan, causing widespread destruction across the island. A 53-mile (85-km) long segment of the Cher-Lung Pu thrust fault ruptured at 1:47 A.M. when most people were sleeping, moving laterally up to 33 feet (10 m) and vertically up to 32 feet (9.8 meters) within 60 seconds. The earthquake released an amount of energy roughly equivalent to 30 times that released by the atomic bomb dropped on Hiroshima, and was the largest earthquake to strike the island in a century. There were 2,333 documented deaths and more than 10,000 people were injured, plus more than 100,000 homes destroyed, with total economic losses topping $14 billion.

The Chi Chi earthquake was associated with many ground ruptures and surface displacement features, even though the epicenter was located five miles (8 km) below the western foot of the Western Foothill Mountains of Taiwan. New waterfalls were created where the fault rupture crossed rivers, and bridge spans collapsed from movement on unstable riverbanks, and buildings on one side of the scarps were suddenly raised several tens of feet about their neighbors. Landslides and slumps tilted and destroyed many other buildings, typically sending riverfront buildings crashing into the riverbeds. In one example more than 1,060,000,000 cubic feet (30 million m³) of a mountain slope slid in the Jiu-Feng Er Shan slide, killing 42 people. Miraculously a woman and two children were carried more than half a mile (1 km) downhill in the landslide but survived unscratched. In many places sand and even gravel boiled up from liquefaction of buried sediments, forming ridges and sand volcanoes. Near the Ta-An River along the front of the mountains, the slopes of the mountains were changed, forming a new anticline above the thrust fault. The Shih-Kang Dam along the Ta-Chia River collapsed where a strand of the fault crossed the spillways, and the southern side of the dam was raised more than 30 feet (9.8), destroying the water-supply system for Tai Chung County. Numerous buildings collapsed, but often where one building crumbled, the one adjacent to it was barely damaged. Earthquake engineers have studied the structural differences between buildings to improve building codes in the region.

Bam, Iran, 2003 (magnitude 6.7)

Iran sits in the zone of convergence between the Arabian and Asian plates and has numerous mountain ranges that formed by folding and faulting of the rocks in the collision zone. There are many earthquakes in Iran, some of which are extremely destructive and have killed many people. For instance, in 893 an earthquake in Ardabil Iran killed an estimated 150,000 people, and other deadly earthquakes have stricken most regions of Iran, including the capital, Tehran. On December 26, 2003, the ancient Silk Road walled city and citadel of Bam, Iran, was leveled by a magnitude 6.7 earthquake that struck at 5:27 A.M., killing an estimated 27,000–50,000 people and injuring more than 20,000. Bam was in a region characterized by high seismicity and many earthquakes. In fact Bam had survived larger earthquakes in the past during its 2,000-year history without being destroyed, so many scientists were puzzled why a moderate-sized earthquake would totally destroy the city.

The earthquake was preceded by foreshocks on the afternoon of December 25, but since this area is characterized by high seismicity, residents were not alarmed and did not prepare for what was to come.

Many of Bam's residents returned home on Thursday evening, December 25, for the Friday holiday, and were woken at 4:00 A.M. on Friday morning by a strong foreshock that sent many residents into the streets. All seemed calm after a short while, so the residents returned indoors, and most were sleeping at 5:27 A.M., when the main earthquake hit, releasing most of its energy directly below Bam, leveling most of the ancient city that had withstood many earthquakes, drought, and seizure by roving warriors including Genghis Khan. Bam was the oldest walled city, originally founded during the Sassanian period (250 B.C.E.), but much was built in the 12th century, and more in the Safavid period between 1502 and 1722. The walled city included about 4 square miles (6 km²), including more than 10,000 buildings, and was surrounded by 38 towers. Most of the buildings were made of mud bricks, clay, and straw and were not reinforced; hence when the shaking was most intense, these buildings collapsed, burying the residents inside.

Why was Bam destroyed by a magnitude 6.7 earthquake, when it had survived larger earthquakes in the past? This example illustrates that every earthquake is different in terms of where and how its energy is released, and Bam had been fortunate in the past. This event was relatively shallow, and the earthquake focus (the point where the energy was first released) was below Bam. The energy was focused, or directed, by surrounding rock structure directly at the old city, much like sound can be focused or directed by cupping one's hand around the mouth, or by walls in a city or canyons in the wilderness. This earthquake released most of its energy directly toward Bam, destroying the city.

Kobe, Japan, 1995 (magnitude 6.9)

The industrial port city of Kobe, Japan, was hit by history's costliest earthquake ($100 billion in property damage) at 5:46 A.M. on January 17, 1995. The 30-mile (50-km) long fault rupture passed directly through the world's third-busiest port and home to 1.5 million people. With little warning 6,308 died in Kobe before sunrise on that cold January morning. The rupturing event lasted 15 seconds, moved each side of the fault more than six feet (1.7 m) horizontally relative to the other side, and uplifted the land by three feet (1 m). The many areas of unconsolidated sediment in and around Kobe saw some of the worst damage, and shook for as long as 100 seconds because of the natural amplification of the seismic waves. Liquefaction was widespread and caused much of the damage, including collapse of buildings and port structures and destruction of large parts of the transportation network. Water, sewer, gas, and electrical systems were rendered useless. More than 150,000 buildings were destroyed in the initial quake, and a huge fire that started from ruptured gas lines consumed the equivalent of 70 square blocks.

These examples of convergent margin earthquakes show that the strongest, deadliest earthquakes tend to occur at convergent margins. Areas thousands of miles (km) long can suddenly slip in one large earthquake event, generating fast-moving seismic waves, giant tsunamis, landslides, shifts of land level, and indirect effects such as fires, disease, and loss of livelihood for millions. Convergent margin earthquakes are capable of releasing more energy than any other catastrophic Earth event, and are therefore among the most destructive forces of nature. When considered with the volcanic eruptions that characterize many sections of convergent margins it is clear that these beautiful, mountainous areas are among the most hazardous on Earth.

See also ACCRETIONARY WEDGE; CONTINENTAL CRUST; DEFORMATION OF ROCKS; GRANITE, GRANITE BATHOLITH; ISLAND ARCS, HISTORICAL ERUPTIONS; METAMORPHISM AND METAMORPHIC ROCKS; PLATE TECTONICS; STRUCTURAL GEOLOGY; VOLCANO.

FURTHER READING

Moores, Eldridge, and Robert Twiss. *Tectonics*. New York: W.H. Freeman, 1995.

U.S. Geological Survey. "Earthquake Hazards Program." Available online. URL: http://earthquake.usgs.gov/. Accessed October 31, 2008.

Copernicus, Nicolaus (1473–1543) *Prussian (Polish or German, disputed) Astronomer, Mathematician, Physician, Economist, Military Leader, Diplomat* Nicolaus Copernicus is credited with being one of the earliest scientists to propose a scientifically sound model with the Sun as the center of the universe, displacing the Earth from this role in earlier models. His heliocentric model was described in his book *De revolutionibus orbium coelestium* (*On the Revolutions of the Celestial Spheres*, 1939), which many regard as the starting point of modern astronomy and the beginning of a revolution in science known as the Copernican revolution.

EARLY YEARS

Nicolaus Copernicus was born on February 19, 1473, in Toruń, a town on the Vistula River in what is now Poland. His father was a copper trader from Kraków, and his mother, Barbara Watzenrode, was from a wealthy merchant family in Toruń. Nicolaus's father died when he was 10 or 12, at which point

his maternal uncle, Lucas Watzenrode the Younger, who became archbishop of Warmier, raised Nicolaus through his school years.

In 1491 Copernicus began his college years by enrolling in the Kraków Academy (the contemporary Jagiellonian University), where he began his studies in astronomy. After four years of study at Kraków, Copernicus returned to Toruń, then moved to the Universities of Bologna and Padua, where he studied law and medicine, supported financially by his uncle Watzenrode the Younger.

SCIENTIFIC CONTRIBUTIONS

After completing his studies, Nicolaus Copernicus returned to Prussia and took on the position of secretary to his uncle Lucas Watzenrode, who was at the time the bishop of Warmia. During this time he lived at the bishop's castle at Lidzbark Warminski (Heilsberg) and started his research on the heliocentric model of the universe. Copernicus obtained a position as a burgher of Warmia in the Collegiate Church of the Holy Cross in Wroclaw (Breslau) in Bohemia, and he kept this position for most of his life while carrying out his studies as an amateur astronomer. Copernicus was a polymath, serving as economic administrator of Warmia from 1516 to 1521, and as head of the Royal Polish forces for Olsztyn (Allenstein) castle when it was besieged by the Teutonic Knights during the Polish-Teutonic War of 1519–21. He also worked as a diplomat on behalf of the bishop of Warmia and adviser to Duke Albert of Prussia, especially in the fields of monetary reforms, where he was charged with determining and negotiating who had the right to mint coins. Copernicus was also called on for his medical skills, even diagnosing and saving the life of one of Duke Albert's counselors.

Copernicus is most famous for proposing the heliocentric model of the universe. He first formulated it in a six-page, handwritten text, the "Commentariolus" (Little Commentary), preceding his famous six-volume work, *De revolutionibus orbium coelestium*, published in 1543 over three decades or more, although the exact date of the "Commentariolus" is not known. "Commentariolus" contained seven main assumptions:

- There is no one center for all the celestial circles or spheres.
- The center of the Earth is not the center of the universe, but only of gravity and of the lunar sphere.
- All the spheres revolve about the Sun as their midpoint, and therefore the Sun is the center of the universe.

- The ratio of the Earth's distance from the Sun to the height of the firmament (the heavens) is so much smaller than the ratio of the Earth's radius to its distance from the Sun that the distance from the Earth to the Sun is imperceptible in comparison with the height of the firmament.

- Whatever motion appears in the firmament arises not from any motion of the firmament, but from the Earth's motion. The Earth together with its circumjacent elements performs a complete rotation on its fixed poles in a daily motion, while the firmament and highest heaven abide unchanged.

- What appear to observers on Earth as motions of the Sun arise not from its motion but from the motion of the Earth and its sphere, which revolves about the Sun like any other planet. The Earth has, then, more than one motion.

- The apparent retrograde and direct motion of the planets arises not from their motion but from the Earth's. The motion of the Earth alone, therefore, suffices to explain so many apparent inequalities in the heavens.

In the period between the publication of his two major works, Copernicus's ideas were discussed among many scholars and clergy of the time, including the archbishop of Capua, Nicholas Shonberg, who encouraged Nicolaus to communicate his work and discoveries to scholars and to share his writings at the earliest possible moment.

While Copernicus was still finishing *De revolutionibus orbium coelestium,* a student of mathematics named Georg Joachim Rheticus came to work with him, as arranged by Phillip Melanchton, a professor from Prussia (later part of Germany). Over a period of two years Rheticus studied with Copernicus and wrote a book about Copernicus's work, *Narratio prima* (First account), which many scholars read who as a result began to appreciate Copernicus's works. Rheticus followed this in 1542 with a well-received book on trigonometry outlining Copernicus's ideas in this field. Having seen his ideas generally well accepted and not criticized by the clergy at the time, Copernicus agreed to publish his major works through the printer Johannes Petreius in Nuremberg. He published *De revolutionibus orbim coelestium* in 1543 in six volumes categorized by the following:

- general vision of the heliocentric theory, and a summarized exposition of his idea of the world

- mainly theoretical, presents the principles of spherical astronomy and a list of stars (as a basis for the arguments developed in the subsequent books)
- mainly dedicated to the apparent motions of the Sun and to related phenomena
- description of the Moon and its orbital motions
- concrete exposition of the new system

Copernicus died after a long illness on May 24, 1543, the same year that his treatise was published. Legend has it Copernicus awoke from a stroke-induced coma and looked at his book as it was placed in his hands, then died peacefully. He was buried at Frombork Cathedral in northern Poland.

Copernicus's ideas were not only revolutionary, but were quite different from those advocated by the Catholic Church. Despite this, the books caused only minor controversy at first, until three years later in 1546, when Giovanni Maria Tolosani, a Dominican priest, denounced the work, stating that it would have been condemned earlier if the chief censor of the Catholic Church at the time (Bartholomeo Spina) had not died while reviewing it. In 1616 the Roman Catholic Church issued a decree that suspended *De revolutionibus orbim coelestium* until it could be corrected, on the basis that it opposed Holy Scripture. The Italian astronomer Galileo Galilei (1564–1642), who supported Copernicus's ideas, was investigated by Cardinal Bellamine on the orders of Pope Paul V and in 1633 was convicted of grave suspicion of heresy for "following the position of Copernicus, which is contrary to the true sense and authority of Holy Scripture" (Papal Condemnation [Sentence] of Galileo, June 22, 1633 [translated from the Latin], in Giorgio de Santillana, *The Crime of Galileo,* University of Chicago Press, 1955). Galileo was placed under house arrest for the rest of his life. The church's influence was strong in this era—another follower of Copernicus's ideas, Giordano Bruno, was condemned and burned at the stake on February 17, 1600, for being a heretic. The censorship of Copernicus's views continued for centuries, with the original *De revolutionibus orbim coelestium* remaining on the Index of Prohibited Books published by the Catholic Church in 1758. The revolutionary book was finally dropped in the 1835 edition of prohibited works, nearly 300 years after its initial publication.

See also ASTRONOMY; GALILEI, GALILEO; SOLAR SYSTEM.

FURTHER READING
Armitage, Angus. *The World of Copernicus.* New York: Mentor Books, 1951.

Bienkowska, Barbara. *The Scientific World of Copernicus: On the Occasion of the 500th Anniversary of His Birth, 1473–1973.* New York: Springer, 1973.

Koyre, Alexandre. *The Astronomical Revolution: Copernicus-Kepler-Borelli.* Ithaca, N.Y.: Cornell University Press, 1973.

Rosen, Edward. *Copernicus and His Successors.* London: Hambledon Press, 1995.

coral Corals are invertebrate marine fossils of the phylum Cnidaria characterized by radial symmetry and a lack of cells organized into organs. They are related to jellyfish, hydroids, and sea anemones, all of which possess stinging cells. Corals are the best preserved of this phylum because they secrete a hard, calcareous skeleton. The animal is basically a simple sac with a central mouth, surrounded by tentacles, that leads to a closed stomach. Cnidarians are passive predators, catching food that wanders by in their tentacles. Corals and other cnidarians produce alternating generations of two body forms. Medusae are forms that reproduce sexually to form polyps, the asexual forms from which the medusae may bud. Corals belong to a subclass of the anthozoan cnidarians known as the Zooantharia. The jellyfish belong to the Scyphozoa class, and the Hydrozoa class includes both fresh- and saltwater cnidarians dominated by the polyp stage.

Corals can live in a range of conditions from shallow tidal pools to 19,700 feet (6,000 m) depth. They have a cylindrical or conical skeleton secreted by the polyp-stage organism, which lives in the upper exposed part of the structure. The skeleton is characterized by radial ridges known as septa that join the skeleton's outer wall (the theca), and may have flat floors that were periodically secreted by the polyp.

Corals range from the Early Ordovician Tabulata forms, joined in the Middle Ordovician by the rugose corals. They both experienced a major extinction in the Late Devonian, from which the rugose forms recovered stronger. Both forms became extinct in the Early Triassic and were replaced by modern coral forms known as Scleractinia, which apparently arose independently from different soft-bodied organisms.

Most corals grow in colonial communities and form reefs that provide numerous advantages, including shelter for larvae and young stages. Reefs are framework-supported carbonate mounds built by carbonate-secreting organisms, or in some instances

Coral reef in the Red Sea, Egypt *(Specta, Shutterstock, Inc.)*

any shallow ridge of rock lying near the surface of the water. Reefs contain a plethora of organisms that together build a wave-resistant structure that reaches up to just below the low-tide level in the ocean waters and provide shelter for fish and other organisms. The spaces between the framework are typically filled by skeletal debris, which together with the framework become cemented together to form a wave-resistant feature that shelters the shelf from high-energy waves. Modern corals can survive only in shallow waters that range in temperature from 77 to 84°F (25–29°C), at depths of fewer than 300 feet (90 m). Reef organisms (presently consisting mainly of zooxanthellae) can survive only in the photic zone, so reef growth is restricted to the upper 328 feet (100 m) of the seawater. Reefs have various forms including fringing, barrier, and atoll reefs.

Reefs are built by a wide variety of organisms, including red algae, mollusks, sponges, and cnidarians (including corals). The colonial scleractinia corals are presently the principal reef builders, producing a calcareous external skeleton characterized by radial partitions known as septa. Inside the skeleton are soft-bodied animals called polyps, containing symbiotic algae that are essential for the life cycle of the coral and for building the reef structure. The polyps contain calcium bicarbonate that is broken down into calcium carbonate, carbon dioxide, and water. The calcium carbonate is secreted to the reefs building its structure, whereas the algae photosynthesize the carbon dioxide to produce food for the polyps.

There are several different types of reefs, classified by their morphology and relationship to nearby landmasses. Fringing reefs grow along and fringe the coast of a landmass and are often discontinuous. They typically have a steep outer slope, an algal ridge crest, and a flat, sand-filled channel between the reef and the main shoreline. Barrier reefs form at greater distances from the shore than fringing reefs and are generally broader and more continuous than fringing reefs. They are among the largest biological structures on the planet—for instance, the Great Barrier Reef of Australia is 1,430 miles (2,300 km) long. A wide, deep lagoon typically separates barrier reefs from the mainland. All these reefs show a zonation from a high-energy side on the outside or windward side of the reef, and typically grow fast, and have a smooth outer boundary. In contrast the opposite side of the reef receives little wave energy and may be irregular, poorly developed, or grade into a lagoon. Many reefs also show a vertical zonation in the types of organisms present, from deep water to shallow levels near the sea surface.

Atolls or atoll reefs form circular-, elliptical-, or semicircular-shaped islands made of coral reefs that rise from deep water; atolls surround central lagoons, typically with no internal landmass. Some atolls do have small central islands, and these, as well as parts of the outer circular reef, are in some cases covered by forests. Most atolls range in diameter from half a mile to more than 80 miles (1–130 km) and are most common in the western and central Pacific Ocean basin and in the Indian Ocean. The outer margin of the semicircular reef on atolls is the most active site of coral growth, since it receives the most nutrients from upwelling waters on the margin of the atoll. On many atolls coral growth on the outer margin is so intense that the corals form an overhanging ledge from which many blocks of coral break off during storms, forming a huge *talus* slope at the base of the atoll. Volcanic rocks, some of which lie more than half a mile (1 km) below current sea level, underlie atolls. Since corals can grow only in shallow water fewer than 65 feet (20 m) deep, the volcanic islands must have formed near sea level, grown coral, and subsided with time, with the corals growing at the rate that the volcanic islands were sinking.

Charles Darwin proposed such an origin for atolls in 1842 based on his expeditions on the *Beagle* from 1831 to 1836. He suggested that volcanic islands were first formed with their peaks exposed above sea level. At this stage coral reefs were established as fringing reef complexes around the volcanic island. He suggested that with time the volcanic islands subsided and were eroded, but that the growth of the coral reefs kept up with the subsidence. In this way as the volcanic islands sank below sea level, the coral reefs continued to grow and eventually formed a ring circling the location of the former volcanic island. When Darwin proposed this theory in 1842, he did not know that ancient eroded volcanic mountains underlay the atolls he studied. More than 100 years later, drilling confirmed his prediction that volcanic rocks would be found beneath the coralline rocks on several atolls.

With the advent of plate tectonics in the 1970s, the cause of the subsidence of the volcanoes became apparent. When oceanic crust is created at midocean ridges, it is typically about 1.7 miles (2.7 km) below sea level. With time, as the oceanic crust moves away from the midocean ridges, it cools and contracts, sinking to about 2.5 miles (4 km) below sea level. In many places on the seafloor small volcanoes form on the oceanic crust a short time after the main part of the crust formed at the midocean ridge. These volcanoes may stick above sea level a few hundred meters. As the oceanic crust moves away from the midocean ridges, these volcanoes subside below sea level. If the volcanoes happen to be in the tropics where corals can grow, and if the rate of subsidence is slow enough for the growth of corals to keep up with subsidence, then atolls may form where the volcanic island used to be. If corals do not grow or cannot keep up with subsidence, then the island subsides below sea level and the top of the island gets scoured by wave erosion, forming a flat-topped mountain that continues to subside below sea level. These flat-topped mountains are known as guyots, many of which were mapped by military surveys during exploration of the seafloor associated with military operations of World War II.

Reefs are extremely sensitive and diverse environments, and cannot tolerate large changes in temperature, pollution, turbidity, or water depth. Reefs have also been subject to mining and destruction for navigation, and have even been sites of testing nuclear bombs in the Pacific. Thus, human-induced and natural changes in the shoreline environment pose a significant threat to the reef environment.

Reefs are rich in organic material and have high primary porosity, so they are a promising target for many hydrocarbon exploration programs. Reefs are well represented in the geological record, with examples including the Permian reefs of west Texas; the Triassic of the European Alps, the Devonian of western Canada, Europe, and Australia; and the Precambrian of Canada and South Africa. Organisms that produced the reefs have changed dramatically with time, but, surprisingly, the gross structure of the reefs has remained broadly similar.

See also DIVERGENT PLATE MARGIN PROCESSES.

FURTHER READING
Botkin, D., and E. Keller. *Environmental Science.* Hoboken, N.J.: John Wiley & Sons, 2003.

Davis, R., and D. Fitzgerald. *Beaches and Coasts.* Malden, Mass.: Blackwell, 2004.

Coriolis, Gustave (1792–1843) French *Mathematician, Engineer, Scientist* Gustave Coriolis, also known as Gaspard-Gustave de Coriolis, was born on May 21, 1792, son of Jean-Baptiste-Elzéar Coriolis and Marie-Sophie de Maillet. He is best known for his work on the Coriolis force caused by the Earth's rotation, but also was the first person to define work as the product of force times distance, and he defined kinetic energy as it is used in the current scientific meaning. He died on September 19, 1843, at the age of 51, in Paris, while he was a professor at the École Centrale Paris.

Gustave's father was a military officer who served with Louis XVI in 1790. This caused difficulties for the family during the French Revolution when the king was caught while attempting to flee Paris and was returned to the capital, where he was guillotined

in January 1793. Gustave's family fled to Nancy, where his father became an industrialist, while the son attended schools. In 1808 he entered the École Polytechnique and on graduating entered the École des Ponts et Chaussées in Paris. For the next few years Coriolis worked with the engineering corps in the Meurthe-et-Moselle district in the Vosges Mountains, and after his father died he worked long hours to raise money to support his family, even though his health was failing.

Coriolis became a tutor at the École Polytechnique in 1816, where he experimented on friction and hydraulics, publishing a textbook in 1829, *Calcul de l'effet des machines* (Calculation of the effect of machines), describing the science of mechanics in a way that industry could apply. The same year Coriolis took the position of professor of mechanics at the École Centrale des Arts et Manufactures in Paris, and in 1832 he took on a position at the École des Ponts and Chaussées and was elected to the Academie des Sciences. Coriolis spent much of his time over the next years working on the principles of kinetic energy as applied to rotating systems and eventually published his famous paper in 1835 "Sur les équations du mouvement relatif des systèmes de corps" (On the equations of relative motion of a system of bodies), relating to the transfer of energy in rotating systems such as gears and waterwheels. In 1838 Coriolis ended his teaching career and became director of studies, but in spring 1843 his poor health took a dramatic turn for the worse, and he died in early fall of that year.

Gustave Coriolis did not work directly with the forces in the atmospheric system, yet by the end of the 19th century his work was being applied to ideas about the ceneral circulation of the atmosphere and relationships between atmospheric pressure and winds. He is recognized for this contribution since he showed that the laws of motion could be applied to a rotating frame of reference if an extra force, now called the Coriolis acceleration, is added to the equations of motion.

See also ATMOSPHERE; CORIOLIS EFFECT.

FURTHER READING
Coriolis, G.-G. "Calcul de l'effet des machines." 1829. Reprinted as "Traité de la Mécanique des corps solides." 1844.

———. "Sur le principe des forces vives dans les mouvements relatifs des Machines." *Journal de l'Ecole royale polytechnique* 13 (1832): 268–302.

———. "Sur les équations du mouvement relatif des systèmes de corps." *Journal de l'Ecole royale polytechnique* 15 (1835): 144–154.

———. *Theorie mathématique des effets du jeu de billard.* Paris: Carilian-Goeury, 1835. Reprint. Paris: J. Gabay, 1990.

Coriolis effect   The Coriolis effect, or force, produces a deflection of moving objects and currents to the right in the Northern Hemisphere and to the left in the Southern Hemisphere. The force ranges from zero at the equator to a maximum at the poles and can be understood by considering the rotation of the Earth from a position above the poles. If we look down from above the North Pole, the rotation appears to be counterclockwise, and from above the South Pole it appears to be clockwise. Points on or very near the poles do not have to travel far during one rotation of the Earth about its axis because the circular path traveled has a relatively small diameter, whereas points that lie on the equator must speed through space at 1,000 miles per hour (1,600 km/hr) to complete a rotation, traveling a distance approximately the length of the Earth's diameter within 24 hours.

As air and water move poleward from the equatorial regions, they bring with them the higher velocity they acquired while traveling closer to the equator. The slower speeds of rotation of the Earth under the moving air and water cause these fluids to move a greater distance per unit time than the underlying Earth, and this results in a deflection to the right in the Northern Hemisphere, and to the left in the Southern Hemisphere. Likewise, air and water moving from the poles to the equator will be moving slower than the underlying Earth, this time causing the Earth to move more per unit time than the air or water. This again causes a deflection to the right in the Northern Hemisphere, and to the left in the Southern Hemisphere.

See also ATMOSPHERE; CORIOLIS, GUSTAVE.

cosmic microwave background radiation
In 1964 two American scientists, Arno Penzias and Robert Wilson, were working on a project at Bell Laboratories in New Jersey to identify and eliminate sources of interference with satellite communications. In their work they accidentally stumbled on one of the most important finds in astronomy and astrophysics of the century. While Penzias and Wilson were examining the radiowave emissions from the Milky Way Galaxy using microwave wavelengths, they discovered a background "hiss" that would not go away, no matter which direction they pointed their receiving antenna or when they took the measurements. They then showed that this background hiss persisted throughout the year and was isotropic, meaning it had the same intensity in all directions.

Penzias and Wilson labored over ways to explain this noise in the radio signal, seeking explanations ranging from short circuits and equipment malfunction, to ground interference, atmospheric storms, and everything else they and their colleagues could think of as possible causes. All of these simple explanations failed, and in the end Penzias and Wilson concluded that this low-level noise signal was a remnant of the radiation left over from the initial big bang and the creation of the universe. They presented the results of their findings and analysis, and won the 1978 Nobel Prize in physics.

The background hiss at microwave wavelengths that Penzias and Wilson discovered became known as the cosmic microwave background radiation. Such a signal had been predicted to exist by physicists since the 1940s, and more specifically by Princeton University researchers in the early 1960s. These predictions were based on the idea that the early universe after the big bang must have been extremely dense and hot, and filled with high-energy thermal radiation including short-wavelength gamma rays. In the papers published before Penzias and Wilson's discovery, scientists predicted that as the universe expanded and cooled, the frequency of this primordial background radiation would have shifted from gamma rays, to X-rays, to ultraviolet rays, and eventually to radio waves. This became known as theoretical black body radiation, with its frequency decreasing as the body (the universe) cooled and expanded. As other scientists, including the Princeton group, examined the discovery of Penzias and Wilson, they determined that the temperature of the present-day cosmic background radiation corresponds to a black body with a temperature of about 3 kelvin.

Some wavelengths of the cosmic background radiation are difficult to measure from the Earth. Scientists spent years of research trying to make better measurements of this radiation, using observation platforms including high-flying balloons, and designed other experiments that were never completed. In 1989, however, the satellite *Cosmic Background Explorer* (COBE) was launched, and it measured the intensity of the cosmic background radiation at different wavelengths, verifying that it corresponded to a black body radiation curve for a body with a temperature of 2.7 kelvin. COBE also verified that the background radiation is highly isotropic, or uniform in different directions. But COBE detected a very slight shift of the radiation toward the blue in the direction of the constellation Leo, and toward the red in the exact opposite direction. These shifts were equated with Doppler shifts of the background radiation caused by the motion of the Earth in space. Motion causes radiation to shift toward shorter and hotter wavelengths (blue) toward the direction the body (Earth) is moving, caused by the contraction of the incoming radiation, and to the red (cooler) in the opposite direction, as the radiation has to travel a little farther to get to the new position of the Earth as it moves. The Doppler shifts (0.0034 kelvin) of the cosmic background radiation correspond to motion of the Earth through space of 248 miles/sec (400 km/sec) toward the constellation Leo.

Since radiation cannot travel faster than the speed of light, and the cosmic background radiation is coming from so far away, the 3-kelvin radiation now observed from Earth (or Earth orbit) was emitted far back in time, almost from the time of the beginning of the universe. The photons now being received on Earth as this cosmic background radiation have not interacted with any matter since the universe was only about 100,000 years old (some 13 or 14 billion years ago), and when the universe was less than 1/1,000 of its present size. At present this is about the furthest back in time and space that can be observed without entering the fields of nuclear and particle physics. The cosmic background radiation represents the oldest snapshot of how the universe looked when it was young.

See also ASTRONOMY; ASTROPHYSICS; ORIGIN AND EVOLUTION OF THE UNIVERSE.

FURTHER READING

Chaisson, Eric, and Steve McMillan. *Astronomy Today.* 6th ed. Upper Saddle River, N.J.: Addison-Wesley, 2007.

Comins, Neil F. *Discovering the Universe.* 8th ed. New York, W. H. Freeman, 2008.

National Aeronautics and Space Administration. "Universe 101, Our Universe, Tests of Big Bang: The CMB web page." Available online. URL: http://map.gsfc.nasa.gov/universe/bb_tests_cmb.html. Last updated April 28, 2008.

Snow, Theodore P. *Essentials of the Dynamic Universe: An Introduction to Astronomy.* 4th ed. St. Paul, Minn.: West, 1991.

cosmic rays Extremely energetic particles that move through space at close to the speed of light are known as cosmic rays. Some cosmic rays are so energetic that they can have energies of about 10^{20} electron volts, much greater than the 10^{13} electron volts that can be produced by particle accelerators on Earth. Most cosmic rays are made of atomic nuclei, about 90 percent of which are bare hydrogen nuclei (protons), 9 percent are helium nuclei (alpha particles), and about 1 percent are electrons (beta minus particles), but cosmic rays include a whole range of particles that spans the entire periodic table of the elements. The name *ray* as applied to cosmic rays is a misnomer, since these particles act individually, not as rays or beams.

The Austrian-American physicist Victor Hess discovered cosmic rays in 1912. There are several different classes of sources for cosmic rays. Galactic cosmic rays come from outside the solar system, with most originating from explosive processes in other stars of the Milky Way Galaxy. These sources can include neutron stars, supernovas, and black holes. Other galactic cosmic rays come from the most distant parts of the visible universe, particularly the cosmic rays with very high energy. Supernovas can produce cosmic rays with energies up to 10^{14} electron volts, but other, more energetic processes must be the source of the very high-energy particles. The source of these very high-energy particles was unknown until observations by a team from 17 countries from the Pierre Auger International Cosmic Ray Observatory in Mendoza, Argentina, identified the most likely source for these particles to be from active galactic nuclei. Solar cosmic rays, which consist mostly of protons and alpha particles, are ejected from the Sun during solar flare events and generally have lower energy than galactic cosmic rays. Also known as solar energetic particles, they have a composition similar to that of the Sun and overlap in their phases with the solar wind.

Cosmic rays constantly bombard the surface of the Earth, and the flux, or flow rate, of these subatomic particles is controlled by the background amount of particles, the solar wind, and the Earth's magnetic field. The magnetized plasma of the solar wind decelerates incoming cosmic ray particles and partially excludes those with energies of less than 1 giga electron volt (GeV, equal to 10^9 electron volts, about equal to the mass equivalent of one atomic unit). The 11-year cycle of solar activity means that the solar wind is not constant over time—the modulating effect of the solar wind on the flux of cosmic rays changes with time. The Earth's magnetic field also modifies and deflects cosmic rays, as confirmed by the fact that the flux shows variation in intensity based on latitude, longitude, and azimuth (direction of approach), and with polarity of the magnetic field. More cosmic rays hit the Earth at the magnetic poles than at the equator, since the magnetic field lines tend to move the charged particles in the direction of the field and not across them, as would happen at the equator.

Cosmic rays can be detected on the Earth by tracing their tracks in specially designed liquid-filled devices called bubble chambers or with a number of other detection instruments. Bubble chambers work on the principle that the cosmic rays are electrically charged; their passage through a special liquid-filled container leaves a trail of bubbles whose length and shape are related to the charge and speed of the particles. Cosmic rays travel at close to the speed of light in the low density of interstellar space. When they approach the Earth, they interact with matter such as atmospheric gases, forming collisions that then form pions, kaons, and mesons, all of which are subatomic particles that rapidly decay into muons, which may reach the Earth's surface. Muons are ionizing radiation, and they are detected by devices such as the bubble chambers and scintillation detectors. They can also be measured in high-altitude particle detectors where the muons interact with special plastics that are then etched to reveal their tracks.

Cosmic rays can interact with some electronics, even changing the states of elements in integrated circuits, leading to corrupted data and transient errors in electronic devices, especially in high-altitude devices such as satellites. While the radiation received from cosmic rays by people on Earth is very small, this would increase for any long-term space travelers, posing a potentially serious obstacle to long-distance space travel for humans.

See also ASTRONOMY; BLACK HOLES; ORIGIN AND EVOLUTION OF THE UNIVERSE; SUPERNOVA.

FURTHER READING

Chaisson, Eric, and Steve McMillan. *Astronomy Today.* 6th ed. Upper Saddle River, N.J.: Addison-Wesley, 2007.

Comins, Neil F. *Discovering the Universe.* 8th ed. New York: W. H. Freeman, 2008.

Snow, Theodore P. *Essentials of the Dynamic Universe. An Introduction to Astronomy.* 4th ed. St. Paul, Minn.: West, 1991.

cosmology Cosmology is the study of the structure and evolution of the universe. Cosmologists seek answers to questions such as how big and how old is the universe, what is its structure on small and large scales, and what is its history and fate—will it expand forever, or collapse back on itself? Cosmology encompasses topics including understanding the life cycles of stars and galaxies, the formation of matter, and the natural laws that hold the universe together.

To study the universe and its history, it is first necessary to make some basic observations. The first is that the universe has structure and order, at every scale observable, from subatomic particles to the distribution of galactic clusters. This order means that laws of nature are being followed that control this structure (an alternative way to view this is that the laws of nature that we construct are our best attempt to explain the observable universe). One interesting observation about the order of the universe is that matter is clustered at scales ranging from subatomic particles to superclusters of galaxies. At the largest scales of observation, however, covering huge expanses of the universe, it becomes clear that the distribution of matter becomes homogeneous at

huge distances. Said another way, each area of universe that is about 200–300 megaparsecs across ([1 parsec = about 3.26 light years, 19.174×10^{12} miles [30.857×10^{12} km]; mega parsec = 1 million parsecs, or 3,262,000 light years, or 2×10^{19} miles) is roughly homogenous.

Cosmology is based on a foundation of some observationally and theoretically supported assumptions about the structure of the universe. The cosmological principle states that the large-scale structure of the universe is homogenous and isotropic, meaning that at large scales the distribution of mass is the same in all parts of the universe, and that the universe looks the same in all directions of observation. The cosmological principle has important implications. For instance, the concept of homogeneity means that there is no edge to the universe, and the idea of isotropy means that there is no center of the universe. This seems to be confirmed by the observation that all galaxies in the universe can be measured to be rushing away from our galaxy, and that all galaxies appear to be moving relatively away from one another, in an expanding universe.

Accepting the cosmological principle of an isotropic and homogenous infinite universe leads to problems with basic observations from Earth. If the universe is infinite, homogenous, and isotropic, then a line of sight in any direction from Earth (or anywhere else) should eventually intersect with a star, and therefore there should be stars visible in all places in the sky, and the sky should be uniformly bright at night. The fact that the sky is mostly dark is known as Olber's paradox, after the German astronomer Heinrich Olber, who popularized this idea in the 19th century.

How can one reconcile Olber's paradox with the cosmological principle and with the observations of an expanding universe? Hubble's law (formulated by two American astronomers from the Midwest, Edwin Hubble [1889–1953] and Milton Humason [1891–1972] in 1929) states that the redshift of light coming from distant galaxies is proportional to their distance. This can be expressed as follows:

$$\text{recession velocity} = H_0 \times \text{distance}$$

where H_0 is Hubble's constant, which is equal to 46.5 miles (75 km)/sec/Megaparsec. Hubble's law can be used to estimate the age of the universe. Since the rate of recession of different galaxies from one another is known, and the recession rate increases with increasing distance, the time backward since the galaxies were all together in one place can be calculated. Since the rate increases with increasing distance, this number turns out to be the same regardless of which galaxies are used, and the date at which all galaxies were together turns out to be roughly 13 billion years ago. At that time everything in the universe, all matter, radiation, dark matter, dark energy—everything—was confined to a singular point, and from that instant exploded outward in a massive primordial expansion known as the big bang. One must remember, however, that the rate of expansion is assumed to be constant in this calculation, and most models for the universe suggest that the initial expansion may have been faster than that at present, which would make the true age of the universe younger than these estimates.

The concept of the big bang, where the universe started at a point and has been expanding with a certain velocity since then, implies that the universe is finite. This idea seems at odds with the cosmological principle, stating that the universe is infinite. The reconciliation can be understood by realizing that in the night sky only a finite part of the universe is visible. The light that reaches Earth is only from the part of the universe that is fewer than 13 billion years old; light from more distant parts, both in time and in space, has not reached Earth yet. Further, the idea of the big bang seems to indicate that the universe is finite, and that light from more than 13 billion years ago does not exist. The paradox still exists.

To understand this paradox more deeply, it is necessary to consider the nature of the big bang. The big bang was not matter suddenly exploding into space, expanding from a point. The big bang instead was the sudden expansion of everything, including space itself. The only known explanation for these observations is that the big bang was the sudden creation and expansion of space into nothing, and the galaxies, and everything else in the universe, is just along for the ride, expanding and moving away from one another, as the space of the universe expands. Therefore although the galaxies are moving away from one another, they are not moving (except for the smaller, and internal motions) with respect to the fabric of space. The space is actually expanding at the same rate the galaxies are moving apart from one another. Understanding the universe this way makes it clear that the big bang did not happen "somewhere," with galaxies exploding from a distant point into space, but instead, the big bang happened "everywhere"—before the big bang, the infinitesimally small point that space expanded from was the entire universe, and the big bang happened everywhere in the whole universe at the same instant. Since then all space and all points in the universe have been moving away from one another.

It is difficult for the human mind to comprehend what existed before the big bang. The big bang itself was a singularity, a moment when according to the laws of physics, all of the mass and energy and space of the universe had zero size, and infinite density and temperature. This idea is based on the laws of physics as deciphered for a finite universe, however, and these laws do not apply to states such as singularities. The big bang represents the start of space, the beginning of mass, and the start of time. Therefore there is no concept such as what there was before the big bang, since before the time of the big bang, there was no time, there was no "before."

The best descriptions of what the universe may have looked liked, and how it behaved, in the instant of the big bang are found in Einstein's general theory of relativity, where the presence of extremely dense mass warps the space and time around it—and conversely, the curvature of space dictates how matter moves through that space. This leads to a different interpretation of the redshifts and inferred increasing velocity of distant galaxies as the distance between them increases. According to Hubble's law, described above, this is due to the increasing relative velocities with increasing distance. Now according to Einstein's general theory of relativity, it is necessary to think of this not as an increase in velocity, but instead as an expansion of the space in the universe. As the space through which the light is traveling expands, the wavelengths of the photons of light are influenced by this expansion, and they shift to longer wavelengths, causing the redshift. The amount of the redshift therefore corresponds to how much the universe has expanded since the photon was emitted—the further the photon came from, the more the expansion and the greater the redshift. The redshift really measures how much the universe has expanded since the photon was emitted, not the relative velocity of the distant galaxy from the observer. For most situations, however, the more familiar Newtonian laws of motion work quite well for objects and forces in the universe.

The shape of the universe depends on how much mass the universe contains, according to these relativistic concepts. Since mass warps space and time, the amount of mass in the universe actually determines the geometry of space itself. If the mass is above a critical value, then the space in the universe is curved back on itself and is said to be a closed universe, with a positive curvature—much like the inside of a sphere. In this kind of space, anything that moves in a straight line could theoretically end up back at the same position it started from, much like a small ball rolling inside a larger sphere. If the mass is less than the critical value, the shape of space is said to be open and has a negative curvature—much like the shape of a saddle. The third possibility is of a flat universe, where the density is exactly equal to the critical value.

COSMOLOGICAL MODELS FOR THE EVOLUTION OF THE UNIVERSE

Observations on the current state of the universe such as the relative velocities of galaxies show that it is expanding. A range of models suggests different fates for the universe from this point. Some models suggest that the universe will continue to expand forever, some that it will slow down and eventually collapse back on itself, perhaps repeating expansion/contraction cycles many times. Still other models suggest that the universe will reach a steady state, with the expansion slowing until it just stops. The factor deciding which of these models will occur depends primarily on how much mass is present in the universe. With a sufficient quantity, gravity may take over and cause the universe to collapse eventually; if the universe contains too little mass, then the expansion can continue indefinitely. A thin critical mass in between would suggest a steady-state universe that would eventually slow its expansion, attaining a steady state. The amount of matter per volume of space it takes to determine the difference among these different possible scenarios depends on the rate of expansion and is known as the critical density.

The future of the universe is vitally dependent on this critical density. If the density of matter is above the critical value, then the gravitational attraction will be enough to halt the expansion, and gravity would start to pull all the matter of the universe back inward, with all points moving toward one another as space contracted just in the opposite way from when it was expanding. This would take precisely the same amount of time that it took for the universe to expand, and everything would happen in reverse. Observers would see that nearby galaxies were blue-shifted—but distant galaxies could still show red-shifts, since the time it takes for the light to reach the observer could be greater than the time since the contraction started.

As the universe and space itself contract in this model, the frequency of collisions between galaxies increases, and the whole universe heats up and becomes denser. The process continues until the temperature is so high the whole universe is hotter than typical stars, and the contraction will continue to the point of a superhot, superdense, supersmall singularity, very similar or identical to that from which the universe first expanded. At this point the normal laws of physics break down again, and understanding the meaning of the universe that has no space and no time and infinite mass is beyond comprehension. Nonetheless, many cosmologists speculate that at this point the universe may suddenly go through another big bang and expansion phase, and that the universe may experience an infinite number of expansions and contractions, in an oscillating or bouncing universe model.

If the density of the universe is below the critical value, then its fate and future will be dramatically different. A universe with a density below the critical value will expand forever, and the radiation and light received from distant galaxies will eventually fade and disappear. Eventually even the light from the stars in the local Milky Way would disappear as they use up their fuel and go dark. The fate of this universe is a cold, dark death. At present it is difficult to tell which of these models may be closer to the real fate of the universe, but the critical measurement is being able to tell the true density of the universe.

Determining the density of the universe is not a simple task, since only about 4 percent of matter in the universe is visible and about 22 percent is dark matter, and about 74 percent of the rest of the universe is estimated to be dark energy. When astronomers calculate the density of the universe, using all the matter that they can see, the density of this luminous matter comes to about $10–28$ kg/m^3, or about 1 percent of the mass needed to be the critical mass of the universe. Even if we include the present estimates for the dark matter in the universe, however, most models for the density suggest that the universe has only 20–30 percent of the mass of the critical density, implying that the universe will expand forever, and not retract upon itself.

Large uncertainties about the amount of dark matter in the voids in space and on very large scales may cause the amount of mass in the universe to be grossly underestimated. Some estimates of the amount of dark matter in these areas place the density of the universe very close to the critical value, so the present limits of detection cannot tell the future of the universe, or whether the future is dark and cold, or bright and very hot.

See also ASTRONOMY; ASTROPHYSICS; DARK MATTER; GALAXIES; ORIGIN AND EVOLUTION OF THE UNIVERSE; UNIVERSE.

FURTHER READING

Chaisson, Eric, and Steve McMillan. *Astronomy Today.* 6th ed. Upper Saddle River, N.J.: Addison-Wesley, 2007.

Comins, Neil F. *Discovering the Universe.* 8th ed. New York: W. H. Freeman, 2008.

National Aeronautics and Space Administration. "Universe 101, Our Universe, Big Bang Theory. Cosmology: The Study of the Universe web page." Available online. URL: http://map.gsfc.nasa.gov/universe/. Updated May 8, 2008.

Snow, Theodore P. *Essentials of the Dynamic Universe: An Introduction to Astronomy.* 4th ed. St. Paul, Minn.: West, 1991.

craton Cratons are large areas of relatively thick continental crust that have been stable for long periods of geological time, generally since the Archean. Most cratons are characterized by low heat flow and few or no earthquakes, and many have a thick mantle root or tectosphere that is relatively cold and refractory, having had a basaltic melt extracted from it during the Archean.

Understanding the origin of stable continental cratons hinges on recognizing which processes change the volume and composition of continental crust with time, and how and when juvenile crust evolved into stable continental crust. The evidence from the preserved record suggests that the continental landmass has been growing since the early Archean, although the relative rates and mechanisms of crustal recycling and crustal growth are not well known and have been the focus of considerable geological debate. The oldest rocks known on the planet are the circa 4.0-Ga Acasta gneisses from the Anton terrane of the Slave Province. The Acasta gneisses are chemically evolved and show trace and REE patterns similar to rocks formed in modern suprasubduction zone settings. Furthermore, the 3.8-billion-year-old Isua sequence from Greenland, the oldest known sedimentary sequence, is an accretionary complex. A few circa 4.2-Ga zircon grains have been found, but it is not clear whether these were ever parts of large continental landmasses. Approximately half of the present mass of continental crust was extracted from the mantle during the Archean.

Exposed portions of Archean cratons are broadly divisible into two main categories. The first are the "granite-greenstone" terranes, containing variably deformed assemblages of mafic volcanic/plutonic rocks, metasedimentary sequences, remnants of older quartzo-feldspathic gneissic rocks, and abundant late granitoids. The second main class of preserved Archean lithosphere is found in the high-grade quartzo-feldspathic gneiss terranes. Relatively little deformed and metamorphosed cratonic cover sequences are found over and within both types of Archean terrain, but they are especially abundant on southern Africa's Kaapvaal craton. Also included in this category are some thick and laterally extensive carbonate platforms similar in aspect to Phanerozoic carbonate platforms, indicating that parts of the Archean lithosphere were stable, thermally subsiding platforms.

Although the rate of continental growth is a matter of geological debate, most geological data indicate that the continental crust has grown by accretionary and magmatic processes taking place at convergent plate boundaries since the early Archean. Arclike trace element characteristics of continental crust suggest that subduction zone magmatism has played an important role generating the continental crust. Convergent margin accretionary processes that contribute to the growth of the continental crust can be divided into five major groups: (1) oceanic plateau accretion, (2) oceanic island arc accretion, (3) normal ocean crust (midocean ridge) accretion/ophiolite obduction, (4) back arc basin accretion, and (5) arc-trench migration/Turkic-type orogeny accretion. These early accretionary processes are typically followed by intrusion of late-stage anatectic granites, late gravitational collapse, and late strike-slip faulting. Together these processes release volatiles from the lower crust and mantle, and help to stabilize young accreted crust and form stable continents.

JUVENILE ISLAND ARC ACCRETION

Many Archean granite-greenstone terranes are interpreted as juvenile island arc sequences that grew above subduction zones and later amalgamated during collisional orogenesis to form new continental crust. The island arc model for the origin of the continental crust is supported by geochemical studies that show the crust has a bulk composition similar to arcs. Island arcs are extremely complex systems that may exhibit episodes of distinctly different tectonics, including accretion of ophiolite fragments, oceanic plateaux, and intra-arc extension with formation and preservation of back arc and intra-arc basins. Many juvenile arcs evolve into mature island arcs in which the magmatic front has migrated through its own accretionary wedge, and many evolve into continental margin arcs after they collide with other crustal fragments or continental nuclei.

Although accretion of immature oceanic arcs appears to have been a major mechanism of crustal growth in Archean orogens, some people argue that oceanic arc accretion alone is insufficient to account for the rapid crustal growth in Precambrian shields. Furthermore, most oceanic arcs are characterized by mafic composition, whereas the continental crust is andesitic.

OPHIOLITE ACCRETION

Ophiolites are a distinctive association of allochthonous rocks interpreted to form in a variety of plate tectonic settings such as oceanic spreading centers, back arc basins, forearcs, arcs, and other extensional magmatic settings including those in association with plumes. A complete ophiolite grades downward from pelagic sediments into a mafic volcanic complex generally made of mostly pillow basalts, underlain by a sheeted dike complex. These are underlain by gabbros exhibiting cumulus textures, then tectonized peridotite, resting above a thrust fault that marks the contact with underlying rock sequences. The term *ophiolite* refers to this distinctive rock association and should not be used in a purely genetic way to refer to allochthonous oceanic lithosphere rocks formed at midocean ridges.

Very few complete Phanerozoic-like ophiolite sequences have been recognized in Archean greenstone belts. However, the original definition of ophiolites includes "dismembered," "partial," and "metamorphosed" varieties, and many Archean greenstone belts contain two or more parts of the full ophiolite sequence. Archean oceanic crust was possibly thicker than Proterozoic and Phanerozoic counterparts, and this resulted in accretion predominantly of the upper section (basaltic) of oceanic crust. The thickness of Archean oceanic crust may in fact have resembled modern oceanic plateaux. If this were the case complete Phanerozoic-like, midocean ridge basalt (MORB)–type ophiolite sequences would have

been very unlikely to be accreted or obducted during Archean orogenies. In contrast, only the upper, pillow lava–dominated sections would likely be accreted.

Portions of several Archean greenstone belts have been interpreted to contain dismembered or partial ophiolites. Accretion of MORB-type ophiolites has been proposed as a mechanism of continental growth in a number of Archean, Proterozoic, and Phanerozoic orogens. Several suspected Archean ophiolites have been particularly well documented. One of the most disputed is the circa 3.5-Ga Jamestown ophiolite in the Barberton greenstone belt of the Kaapvaal craton of southern Africa. The Jamestown sequence contains a 1.8-mile (3-km) thick sequence including a basal peridotite tectonite unit with chemical and textural affinities to Alpine-type peridotites, overlain by an intrusive-extrusive igneous sequence, and

DIAMONDS ARE FOREVER

Diamonds are the most precious of all stones, adorning many engagement rings, necklaces, and other jewelry. They are admired for their hardness, clarity, beauty, and ability to divide light into its component colors. Diamonds, it is said, are forever. Diamonds are stable crystalline forms of pure carbon that form only at high pressures in cool locations in the Earth's mantle. Their origin is restricted, therefore, to places in the subcontinental mantle where these conditions exist, between 90 and 125 miles (150–200 km) below the surface. The vast majority of diamonds that make their way back to the surface are brought up from these great depths by rare and strange explosive volcanic eruptions known as kimberlites.

Kimberlites and related rocks found in diatremes are rare types of continental volcanic rock, produced by generally explosive volcanism with an origin deep within the mantle. They form pipelike bodies extending vertically downward and are the source of most of the world's diamonds. Kimberlites were first discovered in South Africa during diamond exploration and mining in 1869, when the source of many alluvial diamonds on the

Vaal, Orange, and Riet Rivers was found to be circular mud "pans," later appreciated to be kimberlite pipes. In 1871 two diamond-rich kimberlite pipes were discovered on the Vooruitzigt Farm in South Africa, owned by Nicolas de Beer. These discoveries led to the establishment of several large mines and one of the most influential mining companies in history.

Kimberlites are complex volcanic rocks with mixtures of material derived from the upper mantle and water-rich magma of several different varieties. A range of intrusive volcanic styles, including some extremely explosive events, characterizes kimberlites. True volcanic lavas are only rarely associated with kimberlites, so volcanic styles of typical volcanoes are not typical of kimberlites. Most near-surface kimberlite rocks are pyroclastic deposits formed by explosive volcanism filling vertical pipes, and they are surrounded by rings of volcanic tuff and related features. The pipes are typically a couple hundred yards wide, with the tuff ring extending another hundred yards or so beyond the pipes. The uppermost part of many kimberlite pipes includes reworked pyroclastic rocks,

deposited in lakes that filled the kimberlite pipes after the explosive volcanism blasted much of the kimberlite material out of the hole. Geologic studies of kimberlites have suggested that they intrude the crust suddenly and behave differently from typical volcanoes. Kimberlites intrude violently and catastrophically, with the initial formation of a pipe filled with brecciated material from the mantle, sometimes including diamonds, reflecting the sudden and explosive character of the eruption. As the eruption wanes, a series of tuffs fall out of the eruption column, forming the tuff ring around the pipes. Unlike most volcanoes, kimberlite eruptions are not followed by the intrusion of magma into the pipes. The pipes simply get eroded by near-surface processes, lakes form in the pipes, and nature tries to hide the very occurrence of the explosive event.

Below these upward-expanding craters are deep vertical pipes known as diatremes that extend down into the mantle source region of the kimberlites. Many diatremes have features that suggest the brecciated mantle and crustal rocks were emplaced at low temperature capped by a chert-shale sequence. This partial ophiolite is pervasively hydrothermally altered and shows chemical evidence for interaction with seawater with high heat and fluid fluxes. Silicon dioxide (SiO_2) and magnesium dioxide (MgO) alteration and black smokerlike mineralization is common, with some hydrothermal vents traceable into banded iron formations and subaerial mudpool structures. These features led Maarten de Wit and others in 1992 to suggest that this ophiolite formed in a shallow sea and was locally subaerial, analogous to the Reykjanges ridge of Iceland. In this sense Archean oceanic lithosphere may have looked very much like younger oceanic plateaux lithosphere.

Several partial or dismembered ophiolites have been described from the Slave Province of northern Canada. A fault-bounded sequence on Point Lake grades downward from shales and chemical sediments (umbers) into several kilometers of pillow lavas intruded by dikes and sills, locally into multiple dike/sill complexes, then into isotropic and cumulate-textured layered gabbro. The base of this partial Archean ophiolite is marked by a 3,000-foot (1-km) thick shear zone composed predominantly of mafic and ultramafic mylonites, with less deformed domains including dunite, websterite, wherlite, serpentinite, and anorthosite. Syn-orogenic conglomerates and sandstones were deposited in several small foredeep basins, and are interbedded with mugearitic lavas (and associated dikes), all deposited/intruded in a foreland basin setting.

A complete but dismembered and metamorphosed 2.5-billion-year-old ophiolite complex from the North China craton has been described. This

---

nonviolently, presenting a great puzzle to geologists. How can a deep source of broken mantle rocks passively move up a vertical pipe to the surface, suddenly explode violently, then disappear beneath a newly formed lake?

Early speculations on the intrusion and surface explosion of the diamond-containing kimberlites suggested that they rose explosively and catastrophically from an origin in the mantle. Subsequent studies revealed that the early deep parts of their ascent did not seem to be explosive. It is likely that kimberlite magma rises from deep in the upper mantle along a series of cracks and fissures until it gets to shallow levels, where it mixes with water and becomes extremely explosive. Other diatremes may be more explosive from greater depths, and they may move as gas-filled bodies rising from the upper mantle. As the gases move into lower-pressure areas, they would expand and the kimberlite would move faster until it explodes at the surface. Still other ideas for the emplacement of kimberlites and diatremes invoke hydrovolcanism, or the interaction of the deep magma with near-surface water. Magma may rise slowly from depth until it encounters groundwater in fractures or other voids, leading to an explosion when the water mixes with the magma. The resulting explosion could produce the volcanic features and upward-expanding pipe found in many kimberlites, spewing the kimberlite magma and diamonds across a wide area on the surface.

It is likely that some or all of the processes discussed here play a role in the intrusion of kimberlites and diatremes, the important consequence being a sudden, explosive volcanic eruption at the surface, far from typical locations of volcanism, and the relatively rapid removal of signs of this volcanism. The initial explosions are likely to be so powerful that they may blast material into the stratosphere, though other kimberlite eruptions may form only small eruptions and ash clouds.

Diamonds are the hardest substance known and are widely used as gemstones. Uncut varieties may show many different crystal shapes, and many show striated crystal faces. They crystallize in isometric tetrahedral forms, exhibit concoidal fracture, have a greasy luster, and may be clear, yellow, red, orange, green, blue, brown, or even black. Triangular depressions are common on some crystals, and others may be preserved as elongate or pear-shaped forms. Diamonds have been found in alluvial deposits such as gravel, and some mines have been located by tracing the source of the gravel to the kimberlite pipe from where the diamonds were brought back to the surface. Some diamond-mining operations such as those of the Vaal River, South Africa, proceeded for many years before it was recognized that the source was in nearby kimberlites.

Dating the age of formation of small mineral inclusions in diamonds has yielded important results. All diamonds from the mantle appear to be Precambrian, with one type being up to 3.2 billion years old, and another 1.0 to 1.6 billion years old. Since diamonds form at high pressures and low temperatures, their very existence shows that the temperature deep in the Earth beneath the continents in the Precambrian was not much hotter than today. The diamonds were stored deep beneath the continents for billions of years before being erupted in the kimberlite pipes. Diamonds really are forever.

FURTHER READING

Dawson, J. B. *Kimberlites and their Xenoliths.* New York: Springer-Verlag, 1980.

Gemological Institute of America homepage. Available online. URL: www.gia.edu. Accessed January 14, 2009.

Mitchell, R. H. *Kimberlites, Mineralogy, Geochemistry, and Petrology.* New York: Plenum Press, 1989.

ophiolite has structurally disrupted (faulted) pillow lavas, mafic flows, breccia, and chert overlying a mixed dike and gabbro section that grades down into layered gabbro, cumulate ultramafics, and mantle peridotites. High-temperature mantle fabrics and ophiolitic mantle podiform chromitites have also been documented from the Dongwanzi ophiolite, and it has ophiolitic mélange intruded by arc magmas.

Dismembered ophiolites appear to be a widespread component of greenstone belts in Archean cratons, and many of these apparently formed as the upper parts of Archean oceanic crust. Most of these are interpreted to have been accreted within forearc and intra-arc tectonic settings. The observation that Archean greenstone belts have such an abundance of accreted ophiolitic fragments compared to Phanerozoic orogens suggests that thick, relatively buoyant, young Archean oceanic lithosphere may have had a rheological structure favoring delamination of the uppermost parts during subduction and collisional events.

OCEANIC PLATEAUX ACCRETION

Oceanic plateaux are thicker than normal oceanic crust formed at midocean ridges; they are more buoyant and relatively unsubductable, forming potential sources of accreted oceanic material to the continental crust at convergent plate boundaries. Accretion of oceanic plateaux has been proposed as a mechanism of crustal growth in a number of orogenic belts, including Archean, Proterozoic, and Phanerozoic examples. Oceanic plateaux are interpreted to form from plumes or plume heads that come from the lower mantle (D″) or the 415-mile (670-km) discontinuity, and they may occur either within the interior of plates or interact with the upper mantle convective/magmatic system and occur along midocean ridges. Oceanic plateaux may be sites of komatiite formation preserved in Phanerozoic through Archean mountain belts, based on a correlation of allochthonous komatiites and high-MgO lavas of Gorgona Island, Curaçao, and the Romeral fault zone, with the Cretaceous Caribbean oceanic plateau.

Portions of several komatiite-bearing Archean greenstone belts have been interpreted as pieces of dismembered Archean oceanic plateaux. For instance, parts of several greenstone belts in the southern Zimbabwe craton are allochthonous and show a similar magmatic sequence, including a lower komatiitic unit overlain by several kilometers of tholeiitic pillow basalts. These may represent a circa 2.7-Ga oceanic plateau dismembered during a collision between the passive margin sequence developed on the southern margin of the Zimbabwe craton and an exotic crustal fragment preserved south of the suturelike Umtali line.

The accretion of oceanic plateaux and normal oceanic crust in arc environments may cause a backstepping of the subduction zone. As the accretionary complex grows, it is overprinted by calc-alkaline magmatism as the arc migrates through the former subduction complex. Further magmatic and structural events can be caused by late-ridge subduction and strike-slip segmentation of the arc. Average geochemical compositions of the continental crust, however, are not consistent with ocean plateau accretion alone.

Parts of many Archean, Proterozoic, and Phanerozoic greenstone belts interpreted as oceanic plateau fragments are overprinted by arc magmatism, suggesting that they either formed the basement of intraoceanic island arcs or they have been intruded by arc magmas following their accretion. Perhaps the upper and lower continental crusts have grown through the accretion of oceanic island arcs and ocean plateaus, respectively. Accreted oceanic plateaux may form a significant component of the continental crust, although most are structurally disrupted and overprinted by arc magmatism.

BACK ARC BASIN ACCRETION

The formation, closure, and preservation of back arc basin sequences has proven to be a popular model for the evolution of some greenstone belts. Paradoxically, the dominance of buoyant subduction styles in the Archean should have led to dominantly compressional arc systems, but many workers suggest back arc basins (which form in extensional arcs) as a modern analog for Archean greenstone belts.

ARC-TRENCH MIGRATION AND ACCRETIONARY OROGENS; A PARADIGM FOR THE ARCHEAN

Turkic-type accretionary orogens are large, subcontinent-size accretionary complexes built on one or two of the colliding continents before collision, through which magmatic arc axes have migrated, and are later displaced by strike-slip faulting. These accretionary wedges are typically built of belts of flysch, disrupted flysch and mélange, and accreted ophiolites, plateaux, and juvenile island arcs. In 1996 A. M. Celal Sengör and Boris Natal'n reviewed the geology of several Phanerozoic and Precambrian orogens and concluded that Turkic or accretionary-type orogeny is one of the principal builders of continental crust with time. The record of Archean granite-greenstone terranes typically shows important early accretionary phases followed by intrusion by arc magmatism, possibly related to the migration of magmatic fronts through large accretionary complexes. In examples like the Superior Province, many subparallel belts of accreted material are located between continental fragments separated by many hundreds of miles, and thus may represent large accretionary complexes that formed prior to a "Turkic-type" collision. Late-stage strike-slip faulting is important in these Archean orogens, as in the Altaids and Nipponides, and may be partly responsible for the complexity and repetition of belts of similar character across these orogens.

Turkic or accretionary-type of orogeny provides a good paradigm for continental growth. These orogenic belts possess very large sutures (up to several hundred kilometers wide) characterized by subduction-accretion complexes and arc-derived granitoid intrusions, similar to the Circum-Pacific accreted terranes (e.g., Alaska, Japan). These subduction-accretion complexes are composed of tectonically juxtaposed fragments of island arcs, back arc basins, ocean islands/plateaux, trench turbidites, and microcontinents. Turkic or accretionary-type orogens may also experience late-stage extension associated with gravitational collapse of the orogen, especially in association with late collisional events that thicken the crust in the internal parts of the orogen. In the Archean slightly higher mantle temperatures may have reduced the possible height that mountains would have reached before the strength of deep-seated rocks was exceeded, so that extensional collapse would have occurred at crustal thickness lower than those of the younger geological record. Another important feature of these orogens is the common occurrence of orogen parallel strike-slip fault systems, which resulted in lateral stacking and bifurcating lithological domains. In these respects the accretionary-type orogeny may be considered as

a unified accretionary model for the growth of the continental crust.

LATE-STAGE GRANITES AND CRATONIZATION

Archean cratons are ubiquitously intruded by late- to post-kinematic granitoid plutons, which may play a role in or be the result of some process that has led to the stabilization or "cratonization" of these terranes and their preservation as continental crust. Most cratons also have a thick mantle root or tectosphere, characterized by a refractory composition (depleted in a basaltic component), relatively cold temperatures, high flexural rigidity, and high shear wave velocities.

Outward growth and accretion in granite-greenstone terranes provides a framework for the successive underplating of the lower parts of depleted slabs of oceanic lithosphere, particularly if some of the upper sections of oceanic crust are offscraped and accreted, to be preserved as greenstone belts or eroded to form belts of greywacke turbidites. These underplated slabs of depleted oceanic lithosphere will be cold and compositionally buoyant compared with surrounding asthenosphere (providing that the basalt is offscraped and not subducted and converted to eclogite) and may contribute to the formation of cratonic roots. One of the major differences between Archean and younger accretionary orogens is that Archean subducted slabs were dominantly buoyant relative to the dense mantle, whereas younger slabs were not. This may be a result of the changing igneous stratigraphy of oceanic lithosphere, resulting from a reduction in heat flow with time,

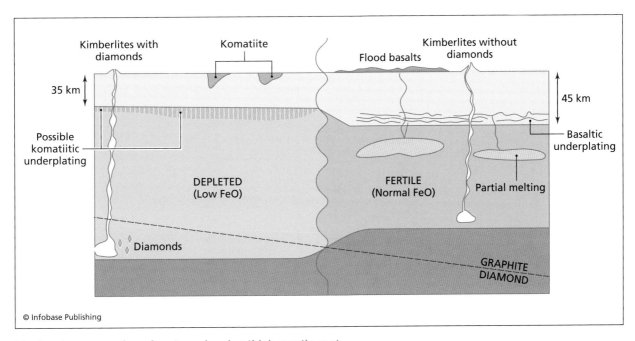

Idealized cross section of craton, showing thick mantle root

perhaps explaining why Archean cratons have thick roots and are relatively undeformable compared with their younger counterparts. Geometric aspects of underplating these slabs predict that they will trap suprasubduction mantle wedges of more fertile and hydrated mantle, from which later generations of basalt can be generated.

Many granites in Archean terranes appear to be associated with crustal thickening and anatexis during late stages of collision. However, some late-stage granitoids may directly result from decompressional melting associated with upper-crustal extensional collapse of Archean orogens thickened beyond their limit to support thick crustal sections, as determined by the strength of deep-seated rocks. Decompressional melting generates, from the trapped wedges of fertile mantle, basaltic melts that intrude and partially melt the lower crust. The melts assimilate lower crust, become more silicic in composition, and migrate upward to solidify in the mid to upper crust, as the late to postkinematic granitoid suite. In this model the tectosphere (or mantle root) becomes less dense (compositionally buoyant) and colder than surrounding asthenosphere, and this makes it a stable cratonic root that shields the crust from further deformation.

Late-stage strike-slip faults that cut many Archean cratons may also play an important role in craton stabilization. Specifically the steep shear zones may provide conduits for massive fluid remobilization and escape from the subcontinental lithospheric mantle, which would both stabilize the cratonic roots of the craton and initiate large-scale granite emplacement into the mid and upper crust.

See also ARCHEAN; GREENSTONE BELTS; OROGENY; PRECAMBRIAN.

FURTHER READING

de Wit, Maarten J., Chris Roering, Robert J. Hart, Richard A. Armstrong, C. E. J. de Ronde, Rod W. E. Green, Marian Tredoux, Ellie Peberdy, and Roger A. Hart. "Formation of an Archean Continent." *Nature* 357 (1992): 553–562.

Kusky, Timothy M. "Collapse of Archean Orogens and the Generation of Late- to Post-Kinematic Granitoids." *Geology* 21 (1993): 925–928.

Kusky, Timothy M., Jianghai Li, and Robert T. Tucker. "The Archean Dongwanzi Ophiolite Complex, North China Craton: 2.505 Billion Year Old Oceanic Crust and Mantle." *Science* 292 (2001): 1,142–1,145.

Kusky, Timothy M., and Ali Polat. "Growth of Granite-Greenstone Terranes at Convergent Margins and Stabilization of Archean Cratons." *Tectonophysics* 305 (1999): 43–73.

Kusky, Timothy M., ed. *Precambrian Ophiolites and Related Rocks, Developments.* In *Precambrian Geology.* Amsterdam: Elsevier, 2003.

Li, Jianghai H., Timothy M. Kusky, and Xiongnan Huang. "Neoarchean Podiform Chromitites and Harzburgite Tectonite in Ophiolitic Mélange, North China Craton, Remnants of Archean Oceanic Mantle." *GSA Today* 12, no. 7 (2002): 4–11.

Sengor, A. M. Celal, and Boris A. Natal'n. "Turkic-Type Orogeny and Its Role in the Making of the Continental Crust." *Annual Review of Earth and Planetary Sciences* 24 (1996): 263–337.

Cretaceous The youngest of three periods of the Mesozoic, during which rocks of the Cretaceous System were deposited, the Cretaceous ranges from 144 million years (Ma) ago until 66.4 Ma and is divided into the Early and Late Epochs and 12 ages. The name derives from the Latin *creta,* "chalk," in reference to the chalky terrain of England of this age.

Pangaea was dispersing during the Cretaceous, and the volume of ridges plus apparently the rate

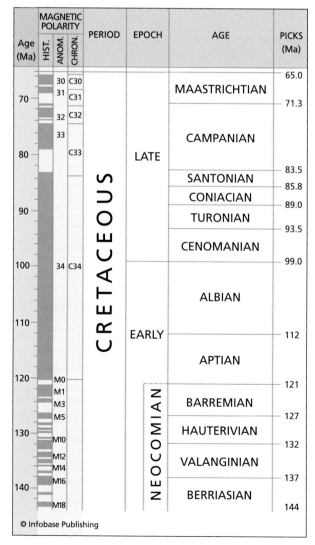

Cretaceous timescale

Artwork of Cretaceous period scene, showing many dinosaur and other species, including pterosaurs, nautoloid mesosaur, pleisiosaur, hesperomis, belemnite, and other life-forms. *(Publiphoto/Photo Researchers, Inc.)*

of seafloor spreading were dramatically increased. The consequential displacement of seawater caused global sea levels to rise, so the Late Cretaceous was marked by high sea levels and the deposition of shal- low water limestones in many epicontinental seas around the world. On the North American craton, the Zuni Sequence was deposited across wide parts of the craton during this transgression. Increased magmatic activity in the Cretaceous may reflect more rapid mantle convection or melting, as marked by a number of igneous events worldwide. The South American Cordillera and the western United States saw unusual amounts of intrusive and volcanic activity. The giant flood basalt provinces of Paraná in South America and the Deccan of India were formed, and kimberlite pipes punctured the lithosphere of South Africa and Greenland. The dispersal of Pangaea was associated with the opening of the Atlantic Ocean. Africa rotated counterclockwise away from South America, closing the Tethys Ocean in the process of opening the Atlantic. The closure of Tethys was associated with the emplacement of many ophiolites onto continents, including the giant Oman (Semail) ophiolite that was thrust to the south onto the Arabian continental margin.

Cretaceous sedimentary patterns suggest that the climate was warming through the period and was more varied and seasonal than in the earlier Mesozoic. The famous Cretaceous chalks were formed by the accumulation of tests of calcareous marine algae known as coccoliths, which thrived in the warm shallow seas. The chalks are in many places interbedded with fossiliferous limestones with abundant brachiopods and rudist coral fragments.

Life on the Cretaceous continents saw the development of the angiosperms, which became the planet's dominant flora by the middle of the period. Invertebrate and vertebrate animals were abundant and included many species of dinosaurs, giant flying pterosaurs, and giant marine reptiles. Dinosaurs occupied many different geological niches, and most continents have fossil dinosaurs, including herbivores, carnivores, and omnivores. Birds had appeared, both flying and swimming varieties. Mammals remained small, but their diversity increased. Many life-forms began a dramatic, progressive disappearance toward the end of the period. These marine and land extinctions seem to be a result of a combination of events including climate change and exhalations from the massive volcanism in the Indian Deccan and South American Paraná flood basalt provinces, coupled with an impact of a six-mile (10-km) wide meteorite that hit the Yucatán Peninsula of Mexico. The extinctions were not all abrupt—many of the dinosaur and other genera had gone extinct, probably from climate stresses, before the meteorite hit the Yucatán Peninsula. When the impact occurred, a 1,000-mile (1,600-km) wide fireball erupted into the upper atmosphere, and tsunami hundreds or thousands of feet (hundreds of meters) high washed across the Caribbean, southern North America, and much of the Atlantic. Huge earthquakes accompanied the explosion. The dust blown into the atmosphere immediately initiated a dark global winter, and as the dust settled months or years later, the extra carbon dioxide in the atmosphere warmed the Earth for many years, creating a greenhouse condition. Many forms of life could not tolerate these rapid changes and perished. The end Cretaceous extinction, commonly referred to as the K-T event, is one of the most significant mass extinction events known in the history of life.

See also Cenozoic; mass extinctions; Mesozoic.

FURTHER READING

Alvarez, Walter. *T Rex and the Crater of Doom.* Princeton, N.J.: Princeton University Press, Princeton. 1997.

Eldredge, N. *Fossils: The Evolution and Extinction of Species.* Princeton, N.J.: Princeton University Press. 1997.

Stanley, Steven M. *Earth and Life Through Time* New York: W. H. Freeman 1986.

crust Thin, low-density rock material called crust makes up the outer layer of the solid Earth, ranging in thickness from about three miles (five km) and less near the midocean ridges, to more than 50 miles (70 km) beneath the tallest mountain ranges. This is followed inward by the mantle, a solid rocky layer extending to 1,802 miles (2,900 km). The outer core is a molten metallic layer extending to 3,169-mile (5,100-km) depth, and the inner core is a solid metallic layer extending to 3,958 miles (6,370 km).

The temperature increases with depth with a gradient of 30°C per kilometer (139°F per mile) in the crust and upper mantle, and with a much smaller gradient deeper within the Earth. The heat of the Earth comes from residual heat trapped from initial accretion, radioactive decay, latent heat of crystallization of outer core, and dissipation of tidal energy of the Sun-Earth-Moon system. Heat flows from the interior of the Earth toward the surface through convection cells in the outer core and mantle. The top of the mantle and the crust compose a relatively cold and rigid boundary layer, or lithosphere, which is about 65 miles (100 km) thick. Heat escapes through the lithosphere largely by conduction or transport in igneous melts, and in convection cells of water through midocean ridges.

The Earth's crust is divisible broadly into continental crust of granodioritic composition and oceanic crust of basaltic composition. Continents make up 29.22 percent of surface, whereas continental crust underlies 34.7 percent of the Earth's surface, with continental crust under continental shelves accounting for the difference. The continents are in turn divided into orogens, made of linear belts of concentrated deformation, and cratons, making the stable, typically older interiors of the continents.

The distribution of surface elevation is strongly bimodal, as reflected in the hypsometric diagrams. Continental freeboard is the difference in elevation between the continents and ocean floor and results from difference in thickness and density between continental and oceanic crust, tectonic activity, erosion, sea level, and strength of continental rocks.

See also CONTINENTAL CRUST; CRATON; LITHO-SPHERE; OCEAN BASIN.

crystal, crystal dislocations Crystals are homogeneous solid structures composed of chemical elements or compounds having a regularly repeating arrangement of atoms, as well as a large number of defects in the crystal lattice, such as dislocations. All

minerals are solids, and in all minerals the atoms are arranged in a very regular geometric form that is unique to that mineral. Every mineral of that species has an identical crystalline structure. This regular structure gives each mineral its characteristic color, chemistry, hardness, and crystal form.

Many minerals may not have a well-developed external crystal form, but still must have a regularly repeating crystal lattice composed of the constituent atoms. Crystal forms and the internal atomic lattice exhibit symmetry, which may be of several different varieties. Crystals exhibit four main types of symmetry. In mirror plane symmetry, the simplest, the crystal can be divided by an imaginary mirror plane that would result in two halves that are mirror images of one another. Crystals may have symmetry about

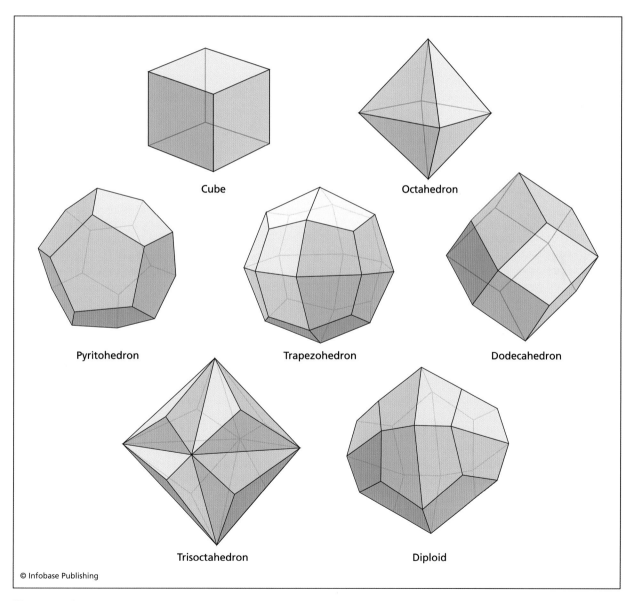

Cube

Octahedron

Pyritohedron

Trapezohedron

Dodecahedron

Trisoctahedron

Diploid

© Infobase Publishing

The seven basic types of crystal forms

an axis that runs through the center of the crystal, in which the crystal lattice would be rotated into an identical configuration two, three, four, or six times in a 360° circuit. These symmetry systems are known as diads, triads, tetrads, and hexads. A more complex form of symmetry is known as roto inversion, characterized by a rotational operation followed by inversion of the lattice across its center point. Finally, a simple inversion across the center of the crystal leads to a crystal face diametrically opposite to every other crystal face. In various combinations of these symmetry operations, all crystals belong to one of seven crystal systems. These include cubic, tetragonal, orthorhombic, monoclinic, triclinic, hexagonal, and trigonal.

Crystals may appear to be close to perfect, but they always have millions of atomic defects. These include vacancies in the crystal lattice, various types of defects in the arrangement of the atoms and lattice, and replacements of one type of atom or ion by another with a similar charge and size.

CRYSTAL DISLOCATIONS AND OTHER CRYSTAL DEFECTS

Crystals are regularly ordered symmetrical arrays of atoms, but like anything else in nature they are not perfect and have many defects. Some defects are acquired during growth of the crystals when they first form, and others develop in the crystals during deformation. These defects determine the strength of crystals, minerals, and rocks. The motion of these defects accommodates the strain of crystals.

The motion of crystal defects is an important deformation mechanism, with different types of motion of different types of defects operating at different temperatures, pressures, and applied stresses. There are two main types of crystal defects: point defects and line defects. Point defects include vacancies, impurities, and interstitials, whereas line defects are known as dislocations.

Point defects are considered irregularities, or defects, that affect one point in a crystal lattice. Several types include impurities, in which the wrong type of atom is present in the crystal lattice in the place of another; vacancies, in which an atom is missing from the atomic lattice; and interstitials, in which an atom occupies a site that is not normally occupied. Other types of point defects are more complex and involve more than one atom at a time.

In a regularly ordered crystal lattice most electric charges are satisfied by bonding and balancing positive and negative charges. Crystals having point or other defects have more internal energy because many bonds are broken or unsatisfied and the electric charges are not neutralized. Crystals are more apt

Cubic crystal of fluorite, about two inches (5 cm) on each face, on quartz with barite from Frazer's Hush Mine, Weardale, England *(Mark A. Schneider/Photo Researchers, Inc.)*

to react or deform when they have a higher internal energy. Temperature causes the number of vacancies to increase, whereas pressure causes the number of vacancies to decrease.

When a crystal is stressed, vacancies tend to move, or diffuse, in an orderly manner related to the stress field, migrating toward the crystal face with the highest stress, whereas atoms tend to migrate in the opposite direction toward the crystal faces with the lowest stress. This process of migration of vacancies is called Nabarro-Herring creep. This kind of diffusion can accommodate a general shape change of a crystal, such as occurs during regional deformation in many mountain belts worldwide.

Twinning, the misorientation of a crystallographic plane across a plane in the crystal lattice, can be caused by mismatched growth or by deformation. Mechanical or stress-induced twinning differs from growth twinning in that shear across a crystal plane changes the lattice orientation. The shear occurs across a crystal plane, which must be a symmetry plane of the crystal. The amount of strain is limited by the crystallographic relationship for each type of crystal, though it typically falls in the range of 20°–45°.

Translation gliding is a deformation mechanism in crystals whereby the crystal lattice structure slips along some internal crystallographic plane, after some critical value of shear stress is reached. Slip, or translation gliding, usually occurs in very specific crystallographic directions in crystals called slip systems, favoring directions that

- have short distances between equivalent atoms
- occur in directions that do not juxtapose ions of like charge.

Slip on these crystallographic directions begins when the critical resolved shear stress for that crystallographic direction is surpassed.

Most crystals have many slip systems activated at different critical resolved shear stresses, so slip begins on the planes with the lowest critical resolved shear stress. Soon different slip systems start to interfere with each other and deformation will either stop or the stress will rise to enable continued deformation. This increase in stress is known as work hardening.

Crystals need to have five independent slip systems to accommodate any general homogenous strain. If the crystal has fewer than five, it will eventually crack or fracture. Different slip systems may be activated at different temperatures, or strain rates, so the relationships of which slip systems are active under different deformation conditions are complex.

Dislocations are best thought of as an extra plane of atoms that terminates somewhere in the crystal lattice. They are often referred to as an extra half-plane because of this geometry. The presence of dislocations weakens the crystal structure. Their motion accounts for much of the strain in crystals. Dislocations move through a crystal lattice much like a ridge can be moved across a carpet, slowly moving the carpet across a floor without moving the whole carpet at one time. The stress required to break one bond at a time to move a dislocation from one place to another in a crystal lattice is much less than breaking all the bonds along a specific crystallographic direction simultaneously.

There are two basic types of dislocations: edge dislocations and screw dislocations. Edge dislocations are simpler, and are basically an extra half-plane of atoms that extends partway across the crystal lattice structure. Screw dislocations, however, are more complex, and have a twisted shape that resembles a parking garage, where one layer of the lattice is offset by a twisted, coil-like motion about an axis perpendicular to the planes. Most dislocations in crystals have both edge and screw components and form complex geometrical shapes.

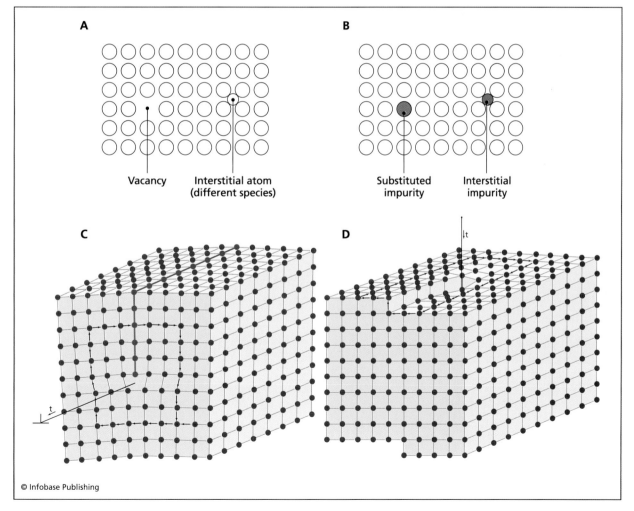

© Infobase Publishing

Types of point and line defects in crystals, including vacancies, interstitials, edge dislocation, and screw dislocation

Dislocations form loops within crystals, marking boundaries between slipped and unslipped portions. This type of motion and slipping in crystals is exactly analogous to motions on faults.

Interactions of Dislocations

During deformation there are so many dislocations moving around in a crystal that they are bound to interact in several ways. Dislocation annihilation occurs when two dislocations of opposite signs move toward each other on the same slip plane and their extra half planes meet and form a complete crystal lattice, effectively canceling each other out of existence. Each dislocation in a crystal induces a stress field known as the self-stress field, since they disrupt the normal crystallographic structure. These stress fields interact for large distances within the crystals, causing either repulsions or attractions between dislocations, much as magnets repel or attract each other depending on charge. In the case of dislocations, however, attractions are deadly because if two dislocations are on the same slip plane and are attracted, they come together and annihilate each other. If they repel each other, the number of dislocations increases and higher and higher stresses are needed to make the dislocations move as the number of repelled dislocations increases during deformation.

Dislocations often encounter immobile, tightly bounded impurities, such as interstitials, which pin dislocations behind them. As more dislocations move toward the region with the impurities, they too become stuck and are repelled by each other's stress fields. This causes dislocation pile-ups. Dislocations can also interact with other dislocations, also causing pile-ups.

When the temperature of the deforming crystal is high, vacancies can diffuse toward the obstacle, or atoms can move away from the half-plane, enabling dislocations to climb over obstacles. Thus there are two main mechanisms by which dislocations move through crystals: gliding and climbing.

When dislocations of different slips systems move through each other, they offset the other slip plane, forming a dislocation jog, which is basically a step in the slip system that one of the dislocations moved along. Once there is a jog in the slip plane, it becomes more difficult for dislocations to move along that slip plane, and they must climb over the jog to progress. Dislocation jogs can be made to disappear, or evaporate, by diffusion of vacancies toward them or movement of atoms away from the jog.

Work hardening is any process that makes increased deformation harder, requiring more stress to do the same amount of deformation. Work hardening can occur by

- formation of dislocation jogs
- dislocation pile-ups
- interaction of stress fields
- increase in dislocation density

Work hardening is more common at low temperatures, as high temperatures cause increased diffusion of vacancies and climb to occur.

Annealing is any process that tends to return a crystal lattice to a less deformed state, such as through a reduction in the number of dislocations. A lattice with fewer dislocations has lower energy and is more stable that one with a high dislocation density. There are several different common ways that a crystal anneals:

- group dislocations in a more stable configuration
- migration of dislocations to an edge of crystal
- recrystallization of grains

Diffusion helps all of these processes, so annealing is faster at high temperatures. Annealing mechanisms are also diverse and include the following:

- Dislocations of opposite signs can climb to the same slip plane and annihilate each other.
- Dislocations can glide and climb to grain boundaries.
- Formation of subgrain boundaries by dislocation motion concentrates dislocations into planes, or walls that bound domains of low-dislocation density.
- Recrystallization, or regrowth, of the entire crystal lattice, with new grains having low-dislocation density. Often this process starts in regions of high-dislocation density or along grain boundaries. Prolonged heating leads to grain growth, and some grains grow at the expense of neighbors. Recrystallization forms grains that are equant, have 120° grain boundaries at triple junctions.

At temperatures lower than those where dislocation glide and climb operate, rocks flow by other mechanisms. Pressure solution, or grain boundary diffusion (also called Coble creep), is where crystals are flattened and dissolved along their edges. The extra dissolved material is either precipitated at the ends of the grains or moved away to be precipitated in veins or pores, or far away. This deformation mechanism can easily accommodate large bulk shortening and stretching. Pressure solution often produces seams, or stylolites, which are leftover concentrations of insoluble material from where the rock dissolved much of the other material. Stylolites are interesting because in three dimensions they form irregular surfaces with cones or teeth on them, pointed to the maximum compressive stress. Pressure solution works much like squeezing an ice cube. Grain-grain boundaries that are initially touching have the highest stresses on them and are the first to be dissolved.

Compaction is also an important deformation mechanism in sedimentary basins due to the weight of overlying rocks. Compaction often involves dewatering, or removal of fluids, from the pore spaces of a rock. The weight of newly deposited sediments and overlying rocks adds pressure and pushes the grain-to-grain contacts between crystals or grains closer together, expelling the fluids. Some muds begin with a porosity of 80 percent and end with 10 percent on burial. In these cases a large quantity of water is expelled from the system during compaction. Sand-stones have initial porosities of up to 45 percent, reduced to about 10–30 percent depending on the rock, pressure, and fluid. Porosity decreases with increasing burial depth.

See also DEFORMATION OF ROCKS; MINERAL, MINERALOGY; STRUCTURAL GEOLOGY.

FURTHER READING

Hull, D., and D. J. Bacon. *Introduction to Crystal Dislocations.* 3rd ed. Oxford: Pergamon Press, 1984.

Kosevitch, Arnold M. *The Crystal Lattice: Phonons, Solitons, Dislocations, Superlattices.* 2nd ed. New York: John Wiley & Sons, 2005.

Nabarro, F. R. N. *Theory of Crystal Dislocations.* New York: Dover, 1987.

Shelly, David. *Manual of Optical Mineralogy.* Amsterdam: Elsevier, 1980.

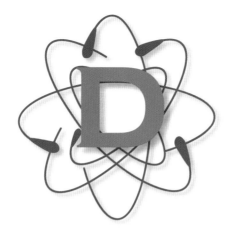

Dana, James Dwight (1813–1895) American *Geologist, Mineralogist* James Dana was born on February 12, 1813, in Utica, New York, in a region surrounded by Paleozoic sedimentary rocks and Precambrian crystalline rocks of the Adirondack Mountains. His teacher at Utica High School, Fay Edgerton, noticed the young Dana's keen interest in science and helped him gain entrance to Yale College, where he received scientific training by Benjamin Silliman, the prominent scientist and founder of the *American Journal of Science*. In terms of enduring scientific achievement James Dwight Dana is one of Yale's most notable scientific figures. His contributions to geology, mineralogy, and zoology formed the basis of classification systems still in use today by scientists in these fields of study.

After Dana graduated from Yale in 1833, he became a mathematics teacher to midshipmen in the U.S. Navy and sailed to the Mediterranean from 1833 to 1835. Dana returned to Yale in 1836 and 1837 and worked as an assistant to Silliman, helping with chemistry laboratories. In 1836 Dana was invited to be a scientific participant of the United States Exploring Expedition, due to sail to the South Seas in 1838. Originally invited on the expedition as its geologist, he assumed the role of zoologist after the departure of James Couthouy in 1840. Dana produced two important monographs based on his study of animals collected during the exploring expedition. These monographs, one on corals and anemones and the other on crustaceans, were extraordinary for their sheer size, scope, and detail. Virtually no modern coral or crustacean researcher today can undertake significant systematic research without encountering by James Dana's legacy.

Dana returned to America in 1842 and spent the next 13 years writing reports from his four years at sea, including detailed reports of the geology and mineralogy of the Mount Shasta, California, area, a poorly studied region at the time. Dana returned to New Haven, Connecticut, and Yale University in 1844, where he married Silliman's daughter Henrietta. He then succeeded Silliman by taking the Silliman professorship of natural history and geology and stayed in that position until 1892.

James Dana is best known for his books on mineralogy, including his *System of Mineralogy*, (1837), *Manual of Mineralogy* (1848), and *Manual of Geology* (1863). His other papers and books number more than 200, on topics ranging from crustaceans to volcanoes, to texts reconciling science and religion, and covering geographic areas from California to the South Pacific.

See also MINERAL, MINERALOGY; NORTH AMERICAN GEOLOGY.

FURTHER READING

Dana, James Dwight. "On the Areas of Subsidence in the Pacific as Indicated by the Distribution of Coral Islands." *American Journal of Science* 45 (1943): 131–135.

———. "Certain Parallel Relations between the Classes of Vertebrates and Some Characteristics of the Reptilian Birds." *American Journal of Science* 36 (1863): 315–321.

———. "A New Mineralogical Nomenclature." *Annals of the Lyceum of Natural History of New York* 8 (1837): 9–34.

———. "Manual of Geology: *Treating of the Principles of the Science with Special Reference to American Geological History, for the use of Colleges, Academies, and Schools of Science.*" Philadelphia: T. Bliss and Co., 1863.

———. *A System of Mineralogy*. New York: John Wiley & Sons, 1868.

dark matter The mass in galaxies and galactic clusters inferred to exist by the rotational properties of galaxies, the bending of light, and other techniques but has not been confirmed to exist by observations at any electromagnetic wavelength is known as dark matter. Dark matter is mysterious; with an unknown composition it interacts only very weakly with normal matter and has been decoupled from the rest of the universe since before the time primordial nucleosynthesis began. Scientists believe that dark matter has experienced large fluxuations in its density distribution since early in the history of the universe, without affecting the background radiation of the universe, but effectively forming large-scale clumping and other mass distributions in the universe. In this way dark matter is able to control the overall mass distribution in the universe without affecting the microwave background radiation or any other observational constraints on the universe.

The large gravitational attraction of dark matter is theorized to have drawn gas and other matter into the vicinity of peaks in its distribution over the history of the universe, accounting for the distribution of galaxies and clusters observed today. One of the shocking features of dark matter is that, although it cannot be seen or directly observed, it is thought to make up the bulk of the universe. About 4 percent of the universe is thought to be made of visible matter, about 22 percent is estimated to be dark matter, and a remarkable 74 percent of the universe is thought to be dark energy. Dark energy permeates the entire universe and is thought to be the cause of the recently detected increase in the rate of expansion of the universe. Dark energy has been proposed to consist of two forms, including the cosmological constant, a constant form of energy that fills space homogeneously, and other more exotic forms of energy (known as scalar fields with names such as moduli and quintessence) that vary in time and in space.

Astrophysicists have classified dark matter into two basic theoretical types, hot and cold, based on its temperature at the time the galaxies began to form. Whether the dark matter was hot or cold at the time the galaxies formed results in vastly different structures for the universe in later times. Astrophysicists and cosmologists have used this variation to model the evolution and structure of the universe using different combinations of hot and cold dark matter as gravitational building blocks. This work is purely theoretical, carried out by simulations in supercomputers, since dark matter has never been directly observed.

Hot dark matter is thought to be made of very lightweight particles, even lighter than electrons. Some astrophysicists think that hot dark matter may be composed of neutrinos. Models for the evolution of the universe using hot dark matter account for the development of very large-scale structures, such as superclusters, and vast empty regions called voids, but they do not explain the smaller-scale structures very well. This is because small amounts of hot material tend to disperse, not to group together to help form smaller-scale structures. Therefore most astrophysicists suggest that models for the evolution of the universe that rely only on hot dark matter are not feasible.

Cold dark matter is thought to consist of heavy particles that formed in the earliest microseconds (10^{-43} second after the big bang) at a time when the strong, weak, and electromagnetic forces were still unified (a time known as the grand unified theory time). Unlike the (theoretical) hot dark matter, computer models for the origin of the universe that use cold dark matter can explain the formation of both large-scale and small-scale structures in the universe. Most astrophysicists and cosmologists therefore prefer models of dark matter that suggest it consists of heavy cold particles that formed very soon after the big bang.

Some new models for the universe suggest that dark matter may consist of both hot and cold particles. Supercomputer simulations can match the theoretical evolution of the universe and its current structure with what might have happened by specific mixtures of hot and cold dark matter.

See also ASTRONOMY; ASTROPHYSICS; COSMIC MICROWAVE BACKGROUND RADIATION; COSMOLOGY; ORIGIN AND EVOLUTION OF THE UNIVERSE.

FURTHER READING

Chaisson, Eric, and Steve McMillan. *Astronomy Today.* 6th ed. Upper Saddle River, N.J.: Addison-Wesley, 2007.

Cline, David B. "The Search for Dark Matter." *Scientific American* 288 (February 2003): 28–35.

Comins, Neil F. *Discovering the Universe.* 8th ed. New York: W. H. Freeman, 2008.

National Aeronautics and Space Administration. "Universe 101, Our Universe, Big Bang Theory. What is the Ultimate Fate of the Universe Web page." Available online. URL: http://wmap.gsfc.nasa.gov/universe/uni_fate.html. Updated April 17, 2008.

Snow, Theodore P. *Essentials of the Dynamic Universe: An Introduction to Astronomy.* 4th ed. St. Paul, Minn.: West, 1991.

Darwin, Charles (1809–1882) British *Natural Historian, Geologist, Evolutionist* Charles Darwin was a British naturalist well known for his theory of evolution by means of natural selection. He proposed that all species of life have evolved over time from common simple ancestors. His ideas on the transmutation of species and natural selection were conceived in 1838 but not published until 1859. These ideas were used to build the modern concept of evolution and form some of the basic foundations of biology by offering viable explanations for the diversity of life, expressed in his famous book *On the Origin of Species,* published in 1859. His ideas are widely accepted by the scientific community but are often still attacked by religious fundamentalist communities. Darwin, who ranks as one of the most influential natural scientists of the 19th century, also made many observations and published several books on the geological sciences. His notes from a five-year-long voyage from 1831 to 1836 on the voyage of the HMS *Beagle* demonstrated his prowess as a geologist, following the uniformitarian principles of Scottish geologist Charles Lyell. Darwin also wrote about the origin of humans in his *The Descent of Man, and Selection in Relation to Sex,* published in 1871. Charles Darwin's contributions and influence in science were so great that when he died in 1882 he was honored with a state funeral in London and buried at Westminster Abbey.

EARLY YEARS

Charles Robert Darwin was born on February 12, 1809, in Shrewsbury, England, the fifth of six children born to the wealthy doctor and financier Robert Darwin and Susannah Wedgwood Darwin. Susannah died when Charles was only eight, and he then joined his older brother Erasmus as a border at the Anglican Shrewsbury School.

At the age of 16 Charles spent the summer of 1825 as an apprentice doctor helping his father treat the poor of Shropshire in the West Midlands region of England. He then returned to medical school in Edinburgh, but he was not interested in surgery, so instead he learned taxidermy from John Edmonston, a freed slave who had worked for English naturalist Charles Waterton (1782–1865) in the South American rain forest. In 1826 in the second year of his studies at Edinburgh, Darwin joined the Plinian Society, a student-run organization dedicated to the study of natural history under the guidance of Dr. Robert Grant. In 1827 Darwin made a presentation to this group about his studies that black "spores" found in oyster shells in the Firth of Forth were the eggs of a skate leech species. He spent much time during these years studying the collections of plants at the University Museum, while neglecting his course work in the geology course of Robert Jameson.

Darwin's father was worried about the direction of his son's studies, drifting further from the medical field, and he enrolled him in a bachelor of arts program at Christ's College in Cambridge, with

hopes that he would become a clergyman with a steady income. Darwin instead sought company riding horseback and shooting in the countryside, while becoming engrossed in beetle collecting that eventually led him to publish his work in *Stevens' Illustrations of British Entomology.* These investigations led him to become friends with botany professor John Stevens Henslow, who helped Darwin investigate natural sciences, mathematics, religious studies, and physics, passing his exams as 10th in his class of 178 in 1831.

Darwin continued to study theology and science at Cambridge, where he read William Parley's books on natural theology and his arguments for divine design in nature. His courses and studies inspired Darwin to travel and contribute to science, and he planned to visit Tenerife in the tropical Canary Islands in the Atlantic Ocean to study the natural history of the region. He took the geology course of Adam Sedgwick, traveling to Wales for fieldwork, and when he returned to Cambridge he found that his friend John Stevens Henslow had recommended him to the unpaid position of naturalist on a voyage of the HMS *Beagle* under Captain Robert Fitzroy. Darwin's father initially objected to his participation on the journey but eventually relented, and Darwin left on the historic voyage four weeks later in 1831.

VOYAGE OF THE HMS *BEAGLE*

Darwin left England aboard HMS *Beagle* on December 27, 1831, under the command of Captain Robert Fitzroy. The *Beagle* voyaged across the Atlantic Ocean, while the crew carried out hydrographic surveys around the coasts of southern South America, then sailed to Tahiti and Australia, returning nearly five years later on October 2, 1836. Throughout this voyage, the second for the *Beagle,* Darwin spent most of his time, 39 months out of the 57-month-long voyage, exploring on land. This time was dedicated to his studies of geology, collecting fossils, and making detailed observations of plants and animals that would later form the basis for his theory of evolution by natural selection.

Several times during the voyage of the *Beagle* Darwin sent his notes and samples back to Cambridge, along with letters to his family. These materials focused on his areas of expertise in geology, beetle collecting, and marine invertebrates, but also increasingly on zoology and related biological subjects in which he was a novice. In the seacliffs of Patagonia he discovered extinct fossil species including skeletons of giant mammals, which he named *Megatherium* (now known to be a type of giant ground sloth that lived from 2 million to 8,000 years ago) and shipped back to England. Also in Patagonia Darwin noted that some areas showed stepped terraces along the coast, and each had seashells along the flat surfaces. He correctly interpreted the terraces to reflect that the land was rising relative to the sea, and these shelves represented former shorelines. While in Chile Darwin experienced a major earthquake and noticed that many mussel beds were stranded above the high-water mark, reminiscent of the raised beaches he saw in Tierra del Fuego. While high in the Andes Darwin found beds of marine shelly fossils and beach deposits; he inferred that these ranges had been uplifted from the sea.

The *Beagle* visited the Galápagos Islands in September and October 1835, then traveled to other islands including atolls in the Pacific. Darwin investigated the coral reefs around these islands and hypothesized that the volcanic islands gradually sank below sea level, and as they did the corals grew upward, eventually forming a ring of coral reefs surrounding a sunken island. While in the Galápagos, Darwin found that different types of mockingbirds were on different islands, and from this he inferred, nearly a quarter century later, that the different species evolved separately on each island, from a common ancestor. While the *Beagle* was at the Galápagos, Darwin also collected many samples of finches. At the time, however, he did not appreciate the differences between them, and only after he returned to England and shared the finches with the ornithologist John Gould on January 4, 1837, did he recognize that the finches represented 12 separate species. After more work Darwin appreciated that the finches, mockingbirds, and also tortoises had similar ancestors on the South American mainland and had all evolved into distinct species on each of the separate islands of the Galápagos. The term *Darwin's finches* was coined by English ornithologist Percy Lowe for these birds in 1936 and popularized by British ornithologist David Lack (1910–73) in 1947.

DARWIN'S THEORY OF EVOLUTION
When Darwin returned to London he was already a celebrity in some scientific circles, since his mentor Henslow had shared many of his geological notes and biological findings with colleagues. After visiting family and friends Darwin returned to Cambridge and studied his notes, data, and samples with the help of many colleagues and scientists recommended by Henslow. Together they cataloged his collections from around the world and discussed many of the possibilities suggested by his geological findings and, most famously, discussed ideas about the variety of different species in the plant and animal kingdoms and whether or not species were immutable, as commonly assumed in those times. Darwin met British geologist Charles Lyell (1797–1875), with whom he

discussed uniformitarianism. Lyell introduced Darwin to the young anatomist Richard Owen, and together they identified several extinct species that closely resembled living species in South America.

In December 1836 Darwin began writing up much of his work. He surmised that the South American continent was slowly rising and presented this idea to the Geological Society of London on January 4, 1837, the same day he presented his collections of mammals and birds to the zoological society, where ornithologist John Gould identified the Galápagos birds as 12 separate species of finches. Darwin moved to London and continued his work and discussions with other scientists. By spring and summer 1837 Darwin's notebooks showed that he had derived his ideas that different species transmuted into other species and formed a branching evolutionary tree in which species merged backward in time to common ancestors. Darwin worked long hours and formally recorded his ideas in his journal in summer 1837. He had classified his numerous biological collections and derived the following ideas:

- Evolution did occur.
- Evolutionary change to form new species was gradual, requiring up to 3 million years.
- Natural selection was the main driving force for evolution.
- All the species of life that are present today came from a single unique life-form.

Darwin's theory states that within each species nature randomly selects which animal or plant will survive and which will die out. Survival depends on how adaptable the species is to its surrounding dynamic environment.

By September 1837 Darwin developed health problems, including heart palpitations, and he rested for a while in the country. While there he observed the work of earthworms, then delivered a talk on the subject to the Geological Society in November, and in March 1838 Darwin became secretary of the society. His long hours of work took their toll, and Darwin developed more health problems such as stomach ailments, headaches, and heart symptoms, which caused him to be laid up for days at a time.

On November 11, 1838, Darwin proposed marriage to his cousin Emma Wedgwood, and she accepted. They were married on January 29, 1839, five days after the Royal Society of London elected him a fellow.

The Darwins returned to London, and for the next decade he prepared the results of his research for scientific publication. These works included his Journal and Remarks (*The Voyage of the Beagle*) published in 1839, and a book on coral reefs published in 1842. After this Darwin and his family moved to Down House outside London, so he could work in a more peaceful setting. Darwin's health continued to fluctuate, and in 1851 his daughter Annie became gravely ill and died. This loss led Darwin to abandon his faith in religion.

Darwin continued working to publish his main ideas, sometimes fearing he would die before his work was complete. In June 1858 a correspondence with the British naturalist Alfred Russel Wallace about the introduction of species and natural selection shocked Darwin, as it included a paper Wallace had written describing natural selection. Wallace and Darwin decided to present their work together at a meeting of the Linnean Society on July 1, 1858, but, tragically, just before the meeting, Darwin's baby son died of scarlet fever so he was unable to attend. Darwin's health continued to decline, and his works were not completed.

After another 13 months in preparation, Darwin arranged for his book on natural selection to be published in 1859 through British publisher John Murray (1745–93). The book, *On the Origin of Species by Means of Natural Selection or the Preservation of Favored Races in the Struggle of Life*, proved to be immensely popular as soon as it was published, with all 1,250 copies sold before publication. The text caused great controversy, especially with various religious institutions. Even some of Darwin's former tutors at Cambridge, including Sedgwick and Henslow, opposed the ideas in his book. It was said that his theories went against the teachings of the church, although perhaps some of his colleagues had merely been afraid to speak out against the church. Some liberal clergymen, however, spoke in favor of Darwin's ideas, calling them noble conceptions of deity. Darwin did not discuss religious views in his works, though he was in fact a very religious man (before his daughter's death), but other scientists after him have used his work as a basis of their own theories that there is no room for religion in science.

Darwin continued to work in the fields of botany, geology, and zoology, publishing works including *Variation in Plants and Animals under Domestication* (1868), followed by *The Descent of Man, and Selection in Relation to Sex* (1871). In 1872 Darwin published *Expression of the Emotions in Man and Animals*, followed by several books on plants, including insectivorous plants, and an insightful volume on *The Power of Movement in Plants,* describing heliotropism and phototropism in plants, published in 1880, and a final book on earthworms.

Darwin died on April 19, 1882, and was buried in a state funeral at Westminster Abbey, near Sir Issac Newton and John Herschel. Charles and Emma Darwin had 10 children, three of whom (Annie, Mary, and Charles Waring) died in childhood, and seven of whom survived Charles. These included Charles Erasmus, Henrietta Emma, George Howard, Elizabeth Bessy, Francis, Leonard, and Horace.

See also EVOLUTION; HISTORICAL GEOLOGY; LYELL, SIR CHARLES.

FURTHER READING

Darwin, Charles. *The Origin of Species by Means of Natural Selection, or the Preservation of Favoured Races in the Struggle for Life.* 6th ed., vii, edited by C. and W. Irvine. New York: Frederick Ungar Co., 1956.

———. *Voyage of a Naturalist, or Journal of Researches into the Natural History and Geology of the Countries Visited during the Voyage of HMS* Beagle *Round the World, under the Command of Capt. Fitz Roy, R. N.* New York: Harper and Brothers, 1846.

———. *The Variation of Animals and Plants under Domestication.* London: John Murray, 1868.

———. *The Descent of Man, and Selection in Relation to Sex.* London: John Murray, 1871.

———. *The Expression of the Emotions in Man and Animals.* London: John Murray, 1872.

deformation of rocks Deformation of rocks is measured by three components: strain, rotation, and translation. Strain measures the change in shape and size of a rock, rotation measures the change in orientation of a reference frame in the rock, and translation measures how far the reference frame has moved between the initial and final states of deformation.

The movement of the lithospheric plates causes rocks to deform, creating mountain belts and great fault systems like the San Andreas. The terms *strain* and *stress* describe how rocks are deformed. Stress, a measure of force per unit area, is a property that has directions of maximum, minimum, and intermediate values. *Strain* describes the changes in the shape and size of an object, and it is a result of stress.

There are three basic ways by which a solid can deform. The first is by elastic deformation, which is a reversible deformation exemplified by a stretching rubber band or the rocks next to a fault that bend and then suddenly snap back to place during an earthquake. Most rocks can undergo only a small amount of elastic deformation before they suffer permanent irreversible, nonelastic strain. Elastic deformation obeys Hooke's law, which simply states that for elastic deformation, a plot of stress versus strain yields a straight line. In other words strain is linearly proportional to the applied stress. So for elastic deformation, the stressed solid returns to its original size and shape after removal of the stress.

Solids may deform through fracturing and grinding processes during brittle failure or by flowing

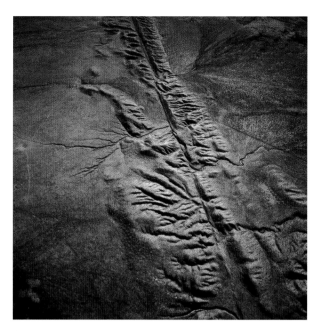

San Andreas Fault crossing Carrizo Plain in California
(Bernhard Edmaier/Photo Researchers, Inc.)

during ductile deformation processes. Fractures form when solids are strained beyond the elastic limit and the rock breaks; they are permanent, or irreversible, strains. Ductile deformation is also irreversible, but the rock changes shape by flowing, much like toothpaste squeezed out of a tube.

When compressed, rocks first experience elastic deformation, then as the stress increases they hit the yield point, at which ductile flow begins, and eventually the rock may rupture. Many variables determine why some rocks deform by brittle failure and others by ductile deformation. These variables include temperature, pressure, time, strain rate, and composition. The higher the temperature of the rock during deformation, the weaker and less brittle the rock will be. High temperature therefore favors ductile deformation mechanisms. High pressures increase the strength of the rock, leading to a loss of brittleness, and therefore hinder fracture formation. Time is also an important factor determining which type of deformation mechanism may operate. Fast deformation favors the formation of brittle structures, whereas slow deformation favors ductile deformation mechanisms. Strain rate is a measure of how much deformation (strain) occurs over a given time. Slow deformation rates favor ductile deformation, whereas fast deformation rates favor brittle deformation. Finally, the composition of the rock is also important in determining what type of deformation will occur. Some minerals (like quartz) are relatively strong, whereas others (such as calcite) are weak. Strong minerals or rocks may deform by brittle

mechanisms under the same (pressure, temperature) conditions that weak minerals or rocks deform by ductile flow. Water reduces the strength of virtually all minerals and rocks; therefore the presence of even a small amount of water can significantly affect the type of deformation that occurs.

BENDING OF ROCKS

The bending or warping of rocks is called folding. Monoclines are folds in which both sides are horizontal, which often form over deeper faults. Anticlines are upward-pointing arches that have the oldest rocks in the center, and synclines are downward-pointing arches, with the oldest rocks on the outside edges of the structure. Though many other geometric varieties of folds exist, most are variations of these basic types. The fold hinge is the region of maximum curvature on the fold, whereas the limbs are the regions between the fold hinges. Folds may be further classified according to how tight the hinges are, which can be measured by the angle between individual fold limbs. Gentle folds have interlimb angles between 180° and 120°, open folds have interlimb angles between 120° and 70°, close folds between 70° and 30°, tight folds have interlimb angles of fewer than 30°, and isoclinal folds have interlimb angles of 0°. Folds may be symmetrical, with similar lengths of both fold limbs, or asymmetrical, in which one limb is shorter than the other limb. Fold geometry may also be described by using the orientation of an imaginary surface (the axial surface), that divides the fold limbs into two symmetric parts, and the orientation of the fold hinge. Folds with vertical axial surfaces

Folded rock strata in Austrian Alps *(Bernhard Edmaier/Photo Researchers, Inc.)*

and subhorizontal hinges are known as upright gently plunging folds, whereas folds with horizontal hinges and axial surfaces are said to be recumbent.

BREAKING OF ROCKS

Brittle deformation results in the breaking of rock along fractures. Joints are fractures along which no movement has occurred. These may be tectonic structures formed in response to regional stresses or formed by other processes such as cooling of igneous rocks. Columnar joints are common in igneous rocks, forming six-sided columns when the magma cools and shrinks.

Fractures along which relative displacement has occurred are known as faults. Most faults are inclined surfaces. The block of rock above the fault is the hanging wall, and the block beneath the fault is the footwall, after old mining terms. Faults are classified according to their dip and the direction of relative movement across the fault. Normal faults are faults along which the hanging wall has moved down relative to the footwall. Reverse faults are faults along which the hanging wall moves up relative to the footwall. Thrust faults are a special class of reverse faults that dip fewer than 45°. Strike-slip faults are steeply dipping (nearly vertical) faults along which the principal movement is horizontal. The sense of movement on strike-slip faults may be right lateral or left lateral, determined by standing on one block and describing whether the block across has moved to the right or to the left.

REGIONAL DEFORMATION OF ROCKS

Deformation of rocks occurs at a variety of scales, from the atomic to the scale of continents and entire tectonic plates. Deformation at the continental-to-plate scale produces distinctive regional structures. Cratons, large stable blocks of ancient rocks that have been stable for a long time (since 2.5 billion years ago), form the cores of many continents and represent continental crust that was formed in the Archean Era. Most cratons are characterized by thick continental roots made of cold mantle rocks, by a lack of earthquakes, and by low heat flow.

Orogens, or orogenic belts, are elongate regions that represent eroded mountain ranges, and they typically form belts around older cratons. Characterized by abundant folds and faults, they typically show shortening and repetition of the rock units by 20–80 percent. Young orogens are mountainous—for instance, the Rocky Mountains have many high peaks, and the slightly older Appalachians have lower peaks.

Continental shields are places where ancient cratons and mountain belts are exposed at the surface, whereas continental platforms are places where younger, generally flat-lying sedimentary rocks overlie the older shield. Many orogens contain large portions of crust that have been added to the edges of the continental shield through mountain-building processes related to plate tectonics. Mountain belts may be subdivided into three basic types: fold and thrust belts, volcanic mountain chains, and fault block ranges.

FOLD AND THRUST BELTS

Fold and thrust mountain chains are contractional features, formed when two tectonic plates collide, forming great thrust faults and folding metamorphic rocks and volcanic rocks. By examining and mapping the structure in the belts we can reconstruct their history and essentially pull them back apart in the reverse of the sequence in which they formed. By reconstructing the history of mountain belts in this way, we find that many of the rocks in the belts were deposited on the bottom of the ocean or on the ocean margin deltas and continental shelves, slopes, and rises. When the two plates collide, many of the sediments get scraped off and deformed, creating the mountain belts; thus fold and thrust mountain belts mark places where oceans have closed.

The Appalachians of eastern North America represent a fold and thrust mountain range. They show a detachment surface, or decollement, folds, and thrust faults. The sedimentary rocks in the mountain belt are like those now off the coast, so the Appalachians are interpreted to represent a place where an old ocean has closed.

VOLCANIC MOUNTAIN RANGES

Volcanic mountain ranges represent thick segments of crust that formed by addition of thick piles of volcanic rocks, generally above a subduction zone. Examples of volcanic mountain chains include the Aleutians of Alaska, the Fossa Magna of Japan (including Mount Fuji), and the Cascades of the western United States (including Mount Saint Helens). These mountain belts are not formed primarily by deformation but by volcanism associated with subduction and plate tectonics. Many do have folds and faults, however, showing that there is overlap between fold and thrust types of mountain chains and volcanic ranges.

FAULT-BLOCK MOUNTAINS

Fault-block mountains generally form by extension of the continental crust. The best examples include the Basin and Range Province of the western United States, and parts of the East African Rift System, including the Ethiopian Afar. These mountain belts are formed by the extension or pulling apart of the continental crust, forming basins between individual tilted fault-block mountains. These types of ranges are associated with thinning of the continental crust, and some have active volcanism as well as active extensional deformation.

See also OROGENY; PLATE TECTONICS; STRUCTURAL GEOLOGY.

FURTHER READING

Hatcher, Robert D. *Structural Geology, Principles, Concepts, and Problems.* 2nd ed. Englewood Cliffs, N.J.: Prentice Hall, 1995.

van der Pluijm, Ben A., and Stephen Marshak. *Earth Structure: An Introduction to Structural Geology and Tectonics.* Boston: WCB-McGraw Hill, 1997.

deltas Found at the mouths of streams and rivers, deltas are low, flat deposits of alluvium that form broad triangular or irregular-shaped areas that extend into bays, oceans, or lakes. They are typically crossed by many distributaries from the main river and may extend for a considerable distance underwater. Deltas are extremely sensitive coastal environments and are particularly susceptible to the effects of rising sea level and human activities. Since deltas are the sites of rich oil deposits, there is currently a sensitive interplay between meeting the world's energy needs by extracting oil from beneath the fragile delta environment and the environmental concerns about preserving the delta ecosystem.

The velocity of the water and capacity of a river or stream to hold sediment in suspension suddenly drop when it enters the relatively still body of water such as a lake or the ocean. Thus the stream dumps its sediment load here, and the resulting deposit is known as a delta. The term *delta* was first used for these deposits by Herodotus in the fifth century B.C.E. for the triangular-shaped alluvial deposits at the mouth of the Nile River. The stream first drops the coarsest material, then progressively finer material farther out, forming a distinctive sedimentary deposit. In a study of several small deltas in ancient Lake Bonneville in Utah, Idaho, and Nevada, American geologist Grover Karl Gilbert in 1890 recognized that the deposition of finer-grained material farther away from the shoreline also created a distinctive vertical sequence in delta deposits. The resulting foreset layer is thus graded from coarse nearshore to fine offshore. The bottomset layer consists of the finest material, deposited far out. As this material continues to build outward, the stream must extend its length and forms new deposits, known as topset layers, on top of all this. Topset beds may include a variety of subenvironments, both subaqueous and subaerial, formed as the delta progrades seaward.

Most of the world's large rivers, such as the Mississippi, the Nile, and the Ganges, have built enormous deltas, yet all of these are different in detail. Deltas may have various shapes and sizes or may even be completely removed, depending on the relative amounts of sediment deposited by the stream, the erosive power of waves and tides, the climate, and the tectonic stability of the coastal region. Most deltas are located along passive or trailing continental margins, and few are found along convergent boundaries (exceptions include the Copper River in Alaska and the Fraser River in British Columbia). This is largely because river systems on passive margins tend to be long and to drain huge areas composed of easily eroded soil, carrying large sediment loads. Rivers along active margins tend to be much shorter and cut through bedrock, which is not eroded as easily so yields smaller sediment loads. Additionally, convergent margins do not contain wide continental shelves needed for the delta to be deposited on, but instead are marked by deep-sea trenches where sediments are rapidly deformed and buried.

Most deltas are quite young, having formed since the glaciers melted 18,000–10,000 years ago and sea levels rose onto the continental shelves. During the last glacial maximum when glaciers were abundant for much of the period from 2.5 million years ago until about 18,000 years ago, sea levels were about 395 feet (120 m) lower than at present. During the glacial maximum, most rivers eroded canyons across the continental shelves and carried their sedimentary

False-color composite of Mississippi River delta from ASTER instrument on NASA's *Terra* satellite, May 24, 2001 *(USGS EROS Data Center Satellite Systems Branch as part of "Earth as Art II" image series, NASA)*

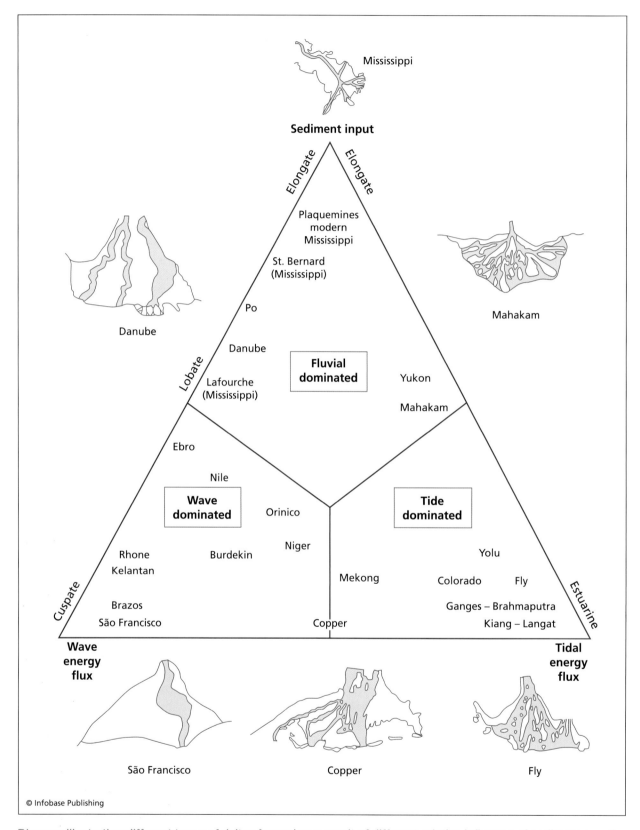

Diagram illustrating different types of deltas formed as a result of different relative influence of sediment supply, tidal energy flux, and wave energy flux. The shapes of deltas characteristic of each are shown on the edges of the diagram, and names of other deltas are plotted in the space inside the diagram, in positions that reflect the relative strength of each component for each delta.

load to the deep oceans. As the glaciers melted, sea level rose onto the broad continental shelves of many continents, where wide and thick delta sediments have space to accumulate. Nearly all of the active parts of deltas are younger than 18,000 years, but many have older, deeper parts that formed during older sea-level high stands (some from interglacial periods) that have subsided deep below sea level. As sea levels were initially rising fast as the glaciers were undergoing rapid melting, the river mouths were moving so rapidly inland that deltas did not have time to form. The rate of sea-level rise slowed significantly around 6,000 years ago, and most of the world's deltas began to grow significantly since that time. This history of sedimentation is reflected in the Mississippi River delta, which has components that are older than several million years, but the active lobes only began forming about 6,000 years ago.

Deltas exhibit a range in conditions and environments from terrestrial and river–dominated at their landward boundaries to marine and wave and tide–dominated at their fronts. The mere presence of a delta along a coast indicates that the amount of sediment input by the river is greater than the amount of sediment that can be removed by the action of waves, tides, currents, wind, and submarine slumping. The distributaries and main channel of the rivers forming deltas typically move to find the shortest route to the sea, and this causes shifting of the active locus of deposition on deltas. Inactive areas, which may form lobes or just parts of the delta, typically subside and are reworked by tidal currents and waves. The processes involved in the growth or seaward progradation of deltas result in the formation of many environments, including those influenced by subaerial, intertidal, and subaqueous processes, and include freshwater, brackish, and saltwater conditions. Most deltas can be divided into three main parts: the landward delta plain, the delta front, and the prodelta in the subtidal to deep continental shelf environment.

The delta plain is really a coastal extension of the river system. It comprises river and overbank sedimentary deposits in a flat, meandering stream-type of setting. These environments are at or near (or in some cases below) sea level, and it is essential that the overbank regions receive repeated deposits of muds and silts during flood stages to build up the land surface continuously as the entire delta subsides below sea level by tectonic processes. Deltas deprived of this annual silt by the construction of levees gradually sink below sea level. If homes were built on delta flood plains without levees, however, they would gradually be buried in mud, as opposed to sinking below sea level behind the false protection of a levee.

False-color image of Lena delta in Russia acquired by Landsat 7's Enhanced Thematic Mapper plus sensor on February 27, 2000 *(USGS EROS Data Center Satellite Systems Branch)*

The stream channels are bordered by natural levee systems that may rise several feet (1–2 m) above the floodplain; these areas are often the only places above water level during river flood stages. In many outer delta plains the only places above sea level are the natural levees. During floods the levees sometimes break, creating a crevasse splay that allows water and muddy sediment to flow rapidly out of the channel and cover the overbank areas, plus any homes or other human infrastructure built in this sensitive area.

The delta front environment, located on the seaward edge of the delta, is an extremely sensitive environment. It is strongly affected by waves, tides, changing sea level, and changes in the flux or amount of sediment delivered to the delta front. Many delta fronts have an offshore sandbar, called a distributary mouth bar, or barrier island system, parallel to the coast along the delta front. Some deltas, such as the Mississippi, are losing huge areas of delta front to subsidence below sea level, due to combined effects of a decrease in sediment supply to the delta front, tectonic subsidence, sea-level rise, human activities such as oil drilling and building levees, and severe erosion from storms such as Hurricanes Hugo, Katrina, and Ike.

Deltas have been classified various ways over time, including by schemes based on their shapes and on the processes involved in their construction. High-constructive deltas form where the fluvial transport dominates the energy balance on the delta. These deltas dominated by riverine processes are typically elongate, such as the modern delta at the mouth of the Mississippi, which has the shape of a bird's foot, or they may be lobate, such as the older Holocene lobes of the Mississippi that have now largely subsided below sea level.

High-destructive deltas form where the tidal and wave energy is high and much of the fluvial sediment gets reworked before it is finally deposited. In wave-dominated, high-destructive deltas sediment typically accumulates as arcuate barriers near the mouth of the river. Examples of wave-dominated deltas include the Nile and the Rhône. In tide-dominated, high-destructive deltas, tides rework the sediment into linear bars that radiate from the mouth of the river, with sands on the outer part of the delta sheltering a lower-energy area of mud and silt deposition inland from the segmented bars. Examples of tide-dominated deltas include the Ganges, and the Kikari and Fly River deltas in the Gulf of Papua New Guinea. Other rivers drain into the sea in places where the tidal and wave current is so strong that these systems completely overwhelm the fluvial deposition, removing most of the delta. The Orinoco River in South America has had its sediment deposits transported southward along the South American coast, with no real delta formed at the mouth of the river.

Where a coarse sediment load of an alluvial fan dumps its load in a delta, the deposit is known as a fan-delta. Braid-deltas are formed when braided streams meet local base level and deposit their coarse-grained load.

Deltas create unique, diverse environments where fresh and saltwater ecosystems meet, and swamps, beaches, and shallow marine settings are highly varied. They contain some of the most productive ecological areas in the world. Deltas also form some of the world's greatest hydrocarbon fields, however, as the muds and carbonates make good source rocks and the sands make excellent trap rocks. Thus there is a delicate struggle between preserving natural ecosystems and using the planet's resources that must be maintained on the deltas of the world. Resting at sea level, delta environments are also the most susceptible to disaster from hurricanes and coastal storms.

See also BASIN, SEDIMENTARY BASIN; BEACHES AND SHORELINES; SEDIMENTARY ROCK, SEDIMENTATION; SUBSIDENCE.

FURTHER READING

Burkett, Virginia R., D. B. Zikoski, and D. A. Hart. "Sea-Level Rise and Subsidence: Implications for Flooding in New Orleans, Louisiana." In *U.S. Geological Survey Subsidence Interest Group Conference, Proceedings for the Technical Meeting.* Reston, Va.: U.S. Geological Survey. USGS Water Resources Division, Open File Report Series 03-308, 2003: 63–70.

Davis, R., and D. Fitzgerald. *Beaches and Coasts.* Malden, Mass.: Blackwell, 2004.

Leatherman, Stephen P., ed. *Barrier Islands, from the Gulf of St. Lawrence to the Gulf of Mexico.* New York: Academic Press, 1979.

National Research Council. *Drawing Louisiana's New Map: Addressing Land Loss in Coastal Louisiana.* Washington, D.C.: National Academies Press, 2005.

Salvador, A., ed. *The Gulf of Mexico Basin, Geology of North America.* Boulder, Colo.: Geological Society of America, 1991.

U.S. Geological Survey, Delta Subsidence in California. "The Sinking Heart of the State. U.S. Geological Survey Fact Sheet." Available online. URL: http://ca.water.usgs.gov/archive/reports/fs00500/fs00500.pdf. Accessed October 10, 2008.

deserts The driest places on Earth, deserts by definition receive less than one inch (250 mm) of rain per year. Most deserts are so dry that more moisture is able to evaporate than falls as precipitation. At present about 30 percent of the global landmass is desert, and the United States has about 10 percent desert areas. With changing global climate patterns and shifting climate zones, much more of the planet is in danger of becoming desert.

Most deserts are also hot, with the highest recorded temperature on record being 136°F (58°C) in the Libyan Desert. With high temperatures the evaporation rate is high, and in most cases deserts evaporate more than the amount of precipitation that falls as rain. Many deserts evaporate 20 times the amount of rain that falls, and some places, like much of the northern Sahara, are capable of evaporating 200–300 times the amount of rain that falls in rare storms. Deserts are also famous for large variations in daily temperature, sometimes changing as much as 50–70°F (28–39°C) between day and night (called a diurnal cycle). These large temperature variations can be enough to shatter boulders. Deserts are also windy and are prone to sand and dust storms. The winds arise primarily because the heat of the day causes warm air to rise and expand, and other air must rush in to take its place. Airflow directions also shift frequently between day and night (in response to the large temperature difference between day and night), and between any nearby water bodies, which tend to remain at a constant temperature over a 24-hour period.

There are many different types of deserts located in all different parts of the world. Some deserts are associated with patterns of global air circulation, and others form because they are in continental interiors far from any sources of moisture. Deserts can form on the "back," or leeward, side of mountain ranges,

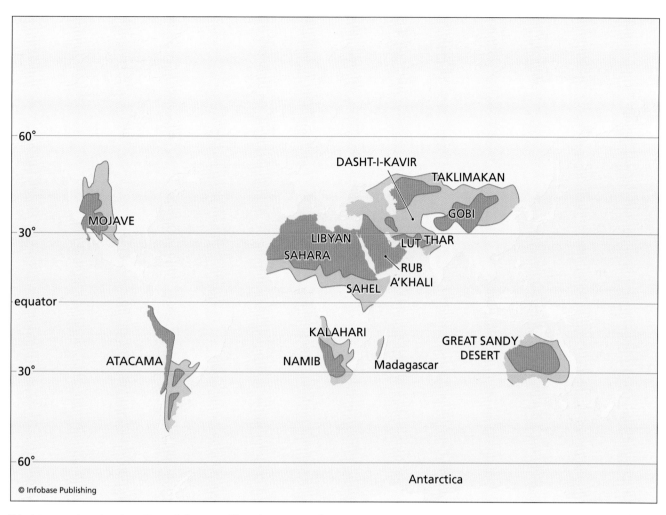

World map showing location of deserts: Note how most deserts are concentrated between 15° and 30° latitude.

where downwelling air is typically dry, or they can form along coasts where cold, upwelling ocean currents lower the air temperature and lower its ability to hold moisture. Deserts can also form in polar regions, where extremely dry and cold air can evaporate (or sublimate) much more moisture than falls as snow in any given year. Parts of Antarctica have not had any significant ice or snow cover for thousands of years.

Deserts have a distinctive set of landforms and hazards associated with these landforms. The most famous desert landform is a sand dune, a mobile accumulation of sand that shifts in response to wind. Deserts tend to be very windy, and some of the hazards in deserts are associated with sand and dust carried by the wind. Dust eroded from deserts can be carried around the world and is a significant factor in global climate and sedimentation. Some sandstorms are so fierce that they can remove paint from cars or skin from an unprotected person. Other desert hazards are associated with flash floods, debris flows, avalanches, extreme heat, and extreme temperature fluctuations.

Droughts are different from deserts—a drought is an extended lack of rainfall across a region that typically gets more rainfall. If a desert normally receives a small amount of rainfall and it still is getting little rainfall, then it is not experiencing a drought. In contrast, a different area that receives more rainfall than the desert may be experiencing a drought if it normally receives significantly more rainfall than it is at present. A drought-plagued area may become a desert if the drought is prolonged. Droughts can cause widespread famine, loss of vegetation, loss of life, and eventual death or mass migrations of entire populations.

Desertification is the conversion of previously productive lands to desert through a prolonged drought. Desertification may occur if the land is stressed before or during the drought, typically from poor agricultural practices, overuse of ground and surface water resources, and overpopulation. Global climate goes through several different variations that can cause belts of aridity to shift back and forth with time. The Sahel region of northern

DRYING OF THE AMERICAN SOUTHWEST

A drought is a prolonged lack of rainfall in a region that typically experiences a significant amount of precipitation. If a desert normally receives a small amount of rainfall, and it still is getting little rainfall, then it is not experiencing a drought. In contrast, a different area that normally receives more rainfall than the desert may be experiencing a drought if it normally receives significantly more rainfall than it is at present, even if it still experiences more rainfall than the desert. A drought-plagued area may become a desert if the drought is prolonged. Droughts are the most serious natural hazard in terms of their severity, area affected, loss of livelihood, social impact, and other long-term effects. Droughts can cause widespread famine, loss of vegetation, loss of life, and eventual death or mass migrations of entire populations.

Droughts may lead to the conversion of previously productive lands to desert. This process, called desertification, may occur if the land is stressed before or during the drought, typically from poor agricultural practices, overuse of ground and surface water resources, and overpopulation. Global climate undergoes several different variations that can cause belts of aridity to shift back and forth with time. The Sahel region of Africa has experienced some of the more severe droughts in recent times. The Middle East and parts of the desert southwest of the United States are overpopulated and the environment there is stressed. If major droughts occur in these regions, major famines could result and the land may be permanently converted to desert.

Much of the desert southwest region of the United States was settled in the past century following a century of historically high rainfall. Towns and cities grew, and the Bureau of Land Management diverted water from melting snows, rivers, and underground aquifers to meet the needs of growing cities. Some of the country's largest and newest cities, including Phoenix, Tucson, Denver, Las Vegas, Los Angeles, San Diego, and Albuquerque, have grown out of the desert using water from the Colorado River system. Even though the temperatures can be high, the air is good, and many people have chosen to move to these regions to escape crowded, polluted, or allergen-rich cities and air elsewhere. The surge in population has been met with increases in the water diverted to these cities, and fountains, swimming pools, resorts, golf courses, and green lawns have sprung up all over. In general the life can be comfortable.

In the past decade the water supply seems to be diminishing. Lake Powell in Arizona has shrunk to half its capacity, and the Colorado River flow shrunk to a quarter of its typical rates. The Colorado River typically supplies 30 million people with water and irrigates 4 million acres of fertile farmland, producing billions of dollars worth of crops. The massive water work systems across seven states in the Southwest were all built using river-flow data for the Colorado River based on 20th-century flow records. Now studies of the ancient climate history in the region going back thousands of years indicate that the 20th century may have been one of the wettest on record for the region. The Hoover Dam, the California aqueduct, and cities across the region were all built during this high-flow stage of the Colorado River, and water budgets for the region were calculated assuming these flows would continue. Now precipitation is decreasing, and the historical records show that the region regularly experiences droughts where the flow decreases to 80 percent and even 50 percent of the 20th-century values used for building the civilization in the desert southwest. Now that more than 80 percent of the water from the river is used for human consumption, droughts of this magnitude have severe implications for any community, and the water wars of the southwest may eventually start again. Historical records show that past civilizations such as the Anasazi in the region disappeared at the end of the 13th century during a similar drought period, and similar trends are expected by climate modelers for the future in the region.

Climate-change models released by the National Ocean and Atmospheric Administration show that the flow of the Colorado River may decrease to half of its 20th-century values by the middle of this century, and that these lower flow values will persist into the foreseeable future. The region is already experiencing rapid changes, with wildfires burning huge tracts of vegetation and occasional storms initiating mudflows and other desert processes. Climate models predict a likely descent of the region into dust bowl conditions, and these changes have already begun. The region saw many mega-droughts in medieval times and throughout history, and states of the region need to prepare for the likelihood of many years of water shortage and increasing drought conditions.

FURTHER READING
Kunzig, Robert. "Drying of the West." *National Geographic Magazine* 213, no. 2 (2008): 90–113.

Kusky, T. M. *Climate Change: Shifting Glaciers, Deserts, and Climate Belts.* New York: Facts On File, 2009.

Reisner, M. *Cadillac Desert: The American West and Its Disappearing Water.* New York: Penguin, 1986.

Africa has experienced some of the more severe droughts in recent times. The Middle East and parts of the desert southwest of the United States are overpopulated and the environment is stressed. If major droughts occur in these regions, major famines could result and the land may be permanently desertified.

LOCATION AND FORMATION OF DESERTS

More than 30 percent of the planet's land area is arid or semiarid, and these deserts form an interesting pattern that reveals clues about how they develop. There are six main categories of desert, based on geographic location with respect to continental margins, oceans, and mountains: trade wind, or Hadley cell, deserts; continental interior/midlatitude deserts; rain-shadow deserts; coastal deserts; monsoon deserts; and polar deserts.

TRADE WIND, OR HADLEY CELL, DESERTS

Many of the world's largest and most famous deserts are located in two belts between 15° and 30° North and South latitude. Included in this group of deserts are the Sahara, the world's largest desert, and the Libyan Desert of North Africa. Other members of this group include the Syrian Desert, Rub' al-Khali (Empty Quarter), and Great Sandy Desert of Arabia; the Dasht-e-Kavir, Lut, and Sind of southwest Asia; the Thar Desert of Pakistan; and the Mojave and Sonoran Deserts of the United States. In the Southern Hemisphere deserts that fall into this group include the Kalahari Desert of Africa and the Great Sandy Desert of Australia. The formation of the Atacama Desert in South America, the world's driest place, can be partly attributed to its location between these latitudes.

The location of these deserts is controlled by a large-scale atmospheric circulation pattern driven by energy from the Sun. The Sun heats equatorial regions more than high-latitude areas, which causes large-scale atmospheric upwelling near the equator. As this air rises, it becomes less dense and can hold less moisture, a condition leading to the formation of large thunderstorms in equatorial areas. This drier air then moves away from the equator at high altitudes, cooling and drying more as it moves, until it eventually forms two circumglobal downwelling belts between 15–30°N and S latitude. This cold downwelling air is dry and can hold much more water than it has brought with it on its circuit from the equator. These belts of circulating air, known as Hadley cells, are responsible for the formation of many of the world's largest, driest deserts. As this air completes its circuit back to the equator, it forms dry winds that heat up as they move toward the equa-tor. The dry winds dissipate existing cloud cover and allow more sunlight to reach the surface, which consequently warms even more.

Deserts formed by global circulation patterns are particularly sensitive to changes in global climate; seemingly small changes in the global circulation can lead to catastrophic expansion or contraction of some of the world's largest deserts. For instance, the sub-Saharan Sahel has experienced several episodes of expansion and contraction, displacing or killing millions of people in this vicious cycle. When deserts expand, croplands dry up and livestock and humans cannot find enough water to survive. Desert expansion is the underlying cause of some of the world's most severe famines.

CONTINENTAL INTERIOR/ MIDLATITUDE DESERTS

Some places on Earth are so far from ocean moisture sources that by the time weather systems reach them, most of the moisture they carry has already fallen. This effect is worsened if the weather systems must rise over mountains or plateaus to reach these areas, because cloud systems typically lose moisture as they rise over mountains. These remote areas therefore have little chance of receiving significant rainfall. The most significant deserts in this category are the Taklimakan-Gobi region of China, resting south of the Mongolian steppe on the Alashan plateau, and the Karakum of western Asia. The Gobi is the world's northernmost desert, and it is characterized by 1,000-foot (305-m) high sand dunes made of coarser than normal sand and gravel, built up layer by layer by material moved and deposited by the wind. It is a desolate region, conquered successively by Genghis Khan, warriors of the Ming dynasty, then the People's Army of China. The sands are still littered with remains of many of these battles, such as the abandoned city of Khara Khoto. In 1372 Ming dynasty warriors conquered this walled city by cutting off its water supply, consisting of the Black River, waiting, then massacring anyone remaining in the city.

RAIN-SHADOW DESERTS

A third type of desert is found on the leeward, or back, side of some large mountain ranges, such as the sub-Andean Patagonian Gran Chaco and Pampas of Argentina, Paraguay, and Bolivia. A similar effect is partly responsible for the formation of the Mojave and Sonoran Deserts of the United States. These deserts form because as moist air masses move toward the mountain ranges, they must rise to move over the ranges. As the air rises it cools, and cold air cannot hold as much moisture as warm air. The clouds thus drop much of their moisture on the windward side of the mountains, explaining why places like the western Cascades and western Sierras of the United States are extremely wet, as are the western Andes in Peru. But the eastern lee, or back, sides of these mountains are extremely dry. This is because as the air rose over the fronts, or windward, sides of the mountains, it dropped its moisture. As the same air descends on the lee side of the mountains, it warms and can hold more moisture than it has left in the clouds. The result is that the air is dry and it rarely rains. This explains why places like the eastern sub-Andean region of South America and the Sonoran and Mojave Deserts of the western United States are extremely dry.

Rain-shadow deserts tend to be mountainous because of the way they form, and they are associated with a number of mass wasting hazards such as landslides, debris flows, and avalanches. Occasional rainstorms that make it over the blocking mountain ranges can drop moisture in the highlands, leading to flash floods coming out of mountain canyons into the plains or intermountain basins on the lee side of the mountains.

COASTAL DESERTS

Some deserts are located along coastlines, where intuition would seem to indicate that moisture should be plentiful. The driest place on Earth is the Atacama Desert, however, located along the coast of Peru and Chile. The Namib Desert of southern Africa is another coastal desert, known legendarily as the Skeleton Coast, because it is so dry that many of the large animals that roam out of the more humid interior climate zones perish there, leaving their bones sticking out of the blowing sands.

How do these coastal deserts form adjacent to such large bodies of water? The answer lies in the ocean currents, for in these places cold water upwells from the deep ocean and cools the atmosphere. The effect is similar to that of rain-shadow deserts, where cold air can hold less moisture, and the result is no rain.

MONSOON DESERTS

In some places seasonal variations in wind systems bring alternating dry and wet seasons. The Indian Ocean is famous for its monsoonal rains in the summer, as the southeast trade winds bring moist air onshore. As the clouds move across India, however, they lose moisture and must rise to cross the Aravalli Mountain range. The Thar Desert of Pakistan and the Rajasthan Desert of India are located on the lee side of these mountains and do not generally receive this seasonal moisture.

POLAR DESERTS

A final class of deserts is the polar desert, found principally in the Dry Valleys and other parts of Antarctica, parts of Greenland, and northern Canada. Approximately 3 million square miles (7.8 million km²) on Earth consists of polar desert environments. In these places cold downwelling air lacks moisture, and the air is so dry that the evaporation potential is much greater than the precipitation. Temperatures do not exceed 50°F (10°C) in the warmest months, and precipitation is less than one inch (2.5 cm) per year. There are places in the Dry Valleys of Antarctica that have not been covered in ice for thousands of years.

Polar deserts are generally not covered in sand dunes but are marked by gravel plains or barren bedrock. The landforms of polar deserts are shaped by frost wedging, where alternating freeze-thaw cycles allow small amounts of water to seep into cracks and other openings in rocks. When the water freezes it expands, pushing large blocks of rock away from the main mountain mass. In polar deserts and other regions affected by frost wedging, large talus slopes may form adjacent to mountain fronts, and these are prone to frequent rock falls from frost wedging.

LOESS

Loess, silt and clay deposited by wind, forms a uniform blanket that covers hills and valleys at many altitudes, distinguishing it from deposits of streams. Strong winds that blow across desert regions sometimes pick up dust made of silt and clay particles and transport them thousands of miles (thousands of km) from their source. For instance, dust from China is found in Hawaii, and the Sahara Desert commonly drops dust in Europe. This dust is a nuisance, has a significant influence on global climate, and has at times, as in the dust bowl days of the 1930s, been known nearly to block out the Sun.

Recently scientists have recognized that wind-blown dust contributes significantly to global climate. Dust storms that come out of the Sahara can be carried around the world and partially block out some of the Sun's radiation. The dust particles may also act as nuclei for raindrops to form around, perhaps acting as a natural cloud-seeding phenomenon. One interesting point to ponder is that as global warming increases global temperatures, the amount and intensity of storms increase, and some of the world's deserts expand. Dust storms may reduce global temperatures and increase precipitation. In this way dust storms may be some kind of self-regulating mechanism whereby the Earth moderates its own climate.

DESERT LANDFORMS

Desert landforms are some of the most beautiful on Earth, often presenting bizarre sculpted mountains, steep walled canyons, and regional gravel plains. They can also be some of the most hazardous landscapes on the planet. The regolith, or mixture of soil and altered bedrock in deserts is thin, discontinuous, and much coarser-grained than in moist regions, and is produced predominantly by mechanical weathering. Chemical weathering is only of minor importance because of the rare moisture. Also the coarse size of particles produced by mechanical weathering causes steep slopes, eroded from steep cliffs and escarpments.

Much of the regolith that sits in deserts is coated with a dark layer of manganese and iron oxides, known as desert varnish, produced by a combination of microorganism activity and chemical reactions with fine manganese dust that settles from the wind.

DESERT DRAINAGE SYSTEMS

Most streams in deserts evaporate before they reach the sea. Most are dry for long periods of time and subject to flash floods during brief but intense rains. These flash floods transport most of the sediment in deserts and form fan-shaped deposits of sand, gravel, and boulders found at the bases of many mountains in desert regions. These flash floods also erode deep, steep-walled canyons through the upstream mountain regions, which is the source of the boulders and cobbles found on the mountain fronts. Intermountain areas in deserts typically have finer-grained material, deposited by slower-moving currents that represent the waning stages of floods as they expand into open areas between mountains after they escape out of mountain canyons.

Flash floods can be particularly hazardous in desert environments, especially when the floods are the result of distant rains. More people die in deserts from drowning in flash floods than die from thirst or dehydration. In many cases rain in faraway mountains occurs while people in downstream areas are not aware it is raining upstream. Rain in deserts is typically a brief but intense thunderstorm, which can drop a couple of inches (> 5 cm) of rain in a short time. The water may then quickly move downstream as a wall of water in mountain canyons, sweeping away all loose material in its path. People or vehicles caught in such a flood are likely to be swept away by the swiftly moving torrent.

Saguaro cactus in Arizona desert *(Paul B. Moore, from Shutterstock, Inc.)*

Dry lake beds in low-lying flat areas, which may contain water only once every few years, characterize many deserts. These playas, or hardpans, typically have deposits of white salts that formed when water from storms evaporated, leaving the lakes dry. There are more than 100 playas in the American southwest, including Lake Bonneville, which formed during the last ice age and now covers parts of Utah, Nevada, and Idaho. When there is water in these basins, they are known as playa lakes. Playas are flat surfaces that make excellent racetracks and runways. The U.S. space shuttles commonly land on Rogers Lake playa at Edwards Air Force Base in California.

Alluvial fans are coarse-grained deposits of alluvium that accumulate at the fronts of mountain canyons. Alluvial fans are very common in deserts, where they are composed of both alluvium and debris-flow deposits. Alluvial fans are quite important for people in deserts, because they are porous and permeable and they contain large deposits of groundwater. In many places alluvial fans so dominate the land surface that they form a *bajada,* or slope, along the base of the mountain range, formed by fans that have coalesced to form a continuous broad alluvial apron.

Pediments represent different kinds of desert surfaces. They are surfaces sloping away from the base of a highland and are covered by a thin or discontinuous layer of alluvium and rock fragments. These erosional features are formed by running water and are typically cut by shallow channels. Pediments grow as mountains are eroded.

Inselbergs are steep-sided mountains or ridges that rise abruptly out of adjacent, monotonously flat plains in deserts. Ayres Rock in central Australia is perhaps the world's best-known inselberg. These are produced by differential erosion, leaving behind as a mountain rocks that for some reason are more resistant to erosion.

WIND IN DESERTS

Wind plays a significant role in the evolution of desert landscapes. Wind erodes in two basic ways. Deflation is a process whereby wind removes material from an area, reducing the land surface. The process is akin to deflating a balloon below the surface of the ground, hence its name. Abrasion is a different process that occurs when particles of sand and other size grains are blown by the wind and impact one another. Exposed surfaces in deserts are subjected to frequent abrasion, which is similar to sandblasting.

Yardangs are elongate (several miles [kms] long), streamlined wind-eroded ridges that resemble an overturned ship's hull sticking out of the water. These unusual features are formed by abrasion, by long-term sand blasting along specific corridors. The sandblasting leaves erosionally resistant ridges but removes the softer material, which itself will contribute to sandblasting in the downwind direction and eventually to the formation of sand, silt, and dust deposits.

Deflation is important on a large scale in places where there is no vegetation. In some places wind has excavated large basins known as deflation basins. Deflation basins are common in the United States from Texas to Canada, as elongate depressions, typically only 3–10 feet (1–3 m) deep. In some places like the Sahara, however, deflation basins may be as many as several hundred feet (100 m) deep.

Deflation by wind can move only small particles away from the source, since the size of the particle that can be lifted is limited by the strength of the wind, which rarely exceeds a few tens of miles per hour (~50 km/h). Deflation therefore leaves boulders, cobbles, and other large particles behind. These get concentrated on the surface of deflation basins and other surfaces in deserts, leaving a surface concentrated in boulders known as desert pavement.

Desert pavements are a long-term, stable desert surface and are not particularly hazardous. When the desert pavement is broken, however, for instance, by being driven across, the coarse cobbles and pebbles get pushed beneath the surface and the underlying sands get exposed to wind action again. Driving across a desert pavement can raise a considerable amount of sand and dust, and if many vehicles drive across the surface, then it can be destroyed and the whole surface becomes active.

A striking, large-scale example of this process was provided by events in the Gulf War of 1991. After Iraq invaded Kuwait in 1990, U.S. and allied forces massed hundreds of thousands of troops on the Saudi Arabian side of the border with Iraq and Kuwait and eventually mounted a multipronged counterattack on Kuwait City that led to its liberation. Several of the prongs circled far to the north then turned around and returned southward to Kuwait City. These prongs took many thousands of heavy tanks, artillery, and other vehicles across a region of stable desert pavement, and the weight of these military vehicles destroyed the pavement to free Kuwait. Since the liberation, the steady winds from the northwest have continued, and this area that was once stable desert pavement and stable dune surfaces (covered with desert pavement and minor vegetation) has been remobilized. Large sand dunes have now formed from the sand previously trapped under the pavement. Other dunes that were stable have been reactivated. Now Kuwait City residents are bracing for what they call the second invasion of Kuwait, but this time the invading force is sand and dust, not a foreign army.

Several measures have been considered to try to stabilize the newly migrating dunes. One is to try to reestablish the desert pavement by spreading cobbles across the surface, but this is unrealistic because of the large area involved. Another proposition, being tested, is to spray petroleum on the migrating dunes to effectively create a blacktop or tarred surface that would be stable in the wind. This is feasible in oil-rich Kuwait but not particularly environmentally friendly.

In China's Gobi and Taklimakan Deserts a different technique to stabilize dunes has proven rather successful. Bales of hay are initially placed in a grid pattern near the base of the windward side of dunes, which decreases the velocity of the air flowing over the dune and reduces the transportation of sand grains over the slip surface. Drought-resistant vegetation is planted between the several-foot (1-m) wide units of hay bales, and then when the dune is more stabilized, vegetation is planted along the dune crest. China is applying this technique across much of the Gobi and Taklimakan Deserts, protecting railways and roads. In northeastern China this technique is being applied in an attempt to reclaim lands that became desert through human activity, and the Chinese are constructing a 5,700-mile (9,000-km) long line of hay bales and drought-resistant vegetation. China is said to be building a new "Green Wall" that will be longer than the famous Great Wall of China, and will, the Chinese hope, prove more effective at keeping out invading forces (in this case, sand) from Mongolia.

WINDBLOWN SAND AND DUST

Most people think of deserts as areas with lots of big sand dunes and continual swirling winds of dust storms. But really dunes and dust storms are not as common as depicted in popular movies, and rocky deserts are more common than sandy deserts. For instance, only about 20 percent of the Sahara is covered by sand; the rest is covered by rocky, pebbly, or gravel surfaces. Sand dunes are locally very important in deserts, however, and wind is one of the most important processes in shaping deserts worldwide. Shifting sands are one of the most severe geologic hazards of deserts. In many deserts and desert border areas the sands are moving into inhabited areas, covering farmlands, villages, and other useful land with thick accumulations of sand. This is a global problem, as deserts are currently expanding worldwide. The Institute of Desert Research in Lanzhou, China, recently estimated that in China alone, 950 square miles (2,460.5 km²) are encroached on by migrating sand dunes from the Gobi Desert each year, costing the country $6.7 billion per year and affecting the lives of 400 million people.

Wind moves sand by saltation, in arced paths, in a series of bounces or jumps. The surface of dunes on beaches or deserts is typically covered by a thin moving layer of sand particles that are bouncing and rolling along it in this process of saltation.

Wind typically sorts different sizes of sedimentary particles, forming small elongate ridges known as sand ripples, very similar to ripples found in streams. Sand dunes are larger than ripples, up to 1,500 (~450 m) feet high, made of mounds or ridges of sand deposited by wind. These may form where an obstacle distorts or obstructs the flow of air, or they may move freely across much of a desert surface. Dunes have many different forms, but all are asymmetrical. They have a gentle slope that faces into the wind and a steep face that faces away from the wind. Sand particles move by saltation up the windward side, and fall out near the top where the pocket of low-velocity air cannot hold the sand anymore. The sand avalanches, or slips down the leeward slope, known as the slip face. This keeps the slope at the angle of repose, 30–34°. The asymmetry of old dunes indicates the directions ancient winds blew.

The steady movement of sand from one side of the dune to the other causes the whole dune to migrate slowly downwind, typically about 80–100 feet (28–30 m) per year, burying houses, farmlands, temples, and entire towns. Rates of dune migration of up to 350 feet (107 m) per year have been measured in the Western Desert of Egypt and the Ningxia Autonomous Region of China.

A combination of many different factors leads to the formation of very different types of dunes, each with a distinctive shape, potential for movement, and hazards. The main variables that determine a dune's shape are the amount of sand available for transportation, the strength (and directional uniformity) of the wind, and the amount of vegetation that covers the surface. If there is a lot of vegetation and little wind, no dunes will form. In contrast, if there is very little vegetation, a lot of sand, and moderate wind strength (conditions that might be found on a beach), then a group of dunes known as transverse dunes form, with the dune crests aligned at right angles to the dominant wind.

Barchan dunes have crescent shapes and horns pointing downwind; they form on flat deserts with steady winds and a limited sand supply. Parabolic dunes have a U-shape with the U facing upwind. These form where there is significant vegetation that pins the tails of migrating transverse dunes, with the dune being warped into a wide U-shape. These dunes look broadly similar to barchans, except the tails point in the opposite direction. They can be distinguished because, in both cases, the steep side of the dune points away from the dominant wind direction.

Linear dunes are long, straight, ridge-shaped dunes that elongate parallel to the wind direction. These occur in deserts with little sand supply and strong, slightly variable winds. Star dunes form isolated or irregular hills where the wind directions are irregular.

Strong winds that blow across desert regions sometimes pick up dust made of silt and clay particles and transport it thousands of miles (kilometers) from their source. For instance, dust from China is found on Pacific islands, and the Sahara commonly drops dust in southern Europe. This dust is a nuisance, has a significant influence on global climate, and has at times, as in the dust bowl days of the 1930s, been known nearly to block out the sun.

Loess is the name for silt and clay deposited by wind. It forms a uniform blanket that covers hills and valleys at many altitudes, which distinguishes it from deposits of streams. In Shaanxi Province, China, an earthquake that killed 830,000 in 1556 had such a high death toll in part because the inhabitants of the region built their homes out of loess. The loess formed an easily excavated material that hundreds of thousands of villagers cut homes into, essentially living in caves. When the earthquake struck, the loess proved to be a poor building material, and large-scale collapse of the fine-grained loess was directly responsible for most of the high death toll.

Recently it has been recognized that windblown dust contributes significantly to global climate. Dust storms that come out of the Sahara can be carried around the world and partially block some of the sun's radiation. The dust particles may also act as small nucleii for raindrops to form around, perhaps acting as a natural cloud-seeding phenomenon. One interesting phenomenon to consider is that as global warming increases global temperatures, the amount and intensity of storms increase, and some of the world's deserts expand. Dust storms may counter this effect and reduce global temperatures, and increase precipitation.

THE SAHARA

The Sahara is the world's largest desert, covering 5,400,000 square miles (8,600,000 km²) in northern Africa, including Mauritania, Morocco, Algeria, Tunisia, Libya, Egypt, Sudan, Chad, Niger, and Mali. The desert is bordered on the north and northwest by the Mediterranean Sea and Atlas Mountains, on the west by the Atlantic Ocean, and on the east by the Nile River. But the Sahara is part of a larger arid zone that continues eastward into the Eastern Desert of Egypt and Nubian Desert of Sudan, the Rub' al-Khali of Arabia, and the Lut, Tar, Dasht-I-Kavir, Takla Makan, and Gobi Deserts of Asia. Some classifications include the Eastern and Nubian Deserts as part of the Sahara and call the region of the Sahara west of the Nile the Libyan Desert, whereas other classifications consider them separate entities. The southern border of the Sahara is less well defined, but is generally taken as about 16° latitude where the desert grades into transitional climates of the Sahel steppe.

Rocky- and stone- or gravel-covered denuded plateaus known as hammada cover about 70 percent of the Sahara, and sand dunes cover about 15 percent. High mountains, rare oases, and transitional regions occupy the remaining 15 percent. Major mountain ranges in the eastern Sahara include the uplifted margins of the Red Sea that form steep escarpments dropping more than 6,000 feet (2,000 m) from the Arabian Desert into the Red Sea coastal plain. Rocks in these mountains include predominantly Precambrian granitic gneisses, metasediments, and mafic schists of the Arabian shield, and are rich in mineral deposits, including especially gold that has been exploited by the Egyptians since Pharaonic times. The highest point in the Sahara is Emi Koussi in Chad, which rises to 10,860 feet (3,415 m), and the lowest point is the Qattara Depression in the northwest Desert of Egypt.

High, isolated mountain massifs rise from the plains in the central Sahara, including the massive Ahagger (Hogger) in southern Algeria, Tibesti in northern Chad, and Azbine (Air Mountains) in northern Niger. Ahagger rises to more than 9,000 feet (2,740 m) and includes a variety of Precambrian crystalline rocks of the Ouzzalian Archean craton and surrounding Proterozoic Shield. The Air Mountains, rising to more than 6,000 feet (1,830 m), are geologically a southern extension of the Ahagger to the north, containing metamorphosed Precambrian basement rocks. Tibesti rises to more than 11,000 feet (3,350 m) and also includes a core of Precambrian basement rocks, surrounded by Paleozoic and younger cover. Northeast of Tibesti near the Egypt-Libya-Sudan border, the lower Oweineat (Uwaynat) Mountains form a similar dome, rising to 6,150 feet (1,934 m), and have a core of Precambrian igneous rocks.

The climate in the Sahara is among the harshest on the planet, falling in the trade wind belt of dry descending air from Hadley circulation, with strong constant winds blowing from the northeast. These winds have formed elongate linear dunes in specific corridors across parts of the Sahara, with individual sand dunes continuous for hundreds of miles, and virtually no interdune sands. These linear dunes reach heights of more than 1,100 feet (350 m) and may migrate tens of feet (a couple of meters) or more per year. When viewed on a continental scale (as from space or on a satellite image) these linear dunes display a curved trace, formed by the Coriolis force deflecting the winds and sand to the right of the movement direction (northeast to southwest). Most parts of the Sahara receive an average of fewer than five inches (12 cm) of rain a year, and this typically comes in a single downpour every few years, with torrential rains causing flash flooding. Rains of this type run off quickly, and relatively little is captured and returned to the groundwater system for future use. The air is extremely dry, with typical relative humidities ranging from 4 percent to 30 percent. Temperatures can be extremely hot, and the diurnal variation is high. The world's highest recorded temperature is from the Libyan Desert, 136°F (58°C) in the shade, during fall 1922. The temperature drop at night can be up to 90°F (30°C), even dropping below freezing after a scorching hot day.

Most of the Sahara is sparsely vegetated, with shrub brushes being common, along with grasses, and trees in the mountains. Some desert oases and sections along the Nile River are extremely lush, however, and the Nile valley has extensive agricultural development. Animal life is diverse, including gazelles, antelopes, jackals, badgers, hyenas, hares, gerbils, sheep, fox, wild ass, weasel, baboon, mongoose, and hundreds of species of birds.

A variety of minerals are exploited from the Sahara, including major deposits of iron ore from Algeria, Mauritania, Egypt, Tunisia, Morocco, and Niger. Uranium deposits are found throughout the Sahara, with large quantities in Morocco. Manganese is mined in Algeria, and copper is found in Mauritania. Oil is exported from Algeria, Libya, and Egypt.

Vast groundwater reservoirs underlie much of the Sahara, both in shallow alluvial aquifers and in fractured bedrock aquifers. The water in these aquifers fell as rain thousands of years ago, reflecting a time when the climate over North Africa was much different. In the Pleistocene much of the Sahara experienced a wet, warm climate, and more than 20 large lakes covered parts of the region. The region experienced several alternations between wet and dry climates in the past couple of hundred thousand years, and active research projects aim to correlate these climate shifts with global events such as glacial and interglacial periods, sea surface and current changes (such as the El Niño–Southern Oscillation). The implications for understanding these changes is enormous, with millions of people affected by expansion of the Sahara, and undiscovered groundwater resources that could be used to sustain agriculture and save populations from decimation. Many of the present drainage and wadi networks in the Sahara follow a drainage network established during the Pleistocene. In the Pliocene the Mediterranean shoreline was about 60 miles (40 km) south of its present location, when sea levels were about 300 feet (100 m) higher than today. Sand sheets and dunes, which are currently moving southward, have been active for only the past few thousand years. These are known to form local barriers to wadi channels in the Sahara, Sinai, and Negev Deserts, and locally block wadis.

The sand of the Sahara and adjacent Northern Sinai probably originated by fluvial erosion of rocks in the uplands to the south, and was transported from south to north by paleo-rivers during wetter climate times, then redistributed by wind. Dry climates such as the present, and low sea levels during glacial maxima, exposed the sediment to wind action that reshaped the fluvial deposits into dunes, whose form depended on the amount of available sand and prevailing wind directions. This hypothesis was developed to suggest the presence of a drainage network to transport fluvial sediments. Indeed, numerous channels incise into the limestone plateau of the central and northern Sahara, and many lead to elongate areas that have silt deposits. Several of these deposits have freshwater fauna and are interpreted as paleolakes and long-standing slack water deposits from floods.

Plio-Pleistocene lakebed sediments have also been identified in many places in the mountains in the Sahara, where erosionally resistant dikes that formed dams in steep-walled bedrock canyons controlled the lakes. The paleolake sediments consist of silts and clays interbedded with sands and gravels, cut by channel deposits. These types of lake beds were formed in a more humid Late Plio-Pleistocene climate, based on fossil roots and their continuity with wadi terraces of that age.

The fluvial history of the region reflects earlier periods of greater effective moisture, as is evident also from archaeological sites associated with remnants of travertines and playa or lake deposits. An Early Holocene pluvial cycle is well documented by archaeological investigations at Neolithic playa sites in Egypt. Late Pleistocene lake deposits with associated early and middle Paleolithic archaeological sites are best known from work in the Bir Tarfawi area of southwest Egypt. Similar associations occur in northwest Sudan and Libya.

An extensive network of sand-buried river and stream channels in the eastern Sahara appears on shuttle-imaging radar images. Calcium carbonate associated with some of these buried river channels is thought to have precipitated in the upper zone of saturation during pluvial episodes, when water tables were high. As documented by radiocarbon dating and archaeological investigations, the eastern Sahara experienced a period of greater effective moisture during Early and Middle Holocene time, about 10–5 thousand years ago. Uranium-series dating of lacustrine carbonates from several localities indicated that five paleolake-forming episodes occurred at about 320–250, 240–190, 155–120, 90–65, and 10–5 thousand years ago. These five pluvial episodes correlate with major interglacial stages.

These results support the contention that past pluvial episodes in North Africa correspond to the interglacial periods. Isotopic dating results and field relationships suggest that the oldest lake- and groundwater-deposited carbonates were more extensive than those of the younger period, and the carbonates of the late wet periods were geographically localized within depressions and buried channels.

This archaeological evidence of previous human habitation, coupled with remains of fauna and flora, suggests the presence of surface water in the past. Remains of lakes and segments of dry river and stream channels occur throughout the Sahara. Archaeological evidence of human habitation during the Early Holocene was recently uncovered in the northeast Sinai Peninsula where an Early Middle Paleolithic site shows evidence for habitation at 33,800 years before present (BP).

ATACAMA DESERT

The Atacama Desert is an elevated arid region located in northern Chile, extending more than 384 square miles (1,000 km^2) south from the border with Peru. The desert is located 2,000 feet (600 m) above sea level and is characterized by numerous dry salt basins (playas), flanked on the east by the Andes and on the west by the Pacific coastal range. The Atacama is the driest place on Earth, with no rain ever recorded in many areas, and practically no vegetation. Nitrate and copper are mined extensively in the region.

The Atacama is first known to be crossed by the Spanish conquistador Diego de Almagro in 1537, but was ignored until the middle 19th century, when mining of nitrates in the desert began. After World War I, however, synthetic nitrates were developed and the region has experienced economic decline.

GOBI DESERT

One of the world's great deserts, the Gobi is located in central Asia encompassing more than 500,000 square miles (1,295,000 km^2) in Mongolia and northern China. The desert covers the region from the Great Khingan Mountains northwest of Beijing to the Tien Shan north of Tibet, but the desert is expanding at an alarming rate, threatening the livelihood of tens of thousands of farmers and nomadic sheepherders every year. Every spring dust from the Gobi covers eastern China, Korea, and Japan, and may extend at times around the world. Northwesterly winds have removed almost all the soil from

land in the Gobi, depositing it as thick loess in eastern China. Most of the Gobi is situated on a high plateau resting 3,000 to 5,000 feet (900–1,500 m) above sea level, and it contains numerous alkaline sabkhas and sandy plains in the west. Regions in the Gobi include abundant steppes, high mountains, forests, and sandy plains. The Gobi has yielded many archaeological, paleontological, and geological finds, including early stone implements, dinosaur eggs, and mineral deposits and precious stones including turquoise and jasper.

NAMIB DESERT AND THE SKELETON COAST

Namibia's Atlantic coastline is known as the skeleton coast, named for the suffering and death that beset many sailors attempting to navigate the difficult waters swept by the cold Benguela current, which moves along the coast, and warm winds coming off the Namib and Kalahari Deserts. The coastline is littered with numerous shipwrecks, testifying to the difficult and often unpredictable nature of shifting winds and ocean currents. Giant sand dunes of the Namib sand sea reach to the coast, and in places these dunes are also covered in bones of mammals that have searched in vain for water. Many dune types are present, including transverse dunes and barchans, and the winds in the region often cause a steady moan to grow from the blowing sands. The desert elephant lives in the region, eating and drinking in generally dry, inland riverbeds, but sometimes venturing to the harsh coast. Oryx, giraffe, hyena, springbok, ostrich, rare rhinos, and lions also roam the area, whereas Cape fur seals populate parts of the coast. Whales and dolphins swim along the coast, and occasionally, giant whale skeletons are washed up and exposed on the shore.

The cold Benguela current breaks off from the circum-Antarctica cold current and forms a cold sea breeze that often shrouds the region in mist and fog, especially during winter months. This mist sustains an unusual plant life in the desert and forms an additional navigational hazard for ships.

DRY VALLEYS OF ANTARCTICA

The Dry Valleys are the largest area on Antarctica not covered by ice. Approximately 98 percent of the continent is covered by ice, but the Dry Valleys, located near McMurdo Sound on the side of the continent closest to New Zealand, have a cold desert climate and receive only four inches (10 cm) of precipitation per year, overwhelmingly in the form of snow. The Dry Valleys are one of the coldest, driest places on Earth and are used by researchers from NASA as an analog for conditions on Mars. No vegetation grows in the Dry Valleys, although a number of unusual microbes live in the frozen soils and form cyanobacterial mats in places. In the Southern Hemisphere summer, glaciers in the surrounding Transantarctic Mountains release significant quantities of meltwater so that streams and lakes form over the thick permafrost in the valleys.

See also ARABIAN GEOLOGY; ASIAN GEOLOGY; ATMOSPHERE; CLIMATE; CLIMATE CHANGE.

FURTHER READING

Abrahams, Athol D., and Anthony J. Parsons. *Geomorphology of Desert Environments.* London: Chapman and Hall, 1994.

Bagnold, Ralph A. *The Physics of Blown Sand and Desert Dunes.* London: Methuen, 1941.

Burke, Kevin, and G. L. Wells. "Trans-African Drainage System of the Sahara: Was It the Nile?" *Geology* 17 (1989): 743–747.

El-Baz, Farouk. "Origin and Evolution of the Desert." *Interdisciplinary Science Reviews* 13 (1988): 331–347.

El-Baz, Farouk, Timothy M. Kusky, Ibrahim Himida, and Salel Abdel-Mogheeth. *Ground Water Potential of the Sinai Peninsula, Egypt.* Cairo, Egypt: Ministry of Agriculture and Land Reclamation, 1998.

Guiraud, R. "Mesozoic Rifting and Basin Inversion along the Northern African-Arabian Tethyan Margin: An Overview." In *Petroleum Geology of North Africa,* edited by D. S. MacGregor, R. T. J. Moody, and D. D. Clark-Lowes, 217–229. *Geological Society of London Special Publication* 133 (1998).

Haynes, C. Vance, Jr. "Great Sand Sea and Selima Sand Sheet: Geochronology of Desertification." *Science* 217 (1982): 629–633.

Haynes, C. Vince, Jr., C. H. Eyles, L. A. Pavlish, J. C. Rotchie, and M. Ryback. "Holocene Paleoecology of the Eastern Sahara: Selima Oasis." *Quaternary Science Reviews* 8 (1989).

Klitzsch, E. "Geological Exploration History of the Eastern Sahara." *Geologische Rundschau* 83 (1994): 1437–3254.

Kusky, Timothy M., Mohamed A. Yahia, Talaat Ramadan, and Farouk El-Baz. "Notes on the Structural and Neotectonic Evolution of El-Faiyum Depression, Egypt: Relationships to Earthquake Hazards." *Egyptian Journal of Remote Sensing and Space Sciences* 2 (2000): 1–12.

McCauley, J. F., G. G. Schaber, C. S. Breed, M. J. Grolier, C. Vince Haynes, Jr., B. Issawi, C. Elachi, and R. Blom. "Subsurface Valleys and Geoarchaeology of the Eastern Sahara Revealed by Shuttle Radar." *Science* 218 (1982): 1004–1020.

McKee, E. D., ed. *A Study of Global Sand Seas.* United States Geological Survey Professional Paper 1052, 1979.

Pachur, H. J., and G. Braun. "The Paleoclimate of the Central Sahara, Libya, and the Libyan Desert." *Paleoecology Africa* 12 (1980): 351–363.

Sestini, G. "Tectonic and Sedimentary History of the NE African Margin (Egypt/Libya)." In *The Geological Evolution of the Eastern Mediterranean,* edited by J. E. Dixon, and A. H. F. Robertson, 161–175. Oxford: Blackwell Scientific Publishers, 1984.

Szabo, B. J., W. P. McHugh, G. G. Shaber, C. Vince Haynes, Jr., and C. S. Breed. "Uranium-Series Dated Authigenic Carbonates and Acheulian Sites in Southern Egypt." *Science* 243 (1989): 1053–1056.

Walker, A. S. "Deserts: Geology and Resources." *United States Geological Survey Publication* 60 (1996): 421–577.

Webster, D. "Alashan, China's Unknown Gobi." *National Geographic* (2002): 48–75.

Wendorf, F., and R. Schild. *Prehistory of the Eastern Sahara.* New York: Academic Press, 1980.

Devonian The Devonian is the fourth geological period in the Paleozoic Era, spanning the interval from 408 to 360 million years ago. It was named after exposures in Devonshire in southwest England. British geologists Adam Sedgwick (1785–1873) and Roderick I. Murchison (1792–1871) first described the Devonian in detail in 1839. The Devonian is divided into three series and seven stages based on its marine fauna.

Devonian rocks are known from all continents and reflect the distribution of the continents grouped into a large remnant Gondwanan fragment in the Southern Hemisphere, and parts of Laurasia (North America and Europe), Angaraland (Siberia), China, and Kazakhstania in the Northern Hemisphere. The eastern coast of North America and adjacent Europe experienced the Acadian orogeny, formed in response to subduction and eventual collision between Avalonian fragments and ultimately Africa with Laurasia. Other orogenies affected North China, Kazakstania, and other fragments. These mountain-building events shed large clastic wedges, including the Catskill delta in North America and the Old Red sandstone in the British Isles.

The Devonian experienced several eustatic sea-level changes and had times of glaciation. There was a strong climatic gradation with tropical and monsoonal conditions in equatorial regions, and cold water conditions in more polar regions.

Marine life in the Devonian was prolific, with brachiopods reaching their peak. Rugose and tabulate corals, stromatoporoids, and algae built carbonate reefs in many parts of the world including North America, China, Europe, North Africa, and Australia. Crinoids, trilobites, ostracods, and a variety of bivalves lived around the reefs and in other shallow water environments, whereas calcareous foraminifera and large ammonites proliferated in the pelagic

Middle Devonian coral reef construction, including different metazoans such as the coelenterates *(Tom McHugh/Photo Researchers, Inc.)*

realm. The pelagic conodonts peaked in the Devonian, and their great variety, widespread distribution, and rapid changes make them useful biostratigraphic markers and form the basis for much of the biostratigraphic division of Devonian time. Bony fish evolved in the Devonian and evolved into tetrapod amphibia by the end of the period.

The land was inhabited by primitive plants in the Early Devonian, but by the middle of the period great swampy forests with giant fern trees (*Archaeopteris*) and spore-bearing plants populated the land. Insects, including some flying varieties, inhabited these swamps.

The end of the Devonian brought widespread mass extinction of some marine animal communities, including brachiopods, trilobites, conodonts, and corals. The cause of this extinction is not well known, with models including cooling caused by a southern glaciation, or a meteorite impact.

The Devonian saw the climactic development of the Appalachian Mountain belt in eastern North America. The Appalachians extend for 1,600 miles (1,000 km) along the east coast of North America, stretching from the St. Lawrence River valley in Quebec Canada, to Alabama. Many classifications consider the Appalachians to continue through New-

foundland in maritime Canada, and before the Atlantic Ocean opened, the Appalachians were continuous with the Caledonides of Europe. Home to many of America's great universities, the Appalachians are one of the best-studied mountain ranges in the world, and understanding of their evolution was one of the factors that led to the development and refinement of the paradigm of plate tectonics in the early 1970s.

Rocks that form the Appalachians include those that were deposited on or adjacent to North America and thrust on the continent during several orogenic events. For the length of the Appalachians, the older continental crust consists of Grenville Province gneisses, deformed and metamorphosed about 1 billion years ago during the Grenville orogeny. The Appalachians grew in several stages. After Late Precambrian rifting, the Iapetus Ocean evolved and hosted island arc growth, while a passive margin sequence was deposited on the North American rifted margin in Cambrian-Ordovician times. In the Middle Ordovician the collision of an island arc terrane with North America marks the Taconic orogeny, followed by the Mid-Devonian Acadian orogeny, which probably represents the collision of North America with Avalonia, off the coast of Gondwana. This orogeny formed huge molassic fan delta complexes of the Catskill Mountains, and was followed by strike-slip faulting. The Late Paleozoic Alleghenian orogeny formed striking folds and faults in the southern Appalachians, but was dominated by strike-slip faulting in the Northern Appalachians. This event appears to be related to the rotation of Africa to close the remaining part of the open ocean in the southern Appalachians. Late Triassic-Jurassic rifting reopened the Appalachians, forming the present Atlantic Ocean.

See also NORTH AMERICAN GEOLOGY; PALEOZOIC.

FURTHER READING

Condie, Kent C., and Robert Sloan. *Origin and Evolution of Earth, Principles of Historical Geology.* Upper Saddle River, N.J.: Prentice Hall, 1997.

Prothero, Donald R., and Robert H. Dott. *Evolution of the Earth.* 6th ed. Boston: McGraw Hill, 2002.

Stanley, Steven M. *Earth and Life through Time.* New York: W. H. Freeman, 1986.

Windley, Brian F. *The Evolving Continents.* 3rd ed. Chichester, U.K.: John Wiley & Sons, 1995.

Dewey, John F. (1937–) British *Tectonicist, Structural Geologist* John F. Dewey is regarded as one of the founders of the modern plate tectonic paradigm and one of the earliest scientists to apply the plate tectonic model to ancient mountain belts such as the Appalachians. He is also well known for his pioneering work on plate kinematics and using principles of plate kinematics on a sphere to understand complex geological problems. After helping to offer explanations of the Appalachian Mountain belt in terms of plate tectonics, Dewey later expanded his studies to a global scale. He presented plate tectonic concepts in a kinematic framework, clearly describing many phenomena for the first time.

John Dewey grew up in London during World War II, and excelled in athletics, especially boxing, rugby, cricket, gymnastics, high jump, and javelin. Although he was good at sports, he realized he could not make a profession out of it, and when he turned 16 he found another passion—geology. Dewey was inspired by his great uncle, Henry Dewey, a British government geologist, and he convinced his housemaster at Bancroft's School (Essex, United Kingdom) to let him pursue geology as a career. The headmaster at his school, John Hayward, was also an amateur geologist and encouraged Dewey in his studies, which eventually led to Dewey's receiving a first-class degree in geology from Queen Mary's College, University of London, in 1957. He next received a Ph.D. in geology from the University of London in 1960. After this Dewey turned down many opportunities

for careers in the oil and mining industry, deciding instead to pursue an academic career, taking a job as a lecturer at the University of Manchester, then at the Cambridge University.

In the late 1960s the field of geology entered a revolution with the new theory of plate tectonics. Dewey was fascinated by the developments and in 1967 took a three-month sabbatical position at the Lamont Geological Observatory in New York City, where he studied the Appalachian/Caledonian mountain belt and began to apply the principles of plate tectonics to this ancient orogenic belt. While seeing the plate tectonic and kinematic model develop, in part from his colleagues at Lamont, Dewey became one of the early pioneers to apply the same principles to old mountain belts, using the Appalachian-Caledonian belt as the prime example. Dewey and his colleagues John Bird and, later, Kevin Burke recognized that the Appalachians and other mountain belts preserved a history of ocean opening and closing, a cycle they termed the Wilson Cycle, after the Canadian geologist J. Tuzo Wilson, one of the pioneers of the plate tectonic model in the oceans. Dewey stayed in America, taking a position at the State University of New York at Albany, where he remained until the mid-1980s. During that time the university became one of the world's leading research institutions for plate tectonics and its applications to old mountain belts, with numerous faculty and students studying the Appalachians, Alps, Himalayas, Andes, Asia, and Precambrian mountain belts of the world.

After 12 years in Albany Dewey moved back to England to become chair of the department of geological sciences at Durham, then moved to become chair of the earth sciences at the University of Oxford. Finally, he left the old society geological network in England and accepted the position of professor of geology at the University of California at Davis in 2001. He was inducted as a member of the National Academy of Sciences in 2005.

Dewey's basic interests and knowledge remain structural geology and tectonics, from the small-scale materials science of deformed rocks to the large-scale origin of topography and structures. Some of his ongoing field-based research is on the rock fabrics and structures of transpression and transtension, especially in California, New Zealand, Norway, Ireland, and Newfoundland. Evolving interests include the neotectonics of California and Nevada and the relationship between faulting, topography, and sediment provenance, yield, and distribution. Derivative interests are the geohazards of volcanoes, earthquakes, and landslides.

See also CONTINENTAL DRIFT; NORTH AMERICAN GEOLOGY; PLATE TECTONICS.

FURTHER READING

Bird, John M., and John F. Dewey. "Lithosphere Plate-Continental Margin Tectonics and the Evolution of the Appalachian Orogen." *Geological Society of America Bulletin* 81 (1970): 1,031–1,059.

Dewey, John F., and John M. Bird. "Mountain Belts and the New Global Tectonics." *Journal of Geophysical Research* 75 (1970): 2,625–2,647.

Dewey, John F., Michael J. Kennedy, and William S. F. Kidd. "A Geotraverse through the Appalachians of Northern Newfoundland." *Geodynamics Series* 10 (1983): 205–241.

diagenesis Diagenesis is a group of physical and chemical processes that affect sediments after they are deposited but before they undergo deformation and metamorphism. Diagenesis occurs under low temperature (T) and pressure (P) conditions, with its upper PT limit defined as when the first metamorphic minerals appear. Diagenesis typically changes the sediment from a loose, unconsolidated state to a rock that is cemented, lithified, or indurated.

The style of diagenetic changes in sediments is controlled by several factors other than pressure and temperature, including grain size, rate of deposition, sediment composition, environment of deposition, nature of pore fluids, porosity and permeability, and types of surrounding rocks. One of the most important diagenetic processes is dewatering, or the expulsion of water from the pore spaces by the weight of overlying, newly deposited sediments. These waters may escape to the surface or enter other nearby, more porous sediments, where they can precipitate or dissolve soluble minerals. Compaction and dewatering of sediments cause reduction of the thickness of the sedimentary pile. For instance, many muds may contain 80 percent water when they are deposited and compaction rearranges the packing of the constituent mineral grains to reduce the water-filled pore spaces to about 10 percent of the rock. This process results in the clay minerals' being aligned, forming a bedding-plane parallel layering known as fissility. Organic sediments also experience large amounts of compaction during dewatering, whereas other types of sediments including sands may experience only limited compaction. Sands typically are deposited with about 50 percent porosity, and they may retain about 30 percent even after deep burial. The porosity of sandstone is reduced by the pressing of small grains into the pore spaces between larger grains, and the addition of cement.

Chemical processes during diagenesis are largely controlled by the nature of the pore fluids. Fluids may dissolve or more commonly add material to the pore spaces in the sediment, increasing or decreasing pore space, respectively. These chemical changes may occur in the marine realm or in the continental realm with freshwater in the pore spaces. Chemical diagnetic processes tend to be more effective at the higher PT end of the diagenetic spectrum, when minerals are more reactive and soluble.

Organic material experiences special types of diagenesis, as bacteria aid in the breakdown of the organic sediments to form kerogen and release methane and carbon dioxide gas. At higher diagenetic temperatures kerogen breaks down to yield oil and liquid gas. Humus and peat are progressively changed into soft brown coal, hard brown coal, then bituminous coal during the diagenetic process of coalification. This increases the carbon content of the coal and releases methane gas in the process.

Most sandstones and coarse-grained siliciclastic sediments experience few visible changes during diagenesis, but they may experience the breakdown of feldspars to clay minerals and see an overall reduction in pore spaces. The pore spaces in sandstones may become filled with cements such as calcite, quartz, or other minerals. Cements can form at several times in the diagenetic process. Carbonates are susceptible to diagenetic changes and typically see early and late cements, and many are altered by processes such as replacement by silica, dolomitization, and the transformation of aragonite to calcite. Carbonates normally show an interaction of physical and chemical processes, with the weight of overlying sediments forming pressure solution surfaces known as stylolites, where grains are dissolved against one another. These stylolites have characteristic crinkly or wavy surfaces oriented parallel to bedding. The material dissolved along the stylolites is then taken in solution and expelled from the system or, more commonly, reprecipitated as calcite or quartz veins, often at high angles to the stylolites reflecting the stresses induced by the weight of the overburden.

See also HYDROCARBONS AND FOSSIL FUELS; METAMORPHISM AND METAMORPHIC ROCKS; STRUCTURAL GEOLOGY.

divergent plate margin processes Divergent plate margins occur where two tectonic plates are moving apart from each other. The world's longest mountain chain is the midocean ridge system, extending 25,000 miles (40,000 km) around the planet. The midocean ridge system represents places where two oceanic plates are moving apart or diverging, and new material is moving up from the mantle to form new oceanic crust and lithosphere in the space created by the divergence. These midocean ridge systems are mature extensional boundaries, many of which began as immature extensional boundaries in continents, known as continental rifts. Some continental rift systems are linked to the world rift system in the oceans and are actively breaking continents into pieces. An example is the Red Sea-East African rift system. Other continental rifts are accommodating small amounts of extension in the crust and may never evolve into oceanic rifts. Examples of where this type of rifting occurs on a large scale include the Basin and Range Province of the western United States and Lake Baikal in Siberia, Russia.

DIFFERENT STYLES OF EXTENSION AT DIVERGENT PLATE BOUNDARIES

Although divergent boundaries are all similar in that they are places where the crust and entire lithosphere are breaking and moving apart, there are large differences in the processes that allow this extension to occur. Some of these different processes act in different places, while others may work together to produce the extension and associated sinking (subsidence) of the land surface. There are three main end-member models for the mechanisms of extension and subsidence in continental rifts. These are the pure shear model, the simple shear model, and the dike injection model.

In the pure shear model for extension, the lithosphere thins symmetrically about the rift axis, being pulled apart like taffy at depth and along brittle faults near the surface. The base of the lithosphere (defined by the 2,425°F [1,330°C] isotherm, or line of equal temperature) rises to 10–20 miles (15–30 km) below the surface near the center of the rift axis but remains at normal depths of 75 miles (120 km) away from the rift. This causes high heat flow and high temperature gradients with depth (geothermal gradients) in rifts, and is consistent with many measurements of the strength of the Earth's gravity force that suggest an excess mass at depth (this would correspond to the denser asthenosphere near the surface). Stretching mechanisms in the pure shear model include brittle accommodation of stretching on faults near the surface. At about four miles (7 km) depth, the rocks no longer deform by fracturing but begin to flow like silly putty—a transition known as the brittle ductile transition—and extension below this depth is accommodated by shear on ductile shear zones.

In the simple shear model for extension, an asymmetric fault known as a detachment fault penetrates the thickness of the lithosphere, dipping a few degrees, forming a system of asymmetric structures across the rift. A series of rotated fault blocks may form where the detachment is close to the surface, whereas the opposite side of the rift (where the lithosphere experiences the most thinning) may be dominated by the eruption of volcanic rocks. Heating of the crust associated with the lithospheric

thinning typically causes the upward doming of the detachment fault, and since the heating (and uplift) is greatest in a region offset from the center of the rifting zone, the center of the uplifted dome tends to be located on one side of the rift. This model explains differences on either side of rifts, such as faulted and volcanic margins now on opposite continental margins (conjugate margins) of former rifts that have evolved into oceans. A good example of where this can be observed is along the Red Sea of the Middle East, where many volcanic rocks are located on the Arabian side of the sea, and few are found on the African side of the rift.

The dike injection model for extension in rifts suggests that a large number of dense igneous dikes (with basaltic composition) intrude the continental lithosphere in rifts, causing the lithosphere to become denser and to sink or subside. This mechanism does not really explain most aspects of rifts, but it may contribute to the total amount of subsidence in the other two models.

In all of these models for initial extension of the rift, initial geothermal gradients (how the temperature changes with depth) are raised and the temperatures become elevated and compressed beneath the rift axis. After the initial stretching and subsidence phases, the rift either becomes inactive or evolves into a midocean ridge system. In the latter case the initial shoulders of the rift become passive continental margins. Failed rifts and passive continental margins both enter a second, slower phase of subsidence related to the gradual recovery of the isotherms (lines of equal temperature) to their deeper, prerifting levels. This process takes about 60 million years and typically forms a broad basin over the initial rifts, characterized by no active faults, no volcanism, and rare lakes. The transition from initial stretching with coarse clastic sediments and volcanics to the thermal subsidence phase is commonly called the "rift to drift" transition on passive margins.

DIVERGENT PLATE BOUNDARIES IN CONTINENTS

Rifts are elongate depressions formed where the entire thickness of the lithosphere has ruptured in extension. In these places the continents are beginning to break apart as immature divergent boundaries, and if successful, may form new ocean basins. The general geomorphic feature that initially forms is known as a rift valley. Rift valleys have steep, fault-bounded sides, with rift shoulders that typically tilt slightly away from the rift valley floor. Drainage systems tend to be short, internal systems, with streams forming on the steep sides of the rift, flowing along the rift axis, and draining into deep, narrow lakes within the rift. If the rift is in an arid

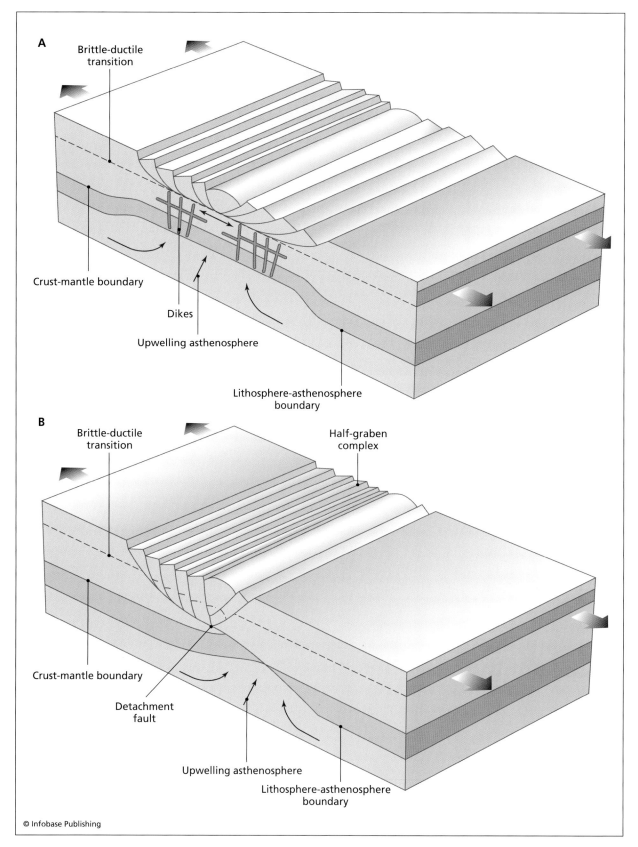

A

Brittle-ductile transition

Crust-mantle boundary

Dikes

Upwelling asthenosphere

Lithosphere-asthenosphere boundary

B

Brittle-ductile transition

Half-graben complex

Crust-mantle boundary

Detachment fault

Upwelling asthenosphere

Lithosphere-asthenosphere boundary

© Infobase Publishing

Modes of extension in rifts. (A) Shows pure shear model, in which the lithosphere extends symmetrically and asthenosphere rises to fill the space vacated by the extending lithosphere. (B) Shows simple shear or asymmetric rifting, where a shallow-dipping detachment fault penetrates the thickness of the lithosphere, and asthenosphere rises asymmetrically on the side of the rift where the fault enters the asthenosphere. Faulting patterns are also asymmetric, with different styles on either side of the rift.

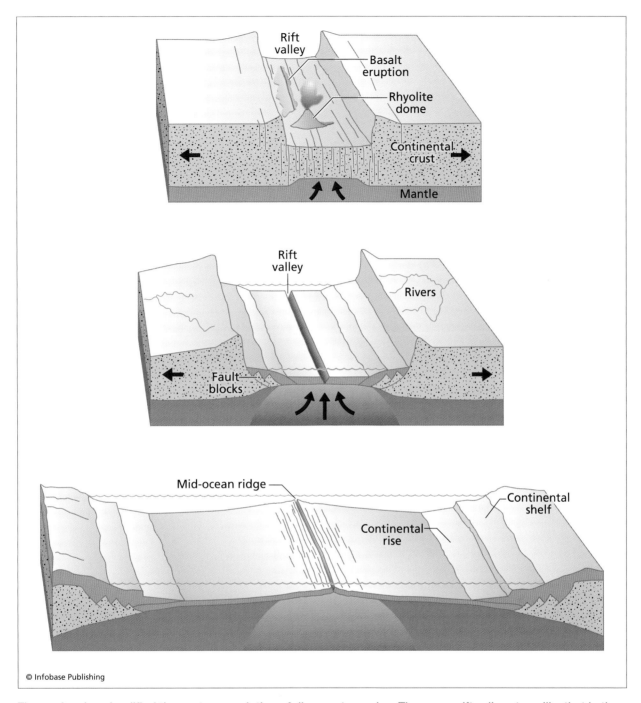

Figure showing simplified three-stage evolution of divergent margins. The young rift valley stage like that in the East African rift system has steep rift shoulders and basaltic and rhyolitic volcanoes. The young ocean stage, similar to the modern Red Sea, has seafloor spreading and steep rift shoulders. Mature ocean stage is like the modern Atlantic Ocean, with thick passive margin sequences developed on continental edges around a wide ocean basin.

environment, such as much of East Africa, the drainage may have no outlet and the water will evaporate before it can reach the sea. This process leaves distinctive deposits of salts and other minerals that form by being left behind during evaporation of seawater (evaporites), one of the hallmark deposits of continental rift settings. Other types of deposits in rifts include lake sediments in rift centers, and conglomerates (cemented gravels) derived from rocks exposed along the rift shoulders. These sediments may be interleaved with volcanic rocks, typically alkaline (having abundant sodium, Na, and other alkali elements) and bimodal in silica content (i.e., basalts and rhyolites).

DIVERGENT PLATE BOUNDARIES IN THE OCEANS: THE MIDOCEAN RIDGE SYSTEM

Some continental rifts may evolve into midocean ridge-spreading centers. The world's best example of where this transition can be observed is in the Ethiopian Afar, where the East African continental rift system meets juvenile oceanic spreading centers in the Red Sea and Gulf of Aden. Three plate boundaries meet in a wide plate boundary zone in the Afar, including the African/Arabian boundary (Red Sea spreading center), the Arabian/Somalian boundary (Gulf of Aden spreading center), and the African/Somalian boundary (East African rift). The boundary is a complex system known as an RRR (rift-rift-rift) triple junction. The triple junction has many complex extensional structures, with most of the Afar near sea level, and isolated blocks of continental crust such as the Danakil horst isolated from the rest of the continental crust by normal faults.

The Red Sea has a young, or juvenile, spreading center similar in some aspects to the spreading center in the middle of the Atlantic Ocean. Geologists recognize two main classes of oceanic spreading centers, based on characteristics of the shapes of their surfaces (geomorphology) and elevation or topography. These different types are formed in spreading centers with different spreading rates, with slow spreading rates, 0.2–0.8 inches per year (0.5–2 cm/yr), on Atlantic-type ridges, and faster rates, generally 1.5–3.5 inches per year (4–9 cm/yr), on Pacific-type ridges.

Atlantic-type ridges are characterized by a broad, 900–2,000-mile (1,500–3,000-km) wide swell in which the seafloor rises 0.6–1.8 miles (1–3 km) from abyssal plains at 2.5 miles (4.0 km) below sea level to about 1.7 miles (2.8 km) below sea level along the ridge axis. Slopes on the ridge are generally less than 1°. Slow, or Atlantic-type, ridges have a median rift, typically about 20 miles (30 km) wide at the top to 0.6–2.5 miles (1–4 km) wide at the bottom of the long, deep medial rift. Many constructional volcanoes are located along the base and inner wall of the medial rift. Rugged topography and many faults forming a strongly block-faulted slope characterize the central part of Atlantic-type ridges.

Pacific-type ridges are generally 1,250–2,500 miles (2,000–4,000 km) wide, and rise 1.2–1.8 miles (2–3 km) above the abyssal plains, with 0.1° slopes. Pacific-type ridges have no median valley but many shallow earthquakes, high heat flow, and low gravity in the center of the ridge, suggesting that magma may be present at shallow levels beneath the surface. Pacific-type ridges have much smoother flanks than Atlantic-type ridges.

The high topography of both types of ridges shows that they are underlain by low-density material and are floating on this hot substrate. Geologists call this mechanism of making mountains isostatic compensation. New magma upwells beneath the ridges and forms small magma chambers along the ridge axis. The magma in these chambers crystallizes to form the rocks of the oceanic crust that gets added (in approximately equal proportions) to both diverging plates. The crust formed at the ridges is young, hot, and relatively light, so it floats on the hot underlying asthenosphere. As the crust ages and moves away from the ridge, it becomes thicker and denser, and subsides; this explains the topographic profile of the ridges. The rate of thermal subsidence is the same for fast- and slow-spreading ridges (a function of the square root of the age of the crust), explaining why slow-spreading ridges are narrower than fast-spreading ridges.

Abundant volcanoes, with vast outpourings of basaltic lava, characterize the centers of the midocean ridges. The lavas are typically bulbous-shaped forms called pillows, as well as tubes and other, more massive flows. The ridge axes are also characterized by high heat flow, with many thermal vents marking places where seawater has infiltrated the oceanic crust and made its way to deeper levels, where it is heated by coming close to the magma, then rises again to vent on the seafloor. Many of these vents precipitate sulfide and other minerals in great quantities, forming chimneys called black smokers that may be many tens of feet (several meters) tall. These chimneys have high-temperature metal- and nutrient-rich water flowing out of them (at temperatures of several hundred degrees Celsius), with the metals precipitating once the temperature drops on contact with the cold seawater outside the vent. These systems may cover parts of the oceanic crust with layers of sulfide minerals. Unusual primitive communities of sulfide-reducing bacteria, tube worms, and crabs have been found near several black smoker vents along midocean ridges. Many scientists believe that similar settings may have played an important role in the early appearance and evolution of life on the planet.

Geophysical seismic refraction studies in the 1940s and 1950s established that the oceanic crust exhibits seismic layering similar in many places in the oceans. Seismic layer one consists of sediments, layer two is interpreted to be a layer of basalt 0.6–1.5 miles (1–2.5 km) thick, and layer three is approximately four miles (6 km) thick and interpreted to be crystal cumulates, underlain by the mantle. Some ridges and transform faults expose deeper levels of the oceanic lithosphere. These typically include a mafic dike complex, thick sections of gabbro, and ultramafic cumulates. In some places rocks of the mantle are exposed, typically consisting of strongly deformed ultramafic rocks that have had a large amount of

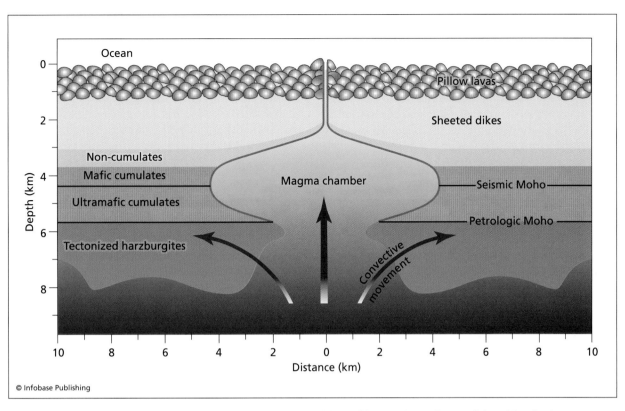

Formation of oceanic crust and lithosphere at midocean ridges. Magma forms by partial melting in the asthenosphere and upwells to make a magma chamber beneath the ridge axis. As the plates move apart, dikes intrude upward from the magma chamber and feed the lava flows on the surface. Heavy crystals settle out of the magma chamber and form layers of crystal cumulates on the magma chamber floor.

magma squeezed out of them. These unusual rocks are called depleted harzburgite tectonites.

As the plates move apart, the pressure on deep underlying rocks is lowered, which causes them to rise and partially melt by 15–25 percent. Basaltic magma is produced by partially melting the peridotitic mantle, leaving a residue-type of rock in the mantle known as harzburgite. The magma produced in this way moves up from deep within the mantle to fill the gap opened by the diverging plates. This magma forms a chamber of molten or partially molten rock that slowly crystallizes to form a coarse-grained igneous rock known as gabbro, which has the same composition as basalt. Before crystallization some of the magma moves up to the surface through a series of dikes and forms the crustal-sheeted dike complex, and basaltic flows. Many of the basaltic flows have distinctive forms, with the magma forming bulbous lobes known as pillow lavas. Lava tubes are also common, as are fragmented pillows formed by the inward explosive collapse (implosion) of the lava tubes and pillows. Back in the magma chamber other crystals grow in the gabbroic magma, including olivine and pyroxene, and are heavier than the magma, so they sink to

the bottom of the chamber. These crystals form layers of dense minerals known as cumulates. Beneath the cumulates the mantle material from which the magma was derived becomes progressively more deformed as the plates diverge and form a highly deformed ultramafic rock known as a harzburgite or mantle tectonite. This process can be seen on the surface in Iceland along the Reykjanes Ridge.

Much of the detailed information about the deep structure of oceanic crust comes from the study of ophiolites, which are interpreted to be on-land equivalents of oceanic crust tectonically emplaced on the continents during the process of convergent tectonics and ocean closure. Studies of ophiolites have confirmed the general structure of the oceanic crust as inferred from the seismic reflection and refraction studies and limited drilling. Numerous detailed studies of ophiolites have allowed unprecedented detail about the structure and chemistry of inferred oceanic crust and lithosphere to be completed, and as many variations as similarities have been discovered. The causes of these variations are numerous, including differences in spreading rate, magma supply, temperature, depth of melting, tectonic setting (arc, forearc, back arc, midocean ridge, etc.), and the presence or absence of water. The ocean floor, however, is still largely unexplored, and scientists know more about many other planetary surfaces than is known about Earth's ocean floor.

The mid-Atlantic ridge rises above sea level on the North Atlantic island of Iceland, lying 178 miles (287 km) off the coast of Greenland and 495 miles (800 km) from the coast of Scotland. Iceland has an average elevation of more than 1,600 feet (500 m) and owes its elevation to a hot spot that is interacting with the midocean ridge system beneath the island. The mid-Atlantic ridge crosses the island from southwest to northeast, and has a spreading rate of 1.2 inches per year (3 cm/yr), with the mean extension oriented toward an azimuth of 103°. The oceanic Reykjanes ridge and sinistral transform south of the island rises to the surface and continues as the Western Rift Zone. Active spreading is transferred to the Southern Volcanic Zone across a transform fault called the South Iceland Seismic Zone, then continues north through the Eastern Rift Zone. Spreading is offset from the oceanic Kolbeinsey ridge by the dextral Tjornes fracture zone off the island's northern coast.

During the past 6 million years the Iceland hot spot has drifted toward the southeast relative to the North Atlantic, and the oceanic ridge system has made a succession of small jumps so that active spreading has remained coincident with the plume of hottest and therefore weakest mantle material. These ridge jumps have caused the active spreading to propagate into regions of older crust that have been remelted, forming unusual alkalic and even silicic volcanic rocks that are deposited unconformably over older oceanic (tholeiitic) basalts. Active spreading occurs along a series of 5–60 mile (10–100 km) long zones of fissures, graben, and dike swarms, with basaltic and rhyolitic volcanoes rising from central parts of fissures. Hydrothermal activity is intense along the fracture zones, with diffuse faulting and volcanic activity merging into a narrow zone within a few miles beneath the surface. Detailed geophysical studies have shown that magma episodically rises from depth into magma chambers located a few miles below the surface, then dikes intrude the overlying crust and flow horizontally for tens of miles to accommodate crustal extension of several to several tens of feet over several hundred years.

Many Holocene volcanic events are known from Iceland, including 17 eruptions of Hekla from the Southern Volcanic zone. Iceland has an extensive system of glaciers and has experienced a number of eruptions beneath them that cause water to infiltrate the fracture zones. The mixture of water and magma induces explosive events including Plinian eruption clouds, phreo-magmatic, tephra-produc-

ing eruptions, and sudden floods known as jokulhlaups induced when the glacier experiences rapid melting from contact with magma. Many Icelanders have learned to use the high geothermal gradients to extract geothermal energy for heating and to enjoy the many hot springs on the island.

See also AFRICAN GEOLOGY; CONVERGENT PLATE MARGIN PROCESSES; PLATE TECTONICS; TRANSFORM PLATE MARGIN PROCESSES.

FURTHER READING

Kious, Jacquelyne, and Robert I. Tilling. "U.S. Geological Survey. This Dynamic Earth: The Story of Plate Tectonics." Available online. URL: http://pubs.usgs.gov/gip/dynamic/dynamic.html. Updated March 27, 2007.

Skinner, Brian, and B. J. Porter. *The Dynamic Earth: An Introduction to Physical Geology.* 5th ed. New York: John Wiley & Sons, 2004.

drainage basin (drainage system)   The total area that contributes water to a stream is called a drainage basin, and the line that divides different drainage basins is known as a divide (such as the continental divide) or interfluve. Drainage basins are the primary landscape units, or systems, concerned with the collection and movement of water and sediment into streams and river channels. They consist of a number of interrelated systems that work together to control the distribution and flow of water within the basin. Hillslope processes, bedrock and surficial geology, vegetation, climate, and many other systems all interact in complex ways that determine where streams will form and how much water and sediment they will transport. A drainage basin's hydrologic dynamics can be analyzed by considering these systems along with how much water enters the basin through precipitation and how much leaves the basin in the discharge of the main trunk channel. Streams are arranged in an orderly fashion in drainage basins, with progressively smaller channels branching away from the main trunk channel. Stream channels are ordered and numbered according to this systematic branching. The smallest segments lack tributaries and are known as first-order streams; second-order streams form where two first-order streams converge, third-order streams form where two second-order streams converge, and so on.

Streams within drainage basins develop characteristic branching patterns that reflect, to some degree, the underlying bedrock geology, structure, and rock types. Dendritic or randomly branching patterns form on horizontal strata or on rocks with uniform erosional resistance. Parallel drainage patterns develop on steeply dipping strata, or on areas with systems of parallel faults or other landforms. Trellis drainage patterns consist of parallel main-stream channels intersected at nearly right angles by tributaries, in turn fed by tributaries parallel to the main channels. Trellis drainage patterns reflect significant structural control and typically form where eroded edges of alternating soft and hard layers are tilted, as in folded mountains or uplifted coastal strata. Rectangular drainage patterns form a regular rectangular grid on the surface and typically form in areas where the bedrock is strongly faulted or jointed. Radial and annular patterns develop on domes, including volcanoes and other roughly circular uplifts. Other, more complex patterns are possible in more complex situations.

Several categories of streams in drainage basins reflect different geologic histories—a consequent stream is one whose course is determined by the direction of the slope of the land. A subsequent stream is one whose course has become adjusted so that it occupies a belt of weak rock or another geologic structure. An antecedent stream has maintained its course across topography that is being uplifted by tectonic forces; these cross high ridges. Superposed streams' courses were laid down in overlying strata onto unlike strata below. Stream capture occurs when headland erosion diverts one stream and its drainage into another drainage basin.

See also ESTUARY; FLUVIAL; GEOMORPHOLOGY; RIVER SYSTEM.

FURTHER READING

Leopold, Luna B. *A View of the River.* Cambridge, Mass.: Harvard University Press, 1994.

Leopold, Luna B., and M. Gordan Wolman. *River Channel Patterns—Braided, Meandering, and Straight.* United States Geological Survey Professional Paper 282-B, 1957.

Parsons, Anthony J., and Athol D. Abrahams. *Overland Flow—Hydraulics and Erosion Mechanics.* London: UCL Press Ltd., University College, 1992.

Ritter, Dale F., R. Craig Kochel, and Jerry R. Miller. *Process Geomorphology.* 3rd ed. Boston: WCB/McGraw Hill, 1995.

Rosgen, David. *Applied River Morphology.* Pasoga Springs, Colo.: Wildland Hydrology, 1996.

Schumm, Stanley A. *The Fluvial System.* New York: Wiley Interscience, 1977.

Du Toit, Alexander (1878–1948) South African Geologist

Alexander Du Toit, known as "the world's greatest field geologist," was an early supporter of the theory of continental drift proposed by German meteorologist Alfred Wegener. Du Toit is credited with extensive mapping of the rocks sequences deposited on the Gondwana supercontinent and was one of the first scientists to knowledgeably correlate sequences between different continents.

Alexander Du Toit was born on March 14, 1878, near Cape Town and attended school at a local diocesan college in Rondebosch and at the University of the Cape of Good Hope. He then spent two years studying mining engineering at the Royal Technical College in Glasgow, United Kingdom, graduating in 1899, and then moved to study geology at the Royal College of Science in London before returning to study surveying and mining in Glasgow. In 1901 he was a lecturer at the Royal Technical College and at the University of Glasgow. He returned to South Africa in 1903, joining the Geological Commission of the Cape of Good Hope, and spent the next several years constantly in the field doing geological mapping. This time in his life was the foundation for his extensive understanding and unrivaled knowledge of South African geology. During his first season he worked with South African geologist Arthur W. Rogers in the western Karoo, where they established the stratigraphy of the Lower and Middle Karoo System. They also recorded the systematic phase changes in the Karoo and Cape Systems. Along with these studies they mapped the dolerite intrusives, their acid phases, and their metamorphic aureoles, publishing numerous papers on the subject. Throughout the years Du Toit worked in many areas including the Stormberg area and the Karoo coal deposits near the Indian Ocean. He was very interested in geomorphology and hydrogeology. The most significant contribution of his work was the theory of continental drift. He was the first to realize that the southern continents had once formed the supercontinent of Gondwana, which was distinctly different from the northern supercontinent Laurasia. In 1927 Du Toit took a position as chief consulting geologist at De Beers Consolidated Mines in South Africa and remained there until he retired in 1941.

Du Toit received many honors and awards. He was the president of the Geological Society of South Africa, a corresponding member of the Geological Society of America, and a member of the Royal Society of London. Some of Du Toit's most famous papers and books that proved influential in the gradual acceptance of the theory of plate tectonics include his *A Geological Comparison of South America with South Africa* (1927) and *Our Wandering Continents*, (1937). In 1933 Du Toit was awarded the Murchison Medal by the Geological Society of London.

See also AFRICAN GEOLOGY; CONTINENTAL DRIFT; GONDWANA, GONDWANALAND; PLATE TECTONICS.

FURTHER READING

Du Toit, Alexander L. "The Origin of the Amphibole Asbestos Deposits of South Africa." *Transactions of* the *Geological Society of South Africa* 48 (1946): 161–206.

———. "The Continental Displacement Hypothesis as Viewed by Du Toit." *American Journal of Science* 17 (1929): 179–183.

———. *Our Wandering Continents; An Hypothesis of Continental Drifting.* London: Oliver & Boyd, 1937.

Du Toit, Alexander, and Reed, F. R. C. *A Geological Comparison of South America with South Africa.* Washington, D.C.: Carnegie Institution of Washington, 1927.

dwarfs (stars) Stellar evolution is strongly related to the mass and size of a star, with normal stars following a stellar evolution curve called the main sequence. The term *dwarf star* oddly refers to stars whose size is normal for their mass, which lie on the main sequence curve, and which are converting hydrogen to helium by nuclear fusion in their cores. Dwarfs are also classified as any star with a radius comparable to or smaller than the Earth's Sun, which is classified as a yellow dwarf. Other types of dwarf stars are more unusual; these include white dwarfs, which are collapsed but still hot and shining stars; black dwarfs, which are cold, dead stars; and brown dwarfs, which are simply not massive enough to fuse hydrogen in their cores.

White dwarfs are small degenerate stars composed of electron-degenerate matter formed by the compression of electrons to positions close to the atomic nucleus in low-energy quantum states during stellar collapse. They are the end state of normal solar mass stars following the main sequence of stellar evolution. Since white dwarfs have masses about equal to the Sun and sizes about equal to the Earth, they are quite dense, yet are only faintly luminous, with the light coming from stored heat. About 6 percent of stars in the vicinity of the Earth's solar system are known to be white dwarfs, yet most stars (perhaps 97 percent) in the solar system may end up as white dwarfs in the end states of their evolution.

Main-sequence stellar evolution shows that after the hydrogen-fusing stage of stellar evolution, main-sequence stars expand into a red giant that fuses helium to carbon and oxygen in its core. If this red giant has sufficient mass to elevate core temperatures above the limit where it can fuse carbon, then spent carbon and oxygen will build up in the stellar core. This will eventually explode, forming a planetary nebula in which the outer layers of the star are shed and the remaining mass becomes a white dwarf star; this process explains why white dwarfs are rich in carbon and oxygen. Some white dwarfs also contain neon and magnesium.

White dwarfs are dead stars that no longer undergo fusion reactions since they have no source of internal energy. The stars collapse to the point at which the electron degeneration pressure is strong enough to stop the gravitational force from compressing the electrons closer to the atomic nuclei, leaving the atoms in a low-energy quantum state. This condition exists for stars with up to 1.4 solar masses, above which collapse may continue, forming a supernova. When white dwarfs initially form by collapse, they are hot, but they gradually cool by radiational transfer to space. With time white dwarfs cool so much that they are no longer visible; they are then called black dwarfs. Cooling to this point by radioactive processes requires more time than the age of the universe, however, so no black dwarfs yet exist, and they are classified as hypothetical stars.

A different variety of dwarf star is known as a brown dwarf. To generate fusion of hydrogen in the star, it must have started on the main sequence with a mass of at least 80 times that of Jupiter. Any star smaller than that would be hard to detect, because it was never massive enough to generate internal fusion of hydrogen. Still, brown dwarfs have convective surfaces and interiors, and are in an intermediate mass range between giant gaseous planets like Jupiter and true stars. There are two known stars with relative sizes and surface temperatures that classify them as brown dwarfs: Tiede 1 and Gliese 229B. Tiede 1 is similar to a yellow dwarf star like Earth's Sun, whereas Gliese 229B is close to a red dwarf category and is the approximate size of Jupiter. Red dwarfs are at the critical mass required for internal fusion, but present observations make it uncertain whether Gliese 229B is just above or just below that limit.

Red dwarf stars are cool, low-mass stars with less than 40 percent of the mass of the Sun, and are characterized by low-energy generation in their cores and relatively low luminosity. The brightest red dwarf has only about 10 percent of the luminosity of the Sun; thus red dwarfs are hard to detect, even though they may constitute the majority of stars in the solar system. Red dwarfs have hydrogen fusion in their cores and transport heat to their surfaces by convection, but the rate of heat generation is so low that they remain dim. Since convection removes the helium produced by hydrogen fusion in the stars' cores, red dwarfs can burn much of their hydrogen fuel before they leave the main sequence of stellar evolution, and they tend to have long life cycles. Stars as massive as the Earth's Sun accumulate helium in their cores and may live only about 10 billion years, but red dwarfs with about 10 percent of the solar mass may live for as long as 10 trillion years.

After the hydrogen in a red dwarf is mostly consumed, the core begins to contract, generating additional heat by gravitational contraction, which is then transferred to the stellar surface by convection. Eventually red dwarfs cool, fusion and contraction cease, and they fade from view. The closest star to the Earth's Sun is Proxima Centauri, which is a red dwarf.

See also ASTRONOMY; ASTROPHYSICS; COSMOLOGY; STELLAR EVOLUTION.

FURTHER READING

Chaisson, Eric, and Steve McMillan. *Astronomy Today.* 6th ed. Upper Saddle River, N.J.: Addison-Wesley, 2007.

Comins, Neil F. *Discovering the Universe.* 8th ed. New York: W. H. Freeman, 2008.

Encyclopedia of Astronomy and Astrophysics. CRC Press, Taylor and Francis Group. Available online. URL: http://eaa.crcpress.com/. Accessed October 24, 2008.

"ScienceDaily: Astrophysics News." Science Daily LLC. Available online. URL: http://www.sciencedaily.com/news/space_time/astrophysics/. Accessed October 24, 2008.

Snow, Theodore P. *Essentials of the Dynamic Universe: An Introduction to Astronomy.* 4th ed. St. Paul, Minn.: West, 1991.

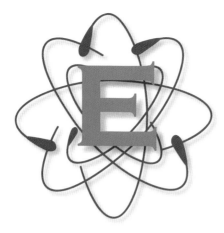

Earth   Earth is the third planet from the center of our solar system, located between Venus and Mars at a distance of 93 million miles (150×10^6 km) from the Sun. It has a mean radius of 3,960 miles (6,371 km), a surface area of 2.04×10^8 square miles (5.101×10^8 km^2), and an average density of 5.5 grams per cubic centimeter. As one of the terrestrial planets (Mercury, Venus, Earth, and Mars), Earth is composed of solid rock, with silicate minerals being the most abundant in the outer layers and a dense iron-nickel alloy forming the core material.

Earth and other planets condensed from a solar nebula about 5 billion years ago. In this process a swirling cloud of hot dust, gas, and protoplanets collided with one another, eventually forming the main planets. The accretion of Earth was a high-temperature process that allowed melting of the early Earth and segregation of the heavier metallic elements such as iron (Fe) and nickel (Ni) to sink to the core, and for the lighter rocky elements to float upward. This process led to the differentiation of Earth into several concentric shells of contrasting density and composition, and was the main control on the large-scale structure of Earth today.

The main shells of Earth include the crust, a light outer shell 3–43 miles (5–70 km) thick. This is followed inward by the mantle, a solid rocky layer extending to 1,802 miles (2,900 km). The outer core is a molten metallic layer extending to 3,170 miles (5,100 km) depth and the inner core is a solid metallic layer extending to 3,958 miles (6,370 km). With the acceptance of plate tectonics in the 1960s, geologists recognized that the outer parts of Earth were also divided into several zones that had very different mechanical properties. The outer shell of Earth was divided into many different rigid plates all moving with respect to each other, with some of them carrying continents in continental drift. This outer rigid layer became known as the lithosphere; it ranges from 45 to 95 miles (75–150 km) thick. The lithosphere is essentially floating on a denser, but partially molten layer of rock in the upper mantle known as the asthenosphere (or weak sphere). The weakness of this layer allows the plates on the surface of Earth to move about.

The most basic division of Earth's surface shows that it is divided into continents and ocean basins, with oceans occupying about 60 percent of the surface and continents 40 percent. Mountains are

Earth seen by *Apollo 17* crew: Africa and Madagascar at center of field of view *(NASA)*

elevated portions of the continents. Shorelines are where the land meets the sea, whereas continental shelves are broad to narrow areas underlain by continental crust, covered by shallow water. The shelves drop off to continental slopes, consisting of steep drop-offs from shelf edges to the deep ocean basin, and at the continental rise the slope flattens to merge with the deep-ocean abyssal plains. Ocean ridge systems are subaquatic mountain ranges where seafloor spreading creates new ocean crust. Mountain belts on Earth are of two basic types. Orogenic belts are linear chains of mountains, largely on the continents, that contain highly deformed, contorted rocks that represent places where lithospheric plates have collided or slid past one another. The midocean ridge system is a 40,000-mile (65,000-km) long mountain ridge that represents vast outpourings of young lava on the ocean floor and places where new oceanic crust is being generated by plate tectonics. After it is formed, it moves away from the ridge crests, and new magmatic plates fill the space created by the plates drifting apart. The oceanic basins also contain long, linear, deep-ocean trenches that are up to several kilometers deeper than the surrounding ocean floor and locally reach depths of seven miles (14 km) below the sea surface. These are places where the oceanic crust is sinking back into the mantle of Earth, completing the plate tectonic cycle for oceanic crust.

External layers of Earth include the hydrosphere, consisting of the ocean, lakes, streams, and the atmosphere. The air/water interface is very active, for here erosion breaks rocks down into loose debris—the regolith.

The hydrosphere is a dynamic mass of liquid, continuously on the move, including all the water in oceans, lakes, streams, glaciers, and groundwater, although most water is in the oceans. The hydrologic cycle describes changes, both long- and short-term, in Earth's hydrosphere. It is powered by heat from the Sun, which causes evaporation and transpiration. This water then moves in the atmosphere and precipitates as rain or snow, which then drains off in streams, evaporates, or moves as groundwater, eventually to begin the cycle over and over again.

The atmosphere is the sphere around Earth consisting of the mixture of gases called air. It is hundreds of kilometers thick and is always moving, because more of the Sun's heat is received per unit area at the equator than at the poles. The heated air expands and rises to where it spreads out, cools and sinks, and gradually returns to the equator. The effects of Earth's rotation modify this simple picture of the atmosphere's circulation. The Coriolis effect causes any freely moving body in the Northern Hemisphere to veer to the right and toward the left in the Southern Hemisphere.

The biosphere is the totality of Earth's living matter and partially decomposed dead plants and animals. It is made up largely of the elements carbon, hydrogen, and oxygen. When these organic elements decay, they may become part of the regolith and are returned through geological processes to the lithosphere, atmosphere, or hydrosphere.

See also ATMOSPHERE; BIOSPHERE; ENERGY IN THE EARTH SYSTEM; LITHOSPHERE; MAGNETIC FIELD, MAGNETOSPHERE; MANTLE.

FURTHER READING

Skinner, Brian J., and Stephen C. Porter. *The Dynamic Earth, an Introduction to Physical Geology.* 5th ed. New York: John Wiley & Sons, 2004.

earthquakes An earthquake occurs when a sudden release of energy causes the ground to shake and vibrate, associated with the passage of waves of energy released at the source. Earthquakes can be extremely devastating and costly events, sometimes killing tens or hundreds of thousands and leveling entire cities in a few seconds. A single earthquake may release the energy equivalent to hundreds or thousands of nuclear blasts and may cost billions of dollars in damage, not to mention the toll in human suffering. Earthquakes are also associated with secondary hazards, such as tsunami, landslides, fire, famine, and disease that also exert their toll on humans and other animals.

Most earthquakes occur along plate boundaries. The lithosphere (or outer rigid shell) of the Earth is broken into about 12 large tectonic plates, each moving relative to the others. There are many other smaller plates. Most earthquakes happen where two of these plates meet and are moving past each other, such as in southern California. Recent earthquakes in China, Turkey, Sumatra (Indonesia), and Mexico have also been located along plate boundaries. A map of plate boundaries of the Earth and earthquakes shows where significant earthquakes have occurred in the past 50 years. Most really big earthquakes occur at boundaries where the plates are moving toward each other (as in Alaska and Japan), or sliding past each other (as in southern California and Turkey). Smaller earthquakes occur where the plates are moving apart, such as along midoceanic ridges where new magma rises and forms oceanic spreading centers.

The area that gets the most earthquakes in the continental United Stated is southern California along the San Andreas Fault. The reason for this high number of earthquakes is that the Pacific plate is sliding north relative to the North American plate along the track of the San Andreas fault. The motion in this

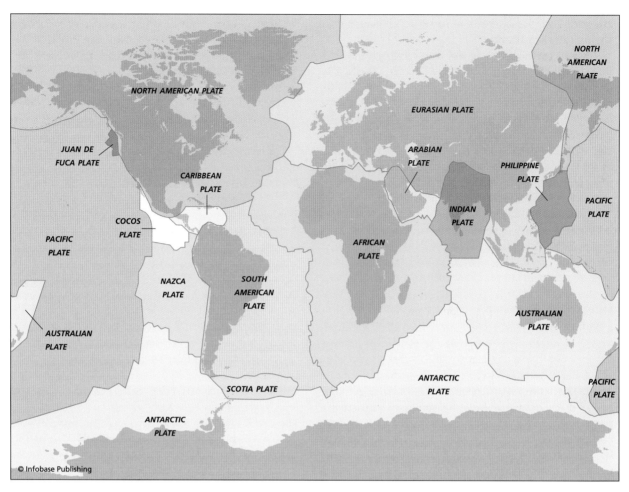

Locations of significant earthquakes and plate boundaries. Shallow-focus earthquakes (in pink) are found at all types of plate boundaries; medium- and deep-focus quakes are found along subduction zones.

area is characterized as a "stick-slip" type of sliding, where the two plates stick to each other along the plate boundary as the two plates slowly move past each other, and stresses rise over tens or hundreds of years. Eventually the stresses along the boundary rise so high that the strength of the rocks is exceeded, and the rocks suddenly break, causing the two plates to move dramatically (slip) up to 20–30 feet (5–7 m) in a few seconds. This sudden motion of previously stuck segments along a fault plane is an earthquake. The severity of the earthquake is determined by how large an area breaks in the earthquake, how far it moves, how deep within the Earth the break occurs, and the length of time that the broken or slipped area along the fault takes to move. The elastic-rebound theory states that recoverable (also known as elastic) stresses build up in a material until a specific level or breaking point is reached. When the breaking point or level is attained, the material suddenly breaks, releasing energy and stresses in an earthquake. In the case of earthquakes, rows of fruit trees, fences, roads, and railroad lines that became gradually bent across

an active fault line as the stresses built up are typically noticeably offset across faults that have experienced an earthquake. When the earthquake occurs, the rocks snap along the fault, and the bent rows of trees, fences, or roads/rail line become straight again, but displaced across the fault.

Some areas away from active plate boundaries are also occasionally prone to earthquakes. Even though earthquakes in these areas are uncommon, they can be very destructive. Places including Boston, Massachusetts; Charleston, South Carolina; and New Madrid, Missouri (near St. Louis) have been sites of particularly bad earthquakes. In 1811 and 1812 three large earthquakes with magnitudes of 7.3, 7.5, and 7.8 were centered in New Madrid and shook nearly the entire United States, causing widespread destruction. Most buildings were toppled near the origin of the earthquake, and several deaths were reported (the region had a population of only 1,000 at the time, but is now densely populated). Damage to buildings was reported from as far away as Boston and Canada, where chimneys toppled, plaster cracked, and church bells were set to ringing by the shaking of the ground.

Many earthquakes in the past have been incredibly destructive, killing hundreds of thousands, like the ones in Wenchuan, China; Iran; Sumatra, Indonesia, and Mexico City in recent years (see the table below). Some earthquakes have killed nearly a million people, such as one in 1556 in China that killed 800,000–900,000, another in China in 1976 that killed an estimated 242,000 to 800,000 people, one in Calcutta, India, in 1737 that killed about 300,000 people, and the earthquake-related Indian Ocean tsunami that killed an estimated 286,000 people in 2004. The 2008 magnitude 8 Wenchuan earthquake in China has an official death toll of about 90,000, but unofficial estimates reach up to 1,000,000.

ORIGINS OF EARTHQUAKES

Earthquakes can originate from sudden motion along a fault, from a volcanic eruption, bomb blasts, landslides, or anything else that suddenly releases energy on or in the Earth. Not every fault is associated with active earthquakes. Most faults are in fact no longer active but were active at some time in the geologic past. Of the active faults, only some are particularly prone to earthquakes. Some faults are slippery, and the two blocks on either side just slide by each other passively without producing major earthquakes. In other cases, however, the blocks stick together and deform

until they reach a certain point at which they suddenly snap, releasing energy in an earthquake event.

Rocks and materials are said to behave in a brittle way when they respond to built-up tectonic pressures by cracking, breaking, or fracturing. Earthquakes represent a sudden brittle response to built-up stress and are almost universally activated in the upper few kilometers of the earth. Deeper than this, the pressure and temperature are so high that the rocks simply deform like silly putty and do not snap, but are said to behave in a ductile manner.

An earthquake originates in one place then spreads out in all directions along the fault plane. The focus is the point in the Earth where the earthquake energy is first released and is the area on one side of a fault that actually moves relative to the rocks on the other side of the fault plane. After the first slip event the area surrounding the focus experiences many smaller earthquakes as the surrounding rocks also slip past one another to even out the deformation caused by the initial earthquake shock. The epicenter is the point on the Earth's surface that lies vertically above the focus.

When big earthquakes occur, the surface of the Earth actually forms into waves that move across the surface, just as in the ocean. These waves can be pretty spectacular and also extremely destructive. When an earthquake strikes, these seismic waves move out in all directions, just like sound waves, or

THE 13 DEADLIEST EARTHQUAKES IN RECORDED HISTORY

Place	Year	Deaths	Estimated Magnitude
Shaanxi, China	1556	830,000	
Calcutta, India	1737	300,000	
Sumatra, Indonesia	2004	286,000	9.0
T'ang Shan, China	1976	242,000 (could be as many as 800,000)	7.8
Port-au-Prince, Haiti	2010	200,000 (preliminary estimate)	7.0
Gansu, China	1920	180,000	8.6
Messina, Italy	1908	160,000	7.5
Tokyo, Japan	1923	143,000	8.3
Beijing, China	1731	100,000	
Chihli, China	1290	100,000	
Naples, Italy	1693	93,000	
Wenchuan, China	2008	90,000 (could be as many as 1,000,000)	7.9
Gansu, China	1932	70,000	7.6

Note: Casualties from the 2008 Wenchuan earthquake and the 2010 Haiti earthquake were still being tabulated at the time of writing.

ripples that move across water after a stone is thrown in a still pond. After the seismic waves have passed through the ground, the ground returns to its original shape, although buildings and other human constructions are commonly destroyed. In really large earthquakes the ground is deformed into waves of rock, several feet high (~1 m), moving at very high speeds.

During an earthquake, several types of seismic waves can either radiate underground from the focus—called body waves—or aboveground from the epicenter—called surface waves. The body waves travel through the whole body of the Earth and move faster than surface waves, whereas surface waves cause most of the destruction associated with earthquakes because they briefly change the shape of the surface of the earth when they pass. There are two types of body waves—P (primary or compressional) waves, and S, or secondary waves. P-waves deform material through a change in volume and density, and these can pass through solids, liquids, and gases. The kind of movement associated with the passage of a P-wave is a back-and-forth–type motion. Compressional (P) waves move with high velocity, about 3.5–4 miles per second (6 km/second), and are thus the first to be recorded by seismographs. This is why

they are called Primary (P) waves. P-waves cause a lot of damage because they temporarily change the area and volume of ground on which humans have built or modified in ways that require the ground to keep its original shape, area, and volume. When the ground suddenly changes its volume by expanding and contracting, many of these constructions break. For instance, if a gas pipeline is buried in the ground, it may rupture and explode when a P-wave passes because of its inability to change its shape along with the earth. Fires and explosions originating from broken pipelines commonly accompany earthquakes. History has shown that fires often do as much damage after the earthquake as the ground shaking did during the quake. This fact is dramatically illustrated by the 1906 earthquake in San Francisco, where much of the city burned for days after the shaking and was largely destroyed. Similarly, much of the damage and loss of life from the 1995 magnitude 7.3 Kobe, Japan, earthquake was from fires ignited by gas lines and home heating systems.

The second kind of body waves are shear waves (S), or secondary waves, because they change the shape of a material but not its volume. Solids can only transmit shear waves. Shear waves move material at right angles to the direction of wave travel,

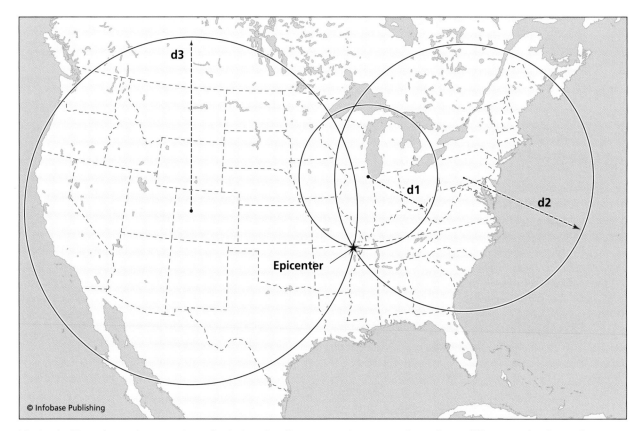

© Infobase Publishing

Method of locating epicenters by calculating the distance to the source from three different seismic stations. The distance to the epicenter is calculated using the time difference between the first arrivals of P- and S-waves. The unique place that the three distance circles intersect is the location of the epicenter.

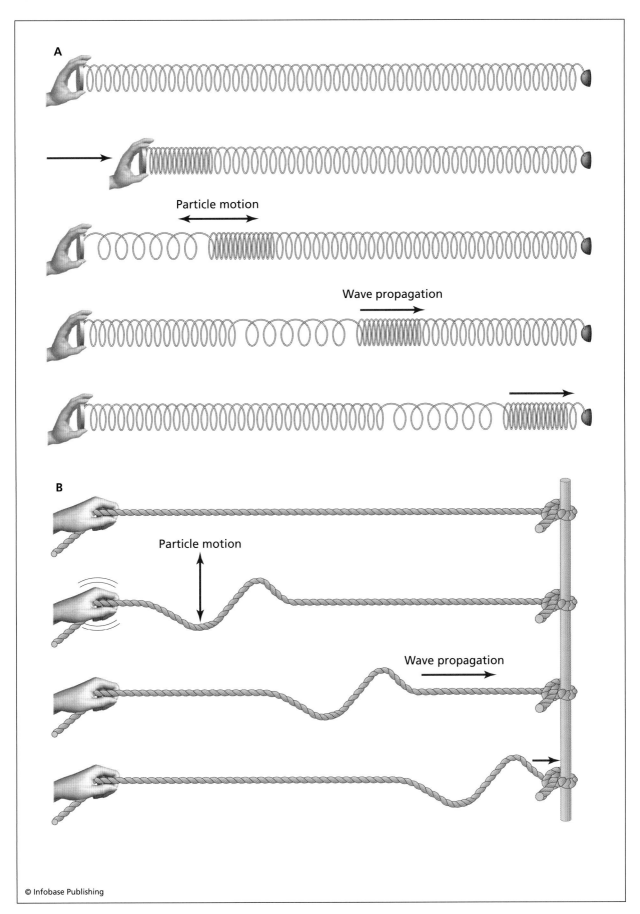

A

Particle motion

Wave propagation

B

Particle motion

Wave propagation

Analogy to seismic P- and S-waves using slinky and rope

and thus they consist of an alternating series of sideways motions. Holding a jump rope at one end on the ground and moving it rapidly back and forth can simulate this kind of motion. Waves form at the end being held, and they move the rope sideways as they approach toward the loose end of the rope. A typical shear wave velocity is two miles per second (3.5 km/second). These kinds of waves may be responsible for knocking buildings off foundations when they pass, since their rapid sideways or back-and-forth motion is often not met by buildings. The effect is much like pulling a tablecloth out from under a set table—if done rapidly, the building (as is the case for the table setting) may be left relatively intact, but detached from its foundation.

Surface waves can also be extremely destructive during an earthquake. These have complicated types of twisting and circular motions, much like the circular motions one might feel while swimming in waves out past the surf zone at the beach. Surface waves travel more slowly than either type of body wave, but because of their complicated motion they often cause the most damage. This is a good thing to remember during an earthquake, because if one realizes that the body waves have just passed one's location, there may be a brief period of no shaking to go outside before the very destructive surface waves hit and cause even more destruction.

MEASURING EARTHQUAKES

To measure the intensity of shaking during an earthquake, geologists use seismographs, which display earth movements by means of an ink-filled stylus on a continuously turning roll of graph paper. Modern seismographs have digital versions of the same design but record the data directly to computer systems for analysis. When the ground shakes, the needle wiggles and leaves a characteristic zigzag line on the paper. Many seismograph records clearly show the arrival of P- and S-body waves, followed by surface waves.

Seismographs are built according to a few simple principles. To measure the shaking of the Earth during a quake, the point of reference must be free from shaking, ideally on a hovering platform. Since building perpetually hovering platforms is impractical, engineers have designed an instrument known as an inertial seismograph, which makes use of the principle of inertia, the resistance of a large mass to sudden movement. When a heavy weight is hung from a string or thin spring, the string can be shaken and the big heavy weight will remain stationary. Using an inertial seismograph, the ink-filled stylus is attached to the heavy weight and remains stationary during an earthquake. The continuously turning graph paper is attached to the ground and moves back and forth

during the quake, recording the zigzag trace of the earthquake motion on the graph paper.

Seismographs are used in series, some set up as pendulums and others as springs, to measure ground motion in many directions. Engineers have made seismographs that can record motions as small as one hundred millionth of an inch, about equivalent to being able to detect the ground motion caused by a car several blocks away. The ground motions recorded by seismographs are very distinctive, and geologists who study them have methods of distinguishing between earthquakes produced along faults, earthquake swarms associated with magma moving into volcanoes, and even between explosions from different types of construction and nuclear blasts. Interpreting seismograph traces has therefore become an important aspect of nuclear test-ban treaty verification. Many seismologists are employed to monitor earthquakes around the world and to verify that countries are not testing nuclear weapons.

EARTHQUAKE MAGNITUDE

Earthquakes vary greatly in intensity, from undetectable ones up to ones that kill millions of people and wreak total destruction. For example, an earthquake in 2008 killed at least 90,000 people in China, yet several thousand earthquakes that do no damage occur every day throughout the world. The energy released in large earthquakes is enormous, up to hundreds of times more powerful than large atomic blasts. Strong earthquakes may produce ground accelerations greater than the force of gravity, enough to uproot trees or send projectiles through buildings, trees, or anything else in their path. Earthquake magnitudes are most commonly measured by the Richter scale.

The Richter scale gives an idea of the amount of energy released during an earthquake. It is based on the amplitudes (half the height from wave-base to wave-crest) of seismic waves at a distance of 61 miles (100 km) from the epicenter. The Richter scale magnitude of an earthquake is calculated by using the zigzag trace produced on a seismograph, once the epicenter has been located by comparing signals from several different, widely separated seismographs. The Richter scale is logarithmic, whereby each step of 1 corresponds to a 10-fold increase in amplitude. This is necessary because the energy of earthquakes changes by factors of more than a hundred million.

The energy released in earthquakes changes even more rapidly with each increase in the Richter scale, because the number of high-amplitude waves increases with bigger earthquakes and also because the energy released is according to the square of the amplitude. Thus it turns out in the end that an

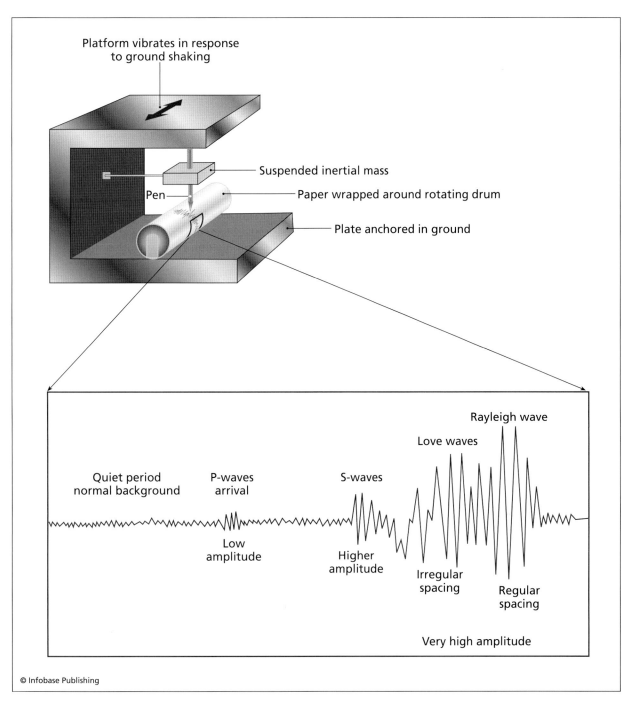

Platform vibrates in response
to ground shaking

Suspended inertial mass

Pen

Paper wrapped around rotating drum

Plate anchored in ground

Rayleigh wave

Love waves

Quiet period
normal background

P-waves
arrival

S-waves

Low
amplitude

Higher
amplitude

Irregular
spacing

Regular
spacing

Very high amplitude

© Infobase Publishing

Schematic diagram of an inertial seismograph showing a large inertial mass suspended from a spring. The mass remains stationary as the ground and paper wrapped around a rotating drum move back and forth during an earthquake, creating the seismogram.

increase of 1 on the Richter scale corresponds to a 30-fold increase in energy released. The largest earthquakes so far recorded are the 9.2 Alaskan earthquake (1964), the 9.5 Chilean earthquake (1960), and the 9.0 Sumatra earthquake (2004), each of which released the energy equivalent to more than 10,000 nuclear bombs the size of the one dropped on Hiroshima.

Before the development of modern inertial seismographs, earthquake intensity was commonly measured by the modified Mercalli intensity scale. This scale, named after Giuseppe Mercalli, was developed in the late 1800s; and it measures the amount of vibration people remember feeling for low-magnitude earthquakes, and measures the amount of damage to buildings in high-magnitude events (see table on

MODIFIED MERCALLI INTENSITY SCALE COMPARED WITH RICHTER MAGNITUDE

Mercalli intensity	Richter magnitude	Description
I–II	< 2	Not felt by most people
III	3	Felt by some people indoors, especially on high floors
IV–V	4	Noticed by most people. Hanging objects swing and dishes rattle
VI–VII	5	Everyone feels. Some building damage (esp. to masonry), waves on ponds
VII–VIII	6	Difficult to stand and people scared or panicked. Difficult to steer cars. Moderate damage to buildings
IX–X	7	Major damage, general panic of public. Most masonry and frame structures destroyed. Underground pipes broken. Large landslides
XI–XII	8 and higher	Near total destruction

Note: This table can be used to compare the relative magnitudes of earthquakes from the historical record for which only information on the Mercalli intensity may be known, with the intensity on the modern Richter magnitude scale.

page 232). One of the disadvantages of the Mercalli scale is that it is not corrected for distance from the epicenter. People near the source of the earthquake therefore may measure the earthquake as an IX or X, whereas those farther from the epicenter might record only a I or II event. The modified Mercalli scale, however, has proven useful for estimating the magnitudes of historical earthquakes that occurred before the development of modern seismographs, since the Mercalli magnitude can be estimated from historical records.

EARTHQUAKE HAZARDS

Earthquakes are associated with a wide variety of specific hazards, including primary effects such as ground motion, ground breaks (or faulting), mass wasting, and liquefaction. Secondary and tertiary hazards are indirect effects, caused by events initiated by the earthquake. These may include tsunami in open ocean and bay waters; waves that rock back and forth in enclosed basins, called seiche waves; fires and explosions caused by disruption of utilities and pipelines; and changes in ground level that may disrupt habitats, change groundwater level, displace coastlines, cause loss of jobs, and displace populations. Financial losses to individuals, insurance companies, and businesses can easily soar into the tens of billions of dollars for even moderate-sized earthquakes.

Ground Motion

One of the primary hazards of earthquakes is ground motion caused by the passage of seismic waves through populated areas. The most destructive waves

Collapsed 10-story apartment building in Islamabad, Pakistan, after earthquake October 8, 2005: The building pancaked as one floor fell, thereby causing each lower floor to collapse. (AP images)

Damage from ground shaking and landslides in Yingxiu, Sichuan Province, China, from May 12, 2008, magnitude 7.9 earthquake *(T. Kusky)*

are surface waves, which in severe earthquakes may visibly deform the surface of the earth into moving waves. Ground motion is most typically felt as shaking; it causes the familiar rattling of objects off shelves reported from many minor earthquakes. The amount of destruction associated with given amounts of ground motion depends largely on the design and construction of buildings and infrastructure according to specific codes.

The amount of ground motion associated with an earthquake generally increases with the magnitude of the quake but depends also on the nature of the substratum—loose, unconsolidated soil and fill tends to shake more than solid bedrock. The Loma Prieta, California, earthquake (1989) dramatically illustrated this phenomenon where areas built on solid rock vibrated the least (and saw the least destruction), and areas built on loose clays vibrated the most. Much of the Bay Area is built on loose clays and mud, including the Nimitz freeway, which collapsed during the event. The area that saw the worst destruction associated with ground shaking was the Marina district. Even though this area is located far from the earthquake epicenter, it is built on loose, unconsolidated landfill, which shook severely during the earthquake, causing many buildings to collapse and gas lines to rupture, initiating fires. More than twice as much damage from ground shaking during the Loma Prieta earthquake was reported from areas

over loose fill or mud than from areas built over solid bedrock. Similar effects were reported from the Mexico City earthquake (1985) because the city is built largely on old lake bed deposits.

Additional variation in the severity of ground motion is noted in the way that different types of bedrock transmit seismic waves. Earthquakes that occur in the western United States generally affect a smaller area than those that occur in the central and eastern parts of the country. This is because the bedrock in the west (California, in particular) is generally much softer than the hard igneous and metamorphic bedrock found in the east. Harder, denser rock generally transmits seismic waves better than softer, less dense rock, so earthquakes of given magnitude may be more severe over larger areas in the east than in the west. From the perspective of ground motion intensity, it is fortunate that more large earthquakes occur in the west than in the east.

Ground motions are measured as accelerations, the rate of change of motion. This type of force is the same as accelerating in a car, where the driver feels pushed gently back against the seat while increasing speed. This is a small force compared with another common force measured as an acceleration, gravity. Gravity is equal to 9.8 meters per second squared, or 1×g (this is what one would feel while jumping out of an airplane). People have trouble standing up and buildings begin falling at one-tenth the acceleration of gravity (0.1×g). Large earthquakes can produce accelerations that greatly exceed—even double or triple—the force of gravity. These accelerations are able to uproot large trees and toss them into the air, shoot objects through walls and buildings, and cause almost any structure to collapse.

Some of the damage typically associated with ground motion and the passage of seismic waves includes swaying and pancaking of buildings. During an earthquake buildings may sway with a characteristic frequency that depends on their height, size, construction, underlying material, and intensity of the earthquake. This causes heavy objects to move rapidly from side to side inside the buildings and can cause much destruction. The shaking generally increases with height, and in many cases the shaking causes concrete floors at high levels to separate from the walls and corner fastenings, causing the floors to fall progressively or, pancake one another, crushing all in between. With greater shaking the entire structure may collapse.

Ground Breaks

Ground breaks, or ruptures, form where a fault cuts the surface, and they may also be associated with mass wasting, or the movements of large blocks of land downhill. These ground breaks may have hori-zontal, vertical, or combined displacements across them and may cause considerable damage. Fissures that open in the ground during some earthquakes are mostly associated with the mass movement of material down slope, not with the fault trace itself breaking the surface. For instance, in the Alaskan earthquake (1964) ground breaks displaced railroad lines by several yards (m); broke through streets, houses, storefronts, and other structures; and caused parts of them to drop by several yards (m) relative to other parts of the structure. Most of these ground breaks were associated with slumping, or movement of the upper layers of the soil downhill toward the sea. Ground breaks during earthquakes are also one of the causes of the rupture of pipelines and communication cables.

Mass Wasting

Mass wasting is the movement of material downhill. In most instances mass wasting occurs by a gradual creeping of soils and rocks downhill, but during earthquakes large volumes of rock, soil, and all that is built on the area affected may suddenly collapse in a landslide. Earthquake-induced landslides occur in areas with steep slopes or cliffs, such as parts of California, Alaska, South America, Turkey, and China. One of the worst recorded earthquake-

A second photograph of damage from ground shaking and landslides in Hongkon, Sichuan Province, China, from May 12, 2008, magnitude 7.9 earthquake *(T. Kusky)*

Damage from ground shaking and liquefaction in Yingxin, Sichuan Province, China, causing buildings to collapse on their sides, from May 12, 2008, magnitude 7.9 earthquake *(T. Kusky)*

induced landslides occurred in the magnitude 7.9 Wenchuan, China, earthquake (2008) when several villages were completely buried under hundreds of feet of debris from mountains that collapsed nearby, killing thousands.

In the 9.2 magnitude Alaska earthquake (1964), landslides destroyed power plants, homes, roads, and railroad lines. Some landslides even occurred undersea and along the seashore. Large parts of Seward and Valdez sat on the top of large submarine escarpments, and during the earthquake extensive areas of these towns slid out to sea in giant submarine landslides and were submerged. Another residential area near Anchorage, Turnagain Heights, was built on top of cliffs with fantastic views of the Alaska Range and Aleutian volcanoes. When the earthquake struck, this area slid out toward the sea on a series of curving faults that connected in a slippery shale unit known as the Bootlegger shale. During the earthquake this shale unit lost all strength and became almost cohesionless, and the shaking of the soil and rock above it caused the entire neighborhood to slide toward the sea along the shale unit and be destroyed.

Liquefaction

Liquefaction is a process in which sudden shaking of certain types of water-saturated sands and muds turns these once-solid sediments into slurry with a liquidlike consistency. Liquefaction occurs where the shaking causes individual grains to move apart, then water moves up in between the individual grains, making the whole water/sediment mixture behave like a fluid. Earthquakes often cause liquefaction of sands and muds. Any structures built on sediments that liquefy may suddenly sink into them as if resting on a thick fluid. Liquefaction caused the Bootlegger shale in the 1964 Alaskan earthquake to become suddenly so weak that it destroyed Turnagain Heights. Liquefaction during earthquakes also causes sinking of sidewalks, telephone poles, building foundations, and other structures. One famous example of liquefaction occurred in the Japan earthquake (1964), where entire rows of apartment buildings rolled onto their sides but were not severely damaged internally. Liquefaction also causes sand to bubble to the surface during earthquakes, forming mounds up to several tens of feet high, known as sand volcanoes, or ridges of sand.

Changes in Ground Level

During earthquakes, blocks of earth shift relative to one another. This may result in changes in ground level, base level, the water table, and high-tide marks. Particularly large shifts have been recorded from some of the historically large earthquakes, such as the magnitude 9.2 Alaskan earthquake (1964) and the Sumatra earthquake (2004). In 1964 an area more than 600 miles (1,000 km) long in south central Alaska recorded significant changes in ground level, including uplifts of up to 12 yards (11 m), downdrops of more than two yards (2 m), and lateral shifts of several to tens of yards. Uplifted areas along the coastline experienced dramatic changes in the marine ecosystem—clam banks were suddenly uplifted out of the water and remained high and dry. Towns built around docks were suddenly located many yards above the convenience of being at the shoreline. Downdropped areas experienced different effects—forests that relied on freshwater for their root systems suddenly were inundated by salt water and were effectively "drowned." Populated areas located at previously safe distances from the high tide (and storm) line became prone to flooding and storm surges, and had to be relocated.

Areas far inland also suffered from changes in ground level—when some were uplifted by many tens of feet (~10 m), the water table recovered to a lower level relative to the land surface, and soon became out of reach of many water wells, which had to be redrilled. Changes in ground level, although seemingly a minor hazard associated with earthquakes, are significant and cause a large amount of damage that may cost millions of dollars to mitigate.

Tsunamis and Seiche Waves

Several types of large waves are associated with earthquakes, including tsunamis and seiche waves. Tsunamis, also known as seismic sea waves, form most usually from submarine landslides that displace a large volume of rock and sediment on the seafloor, which in turn displaces a large amount of water. Tsunamis may be particularly destructive as they travel very rapidly (hundreds of miles per hour), and may reach many tens of yards above normal high-tide levels. The most devastating tsunami in recorded history occurred in 2004, in association with the magnitude 9.0 earthquake in Sumatra, Indonesia. A wave that reached heights locally of 100 feet (30 m) swept across the Indian Ocean, killing about 283,000, mostly in Indonesia, Sri Lanka, and India. Two other particularly devastating examples include a tsunami generated by a magnitude 8.7 earthquake in the Atlantic Ocean (1775) that is estimated to have killed more than 60,000 in Portugal (this number is from Lisbon alone, although the tsunami struck a large section of coastline, and other related tsunamis were reported from North Africa, the British Isles, and the Netherlands). Another tsunami generated in the Aleutian Islands of Alaska (1946) traveled across the Pacific Ocean at 500 miles per hour (800 km/hr) and hit Hilo, Hawaii, with a crest 18 yards (16 m) higher

than the normal high-tide mark, killing 159 people, destroying approximately 500 homes, and damaging 1,000 more structures.

Seiche waves may be generated by the back-and-forth motion associated with earthquakes, causing a body of water (usually lakes or bays) to rock back and forth, gaining amplitude and splashing up to higher levels than normally associated with that body of water. The effect is analogous to shaking a glass of water and watching the ripples suddenly turn into large waves that splash out of the glass. Other seiche waves may be formed when landslides or rockfalls drop large volumes of earth into bodies of water. The largest recorded seiche wave of this type formed suddenly on July 9, 1958, when a large earthquake-initiated rockfall generated a seiche wave 1,700 feet (518 m) tall. The seiche wave raced across Lituya Bay, Alaska, destroying the forest and killing several people, including a geologist who had warned authorities that such a wave could be generated by a large landslide in Lituya Bay.

Damage to Utilities (Fires, Broken Gas Mains, Transportation Network)

Much of the damage and many of the casualties caused by earthquakes are associated with damage to the infrastructure and system of public utilities. For example, much of the damage associated with the San Francisco earthquake (1906) came not from the earthquake itself but from the huge fire that resulted from the numerous broken gas lines, overturned wood and coal stoves, and even from fires set intentionally to collect insurance money on partially damaged buildings. In the Kobe, Japan, earthquake (1995), a large share of the damage was likewise from fires that raged uncontrolled, with fire and rescue teams unable to reach the areas worst affected. Water lines were broken so that even in accessible locations, firefighters were unable to put out the flames.

One of the lessons from these examples is that evacuation routes need to be set up in earthquake hazard zones in anticipation of post-earthquake hazards such as fires, aftershocks, and famine. These routes should ideally be clear of obstacles such as overpasses and buildings that may block access, and efforts should be made to clear these routes soon after earthquake disasters, both for evacuation purposes and for emergency access to the areas worst affected.

See also GEOLOGICAL HAZARDS; PLATE TECTONICS; SEISMOLOGY; TSUNAMI, GENERATION MECHANISMS.

FURTHER READING

Bolt, B. A. *Earthquakes*. 4th ed. New York: W. H. Freeman, 1999.

Coburn, A., and R. Spence. *Earthquake Protection.* Chichester, U.K.: John Wiley & Sons, 1992.

Erikson, Jon. *Quakes, Eruptions, and Other Geologic Cataclysms: Revealing the Earth's Hazards.* Rev. ed. New York: Facts On File, 2001.

Federal Emergency Management Agency home page. Available online. URL: http://www.fema.gov. Accessed October 10, 2008. Updated daily.

Gilbert, Grove Karl, Richard L. Humphrey, John S. Sewell, and Frank Soule. *The San Francisco Earthquake and Fire of April 18, 1906, and their Effects on Structures and Structural Materials.* Washington, D.C.: U.S. Geological Survey Bulletin 324, 1907.

Griggs, G. B., and J. A. Gilchrist. *Geologic Hazards, Resources, and Environmental Planning.* Belmont, Calif.: Wadsworth, 1983.

IRIS consortium (Incorporated Research Institutions for Seismology). Available online. URL: http://www.iris.washington.edu. Accessed October 10, 2008.

Kendrick, T. D. *The Lisbon Earthquake.* London: Metheun, 1956.

Kusky, T. M. *Earthquakes: Plate Tectonics and Earthquake Hazards.* New York: Facts On File, 2008.

———. *Geological Hazards: A Sourcebook.* Westport, Conn.: Greenwood Press, 2003.

Logorio, H. *Earthquakes: An Architect's Guide to Non-Structural Seismic Hazards.* New York: John Wiley & Sons, 1991.

Reiter, L. *Earthquake Hazard Analysis.* New York: Columbia University Press, 1990.

U.S. Geological Survey. "Natural Hazards—Earthquakes. Information on Historic Earthquakes and Real Time Data on Current Events." Available online. URL: http://www.usgs.gov/hazards/earthquakes/. Accessed October 10, 2008.

———. *Lesson Learned from the Loma Prieta Earthquake of October 17, 1989.* Circular 1045, U.S. G.P.O.: Books and Open-File Reports Section, U.S. Geological Survey, 1989.

Verney, P. *The Earthquake Handbook.* New York: Paddington Press, 1979.

Wallace, R. E., ed. *The San Andreas Fault System, California.* Reston, Va.: U.S. Geological Survey Professional Paper 1515, 1990.

economic geology The science of economic geology focuses on the study of earth materials that can be used for economic or industrial purposes. Different subfields include the study of ore deposits and metals; nonmetallic resources such as hydrocarbons and petroleum; gems; and industrial materials for construction and for high-technology fields. Complete analysis of economic resources includes several disciplines of geology, including field geology, geophysics, geochemistry, and structural geology, but also extends to other fields such as investment banking, stock and futures markets analysis, and economic and government planning. The mining of economic resources also brings in environmental scientists and planners who must make sure that the development of economic resources is done with the least detrimental impact on the environment.

MINERAL RESOURCES AND OCCURRENCES AND ORE RESERVES

Geologic processes concentrate many minerals in the Earth's crust, and some of these have economic value of potential benefit to society. When a rare, potentially valuable mineral is concentrated in a location but in quantities too low to be of economic or minable value, it is said to be a mineral occurrence. When the concentration reaches a critical value that makes it minable in current or potentially future economic scenarios, it is classified as a mineral resource. Ore reserves are concentrations of minerals that are economically and technically feasible to extract.

Different types of mineral resources include metallic ores, nonmetallic ores, gems, hydrocarbons, and building materials. Descriptions of the metallic ore deposits are covered here, and links to the nonmetallic ores and resources are cited in the discussions listed at the end of this entry.

THE FORMATION AND CONCENTRATION OF METALLIC ORE DEPOSITS

Most metallic ores form by one of several main processes, including concentration by hydrothermal fluids, crystallization from an igneous magma, metamorphic processes that move fluids and chemical components in rocks from place to place, weathering, sorting by water in streams, or other surficial processes that can remove some elements from a rock or soil while concentrating other elements.

Many of the ores of metallic minerals occur as compounds of the sulfide ion, S^{2-}, with the metals attached as cations. Most of these are soft, resemble metals, and form many of the world's large ore deposits. One of the most common sulfide minerals is pyrite, FeS_2, found as a minor component in many rocks and as an accessory mineral in many ore deposits. Pyrrhotite (Fe_7S_8-FeS) is a less common iron sulfide mineral. Most ore deposits are formed from hydrothermal fluids. Lead (Pb) and Zinc (Zn) are commonly found in sulfide compounds such as galena (PbS) and sphalerite (ZnS), and copper deposits are dominated by the sulfide minerals chalcopyrite ($CuFeS_2$) and bornite (Cu_5FeS_4).

Iron Ore

Iron is one of the most important metalliferous ores used globally for construction, the automobile industry, and a number of other commodities. Most iron ore is found in Precambrian-banded iron formations (BIFs), such as in the Hamersley Basin of Australia, and include finely layered sequences of sedimentary rocks with thin layers of iron oxide minerals, typically interlayered with chert or other silica-rich layers. Most banded iron formations are Archean or Proterozoic in age, and many models suggest that they required special environmental conditions to form, including an oxygen-poor atmosphere, and a setting in which submarine volcanic eruptions were able to help transport the iron to seafloor settings where it was deposited. Many of these deposits were later weathered so that the magnetite in the original deposits was altered by oxidation to hematite, a form of iron more easily mined and used by the steel industry.

Lead-Zinc-Silver Ores

Many deposits of lead and zinc are associated with silver and are found in submarine settings that formed in association with volcanism, forming a type of deposit known as a SEDEX (sedimentary exhalative) deposit. These tend to be associated with the fringes of large volcanic flows or on the margins of subvolcanic plutons, with many examples located in the Superior craton of Canada and others in Australia. Another, quite different type of lead-zinc deposit is found as metalliferous layers that replaced primary carbonate layers; the world's largest deposit of this type is found in the central United States in Missouri, in the Mississippi valley lead-zinc deposit belt.

Gold

Gold is found in a diverse array of deposit types, ranging from concentrations in quartz veins in igneous-metamorphic rocks controlled by the plate tectonic setting, to metamorphic settings, to wide areas called alluvial deposits where streams eroded primary gold sources and deposited them in places where the stream currents slowed and dropped the gold out of suspension. Most of the lode gold deposits are found in quartz veins in intrusions, granites, shear zones, and deformed turbidite sequences. Basalt is a common constituent of many gold provinces, and it appears that metamorphic fluids move through the basalt and leach out the gold and related fluids, depositing them when the chemical and temperature conditions are best suited in the host rocks.

Placer gold is eroded from the lode gold sources and carried by streams and rivers, or reworked by beach processes before reaching the final site of deposition. Most of the the world's placer gold is found in the Archean Witwatersrand Basin in South Africa, but many other placer gold provinces are

Photo of gold-quartz vein *(Layne Kennedy/Corbis)*

known from around the world. One of the most active of these, where the lode-gold sources were identified well after the placer layers, is the gold districts of Alaska.

Historical Note on the Placer Gold Deposits of Alaska

The placer gold deposits of Alaska and the Yukon Territories lured tens of thousands of frontiersmen to the wild and dangerous territories of the north in the late 1800s and early 1900s. Life was extremely tough, the rewards few, and dangers many, but some of these frontiersmen did find gold and managed to settle the north.

Gold typically occurs as a native metal and is found in lode deposits or placer deposits. Lode gold includes primary deposits in hard bedrock or in vein systems in the bedrock. In contrast, placer deposits are secondary, concentrated in stream gravels and soils. Gold is chemically unreactive so it persists through weathering and transportation and concentrates in soils and as heavy minerals in stream gravel deposits known as placers. Placer gold was the sought-after treasure in the great gold rushes of the Fairbanks gold district and the Yukon Territories of Canada, where many placer and lode deposits are still being discovered and mined.

The placer gold mining in Alaska was started after George Washington Carmack and his Native American brothers-in-law Skookum Jim and Tagish Charlie discovered rich deposits of placer gold on a tributary of the Klondike River in the Yukon Territory in 1896. Tens of thousands of would-be gold miners rushed to the Klondike in 1897–98, and many more struggled over the treacherous Chilkoot and White Passes in 1898, only to find that all of the streams and rivers in the area had already been claimed. Many of these enterpreneurs continued moving north into the wilderness of Alaska in their quest for gold. Alaska had been purchased from Russia only in 1867 and offered a new frontier for the United States. Gold had been reported from the Russian River on the Kenai Peninsula in 1834, and in 1886 the first major discovery of gold in interior Alaska was reported from the Fortymile River. Other gold deposits were known from Birch Creek, in what is now the Circle mining district.

The miners who left the Klondike district continued down the Yukon River to the coast on the Seward Peninsula and found gold on the beaches at Nome, starting a new gold rush to the coast. The beaches at Nome were also quickly staked and claimed, and many thousands of explorers and potential gold miners stopped between the Yukon and the Nome beaches, searching for the precious metal in the soils and gravels of central Alaska. In 1902 Italian gold prospector Felix Pedro found gold along a tributary to the Tanana River, at the site of what is now the city of Fairbanks. This became the next gold rush area in Alaska, and has led to many years of gold exploitation along the rivers, as well as the establishment of what has become one of Alaska's biggest cities, founded on the concentration of gold in the regolith.

The early placer mining techniques were very labor intensive, with minors digging gravel from the streams, moving it in wheelbarrows, and washing it in sluices and gold pans in search for the metal. Later, near the turn of the century, new techniques were developed in which miners would build fires to melt the permafrost, tunnel to 20 or 30 feet (6–10 m) into the gravel, and excavate huge piles that were later sluiced in the search for gold. Later techniques saw powerful fire hose–like hydraulic nozzles used to thaw and loosen large quantities of gravel for sluicing, then in the 1930s steam-powered shovels and bucketline dredges rapidly increased the pace of mining. Only in the 1980s did environmental concerns stop these mining methods, by which time the entire environment had been stripped bare, then the barren gravels laid back in the stream channel. Now exploration for gold in the soils and regolith must be done under strict guidelines of the U.S. Environmental Protection Agency.

Platinum Group Elements

Some rare metals known as platinum group elements (PGEs) form economic concentrations in some ultramafic igneous rocks and take several forms. Chromite may occur as layers, typically in continental intrusions or as small pods in ultramafic rock associations. Chromite and platinum group elements are typically associated with sulfide minerals and form when there is enough sulfur in the magma to crystallize these phases while the rock is still in liquid forms. In many cases the metal phases are a result of contamination of the magma by melted country rock, particularly in the continental layered intrusions. Examples of large economic deposits of chromite are found in the Bushveld Complex in South Africa, and the Muskox and Stillwater Complexes of North America.

Nickel

Nickel deposits are typically found as concentrations in lateritic soils or in association with sulfide minerals in ultramafic magmatic rocks. Nickel occurs often with platinum group elements and has geochemical affinities that make it occur with sulfide minerals such as pyrite, chalcopyrite, and pyrrhotite in ultamafic rocks such as komatiites.

Other nickel deposits are found in tropical regions, where lateritic weathering leaches away many elements, leaving just the residual material such as nickel that is not soluble. These deposits, called nickel laterite deposits, are found in parts of Africa including Madagascar and in the Caribbean. In most cases the host rock is ultramafic, but in some examples the tropical weathering is so intense that the nickel and associated metals are concentrated from other host rocks. Gold may also be concentrated in some lateritic weathering profiles.

Copper Deposits

Most economic copper deposits are found in association with volcanic-plutonic arc sequences in porphyry copper deposits, but other economic resources of copper are known from sedimentary deposits. In porphyry copper deposits the copper is carried by the sulfide mineral chalcopyrite, which is enriched and carried upward by the granitic magmas and by hydrothermal fluids associated with the plutons. Copper is also often found in association with nickel, gold, lead, and zinc deposits. Copper can also form in deep oceanic settings, when brine fluids from deeply buried sediments discharge and deposit copper, lead, and zinc directly on the seafloor.

Uranium

Uranium ores generally come from granitic sources, where radioactive minerals such as monazites are leached from the granites, carried by acidic solutions, then deposited where the fluids meet neutralizing conditions such as are found in carbon-bearing sediments and, in some cases, along unconformity surfaces. Other uranium ores are found directly in the granitic host rock, such as Australia's Olympic Dam deposit, containing nearly 35 percent of the world's known sources of economically recoverable uranium.

See also ARCHEAN; BLACK SMOKER CHIMNEYS; FLYSCH; GEOCHEMISTRY; GRANITE, GRANITE BATHOLITH; GREENSTONE BELTS; HYDROCARBONS AND FOSSIL FUELS; IGNEOUS ROCKS; METASOMATIC; MINERAL, MINERALOGY; OPHIOLITES; PETROLEUM GEOLOGY; PLATE TECTONICS; PRECAMBRIAN; SOILS.

FURTHER READING

Evans, A. M. *Ore Geology and Industrial Minerals: An Introduction.* Oxford: Blackwell Science, 1993.

Groves, D. I. "The Crustal Continuum Model for Late-Archaean Lode-Gold Deposits of the Yilgran Block, Western Australia." *Mineralium Deposita* 28 (1993): 366–374.

Jensen, Mead LeRoy, and Alan Bateman. *Economic Mineral Deposits.* New York: John Wiley & Sons, 1979.

ecosystem An ecosystem is an ecological unit that encompasses the total aspect of the physical and biological environment of an area and the connections between the various parts. It is an integrated unit consisting of a community of living organisms, affected by various factors such as temperature, humidity, light, soil, food supply, and interactions with other organisms, as well as the nonliving environments, including matter and energy. Changes in any part of an ecosystem are likely to result in changes in the other parts. Relationships between organisms in an ecosystem depend on changes in the energy input and flow and nutrient flux within the system. The term was coined in 1935 by British ecologist Arthur Tansley (1871–1955).

One of the principal ideas of the ecosystem concept is that living organisms interact in complex ways with their local environments, and change in one part of the system can cause changes in another. There is an overall flow of energy in ecosystems that includes exchange of material between living and nonliving parts of the system. In this way all species are ecologically related to each other, as well as with the abiotic constituents of the environment. Ecosystems are similar to biomes, which are climatically and biologically defined areas with a distinctive community of plants, animals, and soil organisms.

CLASSIFICATION OF ECOSYSTEMS

Ecosystems are classified based on ecological criteria as well as general climate and features recognizable from the field and from satellite imagery. Some are based on the seasonality of changes in the systems, such as changes in leaf characteristics, linked together with information on climate, elevation, humidity, and drainage. These criteria have been modified and adopted by 175 countries in the Convention on Biologic Diversity in Rio de Janeiro in June 1972. This convention had three main goals:

- conservation of biodiversity
- sustainable use of its components
- fair and equitable sharing of benefits arising from genetic resources.

Participants in this conference adopted a new, more encompassing definition of ecosystems as a "dynamic complex of plant, animal, and microorganism communities and their nonliving environment interacting as a functional unit."

Following the criteria and goals of the conference, several different ecological classification systems became widely used. The first is physiognomic-ecological classification of plant formations of the Earth, differentiating between the structures and appearance from above ground and underwater plant systems. The second is a land cover classification system (LCSS) developed by the Food and Agriculture Organization, based mainly on satellite-based observations.

An outcome of the increased attention on ecosystems is the definition of ecosystem services, which are fundamental life-support services on which civilization depends. Such services include pollination, flood control, food for cattle in natural grasslands, wood for the timber industry, erosion, nutrient cycling, natural products for the pharmaceutical industry, and bush meat for indigineous populations. More effort has been made in recent years to assign economic value to ecosystem services, which helps preserve these ecosystems by increasing society's awareness of their inherent value. Secondary services derived from natural ecosystems include natural reserves for populations to enjoy nature, water storage and controls, soil protection, and carbon sequestration. All of these can be assigned commercial values and treated as commodities that can be bartered against pressures to develop threatened ecosystems commercially.

Some of the less concrete values of preserving ecosystems come from the preservation of biodiversity, where preserving the natural environment may help individual organisms in the ecosystem be more resilient to change and avoid extinction, and may eventually contribute to the benefit of humans, for instance, by the discovery of new medicines or ecosystems critical to the stability of the planet's climate.

ECOSYSTEM DYNAMICS

Ecosystems work through the exchange of energy and matter between the various biologic and nonbiologic components of the system. Introduction of new components, or the loss of any component, typically disrupts the system, and some changes can be so severe that they cause a cascading effect and the collapse of the entire ecosystem. In other cases the ecosystem can recover from the introduction of a new toxin, predatory species, or other disruptive agent. The ability of the ecosystem to recover depends on the toxicity of the new element and the resiliency of the original ecosystem.

See also BIOSPHERE; CARBON CYCLE; GAIA HYPOTHESIS; GEOCHEMICAL CYCLES.

FURTHER READING

Christopherson, R. W. *Geosystems: An Introduction to Physical Geography.* Upper Saddle River, N.J: Prentice Hall, 1996.

Ecological Society of America. "Ecosystem Services, A Primer. Ecological Society of America, Fact Sheet." Available Online. URL: http://www.actionbioscience.org/environment/esa.html. Accessed January 19, 2009.

United Nations Environment Programme. Convention on Biological Diversity. June 1992. UNEP Document no. Na.92-78.

Einstein, Albert (1879–1959) German, Swiss *Theoretical Physicist* Albert Einstein was one of the most influential physicists of modern times. He is best known for his theory of relativity and for deriving the equivalence between mass and energy expressed as the following:

$$E = mc^2$$

where E equals energy, m is mass, and c is the speed of light. Einstein won the Nobel Prize in physics in 1921 "for his services to theoretical physics, and especially for his discovery of the photoelectric effect." His most significant contributions include his special theory of relativity, which reconciles mechanics and electromagnetism, and the general theory of relativity, which deals with gravitation and applying the ideas of relativity to nonuniform motion. He also pioneered many contributions in cosmology, mechanics, quantum theory, light and radiation, and unified field theory. Albert Einstein authored more than 300 scientific papers and 150 other works.

EARLY LIFE AND FAMILY

Albert Einstein was born on March 14, 1879, to Hermann Einstein and Pauline (Koch) Einstein, a Jewish family in the kingdom of Württemberg in the German Empire. His father was an engineer and salesman, and the family moved to Munich in 1880, where his father and uncle founded an electrical manufacturing company.

One of the defining moments of Albert's childhood was at age five when his father showed him a compass and the young Einstein realized that there must be something in space that caused the compass needle to move, a realization that proved to be a source of inspiration to him in his early years. Albert developed a hobby of building models and mechanical devices and he developed a talent for mathematics at a young age.

When Albert was 10, a medical student, Max Talmud, introduced Albert to textbooks on classical science, mathematics, and philosophy, including Euclid's *Elements* (referred to by Albert as the "holy little geometry book"), which he mastered by age 12 and moved on to infinitesimal calculus. In his early teens, Einstein grew bored with the regimen of the electrical engineering program his father was encouraging him to pursue at the Luitpold Gymnasium School, and he sought more creative learning. In 1894 his father's business failed and the family moved to Milan, Italy, then on to Pavia. During this time, at the age of 15, Einstein wrote his first scientific paper, which he sent to his uncle Casar Koch, entitled "The Investigation of the State of Aether in Magnetic Fields." Einstein's family left him in Munich to complete his schooling, but missing them, he withdrew from school and traveled to Pavia to rejoin them.

Albert Einstein never finished high school in Munich after that event. He tried to get accepted at the Swiss Federal Institute of Technology (ETH) in Zurich, but failed the entrance exam. His family arranged for him to finish school in Aarau, Switzerland, while boarding with Professor Jost Winteler. While living with the Winteler family Einsten fell in love with their daughter Marie and courted her from 1895 to 1901. He finished his schooling at age 17, renounced his German citizenship to avoid military service, and moved to Zurich, where he finally enrolled in the mathematics program at ETH. There he met his future wife, Serbian Mileva Marič, who was enrolled at ETH as the only woman studying mathematics. Einstein graduated in 1900 with a degree in physics, wrote a prestigious paper first clearly explaining capillary forces, and received Swiss citizenship in 1901.

Einstein married Mileva Marič in 1903, after she gave birth to a mentally handicapped daughter in 1902. The fate of their daughter is unknown, and it is likely that she died of scarlet fever in 1903 or was adopted and raised by a friend of the Marič family, Helene Savič. They also had two sons, Hans Albert, born in Bern, Switzerland, on May 14, 1904, and Eduard, born in Munich, Germany, on July 28, 1910. Einstein and Marič lived apart for five years and divorced on February 14, 1919. On June 2, 1919, he remarried, this time to his cousin Elsa, and though they raised Elsa's daughters from a previous marriage, they had no children together.

EARLY CAREER

After Einstein graduated from ETH he could not find a university teaching position, so he took a job in 1902 at the patent office in Bern, where he evaluated patent applications for electromagnetic devices. During this time he formed a discussion group in which he met with friends to discuss science and philosophy. Einstein's job in the patent office may have influenced his later thinking, since many of the patents he handled had to do with the synchronization of electrical and mechanical signals. This experience helped him establish many of the questions about the fundamental nature of light and connections between space and time.

While he was working in the patent office he published four papers in *Annalen der Physik*, the top German physics journal at the time. These papers have become known as the annus mirabilis (extraordinary year) papers because of their far-reaching implications and importance. His first paper in this group was on the particulate nature of light and how the photoelectric effect could be understood as light behaving as discrete quanta of energy. His next annus mirabilis paper was on Brownian motion, explaining the random movement of small objects as being caused by molecular action, supporting the atomic theory. Einstein's third paper in this series was on special relativity, where he showed that the speed of light was independent of the observer's speed or state of motion. His final annus mirabilis paper was perhaps his most famous; in it he derived the equivalence of mass and energy as described by the equation

$$E = mc^2$$

showing that mass could be converted into energy and predicting the future development of nuclear power. Although modern science recognizes these papers as remarkable achievements and some of the most important works in physics of all time, the physics community barely noticed them, and many actually rejected them as nonsense. At the age of 26 Einstein earned a Ph.D. from ETH under the direction of Alfred Kleiner, after submitting his dissertation, "A New Determination of Molecular Dimensions."

Albert Einstein, shown writing an equation for the density of the Milky Way Galaxy at Carnegie Institution, Mount Wilson Observatory headquarters, Pasadena, California, on January 14, 1931 *(AP Images)*

LATER SCIENTIFIC CONTRIBUTIONS

In 1908 Einstein finally received an academic position at the University of Bern, which gave him the title privatdozent (roughly equivalent to a postdoctoral researcher, granted by some European universities for those who hold a Ph.D. and Habilitation and want to pursue an academic career). Einstein published a paper in 1910 on critical opalescence, describing how light is scattered by molecules in the atmosphere, making the sky appear blue. He also worked more on the quantization of light, showing that light and energy quanta must act as independent pointlike particles. He published this in two papers, including one entitled "The Development of Our Views on the Composition and Essence of Radiation." This work led to the idea of the wave-particle dual nature of light in quantum mechanics.

Einstein took a position as associate professor at the University of Zurich in 1911, then moved quickly to a full professorship at the Charles University of Prague, in the present-day Czech Republic. From there he published a new paper on the effects of gravity on light and the gravitation redshift of light, including a test of the model later confirmed during a solar eclipse.

In 1912 Einstein returned as a full professor at ETH, where he worked on gravitational theory, eventually publishing his general theory of relativity in 1915. The basic idea of general relativity is that gravitation is the distortion of space-time by matter, which affects the inertial motion of other matter.

World War I broke out in 1914 and Marič moved to Zurich while Einstein moved to Berlin, where he became a member of the Prussian Academy of Sciences and professor at the Humboldt University of Berlin, and served as director to the Kaiser Wilhelm Institute for Physics from 1914 to 1932. He also accepted a position as an extraordinary professor at Leiden University, and he traveled to Holland regularly to lecture between 1920 and 1930.

In 1917 Einstein published a paper that added the cosmological constant to his theory of general relativity, in an attempt to explain the behavior of the entire universe. Einstein later abandoned this constant, although new observations in the 1990s suggest that he may have been correct. In 1917 different groups of astronomers also began testing Einstein's prediction of the gravitational redshift of light, but all groups claimed to have disproved his theories until 1919, when the team of British astronomer Arthur Eddington confirmed the gravitational deflection of starlight by the Sun during an eclipse, proving Einstein correct. Scientists around the world then recognized the importance of Einstein's work, with British Nobel laureate Paul Dirac claiming Einstein's theory was "the greatest scientific discovery ever made."

Albert Einstein was awarded the 1921 Nobel Prize in physics "for his service to theoretical physics, and especially for his discovery of the law of the photoelectric effect." The prize was awarded specifically for his paper on the photoelectric effect entitled "On a Heuristic Viewpoint Concerning the Production and Transformation of Light." His theory of relatively was also mentioned but said to be controversial. Earlier that year, while in New York, Einstein was quoted as saying that scientific work proceeds best by examining physical reality and searching for underlying axioms that give consistent explanations that apply in all instances and that do not contradict one another.

After receiving the Nobel Prize and his work on general relativity, Einstein focused on unifying the fundamental laws of physics (including those governing electromagnetism and gravity) into a unified field theory. He was never successful at this (nor was anyone else to date), and as time passed he became progressively more isolated in the physics community, arguing publicly with Danish physicist and Nobel laureate Niels Bohr about scientific determinism and whether or not quantum phenomena are inherently probabilistic or not, and even ignoring many major developments such as the discoveries of the strong and weak nuclear forces. Einstein's drive to find a unifying field theory survives in physics as the current search for a grand unifying theory.

In 1933 Adolf Hitler became chancellor of Germany and immediately removed Jews and other politically suspect employees from office, including from university professorships. Einstein had been active in discussions of science and religion and in the proposed establishment of the State of Israel, and was prudent enough to have left Germany in 1932 and taken up residence in the United States. He spent time at the California Institute of Technology in Pasadena, and at the Institute for Advanced Study in Princeton, New Jersey. After his wife, Elsa, died in 1936, he continued at the Institute for Advanced Study and also became active in helping obtain visas for European Jews trying to flee Nazi persecution and genocide, helping to form the International Rescue Committee.

Meanwhile, in Germany, a campaign by German physicists including Philipp Lenard and Johannes Stark was mounted to try to discredit Einstein's work as "Jewish physics," and there were attempts made to claim his work was done instead by Aryan physicists. Einstein was granted U.S. citizenship, then teamed up with the Hungarian Jewish refugee and physicist Leo Szilard to persuade U.S. President Franklin Roosevelt to develop an atomic weapon before the Germans did, and by 1942 this effort developed into the Manhattan Project. By 1945 the United States had developed operational nuclear weapons and used them on the Japanese cities of Hiroshima and Nagasaki in August 1945, killing 220,000 people and leading to the end of World War II. Einstein made public statements that he did not work on the atomic bomb projects, and that he regretted writing the letter to Roosevelt asking that such research be started.

Einstein was taken to Princeton Hospital on April 17, 1955, for internal bleeding caused by a ruptured aortic aneurysm. He brought along a speech he was working on to commemorate the seventh anniversary of the founding of Israel, but he died the next morning at the age of 76. Before Einstein was cremated the hospital pathologist, without permission, removed Einstein's brain, which has been preserved for science, but the doctor who removed the organ was fired for performing the act without permission of the family. His brain was sliced up and pieces given to various researchers, including Dr. Marian Diamond from the University of California, Berkeley. Other pieces were sent to researchers at Princeton University, and McMaster University in Hamilton, Ontario, Canada.

See also ASTRONOMY; ASTROPHYSICS; COSMOLOGY; GENERAL RELATIVITY; GRAVITY, GRAVITY ANOMALY; ORIGIN AND EVOLUTION OF THE UNIVERSE.

FURTHER READING

Einstein, Albert. "Folgerungen aus den Capillaritätserscheinungen (Conclusions Drawn from the Phenomenon of Capillarity)." *Annalen der Physik* 4 (1901): 513.

———. "On a Heuristic Viewpoint Concerning the Production and Transformation of Light." *Annalen der Physik* 17 (1905): 132–148.

———. "A New Determination of Molecular Dimensions." Ph.D. diss. Swiss Federal Institute of Technology (ETH), Zurich, 1905.

———. "On the Motion—Required by the Molecular Kinetic Theory of Heat—of Small Particles Suspended in a Stationary Liquid." *Annalen der Physik* 17 (1905): 549–560.

———. "On the Electrodynamics of Moving Bodies." *Annalen der Physik* 17 (1905): 891–921.

———. "Does the Inertia of a Body Depend Upon Its Energy Content?" *Annalen der Physik* 18 (1905): 639–641.

———. "Kosmologische Betrachtungen zur Allgemeinen Relativitätstheorie (Cosmological Considerations in the General Theory of Relativity)." *Königlich Preussische Akademie der Wissenschaften* (1917).

———. "Fundamental Ideas and Problems of the Theory of Relativity." *Nobel Lectures, Physics 1901–1921.* Amsterdam: Elsevier, July 11, 1923. Available online. URL: http://nobelprize.org/nobel_prizes/physics/laureates/1921/einstein-lecture.pdf. Accessed November 20, 2008.

———. "Die Ursache der Mäanderbildung der Flussläufe und des sogenannten Baerschen Gesetzes." *Die Naturwissenschaften* 14 (1926): 223–224.

———. "On Science and Religion." *Nature* 146 (1940): 605.

———. "On the Generalized Theory of Gravitation." *Scientific American* 182, no. 4 (1950): 13–17.

The Nobel Foundation. "The Nobel Prize in Physics 1921." Available online. URL: http://nobelprize.org/phsics/laureates/1921/. Accessed November 22, 2008.

electromagnetic spectrum   The electromagnetic spectrum refers to the total range of all possible electromagnetic radiation frequencies, ranging from the smallest gamma rays with wavelengths a fraction the size of an atom, through X-rays, ultraviolet rays, visible radiation, infrared rays, microwave, and radio waves. The spectrum of an object refers to the characteristic distribution of radiation that comes from that object. The wavelengths of the radiation range from gamma rays at 10^{-14} meters, a fraction of an atom, to 10^4 meters, or radio waves that can be thousands of km long. The electromagnetic spectrum is open-ended, so in theory the largest wavelengths of radiation approach the size of the universe, and the smallest are the size of a proton (6.3×10^{-34} inches, or 1.6×10^{-35} m).

Electromagnetic energy is characterized by a specific wavelength λ, at which it has an associated frequency f and photon energy E. The electromagnetic spectrum is therefore expressed by the following three equations:

$$\lambda = c/f$$
$$E = hc/\lambda$$
$$E = hf$$

where c = the speed of light (299,792,458 m/sec), and h = Planck's constant, $6.626068\ 96 \times 10^{-34}$ J·s. These equations mean that high-frequency electromagnetic waves have a short wavelength and high energy, whereas low-frequency waves have a long wavelength and low energy.

The wavelength of electromagnetic radiation is expressed using its wavelength in a vacuum. When the radiation travels though any medium, such as air or water, its wavelength is always decreased. This is important since the behavior of electromagnetic radiation depends on its wavelength. Higher-frequency radiation has shorter wavelengths, and lower-frequency radiation has longer wavelengths. Electromagnetic radiation is also associated with a specific amount of energy, and has a dual nature, as both a wave and a particle, as described by quantum mechanics. When electromagnetic radiation interacts with small particles such as molecules and single atoms, the quantum effects become important, and the behavior of the radiation is then best described as the amount of energy that each quantum contains. In terms of energy levels electromagnetic radiation is divided into octaves, much as sound waves are divided this way. The table "Characteristics of Electromagnetic Radiation" compares the different wave and particle energy levels of the electromagnetic spectrum.

TYPES OF ELECTROMAGNETIC RADIATION

The electromagnetic spectrum is usually divided into several different regions according to wavelength, with the most common classification scheme including radio waves, microwaves, terahertz radiation, infrared radiation, visible light, ultraviolet light, X-rays, and gamma rays.

Radio waves have wavelengths from one millimeter to hundreds of meters and frequencies of about 3 Hz to 300 GHz. They are commonly used to transmit data for television, mobile phones, wireless Internet connections, and many other applications.

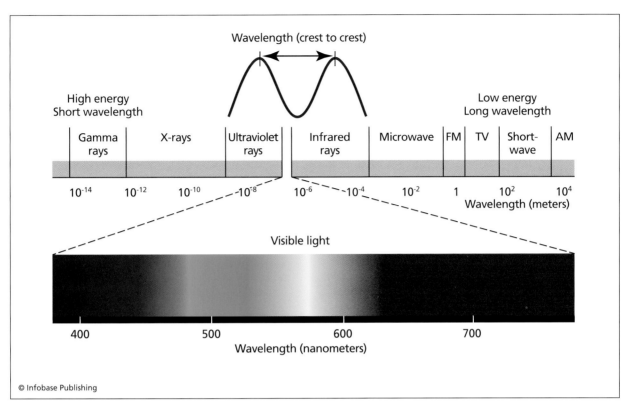

The electromagnetic spectrum

The complex technology to encode radio waves with data involves changing the amplitude and frequency and phase relations of waves within a specific frequency band.

Microwave radiation wavelengths range from one millimeter to one meter and frequencies between 0.3 GHz and 300 GHz. It includes super-high-frequency (SHF) and extremely-high-frequency classes. Microwaves are absorbed by molecules with dipolar covalent bonds, a property used to heat material uniformly, and in rapid amounts of time, in microwave ovens.

Terahertz radiation wavelengths range between the far infrared and microwaves, and frequencies between 300 GHz and 3 terahertz. Radiation in this region can be used for imaging and communications and in electronic warfare to disable electronic equipment.

Infrared radiation has wavelengths between visible light and terahertz radiation, and frequencies of 300 GHz (1 mm) to 400 Terra Hertz (750 nm). Far-infrared radiation (300 GHz to 30 THz) is absorbed by the rotation of many gas molecules, by the molecular motions in liquids, and by phonons (a quantized mode of vibration of a crystal lattice) in solid phases. Most of the far-infrared radiation that enters the Earth's atmosphere is absorbed except for a few wavelength ranges (called windows) where

some energy can penetrate. Mid-infrared radiation has frequencies from 30 to 120 THz, and includes thermal radiation from black bodies (i.e., bodies that absorb all energy at all wavelengths when they are cold). Near-infrared radiation is similar to visible light and has frequencies from 120 to 400 THz.

Higher-frequency radiation (400–790 THz) with wavelengths between 400 and 700 nanometers is detectable by the human eye and is known as visible light. It is also the range of most of the radiation emitted from the Sun and stars. When objects reflect or emit light in the visible range, the human eye and brain process data from these wavelengths into an optical image of the object. The details of how the human brain perceives radiation from these wavelengths and processes it into an image is not completely understood and is actively studied by many molecular biologists, neuroscientists, psychologists, and biophysicists.

Ultraviolet radiation has wavelengths shorter than visible light and longer than X-rays, falling between 400 and 10 nm, and has energies between 3 and 124 electron volts. Ultraviolet radiation is emitted by the Sun and is a highly energetic ionizing radiation that can induce chemical reactions, may cause some substances to glow or fluoresce, and can cause sunburn on human skin. The ultraviolet radiation from the Sun is poisonous to most living organisms but is

CHARACTERISTICS OF ELECTROMAGNETIC RADIATION

Class	Frequency	Wavelength	Energy
Gamma rays	300 EHz	1 pm	1.24 MeV
Hard X-rays	30 EHz 3 EHz	10 pm 100 pm	124 keV 12.4 keV
Soft X-rays	300 PHz 30 PHz	1 nm 10 nm	1.24 keV 124 eV
Extreme ultraviolet	3 PHz	100 nm	12.4 eV
Near ultraviolet	300 THz	1 μm	1.24 eV
Near Infrared	30 THz	10 μm	124 meV
Mid-Infrared	3 THz	100 μm	12.4 meV
Far Infrared	300 GHz	1 mm	1.24 meV
Extremely high frequency	30 GHz	10 mm (1 cm)	124 μeV
Super high frequency	3 GHz	100 mm (1 dm)	12.4 μeV
Ultra high frequency	300 MHz	1 m	1.24 μeV
Very high frequency	30 MHz	10 m	124 neV
High frequency	3 MHz	100 m	12.4 neV
Medium frequency	300 kHz	1 km	1.24 neV
Low frequency	30 kHz	10 km	124 peV
Very low frequency	3 kHz	100 km	12.4 peV
Voice frequency	300 Hz	1 Mm	1.24 peV
Super low frequency	30 Hz	10 Mm	124 feV
Extremely low frequency	3 Hz	100 Mm	12.4 feV

absorbed by the atmospheric ozone layer, preventing significant damage to life on Earth.

X-rays have wavelengths from 10 to 0.01 nanometers with frequencies from 30 petahertz to 30 exahertz (30×10^{15} Hz to 30×10^{18} Hz) and energies from 120 eV to 120 keV. X-rays can "see" through some objects (like flesh) but not others (like bones) and can be used to produce images for diagnostic radiography and crystallography. In the cosmos X-rays are emitted by neutron stars, some nebulae, and the accretion disks around black holes.

Gamma rays are the most energetic photons and have no lower limit to their wavelength. Their frequency is greater than 10^{19} Hz, their energies are more than 100 keV, and their wavelengths are fewer than 10 picometers. Gamma rays are highly energetic and ionizing, so they can cause serious damage to human tissue and are a serious health hazard.

See also ASTRONOMY; ASTROPHYSICS; CORIOLIS EFFECT; COSMIC MICROWAVE BACKGROUND RADIATION; INTERSTELLAR MEDIUM; ORIGIN AND EVOLUTION OF THE UNIVERSE; REMOTE SENSING.

FURTHER READING

Chaisson, Eric, and Steve McMillan. *Astronomy Today*. 6th ed. Upper Saddle River, N.J.: Addison-Wesley, 2007.

Comins, Neil F. *Discovering the Universe*. 8th ed. New York: W. H. Freeman, 2008.

National Aeronautic and Space Administration, Goddard Space Flight Center. "Imagine the Universe! Electromagnetic Spectrum." Available online. URL: http://imagine.gsfc.nasa.gov/docs/science/know_l1/emspectrum.html. Updated August 22, 2008.

Snow, Theodore P. *Essentials of the Dynamic Universe: An Introduction to Astronomy*. 4th ed. St. Paul, Minn.: West, 1991.

El Niño and the Southern Oscillation (ENSO)

El Niño-Southern Oscillation is the name given to one of the better-known variations in global atmospheric circulation patterns. Global oceanic and atmospheric circulation patterns undergo frequent shifts that affect large parts of the globe, particularly those arid and semiarid parts affected by Hadley Cell circulation. Fluctuations in global circulation can account for natural disasters, including the dust bowl days of the 1930s in the midwestern United States. Similar global climate fluctuations may explain the drought, famine, and desertification of parts of the Sahel, and the great famines of Ethiopia and Sudan in the 1970s and 1980s.

The secondary air circulation phenomenon, the El Niño-Southern Oscillation, can also profoundly influence the development of drought conditions and desertification of stressed lands. Hadley cells migrate north and south with summer and winter, shifting the locations of the most intense heating. Several zonal oceanic-atmospheric feedback systems influence global climate, but the most influential is that of the Austral-Asian system. The location of the most intense heating in Austral-Asia in normal Northern Hemisphere summers shifts from equatorial regions to the Indian subcontinent along with the start of the Indian monsoon. Air is drawn onto the subcontinent, where it rises and moves outward to Africa and the central Pacific. In Northern Hemisphere winters the location of this intense heating shifts to Indonesia and Australia, where an intense low-pressure system develops over this mainly maritime region. Air is sucked in, moves upward, and flows back out at tropospheric levels to the east Pacific. High pressure develops off the coast of Peru in both situations, because cold, upwelling water off the coast here causes the air to cool, inducing atmospheric downwelling. The pressure gradient setup causes easterly trade winds to blow from the coast of Peru across the Pacific to the region of heating, causing warm water to pile up in the Coral Sea, off the northeast coast of Australia. This also causes sea level to be slightly depressed off the coast of Peru, and more cold water upwells from below to replace the lost water. This positive feedback mechanism is rather stable—it enhances the global circulation, as more cold water upwelling off Peru induces more atmospheric downwelling, and more warm water piling up in Indonesia and off the coast of Australia causes atmospheric upwelling in this region.

This stable, linked atmospheric and oceanic circulation breaks down and becomes unstable every two to seven years, probably from some inherent chaotic behavior in the system. At these times the Indonesian-Australian heating center migrates eastward, and the buildup of warm water in the western Pacific is no longer held back by winds blowing westward across the Pacific. This causes the elevated warm water mass to collapse and move eastward across the Pacific, where it typically appears off the coast of Peru by the end of December. The ENSO events occur when this warming is particularly strong, with temperatures increasing by 40–43°F (22–24°C) and remaining high for several months. This phenomenon is also associated with a reversal of the atmospheric circulation around the Pacific such that the dry downwelling air is located over Australia and Indonesia, and the warm upwelling air is located over the eastern Pacific and western South America.

The arrival of El Niño is not good news in Peru, since it causes the normally cold upwelling and nutrient-rich water to sink to great depths, and the fish either must migrate to better feeding locations or die. The fishing industry collapses at these times, as does the fertilizer industry that relies on the bird guano normally produced by birds that eat fish and anchovies, which also die during El Niño events. The normally cold dry air is replaced with warm moist air, and the normally dry or desert regions of coastal Peru receive torrential rains with associated floods, landslides, death, and destruction. Shoreline erosion is accelerated in El Niño events, because the warm water mass that moved in from across the Pacific raises sea levels by 4–25 inches (10–60 cm), enough to cause significant damage.

The end of ENSO events also leads to abnormal conditions, in that they seem to turn on the "normal" type of circulation in a much stronger way than is normal. The cold upwelling water returns off Peru with such ferocity that it may move northward, flooding a 1–2° band around the equator in the central Pacific Ocean with water that is as cold as 68°F (20°C). This phenomenon is known as La Niña ("the girl," in Spanish).

The alternation between ENSO, La Niña, and normal ocean-atmospheric circulation has profound effects on global climate and the migration of different climate belts on yearly to decadal timescales and is thought to account for about a third of all the variability in global rainfall. ENSO events may cause flooding in the western Andes and southern California, and a lack of rainfall in other parts of South America, including Venezuela, northeastern Brazil, and southern Peru. It may change the climate, causing droughts in Africa, Indonesia, India, and Australia, and is thought to have caused the failure of the Indian monsoon in 1899, which spread regional famine with the deaths of millions. Recently the seven-year cycle of floods on the Nile has been linked to ENSO events, and famine and desertification in the Sahel, Ethiopia, and Sudan can be attributed to these changes in global circulation as well.

See also ATMOSPHERE; CLIMATE, CLIMATE CHANGE.

FURTHER READING
Ahrens, C. D. *Meteorology Today: An Introduction to Weather, Climate, and the Environment,* 6th ed. Pacific Grove, Calif.: Brooks/Cole, 2000.
Intergovernmental Panel on Climate Change homepage. Available online. URL: http://www.ipcc.ch/index.htm. Accessed January 30, 2008.
Intergovernmental Panel on Climate Change 2007. *Climate Change 2007: The Physical Science Basis. Contributions of Working Group I to the Fourth Assessment Report of the Intergovernmental Panel on Climate Change,* edited by S. Solomon, D. Qin, M. Manning, Z. Chen, M. Marquis, K. B. Averyt, M. Tignor, and H. L. Miller. Cambridge: Cambridge University Press, 2007.
National Oceanographic and Atmospheric Administration, Hazards Research. Available online. URL: http://ngdc.noaa.gov/seg/hazard/tsu.html Accessed January 30, 2008.

energy in the Earth systems *Earth systems* is a term describing all of the interrelated systems on the planet Earth, including those in the geosphere, hydrosphere, biosphere, and atmosphere. Earth systems are driven by energy from internal and external sources. The main external source of energy is the Sun. Internal energy comes from two main sources: the decay of radioactive isotopes and gravitational energy from when the Earth was just forming about 4.5 billion years ago. The outward transfer of heat from these sources inside the Earth is measured as heat flow, and it powers the convection in the Earth's mantle, which in turn drives the motion of the tectonic plates on the planet's surface. Plate tectonics is associated with many of the major Earth surface processes, including volcanic eruptions, earthquakes, and the uplift and erosion of mountain systems. Energy from the Sun heats the Earth's surface and atmosphere and drives the convection in the atmosphere and oceans, producing winds, currents, and storms. Energy transfer from the Sun controls the climate in processes near the Earth's surface, where processes related to internal energy transfer come to the surface. This dynamic transfer of energy in the external energy system includes interaction with the results of internal processes. Cloud formation is influenced by uplifted mountain ranges, their motion is affected by rotation of the Earth, and the energy transfer on the surface of the Earth is dominated by the interaction of systems driven by external energy sources, with such physical features as mountain and volcanic eruptions powered by internal energy sources.

Geological and biological processes on Earth are driven by energy that ultimately comes from either inside the Earth or the Sun. These are intrinsic (or internal) and extrinsic (or external) sources of energy, respectively. Most geological processes, including plate tectonics, the activity of earthquakes, volcanoes, and uplift of mountain ranges, can be attributed to processes associated with the loss of heat from deep in the planet's interior. These intrinsic processes often work together with and interact with extrinsically driven processes, such as rain and water-flow systems that tend to erode the mountains that were uplifted by intrinsic processes.

INTERNAL ENERGY SOURCES: HEAT TRANSFER AND FLOW FROM DEEP IN THE EARTH

In geology crustal heat flow is a measure of the amount of heat energy leaving the Earth from internal energy sources, measured in calories per square centimeter per second. Typical heat-flow values are about 1.5 microcalories per centimeter squared per second, commonly stated as 1.5 heat flow units. Most crustal heat flow is due to heat production in the crust by radioactive decay of uranium, thorium, and potassium. Heat flow shows a linear relationship with heat production in granitic rocks. Some crustal heat flow, however, comes from deeper in the Earth, beneath the crust.

The Earth exhibits a huge variation in temperature, from several thousand degrees in the core to essentially zero degrees Celsius at the surface. The Earth's heat and internal energy were acquired by several mechanisms, including these:

- heat from accretion as potential energy of falling meteorites was converted to heat energy
- heat released during core formation, with gravitational potential energy converted to heat as heavy metallic iron and other elements segregated and sank to form the core soon after accretion
- heat production by decay of radioactive elements
- and heat added by late-impacting meteorites and asteroids, some of which were extremely large in early Earth history

Heat produced by these various mechanisms gradually flows to the surface by conduction, convection, or advection, and accounts for the component of crustal heat flow that comes from deeper than the crust.

Heat flow by conduction involves internal thermal energy flowing from warm to cooler regions, with the heat flux being proportional to the temperature difference, and a proportionality constant k, known as thermal conductivity, related to the material properties. The thermal conductivity of most rocks is low, about one-hundredth that of copper wire.

Advection involves the transfer of heat by the motion of material, such as transport or heat in a magma, in hot water through fractures or pore spaces, and, more important, on a global scale, by the large-scale rising of heated, relatively low-density buoyant material and the complementary sinking of cooled, relatively high-density material in the mantle. The large-scale motion of the mantle, with hot material rising in some places and colder material sinking in other places, is known as convection, an advective heat-transfer mechanism. For convection to occur in the mantle, the buoyancy forces of the heated material must be strong enough to overcome the rock's resistance to flow, known as viscosity. Additionally, the buoyancy forces must overcome the tendency of the rock to lose heat by conduction, since this would cool the rock and decrease its buoyancy. The balance between all of these forces is measured by a quantity called the Raleigh number. Convection in Earth materials occurs above a critical value of the Raleigh number, but below this critical value heat transfer is by conductive processes. Well-developed convection cells in the mantle are very efficient at transporting heat from depth to surface and are the main driving force for plate tectonics.

Heat transfer in the mantle is dominated by convection (advective heat transfer), except in the lower mantle near the boundary with the inner core (the D″ region), along the top of the mantle, and in the crust (in the lithosphere), where conductive and hydrothermal (also advective) processes dominate. The zones where conduction dominates the heat transfer are known as conductive boundary layers, and the lithosphere may be thought of as a convecting, conductively cooling boundary layer.

The main heat-transfer mechanism that takes internal energy from deep in the Earth's mantle to the near-surface region is convection. It is a thermally driven process where heating at depth causes material to expand and become less dense, causing it to rise while being replaced by complementary cool material that sinks. This moves heat from depth to surface in an efficient cycle since the material that rises gives off heat as it rises and cools, and the material that sinks gets heated only to rise again eventually. Convection is the most important mechanism by which the Earth releases heat, with other mechanisms including conduction, radiation, and advection. However, many of these mechanisms work together in the plate tectonic cycle. Mantle convection brings heat from deep in the mantle to the surface, where the heat released forms magmas that generate the oceanic crust. The midocean ridge axis is the site of active hydrothermal circulation and heat loss, forming black smoker chimneys and other vents. As the crust and lithosphere move away from the midocean ridges, they cool by conduction, gradually subsiding (according the square root of their age) from about 1.5–2.5 miles (2.5–4.0 km) below sea level. Heat loss by mantle convection is therefore the main driving mechanism of plate tectonics, and the moving plates can be thought of as the conductively cooling boundary layer for large-scale mantle convection systems.

The heat transferred to the surface by convection is produced by decay of radioactive heat-producing isotopes such as uranium 235, Thorium 232, and Potassium 40, remnant heat from early heat-producing isotopes such as I 129, remnant heat from accretion of the Earth, heat released during core formation, and heat released during impacts of meteorites and asteroids. During the early history of the planet at least part of the mantle was molten, and the Earth has been cooling by convection ever since. Estimating how much the mantle has cooled with time is difficult, but reasonable estimates suggest that the mantle may have been up to a couple of hundred degrees hotter in the earliest Archean.

The rate of mantle convection is dependent on the ability of the material to flow. The resistance to flow is a quantity measured as viscosity, defined as the ratio of shear stress to strain rate. Fluids with high viscosity are more resistant to flow than materials with low viscosity. The present viscosity of the mantle is estimated to be 10^{20}–10^{21} Pascal seconds (Pa/s) in the upper mantle and 10^{21}–10^{23} Pa/s in the lower mantle, viscosities sufficient to allow the mantle to convect and complete an overturn cycle once every 100 million years. The viscosity of the mantle is temperature-dependent, so it is possible that in early Earth history the mantle may have been able to flow and overturn convectively much more quickly, making convection an even more efficient process and speeding the rate of plate tectonic processes.

There is currently on ongoing debate and research relating to the style of mantle convection in the Earth. The relatively heterogeneous upper mantle extends to a depth of 416 miles (670 km), where there is a pronounced increase in seismic velocities. The more homogeneous lower mantle extends to the D″ region at 1,678 miles (2,700 km), marking the transition into the liquid outer core. One school of mantle convection thought suggests that the entire mantle, including both the upper and the lower parts, is convecting as one unit. Another school posits that the mantle convection is divided into two layers, with the lower mantle convecting separately from the upper mantle. A variety of these models, presently held by the majority of geophysicists, is that there is two-layer convection, but that subducting slabs can penetrate the 415-mile (670-km) discontinuity from above, and that mantle plumes that rise from the D″ region can penetrate the 415-mile (670-km) discontinuity from below.

The shapes taken by mantle convection cells include many possible forms reflected to a first order by the distribution of subduction zones and midocean ridge systems. The subduction zones mark regions of downwelling, whereas the ridge system marks broad regions of upwelling. Material is upwelling in a broad cell beneath the Atlantic and Indian Oceans and downwelling in the circum-Pacific subduction zones. There is thought to be a large plumelike "superswell" beneath part of the Pacific that feeds the East Pacific rise. Mantle plumes that originate from the deep mantle punctuate this broad pattern of upper-mantle convection, and their plume tails are distorted by flow in the convecting upper mantle.

The pattern of mantle convection and transfer of internal energy to the surface deep in geological time is uncertain. Some periods such as the Cretaceous seem to have had much more rigorous mantle convection and surface volcanism. More or different types or rates of mantle convection may have helped to allow the early Earth to lose heat more efficiently. Some computer models allow periods of convection dominated by plumes, and others dominated by overturning planiform cells similar to the present Earth. Some models suggest cyclic relationships, with slabs pooling at the 425-mile (670-km) discontinuity, then suddenly all sinking into the lower mantle, causing a huge mantle overturn event. Further research is needed on linking the preserved record of mantle convection in the deformed continents to help interpret the past history of convection.

EXTERNAL ENERGY SOURCES AND VARIATIONS: THE SUN AND CHANGES IN EXTERNAL ENERGY CAUSED BY ORBITAL VARIATIONS

The Sun is the main external contributor of energy to the Earth. The amount of radiation emitted by the Sun is nearly constant on human timescales, but solar emissions vary on 1,500-year timescales. Variations in Earth's orbital parameters around the Sun cause other more significant and systematic changes in the amount of incoming solar radiation. These changes can affect many Earth systems, causing glaciations, global warming, and changes in the patterns of climate and sedimentation. Radiant energy from the Sun drives convection in the atmosphere and oceans, so any changes in the amount of incoming solar radiation affects how these systems work.

Astronomical effects influence the amount of incoming solar radiation; minor variations in the path of the Earth in its orbit around the Sun and the inclination or tilt of its axis cause variations in the amount of solar energy reaching the top of the atmosphere. These variations are thought to be responsible for the advance and retreat of the Northern and Southern Hemisphere ice sheets in the past few million years. In the past 2 million years alone, the ice sheets have advanced and retreated approximately 20 times. The climate record as deduced from ice-core records from Greenland and isotopic tracer studies from deep-ocean, lake, and cave sediments suggest that the ice builds up gradually over periods of about 100,000 years, then retreats rapidly over a period of decades to a few thousand years. These patterns result from the cumulative effects of different astronomical phenomena.

Several orbital variations are involved in changing the amount of incoming solar radiation, or external energy delivered to the Earth. The Earth follows an elliptical orbit around the Sun; the shape of this elliptical orbit is its eccentricity. The eccentricity changes cyclically with a time period of 100,000 years, alternately bringing the Earth closer to and farther from the Sun in summer and winter. This 100,000-year cycle is about the same as the general pattern of glaciers advancing and retreating every 100,000 years in the past 2 million years, suggesting that this is the main cause of variations within the present-day ice age.

The Earth's axis is presently tilting by 23.5°N/S away from the orbital plane, and the tilt varies between 21.5°N/S and 24.5°N/S. The tilt changes by plus or minus 1.5°N/S from a tilt of 23°N/S every 41,000 years. When the tilt is greater, there is greater seasonal variation in temperature.

Wobble of the rotation axis describes a motion much like a top rapidly spinning and rotating with a wobbling motion, such that the direction of tilt toward or away from the Sun changes, even though the amount of tilt stays constant. This wobbling phenomenon is known as precession of the equinoxes; it places different hemispheres closer to the Sun in different seasons. Presently the precession of the equinoxes are such that the Earth is closest to the Sun during the Northern Hemisphere winter. This precession changes with a double cycle, with periodicities of 23,000 years and 19,000 years.

Because each of these astronomical factors acts on different timescales, they interact in a complicated way, known as Milankovitch cycles, after the Yugoslav Milutin Milankovitch who first analyzed them in the 1920s. Understanding these cycles enables one to predict where the Earth's climate is heading, whether the planet heading into a warming or cooling period, and whether civilization needs to plan for sea-level rise, desertification, glaciation, sea-level drops, floods, or droughts.

EXTERNAL ENERGY-DRIVEN PROCESSES IN THE ATMOSPHERE AND OCEANS

The atmosphere constitutes a sphere around the Earth consisting of a mixture of gases held in place by gravity. The atmosphere is divided into several layers, based mainly on the vertical temperature gradients that vary significantly with height. The lower 36,000 feet (11 km) of the atmosphere is the troposphere, where the temperature generally decreases gradually, at about 70°F per mile (21°C per km), with increasing height above the surface. This is because the Sun heats the surface that in turn warms the lower part of the troposphere. External energy received from the Sun drives processes in the atmosphere.

The atmosphere is always moving, because more of the Sun's heat is received per unit area at the equator than at the poles. The heated air expands and rises to where it spreads out, then cools and sinks, and gradually returns to the equator. This pattern of global air circulation forms Hadley cells that mix air between the equator and midlatitudes. Hadley cells are belts of air that encircle the Earth, rising along the equator, dropping moisture as they rise in the Tropics. As the air moves away from the equator at high elevations, it cools, becomes drier, then descends at 15–30°N and S latitude, where it either returns to the equator or moves toward the poles. The locations of the Hadley Cells move north and south annually in response to the changing apparent seasonal movement of the Sun. High-pressure systems form where the air descends, characterized by stable clear skies and intense evaporation, because the air is so dry. Another pair of major global circulation belts is formed as air cools at the poles and spreads toward the equator. Cold polar fronts form where the polar air mass meets the warmer air that has circulated around the Hadley Cells from the Tropics. In the belts between the polar front and the Hadley Cells, strong westerly winds develop. The position of the polar front and extent of the west-moving wind is controlled by the position of the polar jet stream (formed in the upper troposphere), which is partly fixed in place in the Northern Hemisphere by the high Tibetan Plateau and the Rocky Mountains. Dips and bends in the jet stream path are known as Rossby waves; these partly determine the location of high- and low-pressure systems. Rossby waves tend to be semistable in different seasons and have predictable patterns for summer and winter. If the pattern of Rossby waves in the jet stream changes significantly for a season or longer, it may cause storm systems to track to different locations from normal, causing local droughts or floods. Changes in this global circulation may also change the locations of regional downwelling, cold dry air. This can cause long-term drought and desertification. Such changes

may persist for periods of several weeks, months, or years, and may explain several of the severe droughts that have affected Asia, Africa, North America, and elsewhere.

Circulation cells similar to Hadley Cells mix air in middle to high latitudes, and between the poles and high latitudes. The effects of the Earth's rotation modify this simple picture of the atmosphere's circulation. The Coriolis effect describes how any freely moving body in the Northern Hemisphere veers to the right, and toward the left in the Southern Hemisphere. The combination of these effects forms the familiar trade winds, including the easterlies and westerlies, and doldrums.

Like the atmosphere, the ocean is constantly in motion, driven by external energy from the Sun. Ocean currents are defined by the movement paths of water in regular courses, controlled by the wind and thermohaline forces across the ocean basins. Shallow currents are driven primarily by the wind, but are systematically deflected by the Coriolis force to the right of the atmospheric wind directions in the Northern Hemisphere, and to the left of the prevailing winds in the Southern Hemisphere. Shallow water currents therefore tend to be oriented about 45° from the predominant wind directions.

Deep-water currents are driven primarily by thermohaline effects, or the movement of water driven by differences in temperature and salinity. The temperature differences are ultimately controlled by different amounts of solar radiation received by different parts of the global oceans. The Atlantic and Pacific Ocean basins both show a general clockwise rotation in the Northern Hemisphere, and a counterclockwise spin in the Southern Hemisphere, with the strongest currents in the midlatitude sectors. The pattern in the Indian Ocean is broadly similar but seasonally different and more complex because of the effects of the monsoon. Antarctica is bound on all sides by deep water and has a major clockwise current surrounding it, the Antarctic circumpolar current, lying between 40° and 60° south. This strong current moves at 1.6–5 feet per second (0.5–1.5 m/s), and has a couple of major gyres in it at the Ross Ice Shelf and near the Antarctic Peninsula. The Arctic Ocean has a complex pattern, because it is sometimes ice covered and is nearly completely surrounded by land, with only one major entry and escape route east of Greenland, called Fram Strait. Circulation patterns in the Arctic Ocean are dominated by a slow, 0.4–1.6-inch per second (1–4 cm/s) transpolar drift from Siberia to the Fram Strait, and by a thermohaline-induced anticyclonic spin known as the Beaufort gyre that causes ice to pile up on the Greenland and Canadian coasts. Together the two effects in the Arctic Ocean bring numerous icebergs into North Atlantic shipping lanes and send much of the cold deep water around Greenland into the North Atlantic Ocean basin.

See also ATMOSPHERE; BLACK SMOKER CHIMNEYS; CLIMATE; CLIMATE CHANGE; CONVECTION AND THE EARTH'S MANTLE; EARTH; EARTHQUAKES; GEOLOGICAL HAZARDS; HURRICANES; HYDROSPHERE; ICE AGES; MANTLE PLUMES; PLATE TECTONICS; RADIOACTIVE DECAY.

FURTHER READING

Hayes, James D., John Imbrie, and Nicholas J. Shakelton. "Variations in the Earth's Orbit: Pacemaker of the Ice Ages." *Science* 194 (1976): 2,212–2,232.

Schubert, Gerald, Donald L. Turcotte, and Peter Olson. *Mantle Convection in the Earth and Planets.* Cambridge: Cambridge University Press, 2001.

Turcotte, Donald L., and Gerald Schubert. *Geodynamics.* 2nd ed. Cambridge: Cambridge University Press, 2002.

environmental geology Environmental geology is an applied interdisciplinary science focused on describing and understanding human interactions with natural geologic systems. It also includes the field of earth system science, where different systems of the lithosphere, biosphere, hydrosphere, and atmosphere interact, and changes in one system are seen to influence the other systems.

Environmental geology is a diverse field that includes studies of the hydrological system and how humans' use of water resources affects the system. It also encompasses studies of natural resources such as petroleum and other hydrocarbons and how their use affects the natural environment.

Many of the applications of environmental geology have to do with defining and mitigating the effects of exposure to natural hazards, such as floods, earthquakes, volcanoes, tsunamis, landslides, and coastal hazards. In other situations the field of environmental geology is considered more restricted, referring to the environmental issues arising from specific geologic materials such as radon, groundwater contaminants, asbestos, and lead. This discussion focuses on the later aspects of environmental geology, specifically, the processes that concentrate hazardous elements in soils and how these elements make it into homes and human bodies, and harm individuals and entire populations.

HAZARDOUS ELEMENTS, MINERALS, AND MATERIALS

Many of the more than 100 naturally occurring elements are toxic to humans in high doses, and some occur in high concentrations in the soil. The same elements may be beneficial or even necessary in small, dilute doses and pose little or no threat in intermediate concentrations. Most elements show similar toxicity effects on humans, although not all are toxic in high doses. Understanding the effects of trace elements in the environment on human health is the realm of the huge and rapidly growing field of medical geology.

Natural processes in soils in many places on the planet concentrate potentially hazardous geologic materials. The health hazards posed by these elements depend on the way humans interact with their environment, which can vary significantly among different cultures. Primitive cultures that live off the land are more susceptible to hazards and disease associated with contaminated or poor water quality, toxic elements in plants harvested from contaminated soils, and insect- and animal-borne diseases associated with unsanitary environments. In contrast, more developed societies are more likely to be affected by air pollution, different types of water pollution, and indoor pollution such as radon exposure. Some diseases reflect a complex interaction among humans, insects or animals, climate, and the natural concentration of certain elements in the environment. For instance, schistosomiasis-bearing snails are abundant in parts of Africa and Asia where natural waters are rich in calcium derived from soils, but in similar climates in South America, the condition is rare. It is thought that this difference is because the waters in South America are calcium-poor, whereas disease-bearing snails need calcium to build their shells.

All life-forms are composed of a few basic elements, including hydrogen, carbon, nitrogen, oxygen, phosphorus, sulfur, chlorine, sodium, magnesium, potassium, and calcium. Some other elements are important for life, as they play vital roles in controlling how tissues and organs function. Trace element metals are present in very dilute quantities in our bodies; some known to be important for life functions include fluorine, chromium, manganese, iron, cobalt, copper, zinc, selenium, molybdenum, and iodine. Other elements accumulate in tissue as it ages, but their function, and whether they are beneficial or detrimental, is yet to be determined. These age elements include nickel, arsenic, aluminum, and barium.

The distribution of elements in the natural environment is complex and may be changed by many different processes. Geologic processes such as volcanism may concentrate certain elements in some locations even to ore grade or unhealthy levels. When these igneous rocks are weathered, the concentrations of specific elements may be increased or decreased in the soil horizon, depending on the element, climate, and other factors. After this, biological processes may further concentrate elements. Together, leaching and accumulation of elements during soil formation, biological concentration, and many other processes may concentrate or disperse elements that may be harmful to humans.

Some minerals are hazardous when exposed in the natural environment or when extracted in mining operations. In particular, selenium, asbestos, silica, coal dust, and lead can be harmful when inhaled or when present in high concentrations in the environment.

Iodine

Iodine occurs naturally in the geologic environment and is released from rocks by weathering. It is readily soluble in water, so most iodine makes its way to the sea after it is leached from bedrock or soil. A deficiency of iodine in the body can lead to several adverse health effects including thyroid disease and goiter.

There is a strong correlation between the geography of occurrence of thyroid disease and a deficiency of iodine in the environment. Much of the northern half of the conterminous United States has soils low in iodine; this same region yields most of the thyroid disease cases in the United States.

Selenium

Selenium is one of the most toxic elements known in the environment. Like most elements, selenium is needed in small concentrations for normal biological functions. Concentrations of 0.04 to 0.1 parts per million are healthy, but any larger concentration is toxic.

Selenium is produced naturally by volcanic activity, and it is usually ejected as small particles that fall out near volcanoes, causing higher concentrations near volcanic vents. Selenium in natural soils ranges from 0.1 parts per million to more than 12,000 parts in organic-rich soils. Selenium exists in insoluble form in acidic soils and in soluble form in alkaline soils. Biological activity may also concentrate selenium. Some plants take up soluble selenium and concentrate it in their structures. The efficiency of this process depends on the form in which it exists (soluble or insoluble) in the environment. Selenium is also concentrated in human tissue to about 1,000 times the background level in freshwater. It is also concentrated by up to 2,000 times the natural background level in marine fish.

The concentration of selenium in biological material has persisted through geological time; thus many coals and fossil fuels are also rich in selenium. Burning coal releases large amounts of selenium into the atmosphere, and this selenium then rains down on the landscape.

Asbestos

Asbestos was widely used as a flame retardant in buildings through the mid-1970s, and it was present in millions of buildings in the United States. It was also used in vinyl flooring, ceiling tiles, and roofing

Tailings from the Comstock mine of the late 1800s *(Russell Shively, Shutterstock, Inc.)*

LOVE CANAL IS NOT FOR HONEYMOONERS

Love Canal is not a place many people would choose to visit on a honeymoon. Love Canal was a quiet neighborhood in Niagara Falls, New York, that became infamous as one of the most horrific toxic waste dumps in the country. The history of Love Canal began in the 1890s, when entrepreneur William T. Love envisioned building a canal that would connect the two levels of the Niagara River, above and below the falls, for generating electricity and, eventually, as a shipping canal. He dug about a mile (1.6 km) of the canal, with a channel about 15 feet (5 m) wide and 10 feet (3 m) deep, before his scheme failed and the project was abandoned. Eventually his land was sold to the city of Niagara Falls, which used the undeveloped land as a landfill for chemical waste. The canal was thought to be appropriate for this use since the geology consisted of impermeable clay and the area was rural. The area was then acquired by Hooker Chemical, which continued its use as a toxic chemical landfill, dumping more than 22,000 tons of toxic waste into the site from 1942 to 1952, until the canal was full. Then the site was backfilled with four feet (1.2 m) of clay and closed.

As the city of Niagara Falls expanded, land was needed for many purposes, including schools. The local school board attempted to buy the land from Hooker Chemical, but the chemical company initially refused, showing the school board that the site was a toxic waste dump. The board eventually won and purchased the site for one dollar, with a release to Hooker Chemical that the company had explained about the toxic wastes and

would not be liable for deaths or resultant health problems. A school was then built directly on top of the landfill. During construction the contractors broke through the clay seal under the landfill that was intended to prevent leakage of the waste into the local groundwater. Soon after this in 1957, the city constructed sewers for a neighborhood growing around the school, and in doing so broke through the seal of the landfill again, after which chemicals began seeping out of the old canal in more locations. Further construction of roads in the area restricted some of the groundwater flow, and water levels in the old canal rose above ground level, so the site became an elongate pond.

Children in the area began showing health problems, including epilepsy, asthma, and infections, but the source was not known. In 1978 parents in the community united under the leadership of a concerned mother, Lois Gibbs, and discovered that their community was built on top of a huge toxic waste dump. Their complaints of sick children, chemical odors, and strange substances oozing out of the ground were at first ignored by local officials, but were heard in 1979 by the Environmental Protection Agency (EPA). The EPA documented a disturbingly high incidence of miscarriages, nervous disorders, cancers, and strange birth defects. More than half of the children born in Love Canal between 1974 and 1978 were documented as having birth defects, some of which were severe. Many legal and political battles ensued, with the residents unable to sell their homes. Both the city and Hooker

Chemical (by that time a subsidiary of Occidental Petroleum) denied liability, and the health problems persisted. Residents were losing the legal battles against the local government and the chemical company.

On August 7, 1978, President Jimmy Carter declared a federal emergency at Love Canal, and began relocating residents living closest to the canal. Carcinogens such as benzene were discovered in the groundwater around the site, and many residents showed a range of severe health effects, including leukemia. On May 21, 1980, President Carter declared a wider state of emergency and relocated more than 800 families away from the site. This and a similar chemical waste catastrophe at Times Beach, Missouri, led Congress to pass the Comprehensive Environmental Response, Compensation, and Liability Act (CERCLA), commonly known as the Superfund Act. The EPA sued Occidental Petroleum, which paid $129 million in compensation, and a permanent Superfund Act has helped hold many polluters liable for similar negligent acts that have polluted the nation's land and groundwater resources since that time.

FURTHER READING

Finkelman, R. B., H. C. Skinner, G. S. Plumlee, and J. E. Bunnell. "Medical Geology." *Geotimes* (2001): 1–6.

Kusky, T. M. *Landslides: Mass Wasting, Soil, and Mineral Hazards.* New York: Facts On File, 2008.

West, T. R. *Geology Applied to Engineering.* Englewood Cliffs, N.J.: Prentice Hall, 1995.

material. New construction no longer uses asbestos since scientists discovered that it might cause certain types of disease, including asbestosis (pneumoconiosis), a chronic lung disease. Asbestos particles lodge in the lungs, then the lung tissue hardens around the particles, decreasing lung capacity. This decreased lung capacity causes the heart to work harder and

can lead to heart failure and death. Virtually all deaths from asbestosis can be attributed to long-term exposure to asbestos dust in the workplace before environmental regulations governing asbestos were put in place. A less common disease associated with asbestos is mesothelioma, a rare cancer of the lung and stomach linings. Asbestos has become one of the most devastating occupational hazards in U.S. history, costing billions of dollars for cleaning up asbestos in schools, offices, homes, and other buildings. Approximately $3 billion a year is currently spent on asbestos removal in the United States, and many older building still contain large amounts of asbestos in their insulation, panels, and other building materials.

Asbestos is actually a group of six related minerals, all with similar physical and chemical properties. Asbestos includes minerals from the amphibole and serpentine groups that are long and needle-shaped; this makes it easy for them to lodge in the lungs. The Occupational Safety and Health Administration (OSHA) defined asbestos as having dimensions greater than 5 micrometers (0.002 inches) long, with a length-to-width ratio of at least 3:1. The minerals in the amphibole group included in this definition are grunerite (known also as amosite), reibeckite (crocidolite), anthophyllite, tremolite, and actinolite, and the serpentine group mineral that fits the definition is chrysotile. Almost all of the asbestos used in the United States is chrysotile (known as white asbestos), and about 5 percent of the asbestos used was crocidolite (blue asbestos) and amosite (brown asbestos). There is currently considerable debate among geologists, policymakers, and health officials on the relative threats from different kinds of asbestos.

In 1972 OSHA and the U.S. government began regulating the acceptable levels of asbestos fibers in the workplace. The Environmental Protection Agency (EPA) agreed and declared asbestos a Class A carcinogen. The EPA composed the Asbestos Hazard Emergency Response Act, which was signed by President Reagan in 1986. OSHA gradually lowered the acceptable limits from a preregulated estimate of greater than 4,000 fibers per cubic inch (1,600 fibers per cubic centimeter) to four particles per cubic inch (1.6 fibers per cubic centimeter) in 1992. Responding to public fears about asbestosis, Congress passed a law requiring that any asbestos-bearing material that appeared to be visibly deteriorating must be removed and replaced by nonasbestos-bearing material. This remarkable regulation has caused billions of dollars to be spent on asbestos removal, which in many cases may have been unnecessary. Asbestos can be harmful only as an airborne particle, and only long-term exposure to high concentrations leads to disease. In some cases it is estimated that removing the asbestos caused the inside air to become more hazardous than before removal, as the remediation can cause many small particles to become airborne and fall as dust throughout the building.

Asbestos fibers in the environment have led to serious environmental disasters, as the hazards were not appreciated during early mining operations before the late 1960s. One of the worst cases occurred in the town of Wittenoom, Australia. Crocidolite was mined in Wittenoom for 23 years between 1943 and 1966, and the mining was largely unregulated. Asbestos dust filled the air of the mine and the town, and the 20,000 who lived in Wittenoom breathed the fibers daily in high concentrations. More than 10 percent, or 2,300 people, who lived in Wittenoom have since died of asbestosis, and the Australian government has condemned the town and is burying the asbestos in deep pits to rid the environment of the hazard.

In the United States W. R. Grace and Company in Libby, Montana, afflicted hundreds of people with asbestos-related diseases through its mining operations. Vermiculite was mined at Libby from 1963 to 1990 and shipped to Minneapolis to make insulation products, but the vermiculite was mixed with the tremolite (amphibole) variety of asbestos. In 1990 the EPA tested residents of Libby and found that 18 percent who had been there for at least six months had various stages of asbestosis and that 49 percent of the W. R. Grace mine employees had asbestosis. The mine was closed, and Libby is now a Superfund site, where the EPA has determined that toxic wastes were dumped and must be cleaned up. The problem was not limited to Libby, however; 24 workers at the processing plant in Minneapolis have since died from asbestosis, and one resident who lived near the factory also died.

Silica and Coal Dust

Other minerals can be hazardous if made into small airborne particles that can lodge in the lungs. As with asbestos, both silica- and coal-mining operations release large amounts of dust particles into the air, also known, respectively, as quartz dust and coal dust. Workers exposed to these dusts are at risk for diseases broadly similar to asbestosis.

Quartz dust is commonly produced during rock drilling and sandblasting operations. These practices produce airborne particles of various sizes, the largest of which are naturally filtered by hair and mucous membranes during inhalation. Some of the smallest particles can work their way deeply into the lungs, however, and get lodged in the air sacs of the alveoli, where they can do great harm. When small particles get trapped in the air sac, the lungs react by producing fibrolitic nodules and scar tissue around the trapped particles, reducing lung capacity in a disease called silicosis. This disease is easily preventable by simply wearing a respiratory mask when exposed to silica fibers, although this is not yet a common practice.

Coal dust has presented a long-term health problem in the United States and elsewhere, with underground coal miners being at high risk for developing the disease. Mining operations inevitably release fine particles of coal into the air. These particles may lodge in the lungs to cause a myriad of diseases including chronic bronchitis and emphysema, collectively known as black lung disease. The longer a miner works underground, the greater the risk of developing black lung disease. Miners who work underground for fewer than 10 years have about a 10 percent chance of developing these symptoms, whereas miners who have worked underground for more than 40 years have a 60 percent chance of developing black lung disease.

Lead

Lead is a metalliferous element used primarily for pipes, solder, batteries, bullets, pigments, radioactivity shields, and wheel weights. Lead is a known environmental hazard, and ingestion of large amounts can lead to developmental problems in children, including retardation, brain damage, and birth defects. It may also lead to kidney failure, multiple sclerosis, and brain cancer. Some researchers speculate that the fall of the Roman Empire was partly caused by lead poisoning. The Romans drank a lot of wine, and lead was concentrated at several different steps in the process used to make wine then. The upper class also drank from lead cups, and water was pumped into their homes in lead pipes. It is thought that lead poisoning contributed to brain damage, retardation, and the high incidence of birth defects among the Romans. These ideas are supported by the high content of lead measured in the remains of some exhumed Roman citizens. Remarkably, the lead content of ice cores from Greenland representing the Roman Empire period (500 B.C.E–300 C.E.) also preserve about four times the normal level of lead, reflecting the increased mining and use of lead by the Romans.

Lead is present in the natural environment in several different forms. Galena is the most common ore mineral, forming shiny cubes with a silvery "lead" color. Lead is not generally hazardous in its natural mineral form, but it becomes hazardous when mined and released from smelters as particulates, when leached from pipes or other fixtures, or when released into the air from automobile fumes. These processes can lead to high concentrations of native lead in soils, streams, and rivers. Lead may then be taken up by plants or aquatic organisms and thus enter the food chain, where it can do great damage. Lead paint is also a great hazard in many homes in the United States, as lead was used as a paint additive until the 1970s. Paint in many older homes is peeling and ingested by infants, and paint along window frames is turned into airborne dust when windows are opened and closed. Environmental regulations in many states now require the removal of lead paint from homes upon the sale or leasing of properties.

The largest lead smelter in the United States, in Herculaneum, Missouri, brings an example of the legacy of lead mining. Herculaneum is located about 30 miles south of St. Louis, in the heart of the nation's largest lead deposit belt and has been the site of mining operations for generations. The problem in Herculaneum is that the town's smelter releases 34 tons of emissions per year (reduced from 800 tons per year a generation ago), including fine-grained lead dust. This rains down on the local community, and the local street dirt has been tested and found to contain 30 percent lead. Signs on the streets in town warn children not to play in the streets, curbs, or sidewalks, and parents are vigilant in attempting to keep the dust off toys, shoes, and out of the food and water supply. All their efforts were not enough, though, and the State of Missouri has replaced the soil on 535 properties contaminated by lead. Many of the children and adults in the town are suffering the effects of lead poisoning, with retardation, stunted growth, hearing loss, and clusters of brain cancer and multiple sclerosis in town. One-quarter of all the children in the town tested positive for lead poisoning in 2001. Lead contamination had long been suspected in Herculaneum, but it was not until 2002 that the federal government stepped in. In January 2002 the EPA initiated a large-scale relocation program, initially moving 100 families with young children or pregnant women to safer locations. This may be only the beginning of the end, as government officials have been attempting to shut down the Doe Run Lead Smelter and perhaps relocate the 2,800 families remaining in Herculaneum, Missouri.

Radon

Many U.S. homes accumulate radon. Radon is a poisonous gas and a by-product of radioactive decay of the uranium decay series. A heavy gas, radon is a serious indoor hazard in every part of the country. It tends to accumulate in poorly ventilated basements and well-insulated homes built on specific types of soil or bedrock rich in uranium minerals. Radon is known to cause lung cancer; since it is odorless and colorless, it can go unnoticed in homes for years. But the radon hazard is easily mitigated and homes can be made safe once the hazard is identified.

Uranium is a radioactive mineral that spontaneously decays to lighter daughter elements by losing high-energy particles at a predictable rate known as a half-life. The half-life specifically measures how long it takes for half of the original or parent element to decay to the daughter element. Uranium decays to radium through a long series of steps with a cumulative half-life of 4.4 billion years. During these steps intermediate daughter products are produced, and high-energy particles including alpha particles, consisting of two protons and two neutrons, are released. This produces heat. The daughter mineral radium is itself radioactive, and it decays with a half-life of 1,620 years by losing an alpha particle, thus forming the heavy gas radon. Radon escapes from the minerals and ground and makes its way to the atmosphere, where it is dispersed unless it gets trapped in homes. If it gets trapped, it can be inhaled and do damage. Radon is a radioactive gas that decays with a half-life of 3.8 days, producing daughter products of polonium, bismuth, and lead. If this decay occurs while the gas is in someone's lungs, then the solid daughter products become lodged in the lungs. This is how radon damage is initiated. Most of the health risks from radon are associated with the daughter product polonium, which is easily lodged in lung tissue. Polonium is radioactive, and its decay and emission of high-energy particles in the lungs can damage lung tissue, eventually causing lung cancer.

The concentration of radon among geographic regions and in specific places in those regions varies tremendously. There is also a great variation in the concentration of the gas at different levels in the soil, home, and atmosphere. This variation is related to the concentration and type of radioactive elements present at a location. Radioactivity is measured by the picocurie (pCi), which is approximately equal to the amount of radiation produced by the decay of two atoms per minute.

Soils have gases trapped between the individual grains that make up the soil, and these soil gases have typical radon levels of 20 pCi per liter to 100,000 pCi per liter, with most soils in the United States falling in the range of 200–2,000 pCi/L. Radon can also be dissolved in groundwater with typical levels falling between 100–2 million pCi/Liter. Outdoor air typically has 0.1–20 pCi/Liter, and radon inside homes ranges from 1–3,000 pCi/Liter, with 0.2 pCi/Liter being typical.

Formation and Movement of Radon Gas

There are many natural geologic variations that lead to the complex distribution of hazardous radon. One of the main variables controlling radon concentration at any site is the initial concentration of the parent element uranium in the underlying bedrock and soil. If the underlying materials have high concentrations of uranium, it is more likely that homes built in the area may have high concentrations of radon. Most natural geologic materials contain a small amount of uranium, typically about 1–3 parts per million (ppm). The concentration of uranium is typically about the same in soils derived from a rock

as in the original source rock. However, some rock (and soil) types have much higher initial concentrations of uranium, ranging up to and above 100 ppm. Some of the rocks that have the highest uranium content include some granites, some volcanic rocks (especially rhyolites), phosphate-bearing sedimentary rocks, and the metamorphosed equivalents of all of these rocks.

As the uranium in the soil gradually decays, it leaves its daughter product, radium, in concentrations proportional to the initial concentration of uranium. The radium then decays by forcefully ejecting an alpha particle from its nucleus. This ejection is an important step in the formation of radon, since every action has a reaction. In this case the reaction is the recoil of the nucleus of the newly formed radon. Most radon remains trapped in minerals once it forms. But if the decay of radium happens near the surface of a mineral, and if the recoil of the new nucleus of radon is away from the center of the grain, the radon gas may escape the bondage of the mineral. It will then be free to move in the intergranular space between minerals, soil, or cracks in the bedrock, or become absorbed in groundwater between the mineral grains. Less than half (10–50 percent) of the radon produced by decay of radium actually escapes the host mineral. The rest is trapped inside where it eventually decays, leaving the solid daughter products behind as impurities.

Once the radon is free in the open or water-filled pore spaces of the soil or bedrock, it may move rather quickly. The exact rate of movement is critical to whether or not the radon enters homes, because radon does not stay around for long with a half-life of only 3.8 days. The rates at which radon moves through a typical soil depend on how much pore space there is in the soil (or rock), how connected these pore spaces are, and the exact geometry and size of the openings. Radon moves quickly through very porous and permeable soils such as sand and gravel, but moves very slowly through less permeable materials such as clay. Radon also moves very quickly through fractured material, whether bedrock, clay, or concrete.

Considering how the rates of radon movement are influenced by the geometry of pore spaces in a soil or bedrock underlying a home, and how the initial concentration of uranium in the bedrock determines the amount of radon available to move, it becomes apparent that there should be a large variation in the concentration of radon from place to place. Homes built on dry, permeable soils can accumulate radon quickly because it can migrate through the soil quickly. Conversely, homes built on impermeable soils and bedrock are unlikely to concentrate radon beyond their natural background levels.

Radon becomes hazardous when it enters homes and becomes trapped in poorly ventilated or well-insulated areas. Radon moves up through the soil toward places with greater permeability. Home foundations are often built with a porous and permeable gravel envelope surrounding the foundation to allow for water drainage. This also focuses radon movement and brings it close to the foundation, where the radon may enter through small cracks in the concrete, seams, spaces around pipes, sumps, and other openings, as well as through the concrete that may be moderately porous. Most modern homes intake less than 1 percent of their air from the soil. Some homes, however, particularly older homes with cracked or poorly sealed foundations, low air pressure, and other entry points for radon, may intake as much as 20 percent of their internal air from the soil. These homes tend to have the highest concentrations of radon.

Radon can also enter the home and body through groundwater. Homes that rely on well water may be taking in water with high concentrations of dissolved radon. This radon can then be ingested or released from the water by agitation in the home. Radon is released from high-radon water by simple activities such as showers, washing dishes, or running faucets. Radon can also come from municipal water supplies, such as those supplied by small towns that rely on well fields that take the groundwater and distribute it to homes without providing a reservoir for the water to linger in while the radon decays to the atmosphere. Most larger cities, however, rely on reservoirs and surface water supplies, where the radon has had a chance to escape before being used by unsuspecting homeowners.

Radon Hazard Mapping

A greater understanding of the radon hazard risk in an area can be obtained through mapping the potential radon concentrations. This can be done at many scales of observation. Radon concentrations can also be measured locally to determine what kinds of mitigation are necessary to reduce the health risks posed by this poisonous gas.

The broadest sense of risk can be obtained by examining regional geologic maps and determining whether or not an area is located above potential high-uranium-content rocks such as granites, shales, and rhyolites. These maps are available through the U.S. Geological Survey and many state geological surveys. The U.S. Department of Energy has flown airplanes with radiation detectors across the country and produced maps that show the measured surface radioactivity on a regional scale. These maps give a very good indication of the amount of background

uranium concentration in an area and thus are related to the potential risk for radon gas.

More detailed information is needed by local governments, businesses, and homeowners to assess whether or not they need to invest in radon remediation equipment. Geologists and environmental scientists can measure local soil radon gas levels using a variety of techniques, typically involving placing a pipe into the ground and sucking out the soil air for measurement. Other devices may be buried in the soil to measure more passively the formation of the damage produced by alpha particle emission. Via such information, the radon concentrations in certain soil types can be established. This information can be integrated with soil characteristic maps produced by the U.S. Department of Agriculture and by state and county officials to make more regional maps of potential radon hazards and risks.

Most homeowners must resort to private measurements of radon concentrations in their homes by using commercial devices that detect radon or measure the damage from alpha particle emission. The measurement of radon levels in homes has become a standard part of home sales transactions, so more data and awareness of the problem have risen in the past 10 years. The remediation of radon problems in homes or businesses with a radon problem has become relatively simple. An engineer or contractor can be hired simply and cheaply (typically less than $1,000 for an average home) to design and build a ventilation system to remove the harmful radon gas, making the air safe to breathe.

SUMMARY

The formation of soil involves the breakdown of solid rock and the removal of the dissolvable component of the rocks, leaving the residual material behind in the soil. This process concentrates certain elements, and some of these can be harmful to human health. Some of the most hazardous elements common in soils include selenium, arsenic, radon, and lead, while mines may expose workers to other harmful elements such as coal dust, silica dust, and asbestos fibers. A variety of health conditions and ailments around the world, generally among poorer populations, are caused by exposure or ingestion of hazardous elements in the soil. Careful monitoring of the concentrations of these elements in developed nations such as the United States has greatly reduced the health threat from them.

See also BEACHES AND SHORELINES; CLIMATE CHANGE; ECONOMIC GEOLOGY; FLOOD; GEOLOGICAL HAZARDS; GLOBAL WARMING; HURRICANES; HYDROCARBONS AND FOSSIL FUELS; HYDROSPHERE; SEA-LEVEL RISE; SUBSIDENCE.

FURTHER READING

Cothern, C. R., and J. E. Smith, Jr. *Environmental Radon.* New York: Plenum, 1987.

Keller, Edward A. *Environmental Geology.* 8th ed. Upper Saddle River, N.J.: Prentice Hall, 1999.

Raven, Peter, and Linda Berg. *Environment.* New York: John Wiley & Sons, 2008.

Ross, M. *The Health Effects of Mineral Dusts.* Washington, D.C.: Mineralogical Society of America, Reviews in Mineralogy 28, 1993.

United States Environmental Protection Agency. *Consumer's Guide to Radon Reduction: How to Reduce Radon Levels in Your Home.* EPA 402-K92–003, 1992.

United States Environmental Protection Agency and Centers for Disease Control. *A Citizen's Guide to Radon: The Guide to Protecting Yourself and Your Family from Radon.* 2nd ed. EPA 402-K92–001, 1992.

Eocene The Eocene is the middle epoch of the Paleogene (Lower Tertiary) Period and the rock series deposited during this time interval. The Eocene ranges from 57.8 million to 36.6 million years ago, and is divided from base to top into the Ypresian, Lutetian, Bartonian, and Priabonian ages. The Eocene epoch was named by Charles Lyell in 1833 for mollusk-bearing strata in the Paris basin.

The Tethys Ocean was undergoing final closure during the Eocene, and the main Alpine Orogeny occurred near the end of the epoch. Global climates were warm in the Eocene, and shallow-water benthic nummulites and other large foraminifera became abundant. Mammals became the dominant tetrapods on land, and whales were common in the seas about midway through the epoch. Continental flora included broad-leafed forests in low-latitude forests, and conifers at high latitudes.

See also TERTIARY.

eolian Meaning "of the wind" (after Aeolus, Greek god of the winds), eolian refers to sediments deposited by wind. Loess is fine-grained, windblown silt and dust that covers surfaces and forms thick deposits in some parts of the world, such as Shanxi Province in China. Sand dunes and other forms are moved by the wind and form extensive dune terrains, sand sheets, and sand seas in parts of many deserts.

When wind blows across a surface it creates turbulence that exerts a lifting force on loose, unconsolidated sediment. With increasing wind strengths, the air currents are able to lift and transport larger sedimentary grains, which then bump into and dislodge other grains, causing large-scale movement of sediment by the wind in a process called saltation.

When these particles hit surfaces, they may abrade or deflate the surface.

Wind plays a significant role in the evolution of desert landscapes. Wind erodes in two basic ways. Deflation is a process whereby wind picks up and removes material from an area, reducing the land surface. Abrasion occurs when sand and other sizes of particles impact each other. Exposed surfaces in deserts are subjected to frequent abrasion, which is akin to sandblasting.

Yardangs are elongate, streamlined, wind-eroded ridges that resemble an overturned ship's hull sticking out of the water. These unusual features are formed by abrasion, the long-term sandblasting along specific corridors. The sandblasting leaves erosionally resistant ridges but removes the softer material that itself will contribute to sandblasting in the downwind direction, and eventually contribute to the formation of sand, silt, and dust deposits.

Deflation is important on a large scale in places where there is no vegetation, and in some places the wind has excavated large basins known as deflation basins. Deflation basins are common in the United States from Texas to Canada as elongate (several-kilometer-long) depressions, typically only 3–10 feet (1–3 m) deep. In places like the Sahara, however, deflation basins may be as many as several hundred feet deep.

Deflation by wind can move only small particles away from the source since the size of the particle that can be lifted is limited by the strength of the wind, which rarely exceeds a few tens of miles per hour. Deflation leaves behind boulders, cobbles, and other large particles. These get concentrated on the surface of deflation basins and other surfaces in deserts, leaving a surface concentrated in boulders known as desert pavement. Desert pavements are long-term, stable desert surfaces that are not particularly hazardous. The stability of the surface changes when the surface of the desert pavement is broken; for instance, by driving heavy vehicles across the coarse cobbles and pebbles, the gravels get pushed beneath the surface and the underlying sands get exposed to wind action again. Driving across a desert pavement can raise a considerable amount of sand and dust, and if many vehicles drive across the surface, then it can be destroyed, and the whole surface becomes active with sand ripples and dunes forming soon after the surface is disrupted.

Sand dunes are locally important in deserts, and wind is one of the most important processes in shaping deserts worldwide. Shifting sands are one of the most severe geologic hazards of deserts. In many deserts and desert border areas, the sands are moving into inhabited areas, covering farmlands, villages, and other useful land with thick accumulations of sand. This is a global problem, as deserts are currently expanding worldwide.

SAND DUNES

One of the characteristic landforms of deserts is sand dunes, geometrically regular mounds or ridges of sand found in several geological environments, including deserts and beaches. Most people think of deserts as covered with numerous big sand dunes and continual swirling winds of dust storms. Dunes and dust storms are not as common as depicted in popular movies, and rocky deserts are more common than sandy deserts; for instance, only about 20 percent of the Sahara is covered by sand, and the rest is covered by rocky, pebbly, or gravel surfaces. Sand dunes are locally important in deserts, and wind is one of the most significant processes in shaping deserts worldwide. The Institute of Desert Research in Lanzhou, China has recently estimated that in China alone, 950 square miles (2,500 km²) are encroached on by migrating sand dunes from the Gobi Desert each year, costing the country $6.7 billion per year and affecting the lives of 400 million.

Wind moves sand by saltation—an arching path in a series of bounces or jumps. One can see this often by looking close to the surface in dunes on beaches or deserts. Wind sorts different sizes of sedimentary particles, forming elongate small ridges known as sand ripples, similar to ripples found in streams. Sand dunes are larger than ripples (up to 1,500 feet high, or almost 0.5 km), and are composed of mounds or ridges of sand deposited by wind. These may form where an obstacle distorts or obstructs the flow of air, or they may move freely across much of a desert surface. Dunes have many different forms, but all are asymmetrical. They have a gentle slope that faces into the wind and a steep slope that faces away from the wind. Sand particles move by saltation up the windward side, and fall out near the top where the pocket of low-velocity air cannot hold the sand anymore. The sand avalanches, or slips, down the leeward slope, known as the slip face. This keeps the slope at 30°–34°, the angle of repose. The asymmetry of old dunes is used to determine the directions ancient winds blew.

The steady movement of sand from one side of the dune to the other causes the whole dune to migrate slowly downwind (typically about 80–100 feet per year, or 24–30 m/yr), burying houses, farmlands, temples, and towns. Rates of dune migration of up to 350 feet per year (107 m/yr) have been measured in the Western Desert of Egypt and the Ningxia Autonomous Region of China.

Sand rippled by wind in the Qatari Desert *(photoslb, Shutterstock, Inc.)*

A combination of many different factors leads to the formation of very different types of dunes, each with a distinctive shape, potential for movement, and hazards. The main variables that determine a dune's shape are the amount of sand available for transportation, the strength (and directional uniformity) of the wind, and the amount of vegetation that covers the surface. If there is a lot of vegetation and little wind, no dunes will form. In contrast, if there is little vegetation, a lot of sand, and moderate wind strength (conditions that might be found on a beach), then a group of dunes known as transverse dunes form, with their crests aligned at right angles to the dominant wind direction.

Barchan dunes have crescent shapes and horns pointing downwind; they form on flat deserts with steady winds and a limited sand supply. Parabolic dunes have a U-shape with the U facing upwind. These form where there is significant vegetation that pins the tails of migrating transverse dunes, with the dune being warped into a wide U-shape. These dunes look broadly similar to barchans, except the tails point in the opposite direction. They can be distinguished because in both cases, the steep side of the dune points away from the dominant wind direction. Linear dunes are long, straight, and ridge-shaped, elongated parallel to the wind direction. These occur in deserts with little sand supply and strong, slightly variable winds. Star dunes form isolated or irregular hills where the wind directions are irregular.

SAND SEA

Deserts covering vast expanses covered by thick sands, including sand dunes of several types and by an absence of other geographic features, are known as sand seas, or locally as ergs in the North African Sahara. Interdune areas may be covered by relatively flat tabular sand sheets, or even evaporite basins (sabkhas). Sand seas are abundant in parts of the Sahara of North Africa, the Namib of southern Africa, the Rub' al-Khali (Empty Quarter) of Arabia, the Great Sandy Desert of Australia, the Gobi Desert of Asia, and the Nebraska Sand Hills of Nebraska.

Sand seas form where the velocity of the transporting wind decreases, dropping its load. The decreased velocity may be caused by a number of factors, including in topographic lows, or adjacent to topographic barriers, such as mountains that cut across the direction of sand transport. A striking example of this process is found in the Wahiba Sand Sea of Oman. Here the Eastern Hajar Mountains terminate the northward-flowing Wahiba sands, and an intermittent river system at the base of the mountains removes sand that gets close to the mountain front, carrying it to the coast of the Arabian Sea. Longshore transport then carries this sand southward where winds pick it up from beaches and cause it to reenter the Wahiba sand sheet in the south, forming a sort of sand gyre. Sand seas may also form where a large body of water intercepts drifting sand, or where the sand is carried into shifting climate zones where the wind strength decreases.

Surface features in sand seas include bed forms of a variety of scales ranging from several different types of ripples that may be up to an inch (several cm) high, to dunes that are typically up to 300 feet (100 m) tall, to huge bedforms called draa that are giant dunes up to 1,650 feet (500 m) tall, with wavelengths of up to several kilometers. These bedforms are typically superimposed on each other, with dunes migrating over and on top of draa and several different sets of ripples migrating over the dunes. The wind directions inferred from the different sets may also be different, with ripples reflecting the most recent winds, dunes the dominant winds over different seasons, and draa reflecting the very long-term direction of wind in the basin.

See also DESERTS.

FURTHER READING

Abrahams, A. D., and A. J. Parsons. *Geomorphology of Desert Environments.* Norwell, Mass.: Kluwer Academic Publishers for Chapman and Hall, 1994.

Bagnold, R. A. *The Physics of Blown Sand and Desert Dunes.* London: Methuen, 1941.

Blackwell, Major James. *Thunder in the Desert: The Strategy and Tactics of the Persian Gulf War.* New York: Bantam, 1991.

Walker, A. S. *Deserts: Geology and Resources.* Denver: U.S. Department of the Interior, U.S. Geological Survey Publication, 1996.

erosion Erosion encompasses a group of processes that cause Earth material to be loosened, dissolved, abraded, or worn away and moved from one place to another. These processes include weathering, dissolution, corrosion, and transportation. There are two main categories of weathering: physical and chemical processes. Physical processes break down bedrock by mechanical action of agents such as moving water, wind, freeze-thaw cycles, glacial action, forces of crystallization of ice and other minerals, and biological interactions with bedrock such as penetration by roots. Chemical weathering includes the chemical breakdown of bedrock in aqueous solutions. Erosion occurs when the products of weathering are loosened and transported from their origin to another place, most typically by water, wind, or glaciers.

Water is an extremely effective erosional agent, including when it falls as rain and runs across the

Lavaka from severe soil erosion in Ankarafantsika Nature Reserve, Madagascar *(© Pete Oxford/Minden Pictures)*

surface in finger-sized tracks called rivulets, and when it runs in organized streams and rivers. Water begins to erode as soon as raindrops hit a surface—the raindrop impact moves particles of rock and soil, breaking it free from the surface and setting it in motion. During heavy rains, the runoff is divided into overland flow and stream flow. Overland flow is the movement of runoff in broad sheets. Overland flow usually occurs through short distances before it concentrates into discrete channels as streamflow. Erosion performed by overland flow is known as sheet erosion. Streamflow is the flow of surface water in a well-defined channel. Vegetative cover strongly influences the erosive power of overland flow by water. Plants that offer thicker ground cover and have extensive root systems prevent erosion much more than thin plants and crops that leave exposed barren soil between rows of crops. Ground cover between that found in a true desert and in a savanna grassland tends to erode the fastest, while tropical rain forests offer the best land cover to protect from erosion. The leaves and branches break the force of the falling raindrops and the roots form an interlocking network that holds soil in place.

Under normal flow regimes streams attain a kind of equilibrium, eroding material from one bank and depositing it onto another. Small floods may add material to overbank and floodplain areas, typically depositing layers of silt and mud over wide areas. During high-volume floods streams may become highly erosive, even removing entire floodplains that may have taken centuries to accumulate. The most severely erosive floods are found in confined channels with high flow, such as where mountain canyons have formed downstream of many small tributaries that have experienced a large rainfall event. Other severely erosive floods have resulted from dam failures, and in the geological past from the release of large volumes of water from ice-dammed lakes about 12,000 years ago. The erosive power of these floodwaters dramatically increases when they reach a velocity known as supercritical flow, at which time they are able to cut through alluvium like butter and even erode bedrock channels. Supercritical flow cannot be sustained for long periods, as the effect of increasing the channel size causes the flow to self-regulate and become subcritical.

Cavitation in streams can also cause severe erosion. Cavitation occurs when the stream's velocity is so high that the vapor pressure of water is exceeded and bubbles begin to form on rigid surfaces. These bubbles alternately form, then collapse with tremendous pressure, and form an extremely effective erosive agent. Cavitation is visible on some dam spillways, where bubbles form during floods and high-discharge events, but it is different from the more common and significantly less erosive phenomenon of air entrapment by turbulence, which accounts for most air bubbles observed in white-water streams.

Wind is an important but less effective erosional agent than water, particularly in desert or dry environments with exposed soil-poor regolith. Glaciers are powerful agents of erosion, and are thought to have removed hundreds of feet (meters) from the continental surfaces during the last ice ages. These moving masses of ice carve deep valleys into mountain ranges and transport eroded sediments on, within, and in front of glaciers in meltwater stream systems. Glaciers with layers of water along their bases, known as warm-based glaciers, are more effective erosional agents than cold-based glaciers that have no liquid water near their bases. Cold-based glaciers are known from Antarctica.

Mass wasting is considered an erosional process in most definitions, whereas others recognize that mass wasting significantly denudes the surface but classify these sudden events separately. These rapid processes include the transportation of material from one place to another, so they are included here with erosional processes. Most mass wasting processes are related to landslides, debris flows, and rock slides and can significantly reduce the elevation of a region, typically occurring in cycles with intervals ranging from tens to tens of thousands of years.

Humans are drastically altering the planet's landscape, leading to enhanced rates of erosion. Cutting down forests has caused severe soil erosion in Madagascar, South America, the United States, and many other parts of the world. Many other changes are difficult to quantify. Urbanization reduces erosion in some places but enhances it elsewhere. Damming of rivers decreases the local gradient slowing erosion in upland areas, but prevents replenishment of the land in downstream areas. Agriculture and the construction of levees have changed the balance of floodplains. Although difficult to quantify, estimates suggest that human activities in the past couple of centuries have increased erosion rates on average from five times to 100 times previous levels.

See also DESERTS; GLACIER, GLACIAL SYSTEMS; MASS WASTING; WEATHERING.

FURTHER READING

Ritter, D. F., R. C. Kochel, and J. R. Miller. *Process Geomorphology.* 3rd ed. Boston: WCB-McGraw Hill, 1995.
Skinner, Brian J., and Stephen C. Porter. *The Dynamic Earth: An Introduction to Physical Geology.* 5th ed. New York: John Wiley & Sons, 2004.

Eskola, Pentti (1883–1964) *Finnish Geologist* Pentti Eelis Eskola was born in Lellained, Finland, to a farming family. After growing up on the farm he studied at the University of Helsinki, then in 1922 moved to the Carnegie Institution in Washington, D.C., for a postdoctoral position where he conducted experimental studies on the chemical behavior of rock and mineral systems. After two years in Washington Eskola moved back to the University of Helsinki, where he became a chemist before specializing in petrology. He remained a professor for the next 30 years.

Eskola was one of the first geologists to apply physicochemical ideas to the study of metamorphism, and he developed the concept of metamorphic facies. He laid down the foundations for later studies in metamorphic petrology. Throughout his life Eskola was interested in the study of metamorphic rocks, taking an early interest in the Precambrian rocks of Scandinavia and England. Relying heavily on Scandinavian studies, he wanted to define the changing pressure and temperature conditions under which metamorphic rocks were formed. His approach allowed for the comparison of rocks of widely differing compositions in respect of the pressure and temperature under which they had originated.

Pentti Eskola was awarded the Wollaston Medal in 1958, the highest award given by the Geological Society of London, and was honored by many other prizes, including the Penrose and Steinbock medals, and the Vetlesen Prize. Upon his death in 1964 Eskola was given a state funeral.

See also METAMORPHISM AND METAMORPHIC ROCKS; PETROLOGY AND PETROGRAPHY.

FURTHER READING

Eskola, Pentti. "The Mineral Facies of Rocks." *Norsk geologisk tidsskrift* 6 (1920).
———. "Glimpses of the Geology of Finland." *Manchester Geological Association Journal* 2 (1950): 61–79.
———. "The Nature of Metasomatism in the Processes of Granitization." *International Geological Conference, London* (1950): 5–13.

estuary Estuaries are coastal embayments influenced by tides and waves that also have significant freshwater influence derived from a river system that drains into the head of the bay, representing transitional environments between rivers and the sea. Most were formed when sea levels were lower, and rivers carved out deep valleys now flooded by water. They are typically bordered by tidal wetlands and are sensitive ecological zones prone to disturbances by pollution, storms, and overuse by humans. Estuaries accumulate sediment from the river systems from the mainland as well as from the coast, and tend therefore gradually to fill in over time. Each estuary is a unique environment that also preserves a range of water chemistry, reflecting a gradual mixing of the fresh river water from the land and saltwater from the ocean. Saltwater is denser than freshwater, however (having 3.5 percent dissolved salt), and tends in many cases to form a lens that underlies a freshwater cap across the surface of the estuary. The exact nature of the mixing depends on seasonal changes in freshwater influx, basin shape, depth, wave energy, tidal range, and climate. Some estuaries preserve stratified water with saltwater below and freshwater on top, whereas others show complex mixing between the different water types. The biota of estuaries are diverse, reflecting the large range in environments available for different species.

Different parts of estuaries are dominated by river and others by tidal processes. Rivers that enter large estuaries tend to form bayhead deltas that prograde into and may eventually fill the estuary. In other examples, such as Chesapeake Bay, many small rivers may enter the estuary and few have any significant delta, since the river valleys in this system trap most of the stream sediment. Estuaries bordered seaward by barrier islands have much less tidal influence than those with mouths open to the sea, although the tidal range and size and the shape of the estuary also play large roles in determining the strength of tidal versus riverine processes. Tidal-dominated estuaries tend to lack barrier islands at their mouths and exhibit funnel-shaped shorelines that amplify the tides by forcing the incoming tides into progressively more confined spaces. Estuaries that are tidal-dominated with strong tidal currents tend to have well-mixed waters and sandy bottoms, whereas river-dominated estuaries often have stratified water columns and muddy bottoms. Most estuaries exhibit a range of different conditions in different parts—river conditions predominate at the head of the bay, tidal processes dominate at the mouth, and a mixed zone occurs in the middle. An extremely diverse biota inhabits estuaries, with organisms that prefer salty high-energy conditions situated at the mouth of the estuary and freshwater species found near the bayhead. Organisms that tolerate brackish conditions and can adapt to changing salinities and energy are typically found in between in the zone of mixing.

See also BEACHES AND SHORELINES; CONTINENTAL MARGIN; OCEAN BASIN; OCEANOGRAPHY.

FURTHER READING

Davis, R., and D. Fitzgerald. *Beaches and Coasts*. Malden, Mass.: Blackwell, 2004.

Kusky, T. M. *The Coast: Hazardous Interactions within the Coastal Zone*. New York: Facts On File, 2008.

U.S. Environmental Protection Agency. "National Estuary Program." Available online. URL: http://www.epa.gov/nep/about1.htm. Updated January 16, 2008.

U.S. Geological Survey. "Estuaries." Available online. URL: http://www.usgs.gov/science/science.php?term=361&type=feature. Accessed October 10, 2008.

European geology The geology of Europe is dominated by mountains, structures, and basins of the Alpine system in the south, stretching across southern France, northern Italy, then eastward through the Adriatic and Aegean arcs to the Black Sea and finally extending through Turkey to connect with the Himalayan system. This mountain chain formed during the Paleozoic-Cenozoic formation and closure of the Tethys Ocean. The Caspian Sea, on the northern side of the Alpine-Himalayan chain, may represent a piece of oceanic crust trapped during the closure of the Tethys Ocean. Continued closure of the remnants of the Tethys—the Mediterranean—is represented by the active plate margins in southeastern Europe, stretching from the Calabrian arc in Italy, the Hellenic arc in Greece, into the Cyprian arc, which merges with the northern extension of the Dead Sea transform fault. Turkey is moving westward, with the North Anatolian fault as its northern boundary, in response to its escaping from the active collision between Arabia and Asia.

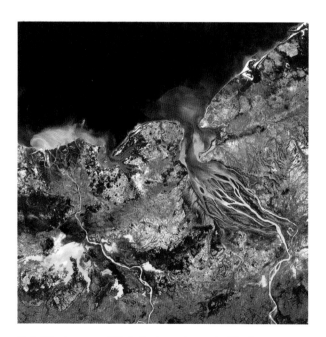

Satellite image of Betsiboka estuary in Madagascar: Waters of estuary run yellow and red from soil washed down by heavy rains. *(M-Sat Ltd./Photo Researchers, Inc.)*

Landsat 5 satellite image of western and central Europe *(M-Sat/Photo Researchers, Inc.)*

The geology of northwestern Europe is delineated largely by the Caledonide system, cutting through the British Isles, well exposed in the Scottish Highlands, and along the northwestern Scandinavian coast. This mountain range formed during the Paleozoic evolution and closure of the Iapetus Ocean. The Baltic shield forms a Precambrian craton in northeastern Europe. It is covered by thick continental and shallow marine deposits, deformed around its edges. The Baltic shield extends through Svalbard Island (Norway) and the Kola Peninsula (Russia) in the north. Geologically the eastern boundary of Europe is considered to be the Ural Mountains, which form a striking north-south line at the edge of the vast East European Plain, which covers deep sections of the Baltic shield and its correlatives.

THE ALPINE MOUNTAIN CHAIN

The Alps form an arcuate or curved mountain system of south-central Europe, about 497 miles (800 km) long and 93 miles (150 km) wide, stretching from the French Riviera on the Mediterranean coast, through southeastern France, Switzerland, southwestern Germany, Austria, and former Yugoslavia (Serbia). The snow line in the Alps is approximately 8,038 feet (2,450 m), with many peaks above this permanently snowcapped or hosting glaciers. The longest glacier in the Alps is the Aletsch, but many landforms attest to a greater extent of glaciation in the Pleistocene. These include famous landforms such as the Matterhorn and other horns, arêtes, U-shaped valleys, erratics, and moraines.

The Alps were formed by plate collisions related to the closure of the Tethys Ocean in the Oligocene and Miocene, but the rocks record a longer history of deformation and events extending back at least into the Mesozoic. Closure of the Tethys Ocean was complex, involving contraction of the older Permian-Triassic Paleo-Tethys Ocean at the same time that a younger arm of the ocean, the Neo-Tethys, was opening in the Triassic and younger times. In the late Triassic carbonate platforms covered older evaporites, and these platforms began foundering and were buried under deep-water pelagic shales and cherts in the early Jurassic. Cretaceous flysch covered convergent margin foreland basins, along with felsic magmatism and high-grade blueschist facies metamorphism. Continent-continent collision-related events dominate the Eocene-Oligocene, with the formation of giant fold nappes, thrusts, and deposition of synorogenic flysch. Late Tertiary events are dominated by late orogenic uplift, erosion, and deposition of post-orogenic molasse in foreland basins. Deformation continues through the present, mostly related to postcollisional extension.

CASPIAN SEA, TRAPPED OCEANIC CRUST

The Caspian is a large, shallow, salty inland sea, located between southern Russia, Kazakhstan, Turkmenistan, Iran, and Azerbaijan. It is 144,444 square miles (373,000 km^2), and its surface rests 92 feet (28 m) below sea level. It has a maximum depth of only 3,280 feet (1,000 m) in the south and is very shallow in the north, with an average depth of only 16.5 feet (5 m). Thus changes in the sea level bring large changes in the position of the shoreline. These historical changes in shoreline position are evident in the lowland continuation of the Caspian depression in the Kalymykiya region to the northwest of the sea. More than 75 percent of the water flowing into the Caspian is from the Volga River in the north, draining the western side of the Urals and the European plains. Other rivers that flow into the Caspian include the Ural, Emba, Kura, and Temek, but there is no outlet. The Caucasus Mountains strike into the sea on the west, and the Elbruz Mountains line its southern border.

The Caspian is mineral-rich and blessed with large oil and gas deposits in several regions, and it is one of the most active exploration areas in the world. The Caspian holds an estimated 200 billion barrels of oil, as much as Iraq and Iran combined. Rich petroleum deposits off the Apseran Peninsula on the west led to the development of Baku, Azerbaijan, where the Nobels (including Alfred Nobel, inventor of dynamite and originator of the Nobel prizes) made their fortune at the end of the 19th century. Unfortunately, decades of careless environmental practices associated with Soviet state-run oil extraction has

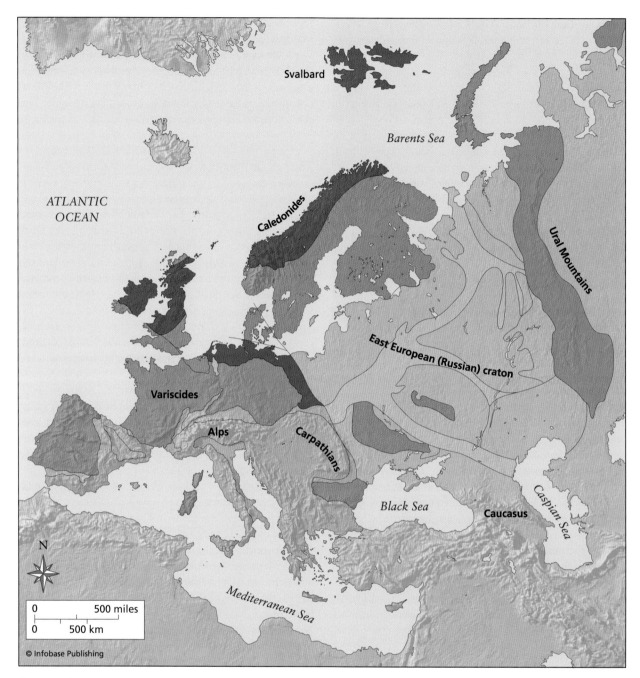

Simplified geological map of Europe showing the location of the Alps, Caledonides, East European (Russian) craton, Urals, Svalbard, and Caspian Sea

led to widespread pollution and contamination, only recently being cleaned up.

The origin of the Caspian depression is somewhat controversial, but many geologists believe that much of the basin is ocean crust trapped during closure of the Tethys Ocean, then deeply buried by sedimentary sequences that host the many petroleum deposits in the area. The sea is also rich in salt deposits and is extensively fished for sturgeon, although the catches have declined dramatically since the early

1990s. The reasons for the fish decline include loss of spawning grounds, extensive poaching, overfishing, and pollution. A single large (typically up to 15 feet) female beluga sturgeon can weigh 1,300 pounds and carry 200 pounds of roe, which retailers can sell as caviar for $250,000 in the United States.

CALEDONIDES

The Caledonides are an early Paleozoic orogenic belt in north and east Greenland, Scandinavia, and the northern British Isles. The Caledonides were continuous with the Appalachian Mountains before the opening of the Atlantic Ocean, together extending more than 4,101 miles (6,600 km). The history of the opening and closing of the Early Paleozoic Iapetus Ocean and the Tornquist Sea is preserved in the Caledonian-Appalachian orogen, which is one of the best-known and studied Paleozoic orogenic belts in the world. The name is derived from the Roman name for the part of the British Isles north of the firths of Clyde and Forth, used in modern times for Scotland and the Scottish Highlands.

The Paleozoic Iapetus Ocean separated Laurentia (proto-North America) from Baltica and Avalonia, and the Tornquist Sea separated Baltica from Avalonia. The eastern margin of Laurentia has Neoproterozoic and Cambrian rift basins overlain by Cambro-Ordovician carbonate platforms, representing a rifting to trailing or passive margin sequence developed as the Iapetus Ocean opened. Similarly Baltica has Neoproterozoic rift basins overlain by Cambro-Ordovician shelf sequences, whereas the Avalonian margin in Germany and Poland records Neoproterozoic volcanism and deformation, overlain by a Cambro-Silurian shelf sequences, with an arc accretion event in the Ordovician. Gondwana sequences include Neoproterozoic orogens overlain by Ordovician shelf rocks deformed in the Devonian and Carboniferous. Significantly, faunal assemblages in Laurentia, Baltica, Gondwana, and Avalonia all show very different assemblages, interpreted to reveal a wide ocean between these regions in the early Paleozoic. Paleomagnetic data support this conclusion. Middle Ordovician ophiolites and flysch basins on Laurentia and Baltica reflect an arc accretion event in the Middle Ordovician, with probable arc polarity reversal leading to volcanism and thin-skinned thrusting preceding ocean closure in the Silurian.

From these and many other detailed studies, a brief tectonic history of the Appalachian-Caledonide orogen is as follows. Rifting of the Late Proterozoic supercontinent Rodinia at 750–600 million years ago led to the formation of rift to passive margin sequences as Gondwana and Baltica drifted away from Laurentia, forming the wide Iapetus Ocean and the Tornquist Sea. Oceanic arcs collided with each other in the Iapetus in the Cambrian and with the margin of Laurentia and Avalon (still attached to Gondwana) in the Ordovician. These collisions formed the well-known Taconic orogeny on Laurentia, ophiolite obduction, and the formation of thick foreland basin sequences. Late Ordovician and Silurian volcanism on Laurentia reflects arc polarity reversal and subduction beneath Laurentia and Gondwana, rifting Avalon from Gondwana and shrinking the Iapetus as ridges were subducted and terranes were transferred

from one margin to another. Avalonia and Baltic collided in the Silurian (430–400 million years ago), and Gondwana collided with Avalon and the southern Appalachians by 300 million years ago, during the Carboniferous Appalachian Orogeny. At this time the southern Rheic Ocean also closed, as preserved in the Variscan Orogen in Europe.

The Caledonides in the Scottish Highlands
Located in the north of the United Kingdom, Scotland includes a variety of generally rugged Lower Ordovician to Archean terranes, dissected by numerous northeast-trending faults that form deep valleys. The coastline of Scotland is highly irregular and has many narrow to wide indented arms of the sea known, respectively, as lochs and firths. The Hebrides, Orkney, and Shetland Islands lie off the coast of northern Scotland. The Southern Uplands form a series of high, rolling, grassy and swampy hills known locally as moors, underlain by a series of strongly folded and faulted Ordovician and Silurian strata. These are separated from the Midland valley by the Southern Uplands fault, an Early Paleozoic strike-slip fault later converted to a normal fault. The Midland Valley includes thick deposits of the Devonian-Carboniferous Old Red Sandstone deposited under continental conditions. The Highland Boundary fault separates the Midlands Valley from the Grampian Highlands, where Precambrian to Early Paleozoic metamorphic and igneous rocks of the Dalradian and Moine Groups are exposed in rugged mountains. The Great Glen fault, a Late Paleozoic left-lateral fault, separates the Grampian Highlands from the Northern Highlands, where Grenvillian-age Moine and Archean Lewisian rocks are exposed. The tallest mountain in Scotland, Ben Nevis (4,406 feet; 1,343 meters) is located in the highlands. The Moine thrust forms the northwestern edge of the Caledonian Orogen, with Archean and Proterozoic rocks of the Lewisian gneisses forming the basement of the orogen.

The oldest rocks exposed in the Scottish Highlands are the Archean (3 billion years old) through Lower Proterozoic Lewisian gneisses formed during the Scourian tectonic cycle, found principally in the Hebrides Islands and the Northern Highlands. The Late Archean gneisses include tonalitic and gabbroic types, with rare ultramafic-mafic plutonic units, probably formed in a volcanic arc setting. Other shallow-shelf metasedimentary rocks including quartzites, limestones, and pelites were metamorphosed to granulite facies at 2.7 billion years ago.

The Invernian tectonic cycle in the Early Proterozoic deformed large tracts of the Scourian gneisses into steep-limbed, west-northwest trending linear structures, accompanied by retrograde amphibolite facies metamorphism. Mafic dike swarms intruded at 2.2 and 1.91 billion years ago, and are not metamorphosed. Post–1.9 billion-year-old Laxfordian cycle events include the formation of shear zones and intrusion of granite plutons near 1.72 billion years ago and a cessation of events by 1.7 billion years ago.

The Moinian Assemblage is a Middle Proterozoic (older than 730 million years, and probably older than 1 billion years) group of pelites and psammites complexly folded into fold interference patterns and metamorphosed to amphibolite facies. Late Proterozoic (970–790 million years old) rocks include two sequences of red beds including the Stoer, Sleat, and Torridon Groups. These groups include conglomerates, siltstones, and sandstones more than a mile (several kilometers) thick in most places, and up to four miles (6 km) thick in a few places. Most of these Late Proterozoic rocks were probably deposited in fluvial or deltaic environments, perhaps in fault-bounded troughs along a continental margin.

The Dalradian Supergroup is found within the Caledonian Orogen south of the Great Glen Fault and north of the Highland Boundary Fault. The Dalradian is more than 12 miles (20 km) thick and is divided into four groups. The lowermost Grampian Group includes shallow to deepwater sandstones and graywackes, overlain by shallow shelf rocks including limestones, shales, and sandstones of the Appin Group. The succeeding Lochaber Group includes sandstones, siltstones, and carbonates deposited in a deltaic environment. The top of the Dalradian consists of the Argyll Group, including a glacial tillite, limestones, and deeper-water graywackes, interbedded with Late Proterozoic (595 million-year-old) basalts. The Dalradian rocks were deformed into large nappe structures in the Late Proterozoic Grampian Orogeny and metamorphosed to the amphibolite facies.

The Paleozoic Era in the Scottish Highlands is marked by a basal transgression of the Durness sequence of shallow-marine quartzites with peculiar trace fossils called skolithos, and limestones onto the Torridonian and Lewisian gneisses. The basal transgressive sequence is about 1,100 feet (350 m) thick, overlain by Lower Cambrian through Ordovician shelf limestones. This sequence is correlated with the basal Cambrian-Ordovician shelf sequence in the Appalachian Mountains, as the Scottish Highlands was linked with Greenland and the Laurentian margin in the Early Paleozoic. However, there is no correlation of rocks of the Scottish Highlands with rocks south of the Highland Boundary Fault, supporting tectonic models that suggest that the southern British Isles were separated from Scotland by a major ocean, known as Iapetus. In the latest Cambrian or Early Ordovician times the region was affected by

main phases of the Caledonian Orogeny, known as the Athollian Orogeny, in the Scottish Highlands. Several generations of folds and regional metamorphism are related to the closure of the Iapetus Ocean along the Highland Boundary Fault, where an oceanic assemblage of cherts, pillow lavas, serpentinites, and Cambro-Ordovician limestones are preserved in a complexly deformed wedge of rock known as a mélange. These events associated with the Early Paleozoic closure of the Iapetus Ocean are correlated with the Taconic and Penobscottian Orogenies in the northern Appalachians. In the Southern Uplands a tectonically complex wedge of imbricated slivers of Ordovician-Silurian deepwater turbidites, shales, and slivers of pillow lavas may represent an oceanic accretionary wedge associated with continued closure of additional segments of the Iapetus Ocean.

The Moine thrust zone in the Northern Highlands formed at the end of the Silurian and places the Caledonian orogenic wedge over the foreland rocks of the Lewisian and Dalradian sequences to the northwest. The Moine thrust is one of the world's classic zones of imbricate thrust tectonics, clearly displaying a sole thrust and imbricate splays, thrust-related folds, klippen, and windows. These structures formed as a Late-Caledonian effect of convergence and shortening between the formerly separated margins of the Iapetus Ocean and placed the orogen wedge allochthonously over basement rocks of the Laurentian margin.

The Old Red Sandstone is a Silurian-Devonian sequence of conglomerates, sandstones, siltstones, shales, and bituminous limestones that is up to 10,000 feet (3,048 m) thick. These rocks represent fluvial-deltaic to lacustrian deposits eroded from the southeast and are interpreted as a molasse sequence representing denudation of the Caledonides. The Old Red Sandstone is loosely correlated with the Devonian Catskill Mountains deltaic complex in the Appalachians, representing erosion of the Appalachian Mountains after the Devonian Acadian Orogeny. Carboniferous deposits in Scotland include shales, coal measures, basalts, and limestones, deposited in deltaic environments mostly in the Midland Valley. Devonian through Carboniferous sinistral strike-slip faults cut many parts of the Scottish Highlands and are associated with Hercynian tectonic events in Europe, and the Acadian-Appalachian Orogenies in the Appalachians.

BALTIC SHIELD

The Baltic (Fennoscandian) shield is an Archean craton divided into three distinct parts. The northern, Lapland-Kola province consists mainly of several previously dispersed Archean crustal terranes that together with the different Paleoproterozoic belts have been involved in a collisional-type orogeny at 2.0 to 1.9 billion years ago. A central, northwest-trending segment known as the Belomorian mobile belt is occupied by assemblages of gneisses and amphibolites. This part of the Baltic shield has experienced two major orogenic periods, in the Neoarchean and Paleoproterozoic. The Neoarchean period included several crust-forming events between 2.9 and 2.7 billion years ago that can be interpreted in terms of first subduction-related and later collisional orogeny. In the end of the Paleoproterozoic at 1.9–1.8 billion years ago, strong structural and thermal reworking occurred during an event of crustal stacking and thrusting referred to as the Svecofennian Orogeny, caused by overthrusting of Lapland granulite belt onto the Belomorian belt. Although the Svecofennian high-grade metamorphism and folding affected all of the belt, its major Neoarchean crustal structure reveals that early thrust and fold nappes developed by 2.74–2.70 billion years ago. In contrast, the Karelian Province displays no isotopic evidence for strong Paleoproterozoic reworking. The Karelian craton forms the core of the shield and largely consists of volcanic and sedimentary rocks (greenstones) and granites/gneisses that formed between 3.2 and 2.6 billion years ago and were metamorphosed at low-grade. Local synformal patches of Paleoproterozoic 2.45 to 1.9 billion-year-old volcano-sedimentary rocks unconformably overlie the Karelian basement. To the southwest of the Archean Karelian craton the Svecofennian domain represents a large portion of Paleoproterozoic crust developed between 2.0 and 1.75 billion years ago.

Although tectonic settings of the Karelian Archean greenstone belts are still a matter of debate, there are some indications that subduction-accretion processes similar to modern-day convergent margins operated at least since 2.9 billion years ago. But a large involvement of deep mantle–plume derived oceanic plateaus in Archean crustal growth processes remains questionable in respect to subduction style.

Baltic Shield on the Kola Peninsula

The Kola Peninsula occupies 50,000 square miles (129,500 km²) in northwestern Russia as an eastern extension of the Scandinavian peninsula, on the shores of the Barents Sea, east of Finland and north of the White Sea. Most of the peninsula lies north of the Arctic Circle. The peninsula is characterized by tundra in the northeast, and taiga forest in the southwest. Winters are atypically warm and snowy for such a northern latitude because of nearby warm Atlantic Ocean waters, and warm summers are filled with long daylight hours.

The Kola Peninsula is part of the Archean Baltic shield, containing medium to high-grade mafic and granitic gneisses including diorite, tonalite, trondhjemite, granodiorite, and granite. Metasedimentary schist, metapelitic gneiss, quartzite, and banded-iron formation known as the Keivy Assemblage form linear outcrop belts in the eastern part of the Kola Peninsula. Mafic/ultramafic greenstone belts and several generations of intrusions are found on the peninsula; these may correlate with ophiolitic rocks of the North Karelian greenstone belts farther south in the Baltic shield. Metamorphism is mostly at amphibolite facies but locally reaches granulite facies, and deformation is complex with abundant fold interference patterns and early isoclinal folds possibly associated with early thrust faults. The Kola schist belts are intruded by several generations of mafic to granitic intrusions.

Baltic Shield and Caledonides on Svalbard and Spitzbergen Island

Spitzbergen is the largest island (15,000 square miles; 40,000 km²) of Svalbard, a large island territory of Norway located in the Arctic Ocean. The islands are on the Barents Shelf, bounded by the Greenland Sea on the west and the Arctic Ocean on the north. The entire Svalbard archipelago was originally referred to as Spitzbergen, but in 1940 the name was changed to Svalbard, and "Spitzbergen" was reserved for the largest island of the archipelago that also includes the islands of Nordaustlandet, Edgeoya, Barentsoya, Prins Karls Forland, and many smaller islands. About half of the island of Spitzbergen is covered by permanent ice and glaciers, and many deeply incised fjords rise to a level of about 3,200 feet (1,000 m), the present height of a flat erosion surface known as a peneplain that has rebounded since the Cenozoic. Since the entire archipelago lies so far north between 76°–81°N, the Sun remains above the horizon from late April through late August but remains below the horizon in winter months. The warm Gulf Stream current has a moderating effect on the climate.

The Svalbard archipelago is well exposed and preserves a complex history of Archean and younger events. The island chain is broken into three main terranes separated by north-south striking faults. The eastern terrane has a basement of Archean through Proterozoic gneisses and amphibolites overlain by psammitic and pelitic schists and marbles that are approximately 1,750 million years old, related to the Baltic shield. These are overlain by pelites, psammites, and felsic volcanics that are about 970 million years old, overlain by 900–800 million-year-old quartzites, silts, and limestones. A Vendian group of pelites and glacial tillites formed during the Varanger glaciation. These are overlain by Cambro-Ordovician carbonates, correlated with similar rocks of eastern Greenland. Mid-Paleozoic tectonism is related to the closure of the Iapetus Ocean during the Caledonian Orogeny, known locally as the Friesland Orogeny. West-vergent fold and thrust structures formed in the Middle and Late Ordovician, whereas late tectonic batholiths intruded in the Silurian through Early Devonian. North-south–striking mylonite zones are concentrated on the western side of the terrane and indicate sinistral transpressive strains.

The central terrane contains a basement of mainly Proterozoic and possible Archean igneous gneisses, overlain by dolostones and Varanger tillites, overlain by Ediacarian phyllites. These are followed by Cambro-Ordovician carbonates. Devonian strata on Svalbard are exposed only in the central terrane and include Old Red Sandstone facies dated by identification of fossil fish remains, similar to that of Scotland. These beds are associated with sinistral transpressive tectonics with the opening of pull-apart basins and the deposition of conglomerates, sandstones, and shales in fluvial systems in these basins. Devonian and Mesozoic strata are folded and show eastward vergence.

The western terrane has a gneissic Proterozoic basement, overlain by Varanger tillites interbedded with mafic volcanics, and overlain by Ediacarian fauna. It is thought that this terrane correlates more with sequences on Ellsemere Island than in the rest of Svalbard or Greenland, so it was probably brought in later by strike-slip faulting. Deformation in Early Ordovician times in the western terrane is linked with subduction tectonics, which may have continued to the Late Ordovician. Later deformation occurred in the Devonian, possibly associated with the Ellsemerian Orogeny.

Some models for the tectonic evolution of Svalbard invoke more than 600 miles (1,000 km) of sinistral strike-slip displacements in the Silurian-Late Devonian on the north-south faults, bringing the eastern terrane into juxtaposition with central Greenland. This motion is associated with the formation of the pull-apart basins filled by the Old Red Sandstone in the central terrane.

In the Carboniferous through Early Eocene most of Svalbard was relatively stable and experienced platform sedimentation, continuous with that of northern Greenland and the Sverdrup basin of northern Canada. Early Carboniferous anhydrites, breccias, conglomerates, sabkha deposits, and carbonates form the basal 3,000 feet (1,000 m) of the section, and these grade up into 1,500 feet (450 m) of fine-grained siliciclastic rocks, cherts, and glauconititic sandstones. Mesozoic strata include more than 8,000 feet (2,500 m) of interbedded deltaic and marine deposits. A Late Cretaceous period of nondeposition was followed by the deposition of nearly 5,000 feet (1,500 m) of deltaic sandstones, shales, and marine beds in the Paleocene and Early Eocene.

In the Eocene western Spitzbergen collided in a dextral transpressional event with the northeast margin of Greenland, forming folds, thrusts, and, later, normal faults along the western coast of the island. Small pull-apart basins formed during this event and are filled by sediments derived from the contemporaneous uplifted fold belt. Erosion and peneplaination in the Oligocene through Holocene formed the flat surface evident across the archipelago today, with Quaternary glaciations depressing the crust. Postglacial rebound, plus thermal uplift associated with the opening of the Arctic Ocean, and the Norwegian and Greenland basins has uplifted the land surface in recent times. Quaternary flood basalts in the northern part of Svalbard are associated with these extensional basin-forming events.

See also ACCRETIONARY WEDGE; CAMBRIAN; CARBONIFEROUS; CONTINENTAL MARGIN; CONVERGENT PLATE MARGIN PROCESSES; DEFORMATION OF ROCKS; DIVERGENT PLATE MARGIN PROCESSES; METAMORPHISM AND METAMORPHIC ROCKS; OPHIOLITES; ORDOVICIAN; OROGENY; PALEOZOIC; PLATE TECTONICS; PRECAMBRIAN; SILURIAN; STRUCTURAL GEOLOGY; SUBDUCTION, SUBDUCTION ZONE; TRANSFORM PLATE MARGIN PROCESSES.

FURTHER READING

Blundell D., R. Freeman, and S. Mueller, eds. *A Continent Revealed: The European Geotraverse.* Cambridge: European Science Foundation, 1992.

Coward, M., and D. Dietrich., "Alpine Tectonics: An Overview." In: *Alpine Tectonics,* edited by M. P. Coward, D. Dietrich, and R. G. Park. London Geological Society Special Publication 45 (1989): 1–29.

Craig, G. Y. *Geology of Scotland,* 3rd ed. Bath, U.K.: Geological Society of London, 1991.

Dewey, J. F., M. L. Helman, E. Turco, D. H. W. Hutton, and S. D. Knott. "Kinematics of the Western Mediterranean." In *Alpine Tectonics,* edited by M. P. Coward, D. Dietrich, and R. G. Park, London Geological Society Special Publication 45 (1989): 265–283.

Laubscher, H. P., and D. Bernoulli. "History and Deformation in the Alps." In *Mountain Building Processes,* edited by K. J. Hsu, 169–180. London: Academic Press, 1982.

Park, R. G., and John Tarney. *Evolution of the Lewisian and Comparable Precambrian High Grade Terranes.* Oxford: Blackwell Scientific Publications, Geological Society of London Special Publication 27, 1987.

Shchipansky, Andrey, Andrei V. Samsonov, E. V. Bibikova, Irena I. Babarina, Alexander N. Konilov, K. A. Krylov, Aleksandr I. Slabunov, and M. M. Bogina. "2.8 Ga Boninite-Hosting Partial Suprasubduction Zone Ophiolite Sequences from the North Karelian Greenstone Belt, NE Baltic Shield, Russia." In *Archean Ophiolites and Related Rocks: Developments in Precambrian*

Geology, edited by Tim Kusky, 425–486, Amsterdam: Elsevier, 2004.

Windley, B. F. *The Evolving Continents*. New York; John Wiley & Sons, 1995.

Ziegler, P. A. *Evolution of the Arctic-North Atlantic and the Western Tethys*, Tulsa, Okla.: American Association of Petroleum Geologists Memoir 43, 1988.

evolution The fossil record indicates that organisms have changed through time, and these changes are best explained by the theory of evolution. Organic evolution is the cumulative and irreversible change of organisms through time, and results of this process explain the distribution and diversity of life throughout Earth history.

Species are a group of interbreeding populations reproductively isolated from other groups. An adaptation of a species is a change that occurs in one species to make the organism better able to cope with its environment. Genetic mutations can result in positive or negative changes to species, but negative changes tend to cause extinction, whereas positive changes tend to help species survive. The theory of evolution explains how these changes occur, the role of the environment, and how these changes help the organism to survive. To understand evolution and the history of theories of evolution, it is necessary to also understand the differences between ontogeny, the development of an individual organism from young to old age (as from a tadpole to a frog), and phylogeny, the evolutionary history of an organism.

In his 1809 treatise *Philosophie Zoologique* Jean-Baptiste Lamarck (1744–1829), a French naturalist, soldier, and academic, proposed a theory of evolution, stating that the fundamental aspect of nature is change, and "life is a stream of gradual complication." Lamarck was one of the first scientists to suggest that the environment can give rise to change in animals, driving evolution. He suggested that if the environment changes, this would force habits of organisms to change, and if these habits persist they would give rise to new characteristics. The most famous example he used to illustrate his case was the giraffe, which, he suggested, evolved from a horse, the descendants of whose necks became gradually stretched to reach leaves high on trees. Lamarck believed that the natural system was constantly replenished by spontaneous generation from inorganic material, and that the youngest organisms were the simplest, and the oldest were the most complex. Lamarckian evolution is not believed by scientists today, since no experimental or other evidence was ever found to support this type of cause-and-effect evolution. But Lamarck is remembered for his contribution of the idea of the inheritance of acquired characteristics.

Charles Robert Darwin (1809–82) was an English naturalist widely regarded as the father of the modern theory of evolution. Some of his most important work stems from his voyage on the research vessel HMS *Beagle* from 1831 to 1836, when he traveled around the world studying and collecting plants, animals, rocks and minerals. In his studies of modern and ancient biological assemblages Darwin noted that the number of individuals in a given population remains relatively constant and that predator-prey relationships within the environment ensure that only those individuals best suited for life in that environment survive. Many individuals within a population are born with inherited characteristics that prevent them from surviving in that environment, and they die. Darwin called this process natural selection, and noted that with time the process makes the entire population of that species better suited for its environment. Darwin noted that as the environment changes, "only the fittest will survive." However, as with Lamarckian evolution, the mechanism for Darwinian evolution was lacking, and no explanation for acquiring advantageous adoptions was known.

In 1869 the Austrian scientist and priest Gregor Johan Mendel (1822–84) described a system of genes, or heritable units, by which characteristics are transmitted from parents to offspring. Mendel became known as the father of genetics, and led to the study of heredity by 1910, after a period of 40 years in which the significance of his work was not appreciated. After this delay it was realized that new characteristics, or mutations, arise completely at random, and are related to chemical changes in DNA. These were thought to be caused by one of the three following mechanisms:

- imperfect replication of DNA during cell division
- physical alteration of segments of DNA by twisting, breaking, or reversed strands
- alteration by an external force such as a virus or radiation

Many mutations turn out to be harmful, causing death, while others cause very small, almost imperceptible changes.

The next major step in understanding evolution was made in 1869 by Ernst Haeckel (1834–1919), a German biologist, naturalist, physician, and professor who documented a direct relationship between the development of an embryo and the history of the group to which it belonged. Haeckel suggested that the "ontogeny [development of the individual] is a short history of the phylogeny [history of the race]." Haeckel further postulated that evolution proceeded by adding stages to the end of an individual's life. But scientists now know that evolutionary stages can be added at any time during ontogeny but are more typically added during the early stages.

EVOLUTION IN THE FOSSIL RECORD

The vast expanses of time needed to test models of evolution are provided by the fossil record, which extends back hundreds of millions of years for complex organisms, and billions of years for simple organisms. The first example of evolution described from the geologic record was in 1869, when German geologist Wilhelm Heinrich Waagen (1841–1900), who studied Jurassic ammonites, published his classic *Die Formenreihe des Ammonites subradiatus* (The Sequence of Form of the Ammonite's Subradiatus), where he showed a series of very small changes gradually accumulated to make much larger changes that contributed to a gradual evolution of the species. Also in 1869 British biologist Thomas H. Huxley (1825–95) proposed a model for the linear evolution of horses, suggesting that smaller forms evolved into larger forms, a model that was later proven too simple and incorrect. Huxley proclaimed himself an *agnostic* (meaning that claims of metaphysical relationships and proving the existence of and establishing relationships with God are unattainable ideals), a term that has stuck to this day, and that caused him to have many debates with the religious community.

The period 1870–80 saw intense fossil collecting and description in attempts to document evolution in the fossil record. Many collectors, biologists, and naturalists were attempting to test the idea of phyletic gradualism, that many groups of organisms began as simple, unspecialized forms, and gradu-

ally became more specialized and larger. Huxley's model for the evolution of the horse was a prime example. The fossil record showed the opposite to be true, however, and revealed that most major groups appear suddenly in the record and many are already highly advanced with their first appearance. The model of phyletic gradualism obviously needed to be replaced with a theory that could explain the sudden appearance of many highly specialized species.

Ideas of evolution experienced a new revolution in 1972, with the publication of a landmark paper by American paleontologists Niles Eldredge (b. 1943–) and Stephen Jay Gould (1941–2002) proposing an alternative mechanism for evolution called punctuated equilibrium. The basic idea of this model is that new characteristics (mutations) may be found in small populations of the main group of a species, and these tend to become isolated near the geographic periphery of the main species range. These can eventually become completely isolated and evolve into a new species reproductively isolated from the original group. Yet the chances of survival and of preserving this new group in the geological record are small. In the case that they do survive, environmental conditions may be such that they could (in rare circumstances) be adapted to fill suddenly an ecological niche, especially if the original group becomes extinct or less able to survive in the same ecological niche under the changing environmental conditions.

Understanding evolution requires a systematic method to classify and describe organisms and fossils. All life-forms are classified into the hierarchy of Kingdom-Phyla-Class-Order-Families-Genera-Species. Most organisms are classified based on their morphology, or general physical appearance. For this it is common to search for homologous features, which are the same on two different organisms or species, and suggest that the two had a common ancestor. In most cases organisms diverge in their characteristics with evolution, but in some cases there is a convergence of characteristics between different species with different ancestors, resulting in similar or analogous features. This phenomenon can happen from similar environmental stimuli, such as the development of advantageous wings in birds and in insects.

Cladistics is the hierarchical classification of species based on the evolutionary ancestry of the species, and uses cladograms (family trees) to show the relationships of organisms through the evolutionary chain. If there is information on the cladogram about the time or age of the different species or their branching, then this cladogram becomes a phylogenetic tree. The general principles of cladistics operate

Fossil Neanderthal skull (found at La Ferrassie) and Cro-Magnon skull of similar antiquity *(John Reader/ Photo Researchers, Inc.)*

by comparing specialized characteristics of organisms and placing organisms with the same derived characteristics on the branch (clade).

Studies of phylogenetic trees and fossil assemblages have shown that at certain times in the geological past large numbers of new species and genera have suddenly appeared and rapidly filled ecological niches. These intervals are called adaptive radiations, and are thought to occur in response to rapid changes in external factors such as environmental differences caused by plate tectonic upheavals and supercontinent rearrangements. One famous example of such an adaptive radiation is the Cambrian explosion, during which large numbers of complex new organisms suddenly appeared in the geologic record, filling many ecological niches, soon after the breakup of the supercontinent Gondwana. Adaptive radiations are often terminated by periods of mass extinction.

See also Darwin, Charles; historical geology; mass extinctions; stratigraphy, stratification, cyclothem.

FURTHER READING

Evolution: A Journey Into Where We're From and Where We're Going. A coproduction of the WGBH/NOVA Science Unit and Clear Blue Sky Productions. Available online. URL: http://www.pbs.org/wgbh/evolution/. Accessed October 10, 2008.

Prother, Donald R. *Bringing Fossils to Life: An Introduction to Paleobiology.* New York: W. H. McGraw-Hill, 2004.

Stanley, Steven M. *Earth and Life through Time.* New York: W. H. Freeman, 1986.

flood A flood occurs when too much water is in one place at one time. When rains, heavy snowmelts, or combinations of these events bring more water than normal into populated areas, floods result. Many floods cause significant damage and destruction because over the past couple of centuries many cultures have moved large segments of their populations onto floodplains, the flat areas adjacent to rivers that naturally flood. Ancient cultures used these floods and the rich organic mud that covered the floodplains during the floods as natural fertilizers for farmlands. Now that many towns, cities, and other population centers have been built on floodplains, people regard these natural flood cycles as disasters. In fact, about 9 out of every 10 disaster proclamations by the president of the United States is for flood disasters, typically to provide funds to those who have built on floodplains. Additionally, many types of natural vegetation have been removed from hillsides, particularly in urban areas. This reduces the amount of infiltration of water into the hillsides and increases the amount and rate of surface runoff, amplifying the danger of floods.

Floods come in many forms, including flash floods where huge volumes of water come rushing out of mountain canyons, carrying mud, boulders, and every other kind of debris washing into valleys and lowlands. Some floods are associated with coastal storms that bring high tides into coastal lowlands and back up river systems far across deltas and coastal plains. Many regions experience slowly rising, long-lasting regional floods associated with spring snowmelts and unusually heavy rains that can last for weeks or months. Some high-latitude climate zones also experience floods in association with the spring breakup of ice on rivers. As the ice melts, blocks move downriver, occasionally jamming and forming ice dams that can cause rapidly rising ice-cold floodwater to cover the river floodplains.

RIVER FLOODS AND FLOOD FREQUENCY

Seasonal variations in rainfall cause stream discharge to rise and sometimes overflow the stream's banks. Both stream discharge and velocity increase during floods, so at floods the streams carry larger particles. Many of the most dramatic changes in river channels occur during floods—meander channels may be cut off, new channels may form, natural levees may be breached, and, occasionally, the river may abandon one channel altogether in favor of another. Floods have a probable interval of recurrence for floods of specific magnitude. Small floods occur quite often, typically every year. Larger floods occur less frequently, and the largest floods occur with the longest time interval between them. The time interval between floods of a specific discharge is known as the recurrence interval, and this is commonly cited using statistics for the 50-year flood, 100-year flood, 500-year flood, and so on.

Curves of the discharge versus recurrence interval can be drawn for every stream and river to determine its characteristic flooding frequency. Knowing how likely it is that a flood of a certain height will recur within a certain time frame is important information for everyone living near a stream or on a river floodplain. For instance, if a flood of 150 cubic feet/second (4.25 m³/s) covered a small town with 10 feet (3 m) of water 30 years ago, is it safe to build a new housing development on the floodplain on the outskirts of town? Using the flood frequency curve for that river,

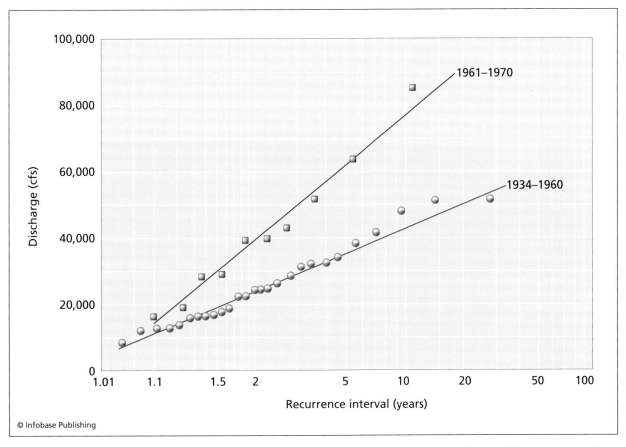

Flood frequency curve for two different periods of different climates, along a river in Africa. The recurrence time (horizontal axis) represents how often a flood with a specific volume of water (vertical axis, showing discharge in cubic feet per second) occurs at a location along a river. Note that the river had a dry climate from 1934 to 1960, and floods were not as large or as frequent as in the wetter period from 1961 to 1970. Changes in climate can severely impact the predicted frequency of floods. *(Data from the Ministry of Water Development, Kenya)*

planners could determine that floods of 150 cubic feet/second (4.25 m³/s) are expected on average every 40 years, and floods of two times that magnitude are expected every 100 years. Planners and insurers might (in the best of situations) conclude from that information that it is unwise to build extensively on the floodplain—the new community should be located on higher ground.

Understanding flood frequency and the chances of floods of specific magnitude occurring along a river is also essential for planning many other human activities. Engineers must determine how much water bridges and drainage pipes must be built to handle and how to plan for land use across the floodplain. In many cases bigger, more expensive bridges should be built, even if it seems unlikely that a small stream will ever rise high enough to justify such a high bridge. In other cases structures are built with a short lifetime of use expected, and planners must calculate whether the likelihood of flood warrants the extra cost of building a flood-resistant structure.

EXAMPLES OF DIFFERENT TYPES OF FLOODS

Floods are the most common natural hazard and have also proven to be the deadliest and costliest of all natural disasters in history. Individual floods have killed upwards of a million people in China on several occasions, and cause billions of dollars of damage annually in different parts of the world. The risk of flooding increases with time as many countries are allowing settlements on floodplains and even encouraging commercial and residential growth on floodplains known to experience floods at frequencies of every several to every couple of hundred years. As world population continues to grow and people move into harm's way on floodplains, this problem will only worsen. Further, as the climate changes, some areas will experience more rainfall while others experience drought, so areas that may be relatively safe on floodplains now may be frequently inundated with floodwaters in the near future. Development of floodplains should not proceed without proper scientific analysis of the risks. In most cases floodplains should be preserved as natural areas or used for farming, but should not be the sites of major commercial or residential development.

Several kinds of floods affect different areas, act on different timescales, and present different types of hazards. Floods associated with hurricanes and tidal surges in coastal areas can cause extreme damage during these coastal storms. Rare, large thunderstorms in mountains and canyon territory typically cause a second type of flood, known as a flash flood. These floods can move into areas as a mud- and debris-laden wall of water, wiping out buildings and towns within a few seconds to minutes. Prolonged rains over large drainage basins cause a third category of floods, called regional floods. A final type of flood occurs in areas where rivers freeze over. In cold climate zones the annual spring breakup can cause severe floods, initiated when blocks of ice jam up behind islands, beneath bridges, or along river bends. These ice dams can create severe floods, causing the high spring waters to rise quickly, bringing the ice-cold waters into low-lying villages on floodplains. When ice dams break up, the force of the rapidly moving ice is sometimes enough to cause severe damage, knocking out bridges, roads, and homes. Ice-dam floods are fairly common in parts of New England, including New Hampshire, Vermont, and Maine. They are also common across much of Alaska and Canada, but the floodplains in these areas tend to be less developed so the floods pose fewer hazards to humans.

COASTAL STORMS AND STORM SURGES

Coastal areas affected by cyclones and hurricanes are prone to flooding by storm surges associated with these storms. Storm surges, formed by water pushed ahead of storms, typically move on land as exceptionally high tides in front of these severe ocean storms. Storm surges are one of the major, most unpredictable hazards to people living along coastlines.

When hurricanes, cyclones, or extratropical lows (also known as coastal storms and northeasters) form, they rotate and the low pressure at the centers of the storms raises the water several to several tens of feet (<1–10 m). This extra water moves ahead of the storms as a storm surge that represents an additional height of water above the normal tidal range. The wind from the storms adds further height to the storm surge, with the total height of the storm surge being determined by the length, duration, and direction of wind, plus how low the pressure gets in the center of the storm. The most destructive storm surges are those that strike low-lying communities at high tide, as the effects of the storm surge and the regular astronomical tides are cumulative.

Like many natural catastrophic events it is possible to predict the statistical probability of a storm surge of a specific height hitting a section of coastline in a specific time interval. If the height of the storm surge is plotted on a semilogarithmic plot, with the height in a linear interval and the frequency (in years) on a logarithmic scale, then a linear slope results. This means that statistically some coastal communities can plan for storm surges of certain height to occur about once in a specified interval, typically calculated as every 50, 100, 300, or 500 years, although there is no way to predict when the actual storm surges will occur. This is a long-term statistical average; one, two, three, or more 500-year events may occur over a relatively short period, but over a long time, the events average out to once every 500 years.

During some hurricanes and coastal storms the greatest destruction and largest number of deaths are associated with inundation by the storm surge. The waters can rise and cover large regions, staying high for many hours during intense storms, drowning victims in low-lying areas, and continuously pounding structures with the waves that move in on top of the storm surges.

Storm Surges and Bangladesh

The area that seems to be hit by the most frequent and most destructive storm surges is Bangladesh. A densely populated, low-lying country, Bangladesh sits mostly at or near sea level between India and Myanmar. It is a delta environment, built where the Ganges and Brahmaputra Rivers drop their sediment eroded from the Himalaya Mountains. Bangladesh is frequently flooded from high river levels, with up to 20 percent of the low-lying country being under water in any year. It also sits directly in the path of many Bay of Bengal tropical cyclones (another name for a hurricane), and has been hit by eight of the 10 most deadly hurricane disasters in world history, including Typhoon Sidr in late 2007.

On November 12 and 13, 1970, a category 3 typhoon known as the Bhola cyclone hit Bangladesh with 115-mile-per-hour (185 km/hr) winds, and a 23-foot (7-m) high storm surge that struck at the astronomically high tides of a full moon. The result was devastating, with about 500,000 human deaths and a similar number of farm animals perishing. The death toll is hard to estimate in this rural region, with estimates ranging from 300,000 to 1 million people lost in this one storm alone. Most perished from flooding associated with the storm surge that covered most of the low-lying deltaic islands on the Ganges River. The most severely hit area was in Tazmuddin Province, where nearly half the population of 167,000 in the city of Thana were killed by the storm surge. Again in 1990 another cyclone hit the same area, this time with a 20-foot (6.1-m) storm surge and 145-mile-per-hour (233.4 km/hr) winds, killing another 140,000 people and another half-million farm animals. In November 2007 Bangladesh was hit by a powerful category 5 cyclone, with 150 mph (242 km/hr) winds and was inundated with a 20-foot (6-m) high storm surge. Since the 1990 storm the area had a better warning system in place, so many more people evacuated low-lying areas before the storm. Still, it is estimated that 5,000–10,000 people perished during Typhoon Sidr, most from the effects of the storm surge.

FLASH FLOODS

Flash floods result from short periods of heavy rainfall and are common near warm oceans, along steep mountain fronts in the path of moist winds, and in areas prone to thunderstorms. They are well known from the mountain and canyon lands of the U.S. desert Southwest and many other parts of the world. Some of the heaviest rainfalls in the United States have occurred along the Balcones escarpment in Texas. Atmospheric instability in this area often forms along the boundary between dry desert air masses to the northwest and warm moist air masses rising up the escarpment from the Gulf of Mexico to the south and east. Up to 20 inches (0.5 m) of rain have fallen along the Balcones escarpment in as few as three hours from this weather situation. The Balcones escarpment also seems to trap tropical hurricane rains, such as those from Hurricane Alice, which dumped more than 40 inches (102 cm) of rain on the escarpment in 1954. The resulting floodwaters were 65 feet (20 m) deep, one of the largest floods ever recorded in Texas. Approximately 25 percent of the catastrophic flash-flooding events in the United States have occurred along the Balcones escarpment. On a slightly longer timescale tropical hurricanes, cyclones, and monsoonal rains may dump several feet of rain over periods of a few days to a few weeks, causing fast, but not quite flash, flooding.

The national record for the highest, single-day rainfall is held by the south Texas region, when Hurricane Claudette dumped 43 inches (1.1 m) of rain on the Houston area in 1979. The region was hit again by devastating floods during June 8–10, 2001, when an early-season tropical storm suddenly grew off the coast of Galveston, dumping 28–35 inches (0.7–0.9 m) of rain on Houston and surrounding regions. The floods were among the worst in Houston's history, leaving 17,000 homeless and 22 dead. More than 30,000 laboratory animals died in local hospital and research labs, and the many university and hospital research labs experienced hundreds of millions of dollars in damage. Fifty million dollars were set aside to buy out the properties of homeowners who had built on particularly hazardous floodplains. Total

damages exceeded $5 billion. The standing water left behind by the floods became breeding grounds for disease-bearing mosquitoes, and the humidity led to a dramatic increase in the release of mold spores, which cause allergies in some people and are toxic to others.

The Cherrapunji region in southern India at the base of the Himalaya Mountains has received the world's highest rainfalls. Moist air masses from the Bay of Bengal move toward Cherrapunji, where they begin to rise over the high Himalayas. This produces a strong orographic effect, where the air mass cannot hold as much moisture as it rises and cools, so heavy rains result. Cherrapunji has received as many as 30 feet (9 m) of rain in a single month (July 1861) and more than 75 feet (23 m) of rain for all of 1861.

Flash floods typically occur in localized areas where mountains cause atmospheric upwelling, leading to the development of huge convective thunderstorms that can pour several inches of rain per hour onto a mountainous terrain, which focuses the water into steep-walled canyons. The result can be frightening, with floodwater thundering down canyons in steep walls that crash into and wash away all in their paths. Flash floods can severely erode the landscape in arid and sparsely vegetated regions but do much less to change the landscape.

Many canyons in mountainous regions have fairly large upriver parts of their drainage basins. Sometimes the storm that produces a flash flood with a wall of water may be located so far away that people in the canyon do not even know it is raining somewhere, or that they are in immediate grave danger. Such was the situation in some of the examples described below.

A number of factors other than the amount of rainfall determine the severity of a flash flood. The shape of the drainage basin is important, because it determines how quickly rainfall from different parts of the basin converges at specific points. The soil moisture and previous rain history are important, as are the amounts of vegetation, urbanization, and slope.

Big Thompson Canyon, Colorado, 1976

Big Thompson Canyon is a popular recreation area about 50 miles (80 km) northwest of Denver, in the Front Ranges of the Rocky Mountains. On July 31, 1976, a large thunderhead cloud had grown over the front ranges, and it suddenly produced a huge cloudburst (rainfall) instead of blowing eastward over the plains as it normally does. Approximately 7.5 inches (0.2 m) of rain fell in a four-hour period, an amount approximately equal to the average yearly rainfall in the area. The steep topography focused the water into Big Thompson Canyon, where a flash flood with a raging 20-foot (6-m) high wall of water rushed through the canyon narrows at 15 mph (24 km), killing 145 people who were driving into or out of the canyon. As the wall of water roared through the canyon, many abandoned their cars and scrambled up the canyon walls to safety, only to watch their cars wash away in the floods. Those who climbed the canyon walls to escape the flash flood survived, but others perished in the flood. In addition to the deaths, this flash flood destroyed 418 homes, wrecked 52 businesses, and washed away 400 cars. Damage totals are estimated at $36 million.

Flash Floods in the Northern Oman Mountains

The Northern Oman (Hajar) Mountains are a steep, rugged mountain range on the northeastern Arabian Peninsula, with deep, long canyons that empty into the Gulf of Oman and the Arabian Sea. These are normally dry canyons or wadis, and the local villagers dig wells in the wadi bottoms to reach the groundwater table for use in homes and agriculture. The region is normally very dry, but infrequent thunderstorms grow and explode over parts of the mountains. Occasionally a typhoon works its way from the Indian Ocean across the Arabian Peninsula and may also dump unusual amounts of rain on the mountains. In either situation the canyons become extremely unsafe, and local villagers have tales of flash floods with hundred-foot (30-m) tall walls of water wiping away entire settlements, leaving only coarse gravel in their place. The inhabitants of this region have learned to build their villages on high escarpments above the wadis, out of reach of the rare but devastating flash flood. Older destroyed villages are visible in some wadi floors, but the wisdom acquired from experiencing a devastating flash flood has encouraged these people to move to higher ground. The inconvenience of being located a hundred feet or more above their water source is avoided by building long aqueduct-like structures known as *falaj* from water sources located at similar elevations far upstream, and letting gravity bring the water to the elevated village. In August 2007 Typhoon Gonu caused severe flooding across northern Oman, with floods ripping through mountain canyons and inundating the streets of Muscat with 5–10 feet (1–3 m) of water. Damage was severe with many homes and shops destroyed, and people needed to be rescued from the desert streets by motorboats.

Flash Floods in the Southern Alps and Algeria, 2000 and 2001

In November 2001 parts of Algeria in North Africa received heavy rainfall over a period of two days that led to the worst flooding and mudslides in the capital city, Algiers, in more than 40 years. An estimated 1,000 died in Algiers, being buried by fast-moving mudflows that swept out of the Atlas Mountains to the south and moved through the city, hitting some of the poorest neighborhoods with the worst flooding. The Bab El Oued District, one of the poorest in Algiers, was hit the worst, where 600 people were buried under mud flows several feet (1 m) thick.

These floods followed similar heavy rains and mudslides that devastated parts of southern Europe in October 2000. Northern Italy and Switzerland were among the worst hit, where water levels reached their highest in 30 years, killing about 50 people. In Switzerland the southern mountain village of Gondo was devastated when a 120-foot (37-m) wide mudflow ripped through the town center, removing 10 homes (one-third of the village) and killing 13 people. Numerous roads, bridges, and railroads were washed away throughout the region, stretching from southern France, through Switzerland and Italy, to the Adriatic Sea. Crops were destroyed on a massive scale. Tens of thousands of people had to be evacuated from the region, and total damage estimates are in the range of many billions of dollars.

REGIONAL FLOOD DISASTERS

Some flooding events are massive, covering with water hundreds of thousands of acres along the entire floodplain of a river system. These floods tend to rise slowly and may have high water for weeks or even months. History has shown that many levees fail during regional long-term flooding events, because most levees are designed to hold back high water for only a short time. The longer the water remains high, the more the water pressure acts on the levee, slowly forcing the water into the pores, letting the water seep into and under it. This often causes levee failure and explains why so many levees fail in long-term regional floods. Long-term flooding events can affect hundreds of thousands or millions of people, and cause widespread disease, famine, loss of jobs, and displacement of populations as a result of the disaster. Floods of this magnitude are among the costliest of all natural disasters.

Mississippi River Basin and the Midwest of the United States

The Mississippi River is the largest river basin in the United States, and the third-largest river basin in the world. It is the site of frequent, sometimes devastating floods. All of the 11 major tributaries of the Mississippi have also experienced major floods, including events that have at least quadrupled the normal river discharge in 1993, 1973, 1927, 1909, 1903, 1892, and 1883. Three of the major rivers (Mississippi, Missouri, and Illinois) meet in St. Louis,

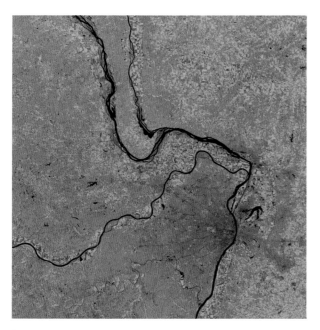

Satellite image of the St. Louis, Missouri, area that shows the Mississippi, Missouri, and Illinois Rivers at normal flow stages *(Image taken August 14, 1991; NASA images created by Jesse Allen, Earth Observatory, using data provided courtesy of Landsat Project Science Office)*

which has seen some of the worst flooding along the entire system.

Floods along the Mississippi in the 1700s and 1800s prompted the formation of the Mississippi River Commission, which oversaw the construction of high levees along much of the length of the river from New Orleans to Iowa. By 1926 more than 1,800 miles (2,896.8 km) of levees had been constructed, many of them higher than 20 feet (6.1 m). The levees imparted a false sense of security against the floodwaters of the mighty Mississippi, and they restricted the channel, causing floods to rise more quickly and forcing the water to flow faster.

Many weeks of rain in late fall 1926 followed by high winter snowmelts in the upper Mississippi River basin caused the river to rise to alarming heights by spring 1927. Worried residents all along the Mississippi strengthened and heightened the levees and dikes along the river, hoping to avert disaster. The crest of water was moving through the upper Midwest and had reached central Mississippi, and the rains continued. In April levees began collapsing along the river, sending torrents of water over thousands of acres of farmland, destroying homes, killing livestock, and leaving 50,000 people homeless. One of the worst-hit areas was Washington County, Mississippi, where an intense late April storm dumped an incredible 15 inches (457.2 cm) of rain in 18 hours, causing additional levees along the river to collapse. One of the

most notable was the collapse of the Mounds Landing levee, which caused a 10-foot-deep lobe of water to cover the Washington County town of Greenville on April 22. The river reached 50 miles in width and had flooded approximately 1 million acres, washing away an estimated 2,200 buildings in Washington County alone. Many perished trying to keep the levees from collapsing and were washed away in the deluge. The floodwaters remained high for more than two months, and people were forced to leave the area (if they could afford to) or to live in refugee camps on the levees, which were crowded and unsanitary. An estimated 1,000 people perished in the floods of 1927, some from the initial deluge, more from famine and disease in the months following the initial inundation by the floodwaters. More than 1 million people were displaced from their homes, and a total of 27,000 square miles (43,450 km², or 16.6 million acres) were flooded. Crop losses amounted to $102.6 million, and 162,000 homes were inundated.

Another wet year along the Mississippi was 1972, with most tributaries and reservoirs filling by the end of summer. The rains continued through winter 1972–73, and the snowpack thickened over the northern part of the Mississippi basin. The combined snowmelts and continued rains caused the river to reach flood levels at St. Louis in early March, before the snow had even finished melting. Heavy rain continued throughout the Mississippi basin, and the

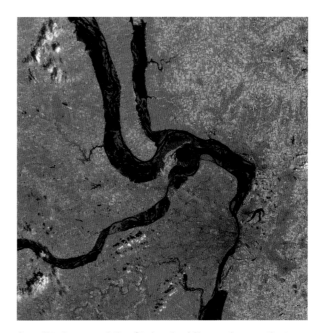

Satellite image of the St. Louis, Missouri, area that shows the Mississippi, Missouri, and Illinois Rivers at stages at the height of the 1993 flood *(Image taken August 19, 1993; NASA images created by Jesse Allen, Earth Observatory, using data provided courtesy of Landsat Project Science Office)*

The Blackfoot River in Montana, showing natural meanders, oxbow lakes, and floodplain *(James Steinberg/ Photo Researchers, Inc.)*

river continued to rise through April and May, spilling into fields and low-lying areas. The Mississippi was so high that it rose to more than 50 feet above its average levels for much of the lower river basin, and these heights caused many of the smaller tributaries to back up until they too reached this height. The floodwaters rose to levels not seen for 200 years. At Baton Rouge, Louisiana, the river nearly broke through its banks and established a new course to the Gulf of Mexico, which would have left New Orleans without a river.

The floodwaters began peaking in late April, causing 30,000 to be evacuated in St. Louis by April 28, and close to 70,000 in the region. The river remained at record heights throughout the lower drainage basin through late June. Damage estimates exceeded $750 million (1973 dollars).

In late summer 1993 the Mississippi and its tributaries in the upper basin rose to levels not seen in more than 130 years. The discharge at St. Louis was measured at more than 1 million cubic feet (28,320 m³) per second. The weather situation that led to these floods was remarkably similar to that of the floods of 1927 and 1973, only worse. High winter snowmelts were followed by heavy summer rainfalls caused by a low-pressure trough that stalled over

the Midwest, because it was blocked by a stationary high-pressure ridge that formed over the East Coast of the United States. The low-pressure system drew moist air from the Gulf of Mexico that met the cold air from the eastern high-pressure ridge, initiating heavy rains for much of the summer. The rivers continued to rise until August, when they reached unprecedented flood heights. The discharge of the Mississippi was the highest recorded, and the height of the water was even greater because all the levees that had been built restricted the water from spreading laterally and caused it to rise more rapidly than it would have without the levees in place. More than two-thirds of all the levees in the upper Mississippi River basin were breached, overtopped, or damaged by the floods of 1993. Forty-eight people died in the 1993 floods, and 50,000 homes were damaged or destroyed. Total damage costs are estimated at more than $20 billion.

The examples of the floods of 1927 and 1993 on the Mississippi reveal the dangers of building extensive levee systems along rivers. Levees adversely affect the natural processes of the river and may actually make floods worse. Their first effect is to confine the river to a narrow channel, causing the water to rise faster than if it were able to spread across its floodplain. Additionally, since the water can no longer flow across the floodplain, it cannot seep into the ground as effectively, and a large amount of water that the ground would normally absorb must now flow through the confined river channel. The floods are therefore larger because of the levees. A third hazard of levees is associated with their failure. When a levee breaks, it does so with the force of hundreds or thousands of acres of elevated river water pushing it from behind. The force of the water that broke through the Mounds Landing Levee in the 1927 flood is estimated to be equivalent to the force of water flowing over Niagara Falls. If the levees were not in place, the water would have risen gradually and would have been much less catastrophic when it eventually entered the farmlands and towns along the Mississippi River basin.

The U.S. Army Corps of Engineers is mitigating another hazard and potential disaster where the Atchafalaya branches off the Mississippi. Over geological time the Mississippi River has altered its course so that its mouth has migrated east and west by hundreds of miles. Each course of the river has produced its own delta, which subsides below sea level after the river migrates to another location. Subsidence of the delta deposit occurs primarily because the river no longer replenishes the top of the delta, and the buried muds gradually compact as the weight of the overlying sediments expels water from the pore spaces. As the delta subsides to sea level, waves add to the erosion, keeping the delta surface below sea level. At the present time the lower Mississippi River follows a long, circuitous course from where the Atchafalaya branches off from it, past New Orleans, to its mouth near Venice. The Mississippi is ready to switch its course back to its earlier position, following the Atchafalaya, which would offer it a shorter course to the sea, and would take less energy to transport sediment to the Gulf of Mexico. If this were to occur, it would be devastating to the lower delta, which would quickly subside below sea level. The city of New Orleans is currently below sea level and protected from the river, storms, and the Gulf of Mexico only by high levees built around the city. To prevent this disaster the Army Corps has constructed an extensive system of diversions, levees, and dams at the Mississippi/Atchafalaya junction to keep the Mississippi in its channel.

Yellow River, China

More people have been killed from floods along the Yellow River in China than from any other natural feature, whether river, volcano, fault, or coastline. An estimate of millions of people have died as a result of floods and famine generated by the Yellow River, which has earned it the nickname River of Sorrow in China.

The Yellow River flows out of the Kunlun Mountains across much of China into the wide lowland basin between Beijing and Shanghai. The river has switched courses in its lower reaches at least 10 times in the last 2,500 years. It currently flows into Chihli (Bohai) Bay, then into the Yellow Sea.

The Chinese have attempted to control and modify the course of the Yellow River since dredging operations in 2356 B.C.E. and the construction of levees in 602 B.C.E. One of the worst modern floods along the Yellow River occurred in 1887, when the river rose over the top of the 75-foot (22-meter) high levees and covered the lowlands with water. More than 1 million people died from the floods and subsequent famine. Crops and livestock were destroyed, and sorrow returned to the river.

The Yellow River was also the site of a mixed natural and unnatural disaster in 1938. As part of the war effort, in 1938 Chiang Kai-shek (Jiang Jieshi, Nationalist Chinese leader and later, president of Taiwan) is said to have attacked and bombed the levees along the Yellow River to trap the advancing Japanese army. The Japanese had been brutally advancing inland from the coast, and the Chinese adopted a scorched earth policy, burning towns and villages before retreating to leave nothing for the Japanese. The war was causing more than a thousand deaths a day for the Japanese, and more for the Chinese, in some of history's largest military battles. When the Japanese army arrived in Suchow (now Xuzhou, Jiangsu Province), the area was deserted. They captured the empty city on May 20, 1938, and the Japanese were preparing to move farther inland. But torrential rains were causing flooding along the Yellow and Yangtze Rivers, and progress was slow. Then, in June, the levees along the Yellow River were apparently cut (some historians say they broke naturally), and one of the river's greatest floods ensued. The river escaped, initially inundating 500 square miles (1,295 km^2), and took another 1 million lives as the flooding spread throughout June and July while rains continued. Despite the enormous loss of life and destruction, the massive floods of the Yellow River, and the Yangtze, the Japanese began a disorganized retreat in rafts and boats. They next tried to advance up the Yangtze, and with great loss of life made progress and captured towns up to Kiukang (now Jiujiang, in Jiangxi Province). On August 3 the Chinese army cut the dikes on the Yangtze, flooding and killing many more people, but effectively ending the two-month-long drive by the Japanese army up the Yangtze. By the middle of August the Japanese were retreating from their drive into central China up the Yangtze, following the floodwaters to the sea.

The Yellow River is continuing its natural process of building up its bottom, and the people along the river continue to raise the level of the levees to keep the river's floods out of their fields. Today, the river bottom rests an astounding 65 feet (20 meters) above the surrounding floodplain, a testament to the attempts of the river to find a new, lower channel and to abandon its current channel in the process of avulsion. What will happen if heavy rains cause another serious flood along the River of Sorrow? Will another million people perish?

URBANIZATION AND FLASH FLOODING

Urbanization is the process of building up and populating a natural habitat or environment, such that the habitat or environment no longer responds to input the way it did before being altered by humans. When heavy rains fall in an unaltered natural environment, the land surface responds to accommodate the additional water. Desert regions may experience severe erosion in response to the force of falling raindrops that dislodge soil and also by overland flow during heavy rains. This causes upland channel areas to enlarge to accommodate larger floods. Areas that frequently receive heavy rains may develop lush vegetative cover, which helps to break the force of the raindrops and reduce soil erosion, and the extensive root system holds the soil in place against erosion by overland flow. Stream channels may be large so that they can accommodate large-volume floods.

When the natural system is altered in urban areas, the result can be dangerous. Many municipalities have paved over large parts of drainage basins and covered much of the recharge area with roads, buildings, parking lots, and other structures. The result is that much of the water that used to seep into the ground and infiltrate into the groundwater system now flows overland into stream channels, which may themselves be modified or even paved over. The net effect of these alterations is that flash floods may occur much more frequently than in a natural system, since more water flows into the stream system than before the alterations. The floods may occur with significantly lower amounts of rainfall as well, and since the water flows overland without slowly seeping into the ground, the flash floods may reach urban areas more quickly than the floods did before the alterations to the stream system. Overall the effect of urbanization is faster, stronger, bigger floods that have greater erosive power and do more damage. It is almost as if the natural environment responds to urban growth by increasing its ability to return the environment to its natural state.

Urbanization and Changes in the Missouri River Floodplain

The Missouri River stretches more than 2,300 miles and drains one-sixth of the United States. It was once one of the wildest stretches of rivers in the American Midwest. During the past two centuries the Missouri, along with its adjacent wetlands and floodplains, has been dramatically modified in various attempts to promote transportation, agriculture, and development. These modifications have included draining wetlands for cultivation, straightening stream channels to facilitate navigation, stabilizing banks to prevent erosion, and constructing agricultural levees, dams, reservoirs, and flood-control levees to control water flow and exclude floodwaters from the floodplain. These modifications have resulted in a severe loss of wetlands.

Historically the Missouri River floodplain below Sioux City, Iowa, covered 1.9 million acres. According to the Sierra Club, modifications in the river-floodplain system described above have resulted in the loss of approximately 168,000 acres of natural channel, 354,000 acres of meander belt habitat, and 50 percent of the river's surface. In addition, shallow-water habitat has been reduced by up to 90 percent in some areas, while sandbars, islands, oxbows, and backwaters have been virtually eliminated. Forested floodplains along the Missouri have decreased from 76 percent in the 19th century to 13 percent in 1972, and cultivated lands have increased from 18 to 83 percent.

By the late 1970s the lower Missouri River had been totally channelized and its natural floodplain ecosystems almost completely converted to agricultural or other uses. Today levees and other flood-control structures flank the lower Missouri River for most of its length. Environmental groups such as the Sierra Club, Great Rivers Habitat Alliance, and Ducks Unlimited have been fighting further development of the floodplain to prevent the complete loss of this habitat and reduce the risks of hazardous floods along the system.

SUMMARY

Floods are the costliest, deadliest, most common natural disaster to affect humans. Individual floods have killed on the order of a million people at a time, and other floods cause billions of dollars in damages and ruin entire towns, disrupt livelihoods of hundreds of thousands of people, and bring disease and famine to affected regions.

There are many types of floods, ranging from isolated flash floods that sweep down isolated mountain canyons, to coastal floods associated with tropical cyclones and other large ocean storms. Bangladesh has experienced the most frequent and most deadly storm surge–related flooding of anywhere in the world, with some storms killing hundreds of thousands of people. Some of the most devastating floods in history have been large, slowly rising floods that cover entire regions, with the Yellow and Yangtze Rivers of China having the dubious distinction of recording the two deadliest floods of all time, each flood claiming more than 1 million lives. Floods along the Mississippi-Missouri-Ohio River basins in the United States have been frequent and long-lasting, and have caused great damage to areas on the floodplains that have been built up for commercial or residential uses. Had these areas remained natural or been used for agriculture, the damage would have been much less, and the river floods would have been lower in magnitude. Levees along these rivers have constricted the rivers with time, raising their base and in turn raising the river flood stages, leading to more disastrous floods.

People have modified rivers and floodplains for navigation and flood control for thousands of years, often with disastrous results. In the United States construction of levees along the Mississippi River began essentially as soon as western settlers arrived in the New Orleans area. Time and again levees were built, floods seemed to become larger, and the levees were breached or collapsed through processes of underseepage, piping, scouring, or liquefaction. Some scientists noticed that as the levees were built higher, the base of the river seemed to aggrade or rise as well. However, it took 250 years before quantitative evidence demonstrated that the construction of levees and other flood-control measures constricted the river and caused flood stages to rise higher and faster, and become more dangerous, with increasing constriction of the river by levees. This understanding has not reached the level of policy in the United States, as rivers are still being actively constricted by levees, and floodplains are being widely developed.

Urbanization of floodplains also causes floods to rise faster, be more powerful, and do more damage than in natural settings. Some places such as the American Southwest and parts of the Middle East have extensively altered the natural drainage network to provide drinking and irrigation water in arid and semiarid climates. These water resources are presently extremely stressed and reaching their limits, yet the population keeps on expanding at alarming rates. New sources of water must be sought to meet the demands of a growing global population.

See also DESERTS; DRAINAGE BASIN (DRAINAGE SYSTEM); FLUVIAL; GEOLOGICAL HAZARDS; GEOMORPHOLOGY; GROUNDWATER.

FURTHER READING

Arnold, J. G., P. J. Boison, and P. C. Patton. "Sawmill Brook—An Example of Rapid Geomorphic Change Related to Urbanization." *Journal of Geology* 90 (1982): 115–166.

Baker, Victor R. "Stream-Channel Responses to Floods, with Examples from Central Texas." *Geological Society of America Bulletin* 88 (1977): 1057–1071.

Belt, Charles B., Jr. "The 1973 Flood and Man's Constriction of the Mississippi River." *Science* 189 (1975): 681–684.

Junk, Wolfgang J., Peter B. Bayley, and Richard E. Sparks. "The Flood Pulse Concept in River-Floodplain Systems." *Canadian Special Publication Fisheries and Aquatic Sciences* 106 (1989): 110–127.

Kusky, T. M. *Floods; Hazards of Surface and Groundwater Systems. The Hazardous Earth Set.* New York: Facts On File, 2008.

Leopold, L. B. *A View of the River.* Cambridge, Mass.: Harvard University Press, 1994.

Maddock, Thomas, Jr. "A Primer on Floodplain Dynamics." *Journal of Soil and Water Conservation* 31 (1976): 44–47.

Noble, C. C. *The Mississippi River Flood of 1973.* In *Geomorphology and Engineering,* edited by D. R. Coates. London: Allen and Unwin, 1980.

United States Geological Survey. "Water Resources." Available online. URL: http://water.usgs.gov/. Accessed December 10, 2007.

fluvial Rivers are the main geological instruments that shape the surface of the land, carrying pieces of the continents grain by grain, steadily to the sea. The term *fluvial* refers to deposits and landforms created by the action of flowing rivers and streams, and also the processes that occur in these rivers and streams. Fluvial systems slowly erode mountains and fill deep valleys with alluvium, and serve as passageways for people, aquatic fauna and flora, sediment, and dissolved elements from one place to another. River systems are not simply channels, but are intricately linked to associated floodplains and deltas, and they are affected by processes that occur throughout the entire drainage basin. Rivers transport water in a critical step in the hydrological cycle, and bring freshwater to even the driest places on Earth. Nearly every city and town in the world is built with a river flowing through it or near it, so vital is water for drinking, agriculture, and navigation. Rivers have controlled history, bringing life to some areas, but they are also prone to floods, sometimes bringing disaster from the same source that has fed populations for ages.

GEOMETRY OF FLUVIAL SYSTEMS

Fluvial systems including streams and rivers are dynamic, ever-changing systems that represent a balance between driving and resisting forces. The ability of a stream to erode and transport sediments depends on how much energy is in the water as it flows, versus how much is consumed by the resistance to flow. As the velocity of the water in the stream increases, the resistance to flow provided by the stream banks, boulders, and material carried by the stream also increases. Therefore at any point in the stream the velocity of the water and the shape of the stream channel represent a balance between the energy causing the flow of water and the energy consumed by resistance to flow.

The water in the stream channel may exhibit one of two main types of flow, laminar or turbulent. In a laminar flow pattern the paths of water particles are parallel and smooth, and the flow is not very erosive. Resistance to flow in laminar systems is provided by internal friction between individual water molecules, and the resistance is proportional to flow velocity. The frictional resistance in laminar flow systems increases from the top of the water surface to the base of the streambed. In contrast, in turbulent flow the direction of flow and the velocity vary in all directions within the stream, and water is being continuously exchanged between adjacent flow zones. In turbulent flows the water may move in different directions and often forms zones of sideways or short backward flows called eddies. These significantly increase the resistance to flow. In turbulent flows the resistance is proportional to the square of the flow velocity. Many zones of turbulence and turbulent eddies are generated along channel margins where the water velocity is reduced by frictional resistance from the bed material and riverbanks.

Streams are defined primarily by their channels, which are the elongate depressions where the water flows. Several different types of stream banks separate the stream channels from the adjacent flat floodplains, including low-profile point bars, steep-cut banks, and eroded cliffs. The shape of the channel and its pattern in map view represents the balance between the driving and resisting forces in the different conditions or environments through which the stream passes. Streams create their broad, flat floodplains by erosion and redeposition during floods, and these plains serve as the stream bottom during large floods. Even though floodplains may have no water over them for many tens of years, they are part of the stream system and the stream will return. Many communities in the United States and elsewhere have built extensively on the floodplains, and these communities will eventually be flooded.

Stream channels are self-adjusting features—they modify their shapes and sizes to best accommodate the amount of water flowing in the stream. A stream's discharge is a measure of the amount of water passing a given point per unit time. During floods the discharge may be two, three, ten, or more times normal levels. The stream channel may then overflow, causing the water to spread across the adjacent floodplain, inundating towns and farms. The cross-sectional shape of streams changes with time and amount of water flow through the channel. The shape of a stream channel is also different in the upstream and downstream parts of the system, as the slope and volume of water changes along the course of the river. Small, narrow streams are typically as deep as they are wide, whereas large streams and rivers are much wider than they are deep.

The gradient or slope of a stream is a measure of the vertical drop over a given horizontal distance, and the average gradient decreases downstream. Going downstream, several changes also occur. First, the discharge increases, which in turn causes both the width and the depth of the channel to increase. Yet downstream, as the gradient decreases, the velocity increases. Although one might expect the velocity of a stream to decrease with a decrease in slope (gradient), anyone who has seen the Mississippi at New Orleans or the Nile at Cairo can testify to their great velocity as compared to their upstream sources. Two reasons explain this increase in velocity. First, the upstream portions of these mighty rivers have courses with many obstacles and more friction per stream volume, reducing velocity. Second, more water flows in the downstream portions of the streams, and this has to move quickly to allow the added discharge from the various tributaries that merge with the main stream.

The base level of a stream is the limiting level below which a stream cannot erode the land. The ultimate base level is sea level, but in many cases streams entering a lake or dammed region form a local base level.

EROSION, SEDIMENT TRANSPORT, AND DEPOSITION IN FLUVIAL SYSTEMS

Most energy in streams is dissipated by turbulent flow, but a small part of a stream's energy is used to erode and transport sediments downstream. Streams carry a variety of materials as they make their way to the sea, and the way this material is eroded and transported depends on the energy balance in the stream. These materials range from minute dissolved particles and pollutants to giant boulders moved only during the most massive floods. The bed load consists of the coarse particles that move along or close to the bottom of the streambed. Particles move more slowly than the stream by rolling, bouncing, or sliding. Saltation is the movement of a particle by short, intermittent jumps caused by the current lifting the particles. Bed load typically constitutes 5–50 percent of the total load carried by the stream, with a greater proportion carried during high-discharge floods. The suspended load consists of the fine particles suspended in the stream. This makes many streams muddy, and the suspended load consists of silt and clay that moves at the same velocity or slightly lower than the stream. The suspended load generally accounts for 50–90 percent of the total load carried by the stream. The dissolved load of a stream consists of dissolved chemicals, such as bicarbonate, calcium, sulfate, chloride, sodium, magnesium, and potassium. The dissolved load tends to be high in streams fed by groundwater. Pollutants, such as fertilizers and pesticides from agriculture, and industrial chemicals also tend to be carried as dissolved load in streams.

Most of the larger particles in streambeds are usually not moving, but move only for short distances at times of high-flow velocity and discharge of the stream. The picking up of particles from the bed load of a stream and the erosion of material from the banks is known as entrainment, a process that depends on the erosive power of the flow and the resistance of the particles. There is a wide range in the sizes and amounts of material that can be entrained and transported by a stream. The competence of a stream refers to the maximum size of particles that can be entrained and transported by a stream under a given set of hydraulic conditions, measured in diameter of the largest bed load. A stream's capacity is the potential load it can carry, measured in the amount (volume) of sediment passing a given point in a set amount of time. The amount of material carried by streams depends on a number of factors. Climate studies show erosion rates are greatest in climates between a true desert and grasslands. Topography affects stream load as rugged topography contributes more detritus, and some rocks are more erodable. Human activity, such as farming, deforestation, and urbanization, all strongly affect erosion rates and stream transport. Deforestation and farming greatly increase erosion rates and supply more sediment to streams, increasing their loads. Urbanization has complex effects, including deceased infiltration and decreased times between rainfall events and floods, as discussed in detail below.

Erosion and Deposition along Stream Banks

The process of entrainment determines how a stream erodes its bank and bed, and the type of sedimentary load the stream can carry. The lateral, or sideways, erosion of a stream bank is an important process that strongly influences other stream processes. The erosion of the stream banks is accomplished through a combination of events, including weathering of the material on the stream bank, mass wasting that may cause the bank to collapse into the stream, and the actual entrainment of the sedimentary particles into the bed load of the stream.

The weathering of the bank material, typically loose sediment deposited by the stream, makes it weaker and more susceptible to mass movement and collapse into the river. The amount of moisture in the soil on the banks is important in this stage, as increased moisture, such as during rain or flooding events, decreases the frictional resistance between the bank sediments, and that is partly why many banks collapse during rains and flooding events. In areas prone to freeze-thaw cycles, stream banks are also susceptible to collapse from the action of frost wedging in small cracks, pushing blocks of sediment into the river. Many stream banks have layers of sand, gravel, and mud deposited on floodplains during earlier stages of the stream development. In these cases groundwater may move along the gravel and sand layers, seeping out along the river bank. This movement of groundwater can actually carry sediment away from the bank into the stream in a process called sapping. This groundwater sapping creates overhanging banks along the river, which are then prone to collapse. The water along these layers may also reduce the friction on this layer, creating a plane along which overlying layers often slide into the river along, forming planar slides. Planar slides are recognized as important mass wasting processes along many riverbanks, including the Mississippi River.

River and stream banks collapse with many different styles of mass wasting, including slipping of large sections of the bank on rotational slides, slumps into the river, and wholesale collapse of large slabs, especially where the stream has undercut the riverbank. Many factors determine which kind of collapse occurs, including the layering in the riverbank/floodplain sediment, the pore fluid pressure.

After the bank materials collapse into the river, the current begins to entrain the finer-grained particles, and to move the coarser material as bed load. This carries the material away and prepares the bank for the next failure, in the steady process of the river migrating across the floodplain.

After the sediments are carried away by the river current, at some point they are deposited again. Fine particles may be carried in suspension all the way to the ocean or local reservoir, whereas most of the coarser particles move by bouncing or rolling along the stream bed. Where these sediments get deposited next depends on the interactions of the type of current in the river and the size, shape, and density of the sedimentary particles. The river channel is a dynamic environment, and the flow velocity and style, whether laminar or turbulent, can vary significantly over short distances. These local variations often determine where a sedimentary particle will be deposited, and whether the current is scouring one place or filling in another. Typically as the river is eroding a bank or scouring its base in one location, moving the material from that location downstream, the current is simultaneously depositing other sedimentary material nearby that was carried from further upstream. For instance, along one bend of a river the outer or cut bank may be eroding, whereas the inner bank of the bend may be experiencing deposition. In this way the river effectively moves its location, filling in the old channel as it cuts a new one step by step. River floodplains are naturally dynamic environments, where the natural forces in the river keep a balance by maintaining this lateral, back-and-forth type of movement of the channel as the river transports the bed and suspended load downstream.

If the river cannot move laterally, such as if confined by bedrock or by levees, it must respond by changing the level of its base. Rivers may downcut through alluvium or bedrock in response to tectonic uplift, or may rise through depositing sediments along their beds in a process called aggradation. If the river is transporting a large bed load, it naturally responds to this by moving it downstream and moving sideways. When the bank is confined, it can move only upward and deposits these extra bed load sediments along its base, causing unnatural aggradation.

CHANNEL PATTERNS

River channels represent a quasi-equilibrium condition between the river discharge, flow regime (whether laminar or turbulent), amount of sediment being transported, and slope of the river channel. The river can respond to these variables by finding an equilibrium or quasi-equilibrium condition by adjusting the channel shape (width and depth), the velocity of the flow, the roughness of the bed and bank, and the slope of the bed. The slope of a riverbed can be adjusted by the river by increasing or decreasing the number of its bends, or meanders. If the river needs to lower the slope to maintain a quasi-equilibrium condition, then it can increase the number of bends and flow more parallel to the contours. If the river needs to increase the slope, it can cut through the banks and flow straight downhill, attaining a slope equal to the regional gradient. This is one of the reasons rivers have so many different forms, from straight to wildly meandering channels.

Straight Channels

Stream channels are rarely straight, and a stream is said to have a straight channel if the ratio of the stream length to valley length is 1.5. Although this ratio, called the sinuosity, seems to have no particular mechanical significance, this measure is useful to describe the shape of stream channels. Many stream

channels are straight because they inherit their path from incision into an underlying bedrock fracture, whereas others are relatively straight for short distances. In either case the velocity of flow changes in different places and, internally, the water in the channel naturally starts to develop some complex flow patterns. Friction makes the flow slower on the bottom and sides of the channel, and the water develops curving traces of highest velocity in plain view, and also begins to circulate in loops from the surface, to the streambed, and back to the surface as the water moves downstream. The changing currents cause sand bars to be deposited alternately on either sides of the straight channel and for deep pools to develop between the bars. These internal bends in the river make the zone of fastest flow swing from side to side.

Straight channels are very rare, and those that do occur have many properties of curving streams. The thalweg is a line connecting the deepest parts of the channel. In straight segments the thalweg typically meanders from side to side of the stream. In places where the thalweg is on one side of the channel, a bar may form on the other side. A bar (for example, a sandbar) is a deposit of alluvium in a stream.

Meandering Streams

Most streams move through a series of bends known as meanders. The main components of meandering channels are similar to the straight channels, with greater curvature. The outer bends of meanders are typically marked by steep-cut banks, with active slumping and mass wasting into the channel, whereas the inner bends of the meanders are marked by deposition of sand and gravel. Meanders are always migrating across the floodplain by the process of the deposition of the point bar deposits and the erosion of the bank on the opposite side of the stream with the fastest flow. The thalweg, the line of fastest flow, bounces into the outer cut bank, and some of the flow moves down along this steep wall, then more slowly upward along the slope of the point bar as the water moves downstream. This results in a twisting helical flow of water in the stream channel, and keeps the outer banks erosive, with the fastest currents, and the inner point bars have slower velocity currents, and receive deposits of sand and gravel. Meanders typically migrate back and forth, and also down-valley at a slow rate. If the downstream portion of a meander encounters a slowly erodable rock, the upstream part may catch up and cut off the meander. This forms an oxbow lake, an elongate and curved lake formed from the former stream channel.

Studies on the mechanics of stream flow have revealed quantitative relationships between the wavelength of a meander (the distance from one cut bank to the next one of similar curvature), the discharge of the stream, the radius of curvature, and other stream parameters. By changing any of these variables the current in the stream will change to attempt to restore the system to an equilibrium state. Therefore it is clear that streams need to be able to maintain their migrating meandering pattern across floodplains to be in equilibrium. Any unnatural changes, such as straightening channels, narrowing channels, and the like, will naturally be met by the river with changes in other parameters that may be unexpected and potentially hazardous.

Braided Stream Channels

Braided streams consist of two or more adjacent but interconnected channels separated by bars or islands, commonly known as braid bars. Braided streams have constantly shifting channels, which move as the bars are eroded and redeposited during large fluctuations in discharge. Most braided streams have highly variable discharge in different seasons, and they carry more load than meandering streams.

Braided streams tend to be wider and shallower, and have steeper gradients than streams with undivided channels. Several factors seem to play significant roles in determining whether a stream channel becomes braided. First, the backs of the stream must be easily erodible, letting the channels migrate and contributing bedload to the channel. Second, the load must be large, as all braided streams carry high-sediment loads. Third, braided streams are characterized by rapid changes in discharge. Braided streams are common in areas such as on glacial outwash plains, where the fluctuation in discharge is large, there is abundant sediment supply, and the river banks are easily erodible.

DYNAMICS OF STREAM FLOW

Streams are dynamic systems and constantly change their channel patterns and the amount of water (discharge) and sediment being transported in the system. Streams may transport orders of magnitude more water and sediment in times of spring floods, as compared with low-flow times of winter or drought. Since streams are dynamic systems, as the amount of water flowing through the channel changes, the channel responds by changing its size and shape to accommodate the extra flow. For instance, in a gradually changing climate scenario, the discharge and load of a river may gradually change, and the river may be able to make small changes accordingly to account for these variables. At some point, however, the balance of controlling forces in the river may exceed a critical threshold value, and the channel may suddenly make a dramatic change into a completely different configuration. In another scenario a river may gradually downcut its gradient through a

mountain range, starting as a juvenile high-gradient stream, and over the course of many years gradually decrease its gradient (slope) as the bed is eroded. At different stages in this evolution, the stream may make transitions, perhaps rapidly, through different channel types and flow regimes. The following five factors control how a stream behaves:

- width and depth of channel, measured in feet (meters)
- gradient, measured as change in elevation in feet per mile (m/km)
- average velocity, measured in feet per second (m/sec)
- discharge, measured in cubic feet per second (m^3/s)
- load, measured as tons per cubic yard (metric tons/m^3)

All of these factors are continually interplaying to determine how a stream system behaves. As one factor, such as discharge changes, so do the others, expressed as

$$Q = w \times d \times v$$

where Q represents discharge, w represents width, d represents depth, and v represents velocity. Other factors may also play a role, though are less important. These include the mean annual flood, meander wavelength, width-depth ratio, and sinuosity. These secondary variables are not totally independent; for instance, sinuosity and gradient are related, the mean annual flood and discharge are related, and so on. The main point is that the variables are all interrelated, and changing one can lead to changes in the others.

All factors vary across the stream, so they are expressed as averages. If one term changes, then all or one of the others must change too. For example, with increased discharge, the stream erodes, widens, and deepens its channel. With increased discharge, the stream may also respond by increasing its sinuosity through the development of meanders, effectively creating more space for the water to flow in and occupy by adding length to the stream. The meanders may develop quickly during floods because the increased stream velocity adds more energy to the stream system, and this can rapidly erode the cut banks, enhancing the meanders.

The amount of sediment load available to the stream is also independent of the stream's discharge, so different types of stream channels develop in response to different amounts of sediment load availability. If the sediment load is low, streams tend to have simple channels, whereas braided stream channels develop where the sediment load is greater. If a large amount of sediment is dumped into a stream, the stream will respond by straightening, thus increasing the gradient and stream velocity, and increasing the stream's ability to remove the added sediment.

When streams enter lakes or reservoirs along their path to the sea, the velocity of the stream suddenly decreases. This causes the sediment load of the stream or river to be dropped as a delta on the lake bottom, and the stream attempts in this way to fill the entire lake with sediment. The stream is effectively attempting to regain its gradient by filling the lake, then eroding the dam or ridge that created the lake in the first place. When the water of the stream flows over the dam, it does so without its sediment load and therefore has greater erosive power and can erode the dam more effectively.

The concept of a graded stream is widely used by geomorphologists to describe how a river may adjust its environment to transport its sedimentary load with the least energy required. In this concept the stream gradually (over many years) erodes its bed to attain an equilibrium gradient just right for transporting the sedimentary load when balanced with the types of channels characteristics and velocity available in the area. This graded profile is typically concave up, steeper in the headwaters, and with a low slope near the mouth of the river. Graded streams are thought to be in a state of relative equilibrium; changes in one variable will be accommodated by changes in the other to keep the balance of forces.

FLUVIAL DEPOSITIONAL FEATURES: FLOODPLAINS, TERRACES, AND DELTAS

During great floods, streams flow way out of their banks and fill the adjacent floodplain. During these times, when the water flows out of the channel, its velocity suddenly decreases and it drops its load, forming levees and overbank silt deposits on the floodplain.

Floodplains

Floodplains are relatively flat areas that occupy valley bottoms, and generally comprise unconsolidated sediments. Most of these sediments are deposited by the river, but some may come from other processes, such as from slopes along the margins of the valley and even from wind or Aeolian processes. In natural river systems (ones not disturbed by levees, etc.) the river will periodically rise out of its banks and cover the floodplain with water and fine sediments. Different flood levels and different parts of the floodplain may be reached with different frequencies of floods. In most natural rivers in humid climates the river rises out of the banks every year or two. Higher levels of the floodplain may be reached only during higher floods, such as *the 100-year flood,* a term that describes the height of water statistically expected to be reached only once every hundred years.

Floodplains are an essential part of the river system, as they allow the river to adjust to changing conditions. During floods the floodplains hold water, reducing the speed and height of floods in downstream areas, and the unconsolidated sediments in the floodplain also absorb large quantities of the water, reducing the amount that flows downstream. The floodplain also serves as a large temporary and mobile storage area for the sediments that have been eroded from throughout the watershed. This storage is important for maintaining the river's ability to respond to changes in discharge, climate, and other variables, so it needs to remain in contact with the river. Attempts to isolate the floodplain from the river by construction of levees and artificial canals disrupt the natural flow and separate different components of the system, setting the stage for disasters.

Stream Terraces

Terraces are abandoned floodplains formed when a stream flowed above its present channel and floodplain level. These form when a stream erodes downward through its deposits, to a new lower level. Paired terraces are terrace remnants that lie at the same elevation on either side of the present floodplain. Nonpaired terraces form at different levels on either side of the current floodplain, and imply several episodes of erosion. Rivers and streams may downcut through older terraces for a variety of reasons, including climatically influenced changes in discharge, or uplift of the river valley and slopes, causing a change in the river profile.

Deltas

When a stream enters the relatively still water of a lake or the ocean, its velocity and capacity to hold sediment drop suddenly. Thus the stream dumps its sediment load here, and the resulting deposit is known as a delta. Where a coarse sediment load of an alluvial fan dumps its load in a delta, the deposit is known as a fan-delta. Braid-deltas are formed when braided streams meet local base level and deposit their coarse-grained load. When a stream deposits its load in a delta, it first drops the coarsest material, then progressively finer material farther out, forming a distinctive sedimentary deposit. The resulting foreset layer is thus graded from coarse nearshore to fine offshore. The bottomset layer consists of the finest material, deposited far out. As this material continues to build outward, the stream must extend its length and forms new deposits, known as topset layers, on top of all this. Most of the world's large rivers—the

Mississippi, Nile, and Ganges—have built huge deltas at their mouths, yet all of these differ in detail.

DRAINAGE SYSTEMS

A drainage basin is the total area that contributes water to a stream, and the line that divides different drainage basins is known as a divide (such as the continental divide) or interfluve. Drainage basins are the primary landscape units or systems concerned with the collection and movement of water and sediment into streams and river channels. Drainage basins consist of a number of interrelated systems that work together to control the distribution and flow of water within the basin. Hillslope processes, bedrock and surficial geology, vegetation, climate, and many other systems all interact in complex ways that determine where streams will form and how much water and sediment they will transport. A drainage basin's hydrologic dynamics can be analyzed by considering these systems along with how much water enters the basin through precipitation and how much leaves the basin in the discharge of the main trunk channel. Streams are arranged in an orderly fashion in drainage basins, with progressively smaller channels branching away from the main trunk channel. Stream channels are ordered and numbered according to this systematic branching. The smallest segments lack tributaries and are known as first-order streams; second-order streams form where two first-order streams converge; third-order streams form where two second-order streams converge, and so on.

Streams within drainage basins develop characteristic branching patterns that reflect, to some degree, the underlying bedrock geology, structure, and rock types. Dendritic or randomly branching patterns form on horizontal strata or on rocks with uniform erosional resistance. Parallel drainage patterns develop on steeply dipping strata, or on areas with systems of parallel faults or other landforms. Trellis drainage patterns consist of parallel main stream channels intersected at nearly right angles by tributaries, in turn fed by tributaries parallel to the main channels. Trellis drainage patterns reflect significant structural control, and typically form where eroded edges of alternating soft and hard layers are tilted, as in folded mountains or uplifted coastal strata. Rectangular drainage patterns form a regular rectangular grid on the surface, and typically form in areas where the bedrock is strongly faulted or jointed. Radial and annular patterns develop on domes including volcanoes and other roughly circular uplifts. Other, more complex patterns are possible in more complex situations, as illustrated by multibasinal and contorted styles of drainage patterns.

Several categories of streams in drainage basins reflect different geologic histories. A consequent stream is one whose course is determined by the direction of the slope of the land. A subsequent stream is one whose course has become adjusted so that it occupies a belt of weak rock or another geologic structure. An antecedent stream is one that has maintained its course across topography being uplifted by tectonic forces; these cross high ridges. Superposed streams are those whose courses were laid down in overlying strata onto unlike strata below. Stream capture occurs when headland erosion diverts one stream and its drainage into another drainage basin.

EFFECTS OF RIVER MODIFICATIONS ON RIVER DYNAMICS

The long history of flooding and attempted flood-control measures along the Mississippi River basin had taught engineers valuable lessons on how to manage flood control on river basins. Levees are commonly built along riverbanks to protect towns and farmlands from river floods. These levees usually succeed at the job they were intended to do, but they also cause other collateral effects. First, the levees do not allow waters to spill onto the floodplains, so the floodplains do not receive the annual fertilization by thin layers of silt, and they may begin to deflate and slowly degrade as a result of this loss of nourishment by the river. The ancient Egyptians relied on such yearly floods to maintain their fields' productivity, which has declined since the Nile has been dammed and altered in recent times. Another effect of levees is that they constrict the river to a narrow channel, so that floodwaters that once spread slowly over a large region are now focused into a narrow space. This causes floods to rise faster, reach greater heights, have a greater velocity, and reach downstream areas faster than rivers without levees. The extra speed of the river is in many cases enough to erode the levees and return the river to its natural state.

One of the less appreciated effects of building levees on the sides of rivers is that they sometimes cause the river to slowly rise above the height of the floodplain. Many rivers naturally aggrade or accumulate sediment along their bottoms. In a natural system without levees this aggradation is accompanied by lateral or sideways migration of the channel so that the river stays at the same height with time. If a levee is constructed and maintained, however, the river is forced to stay in the same location as it builds up its bottom. As the bottom rises, the river naturally adds to the height of the levee, and people will also build up the height of the levee as the river rises to prevent further flooding. The net result is that the river may gradually rise above the floodplain, until some catastrophic flood causes the levee to break, and the river establishes a new course.

Breaking through a levee happens naturally as well and is known as avulsion. Avulsion has occurred seven times in the last 6,000 years along the lower Mississippi River. Each time the river has broken through a levee a few hundred miles from the mouth of the river and has found a new, shorter route to the Gulf of Mexico. The old river channel and delta is then abandoned, and the delta subsides below sea level, as the river no longer replenishes it. A new channel is established and this gradually builds up a new delta until it too is abandoned in favor of a younger, shorter channel to the gulf.

The history of constructing levees along the Mississippi River is instructive as it illustrates how the dynamics of the river were not appreciated as the course of the river was being altered, and levees were constricting the flow in efforts to reduce flooding and increase navigability. By the time engineers realized the consequences of constricting the river, a couple of hundred years of river modifications had already taken their toll. Still, further modifications were proposed and implemented, and the floods continue to worsen.

History of Levee Building on the Mississippi River

The Mississippi River is the longest river in the world and encompasses the third-largest watershed, draining 41 percent of the continental United States including an area of 1,245,000 square miles (3,224,550 km^2). The river transports 230 million tons of sediment, including the sixth-largest silt load in the world. Before the Europeans came and began altering the river, this silt used to cover the floodplains with this fertile material during the semiannual floods and carry more downriver to be deposited on the Mississippi River delta. Levee construction along the lower Mississippi River system began with the first settlers who came to the region, and has continued until the present-day levee system, the main parts of which include 2,203 miles (3,580 km) of levees, flood walls, and other control structures. Of this, 1,607 miles (2,586 km) of levees lie along the Mississippi River, and another 596 miles (959 km) are along the banks of the Arkansas and Red Rivers in the Atchafalaya basin. Additional levees are built along the Missouri River.

The first levee along the Mississippi River was built around the first iteration of New Orleans between 1718 and 1727, and consisted of a slightly more than mile-long (5,400 feet; 1,646 m), 4-foot-high earthen mound that was 18 feet (5.5 m) wide at the top, with road along the crown. This levee was meant to protect the residents of the newly founded city from annual floods and pestilence that would last from March until June of each year. New Orleans had only recently been inhabited—Louis XIV of France had commissioned the explorer Pierre Le Moyne, Sieur d'Iberville, to establish a colony near the mouth of the Mississippi River to control the Mississippi valley and the lumber and fur trade moving down the river. D'Iberville's younger brother, Jean Baptiste Le Moyne, Sieur de Bienville, established New Orleans in 1718 in a bend of the river to control the portage between the river and Lake Pontchartrain. The site of New Orleans was surrounded by water on all sides. Lakes Ponchartrain, Maurepas, and Bayou Manchac and the Amite River divide it from higher land on the north, and the Mississippi River wraps around its other sides. The site of New Orleans on the natural levee of the Mississippi on the Isle of Orleans has always been precarious, and the city has been inundated by floods from the river on three sides, and by storm surges from hurricanes on the other side about every 30 years since its founding. The first levee built in 1718–27 did not stop the floods, and the city was destroyed by a hurricane in 1722. On September 23–24 a hurricane almost completely destroyed the newly founded capital city. The storm had 100-mile-per-hour (161 kph) sustained winds and a storm surge of 7–8 feet (2–2.4 m) that overtopped the four-foot (1.2-m) high levee. Almost every building in the city was destroyed or severely damaged. If city planners has taken this warning when the city consisted only of several dozens of buildings, much future damage could have been avoided. Instead, more and higher levees were built, with successive floods by storms destroying or severely damaging the city in 1812, 1819, 1837, 1856, 1893, 1909, 1915, 1947, 1956, 1965, 1969, and 2004. The old levees did not hold in 1722, the new levees did not hold in 2004 during Hurricanes Katrina and Rita, and the levees broke repeatedly during the high-water events in between.

The early river levees along the Mississippi consisted of earthen mounds, generally with a slope of 1:2. The local and state governments made it a policy that local farmers had to build their own levees on the property they owned along the Mississippi. Haul methods for bringing the dirt to make the levees were primitive, typically with horse and carriage, yielding only 10–12 cubic yards (7.5–9 m^3) per day. The federal government became involved in 1820 with legislation that focused mostly on navigation along the river and did not consider flood control. As the levees were built at breakneck pace, the river became constricted, causing the bed of the river to raise itself continuously in a process called aggradation. This happens because if the river is not allowed to migrate laterally, it cannot move out of the way of the sediment it is carrying and depositing, and cannot widen the channel, so therefore it raises the bed as this sediment is deposited. Disastrous floods along the lower Mississippi in 1844, 1849, and 1850 resulted in passage of the Swamp Acts of 1849 and 1850. These acts gave Louisiana, Mississippi, Arkansas, Missouri, and Illinois swamp and overflow lands within their boundaries that were unfit for cultivation. These lands were sold, and the revenues generated were used to construct levees and complete drainage reclamation of the purchased lands. Between 1850 and 1927 the levees along the lower Mississippi had to be continuously heightened because of this river avulsion caused by the construction of the levees.

In 1850 Congress appropriated $50,000 to complete two topographic and hydrographic surveys to promote flood protection along the Mississippi River. One survey was completed by a civilian engineer, Charles Ellet Jr., and the other by army engineers A. A. Humphreys and Henry Abbot. The Humphreys-Abbot report recommended three possible methods for flood control including cutting off the bends in the river, diversion of tributaries creating artificial reservoirs and outlets, and confining the river to its channel using levees. Since the first two options were considered too expensive, the third was enacted, with long-lasting consequences. Their levee design called for freeboards at 3–11 feet (1–3.4 m) above the level of the 1858 flood.

The Civil War (1861–65) saw the levees disregarded and they fell into a state of disrepair, made worse by the large floods of 1862, 1865, and 1867. New floods in 1874 prompted the creation of a Levee Commission to complete a new survey of the state of the levees and recommend how to repair the system and reclaim the floodplain. The Levee Commission made a stark assessment, citing major defects in the system and huge costs to repair and improve it. They documented that previous levees were built in faulty locations, with poor organization, insufficient height, poor construction, and inadequate inspection and guarding. They estimated that it would cost $3.5 million to repair the existing system and $46 million

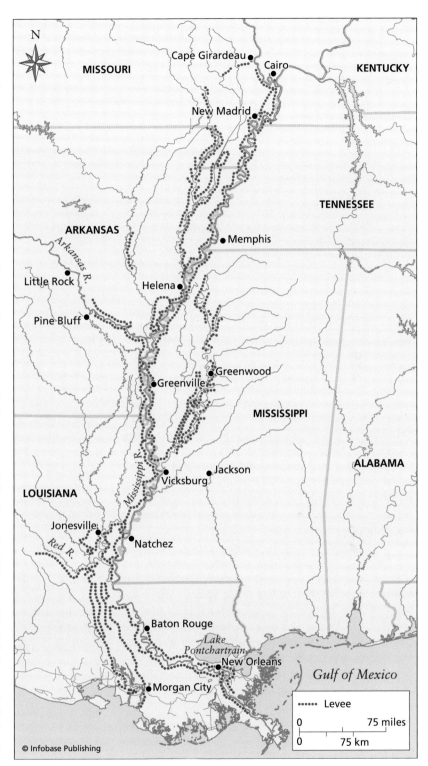

Map of the lower Mississippi River from the mouth of the delta to southern Missouri showing the thousands of miles of levees constructed along the river in the past century

to build a new, complete levee system to reclaim the floodplain from the river.

In 1879 Congress created the Mississippi River Commission (MRC), as organized by James B. Eads.

The commission consisted of three officers from the U.S. Army Corps of Engineers, three civilians, and one officer from the U.S. Coast and Geodetic Survey. The MRC conducted surveys and suggested many modifications and new additions to the flood-control and navigation projects along the river. They made a policy in 1882 to close the breaks along the levee and to construct a line of levees with sufficient height and grade supposedly to contain the frequent floods along the river. They did not have long to wait to see the faults in their model.

The flood of 1890 destroyed 56 miles (90 km) of levees, and the MRC began to raise the levees from 38 to 46 feet (11.5 m to 14 m). During this phase of massive reconstruction the federal government and private citizens added more than 125 million cubic yards of soil to the levees (96 million m³), but much of this was lost to the river by mass wasting processes including slumping and bank caving. Efforts were made to reinforce the banks with various revetments, but then the flood of 1912 destroyed much of the levee system that was meant to protect the adjacent floodplain. The response of the commission was to raise the levees again, to three feet above the 1912 flood line. The lesson was not yet learned that raising the levee and constricting the river causes the bed to aggrade and rise as well.

The first federal flood control act was passed in 1917, authorizing for the first time levees to be built for flood control, along both the Mississippi and its tributaries. The federal government would pay two-thirds of the costs of levees if the local interests would pay the balance. During the 1920s levee construction was stepped up to a higher pace with the mechanization of earth-moving technology, with introduction of large cranes, moving tower machines, and cableway draglines that could move dirt orders of magnitude faster than the traditional horse and cart.

The year 1927 came, and with it, the greatest flood in recorded history along the lower Mississippi River valley. Many of the levees built to the MRC standards failed up and down the river, with enormous consequences in terms of loss of life, displaced people, and loss of property that was supposed to be protected by the levees. The government responded with the 1928 Flood Control Act, passing legislation to improve the grade of the levees and make models of different flood scenarios, including the creation of several large floodways that could be opened to let water out of the river in high flow times. Some of these floodways were quite large, such as the Birds Point–New Madrid floodway, which is about 35 miles (56 km) long, 3–10 miles (5–16 km) wide, and designed to divert 550,000 cubic feet (15,576 m³) per second of flow from the Mississippi during floods. Further downriver the West Atchafalaya floodway was designed to carry half of the modeled projected flood of 1,500,000 cubic feet (139,400 m³) per second. The Bonne Carre floodway was built upriver from New Orleans, designed to restrict the flow to downstream by diverting the water and protecting New Orleans. Levees were redesigned, moved to locations where their projected life span was from 20 to 30 years, and thought to be stronger. As construction on the new levee and floodway system continued, new floods, such as the 1929 flood, disrupted operations, but the construction methods continued to improve, and the levees were built, forming much of the present-day levee system.

In 1937 a large flood emanated from the Ohio River watershed, raising the waters to levels such that the Birds Point–New Madrid floodway was used, opening the floodway by dynamiting the Fuse Plus levee. This released huge volumes of water and eased the flood downstream. One of the lessons from the 1937 flood was that roads should be added to the levees to aid in moving material from place to place during floods. In 1947 the MRC began redesigning levees to be stronger to avoid failure, recognizing the importance of compaction for reducing the chances of levee failure.

Levees fail by three main modes: underseepage of water beneath the levee, where the pressure from the high water opens a channel causing catastrophic failure; hydraulic piping, in which the water finds a weak passage through the levee; and overtopping when the water flows over the top of the levee and erodes the sides. Levees can also fail when the river current scours the base of the levee during high-flow conditions, as happened in many of the Mississippi River floods, and this causes slumping and massive collapse of the levee. Mass wasting is also promoted by long-term floods in which the water gradually saturates the pores of the levee, weakening it, causing massive liquefaction and catastrophic failure, leading large sections of the levee to collapse at the same time. Most levee failures happen during times when the flow has been high for long periods, since this increases the pore pressure, scouring, and liquefaction potential of the levee.

By 1956 the MRC was modeling floods with twice the previous discharge, examining the ability of the river and levee system to handle a discharge of 3,000,000 cubic feet (2,300,000 m³) per second. Then the flood of 1973 hit the Mississippi River basin with one of the highest floods recorded in 200 years. The flood set a record for the number of days the river was out of bank, causing more than $183,756,000 in damages. In terms of flood management, the flood of 1973 brought the realization that building levees, wing dikes, and other navigational and so-called flood-control measures had actually decreased the carrying capacity of the river. This meant that for any given amount of water, the flood levels (called stages) would be higher than before the levees were built.

The catastrophic floods of 1993 provided another test of the levees, and the new system failed massively. The constriction of the river caused by the levees led to numerous cases of levee failure, overtopping, crevasse splays, collapse, and massive amounts of damage as had never been seen along the river. Approximately two-thirds of all the levees in the upper Mississippi River basin collapsed, were breached, or were otherwise damaged by the floods of 1993. Dozens of people died and 50,000 homes were damaged or destroyed, with total damage estimated at most than $15 billion.

SUMMARY

Streams are dynamic systems that represent a balance between the forces that drive the current and those that resist the flow. Channels have many different styles that form in response to a quasi equilibrium between the gradient, or slope, of the streambed, the discharge of the stream, the amount of sediment being transported, the roughness of the streambed, and the resistance of the bank to erosion. The stream may form one of three main types of channels in response to the relative contributions of these variables. Straight channels are the rarest and are usually controlled by incision into a bedrock structure, but within the straight channel the current usually follows a curved path. Meandering streams are most common, with the current actively eroding cut banks and depositing material on the opposite point bars. In this way the meanders move back and forth across the floodplain, maintaining equilibrium through changes in the sinuosity, meander wavelength, width and depth of the channel, and velocity of the current. Meandering channels and their floodplains are different parts of the same dynamic system. Braided streams have multiple channels and are prone to rapid changes; they carry more sediment than meandering or straight channels. They are prone to large fluctuations in discharge and load, and many are found in environments in front of melting glaciers.

Individual stream and river channels are parts of much larger systems, and the patterns of branching and angles between individual streams often define different patterns that reflect underlying processes. Some river systems exhibit control by uniform slopes; some have rectilinear patterns reflecting underlying beds and structures; some are radial, reflecting drainage off domes; and others cut straight through uplifted mountain ranges. The regional pattern of the stream channels reflects control by the underlying geology, and the more local stream channel pattern reflects control by the balance between the forces

driving the current and those opposing it. Streams can deposit thick layers of sand, mud, and gravel on floodplains, cut through them forming terraces, and carry massive amounts of sediment to the sea to deposit them as giant delta complexes. These delta complexes build to sea and have forms that reflect a balance among sediment input, tides, and wave energy. Many delta lobes are active on the order of 1,000 years; then the river switches course and forms a new lobe, as the older one subsides.

See also DRAINAGE BASIN (DRAINAGE SYSTEM); FLOOD; RIVER SYSTEM; SEDIMENTARY ROCK, SEDIMENTATION.

FURTHER READING

Galloway, W. E., and D. K. Hobday. *Terrigineous Clastic Depositional Systems.* New York: Springer-Verlag, 1983.

Gordon, N. D., T. A. McMahon, and B. L. Finlayson. *Stream Hydrology: An Introduction for Ecologists.* New York: John Wiley & Sons, 1992.

Ritter, D. F., R. C. Kochel, and J. R. Miller. *Process Geomorphology.* 3rd ed. Boston: WCB-McGraw Hill, 1995.

Schumm, S. A. *The Fluvial System.* New York, Wiley-Interscience, 1977.

flysch Flysch is a syn-orogenic clastic sedimentary deposit typically marked by interbedded shales and sandstones. The term was first used for sedimentary rocks deposited in the Alps in Cretaceous-Tertiary times, before the main erosional event that shed coarser-grained conglomerates known as molasse. Sedimentary structures in flysch typically include a series of graded and cross-laminated layers in sands forming Bouma sequences, indicating that the sands were deposited by turbidity currents. Flysch is typically deposited in foreland basins and forms regionally extensive clastic wedges, underlain by distal black shales and overlain by fluvial deposits and conglomerates of fluvial origin.

The most common type of sedimentary deposit in flysch sequences are turbidite sequences. A turbidite is a deposit of a submarine turbidity current consisting of graded sandstone and shale, typically deposited in a thick sequence of similar turbidites. Most turbidites are thought to be deposited in various subenvironments of submarine fans, in shallow- to deep-water settings. These form when water-saturated sediments on a shelf or in a shallow water setting are disturbed by a storm, earthquake, or some other mechanism that triggers the sliding of the sediments down slope. The sediment-laden sediment/water mixture then moves rapidly down slope as a density current, and may travel tens or even hundreds of miles at tens of miles per hour until the slope decreases

Alternating sandstone and shale deposits arranged in an anticline fold in flysch sediments in Zumaia, Spain *(Dirk Wiersma/Photo Researchers, Inc.)*

and the velocity of current decreases. As this occurs the ability of the current to hold coarse material in suspension decreases, and the current drops, first, its coarsest load, and then progressively finer material as the current velocity continues to decrease. In this way the coarsest material is deposited closest to the channel or slope that the turbidity current flowed down, and the finest material is deposited further away. The same sequence of coarse to fine material is deposited upward in the turbidite bed as the current velocity decreases with time at any given location. This is how graded beds are formed, with the coarsest material at the base and finer material at the top.

Classical complete turbidite beds consist of a sequence of sedimentary structures divided into a regular A-E sequence known as the Bouma sequence, after the sedimentologist Arnold Bouma, who first described the sequence. The A horizon consists of coarse- to fine-grained graded sandstone beds, representing material deposited rapidly from suspension. The B horizon consists of parallel-laminated sandstones deposited by material that moved in traction on the bed, whereas division C contains cross-laminated sands deposited in the lower-flow regime. The D and E horizons represent the transition from mate-

rial deposited from the waning stages of the turbidity current and background pelagic sedimentation.

Variations in the thickness and presence or absence of individual horizons of the Bouma sequence have been related to where on the submarine fan or slope the turbidite was deposited. Turbidites with more of the A-B-C horizons are interpreted to have been deposited closer to the slope or channel, whereas turbidites sequences with more of the C-D-E horizons are interpreted as more distal deposits.

Many turbidite sequences are deposited in foreland basins and in deep-sea trench settings. These environments have steep slopes in the source areas, a virtually unlimited source of sedimentary material, and many tectonic triggers to initiate the turbidity current.

Molasse sequences overlie many turbidite-bearing flysch sequences, especially those deposited in foreland basins. Molasse consists of thick sequences of coarse-grained postorogenic sandstones, conglomerates shales, and marls that form in response to the erosion of orogenic mountain ranges. The name is derived from the classic Miocene-Oligocene-Pliocene molasse of the European foreland, deposited across much of France, Switzerland, and Germany, and overlying the Alpine flysch sequence. These sediments are up to four miles (7 km) thick on the Swiss Plateau and represent rapid erosion of the Alps. Lower parts of the molasse include shallow marine and tidally influenced sediments, overlain by alluvial fan deltas, alluvial fan complexes, and overbank deposits.

See also BASIN, SEDIMENTARY BASIN; CONVERGENT PLATE MARGIN PROCESSES; OROGENY; SEDIMENTARY ROCK, SEDIMENTATION.

FURTHER READING

Bouma, Arnold H. *Sedimentology of Some Flysch Deposits.* Amsterdam: Elsevier, 1962.

Kuenen, Phillip H., and Carlo I. Migliorini. "Turbidite Currents as a Cause of Graded Bedding." *Journal of Geology* 58 (1950): 91–127.

Walker, Roger G. *Facies Models.* Toronto: Geoscience Canada Reprint Series 1, Geological Association of Canada, 1983.

fossil A fossil is any remains, trace, or imprint of any plant or animal that lived on the Earth. Such remains of past life include body fossils, the preserved record of hard or soft body parts, and trace fossils that record traces of biological activity such as footprints, tracks, and burrows. The oldest body fossils known are 3.4 billion-year-old remnants of early bacteria, whereas chemical traces of life may extend back to 3.8 billion years.

The conditions that lead to fossilization occur so rarely that a mere 10 percent of all species that have ever existed are estimated to be preserved in the fossil record. The record of life and evolution is therefore very incomplete. To be preserved life-forms become mineralized after they die, with organic tissues typically being replaced by calcite, quartz, or other minerals during burial and diagenesis. Fossils are relatively common in shallow marine carbonate rocks where organisms that produced calcium carbonate shells are preserved in a carbonate matrix.

The fossil record has been used to test, modify, and support evolution, a concept traditionally regarded as a slow, gradual process that describes how life has changed with time on the Earth, starting with simple single-celled organisms to the complex biosphere on the planet today. However, a better definition for biological evolution is a sustained change in the genetic makeup of populations over a period of generations leading to a new species. The field of evolution was pioneered by Charles Darwin in his *Origin of Species* (1859) and *The Descent of Man* (1871) and is a multidisciplinary science incorporating geology, paleontology, biology, and, with neo-Darwinism, genetics.

Darwin sailed on the HMS *Beagle* (1831–36) when he made numerous observations of life and fossils from around the world, leading to the development of his theory of natural selection, in which species with favorable traits stand a better chance for survival. The main tenets of his theory are that species reproduce more than necessary, but populations tend to remain stable since there is a constant struggle for food and space, and only the fittest survive. Darwin proposed that the traits that contributed to an individual's survival are passed on to its descendants, hence propagating the favorable traits. But Darwin did not have a good explanation for why some individuals would have favorable traits that others would not. This evidence would not come until much later with the field of genetics and the recognition that mutations can cause changes in character traits. Sequential passing down to younger generations of mutation-induced changes in character traits can lead to changes in the species and, eventually, the evolution of new species. Darwin's process of natural selection works by the gradual elimination of the less successful forms of species, favoring the other forms that had favorable mutations.

More modern variations on evolution recognize two major styles of change. Macroevolution describes changes above the species level and the origin of major groups, whereas microevolution is concerned with changes below the species level and the development of new species. Another major development in the field of evolution over the past century relates to the rate of evolutionary changes as preserved in the fossil record. Darwin thought that evolution pro-gressed slowly, with one species gradually changing into a new species, but the fossil record supports only a few examples of this gradual change (notable examples include changes in trilobites in the Ordovician and changes in horses in the Cenozoic). Fossil evidence demonstrates the persistence of nearly all species with little change for long periods of geologic time, followed by a sudden disappearance and subsequent replacement with entirely new species. In other cases new species suddenly appear without the disappearance of other species. Biologists initially regarded skeptically some of the apparent rapid change in an incomplete fossil record, but many examples of complete records show that these rapid changes are real. A new paradigm of evolution named punctuated equilibrium, advanced in the 1970s by Steven Jay Gould and Niles Eldredge, explains these sudden evolutionary changes. Physical or geographic isolation of some members of a species, such as expected during supercontinent breakup, can separate and decimate the environment of a species and effectively isolate some of its members in conditions that can select for change. This small group may have a mutation that favors their new environment, letting them survive. When supercontinents collide, many species that never encountered one another must compete for the same food and space, and only those best suited to that particular environment will survive to reproduce, leading to extinction of the others.

In other cases major environmental catastrophes such as meteorite impacts and flood basalt eruptions can cause extreme changes in the planetary environment, causing mass extinction. Relatively minor or threatened species that survive can suddenly find themselves with traits that favor their explosion into new niches and their dominance in the fossil record.

See also EVOLUTION; LIFE'S ORIGINS AND EARLY EVOLUTION; MASS EXTINCTIONS; PALEONTOLOGY; SUPERCONTINENT CYCLES.

FURTHER READING

McKinney, Michael L. *Evolution of Life: Processes, Patterns, and Prospects.* Englewood Cliffs, N.J.: Prentice Hall, 1993.

Stanley, Steven M. *Earth and Life Through Time.* New York: W. H. Freeman, 1986.

fracture A general name for a break in a rock or other body that may or may not have any observable displacement. Fractures include joints, faults, and cracks formed under brittle deformation conditions and are a kind of permanent (nonelastic) strain. Brittle deformation processes generally involve the growth of fractures or sliding along existing fractures. Frictional sliding involves the sliding on preexisting fracture surfaces, whereas cataclastic flow includes grain-scale fracturing and frictional sliding producing macroscopic ductile flow over a band of finite width. Tensile cracking involves the propagation of cracks into unfractured material under tensile stress perpendicular to the maximum compressive stress, whereas shear rupture refers to the initiation of fracture at an angle to the maximum principal stress.

Fractures may propagate in one of three principal modes. Mode I refers to fracture growth by incremental extension perpendicular to the plane of the fracture at the tip. In Mode II propagation is where the fracture grows by incremental shear parallel to the plane of the fracture at the tip, in the direction of fracture propagation. Mode III is when the fracture grows by incremental shear parallel to the plane of the fracture at the tip, perpendicular to the direction of propagation.

Joints are fractures with no observable displacement parallel to the fracture surface. They generally occur in subparallel joint sets, and several sets often occur together in a consistent geometric pattern, forming a joint system. Joints are sometimes classified into extension joints or conjugate sets of shear joints, a subdivision based on the angular relationships between joints. Most joints are continuous for only short distances, but in many regions master joints may run for long distances and control geomorphology or form air photo lineaments. Microfractures or joints are visible only under the microscope and affect only a single grain.

Many joints are contained within individual beds and have a characteristic joint spacing, measured perpendicular to the joints. This is determined by the relative strength of individual beds or rock types, the thickness of the jointed layer, and structural position, and is very important for determining the porosity and permeability of the unit. In many regions fractures control groundwater flow; location of aquifers, and migration and storage of petroleum and gas.

Joints and fractures, found in all kinds of environments, form by a variety of mechanisms. The contraction of materials induces the formation of desiccation cracks and columnar joints. Mineral changes during diagenesis that lead to volume changes in the layer produce bedding plane fissility, characterized by fracturing parallel to bedding. Unloading joints form by stress release, such as during uplift, ice sheet withdrawal, or quarrying operations. Exfoliation joints and domes may form by mineral changes, including volumetric changes during weathering, or by diurnal temperature variations. Most joints have tectonic origins, typically forming in response to the last phase of tectonic movements in an area. Other joints seem to be related to regional doming, folding, and faulting.

Many fractures and joints exhibit striated or ridged surfaces known as plumose structures, since they vaguely resemble feathers. Plumose structures develop in response to local variations in propagation velocity and the stress field. The origin is the point at which the fracture originated, the mist is the small ridging on the surface, and the plume axis is the line that starts at the axis and from which individual barbs propagate. The twist hackle refers to the steps at the edge of the fracture plane along which the fracture has split into a set of smaller en echelon fractures.

British geologist E. M. Anderson elegantly explained the geometry and orientation of some fracture sets in 1905 and 1942, and in a now classic work published in 1951, *The Dynamics of Faulting and Dyke Formation*. General acceptance of this model by the scientific community led to Anderson's model being adopted, and many fault and fracture sets are described in terms of Anderson's theory. According to Andersonian theory the attitude of a fracture plane tells a lot about the orientation of the stress field that operated when the fracture formed. Fractures are assumed to form as shear fractures in a conjugate set, with the maximum compressive stress bisecting an acute (60°) angle between the two fractures. In most situations the surface of the Earth may be the maximum, minimum, or intermediate principal stress, since the surface can transmit no shear stress. If the maximum compressive stress is vertical, two fracture sets will form, each dipping 60° toward each other and intersecting along a horizontal line parallel to the intermediate stress. If the intermediate stress is vertical, two vertical fractures will form, with the maximum compressive stress bisecting the acute angle between the fractures. If the least compressive stress is vertical, two gently dipping fractures will form, and their intersection will be parallel to the intermediate principal stress.

Other interpretations of fractures and joints include modifications of Andersonian geometries that include volume changes and deviations of principal stresses from the vertical. Many joints show relationships to regional structures such as folds, with some developing parallel to the axial surfaces of folds and others crossing axial surfaces. Other features on joint surfaces may be used to interpret their mode of formation. For instance, plumose structures typically indicate Mode I or extensional types of formation, whereas the development of fault striations (known as slickensides) indicates Mode II or Mode III propagation. Observations of these surface features, the fractures' relationships to bedding, structures such as folds and faults, and their regional orientation and distribution can lead to a clear understanding of their origin and significance.

FRACTURE ZONE AQUIFERS

Fractures and joints are in many places important aquifers, forming deep spaces in the Earth where water can be stored without evaporation or contamination for centuries or even thousands of years. Faults and fractures develop at various scales from faults that cross continents to fractures that are visible only microscopically. The internal properties of the rock and the external stresses imposed on it determine the location and orientation of these discontinuities in the rock fabric. Fractures at various scales represent zones of increased porosity and permeability. By forming networks they are able to store and carry vast amounts of water.

The concept of fracture zone aquifers explains the behavior of groundwater in large fault-controlled watersheds. Fault zones in this case serve as collectors and transmitters of water from one or more recharge zones with surface and subsurface flow strongly controlled by regional tectonism.

Both the yield and quality of water in these zones are usually higher than average wells in any type of rock. High-grade water for such a region would be 250 gallons (950 liters) per minute or greater. In addition the total dissolved solids measured in the water from such high-yielding wells will be lower than the average for the region.

The fracture zone aquifer concept looks at the variations in groundwater flow as influenced by secondary porosity over an entire watershed. It attempts to integrate data on a basin in an effort to describe the unique effects of secondary porosity on the processes of groundwater flow, infiltration, transmissivity, and storage.

The concept includes variations in precipitation over the catchment area. One example is orographic effects wherein the mountainous terrain precipitation is substantially greater than at lower elevations. The rainfall is collected over a large catchment area, which contains zones with high permeability because of intense bedrock fracturing associated with major fault zones. The multitude of fractures within these highly permeable zones "funnel" the water into other fracture zones that are down gradient from where the water enters the system. These funnels may be in a network of hundreds of square miles (kilometers).

The fault and fracture zones serve as conduits for groundwater and often act as channelways for surface flow. Intersections form rectilinear drainage patterns sometimes exposed on the surface but are also represented below the surface and converge down the hydrologic gradient (at places to which water would flow naturally downhill). In some regions these rectilinear patterns are not always visible on the surface owing to vegetation and sediment cover. The convergence of these groundwater conduits increases the amount of water available as recharge. The increased permeability, water volume, and ratio of water to minerals within these fault/fracture zones help to maintain the quality of water supply. These channels occur in fractured, nonporous media (crystalline rocks) as well as in fractured, porous media (sandstone, limestone).

At some point in the groundwater course, after convergence, the gradient decreases. The sediment cover over the major fracture zone becomes thicker and acts as a water storage unit with primary porosity. The major fracture zone acts as both a transmitter of water along conduits and a water storage basin along connected zones with secondary (and/or primary) porosity. Groundwater within this layer or lens often flows at accelerated rates. The result can be a pressurization of groundwater both in the fracture zone and in the surrounding material. Precipitation can almost instantaneously replenish the rapid flow in the conduit. The surrounding materials are replenished more slowly, but also release the water more slowly and serve as a storage unit to replenish the conduit between precipitation events.

Once the zones are saturated, any extra water that flows into them will overflow, if an exit is available. In a large-area watershed it is likely that this water flows along subsurface channelways under pressure until some form of exit is found in the confining environment. Substantial amounts of groundwater may flow along an extension of the main fault zone controlling the watershed and may vent at submarine extensions of the fault zone, forming coastal or offshore freshwater springs.

The concept of fracture zone aquifers is particularly applicable to areas underlain by crystalline rocks and where these rocks have undergone a multiple deformational history that includes extensional tectonics. This is especially true for areas where recharge is possible from seasonal and/or sporadic rainfall on mountainous regions adjacent to flat desert areas.

Fracture zone aquifers are distinguished from horizontal aquifers in that (a) they drain numerous wadis in extensive areas and many extend for tens of miles (dozens of kilometers); (b) they constitute conduits to mountainous regions where the recharge potential from rainfall is high; (c) some may connect several horizontal aquifers and thereby increase the volume of accumulated water; (d) because the source of the water is at higher elevations, the artesian pressure at the groundwater level may be high; and (e) they are usually missed by conventional drilling because the water is often at the depth of up to 1,000 feet (hundreds of meters).

The characteristics of fracture zone aquifers make them an excellent source of groundwater in arid and semiarid environments. Fracture zone aquifers are located by seeking major faults. The latter are usually clearly displayed in images obtained from spacecraft in Earth's orbit, because they are emphasized by drainage. Thus the first step in evaluating the groundwater potential of any region is to study the structures displayed in satellite images to map the faults, fractures, and linear features of uncertain origin (called lineaments). Such a map is then compared with a drainage map showing wadi locations. The combination of many wadis and major fractures indicates a larger potential for groundwater storage. Furthermore, the intersection between major faults would increase both porosity and permeability and, hence, the water-collection capacity.

Groundwater resources in arid and semiarid lands are scarce and must be properly used and thoughtfully managed. Most of these resources are "fossil," having accumulated under wet climates during the geological past. The present rates of recharge from the occasional rainfall cannot sufficiently replenish the aquifers. Therefore the resources must be used sparingly without exceeding the optimum pumping rates for each water well field.

See also DEFORMATION OF ROCKS; GROUNDWATER; STRUCTURAL GEOLOGY.

FURTHER READING

Anderson, E. M. *The Dynamics of Faulting and Dyke Formation.* London: Oliver and Boyd, 1951.

Bisson, Robert A., and Farouk El-Baz. "Megawatersheds Exploration Model." *Proceedings of the 23rd International Symposium on Remote Sensing of Environment.* Ann Arbor, Mich.: Environmental Research Institute of Michigan, 1990.

El-Baz, Farouk. "Utilizing Satellite Images for Groundwater Exploration in Fracture Zone Aquifers." *International Conference on Water Resources Management in Arid Countries.* Muscat, Oman: Ministry of Water Resources, 1995.

Gale, J. E. "Assessing the Permeability Characteristics of Fractured Rock." In *Recent Trends in Hydrogeology,* edited by T. N. Narasimhan. *Geological Society of America Special Paper* 189 (1982).

Kusky, Timothy M., and Farouk El-Baz. "Structural and Tectonic Evolution of the Sinai Peninsula, Using Landsat Data: Implications for Groundwater Exploration." *Egyptian Journal of Remote Sensing* 1 (1999): 69–100.

National Academy of Sciences. *Rock Fractures and Fluid Flow: Contemporary Understanding and Applications.* Washington, D.C.: National Academy Press, 1996.

Pollard, David D., and Aydin Atilla. "Progress in Understanding Jointing over the Past Century." *Geological Society of America* 100 (1988): 1181–1204.

Ramsay, John G., and Martin I. Huber. *The Techniques of Modern Structural Geology, Volume 2: Folds and Fractures.* London: Academic Press, 1987.

Wright, E. P., and W. G. Burgess. "The Hydrogeology of Crystalline Basement Aquifers in Africa." *Geological Society of London Special Publication* 66 (1992).

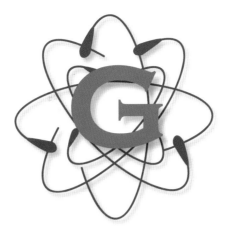

Gaia hypothesis The British atmospheric chemist James Lovecock proposed the Gaia hypothesis in the 1970s, suggesting that Earth's atmosphere, hydrosphere, geosphere, and biosphere interact as a self-regulating system that maintains conditions necessary for life to survive. In this view the Earth acts as if it is a giant self-regulating organism in which life creates changes in one system to accommodate changes in another to keep conditions within the narrow limits that allow life to continue on Earth.

The temperature on the Earth has been maintained at 50°–86°F (10°–30°C) for the past 3.5 billion years, even though the solar energy received by the Earth has increased by 40–330 percent since the Hadean. The temperature balance has been regulated by changes in the abundance of atmospheric greenhouse gases, controlled largely by volcanic degassing and the reduction of carbon dioxide (CO_2) by photosynthetic life. A slight increase or decrease in CO_2 and other greenhouse gases could cause runaway greenhouse or icehouse global climates, yet life has been able to maintain the exact balance necessary to guarantee its survival.

The presence of certain gases such as ammonia at critical levels in the atmosphere for maintaining soil pH near 8, the optimal level for sustaining life, is critical for maintaining atmospheric oxygen levels. This critical balance is unusual, as methane is essentially absent from the atmospheres of Venus and Mars, where life does not exist. The salinity of the oceans has been maintained at around 3.4 percent, in the narrow range required for marine life, reflecting a critical balance between terrestrial weathering, evaporation, and precipitation.

The exact mechanisms that the Earth maintains for these critical balances necessary for life are not well known. As solar luminosity increases, however, the additional energy received by the Earth is balanced by the amount of energy radiated back to space. Changes in the surface reflectance (albedo) can accomplish this through changes in the amount of ice cover, types of plants, and cloud cover. Changes in one Earth system produce corresponding changes in other systems in self-regulation processes known as homeostasis. Critical for Gaia are the links between organisms and the physical environment, such that many proponents of the theory regard the planet as one giant superorganism.

See also ATMOSPHERE; CLIMATE; CLIMATE CHANGE; GREENHOUSE EFFECT; SUPERCONTINENT CYCLE.

galaxies Galaxies are gravitationally bound assemblages of stars, dust, gas, radiation, and dark matter. Most contain vast numbers of star systems and are located at enormous distances from our Milky Way Galaxy, such that the light reaching Earth from these galaxies was generated billions of years ago.

TYPES OF GALAXIES

Telescopic observations of galaxies show that they have a wide range of different shapes first classified by the American astronomer Edwin Hubble in 1924 into four basic types, including spiral galaxies, barred spiral galaxies, elliptical galaxies, and irregular galaxies. Hubble's classification has since been modified and elaborated upon, but astronomers continue to use the same basic scheme.

Spiral galaxies are characterized by a flattened disk shape that exhibits a central bulge and spiral-shaped arms that emerge from this central bulge and extend in variously curved forms to distant reaches of the galaxy. These are surrounded by a galactic halo made of a ball of old faint stars forming a sphere around the other parts of the galaxy. The Milky Way, which contains billions of star and planetary systems, including Earth, is a spiral galaxy. One of the major differences between different types of spiral galaxies is how tightly wrapped the spiral arms are as they circle the bulge in the core of the galaxy. Generally, the larger the central bulge of the galaxy, the more tightly wrapped the arms become. Galaxies with small central bulges in their cores tend to have loosely wrapped spiral arms and more lumpy or knotty distributions of matter within their arms. In the Hubble classification, forms of spiral galaxies are abbreviated by the letter *S*, with small letters *a-d* denoting progressively more open spiral forms, such that Sc galaxies are more open than Sa galaxies.

Most spiral galaxies have galactic disks rich in gas and dust, and halos comprised largely of old dim stars. The spiral arms have many younger stars and newly forming star systems and are the densest parts of these galaxies, providing material for the birth of new star systems.

Barred spirals are a special class of spiral galaxies in which a concentrated "bar" of stellar and interstellar matter passes through the central bulge of the galaxy, and the spiral arms extend from the ends of this bar. Most of these have unusual shapes, resembling giant Z or S shapes, with spiral trails of luminous matter extending around the letter. Barred spiral galaxies are designated by the Hubble classification as SB galaxies, with the small letters *a-c* denoting how open the spiral arms are around the bar.

Elliptical galaxies are circular to highly elliptical concentrations of stellar and interstellar matter whose density increases toward the center of the galaxies. The size of elliptical galaxies and the number of stars contained within elliptical galaxies vary widely. Some are small and known as dwarf ellipticals, being only a kiloparsec across and containing on the order of a million stars. Others are huge, many times the size of the Milky Way Galaxy, and contain trillions of stars in giant elliptical galaxies that can be several megaparsecs across. Elliptical galaxies typically exhibit little internal structure, with no spiral arms and no central bulge. They are designated in the Hubble classification by the letter *E* with the numbers 1–7 indicating a progression from least- to most-elongated varieties.

Elliptical galaxies also differ from spiral galaxies in that they contain little gas and dust, and they seem not to have any young stars or places where star formation is in the process of occurring (when the light

Hubble image of M51 Whirlpool Galaxy dated November 7, 2002 *(NASA Goddard Space Flight Center)*

was formed). Most of the stars in elliptical galaxies appear to be old, relatively cold, reddish low-mass stars, similar to the stars in the halo of the Milky Way and other spiral galaxies. Ellipticals are therefore old systems in which the gas and dust was all swept up by the star systems long ago, and the stars are moving about in irregular paths within the elliptical mass. As with most generalized statements, there are exceptions. Recent observations have shown that some giant elliptical galaxies have smaller areas that contain disks of gas and dust, but some astronomers speculate that these may be other spiral galaxies that collided with the giant ellipticals.

Irregular galaxies include the whole range of other galaxies that do not fit into Hubble's spiral, barred spiral, or elliptical galaxy classes. These galaxies tend to lack systematic structure such as spiral arms or bulges, but they do contain large amounts of interstellar material such as dust and gas. Irr I type irregular galaxies slightly resemble distorted spiral galaxies, with famous examples including the Magellanic Clouds that orbit the Milky Way. The less common Irr II galaxies exhibit explosive or filamentous characteristics. Different models have been advanced to explain these characteristics of Irr II galaxies, including massive explosions inside the galaxies, or effects of close encounters with other galaxies. Most irregular galaxies contain between 1,000,000,000 to 100,000,000,000 stars, with the smaller "dwarf" irregular being more common than the larger elliptical galaxies.

PHYSICAL PROPERTIES OF GALAXIES

The observable universe presents an estimated minimum of 100 billion galaxies, and many of these have billions of stars in them. Most of these galaxies are located far from Earth and the Milky Way Galaxy, and thus are difficult to observe closely. To measure distances to and sizes of these distant galaxies one must use some objects, such as planetary nebulae or certain kinds of supernovae that have known brightnesses, and then use their apparent brightness to measure their distance from Earth. Another relationship that has been exploited to measure the distance to faraway galaxies is to measure their rotational speeds, and then correlate these with a known relationship between the rotational speed of a galaxy, its mass, and its luminosity. The rotational speed can be measured at great distances, and the absolute brightness calculated; then when compared with the apparent brightness, the distance to the galaxy can be calculated. Using these methods to measure distances to galaxies, scientists have found that most lie at vast distances from the Earth, most much greater than 20 megaparsecs away. Furthermore, there is some order to the large-scale structure in the arrangement of galaxies, with many residing in galaxy clusters, superclusters, and other even larger structures.

Determining the masses of distant galaxies can be difficult. For spiral galaxies within about 50 kiloparsecs of Earth, the rotational speed of the different spiral arms can be determined from the Doppler shifts of each arm, and if the distance from the galactic center is known, then Newton's laws of motion can be used to calculate the mass of a galaxy. For more distant galaxies one must depend on less reliable methods to estimate their masses. One way is to search for binary galaxy systems, then measurements of their orbital size and their orbital period enable the calculation of their mass using Kepler's third law. These different methods reveal that most spiral galaxies and large elliptical galaxies have about 10^{11}–10^{12} solar masses in them, while the irregular galaxies tend to be less massive, with 10^8–10^{10} solar masses. Dwarf ellipticals are the least massive, typically containing 10^6–10^7 solar masses.

The rotational properties of most spiral galaxies and many elliptical galaxies indicates that they have excess mass surrounding them, but this mass is not luminous and is thought to be dark matter. The amount of dark matter in many cases is estimated to be 3–10 times the mass of the luminous matter in the galaxies. Galaxy clusters also appear to be associated with massive amounts of dark matter, with calcula-

tions showing that there must be between 10 and 100 times the masses of individual galaxy clusters. These calculations lead to the shocking conclusion that about 90 percent of the universe must be made up of invisible dark matter not detectable at any electromagnetic wavelength but can be observed only by its gravitational effects.

X-ray observations of galaxy clusters have demonstrated that some clusters are associated with strong emissions of X-ray radiation, and these are interpreted to be coming from hot gases that exist as intergalactic gas within the clusters. The mass of this gas is estimated, in some cases, to be about the same as or even more than the mass of the visible matter, but still substantially less than the mass needed to explain the gravitational observations by a factor of 10 to 100.

The motions of galaxies show interesting patterns on different scales of observation. The motion of individual galaxies within clusters of galaxies appears random, but the clusters show very ordered patterns to their motions at some of the largest scales of observation in the universe. Some of these motions have been partly understood for nearly a century. In 1912 Vesto Slipher, an American astronomer working with Percival Lowell (1855–1916), the American astronomer who founded Lowell Observatory and was president of Harvard University, discovered that every spiral galaxy he observed had a redshifted spectrum; Slipher concluded they were all moving away from the Earth. This observation has since been extended to include all known galaxies, which are moving away from the Earth in all directions. Individual galaxies not in clusters are moving away, as are the groups of galaxies in clusters, even though they have some random motions within the clusters. Furthermore, as observations improved, it became clear that the farther away the galaxy is from the Earth, the greater the redshift, and the faster it is receding.

In the 1920s the astronomer Edwin Hubble made a series of plots of the redshifts of galaxies and their calculated recessional velocities with distance from the Earth. He found that there is a straight-line relationship such that velocity increases steadily with distance. This proportionality is known as Hubble's law, and the general picture of all of the galaxies moving apart from one another is known as Hubble flow. Hubble's flow provides clear evidence that the universe is expanding.

Hubble's law can be written as follows:

$$V = H_0 \times D$$

where V is the recessional velocity, D is the distance, and H_0 is the proportionality constant (as Hubble's

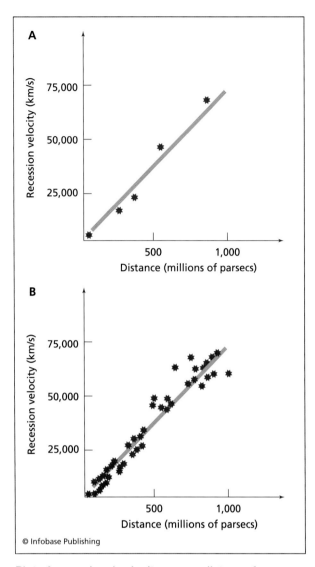

Plot of recessional velocity versus distance for many galaxies within about 1 billion parsecs of Earth, illustrating Hubble's law, that recessional velocity is proportional to distance *(modified from Chaisson and MacMillan)*

© Infobase Publishing

constant) between the recessional velocity and distance. The slope of the straight line on a distance/recessional velocity diagram is equal to Hubble's constant, which turns out to be approximately 75 km/sec per megaparsec (46.5 miles/sec per 3.3 million light years). There is, however, uncertainty in the exact value of the Hubble constant, with nearly all estimates falling in the range of 37–56 miles per second (60–90 km/sec) per megaparsec. Hubble's constant represents the best estimate of the rate of expansion of the universe.

Hubble's law is also extremely useful for measuring distances to faraway objects. Since the recessional velocity is proportional to the distance of the object, it is simple to measure the recessional velocity (from the redshift of the spectrum), then use Hubble's law to estimate the distance directly. This method works well even for very distant objects and is used to calculate the distance to the most distant objects yet known in the universe—object QO51-279, which has a redshift showing a recessional velocity of 93 percent the speed of light, and a distance of 4,000 megaparsecs. The electromagnetic radiation now observable on Earth from QO51-279 was generated about 13 billion years ago, close to the time of the big bang (presently estimated to be 13.73 billion +/- 120 million years ago). Another extremely distant object was discovered in 2004 using the Hubble Space Telescope and an effect of general relativity called gravitational lensing, where massive objects in the foreground of a distant object can bend and magnify the light from the distant object, making it more observable. In 2004 a team of scientists discovered an object magnified by a gravitational lens in a galactic cluster (Abell 2218), and that object is a small, compact system of stars approximately 2,000 light years across and about 13 billion light years away. Using the present estimate of the age of the universe, scientists estimate the light from this object now reaching Earth was generated when the universe was only 750 million years old.

Using powerful telescopes and Hubble's law, scientists can now map the large-scale structure of the universe. It turns out that the universe is not a random collection of star and galaxy systems but rather a patterned distribution of galaxies and clusters of galaxies, which are arranged in a network of string- or filament-like groups, separated by largely empty space known as voids. Astronomers have mapped these stringlike features to be on the surfaces of bubblelike voids, as if the universe were made of a system of empty bubbles with galaxy clusters forming chains along the surfaces of the bubbles. The areas where several bubbles intersect tend to be where the densest galaxy clusters and superclusters are located. The origin of these bubblelike structures is debated but must be related to density fluctuations or ripples in the earliest stages of the formation of the universe that grew during time and the expansion of the universe.

GALACTIC EVOLUTION

Despite years of research, there is still remarkably little known about the processes of galaxy formation and why there is such variety in the structure of different galaxies. Most astronomers suggest that small density fluctuations in the primordial matter led to the formation of many small "pregalactic masses" that were similar to present-day dwarf galaxies, and that collisions and mergers of these galaxies led to the formation of the larger galaxies common in the present universe. As the universe expanded, these merged galaxies grew into the large-scale clusters and voids that now occupy the universe. Much of the evidence for such a history of galaxy formation comes from observations of the most distant objects in the universe, whose light and other electromagnetic radiation that reaches Earth now was generated billions of years ago when the universe was young. It seems that the further back in time one observes, the smaller and less organized individual galaxies appear. Furthermore, there are many examples of galaxies merging, and many stages are observed of irregular galaxies merging to form more complex systems.

Different ideas and models attempt to explain why some galaxies are elliptical, some spiral, and some irregular. This variation may relate to the timing of when the stars in the galaxy formed compared to the galactic formation—if the stars formed early in the galactic evolution, an elliptical galaxy would most likely form, since the gas would be used up, and no central disk would form. In these elliptical galaxies star formation would occur early, and their present state would be dominated by systems with relatively old and cold stars. In contrast, if a lot of gas remained in the galaxy after it formed, then gravity would make this gas tend to collapse into a rotating disk, forming spiral galaxies, with stars able to form throughout the history of these types of galaxies.

The reason why some galaxies may have early star formation and form ellipticals, whereas others have late star formation and form spirals, is still not clearly known. But it is known that spiral galaxies are comparatively rare in parts of the universe that have high galactic density, perhaps because collisions between galaxies are more common in areas of high galactic density, and collisions tend to destroy the spiral arm structures of these galaxies. Observations and computer models suggest that collisions between galaxies tend to leave new galaxies that have elliptical characteristics, and these collisions eject large amounts of gas into intergalactic space. Also observations of deep space show that ellipticals were more common earlier in the universe, and are becoming less common with time, suggesting that collisions may be destroying their spiral structure with time.

Although many galaxies formed early in the history of the universe, many are still forming or being extensively modified through collisions and other interactions that are ongoing in the present-day universe. As galaxies interact, their halos of dark matter first interact and may be transferred from one (relatively smaller) galaxy to another, then the galaxies may spin in toward each other and merge, typically with the larger galaxy absorbing (or cannibalizing) the smaller. Other computer models of interactions

between galaxies show that it is possible for galaxies to come close but not merge, and one possible outcome of these types of interactions is the formation of spiral arms in one galaxy, where none existed before. Other interactions between galaxies produce sudden bursts of new star formation in the affected galaxies, showing that galaxy and stellar formation is an ongoing process in the universe.

See also ASTRONOMY; ASTROPHYSICS; COSMOLOGY; DARK MATTER; GALAXY CLUSTERS; HUBBLE, EDWIN; KEPLER, JOHANNES.

FURTHER READING

Brecher, Kenneth. "Galaxy." World Book Online Reference Center. 2005. World Book, Inc. Available online. URL: http://www.worldbookonline.com/wb/Article?id=ar215080. Accessed October 10, 2008.

Chaisson, Eric, and Steve McMillan. *Astronomy Today.* 6th ed. Upper Saddle River, N.J.: Prentice Hall, 2001.

Comins, Neil F. *Discovering the Universe.* 8th ed. New York: W. H. Freeman, 2008.

National Aeronautics and Space Administration. "Universe 101, Our Universe, Big Bang Theory. Cosmology: The Study of the Universe Web page." Available online. URL: http://map.gsfc.nasa.gov/universe/. Updated May 8, 2008.

Snow, Theodore P. *Essentials of the Dynamic Universe: An Introduction to Astronomy.* 4th ed. St. Paul, Minn.: West, 1991.

galaxy clusters Most galaxies lie at great distances from the Earth and the Milky Way Galaxy, and tend to form groups or clusters held together by their mutual gravitational attraction. These groups or clusters are separated by voids characterized by relatively empty space apparently devoid of luminous matter. The Milky Way Galaxy is part of the Local Group, which also includes the large Andromeda Spiral galaxy, the Large and Small Magellanic Clouds, and about 20 smaller galaxies. Most of the galaxies in the Local Group are elliptical and dwarf irregular systems, and the diameter of the local group is about 1 megaparsec.

Several types of galaxy clusters exist. Regular clusters are spherical with a dense central core, and are classified based on how many galaxies reside within 1.5 megaparsecs of the cluster center. Most regular clusters have a radius of 1–10 megaparsecs and masses of 10^{15} solar masses. In some cases such as the Coma cluster, there may be thousands of galaxies within 1.5 megaparsecs of the center of the cluster, making this one of the densest large-scale regions in the universe.

Irregular clusters generally have slightly lower mass (~10^{12}–10^{14} solar masses) than regular clusters

Cluster of galaxies called Abell 2218, taken by *Hubble Space Telescope* *(NASA, A. Fruchter and the ERO Team, STScI, ST-ECF)*

and have no well-defined center. An example of an irregular cluster is the Virgo cluster.

Galaxies and galaxy clusters are further grouped into even larger structures. Superclusters typically consist of groups or chains of clusters with masses of about 10^{16} solar masses. The Milky Way Galaxy is part of one supercluster, centered on the Virgo cluster, and has a size of about 15 megaparsecs. In contrast, the largest known superclusters, like that associated with the Coma cluster, are about 100 megaparsecs across. Recent advances in astronomers' ability to map the distribution of matter in space reveal that about 90 percent of all galaxies are located within a network of superclusters that permeates the known universe.

The largest-scale structures known in the universe include an understanding that these galaxy clusters and superclusters form a bubbly type of distribution of galaxies and clusters, with voids that are about 25 megaparsecs across separating sheets and filaments of galaxy clusters. The Great Wall is one such structure—a sheet of galaxies about 100 megaparsecs long, located about 100 megaparsecs from Earth.

Studies of the redshifts of galaxies from Earth reveal that many groups of galaxies on scales of 60 megaparsecs across are moving in a relatively coherent way, as if they are linked by some large-scale structure. Consistent with this idea is the determination that the Milky Way Galaxy, along with our local group, is moving at about 600 km/sec toward an area

in space about 45 megaparsecs away in the Centaurus Supercluster, known as the Great Attractor. The mass in this area is calculated to be in excess of 5 X 10^{16} solar masses, and is likely a diffuse concentration of matter about 400 million light years across, perhaps a supercluster itself.

See also ASTRONOMY; ASTROPHYSICS; COSMOLOGY; GALAXIES.

FURTHER READING

Chaisson, Eric, and Steve McMillan. *Astronomy Today.* 6th ed. Upper Saddle River, N.J.: Addison-Wesley, 2007.

Comins, Neil F. *Discovering the Universe.* 8th ed. New York: W. H. Freeman, 2008.

National Aeronautics and Space Administration. "Universe 101, Our Universe, Big Bang Theory. Cosmology: The Study of the Universe" Web page. Available online. URL: http://map.gsfc.nasa.gov/universe/. Updated May 8, 2008.

Snow, Theodore P. *Essentials of the Dynamic Universe: An Introduction to Astronomy.* 4th ed. St. Paul, Minn.: West, 1991.

Galilei, Galileo (1564–1642) Italian *Physicist, Mathematician, Astronomer* Galileo, born on February 15, 1564, in Pisa, Italy, played a major role in the scientific revolution during his life. His main contributions include making dramatic improvements in the telescope and using this as a tool to explore the heavens, providing support for the Copernican helio-centric model for the universe. He worked on the equations of motion for uniformly accelerated bodies and contributed to the science of kinematics, the study of the motion of objects. He became famous for his observations of Venus and sunspots. His views on heliocentrism were opposed by the Roman Catholic Church, which caused him to spend his final years under house arrest. The Church kept this position until 1981, when Pope John Paul II urged the Pontifican Academy of Sciences to reconsider Galileo's case and formed the Galileo Commission. In 1992, after more than a decade of investigation, the commission reversed the Church's earlier conviction of heresy and reversed its earlier verdict, finding Galileo not guilty of heresy. A statue was erected to honor Galileo in the Vatican gardens in 2009 on the 400th anniversary of his invention of the telescope.

PERSONAL LIFE

On February 15, 1564, Galileo Bonaiuti de'Galilei was born as the first of six children of Giulia Ammannati and Vincenzo Galilei, a famous lute player and music theorist of his time. When he was eight his family moved to Florence, and Galileo eventually enrolled at the University of Pisa to pursue a medical degree. During his studies Galileo changed to mathematics, and in 1589 at the age of 25 he was appointed as chair of mathematics at the university. In 1592 he moved to the University of Padua—one of the oldest in Italy, founded in 1222—where he taught geometry, mechanics, and astronomy until 1610. During this interval Galileo made major discoveries in the kinematics of motion, astronomy, and the strength of materials, and made improvements in the telescope.

Galileo had three children with Marina Gamba, but they were not married so their daughters (Virginia and Livia) were considered unmarriageable and spent their lives in the Convent of San Matteo in Arcetri. Their son, Vincenzio, was legitimized by the church and married Sestila Bocchineri.

In 1610 Galileo published a remarkable account of his observations of the moons of Jupiter and used this to argue for a Sun-centered (Copernican) model for the universe. In 1611 he went to Rome to show his telescope and the moons of Jupiter to leading philosophers at the Jesuit Collegio Romano, and at this time he was made a member of the Accademia dei Lincei.

In 1612 Father Tommaso Caccini denounced Galileo's models and ideas as being close to heresy, and when Galileo went to Rome in 1616 to defend his observations, Cardinal Roberto Bellamino admonished him and ordered him to stop teaching

and advocating Copernican astronomy. In 1630 he applied in Rome for a license to print his book, *The Dialogue Concerning the Two Chief World Systems*, which was published in Florence in 1632, but he was ordered instead to appear in the office of the church in Rome. The church put Galileo on trial for heresy, and from that time on, the pope ordered him to remain under house arrest in his country house in Arceti (near Florence). Galileo went blind in 1638 after suffering from insomnia and a hernia. After suffering additional fever and heart palpitations in 1642, he died at the age of 78.

SCIENTIFIC CONTRIBUTIONS

Galileo was one of the first scientists to state clearly that the laws of nature could be explained mathematically. In his book *The Assayer,* published in Rome in 1623, Galileo wrote that "the universe is written in the language of mathematics, and its characters are triangles, circles, and other geometric figures." Galileo was driven by testing assertions by scientists, philosophers, and religious figures through experimentation and mathematics, and this passion and reason in his character led him to reject many positions of authority, especially in the church. He became a major proponent of the need to separate religion and science. Galileo was instrumental in deriving mathematical relationships between curves, the equations to describe the curves, and relating these to the physical world. For instance, he noted that the parabola describes the path of a projectile moving without the effects of friction, and derived equations to explain this motion. He created standards for length and time to be used in laboratory experiments, so that results from different labs and different days could be compared. Many later scientists including German-American physicist Albert Einstein and British physicist Stephen Hawking have suggested that Galileo should be considered the father of modern science.

Galileo is widely credited with inventing the telescope. This invention was based on crude descriptions of a similar device from the Netherlands in 1608. By 1609 Galileo had made a telescope with the ability to magnify objects threefold (3×), soon improving this to a magnification of 30×. He immediately applied this device to study the stars, publishing a short note of his astronomical observations entitled *Sidereus Nuncius* (starry messenger) in March 1610. On January 7, 1610, Galileo observed "three fixed stars, totally invisible by their smallness, all within a short distance of Jupiter, and lying on a straight line through it." On the following nights he determined that these so-called fixed stars were moving in unusual ways that showed they could not be stars, and on January 10 he noted that one of them had disappeared by moving to a position behind Jupiter. A few days later Galileo concluded that he had discovered objects (moons) orbiting Jupiter, and he named them Medicean stars in honor of his patron Cosimo II de'Medici, the grand duke of Tuscany. These moons were later renamed the Galilean satellites (after Galileo), and included Io, Europa, and Callisto. Galileo discovered a fourth moon, Ganymede, on January 13.

Galileo's discovery of moons, or small planets circling another planet, was in direct contradiction to the cosmology of Aristotle and the teachings of the Catholic Church, that all heavenly bodies circle the Earth. Many scholars and theologians were therefore opposed to Galileo's discovery, and many claimed it must be false.

Galileo made more discoveries, including observations that Venus exhibited phases in a manner similar to that of the Moon. He explained this using Copernicus's heliocentric model of the solar system and presented mathematical arguments that if the Earth was the center of the solar system as in the Aristotelian view, then only the crescent phases of Venus would be visible. His observations proved that Venus orbited the Sun and played a major role in convincing the scientific community in the 1600s that the solar system was not purely geocentric. This would stand as one of Galileo's most important contributions to science.

Galileo's observations of Saturn and sunspots also showed changes in both of these heavenly bodies that were making a substantial change in scholars' thoughts about the universe. In contrast to the Aristotelian view of a nonchanging universe, Galileo demonstrated that the universe was dynamic and always changing. This remains one of Galileo's major lasting contributions to science.

Using his telescopes, Galileo Galilei also became the first to describe craters and mountains on the Moon. Using measurements of the lengths of shadows and his skills with trigonometry, Galileo even estimated the heights of these mountains. Using his telescope, Galileo deduced that the Milky Way was not nebulous as previously thought, but actually consisted of a vast number of stars that were so densely concentrated that they appeared nebulous from the Earth when viewed with the naked eye.

In addition to his contributions to pure science Galileo made many contributions to technology. The first of these (1595–98) in his repertoire was a geometric and military compass for surveyors and gunners. It was a remarkable instrument for its time, capable of helping to construct and calculate the area of any polygon or circular section, and for gunners it was helpful to calculate the angle and amount of powder needed to fire cannonballs of different weights to their intended targets. In 1593 Galileo designed and built a thermometer that used the expansion and contraction of an air bubble by heat to move a column of water in a calibrated tube. Galileo's most famous contribution to technology was his construction in 1609 of a refracting telescope that he used to explore the heavens. The following year he turned the telescope earthward and modified the optics so that he could magnify insects, and by 1624 he had perfected the first compound microscope, used the following year to publish his detailed observations of insects. Some of Galileo's inventions turned out to be not so popular, such as his automatic tomato picker, his combination pocket comb and eating utensil, and an early version of a ballpoint pen.

Galileo also contributed significantly to physics and mathematics. Folklore (possibly apocryphal) has it that Galileo demonstrated his ideas of classical mechanics by dropping balls of the same material, but with different masses, from the Leaning Tower of Piza, demonstrating that their rate of descent was not dependent on mass. This was contrary to the ideas of Aristotle, who said that heavier objects fell faster than lighter ones, an idea easily disproved. Galileo appreciated the effects of friction, and suggested that falling bodies would fall with a uniform acceleration as long as the resistance to motion was negligible. Galileo was also the first to state that moving objects retain their velocity unless acted on by an external force, such as friction. This too contradicted Aristotelian physics, which claimed that objects naturally slowed down unless acted on by a force to keep them moving, although this idea had been proposed many centuries earlier by Chinese philosopher Mozi (Mo Tzu, 470–391 B.C.E.). Galileo's principle of inertia specifically stated that "a body moving on a level surface will continue in the same direction at constant speed unless disturbed," an idea later incorporated into Newton's first law of motion. In another prescient tome, Galileo advanced his basic principle of relativity, in which he stated that the laws of physics are the same in any system moving at a constant velocity, regardless of the speed or direction. This principle formed the early basis for Einstein's special theory of relativity.

See also ASTRONOMY; ASTROPHYSICS; BRAHE, TYCHO; COPERNICUS, NICOLAUS; EINSTEIN, ALBERT; IMPACT CRATER STRUCTURES; JUPITER; VENUS.

FURTHER READING

Brodrick, James S. J. *Galileo: The Man, His Work, His Misfortunes.* London: G. Chapman, 1965.

Drabkin, Israel, and Stillman Drake, eds. and trans. *On Motion and On Mechanics.* Madison: University of Wisconsin Press, 1960.

Drake, Stillman, trans. *Dialogue Concerning the Two Chief World Systems.* Berkeley: University of California Press, 1953.

Galilei, Galileo [1638, 1914], Henry Crew, and Alfonso de Salvio, trans. *Dialogues Concerning Two New Sciences.* New York: Dover, 1954.

Galilei, Galileo. *Galileo: Two New Sciences.* (Translation by Stillman Drake of Galileo's 1638 *Discourses and Mathematical Demonstrations Concerning Two New Sciences*) Madison: University of Wisconsin Press, 1974.

Hawking, Stephen. *A Brief History of Time.* New York: Bantam Books, 1988.

Seeger, Raymond J. *Galileo Galilei: His Life and His Works.* Oxford: Pergamon Press, 1966.

Sharratt, Michael. *Galileo: Decisive Innovator.* Cambridge: Cambridge University Press, 1996.

Gamow, George (1904–1968) Russian Empire (Ukrainian) *Theoretical Physicist and Cosmologist* George Gamow was born Georgiy Antonovich Gamov in Odessa, Russian Empire, on March 4, 1904, and was educated at the Novorossiya University in Odessa from 1922 to 1923, and at the University of Leningrad from 1923 to 1929. He is best known for his prediction of the cosmic microwave background radiation, as well as work on the big bang theory and on alpha decay by quantum tunneling.

Gamow's career began after his work on quantum theory landed him a job at the Theoretical Physics Institute at the University of Copenhagen in 1928. He visited other laboratories to collaborate with other leading physicists of the time, including the British physicist Ernest Rutherford, considered the father of nuclear physics. His work at this point included solving a complex problem in radioactivity, in which he described how alpha decay of an atomic nucleus happens when a high-energy particle can tunnel through and escape the high-energy field that keeps the atomic nucleus together, with a specific half-life that is dependent on the type of nucleus. With increasing oppression in Russia, Gamow and his wife, Lyubov Vokhminzeva, attempted to leave the country twice in perilous sea voyages by kayak that both ended in failure because of bad weather. The couple managed to escape in 1933 by getting permission to attend a conference, then defecting to the United States, where they became naturalized citizens in 1940.

Gamow worked in cosmogony, the study of the origin of the universe, publishing a landmark paper in 1948 in the journal *Physical Review* called "The Origin of the Chemical Elements," suggesting that the present abundance of hydrogen and helium in the universe was acquired by reactions in the big bang. In the paper Gamow and his colleagues (one of whom was fictional) predicted the strength of the residual cosmic microwave background radiation that should be left over as an afterglow from the big bang, even though no such radiation had at that time been detected. This radiation was not detected until 16 years later, when Arno Penzias and Robert Wilson of Bell Laboratories in New Jersey detected the cosmic background radiation (earning them the Nobel Prize in physics in 1878) at 2.7 degrees above absolute zero, a couple of degrees lower than Gamow's prediction of 5 Kelvin made in 1948.

See also ASTRONOMY; ASTROPHYSICS; COSMOLOGY.

FURTHER READING

Chaisson, Eric, and Steve McMillan. *Astronomy Today.* 6th ed. New York: Pearson/Addison-Wesley, 2007.

Comins, Neil F. *Discovering the Universe.* 8th ed. New York: W. H. Freeman, 2008.

Snow, Theodore P. *Essentials of the Dynamic Universe: An Introduction to Astronomy.* 4th ed. St. Paul, Minn.: West, 1991.

geochemical cycles *Geochemical cycles* refers to the transportation, cycling, and transformation of the different chemical elements through various reservoirs or spheres in the Earth system, including the atmosphere, lithosphere, hydrosphere, and biosphere. The cycles occur through a great variety of processes, timescales, and different reservoirs or systems within the whole Earth system. Geochemical cycles are characterized by a closed flow in the Earth system; if the system is fully defined, the particular element remains in the same abundance in the cycle but moves from location to location by a series of processes. In other words, there is a material balance in geochemical cycles. The most basic picture of a geochemical cycle is analogous to the rock cycle, where molten magma rises from the deep interior of the Earth, crystallizes to form an igneous rock, then erodes to form a sedimentary rock, which gets buried and becomes a metamorphic rock, which eventually is heated to become a magma that rises back to the surface.

All geochemical cycles have a characteristic time necessary for completion. The longest is the geochemical cycle that brings material from deep within the Earth to form midocean ridges that then form oceanic crust that gets subducted, returned to the deep mantle to eventually rise back to the surface. This cycle takes from several hundred million years to about 4.5 billion years by some estimates.

The material balance of chemical elements in a geochemical cycle can be complex and includes transfer of the element between many different geological, biological, atmospheric, and liquid systems. The cycles are largely controlled by and can act as indicators of past conditions of other factors such as the configuration and elevation of the continents, distribution of landmasses and vegetation, large volcanic eruptions, climate, and biological production. Geochemical cycles are generally divided into two types, those that occur near the Earth's surface (exogenic cycles) and those that occur in the deep interior (endogenic cycles).

WATER CYCLE

The global water cycle is one of the major drivers and reservoirs for other geochemical cycles on the Earth. The water cycle describes the sum of processes operative in the hydrosphere, a dynamic mass of liquid continuously on the move between the different reservoirs on land and in the oceans and atmosphere. The hydrosphere includes all the water in oceans, lakes, streams, glaciers, atmosphere, and groundwater, although most water is in the oceans. The hydrologic, or water, cycle encompasses all of the changes, both long- and short-term, in the Earth's hydrosphere. Heat from the Sun powers the hydrologic cycle, causing water to change its state through evaporation and transpiration. Water is both a means of transport for other chemical components and a reactive agent that removes these other elements from rocks and soils on the continents and moves them into other reservoirs in the oceans.

The water cycle can be thought of as beginning in the ocean, where energy from the Sun causes surface waters to evaporate, changing from the liquid to the gaseous states. Evaporation takes heat from the ocean and transfers this heat into the atmosphere. An estimated 102 cubic miles (425 cubic km^3) of water evaporate from the ocean each year, leaving the salts behind in the ocean. The water vapor then condenses into water droplets in clouds and eventually falls back to the Earth as precipitation. Most (92 cubic miles; 383 cubic km^3) falls directly back into the ocean, but about 26 cubic miles (108 km^3) of precipitation falls as rain or snow on the continents, transforming salty water of the oceans into freshwater on the land. Nearly three-fourths of this water (17 cubic miles, or 71 cubic km/yr) evaporates back to the atmosphere or is aided by the transpiration from plants, returning the water back to the atmosphere. The other estimated 10 cubic miles (42 km^3) per year of water runs across the surface, some merging together to form streams and rivers that flow eventually back into the ocean, and other parts seeping into the ground to recharge the groundwater system. Humans now intercept approximately half of the fresh surface water for drinking, agriculture,

and other uses and make a significant impact on the natural hydrological cycle. Water that seeps into the groundwater system is said to infiltrate, whereas water that flows across the surface is called runoff.

Water in the atmosphere is one of the major greenhouse gases that help to regulate global temperature and climate. Changes in the water content in the atmosphere can change the erosion rate of different chemical elements on land, the evaporation rate from the ocean, and the balance between many other geochemical cycles.

SODIUM CYCLE

One of the most important geochemical cycles is the sodium cycle. Sodium is one of the major constituents of crustal rocks, sediments, and ocean water, and moves from each of these reservoirs to the other over long geological times. Sodium is dissolved from crustal rocks such as granite by rainwater, then streams and rivers carry it in solution to the sea. Sodium (Na) and chlorine (Cl) are the two most abundant elements carried in solution in ocean water. They combine to form the mineral halite (NaCl) that remains following evaporation of sea water. The conversion from dissolved sodium to sodium in the mineral halite is ongoing in many areas of strong evaporation along seashores around the world. At times in the geological past large sections of ocean basins (the Mediterranean Sea, Red Sea, juvenile Atlantic Ocean) have evaporated, leaving thick deposits of salts. Stream waters re-erode some of these deposits and carry them back to the sea, completing one circuit of the geochemical cycle.

When the salt deposits get buried on the seafloor the salt may be interlayered with oceanic muds, then the sodium is removed from the salts and transferred into clay minerals. Replenishing the amount of sodium in the ocean by river flow from the continents takes an estimated 65–100 million years. If the concentration of sodium (or other element in other geochemical cycle) remains the same in one reservoir such as the ocean basins, over time, then there is a balance between the input of that element to the system and its extraction to other systems. The amount of time it takes to replenish that amount reflects this balance, and is known as the residence time, obtained by dividing the mass of the element in the reservoir by the rate of input to the system.

Sodium in the seafloor sedimentary deposits can react with the basalt of the ocean crust, forming veins, and can replace other elements in the basalt. Ultimately these basalts and sediments containing sodium get subducted into the mantle, where some remelt to form igneous rocks that rise to the surface, containing minerals with sodium. These then are prone to erosion by rivers, leaching away sodium to be carried back to the ocean. Other atoms of sodium are carried deeper into the mantle, forming the longest residence time arm of the sodium cycle.

CARBON CYCLE

The carbon cycle preserves a record of many processes on the Earth throughout the planet's history and includes many geologic, biologic, ocean, and atmospheric systems. Many volatile substances including water and carbon dioxide were degassed from the deep interior of the Earth during the early Archean, and some has been added by cometary and meteorite impact. The early atmosphere of the Earth was rich in carbon dioxide (CO_2), and since the Archean this CO_2 has been progressively removed by the precipitation of limestones (with a composition close to $CaCO_3$), and by photosynthesis that converts the CO_2 (along with nitrogen, phosphorus, and sulfur) into organic matter, releasing free oxygen in the process. The development of life on the Earth enhanced the formation of limestones and other carbonates, since many organisms secrete calcium carbonate for their shells and tissues. Inorganic processes since the Archean formed other limestones.

Over long geologic times carbon dioxide returns to the atmosphere by decomposition of limestones subducted to the Earth's deep interior, releasing carbon dioxide through gases dissolved in magmas that rise to the surface. Plate tectonics and the supercontinent cycle also play a large role in cycling carbon between the atmosphere and rock sphere. When many continents collide to form a supercontinent, the passive margins on these continents that contain thick limestone sequences are uplifted above sea level. The tectonic uplifting of carbonate rocks causes them to be exposed to the atmosphere during continental collisions. The calcium carbonate ($CaCO_3$) then combines with atmospheric CO_2, depositing it in the oceans. Thus continental collisions and times of supercontinent formation are associated with drawdown and reduction of CO_2 from the atmosphere, global cooling, and sea-level changes.

The mass of carbon stored in the limestone and organic matter reservoirs on Earth is huge, about 2,000 times greater than all the carbon presently in the atmosphere and oceans combined. Living plants contain about the same amount of carbon as that in the atmosphere, so human activities such as deforestation that change the vegetation balance on the planet may significantly change the balance between atmospheric and living organic reservoirs for the carbon, putting more CO_2 in the atmosphere and altering global climate.

Living plants take CO_2 out of the atmosphere and release one molecule of oxygen for every molecule of carbon dioxide used to make organic matter. When these plants die, much of the organic matter is oxidized and returned into CO_2, but some escapes this process and becomes buried in organic sediments in another reservoir to store carbon. There is a delicate balance between the carbon cycle and the oxygen cycle, as the amount of oxygen released indicated by the mass of the present-day mass of the organic carbon reservoir is 30 times the present atmospheric level, so there is recycling of both carbon and oxygen on geological timescales. Similar relationships exist among biological, geological, and atmospheric processes for the geochemical cycles of nitrogen, phosphorus, and sulfur. Plants absorb these elements in fixed proportions from different environments, store them in organic soils, where groundwater can leach these elements into the hydrological system, bringing them to the ocean, where they form building blocks for new life.

See also ASTHENOSPHERE; ATMOSPHERE; BIOSPHERE; CARBON CYCLE; CLIMATE CHANGE; CONVECTION AND THE EARTH'S MANTLE; ENERGY IN THE EARTH SYSTEM; GAIA HYPOTHESIS; GREENHOUSE EFFECT; HYDROSPHERE; LITHOSPHERE; MAGMA; PASSIVE MARGIN; PHOTOSYNTHESIS; PRECAMBRIAN; SUBDUCTION, SUBDUCTION ZONE; SUPERCONTINENT CYCLES; WEATHERING.

FURTHER READING

Berner, Elizabeth Kay, and Robert Berner. *Global Environment: Water, Air and Geochemical Cycles.* Upper Saddle River, N.J.: Prentice Hall, 1994.

Brantley, Susan, James D. Kubicki, and Art White. *Kinetics of Global Geochemical Cycles.* New York: Springer, 2008.

geochemistry   Geochemistry is the study of the distribution and amounts of elements in minerals, rocks, ore bodies, rock units, soils, the Earth, atmosphere, and by some accounts, other celestial bodies, and the principles that govern the distribution and migration of these elements. This field of earth science includes the study and analysis of the movement of chemical elements, the properties of minerals as related to their distribution and concentrations of specific elements, and the classification of rocks based on chemical composition.

The field of geochemistry began with the discovery of 31 chemical elements by the French chemist Antoine Lavoisier in 1789, with the first mention of the word by German chemist Christian F. Shonbein in 1813. In 1884 the United States Geological Survey (USGS) established a laboratory to investigate the chemistry of the planet and appointed F. W. Clarke as head of the laboratory. Since then the USGS has been one of the world's leaders in the collection and analysis of geochemical data. In 1904 the Carnegie

Institution in Washington, D.C., established the Geophysical Laboratory, which tests physical and chemical properties of minerals and rocks. The Vernadsky Institute in Moscow, Russia, had a similar charge, and both institutions spearheaded a revolution in technologies applied to analyzing the composition of rock materials, leading to the proposition of the concepts of chemical equilibrium, disequilibrium, and amassing of huge databases encompassing the chemistry of rocks of the world. Geochemist Victor M. Goldschmidt from the University of Oslo, Norway, applied the phase rule, explaining metamorphic changes in terms of chemical equilibrium.

Geochemistry is commonly studied by using several different methods:

- chemical analysis, in which rocks are broken down into major and minor (trace) constituents and measured in the laboratory
- analysis of the atomic and chemical structure, physical properties of minerals, and properties of rocks as reflected in their geochemistry
- direct experimentation, under controlled conditions. In experimental geochemistry controlled conditions are created to simulate processes of formation of various materials, and phase relations between the materials are determined.
- study of the dispersion and accumulation of elements under dynamic conditions

Geochemistry is subdivided into several fields:

- isotope geochemistry, which involves the measurement of the concentration of elements and their isotopes in different Earth and planetary systems, uses stable and radiogenic isotopes to understand different mineral and rock systems, and dates geological events using radioactive decay of isotopes
- geochemical cycling, which uses the changes in the distribution of elements in different parts of the Earth to understand these different mineral, rock, water, and biological systems
- cosmochemistry, which analyses the distribution of elements and their isotopes in space
- biogeochemistry, which assesses the role of life and organisms on the chemistry of the Earth and earth systems
- organic geochemistry, which examines the role of processes and compounds by living and once-living organisms
- environmental and exploration geochemistry, including applications of chemistry to environmental, hydrological, and mineral exploration studies

GEOCHEMISTRY AND THE COMPOSITION OF THE EARTH

Geochemists have worked closely with cosmologists to obtain a better understanding of the chemical composition of the universe and Earth-Moon system in the field of cosmochemistry. These cosmochemists have produced models for how the chemical composition of the universe began, initially consisting almost exclusively of hydrogen and helium, and evolved through stellar processes and supernovas to include heavier elements. Studies of meteorites have yielded data on the average composition of the rocky planets in the inner solar system, and these data have been used to derive a model for the average composition of the Earth. The Earth is thought to have an overall composition close to that of a carbonaceous chondrite meteorite. The planet experienced early heating due to the decay of short-lived radioactive isotopes and heat from gravitational compaction and the collection in the core of metallic phases that sink and release heat. This early heating formed a melt phase within the Earth, and the outer core is still molten, as shown by seismic waves. This causes a chemical and density zonation in the Earth.

The chemical and density zonation has broken the Earth into several different shells with different properties. These layers include the crust, upper mantle, transition zone, lower mantle, outer core, and inner core. The Earth's magnetic field is formed by motions in the liquid outer core, where a dynamo effect generates a magnetic field by the motion of electrically charged fluid. The lower mantle has a composition that includes magnesium and iron silicate concentrations, similar to chondritic meteorites.

There are a couple of major polymorphic transitions within the Earth that are of great importance for determining overall Earth structure and behavior. At 250 miles (400 km) depth pyroxene minerals attain a garnet structure by the increase in pressure from higher in the Earth, but they maintain a pyroxene composition. Also the crystal lattice structure of olivine—one of the most abundant elements in the mantle—changes from an olivine to a denser crystal lattice structure known as a spinel structure at the same depth. Deeper, at 435 miles (700 km), the garnet and spinel structures react together and change into progressively denser crystal lattices including ilmenite, then to a perovskite structure, still retaining the olivine and pyroxene composition, although possibly with more iron present.

Geochemists have determined that the Earth is not yet completely chemically fractionated. Volcanoes continue to release gases such as sulfur and carbon dioxide, and deep eruptions such as those

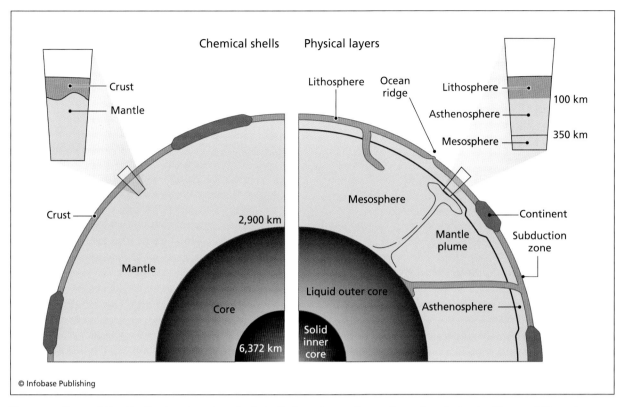

Cross sections of the Earth showing chemical shells (crust, mantle, and core) and physical layers (lithosphere, asthenosphere, mesosphere, outer core, and inner core)

through kimberlites have recently released diamonds, showing that there is still carbon at depth. Carbon dioxide is presently being degassed from the interior of the Earth at high rates, showing that fractionation is an ongoing process. The composition of the Earth and its different layers has been calculated based on density measurements, gravity, deep samples erupted from volcanoes, and models based on the composition of meteorites. The crust has been calculated by geochemists to have a composition that is about:

- 47 percent oxygen (O)
- 28 percent silicon (Si)
- 11 percent iron (Fe), magnesium (Mg), and calcium (Ca)
- 8 percent aluminum (Al)
- 6 percent potassium (K), sodium (Na), and all other elements

In terms of minerals in the Earth's crust this equates with about

- 49 percent feldspar
- 21 percent quartz
- 5 percent pyroboles (includes pyroxene, amphibole, and olivine)

- 8 percent micas
- 7 percent magnetite and all other minerals

The overall distribution of elements in the whole Earth is quite different and resembles a massively differentiate giant meteorite with a crust on top. The whole Earth has a composition that has

- 38 percent of the mass in the core divided between 35 percent iron (Fe) and 2.7 percent nickel (Ni)
- 28 percent oxygen, distributed mostly in the mantle
- 17 percent magnesium (Mg)
- 13 percent silicon (Si)
- 2.7 percent sulfur (S)

See also ASTROPHYSICS; COSMOLOGY; EARTH; IGNEOUS ROCKS; ORIGIN AND EVOLUTION OF THE EARTH AND SOLAR SYSTEM; METEOR, METEORITE; MINERAL, MINERALOGY, MINERALS; STELLAR EVOLUTION.

FURTHER READING
Holland, H. D., and K. K. Turekian. *Treatise on Geochemistry*. 9 Vols. Amsterdam: Elsevier, 2004.
Marshall, C., and R. Fairbridge. *Encyclopedia of Geochemistry*, Berlin: Springer, 2006.

geochronology Geochronology is the study of time with respect to Earth history, and includes both absolute and relative dating systems as well as correlation methods. Absolute dating systems include a variety of geochronometers such as radioactive decay series in specific isotopic systems that yield a numerical value for the age of a sample. Relative dating schemes include cross-cutting features, and discontinuities such as igneous dikes and unconformities, with the younger units being the cross-cutting features or those overlying the unconformity.

During the 19th and early 20th centuries geochronologic techniques were crude. Many ages were estimated by the supposed rate of deposition of rocks, and correlation of units with unconformities with other, more complete sequences. With the development of radioactive dating it became possible to refine precise or absolute ages for specific rock units. Radiometric dating operates on the principle that certain atoms and isotopes are unstable. These unstable atoms tend to decay into stable ones by emitting a particle or several particles. Alpha particles have a positive charge and consist of two protons and two neutrons. Beta particles are physically equivalent to electrons or positrons. These emissions are known as radioactivity. Half-life is the time it takes for half of a given amount of a radioactive element to decay to a stable element. By matching the proportion of original unstable isotope to stable decay product, and knowing the half-life of that element, one can thus deduce the age of the rock. The precise ratios of parent-to-daughter isotopes are measured by a mass spectrometer.

William F. Libby (1908–80) developed radiocarbon, or carbon 14, dating techniques at the University of Chicago in 1946. This major breakthrough in dating organic materials is now widely used by archaeologists, Quaternary geologists, oceanographers, hydrologists, atmospheric scientists, and paleoclimatologists. Cosmic rays entering Earth's atmosphere transform regular carbon (carbon 12) to radioactive carbon (carbon 14). Within about 12 minutes of being struck by cosmic rays in the upper atmosphere, the carbon 14 combines with oxygen to become carbon dioxide that has carbon 14. The radioactive carbon dioxide diffuses through the atmosphere and is absorbed by vegetation (plants need carbon dioxide to make sugar by photosynthesis). Every living thing contains carbon. While it is alive, each plant or animal exchanges carbon dioxide with the air. Animals also feed on vegetation and absorb its carbon dioxide. At death the animal no longer exchanges carbon 14 with the atmosphere, but the radioactive element continues to decay within the organic material. Theoretically analysis of this carbon 14 can reveal the date when the object once lived by the percent of carbon 14 atoms still remaining in the object. The radiocarbon method has subsequently evolved into one of the most powerful techniques to date late Pleistocene and Holocene artifacts and geologic events up to about 50,000 years old.

Dendrochronology is the study of annual growth rings in trees for dating the recent geological past. The ages of trees may be determined most simply by counting the number of annual growth rings that form in the trunk of the tree each year. Sometimes this practice is done in conjunction with carbon 14 or other dating techniques to verify the ages of specific rings. This field is closely related to dendroclimatology, the study of the sizes and relative patterns of tree growth rings to yield information about past climates. Tree rings are most clearly developed in species from temperate forests but not well formed in tropical regions, where seasonal fluctuations are not as great. Most annual tree rings consist of two parts, and early wood consisting of widely spaced thin-walled cells, followed by late wood, consisting of thinly spaced, thick-walled cells. The changes in relative width and density of the rings for an individual species correlates to changes in climate such as soil moisture, sunlight, precipitation, and temperature and also reflects unusual events such as fires or severe drought stress.

The longest dendrochronology record goes back 9,000 years, using species such as the Bristle Cone Pine, found in the southwestern United States, and Oak and Spruce species from Europe. To extend the record from a particular tree, one can correlate rings between individuals that lived at different times in the same microenvironment close to the same location.

Uranium, thorium, and lead isotopes form a variety of geochronometers using different parent/daughter pairs. Uranium 238 decays to lead 206 with a half-life of 4.5 billion years. Uranium 235 decays to lead 207 with a half-life of 0.7 billion years, and thorium 232 decays to lead 208 with a half-life of 14.1 billion years. Uranium, thorium, and lead are generally found together in mixtures, and each one decays into several daughter products (including radium) before turning into lead. The thorium 230/uranium 234 disequilibrium method is one of the most commonly used uranium-series techniques and can be used to date features as old as Precambrian. The fact that uranium is much more soluble than thorium forms the basis for this method, so materials such as corals, mollusks, calcic soils, bones, carbonates, cave deposits, and fault zones are enriched in uranium with respect to thorium.

Uranium-lead dating also uses the known original abundance of isotopes of uranium and the known decay rates of parents to daughter isotopes. This technique is useful for dating rocks up to billions of years old. All naturally occurring uranium contains uranium 238 and uranium 235 in the ratio of 137.7:1. Uranium 238 decays to lead 206 with a half-life of 4,510 Ma through a process of eight alpha-decay steps and six beta-decay steps. Uranium 235 decays to Lead 207 (with a half-life of 713 Ma) by a similar series of stages that involves seven alpha-decay steps and four beta-decay steps. Uranium-lead dating techniques were initially applied to uranium minerals such as uraninite and pitchblende, but these are rare, so geochronologists developed precise methods of measuring isotopic ratios in other minerals with only trace amounts of uranium and lead (zircon, sphene). The amount of radiogenic lead in all these methods must be distinguished from naturally occurring lead; this is calculated using their abundance with lead 204, which is stable. After measuring the ratios of each isotope relative to lead 204, the ratios of uranium 235/lead 207 and uranium 238/lead 206 should give the same age for the sample, and a plot with each system plotted on one axis shows each age.

If the two ages agree, the ages will plot on a curve known as concordia, which tracks the evolution of these ratios in Earth versus time. Ages that plot on concordia are said to be concordant. In many cases, however, the ages determined by the two ratios are different, and they plot off the concordia curve. This occurs when the system has been heated or otherwise disturbed during its history, causing a loss of some of the lead daughter isotopes. Because lead 207 and lead 206 are chemically identical, they are usually lost in the same proportions.

The thorium-lead dating technique is similar to the uranium-lead technique; it uses the decay from thorium 232 to lead 208 (releasing six helium 4), with a half-life of 13,900 years. Minerals used for this method include sphene, zircon, monazite, apatite, and other rare U-Th minerals. The ratio of lead 208/thorium 232 is comparable with lead 207/uranium 235. This not totally reliable method is usually employed in conjunction with other methods. In most cases the results are discordant, showing a loss

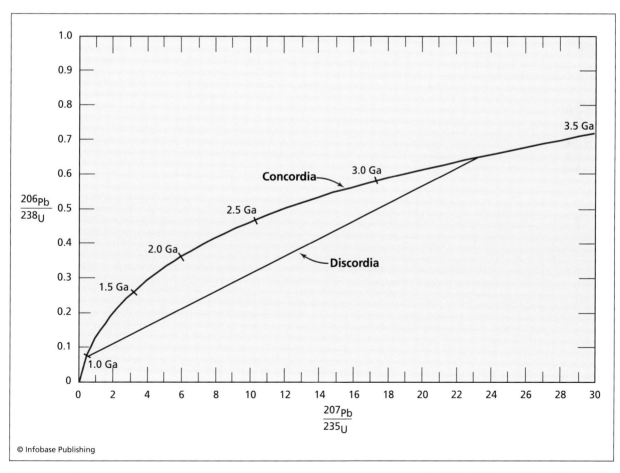

Concordia diagram showing the concordia curve that traces the evolution of the $^{206}Pb/^{238}U$ vs. $^{207}Pb/^{235}U$ ratio with time, from the present to 3.5 billion years (Ga) ago. The discordia curve shows the path that the ratio would follow if the rock example used crystallized at 3.2 billion years ago and lost lead (for example, through metamorphism) at 1.0 billion years ago.

of lead from the system. The Th-Pb method can also be interpreted by isochron diagrams.

Potassium-argon dating is based on the decay of radioactive potassium into calcium and argon gas at a specific rate and is accomplished by measuring the relative abundances of potassium 40 and argon 40 in a sample. The technique is potentially useful for dating samples as old as 4 billion years. Potassium is one of the most abundant elements in Earth's crust (2.4 percent by mass). One out of every 100 potassium atoms is radioactive potassium 40, with 19 protons and 21 neutrons. If a beta particle hits one of the protons, the latter can convert into a neutron. With 18 protons and 22 neutrons, the atom becomes argon 40, an inert gas. For every 100 potassium 40 atoms that decay, 11 become argon 40.

By comparing the proportion of potassium 40 to argon 40 in a sample and knowing the decay rate of potassium 40, one can estimate the age of the sample. The technique works well in some cases but is unreliable in samples that have been heated or recrystallized after formation. Since it is a gas, argon 40 can easily migrate in and out of potassium-bearing rocks, changing the ratio between parent and daughter.

Fission-track dating determines the thermal age of a sample, the time lapsed since the last significant heating event (typically above 215°F, or 102°C). Fission tracks are paths of radiation damage made by nuclear particles released by the spontaneous fission, or radioactive decay, of uranium 238. Fission tracks are created at a constant rate in uranium-bearing minerals, so by determining the density of tracks present, one can determine the amount of time that has passed since the tracks began to form in the mineral. Fission-track dating is used for determining the thermal ages of samples between about 100,000 and 1,000,000 years old; it is also used for estimating the uplift and erosional history of areas by recording when specific points cooled past 215.6°F (102°C).

Thermoluminescence is a chronometric dating method based on the fact that some minerals give off a flash of light when heated. The intensity of the light is proportional to the amount of radiation to which the sample has been exposed and the length of time since the sample was heated. Luminesence results from heating a substance, and thus liberating electrons trapped in its crystal defects. The phenomenon is used as a dating technique, especially for pottery. The number of trapped electrons is assumed to be related to the quantity of ionizing radiation to which the specimen has been exposed since firing, since the crystal defects are caused by ionizing radiation, and therefore the sample's age. Thus, measuring the amount of light emitted upon heating allows one to estimate of the age of the sample.

A number of other isotopic systems can be used for geochronology, but they are less common or less reliable than the methods described above. Geochronologists also incorporate relative and correlation dating techniques, such as stratigraphic correlation of dated units, to explore the wider implications of ages of dated units. A paleomagnetic timescale has been constructed for the past 180 million years, and in many situations one can determine the age of a particular part of a stratigraphic column or location on the seafloor by examining the geomagnetic properties of that position on the column and correlating it with a known geomagnetic period. Finally, geochronologists use structural cross-cutting relationships to determine which parts of a succession are older or younger than a dated sample. Eventually the geochronologist is able to put together a temporal history of a rock terrane by dating several samples and combining these ages with cross-cutting observations and correlation with other units.

See also PALEOMAGNETISM; RADIOACTIVE DECAY; STRATIGRAPHY, STRATIFICATION, CYCLOTHEM.

FURTHER READING

Faure, G. *Principles of Isotope Geology.* 2nd ed. Cambridge: Cambridge University Press, 1986.

Faure, G., and Mensing, D. *Isotopes: Principles and Applications.* 3rd ed. New York: John Wiley & Sons, 2005.

geodesy Geodesy is the study of the size and shape of the Earth, its gravitational field, and the determination of the precise locations of points on the surface. This branch of geology also includes the study of the temporal variations in the shape of the planet and the location of points on the surface as a result of tides, rotation, and plate tectonic movements. Geodetic measurements rely heavily on positional measurements from satellite-based global positioning systems, gravity measurements, and radar altimetry measurements over the oceans. The science of measuring the size of the Earth probably started with Erastosthenes in ancient Greece, who measured the distance from Alexandria to Aswan in Egypt and calculated the curvature of the Earth from his measurements.

One branch of geodesy deals with the measurement of the Earth's gravity field and the geoid, the surface of equal gravitational potential. A person's weight, or the pull of the Earth's gravity on a person, would be the same everywhere on this surface. Since the Earth has internal irregularities in density, this surface is itself irregular. The geoid has a roughly elliptical shape that is slightly flattened at the poles as a result of the planet's rotation, and the shape of this surface is approximated by a reference ellipsoid. Variations in the height of the geoid from the reference ellipsoid are expressed as the geoid height, in many cases reaching 30–50 feet (tens of meters). These variations reflect variations in the mass distribution within the Earth, and smaller, temporary variations may result from tides or winds changing the mass distribution of the oceans.

Geodetic measurements must use some reference frame, typically an astronomical or celestial, or an inertial reference frame. Many geodetic measurements are between different points on the surface, and these terrestrial measurements are useful for determinations of surface deformation such as motion along faults. Regional geodetic measurements rely on the art of triangulation, first developed by the Dutch scientist Gemma Frisius in the 16th century. Triangulation uses precise measurements of the angles and distances between different points in a network or grid to determine the changes in the shape of the grid with time, and hence the deformation of the surface.

Space geodesy uses satellite positioning techniques where GPS satellites emit microwave signals encoded with information about the position of the satellite and the precise time at which the signal left the satellite. The microwave signals are received and decoded by the GPS receivers on the ground, and the distance to the satellite is calculated by knowing the time it took for the signal to travel from the satellite to the receiver using the speed that microwave signals travel through the atmosphere. The resulting calculation gives a result that the receiver may be anywhere on a sphere around the satellite, but the range of possible locations can be narrowed since the receiver is typically on the surface of the Earth. The use of multiple GPS satellite signals allows for the determination of the precise position on the surface of the Earth by finding the unique location where the calculated spheres around the multiple GPS satellites intersect on the surface. The accuracy of GPS positions can be a few feet (1 meter or less) and can be improved by the technique of differential GPS, in which a satellite receiver at a known position is coupled with and emits signals to a roving receiver. Furthermore, multireceiver interferometric and kinematic GPS techniques can improve positional measurements to the submillimeter level. The improved precision for these methods has greatly improved observations of surface deformation needed to predict earthquakes and volcanic eruptions and has aided precise navigation, surveying, and guidance systems.

See also GEOGRAPHIC INFORMATION SYSTEMS; GEOID; GEOPHYSICS; REMOTE SENSING.

FURTHER READING

Turcotte, Donald L., and Gerald Schubert. *Geodynamics*. 2nd ed. Cambridge: Cambridge University Press, 2002.

geodynamics Geodynamics is the branch of geophysical science that deals with forces and physical processes in the interior of the Earth necessary to understand plate tectonics and many other geological phenomena. This field of geology typically involves the macroscopic analysis of forces associated with a process and may include mathematical or numerical modeling. Geodynamics is a quantitative science closely related to geophysics, tectonics, and structural geology with problems including assessment of the forces associated with mantle convection, plate tectonics, heat flow, mountain building, erosion, volcanism, fluid flow, and other phenomena. The aim of many studies in geodynamics is to assess the relationships between different processes, such as to determine the influence of mantle convection on plate movements or to assess plate motions in one area with deformation in another. It is contrasted with many other types of geological studies that tend to be either static, analyzing only present and past states, or kinematic, which analyze the history of motions without a quantitative assessment of the forces involved.

Geodynamics is largely concerned with consideration of the fundamental physical processes that drive plate tectonics and how to interpret signatures of the products of plate interactions. To achieve these goals geodynamics typically takes a continuum mechanical approach to understanding stress and strain in solid materials, and takes a quantitative approach to modeling flexure of materials, and then applies this to studies of the Earth's lithosphere. Studies of heat transfer form a major component of the field of geodynamics. Heat is produced within the Earth and may be transferred by conduction, convection, or advection. Studying heat flow and transfer equations is necessary to understand the role of these different mechanisms in the Earth. Heat transfer and flow and the temperature of Earth materials all play major roles in how the mantle behaves and in how internal processes drive plate tectonics on the surface.

Measurements of gravity and magnetic fields can yield information about the structure and composition of materials at depth. To obtain realistic interpretations of gravity and magnetic anomalies and their causes, it is first necessary to understand such concepts as gravitational acceleration, the geoid, gravity fields of masses at depth, and techniques to model these physical processes.

Fluid mechanics falls in the realm of geodynamics, including flow in the asthenosphere, flow of magma through subvolcanic feeder systems and in lava tubes, flow of material into (and out of) subduction zones, as well for understanding mantle flow associated with glacial rebound. Thermal convection is modeled in fluid dynamics and has obvious applications to mantle convection, the driving forces of plate tectonics, and also to systems such as modeling fluid flow around hot springs, submarine black smoker chimneys, and geological mineral deposits formed by circulating hot fluids.

Geodynamics is concerned with the rheology or mechanical behavior of materials and how strain is accommodated at the crystal lattice and atomic scales, then applies the physics at these scales to the deformation of the mantle, asthenosphere, and lithosphere. Brittle and ductile/brittle deformation mechanisms can be modeled in geodynamics, including the mechanics of thrust-faulted terranes, the geometry of faulting, extensional fault systems, and strike-slip fault systems.

Flow of fluids in porous media is studied in geodynamics, with applications to flow of water in aquifers, petroleum and hydrocarbons in reservoirs, and general flow laws for fluids moving through any porous or fractured medium.

See also CONVECTION AND THE EARTH'S MANTLE; CRYSTAL, CRYSTAL DISLOCATIONS; ENERGY IN THE EARTH SYSTEM; GEOPHYSICS; PLATE TECTONICS.

FURTHER READING

Turcotte, Donald L., and Gerald Schubert. *Geodynamics.* 2nd ed. Cambridge: Cambridge University Press, 2002.

geographic information systems (GIS)

Geographic information systems (GIS) are computer application programs that organize and link information to enable users to manipulate that information constructively. They typically integrate a database management system with a graphics display that shows links between different types of data. For instance, a GIS may show relationships among geological units, ore deposits, and transportation networks. GIS allows users to layer information over other information already in the database. A GIS database allows the storage of information in a particular geographical area no matter what that information may be.

GIS is a powerful tool for environmental data analysis and planning, allowing for better viewing and modeling of changing environmental conditions and the relationships that influence a given critical environmental setting. Many fields have come to rely on GIS for data collection and analysis, including environmental and health science studies concerned with risk assessment and mitigation, environmental modeling, resources exploration, sustainable development, natural resource management, transportation, air pollution and control, and forest fire management. GIS is used widely in science, industry, business, and government to sort out pertinent information for the particular user from the GIS database.

For a GIS to be accurate the data must be entered with precise knowledge of the location of features on the ground, referenced to a map grid or reference frame. Benchmarks are well-defined, uniformly fixed points on the land's surface used for reference points from which other measurements can be made. They are generally marked by circular bronze disks with a diameter of 3.75 inches (10 cm) and are embedded firmly in bedrock or another permanent structure. In the United States benchmarks are installed and maintained by the U.S. Coast and Geodetic Survey and the U.S. Geological Survey. The elevations of many benchmarks were in the past established by the surveying technique of differential leveling. Now it is more common to determine elevations with satellite-based differential global positioning systems. Benchmarks, often identified on topographic maps by the abbreviation B.M., are used for determining elevation and for surveying and construction.

Geographic information systems are typically used in conjunction with global positioning systems to collect spatially accurate location data in the field. Global positioning systems, commonly referred to by their acronym GPS, were developed by the U.S. Department of Defense to provide the U.S. military with a superior tool for navigation, viable at any arbitrary point around the world. The Defense Department pays billions of dollars for the development and maintenance of the GPS program, which has matured a great deal since its conception in the 1960s.

The configuration of the global positioning system includes three main components: the GPS satellites, the control segment, and the GPS receivers. Working together these components provide users of GPS devices with their precise location on the Earth's surface, along with other basic information of substantial use such as time, altitude, and direction. Key to understanding how GPS functions is the understanding of these components and how they interrelate with one another.

GPS satellites, named Navstar Satellites, form the core of the global positioning system. Navstar satellites are equipped with an atomic clock and radio equipment to broadcast a unique signal, called a "pseudorandom code," as well as ephemeris data about the exact position of the satellite relative to Earth and astromical reference frames. This signal distinguishes one satellite from the other and provides GPS receivers accurate information about the exact location of the satellite. These satellites follow a particular orbit around the Earth, and their sum is called a constellation. The GPS Navstar satellite constellation is configured so that at any point on Earth's surface, the user of a GPS receiver should be able to detect signals from at least six Navstar satellites. It is essential to the proper functioning of GPS that the precise configuration of the constellation be maintained.

Geosynchronous satellite orbit maintenance is performed by the control segment (or satellite control centers), with stations in Hawaii, Ascension Island, Diego Garcia, Kwajalein, and Colorado Springs. Should any satellite fall slightly in altitude or deviate from its correct path, the control segment will take corrective actions to restore precise constellation integrity.

Finally, the component most visible to all who use GPS devices is the GPS receiver. GPS receivers are single-direction, asynchronous communication devices; the GPS receiver does not broadcast any information to the Navstar satellites but only receives signals from them. Recent years have seen the miniaturization and mass proliferation of GPS receivers. GPS devices are now so small that they can be found in many other hybrid devices such as cellular phones, some radios, and personal desk accessories. They are also standard on many vehicles for land, sea, and air travel. Accuracy of consumer GPS receivers is typically no better than nine feet (2.7 m), but advanced GPS receivers can measure location to less than a centimeter.

GPSs resolve a location on the surface of Earth through trilaterating, a process that determines the distances to the Navstar satellites and the GPS receiver. To do this two things must be true. The locations of the Navstar satellites must be known and there must be a mechanism for precision time measurements. For very precise measurements there must be a system to reconcile error caused by various phenomena.

GPS receivers are programmed to calculate the location of all the Navstar satellites at any given time. A combination of the GPS receiver's internal clock and trilateration signal reconciliation, performed by the GPS receiver, allow a precise timing mechanism to be established.

When a GPS receiver attempts to locate itself on the surface of the planet, it receives signals from the Navstar satellites. As mentioned previously, these signals are intricate and unique. By measuring the time offset between the GPS receiver's internal pseudorandom code generator and the pseudorandom code signal received from the Navstar satellites, the GPS receiver can use the simple-distance equation to calculate the distance to the Navstar satellite.

To locate a point on Earth's surface accurately, at least three distances must be measured. One measurement is enough only to place the GPS receiver within a three-dimensional arc. Two measurements can place the receiver within a circle. Three measurements place it on one of two points. One possible point location is usually floating in space or traveling at some absurd velocity, so the GPS receiver eliminates this point as a possibility, thus resolving the GPS receiver's location on the surface of the planet. A fourth measurement allows the correct point to be located, as well as provides necessary geometry data to synchronize the GPS receiver's internal clock to the Navstar satellite's clock.

Depending on the quality of the GPS device, error correction may also be performed when calculating location. Errors arise from many sources. Atmospheric conditions in the ionosphere and troposphere cause impurities in the simple-distance equation by altering the speed of light. Weather modeling can help calculate the difference between the ideal speed of light and the likely speed of light as it travels through the atmosphere. Calculations based on the corrected speed of light then yield more accurate results.

As weather conditions rarely fit models, however, other techniques such as dual frequency measurements, in which two different signals are compared to calculate actual speed of the pseudocode signal, are used to reduce atmospheric error.

Ground interference, such as multipath error, which arises from signals bouncing off objects on Earth's surface, can be detected and rejected in favor of direct signals via complicated signal-selection algorithms.

Still another source of error can be generated by the Navstar satellites' being slightly out of position. Even a few meters from the calculated position can throw off a high-precision measurement.

Geometric error can be reduced by using satellites that are far apart rather than close together, as this creates larger distances between satellites, easing certain geometric constraints.

Sometimes precision down to the centimeter is needed. Only advanced GPS receivers can produce precise and accurate measurements at this level. Advanced GPS receivers utilize one of several techniques to pinpoint more precisely locations on Earth's surface, mostly by reducing error or using comparative signal techniques.

One such technique is differential GPS, which involves two GPS receivers. One receiver monitors variations in satellite signals and relates this information to the second receiver. With this information the second receiver is then able to determine more accurately its location through better error correction.

Another method involves using the signal carrier-phase as a timing mechanism for the GPS receiver. As the signal carrier is a higher frequency than the pseudorandom code it carries, carrier signals can be used to synchronize timers more accurately.

Finally, a geostationary satellite can be used as a relay station for transmission of differential corrections and GPS satellite data. Called augmented GPS, this is the basic idea behind the new WAAS system installed in North America. The system encompasses 25 ground-monitoring stations and two geostationary WAAS satellites that allow for better error correction. This sort of GPS is necessary for aviation, particularly in landing sequences.

See also REMOTE SENSING.

FURTHER READING

Demers, Michael N. *Fundamentals of Geographic Information Systems*. 4th ed. New York: John Wiley & Sons, 2007.

Gorr, Wilpen L., and Kristen S. Kurland. *GIS Tutorial: Workbook for ArcView 9*. 3rd ed. Redlands, Calif.: ESRI Press, 2008.

Longley, Paul A. *Geographic Information Systems and Sciences*. 2nd ed. New York: John Wiley & Sons, 2005.

Yeung, Albert K. W. *Concepts and Techniques of Geographic Information Systems*. 2nd ed. Upper Saddle River, N.J.: Prentice Hall, 2007.

geoid The geoid is an imaginary surface near the surface of the Earth, along which the force of gravity is the same and equivalent to that at sea level. This so-called equipotential surface can be thought of as equivalent with sea level and extending through the continents on the Earth, and is often referred to as the figure of the Earth. Theoretically it exists everywhere perpendicular to the direction of gravity (the plumb line) and is used as a reference surface for geodetic measurements. If the Earth were spherically symmetric and not spinning, the gravitational equipotential surfaces would consist of a series of concentric shells with increasing potential energy extending away from the Earth, much like raising a ball to a higher level increases its potential energy. Since the Earth is not perfectly spherical (it is a flattened oblate spheroid) and it is spinning, however, the gravitational potential is modified so that it is an oblate spheroid with its major axis 0.3 percent longer than the minor axis. A best-fit surface to this spheroid is used by geodeticists, cartographers and surveyors, but in many places the actual geoid departs from this simple model shape. Nonuniform distributions of topography and mass with depth cause variations in the gravitational attraction, phenomena known as geoid anomalies. Areas of extra mass, such as mountains or dense rocks at depth, cause positive geoid anomalies known as geoid highs, whereas mass deficits cause geoid lows. The geoid is measured by a

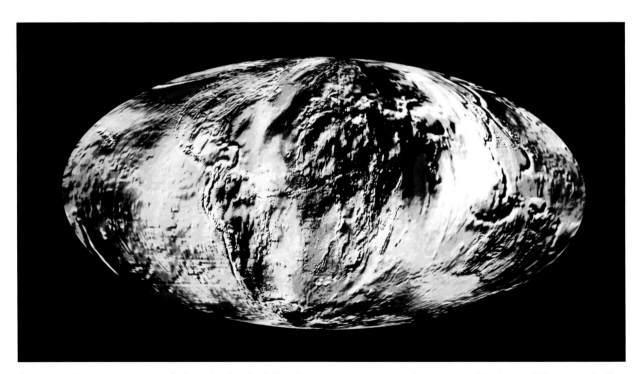

Ocean surface topography. Colors depict deviation from mean ocean surface caused by local differences in the Earth's gravitational field. Purple is up to 280 feet (85 m) above mean, through red, orange, yellow, green, to blue (up to 345 feet [105 m] below mean). *(GFZ/Photo Researchers, Inc.)*

variety of techniques, including direct measurements of the gravity field on the surface, tracking of satellite positions (and deflections due to gravity), and satellite-based laser altimetry that can measure the height of the sea surface a fraction of an inch (the subcentimeter level). Variations in the height of the geoid are typically tens to 50 feet, but range up to 450 feet (tens to even more than 100 meters).

The geoid is considered the baseline figure of the Earth, which can be considered a sea-level surface including local gravitational effects without taking into account topographic features. The geoid surface is continuous over the entire surface of the Earth. Calculating the surface of equal potential energy would yield a map close to that of the sea-surface topography as measured by satellite without the effect of motion of the water, but the true sea-surface topography differs from the geoid from the effects of ocean currents. Oceanographers use this relationship to map ocean circulation patterns, including complex eddies, by removing the height of the sea surface caused by gravity and examining the remaining anomalies caused by motion of the water.

See also GEOPHYSICS; SUPERCONTINENT CYCLES.

FURTHER READING

Turcotte, Donald L., and Gerald Schubert. *Geodynamics*. 2nd ed. Cambridge: Cambridge University Press, 2002.

Vanicek, Petr, and Nikolaos T. Christou. *Geoid and Its Geophysical Interpretations*. New York: CRC Press, 1994.

geological hazards Geological hazards take many shapes and forms, from earthquakes and volcanic eruptions to the slow, downhill creep of material on a hillside and the expansion of clay minerals in wet seasons. Natural geologic processes constantly operate on the planet. They are considered hazardous when they go to extremes and interfere with the normal activities of society. For instance, the surface of the Earth constantly moves through plate tectonics yet is not noticeable until sections of the surface move suddenly, causing an earthquake.

The Earth is a naturally dynamic, hazardous world, with volcanic eruptions spewing lava and ash, earthquakes pushing up mountains and shaking Earth's surface, and tsunamis that sweep across ocean basins at hundred of miles per hour (500 km/hr), rising in huge waves on distant shores. Mountains may suddenly collapse, burying entire villages under massive landslides. Other mountain slopes are gradually creeping downhill, slowly tilting and moving everything built on them farther downslope. Storms sweep coastlines and remove millions of tons of sand

from one place and deposit it in another in a single day. Large parts of the globe are turning into desert. Glaciers that once advanced are rapidly retreating, and sea level is beginning to rise faster than previously imagined. All of these natural phenomena are expected consequences of the way the planet works, and as scientists better understand these geological processes, they are better able to predict when and where natural geologic hazards could become disasters and take preventative measures.

The slow but steady movement of tectonic plates on the surface of the Earth is the cause of many geologic hazards, either directly or indirectly. Plate tectonics controls the distribution of earthquakes and the location of volcanoes, and causes mountains to be uplifted. Other hazards are related to Earth's surface processes, including floods of rivers, coastal erosion, and changing climate zones. Many of Earth's surface processes are parts of natural cycles but are considered hazardous to humans because people did not adequately understand the cycles before building on exposed coastlines and in areas prone to shifting climate zones. A third group of geologic hazards is related to materials such as clay minerals that dramatically expand when wetted, and sinkholes that develop in limestones. Still other hazards are extraterrestrial in origin, such as the occasional impact of meteorites and asteroids with Earth. The exponentially growing human population on Earth worsens the effect of most of these hazards. Species on the planet are now experiencing a mass extinction event, the severity of which has not been seen since the extinction event 66 million years ago that killed the dinosaurs and many other species at that time.

Many geologic hazards are the direct consequence of plate tectonics, associated with the motion of individual blocks of the rigid outer shell of the Earth. With so much energy loss accommodated by plate tectonics, it is clear why plate tectonics is one of the major energy sources for natural disasters and hazards. Most of the earthquakes on the planet are directly associated with plate boundaries, and these sometimes devastating earthquakes account for much of the motion between the plates. Single earthquakes have killed tens and even hundreds of thousands of people, such as the 1976 Tangshan earthquake in China that killed a quarter million people, and the 2008 Sichuan earthquake that killed about 100,000 people in southern China. Earthquakes also cause enormous financial and insurance losses; for instance, the 1994 Northridge, California, earthquake caused more than $14 billion in losses. Most of the world's volcanoes are also associated with plate boundaries. Thousands of volcanic vents are located along the midocean ridge system, and

Mudflow on Whangaehu River after lahar from crater lake of Mount Ruapehu in the central North Island of New Zealand, March 18, 2007: The bridge over river is Tangiwai road bridge. *(Anthony Phelps/Reuters/Landov)*

most of the volume of magma produced on the Earth is erupted through these volcanoes. Volcanism associated with the midocean ridge system is rarely explosive, hazardous, or even noticed by humans. In contrast, volcanoes situated above subduction zones at convergent boundaries are capable of producing tremendous explosive eruptions, with great

Pyroclastic flow on Mayon volcano in Legazpi City, Philippines, March 7, 2000 *(AP Images)*

devastation of local regions. Volcanic eruptions and associated phenomena have killed tens of thousands of people in the 20th century, including the massive mudslides at Nevada del Ruiz, Colombia, that killed 23,000 in 1985. Some of the larger volcanic eruptions cover huge parts of the globe with volcanic ash and are capable of changing the global climate. Plate tectonics is also responsible for uplifting the world's mountain belts, which are associated with their own sets of hazards, particularly landslides and other mass wasting phenomena.

Some geologic hazards are associated with steep slopes and the effects of gravity moving material down these slopes to inhabited areas. Landslides and the slow downhill movement of earth material occasionally kill thousands in large disasters, such as when parts of a mountain collapsed in 1970 in the Peruvian Andes and buried a village several tens of miles away, killing 60,000 people. Downhill movements are typically more localized and destroy individual homes, neighborhoods, roads, or bridges. Some downslope processes are slow and involve the inch-by-inch (cm by cm) creeping of soil and other earth material downhill, taking everything with it during its slide. Creep is one of the costliest of natural

LIQUEFACTION AND LEVEES: POTENTIAL DOUBLE DISASTER IN THE AMERICAN MIDWEST

Liquefaction is a process during which sudden shaking of certain types of water-saturated sands and muds turns these once-solid sediments into a slurry having a liquidlike consistency. Liquefaction occurs when shaking causes individual grains in the soil to move apart, then water moves between the grains, making the whole water/sediment mixture behave like a fluid. Earthquakes often cause liquefaction of sands and muds, and any structures built on soils or sediments that liquefy may suddenly sink into them as if they were resting on a thick fluid. Liquefaction causes sand to bubble to the surface during earthquakes, forming mounds up to several tens of feet (~ 10 m) high, known as sand volcanoes, and ridges of sand to squeeze into cracks in the Earth. Liquefaction is responsible for the sinking of sidewalks, telephone poles, building foundations, and other structures during earthquakes. Famous examples of liquefaction occurred in the 1964 Alaskan earthquake, when entire neighborhoods slid toward the sea on liquefied sand layers, and in the 1964 and 1995 Japan earthquakes, when entire rows of apartment buildings and shipping piers rolled onto their sides but were not severely damaged internally.

Recent earthquakes and floods in the Midwest region of the United States passed with locally huge amounts of damage, but were not catastrophic events for the entire region. But one must consider what could have happened if the earthquakes were slightly larger (which is possible) and occurred while the rivers were at high stages (which happens for several months each year). One real threat is that many apparently stable levees may experience mass failure by liquefaction during an earthquake, potentially causing catastrophic flooding of regions behind these levees.

The U.S. Geological Survey reports that there is a significantly high threat of many of the soils on the floodplains of the Mississippi, Missouri, Ohio, and Illinois Rivers—in a report to the U.S. House of Representatives in 2006, Eugene Schweig (of the U.S. Geological Survey) stated, "If the earthquakes were to occur when the Ohio and Mississippi Rivers were high, loss of levees is likely along with flooding of low-lying communities." When the soils in the levees are saturated with water, such as during flood or high-water events, the potential for liquefaction is much higher. Geological analysis of the banks of many rivers in the region has revealed that a magnitude 6 or 7 earthquake struck about 40 miles (65 km) east of St. Louis some 6,500 years ago, and another event struck about 4,000 years ago. Both events caused massive liquefaction of the thick river sediments on the floodplains. Large earthquakes have continued to hit the region in historic times, as shown by the 1811–12 sequence of magnitude 7–8 events, which also caused liquefaction of huge areas in southern Missouri and surrounding states. The April 18, 2008, earthquake in Illinois and aftershocks remind and warn us that the potential consequences of earthquakes on levees during high water must be considered.

The two earthquake swarms in the Midwest in 2008, although minor, occurred when the local rivers were high. What would have happened if the earthquakes were slightly larger, say a magnitude 6 or 7? Would the levees have experienced mass failure and collapse by liquefaction? Levees fail for several reasons. They may be overtopped by high water, scoured at their bases and sides, or have water seep through the pores between the sand and mud, weakening the structure until the pressure from the high water causes it to collapse. The most catastrophic type of failure can occur when the soils of the levee liquefy from the pressure of the high water, by shaking from earthquakes, or both. In these types of failures, hundreds of linear yards of levee may suddenly collapse, sending torrents of water into the "protected" areas behind the levees. Studies by the U.S. Army Corps of Engineers suggest that many of the levees in the region are not strong enough to withstand shaking during an earthquake and may fail by liquefaction. The problem is especially critical in communities such as those surrounding East St. Louis, Illinois, where the entire levee system is in the process of being decertified, as the levees do not meet modern standards for safety in earthquakes and from other stresses, such as floods. If these levees fail during high water, the force of the Mississippi will surge into East St. Louis with the force of Niagara Falls, pushing into and covering the floodplain with many feet (several m) of water. Approximately 130,000 people live on the floodplain in the Metro East area near East St. Louis and rely on the levees for protection. If an earthquake causes massive liquefaction and failure of the levees, there will be little time to react, and many people will be stranded on rooftops, creating scenes reminiscent of New Orleans and Hurricane Katrina.

What can be done? The U.S. Army Corps of Engineers estimated that it will take $180 million to upgrade the Metro East levees to modern standards, and much of that money would need to come from local communities. If that is too expensive, residents on the floodplain behind the levees need to be aware of the possibility of liquefaction and mass failure of the levee system. Flood-hazard and earthquake-hazard risk maps need to be compared to determine which areas have the greatest threat for liquefaction, and emergency management plans need to be established for the contingency of such a catastrophe. This could save many lives. Homeowners, businesses, local governments, and insurance issuers need to understand these risks and plan accordingly.

FURTHER READING

Kusky, T. M. *Floods: Hazards of Surface and Groundwater Systems.* New York: Facts On File, 2008.

———. *Earthquakes: Plate Tectonics and Earthquake Hazards.* New York: Facts On File, 2008.

Debris in Galle, Sri Lanka, after tsunami, December 27, 2004 *(AP Images)*

hazards, for which American taxpayers pay billions of dollars each year.

Many other geological hazards are driven by energy from the Sun and reflect the interaction of the hydrosphere, lithosphere, atmosphere, and biosphere. Heavy or prolonged rains can cause river systems to overflow, flooding low-lying areas and destroying towns, farmlands, and changing the course of major rivers. There are several types of floods, ranging from flash floods in mountainous areas to regional floods in large river valleys such as the great floods of the Mississippi and Missouri Rivers in 1993. Coastal regions may also experience floods, sometimes the result of typhoons, hurricanes, or coastal storms that bring high tides, storm surges, heavy rains, and deadly winds. Coastal storms can cause large amounts of coastal erosion, including cliff retreat, beach and dune migration, and opening of new tidal inlets and closing of old inlets. All these normal beach processes have become hazardous since so many people have migrated into beachfront homes. Hurricane Andrew caused more than $19 billion of damage to the southern United States in 1992.

Deserts and dry regions are associated with their own natural geologic hazards. Blowing winds and shifting sands hinder agricultural efforts, and deserts have a limited capacity to support large populations. Some of the greatest disasters in human history have been caused by droughts, some associated with the expansion of desert regions into areas that previously received significant rainfall and supported large populations dependent on agriculture. In this century the sub-Saharan Sahel region of Africa has been hit with drought disasters several times, affecting millions of people and animals. This appears to be part of a natural climate cycle of alternating wet and dry periods in the Sahara. Contraction and expansion of the desert fringes has severe consequences for those who try to live in such changing conditions.

Desertification is but one possible manifestation of global climate change. The Earth has fluctuated in climate extremes, from hot and dry to cold and dry or cold and wet, and has experienced several periods when much of the land's surface was covered by glaciers. Glaciers have their own set of local-scale hazards that affect those living or traveling on or near their ice crevasses, which can be deadly if fallen into. Glacial meltwater streams can change in discharge so quickly that encampments on their banks can be washed away without a trace, and icebergs present hazards to shipping lanes. More important, glaciers reflect subtle changes in global climate—when glaciers are retreating, climate may be warming and becoming drier. When glaciers advance, the global climate may be getting colder and wetter. Glaciers have advanced and retreated over northern North America several times in the past 100,000 years. The Earth is currently experiencing an interglacial episode and may see the start of the return of the continental glaciers over the next few hundred or thousand years.

Geologic materials themselves can be hazardous. Asbestos, a common mineral, is being removed from thousands of buildings in the United States because of the threat that certain types of airborne asbestos fibers present to human health. In some cases (for certain types of asbestos fibers), this threat is real and removal of the fibers is necessary. In other cases leaving the asbestos alone is safer than disturbing it and releasing airborne particles. Natural radioactive decay releases harmful gases, including radon, that creep into homes, schools, and offices, and causes numerous cases of cancer every year. Mitigating this hazard is often easy—simple monitoring and ventilation can prevent many health problems associated with radon exposure. Other materials can be hazardous even though they seem inert. For instance, some clay minerals expand by hundredsfolds when wetted. These expansive clays rest under many foundations, bridges, and highways, and cause billions of dollars of damage every year in the United States.

Sinkholes, including the one in Winter Park, Florida, have swallowed homes and businesses in Florida and in other locations in recent years. Sinkhole collapse and other subsidence hazards are more important than many realize. Large parts of southern California near Los Angeles have sunk tens of feet (about 10 meters) in response to pumping of groundwater and oil out of underground reservoirs. Other developments above former mining areas have begun sinking into collapsed mine tunnels. Coastline areas experiencing subsidence face the added risk of having the ocean rise into former living space. Coastal subsidence coupled with gradual sea-level rise is rapidly becoming one of the major global hazards that humans must deal with in the next century, since most of the world's population lives near the coast in the reach of the rising waters. Cities may become submerged and farmlands covered by shallow salty seas. New Orleans and much of the Gulf coast have been sinking at rates of up to one inch per year (2.5 cm per year) in response to natural and human-induced processes, placing some urban areas at high risk for storm surge and hurricane damage. These risks were shown dramatically by Hurricanes Katrina and Rita in 2005. An enormous amount of planning is needed, as soon as possible, to deal with this growing threat.

Occasionally in the Earth's history the planet has been hit with asteroids and meteorites from outer space, and these have completely devastated the biosphere and climate system. Many of the mass extinctions in the geologic record are now thought to have been triggered, at least in part, by large impacts from outer space. For instance, the extinction of the dinosaurs and a huge percent of other species on Earth 66 million years ago is thought to have been caused by a combination of massive volcanism from a flood basalt province preserved in India, coupled with an impact with a six-mile (10-km) wide meteorite that hit the Yucatán Peninsula of Mexico. When the impact occurred, a 1,000-mile (1,610-km) wide fireball erupted into the upper atmosphere, a tsunami hundreds or thousands of feet (hundreds of meters) high washed across the Caribbean, southern North America, and much of the Atlantic, and huge earthquakes accompanied the explosion. The dust blown into the atmosphere immediately initiated a dark global winter, and as the dust settled months or years later, the extra carbon dioxide in the atmosphere warmed the Earth for many years, forming a greenhouse condition. Many forms of life could not tolerate these rapid changes and perished. Similar impacts at several times in the Earth's history have had a profound influence on the extinction and development of life.

The human population is growing at an alarming rate and currently doubling every 50 years. At this rate, there will be only a three-foot by one-foot space (1 square meter) for every person on Earth in 800 years. The unprecedented population growth has put such a stress on other species that we are driving a new mass extinction on the planet. Because details of the relationships between different species are unknown, many fear that destroying so many other life-forms may contribute to our own demise. In response to the population explosion, people are moving into hazardous locations including shorelines, riverbanks, along steep-sloped mountains, and along the flanks of volcanoes. Populations that grow too large to be supported by the environment usually suffer some catastrophe, disease, famine, or other mechanism that limits growth, and society needs to find ways to limit human population growth to sustainable rates. Survival of the planet depends on our ability to maintain these limits.

Advances in science and engineering in recent decades have dramatically changed perceptions of natural hazards. In the past people viewed destructive natural phenomena (including earthquakes, volcanic eruptions, floods, landslides, and tsunamis) as unavoidable and unpredictable. Society's attention to basic scientific research has changed that view dramatically, and the resulting ability to make general predictions regarding the timing, location, and severity of such destructive natural events reduces their consequences significantly. Communities can use this information to plan evacuations, strengthen buildings, and make detailed plans of what needs to be done in natural disasters to such a degree that their costs have been greatly reduced. Increased government responsibility accompanies this greater understanding. Formerly society hardly looked to government for aid in natural disasters. For instance, nearly 10,000 people perished in a hurricane that hit Galveston, Texas, on September 8, 1900, yet since there were no warning systems in place, no one was at blame. In 2001 two feet (0.6 m) of rain with consequent severe flooding hit the same area, and nobody perished, but billions of dollars of insurance claims were filed. When Hurricane Ike hit Galveston in 2008, most people evacuated and the loss of life was minimal.

Public perception of natural hazards and disasters has changed with the development of warning and protection systems, and few disasters occur without blame being assigned to public officials, engineers, or planners. Extensive warning systems, building codes, and increased understanding have certainly prevented the loss of thousands of lives, yet they also have given society a false sense of security. When an earthquake or other disaster strikes, people expect their homes to be safe, yet they were built to withstand only a certain level of shaking. When a natural geological hazard exceeds the expected level, a natural disaster with great destruction may result, and people often blame the government for not anticipating the event or preventing the destruction. However, planning and construction efforts are designed to meet only certain force levels for earthquakes and other hazards; planning for the rare stronger events would be exorbitantly expensive.

Geologic hazards can be extremely costly in terms of price and human casualties. With growing population and wealth, the cost of natural disasters has grown as well. The amount of property damage measured in dollars has doubled or tripled every decade, with individual disasters sometimes costing tens of billions of dollars. A 2000 report to the Congressional Natural Hazards Caucus estimated the costs of some recent disasters: Hurricane Andrew in 1992 cost $23 billion, the 1993 Midwest floods cost $21 billion, and the 1994 Northridge earthquake cost $45 billion. Recovery from Hurricane Katrina cost a staggering $80–200 billion, with some estimates exceeding $1 trillion. In contrast, the entire first Persian Gulf War cost the United States and its allies $65 billion. That the costs of natural geologic hazards are now similar to the costs of warfare demonstrates the importance of understanding their causes and potential effects.

See also EARTHQUAKES; ENERGY IN THE EARTH SYSTEM; HURRICANES; ISLAND ARCS, HISTORICAL ERUPTIONS; MASS WASTING; PLATE TECTONICS; VOLCANO.

FURTHER READING

Abbott, P. L. *Natural Disasters*. 3rd ed. Boston: McGraw Hill, 2002.

Bryant, E. A. *Natural Hazards*. Cambridge: Cambridge University Press, 1993.

Erickson, Jon. *Quakes, Eruptions, and Other Geologic Cataclysms: Revealing the Earth's Hazards*. New York: Facts On File, 2001.

Griggs, Gary B., and J. A. Gilchrist. *Geologic Hazards, Resources, and Environmental Planning*. Belmont, Calif.: Wadsworth, 1983.

Kusky, Timothy M. *Geologic Hazards: A Sourcebook*. Greenwood Press, 2003.

———. *Earthquakes: Plate Tectonics and Earthquake Hazards*. New York: Facts On File, 2008.

——— *Volcanoes: Eruptions and Other Volcanic Hazards*. New York: Facts On File, 2008.

———. *Tsunami: Giant Waves from the Sea*. New York: Facts On File, 2008.

———. *Landslides: Mass Wasting, Soil, and Mineral Hazards*. New York: Facts On File, 2008.

———. *Climate Change: Shifting Deserts, Glaciers, and Climate Belts*. New York: Facts On File, 2008.

———. *The Coast: Hazardous Interactions within the Coastal Zone*. New York: Facts On File, 2008.

———. *Floods: Hazards of Surface and Groundwater Systems*. New York: Facts On File, 2008.

———. *Asteroids and Meteorites: Catastrophic Collisions with Earth*. New York: Facts On File, 2008.

Murck, Barbara W., Brian J. Skinner, and Stephen C. Porter. *Dangerous Earth: An Introduction to Geologic Hazards*. New York: John Wiley & Sons, 1997.

geomagnetism, geomagnetic reversal

The Earth has a magnetic field generated within the core of the planet. The field is generally approximated as a dipole, with north and south poles and magnetic field lines emerging from the Earth at the south pole and extending toward the north pole. The field is characterized at each place on the planet by an inclination and a declination. The inclination is a measure of how steeply inclined the field lines are with respect to the surface, with low inclinations near the surface, and steep inclinations near the poles. The declination measures the apparent angle between the rotational north pole and the magnetic north pole.

The magnetic field originates in the liquid outer core of the Earth and is thought to result from electrical currents generated by convective motions of the iron-nickel alloy that the outer core is made from.

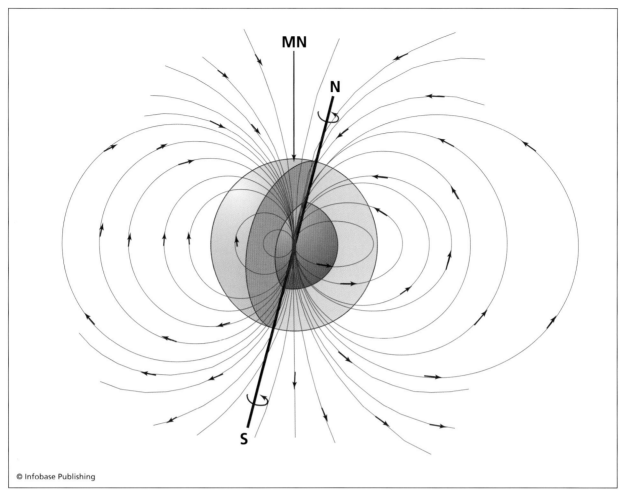

Magnetic field lines of Earth approximate the shape of the field produced by a bar magnet. Magnetic field lines point upward out of the south magnetic pole and form imaginary elliptical belts of equal intensity around Earth and plunge back into Earth at the magnetic north pole. Note how the magnetic poles are not coincident with the rotational poles. The orientation of the magnetic field at any point on Earth can be expressed as an inclination (plunge into Earth) and a declination (angular distance between the magnetic and rotational north poles).

The formation of the magnetic field by motion of the outer core is known as the geodynamo theory, pioneered by the German-American geophysicist and biologist Walter M. Elsasser (1904–91) of Johns Hopkins University in the 1940s. A dynamo generates electrical energy from mechanical energy. The basic principle for the generation of the magnetic field is that mechanical energy from the motion of the liquid outer core, which is an electrical conductor, is transformed into electromagnetic energy of the magnetic field. The convective motion of the outer core, maintained by thermal and gravitational forces, is necessary to maintain the field. If the convection stopped or the outer core solidified, the magnetic field would disappear. Secular variations in the magnetic field have been well documented by examination of the paleomagnetic record in the seafloor, lava flows, and sediments. Every few thousand years the

magnetic field changes intensity and reverses, with the north and south poles abruptly flipping.

See also EARTH; GEOPHYSICS; PALEOMAGNETISM.

geomorphology Geomorphology is the description, classification, and study of the physical properties of and the origin of the landforms of the Earth's surface. Most studies in geomorphology include an analysis of the development of landforms and their relationships to underlying structures, and how the surface has interacted with other Earth systems such as the hydrosphere, cryosphere, and atmosphere. Geomorphologists have become increasingly concerned with the study of global climate change and the development of specific landforms associated with active deformation and tectonics. These relatively new fields of global change and active tectonic

Braided river channel and alluvial terraces on Golmud River in the Kunlun Mountains, Qinghai Province, China *(Fletcher & Baylis/Photo Researchers, Inc.)*

geomorphology represent a significant movement away from classical geomorphology, which is concerned mostly with the evolutionary development of landforms.

Geomorphological phenomena depend on many different processes that operate on the surface of the planet, so the geomorphologist needs to integrate hydrology, climate, sedimentology, geology, forestry, pedology, and many other sciences. This type of research has relevant applications to everyday life—for example, the decomposition of bedrock and the development of soils relate to global climate systems and also can help local engineers solve problems such as determining slope stability for construction sites. Other geomorphologists may study the development and evolution of drainage basins, along with the analysis of fluvial landforms such as floodplains, terraces, and deltas. Desert geomorphology is concerned with the development of desert landforms and climate, whereas glacial geomorphology analyzes the causes and effects of the movement of glaciers. Coastal processes including erosion, deposition, and longshore movement of sand are treated by coastal geomorphologists, whereas the development of various other types of landforms, such as karst,

alpine, seafloor, and other terrains are treated by other specialists.

FURTHER READING
Ritter, Dale, R. Craig Kochel, and Jerry Miller. *Process Geomorphology.* 3rd ed. Boston: WCB/McGraw Hill, 1995.

geophysics Geophysics is the study of the Earth by quantitative physical methods, with different divisions including solid-Earth geophysics, atmospheric and hydrospheric geophysics, and solar-terrestrial physics. The many subdisciplines in the field include seismology, tectonics, geomagnetics, gravity, atmospheric science, ocean physics, and many others. Geophysics also includes the description and study of the origin and evolution of the major Earth systems, including the core, mantle, and crust of the continents and oceans. Geophysical methods include

- the use of tools of reflection and refraction seismology to use and measure the passage of sound waves through the Earth to measure the physical properties of the area in the Earth the waves pass through

- the use of electromagnetic sensors to quantify the electrical and magnetic properties of rocks
- potential field methods, which determine variations in gravity and magnetism across a region

SEISMOLOGY

Seismology is the study of the propagation of seismic, or sound, waves through the Earth, including analysis of earthquake sources, mechanisms, and the determination of the structure of the Earth through variations in the properties of seismic waves. The analysis is quantitative and typically requires high-powered computers.

The structure of the deep parts of the Earth can be mapped by seismology. Seismographs are stationed all over the world. Studying the propagation of seismic waves from natural and artificial sources, such as earthquakes, nuclear explosions, and other seismic events, allows for the calculation of changes in the properties of the Earth in different places. If the Earth had a uniform composition, seismic wave velocity would increase smoothly with depth because increased density is equated with higher seismic velocities. By plotting the observed arrival time of seismic waves, however, seismologists have found that the velocity does not increase steadily with depth, but that several dramatic changes occur at discrete boundaries and in transition zones deep within the Earth.

Seismologists calculate the positions and changes across these zones by noting several different properties of seismic waves. Some are reflected off interfaces, just as light is reflected off surfaces, and other waves are refracted, changing the velocity and path of the rays. These reflection and refraction events happen at specific sites in the Earth, and the positions of the boundaries are calculated by using wave velocities. The core-mantle boundary at 1,802 miles (2,900 km) depth in the Earth strongly influences both P- and S-waves. It refracts P-waves, causing a P-wave shadow and, because liquids cannot transmit S-waves, none gets through, causing a huge S-wave shadow. These contrasting properties of P- and S-waves can be used to map accurately the position of the core-mantle boundary.

Variation in the propagation velocity and direction of seismic waves illustrates several other main properties of the deep Earth. Velocity gradually increases with depth, to about 62 miles (100 km), where the velocity drops slightly between 62–124 miles (100–200 km) depth, in the low velocity zone. The reason for this drop in velocity is thought to be small amounts of partial melt in the rock, corresponding to the asthenosphere, the weak sphere on which the plates move, lubricated by partial melts.

Another seismic discontinuity exists at 248.5 miles (400 km) depth, where velocity increases sharply due to a rearrangement of the atoms within olivine in a polymorphic transition into spinel structure, corresponding to an approximate 10 percent increase in density.

A major seismic discontinuity at 416 miles (670 km) could be either another polymorphic transition or a compositional change. This is the topic of many current investigations. Some models suggest that this boundary separates two fundamentally different types of mantle, circulating in different convection cells, whereas other models suggest that there is more interaction between rocks above and below this discontinuity.

The core-mantle boundary is one of the most fundamental on the planet, with a huge density contrast from 5.5 g/cm^3 above, to 10–11 g/cm^3 below, a contrast greater than that between rocks and air on the surface of the Earth. The outer core consists mainly of molten iron. An additional discontinuity occurs inside the core at the boundary between the liquid outer core and the solid, iron-nickel inner core.

Seismic waves can also be used to understand the structure of the Earth's crust. Andrija Mohorovičić, a Yugoslav seismologist, from Volosko in Croatia, noticed slow and fast arrivals from nearby earthquake source events. He proposed that some seismic waves traveled through the crust, some along the surface, and others were reflected off a deep seismic discontinuity between seismically slow and fast material at about 18.6 miles (30 km) depth. Geologists now recognize this boundary, called the Mohorovicic (or Moho) boundary, to be the base of the crust and use its seismically determined position to measure the thickness of the crust, typically between 6.2–43.5 miles (10–70 km).

GRAVITY ANOMALIES (POTENTIAL FIELD STUDIES)

Gravity anomalies are the difference between the observed value of gravity at a point and the theoretically calculated value of gravity at that point, based on a simple gravity model. The value of gravity at a point reflects the distribution of mass and rock units at depth, as well as topography. The average gravitational attraction on the surface is 32 feet per second squared (9.8 m/s^2), with one gravity unit (g. u.) being equivalent to one ten-millionth of this value. Another older unit of measure, the milligal, is equivalent to 10 gravity units. The range in gravity on the Earth's surface at sea level is about 50,000 g.u., from 32.09–32.15 feet per second squared (9.78–9.83 m/s²). An adult human would weigh slightly more at the poles than at the equator because the Earth has a slightly larger radius at the equator than at the poles.

Geologically significant variations in gravity are typically only a few tenths of a gravity unit, so instruments that measure gravity anomalies must be very sensitive. Some gravity surveys use closely to widely spaced gravity meters on the surface, whereas others use observations of the perturbations of orbits of satellites.

The determination of gravity anomalies involves subtracting the effects of the overall gravity field of the Earth, accomplished by removing the gravity field at sea level (geoid), leaving an elevation-dependent gravity measurement. This measurement reflects a lower gravitational attraction with height and distance from the center of the Earth, as well as an increase in gravity caused by the gravitational pull of the material between the point and sea level. The free-air gravity anomaly is a correction to the measured gravity calculated using only the elevation of the point and the radius and mass of the Earth. A second correction, known as the Bouguer gravity anomaly, depends on the shape and density of rock masses at depth. Sometimes a third correction, known as the isostatic correction, is applied to gravity measurements. This applies when a load such as a mountain, sedimentary basin, or other mass is supported by mass deficiencies at depth, much like an iceberg floating lower in the water. There are several different mechanisms of possible isostatic compensation, however, and it is often difficult to know which mechanisms are important on different scales. Therefore this correction is often not applied.

Different geological bodies are typically associated with different magnitudes and types of gravity anomalies. Belts of oceanic crust thrust on continents (ophiolites) represent unusually dense material and are associated with positive gravity anomalies of up to several thousand g.u. Likewise, massive sulfide metallic ore bodies are unusually dense and are also associated with positive gravity anomalies. Salt domes, oceanic trenches, and mountain ranges all represent an increase in the amount of low-density material in the crustal column, and are therefore associated with increasingly negative gravity anomalies, with negative values of up to 6,000 g.u. associated with the highest mountains on Earth, the Himalayan chain.

GEOPHYSICS AND ISOSTACY

Isostacy is the principle of hydrostatic equilibrium applied to the Earth, referring to the position of the lithosphere essentially floating on the asthenosphere, similar to how low-density ice floats at a certain level on water, depending on the relative densities of the water and ice. Isostatic forces are of major importance in controlling the topography of the Earth's surface. There are several different models for how topography is supported, referred to as isostatic models. The simplest models have blocks of crust that are essentially floating as isolated blocks in a fluid substrate (the asthenosphere). Such blocks are free to migrate vertically and do not interact with neighboring blocks. There are two main variations of these simple isostatic models. In the Pratt model crustal blocks of different density are assumed to extend to a constant depth known as the depth of compensation, and the height of the topography varies inversely with the density of each block. Thus high-density oceanic crust resides at a lower level than lower-density continental crust. In the Airy model the level of isostatic compensation varies for each block, but the crustal layer is assumed to have a constant density. This way thick blocks have high topography and a thick root to compensate the topography, whereas thin crustal blocks have subdued topography. Both of these models are simplistic descriptions of a complex lithosphere, having been derived in the 1700s before an appreciation of plate tectonics. The Airy model is generally more applicable than the Pratt model, but the Airy model does not accommodate known variations in crustal density, such as that between continents and oceans.

Isostatic anomalies are variations in measured gravity values from those expected using an assumed isostatic model and depth of compensation. The anomaly indicates that the model used or the compensation depth assumed needs to be adjusted in the model.

See also ASTHENOSPHERE; CONVECTION AND THE EARTH'S MANTLE; EARTH; EARTHQUAKES; GEODESY; GEODYNAMICS; GEOID; GRAVITY, GRAVITY ANOMALY; MAGNETIC FIELD, MAGNETOSPHERE; MANTLE; PALEO-MAGNETISM; PETROLEUM GEOLOGY; PLATE TECTONICS; REMOTE SENSING; SEISMOLOGY.

FURTHER READING

Keary, P., Keith Klepeis, and Fredick J. Vine. *Global Tectonics*. Oxford: Blackwell, 2008.

Shearer, Peter M. *Introduction to Seismology*. Cambridge: Cambridge University Press, 2009.

Sheriff, Robert E. *Encyclopedic Dictionary of Applied Geophysics*. 4th ed. Tulsa, Okla.: Society of Exploration Geophysicists, 2002.

Turcotte, Donald L., and Gerald Schubert. *Geodynamics*. 2nd ed. Cambridge: Cambridge University Press, 2002.

Vanicek, Petr, and Nikolaos T. Christou. *Geoid and Its Geophysical Interpretations*. New York: CRC Press, 1994.

geyser Geysers are springs in which hot water or steam sporadically or episodically erupts as jets from an opening in the surface, in some cases creating a tower of water hundreds of feet (tens of meters) high. Geysers are often marked on the surface by a cone of siliceous sinter and other minerals that precipitated from the hot water, known as geyser cones. Many also have deposits of thermophilic (heat-loving) bacteria that can form layers and mounds in stromatolitic buildups. Geysers form where water in pore spaces and cracks in bedrock gets heated by an underlying igneous intrusion or generally hot rock, causing it to boil and erupt, then the lost water is replaced by other water that comes in from the side of the system. In this way a circulation system is set up that in some cases is quite regular with a predictable period between eruptions. The most famous geyser in the world is Old Faithful in Yellowstone National Park, Wyoming, which erupts every 20 to 30 minutes.

Thermal springs in which the temperature is greater than that of the human body are known as hot springs. They are found in places where porous structures such as faults, fractures, or karst terrains channel meteoric water (derived from rain or snow) deep into the ground where it warms, and also where it can escape upward fast enough to prevent it from cooling by conduction to the surrounding rocks. Most hot springs, especially those with temperatures above 140°F (60°C), are associated with regions of active volcanism or deep magmatic activity, although some hot springs are associated with regions of tectonic extension without known magmatism. Active faulting is favored for the development of hot springs since the fluid pathways tend to become mineralized and closed by minerals that precipitate out of the hot waters, and the faulting is able to break repeatedly and reopen these closed passageways.

When cold, descending water heats up in a hot spring thermal system, it expands and the density of the water decreases, giving it buoyancy. Typical geothermal gradients increase about 120–140°F per mile (25–30°C per km) in the Earth, so for surface hot springs to attain temperatures of greater than 140°F (60°C), it is usually necessary for the water to circulate to at least two miles (two or three kilometers) depth. This depth may be less in volcanically active areas where hot magmas may exist at shallow crustal levels, reaching several hundred degrees at two miles (three kilometers) depth. Hot springs may boil when the temperatures of the waters reach or exceed 212°F (100°C), and if the rate of upward flow is fast enough to allow decompression. In these cases boiling water and steam may be released at the surface, sometimes forming geysers.

Hot springs are often associated with a variety of mineral precipitates and deposits, depending on the composition of the waters that come from the springs. This composition is typically determined by the type of rocks the water circulates through and is able to leach minerals from, with typical deposits including mounds of travertine, a calcium carbonate precipitate, siliceous sinters, and hydrogen sulfides.

Hot springs are common on the seafloor, especially around the oceanic ridge system where magma is located at shallow levels. The great pressure of the overlying water column on the seafloor elevates the boiling temperature of water at these depths, so that vent temperatures may exceed 572°F (300°C). Submarine hot springs often form 10-foot (several meter) or taller towers of sulfide minerals with black clouds of fine metallic mineral precipitates emanating from the hot springs. These systems, known as black smoker chimneys, host some of the most primitive known life-forms on Earth, some of which derive their energy from the sulfur and other minerals that come out of the hot springs rather than from sunlight.

Geysers and hot springs may be the surface outflows of hot waters that flow from deep within the

Beehive Geyser at Yellowstone National Park, Wyoming *(National Park Service)*

Earth. Many geysers and hot springs contain water that fell as rain, seeped into the Earth where it got heated, then rose again to the surface. Other hot springs and geysers have water that came from deeper levels in the Earth's crust, are known as hydrothermal fluids. Most heated subsurface waters contain dissolved minerals or other substances. Known as hydrothermal solutions, these waters are important because they dissolve, transport, and redistribute many elements in the Earth's crust and are responsible for the concentration and deposition of many ores, including many gold, copper, silver, zinc, tin, and sulfide deposits. These mineral deposits are known as hydrothermal deposits.

Hydrothermal solutions are typically derived from one or more sources, including fresh or saline groundwater, water trapped in rocks as they are deposited, water released during metamorphic reactions, or water released from magmatic systems. The minerals, metals, and other compounds dissolved in hydrothermal solutions often come from the dissolution of the rocks through which the fluids migrate or are released from magmatic systems. Hydrothermal solutions commonly form during the late stages of crystallization of a magmatic body, and these fluids contain many of the chemical elements that do not readily fit into the atomic structures of the minerals crystallizing from the magma. These fluids tend to be enriched in lead, copper, zinc, gold, silver tin, tungsten, and molybdenum. Many hydrothermal fluids are also saline, with the salts derived from leaching of country rocks. Saline solutions are much more effective at carrying dissolved metals than nonsaline solutions, so these hydrothermal solutions tend to be enriched in dissolved metals.

As hydrothermal solutions move up through the crust, they cool from as high as 1,112°F (600°C), and at lower temperatures the solutions cannot hold as much dissolved material. Therefore as the fluids cool, hydrothermal veins and ore deposits form, with different minerals precipitating out of the fluid at different temperatures. Some minerals may also precipitate out when the fluids come into contact with rocks of a certain composition, with a fluid-wall rock reaction.

GEOTHERMAL ENERGY IN REGIONS WITH GEYSERS

Geysers, hot springs, and fumaroles are associated with regions of elevated temperature at depth, and in some cases these high temperatures have been exploited for geothermal energy. Temperatures in the Earth generally increase downward at 90°–250°F (30°–140°C) per mile (68°–212°F [20°–100°C] per km), following the geothermal gradient for the region. Some regions near active volcanic vents and deep-seated plutons have even higher geothermal

gradients and are typified by abundant fumaroles and hot springs. These systems are usually set up when rising magma heats groundwater in cracks and pore spaces in rocks, and this heated water rises to the surface. Water from the sides of the system then moves in to replace the water that rose into hot springs, fumaroles, and geysers, and a natural hydrothermal circulation system is set up. The best natural hydrothermal systems are found in places where there are porous rocks and a heat source such as young magma.

Geothermal energy may be tapped by drilling wells, frequently up to several kilometers deep, into the natural geothermal systems. Geothermal wells that penetrate these systems commonly encounter water and less commonly steam at temperatures exceeding 572°F (300°C). Since water boils at 212°F (100°C) at atmospheric pressure and higher temperatures at higher pressures (being 300°C at a 1-km depth), the water can be induced to boil by reducing the pressure by bringing it toward the surface in pipes. For a geothermal well to be efficient, the temperatures at depth should be greater than 392°F (200°C). Turbines attached to generators are attached to the tops of wells, requiring about two kilograms of steam per second to generate each megawatt of electricity.

Some countries use geothermal energy to produce large amounts of electricity or heated water, with China, Hungary, Iceland, Italy, Japan, Mexico, and New Zealand leading the list. Still, the use of geothermal energy amounts to a very small but growing amount of the total electrical energy used by industrialized nations around the world.

See also BLACK SMOKER CHIMNEYS; ENERGY IN THE EARTH SYSTEM.

Gilbert, Grove K. (1843–1918) *American Geologist, Geomorphologist* Grove Gilbert is well known for his concept of graded streams. This concept maintains that streams always make channels and slopes for themselves either by cutting down their beds or by building them up with sediment so that over a period of time these channels will transport exactly the load delivered into them from above. Gilbert also explained the structure of the Great Basin as a result of extension, that is, individual "basin ranges" are the eroded upper parts of tilted blocks displaced along faults as "comparatively rigid bodies of strata."

Grove Karl Gilbert was born on May 6, 1843, in Rochester, New York, and graduated from the University of Rochester. In 1871 Gilbert joined the geographical survey of George M. Wheeler, a pioneering explorer and cartographer of the western United States. The Wheeler Survey was charged by the U.S.

Congress to map a portion of the United States west of the 100th meridian at a scale of eight miles to the inch, a task the Wheeler Survey and the similar King and Powell Surveys continued until 1879. Gilbert moved to the Powell Survey in 1874, mapping parts of the Rocky Mountain region. In 1879 the Wheeler and the King and Powell surveys were reorganized to form the United States Geological Survey.

In 1877 Gilbert published an important geological monograph *The Geology of the Henry Mountains*. As a result of his studies on the Henry Mountains (1875–76), he was the first to establish that intrusive bodies are capable of deforming the host rock, and he illustrated his monograph with examples of this process. He insisted that the Earth's crust is "as plastic in great masses as wax is in small." Unfortunately he exaggerated the fluidity of magma. Gilbert was conscientious in giving credit to those who deserved it but paid no attention when it was given to him. Gilbert's report also introduced new concepts of erosion and river and stream system development that are commonly used in modern theories of physical geology.

In 1879 Gilbert was a senior geologist with the United States Geological Survey, and from 1889 to 1892, he was the chief geologist. During this time he completed and published several important studies of the American West ranging from the structure and origin of the mountain ranges to the processes involved in the evolution of the stream systems. In 1890 he published his analysis of the Pleistocene Lake Bonneville as the first monograph of the U.S. Geological Survey. A small remnant of Lake Bonneville is preserved as the Great Salt Lake, in Utah. This pluvial lake covered much of the Great Basin region in Utah, Idaho, and Nevada. Lake Bonneville formed about 32,000 years ago, and much of it was drained suddenly through Red Rock Pass in Idaho 16,800 years ago. This catastrophic event occurred when the lake flowed over a natural dam formed by coalesced alluvial fans at Red Rock Pass, and rapidly eroded and downcut the fans as the lake waters rushed through, dropping the lake level by 350 feet (105 m). The flood from this event was a huge geologic event, releasing about 1,000 cubic miles (4,168 km³) of water in the first few weeks and lasting about a year. Before the catastrophic draining, the lake had a surface area similar to that of Lake Michigan (19,691 square miles, or 51,000 km²) and up to 1,000 feet (305 m) deep. The lake dried up further starting 14,000 years ago, as the glaciers retreated from North America and the climate dried, leaving a few remnants including Great Salt Lake, Little Salt Lake, Sevier Lake, and Rush Lake for the past 12,000 years.

Grove Gilbert is most remembered for his major contributions to the field of geomorphology and his observations on landscape evolution, erosion, river incision, and sedimentation. Gilbert was awarded the Wollaston Medal in 1990, the highest award given by the Geological Society of London.

See also GEOMORPHOLOGY; NORTH AMERICAN GEOLOGY; SEDIMENTARY ROCK, SEDIMENTATION.

FURTHER READING
Gilbert, G. K. "Report on the Geology of the Henry Mountains-Reprint of 1880 Paper." *Earth Science* 31 (1976): 68–74.

glacier, glacial systems Any permanent body of ice (recrystallized snow) that shows evidence of gravitational movement is known as a glacier. Glaciers are an integral part of the cryosphere, that portion of the planet where temperatures are so low that water exists primarily in the frozen state. Most glaciers presently reside in the polar regions and at high altitudes. At several times in Earth's history, however, glaciers have advanced deeply into midlatitudes, and the climate of the entire planet was different. Some models suggest that at one time ice covered the entire surface of the Earth, a state referred to as the "snowball Earth." Glaciers are dynamic systems, always moving under the influence of gravity and changing drastically in response to changing global climate systems. Thus, changes in glaciers may reflect coming changes in the environment.

Several types of glaciers exist. Mountain glaciers form in high elevations and are confined by surrounding topography, such as valleys. Specific types of mountain glaciers include cirque glaciers, valley glaciers, and fjord glaciers. Piedmont glaciers are fed by mountain glaciers but terminate on open slopes beyond the mountains. Some piedmont and valley glaciers flow into open water, bays, or fjords, and are known as tidewater glaciers. Ice caps form dome-shaped bodies of ice and snow over mountains and flow radially outward. Ice sheets are huge, continent-sized masses of ice that presently cover Greenland and Antarctica and are the largest glaciers on Earth, making up about 95 percent of all the glacier ice on the planet. If global warming were to continue to melt the ice sheets, sea level would rise by 230 feet (66 m). A polar ice sheet covers Antarctica. This ice sheet consists of two parts that meet along the Transantarctic Mountains. The Antarctic ice sheet contains shelves (thick glacial ice that floats on the sea), which form many icebergs by breaking and separating in a process called calving and which move northward into shipping lanes of the Southern Hemisphere.

Polar glaciers form where the mean average temperature lies below freezing, and these glaciers have little or no seasonal melting because the temperature does not increase sufficiently. Other glaciers, called temperate glaciers, have seasonal melting periods, when the temperature throughout the glacier may be at the pressure melting point (when the ice can melt at that pressure and both ice and water coexist). All glaciers form above snow line, the lower limit at which snow remains year-round, located at sea level in polar regions and at 5,000–6,000 feet (1,525–1,830 m) at the equator (Mount Kilamanjaro in Tanzania has glaciers, although these are melting rapidly).

Glaciers are sensitive indicators of climate change and global warming, shrinking in times of warming and expanding in times of cooling. Glaciers may be thought of as the "canaries in the coal mine" for climate change.

The Earth has experienced at least three major periods of long-term frigid climate and ice ages, interspersed with periods of warm climate. The earliest well-documented ice age is the period of the "snowball Earth" in the Late Proterozoic from about 710–650 million years ago, although there is evidence of several earlier glaciations. Beginning about 350 million years ago the Late Paleozoic saw another ice age lasting about 100 million years until 250 million years ago. The planet entered the present ice age about 55 million years ago. Varied underlying causes of these different glaciations include anomalies in the distribution of continents and oceans and associated currents, variations in the amount of incoming solar radiation, and changes in the atmospheric balance between the amount of incoming and outgoing solar radiation.

FORMATION OF GLACIERS

Glaciers form mainly by the accumulation and compaction of snow, and are deformed by flow under the influence of gravity. When snow falls it is porous, and with time the pore spaces close by precipitation and compaction. When snow first falls, it has a density of about 1/10th that of ice; after a year or more the density is transitional between snow and ice, and it is called firn. After several years the ice reaches a density of 0.9 g/cm^3, and it flows under the force of gravity. At this point glaciers are considered to be metamorphic rocks, composed of the mineral ice.

The mass and volume of glaciers constantly change in response to the seasons and to global climate variations. The mass balance of a glacier is determined by the relative amounts of accumulation and ablation (mass loss through melting and evaporation or calving). Some years see a mass gain leading to glacial advance, whereas some periods have a mass loss and a glacial retreat (the glacial front or terminus shows these effects).

Glaciers have two main zones, best observed at the end of the summer ablation period. The zone of accumulation, found in the upper parts of the glacier, remains covered by the remnants of the previous winter's snow. Below this the zone of ablation is characterized by older, dirtier ice from which the previous winter's snow has melted. An equilibrium line, marked by where the amount of new snow exactly equals the amount that melts that year, separates these two zones.

MOVEMENT OF GLACIERS

When glacial ice becomes thick enough, it begins to flow and deform under the influence of gravity. The thickness of the ice must be great enough to overcome the internal forces that resist movement, which depend on the temperature of the glacier. The thickness at which a glacier starts flowing also depends on the steepness of the slope on which it flows—thin glaciers can move on steep slopes, whereas to move across flat surfaces, glaciers must become very thick. The flow is by creep, or deformation of individual mineral grains. This creep leads to the preferential orientation of mineral (ice) grains, forming foliations and lineations, much the same way as in other metamorphic rocks.

Some glaciers develop a layer of meltwater at their base, allowing basal sliding and surging to occur. Where glaciers flow over ridges, cliffs, or steep slopes, their upper surface fails by cracking, forming large, deep crevasses that can extend to 200 feet (65 m) deep. A thin blanket of snow can cover these crevasses, making dangerous conditions for travelers on the ice.

Ice in the central parts of valley glaciers moves faster than at the sides because of frictional drag against the valley walls on the side of the glacier. Similarly, a profile with depth into the glacier would show that it moves the slowest along its base and faster internally and along its upper surfaces. When a glacier surges, the ice in the base may temporarily move as fast as the ice in the center and on top. This is because during surges, frictional resistance is reduced, and the glacier essentially rides on a cushion of meltwater along the glacial base. During meltwater-enhanced surges, glaciers may advance by as much as several kilometers in a year. Events like this may happen in response to climate changes.

GLACIATION AND GLACIAL LANDFORMS

Glaciation is the modification of the land's surface by the action of glacial ice. When glaciers move over the land's surface, they plow up the soils, abrade and file down the bedrock, carry and transport the sedimentary load, steepen valleys, then leave thick deposits of glacial debris during retreat.

In glaciated mountains a distinctive suite of landforms results from glacial action. Glacial striations are scratches on the surface of bedrock, formed when a glacier drags boulders across the bedrock surface. Rouche moutonnées and other asymmetrical landforms are made when the glacier plucks pieces of bedrock from a surface and carries them away. The step faces in the direction of transport. Cirques are bowl-shaped hollows that open downstream and are bounded upstream by a steep wall. Frost wedging, glacial plucking, and abrasion all work to excavate cirques from previously rounded mountaintops. Many cirques contain small lakes called tarns, which are blocked by small ridges at the base of the cirque. Cirques continue to grow during glaciation, and where two cirques form on opposite sides of a mountain, a ridge known as an arete forms. A steep-sided mountain known as a horn forms where three cirques meet. The Matterhorn of the Swiss Alps is an example of a glacial carved horn.

Valleys that have been glaciated have a characteristic U-shaped profile, with tributary streams entering above the base of the valley, often as waterfalls. In contrast, streams generate V-shaped valleys. Fjords are deeply indented glaciated valleys partly filled by the sea. In many places that were formerly overlain by glaciers, elongate streamlined forms known as drumlins occur. These are both depositional features (composed of debris) and erosional (composed of bedrock).

Terminal moraines are ridgelike accumulations of drift deposited at the farthest point of travel of a glacier's terminus. They may be found as depositional landforms at the bases of mountain or valley glaciers, marking the locations of the farthest advance of that particular glacier, or may be more regional, marking the farthest advance of a continental ice sheet. There are several different categories of terminal moraines, some related to the farthest advance during a particular glacial stage, and others referring to the farthest advance of a group of or all glacial stages in a region. Continental terminal moraines are typically succeeded poleward by a series of recessional moraines marking temporary stops in the glacial retreat or even short advances during the retreat. They may also mark the boundary between glacial outwash terrain toward the equator, and knob and kettle or hummocky terrain toward the pole from the moraine. The knob and kettle terrain is characterized by knobs of outwash gravels and sand separated by depressions filled with finer material. Many of these kettle holes were formed when large blocks of ice were left by the retreating glacier, and the ice blocks melted later leaving large pits where the ice once was. Kettle holes are typically filled with lakes, and many regions characterized by many small lakes have a recessional kettle hole origin.

GLACIAL TRANSPORT

Glaciers transport enormous amounts of rock debris, including large boulders, gravel, sand, and fine silt. The glacier may carry this at its base, on its surface, or internally. Glacial deposits are characteristically poorly sorted or nonsorted, with large boulders next to fine silt. Most of a glacier's load is concentrated along its base and sides, because in these places plucking and abrasion are most effective.

Active ice deposits till as a variety of moraines, which are ridgelike accumulations of drift deposited on the margin of a glacier. A terminal moraine is the farthest point of travel of the glacier's terminus. Glacial debris left on the glaciers' sides forms lateral moraines, whereas where two glaciers meet, their moraines merge into a medial moraine.

Rock flour is a general name for the deposits at the base of glaciers, where they are produced by crushing and grinding by the glacier to make fine silt and sand. *Glacial drift* is a general term for all sediment deposited directly by glaciers, or by glacial meltwater in streams, lakes, and the sea. Till is a type of glacial drift deposited directly by the ice and characterized by a nonsorted random mixture of rock fragments. Glacial marine drift is sediment deposited on the seafloor from floating ice shelves or bergs and may include many isolated pebbles or boulders that were initially trapped in glaciers on land, then floated in icebergs that calved off from tidewater glaciers. These rocks melted out while over open water and fell into the sediment on the sea bottom. These isolated dropstones are often one of the hallmark signs of ancient glaciations in rock layers that geologists find in the rock record. Stratified drift is deposited by meltwater and may include a range of sizes, deposited in different fluvial or lacustrine environments.

Glacial erratics are glacially deposited rock fragments with compositions different from underlying rocks. In many cases the erratics are composed of rock types that do not occur in the area where they are located but are normally found only hundreds or thousands of miles away. Many glacial erratics in the northern United States can be shown to have come from parts of Canada. Some clever geologists have used glacial erratics to help them find mines or rare minerals that they have located in an isolated erratic—they used their knowledge of glacial geology to trace the boulders back to their sources following the orientation of glacial striations in underlying rocks. Recently diamond mines were discovered in northern Canada (Nunavut) by tracing diamonds found in glacial till back to their source.

Sediment deposited by streams washing out of glacial moraines, known as outwash is typically deposited by braided streams. Many of these form on broad outwash plains. When glaciers retreat, the load is diminished, and a series of outwash terraces may form.

ICE CAPS

Glaciers are any permanent body of ice (recrystallized snow) that shows evidence of gravitational movement. Ice caps form dome-shaped bodies of ice and snow over mountains, and flow radially outward. They cover high peaks of some mountain ranges, such as parts of the Kenai and Chugach Mountains in Alaska, the Andes, and many in the Alpine-Himalayan system. Ice caps are relatively small, fewer than 20,000 square miles (50,000 km²), whereas ice sheets are similar but larger. Ice sheets are continent-sized masses of ice that presently cover Greenland and Antarctica. About 95 percent of all the glacier ice on the planet exists as ice sheets. If global warming continues to melt the ice sheets, the sea level will rise by 230 feet (70 m). Antarctica is covered by an ice cap that covers the entire continent, with only peaks of the Transantarctic mountains poking through in the center of the continent. This ice cap is surrounded by many ice shelves that are collapsing and forming many icebergs that move northward into the shipping lanes, where they pose hazards to ships.

Global sea levels are currently rising, partly as a result of the melting of the Greenland and Antarctica ice sheets. The Earth is presently in an interglacial stage of an ice age, and sea levels have risen nearly 400 feet (120 m) since the last glacial maximum 20,000 years ago, and about six inches (15 cm) in the past 100 years. The rate of sea-level rise seems to be accelerating and may presently be as much as an inch (2.5 cm) every 10 years. If all the glaciers melted in Greenland and Antarctica, then sea level would rise globally by 230 feet (70 m), inundating most of the world's major cities and submerging large parts of the continents under shallow seas. The coastal regions of the world are densely populated and experiencing rapid population growth. Approximately 100 million people presently live within 3.2 feet (1 m) of the present-day sea level. If sea level were to rise rapidly and significantly, we would experience an economic and social disaster of a magnitude never experienced by the civilized world. Many areas would flood permanently or become subjected to inundation by storms, beach erosion would accelerate, and water tables would rise.

The Greenland and Antarctic ice sheets have significant differences that cause them to respond differently to changes in air and water temperatures. The Antarctic ice sheet is about ten times as large as the Greenland ice sheet, and since it sits on the South Pole, Antarctica dominates its own climate. The surrounding ocean is cold even during summer,

and much of Antarctica is a cold desert with low precipitation rates and high evaporation potential. Most meltwater in Antarctica seeps into underlying snow and simply refreezes, with little running off into the sea. Antarctica hosts several large ice shelves fed by glaciers moving at rates of up to a thousand feet per year. Most ice loss in Antarctica is accomplished through calving and basal melting of the ice shelves at rates of 10–15 inches (25–38 cm) per year.

In contrast, Greenland's climate is influenced by warm North Atlantic currents and by its proximity to other landmasses. Climate data measured from ice cores taken from the top of the Greenland ice cap show that temperatures have varied significantly in cycles of years to decades. Greenland also experiences significant summer melting and abundant snowfall, and has few ice shelves. Its glaciers move quickly at rates of up to miles per year. These fast-moving glaciers can drain a large amount of ice from Greenland in relatively short time spans.

The Greenland ice sheet is thinning rapidly along its edges, having lost an average of 15–20 feet (4.5–6 m) in the past decade. In addition, tidewater glaciers and the small ice shelves in Greenland are melting at an order of magnitude faster than the Antarctic ice sheets, with rates of melting between 25–65 feet (7–20 m) per year. About half of the ice lost from Greenland is through surface melting that runs off into the sea. The other half is through calving of outlet glaciers and melting along the tidewater glaciers and ice-shelf bases.

These differences between the Greenland and Antarctic ice sheets lead them to play different roles in global sea-level rise. Greenland contributes more to the rapid short-term fluctuations in sea level, responding to short-term changes in climate. In contrast, most of the world's water available for raising sea level is locked in the slowly changing Antarctic ice sheet. Antarctica contributes more to the gradual, long-term sea-level rise.

ICEBERGS AND SEA ICE

Calving is a process in which large pieces of ice break off from the fronts of tidewater glaciers, ice shelves, or sea ice. Typically, the glacier cracks like an explosion, then a large chunk of ice splashes into the water, detaching from the glacier. Calving causes glaciers to retreat rapidly. Ice that has broken off an ice cap, polar sea, or calved off a glacier and is floating in open water is known as sea ice or, more commonly, icebergs. Presenting a serious hazard to ocean traffic and shipping lanes, icebergs have sunk numerous vessels, including the ill-fated RMS *Titanic* in 1912, killing 1,503. Icebergs float on the surface, but between 81 and 89 percent of the ice is submerged. The water level at which sea ice floats depends on the

Calving in Glacier Bay National Park, Alaska *(Scott Kapich, Shutterstock, Inc.)*

exact density of the ice, as determined by the total amount of air bubbles trapped in the ice and how much salt got trapped there during freezing.

Four main categories of sea ice may break off from larger glaciers, ice caps, or ice shelves to form many icebergs. The first comes from ice that formed on polar seas in the Arctic Ocean and around Antarctica. The ice that forms in these regions is typically about 10–15 feet (3–4 m) thick. Antarctica becomes completely surrounded by this sea ice every winter, and the Arctic Ocean is typically about 70 percent covered in the winter. During summer many passages open up in this sea ice, but during winter they reclose, forming pressure ridges of ice that may be 50–100 feet (up to tens of meters) high. Recent observations suggest that the sea ice in the Arctic Ocean is thinning dramatically and rapidly, and may soon disappear altogether. The icecap over the Arctic Ocean rotates clockwise, in response to the spinning of the Earth. This spinning is analogous to putting an ice cube in a glass, and slowly turning the glass. The ice cube will rotate more slowly than the glass, because it is decoupled from the edge of the glass. About one-third of the ice is removed every year by the East Greenland current. This ice then moves south and becomes icebergs, and thus hazards to shipping in the North Atlantic.

A second group of sea ice forms as pack ice in the Gulf of St. Lawrence, along the southeast coast of Canada; in the Bering, Beaufort, and Baltic Seas; in the Seas of Japan and Okhotsk; and around Antarctica. Pack ice builds up especially along the western sides of ocean basins, where cold currents are more common. Occasionally, during cold summers, pack ice persists throughout the summer.

Pack ice is hazardous when it becomes so extensive that it effectively blocks shipping lanes, or when leads (channels) into the ice open and close, forming pressure ridges that become too thick to penetrate with ice breakers. Ships attempting to navigate through pack ice have become crushed when leads close, and the ships are trapped. Pack ice has terminated or resulted in disaster for many expeditions to polar seas, most notably Sir John Franklin's expedition in 1845 in the Canadian arctic and Robert F. Scott's expedition from 1901 to 1904 to Antarctica. Pack ice also breaks up, forming many small icebergs, but because these are not as thick as icebergs of other origins, they do not present as significant a hazard to shipping.

Pack ice also presents hazards when it drifts into shore, usually during spring breakup. With significant winds pack ice can pile up on flat shorelines and accumulate in stacks up to 50 feet (15 m) high. The tremendous force of the ice is enough to crush shoreline wharves, docks, buildings, and boats. Pack ice blown ashore also commonly pushes up high piles of gravel and boulders that may be 35 feet (10.5 m) high in places. These ridges are common around many of the Canadian Arctic islands and mainland. Ice that forms initially attached to the shore presents another type of hazard. If it breaks free and moves away from shore, it may carry with it significant quantities of shore sediment, causing rapid erosion of beaches and shore environments.

Pack ice also forms on many high-latitude lakes, and the freeze-thaw cycle causes cracking of the lake ice. When lake water rises to fill the cracks, the ice cover on the lake expands and pushes over the shoreline, causing damage to any structures built along the shore. This is a common problem on many lakes in northern climates and leads to widespread damage to docks and other lakeside structures.

Icebergs derived from glaciers present the greatest danger to shipping. In the Northern Hemisphere most icebergs calve off glaciers in Greenland or Baffin Island, then move south through the Davis Strait into shipping lanes in the North Atlantic off Newfoundland. Some icebergs calve off glaciers adjacent to the Barents Sea, and others come from glaciers in Alaska and British Columbia. In the Southern Hemisphere most icebergs come from Antarctica, though some come from Patagonia, the southern tip of South America.

Once in the ocean icebergs drift with ocean currents, but the Coriolis force deflects them to the right in the Northern Hemisphere and to the left in the Southern Hemisphere. Most icebergs are about 100 feet to 300 feet (30.5–91.5 m) high, and up to about 2,000 feet (609.5 m) long. In March 2000 a huge iceberg broke off the Ross Ice Shelf in Antarctica. This berg was roughly the size of the state of Delaware, with an area of 4,500 square miles (11,655 km²) and rising 205 feet (62.5 m) out of the water. Icebergs in the Northern Hemisphere pose a greater threat to shipping, as those from Antarctica are too remote and rarely enter shipping lanes. Ship collisions with icebergs have resulted in numerous maritime disasters, especially in the North Atlantic on the rich fishing grounds of the Grand Banks off the coast of Newfoundland.

Satellites now track icebergs, and ships receive updated information about their positions to avoid disastrous collisions. Radio transmitters are placed on larger icebergs to monitor their locations more closely; many ships now carry more sophisticated radar and navigational equipment that helps track the positions of large icebergs and the ship, to avoid collision.

Icebergs also pose a serious threat to oil-drilling platforms and seafloor pipelines in high-latitude seas. Some precautions have been taken, such as building seawalls around near-shore platforms, but not enough planning has gone into preventing an iceberg from colliding with and damaging an oil platform or from one being dragged across the seafloor and rupturing a pipeline.

See also CLIMATE; CLIMATE CHANGE; ICE AGES; SUPERCONTINENT CYCLES.

FURTHER READING

Alley, Richard B., and Michael L. Bender. "Greenland Ice Cores: Frozen in Time." *Scientific American* (February 1998).

Dawson, A. G. *Ice Age Earth*. London: Routledge, 1992.

Douglas, B., M. M. Kearney, and S. Leatherman. *Sea Level Rise: History and Consequence. Vol. 75*. San Diego, Calif.: Academic Press, 2000.

Erickson, Jon. *Glacial Geology: How Ice Shapes the Land*. New York: Facts On File, 1996.

Intergovernmental Panel on Climate Change home page. Available online. URL: http://www.ipcc.ch/index.htm. Accessed January 30, 2008.

Intergovernmental Panel on Climate Change 2007. *Climate Change 2007: The Physical Science Basis. Contributions of Working Group I to the Fourth Assessment Report of the Intergovernmental Panel on Climate Change* (S. Solomon, D. Qin, M. Manning, Z. Chen, M. Marquis, K. B. Avery, M. Tignor, and H. L. Miller, eds.) Cambridge: Cambridge University Press, 2007.

Schneider, D. "The Rising Seas." *Scientific American* (March 1997): 112–118.

Stone, G. "Exploring Antarctica's Islands of Ice." *National Geographic* (December 2001): 36–51.

global warming The term *global warming* generally refers to the phenomenon whereby the planet has been warming dramatically over the past century, whereas the more encompassing term *climate change* refers to longer-term changes. Most scientists suggest that human contributions to climate change have been accelerating to critical levels as the world becomes increasingly industrialized. Much of what is known about these short-term climate changes has been described in a report called *Climate Change 2007: The Physical Science Basis*, issued in late 2007 by the Intergovernmental Panel on Climate Change (IPCC), an international group of hundreds of scientists who have analyzed all available data, assessed the causes of these recent, short-term changes to the global climate, and made predictions of what the climate of the Earth may look like at various times in the future. Much of the information in this report is based on the findings of the IPCC.

Eleven of the 12 years between 1996 and 2006 were the warmest on record since weather-recording instruments were widely used starting in 1850. The temperature rate increase seems to be increasing, with polar areas affected more than equatorial regions. Sea levels are also rising at an increasing rate. Between 1961 and 1993 global sea level was rising at a rate of 0.05–0.09 inches per year (0.13–0.23 cm/yr), and since 1993 they have been rising at 0.09–0.11 inches per year (0.24–0.28 mm/yr). Some of the sea level rise is due to melting glaciers, ice caps, and snow, and some is from thermal expansion of ocean water as the water warms. Glaciers are shrinking in both the northern and southern hemispheres, and the ice caps on the Arctic Ocean and over parts of Antarctica are shrinking rapidly.

Global precipitation patterns are observably changing on the century scale, with much of eastern North and South America, northern Europe, and north and central Asia seeing increased rainfall, but other areas such as the Sahel, Mediterranean, southern Africa, and southern Asia are experiencing decreased precipitation. On a global scale areas experiencing drought or less precipitation are greater than areas receiving greater precipitation.

TEMPERATURE VARIATIONS DURING THE PAST 1,000 YEARS

Understanding changes in the Earth's climate in the past 100–200 years, or the slightly longer interval extending back through the last glacial interval, rely on several types of data. Instrumental records of Earth's climate extend back to about the year 1850, when recording devices were put into widespread use. Long cores of ice obtained from Greenland and other locations are also widely used to measure past climate conditions, with this record extending back for about 650,000 years.

The IPCC issued the following statement in November 2007:

> Warming of the climate system is unequivocal, as is now evident from observations of increases in global average air and ocean temperatures, widespread melting of snow and ice, and rising global average sea level. Global atmospheric concentrations (of greenhouse gases) have increased markedly as a result of human activities since 1750 and now far exceed pre-industrial levels. The global increases in carbon dioxide concentration are due primarily to fossil fuel use and land-use change, while those of methane and nitrous oxide are primarily due to agriculture.

This bold and controversial statement was based on rigorous analysis of data from the past 1,000 years, showing that temperatures remained fairly steady at about 0–0.5 degrees below the 1990 average value from the year 1000 to about 1910, then began a sharp upward turn that flattened off for a short time in the 1950s, and has turned sharply up again since about 1976. Temperatures are now about 0.5–1.0

Mother polar bear and cub, threatened by global warming *(Keith Levit, Shutterstock, Inc.)*

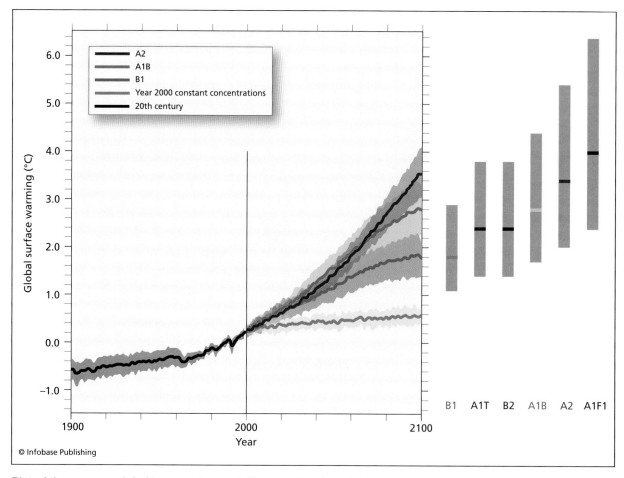

Plot of the average global temperature variations on the planet in the past 1,000 years and what different models predict the temperature will be by 2100. All models show a predicted temperature rise, ranging between 1.5 and 5.5 degrees. *(Data from IPCC 2007)*

degrees above the 1990 value, and expected to rise 2–5 degrees above this value by 2100.

To measure global average temperatures groups of meteorologists, such as the World Meteorological Organization (WMO) and the Global Climate Observing System (GCOS), have a large number of observation points on the continents, which are then gridded into equal areas to assign a temperature for each box in the grid. Observations from satellites are widely used where local observations are not available and for calibrating the models. Local effects, such as any urban heat island effect from cities, are accounted for in these types of model. With rapid improvements in computer modeling it has been possible to make more and more detailed and accurate models for the Earth.

OBSERVED SHORT-TERM CLIMATE CHANGES AND THEIR EFFECTS

Instrumental and ice core records show that several components of the atmosphere and surface have significant changes in the recorded climate history. First, the concentration of greenhouse gases such as carbon

dioxide has increased dramatically since 1850, causing an increase in the atmospheric absorption of outgoing radiation and warming the atmosphere. Aerosols, microscopic droplets or airborne particles, have also increased, and these have reflected and absorbed incoming solar radiation.

The most obvious change in the short-term climate is the increase in temperature of the atmosphere and sea surface. The period between 1995 and 2006 ranks among the hottest on record since instrumental records have been in widespread use since 1850, containing 11 of the 12 hottest years recorded. Moreover, the rate of temperature rise has increased each decade since 1850. The total temperature increase since 1850 is estimated by the IPCC to be 1.4°F (0.76°C). Measurements of the atmospheric water vapor indicate it is increasing with increasing temperature of the atmosphere, although such measurements extend back only to the mid-1980s.

Sea level has been rising at about 0.07 inches per year (0.18 cm/yr) since 1961, and at 0.12 inches per year (0.31 cm/yr) since 1993. The temperature of the oceans to a depth of 1.9 miles (3 km) has been increasing since at least 1961, with seawater absorbing most (~ 80 percent) of the heat energy associated with global warming. This increase in seawater temperature is causing the water to expand, contributing to sea-level rise. Also contributing to sea-level rise is a dramatic melting of mountain glaciers in both the

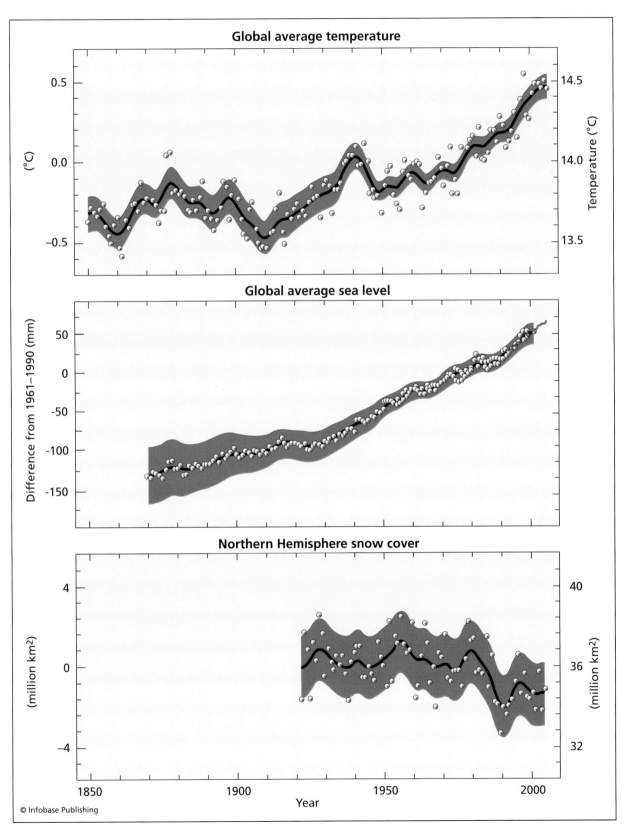

© Infobase Publishing

Graphs of global average temperature, sea level, and snow cover for the past 160 years *(Data from IPCC 2007)*

Spruce killed by spruce bark beetle near Homer, Alaska *(Peter Essick/Aurora/Getty Images)*

Northern and the Southern Hemispheres. Changes in the ice caps on Greenland and Antarctica show an increased outflow of glacial ice and meltwater, so melting of the polar ice caps is very likely contributing to the measured sea-level rise. Both of these ice caps show significant thinning, much due to increased melting, but some (especially on Greenland) due also to decreased snowfall.

Many specific regions of the planet are showing dramatic changes in response to the global average warming surface conditions. For instance, the surface temperatures measured in the Arctic have been increasing at about twice the global rate for the past 100 years, although some fluctuations on a decadal scale have been observed as well. The sea ice that covers the Arctic Ocean may be on the verge of collapse, as the sea ice thins and covers a smaller area each year. Since 1978 the Arctic sea ice has diminished in aerial extent by 2.7 percent each decade. On land in Arctic regions the thick permafrost layer is also warming, by 4–5°F (~3°C), and a total decrease in the area covered by permafrost since 1900 is estimated to be about 7 percent. Permafrost locks a huge amount of peat and carbon into a closed system, so there are fears outlined in the IPCC report that the melting of the permafrost layer may release large amounts of carbon into the atmospheric system. The sea ice around Antarctica shows greater variations on interannual scales and no longer-term trends are yet discernible. Much of the Antarctic region is isolated from other parts of the global climate belt, so overall it shows less change than northern polar regions.

Precipitation patterns across much of the planet are changing as a result of global warming. Observations from 1900 to 2007 show long-term drying and potential desertification over parts of the sub-Saharan Sahel, the Mediterranean region, parts of southern Asia, and much of southern Africa. Deeper and longer droughts have been occurring over larger areas since the 1970s, and some of these conditions can be related to changes in ocean temperature, wind patterns, and loss of snow cover. Westerly winds in the mid latitudes have become stronger in both the Northern and the Southern Hemispheres since the 1960s.

Weather extremes show an increase in frequency, including heavy precipitation events over land, as well as heat waves and extreme temperatures over land. Many studies suggest that oceanic cyclones or hurricanes may also be becoming stronger and more frequent, but some decadal variations in oceanic cyclones may also complicate determination of these trends. Most studies support an increase in tropical cyclone activity since the 1970s over the North Atlantic, and relate this to the increase in sea surface temperatures.

CAUSES OF SHORT-TERM CLIMATE CHANGE

The IPCC issued new reports, *Climate Change 2007,* in 2007 revealing that concentrations of some greenhouse gases have increased dramatically as a result of human activities, mostly starting with the early industrial revolution around 1750 and accelerating in the late industrial revolution around 1850. The greenhouse gases that show the most significant increases are carbon dioxide, methane, and nitrous oxide. Carbon dioxide (CO_2), the most significant anthropogenic greenhouse gas, is produced mainly by burning fossil fuels such as coal, oil, and gasoline. The atmospheric concentrations of CO_2 have increased from a preindustrial revolution level of 280 parts per million (ppm) in the atmosphere to a present (2005) level of 379 ppm, far exceeding the natural range (180–300 ppm) measured over the past 650,000 years, but CO_2 levels have been higher in the geological past for reasons related to global volcanism, supercontinent cycles, and the like that operate on longer timescales than the changes measured since the industrial revolution. Despite significant variations on a year-to-year basis, the rate of CO_2 increase in concentration in the atmosphere has been accelerating over the past 10 years.

Methane in the atmosphere has increased in concentration from a pre–industrial revolution value of about 715 parts per billion (ppb) to 1,774 ppb in 2005. Methane is produced predominantly in agricultural production and also in burning fossil fuels. The rapid increase in atmospheric methane is, like carbon dioxide, well beyond the natural range (320–790 ppm) of the past 650,000 years. Nitrous oxide is released by agricultural activities and is a greenhouse gas. Its concentration has increased in the atmosphere from a pre–industrial revolution level of 270 ppb to 319 ppb in 2005.

Scientists on the IPCC estimate that the total increase in heating of the atmosphere due to anthropogenic increases in greenhouse gases since the start of the industrial revolution in 1750 is greater than other effects, and the rate of planet warming is now faster than any experienced in the past 10,000 years. *Radiative forcing* is the net change in downward minus the upward irradiance at the tropopause, caused by a change in an external driver such as a change in greenhouse gas concentration. The radiative forcing caused by the change in CO_2 between 1995 and 2005 is estimated to be about 20 percent, the largest amount in the past 200 years. The way to counteract this is to enforce climate treaties such as the Kyoto Protocols, which call for a reduction in

CO_2 emissions; through the installation of scrubbers and other cleansing technologies on factories and power plants; and increasing the fuel efficiency of cars.

Some of the warming caused by increases in greenhouse gases may be counteracted by an increase in aerosols, small airborne solid or liquid particles, which may have a cooling effect. Aerosols include particles such as sulfate, organic carbon, black carbon, nitrates, and dust. As the climate warms, more and more dust is being picked up from regions undergoing increased aridity and desertification such as the fringes of the Gobi desert and the Sahara. This dust gets emplaced high into the atmosphere, where it may reside some time and may actually have a small cooling effect.

THE GREENHOUSE EFFECT

The term *greenhouse effect* refers to the Earth's climate as being sensitive to the concentrations of certain gases in the atmosphere. The concept was first coined by French physicist Edme Mariotte (1620–84) in 1681, who noted that light and heat from the Sun easily pass through a sheet of glass, but that heat from candles and other sources does not. This concept was then extended by French mathematician and physicist Joseph Fourier in 1824 to the atmosphere by noting that heat and light from the Sun can pass from space through the atmosphere, but heat radiated back to the atmosphere from Earth may get trapped by some of the atmospheric gases, just like the heat from a candle is partly blocked by the glass pane. Then in 1861 Irish physicist John Tyndall (1820–93) identified that the molecules of water (H_2O) and carbon dioxide (CO_2) were mainly responsible for the absorption of heat radiated back from Earth, and that other atmospheric gases such as nitrogen and oxygen did not play a role in this effect. Tyndall noted that simple changes in the concentrations of CO_2 and H_2O could alternately cool and heat the atmosphere, producing "all the mutations of climate which the researches of geologists reveal." The next step in understanding the greenhouse effect came from the work of Swedish physicist Svante Arrhenius in 1896, who calculated that a 40 percent increase or decrease in the atmospheric concentration of CO_2 could cause the advance and/or retreat of continental glaciers, triggering the glacial and interglacial ages. Much later a change in the atmospheric CO_2 of this magnitude was documented in cores of the Greenland ice sheet, as predicted by Arrhenius. Carbon dioxide can vary naturally in the atmosphere through a variety of driving mechanisms, including changes in volcanism, erosion and plate tectonics, and ocean-atmosphere interactions. The modern concept of linking greenhouse gases with the burning of fossil fuels was formulated by British steam engineer and amateur meteorologist Guy Stewart Callendar (1898–1964) in 1938, who calculated that a doubling of atmospheric CO_2 by burning fossil fuels would cause an average global temperature increase of about 3°F (2°C), with more heating at the poles. Callendar made prescient predictions that humans are changing the composition of the atmosphere at a rate that is "exceptional" on geological timescales, and he sought to understand what effects these changes might have on climate. His prediction was that the "principal result of increasing carbon dioxide will be a gradual increase in the mean temperature of the colder regions of the Earth." These predictions were first confirmed in 1947 when Ahlmann reported a 1–2°F (1.3°C) increase in the average temperature of the North Atlantic sector of the Arctic. However, at this time the nature of the complex interactions of the carbon cycle and exchange of CO_2 in the atmosphere-ocean system was not well understood, and many scientists attributed the entire temperature rise to human production of greenhouse gases. Later studies of ocean-atmosphere relationships, and biogeochemistry, showed more complex relationships. Later in the 1970s effects of aerosols in the atmosphere, principally to reflect solar radiation to space and cooling the Earth, began to be appreciated as another component of the greenhouse effect. The current state of knowledge of the complex physical, chemical, biological, and other processes associated with the greenhouse effect are described in the IPCC's *Climate Change 2007*.

COMPARISON OF SHORT-TERM CLIMATE CHANGES WITH THE MEDIUM-TERM PALEOCLIMATE RECORD

Separating the effects of short-term human-induced climate changes from natural variations on longer-term timescales can be difficult. Present-day global warming is unusual for the climate record of the past 1,300 years, but it has counterparts induced by natural causes about 125,000 years ago and in the older geological record. The last time (125,000 years ago) climates warmed as significantly as the planet is now experiencing, loss of polar ice led to sea-level rise of 13–20 feet (4–6 m), suggesting that the world's coastlines are in grave danger of moving inland to higher ground. Ice core data show that temperatures in Greenland were 4–7°F (3–5°C) hotter than at present, a level that many models predict will be reached by the end of this century. The last 50 years appear to be the hottest in the past 1,300 years, but significant fluctuations have occurred.

The measured increases in anthropogenic greenhouse gases can more than account for the measured temperature rise of the surface of the Earth in the past 50–100 years. The less-than-expected warming is probably related to lowering of the temperature by aerosols from volcanic eruptions and dust from desert environments. These measurements strongly suggest that present-day global warming is being forced by the human-induced injection of greenhouse gases to the atmosphere, not to other long-term climate-forcing mechanisms that have controlled other global warming and cooling events in past geological times.

The measured surface warming is nearly global in scale, with Antarctica the exception; it is sheltered from parts of the global atmosphere/ocean system. Climate models are consistent with the global warming produced by anthropogenic causes. Many local variations exist, such as "warming holes," where local atmospheric effects are stronger than the global changes.

Global warming is also likely affecting wind patterns, the most extreme hot and cold nights, extra-tropical storm patterns, and an increase in heat waves. Effects are stronger in the Northern than in the Southern Hemisphere.

SEA-LEVEL CHANGES

Global sea levels are currently rising as a result of the melting of the Greenland and Antarctica ice sheets and thermal expansion of the world's ocean waters due to global warming. The Earth is presently in an interglacial stage of an ice age, and sea levels have risen nearly 400 feet (130 m) since the last glacial maximum 20,000 years ago, and about six inches (15.25 cm) in the past 100 years. The rate of sea-level rise seems to be accelerating and may presently be as much as an inch (2.5 cm) every 8–10 years. If all the ice on both ice sheets were to melt, global sea levels would rise by 230 feet (70 m), inundating most of the world's major cities and submerging large parts of the continents under shallow seas. The coastal regions of the world are densely populated and are experiencing rapid population growth. Approximately 100 million people presently live within 3 feet (1 m) of the present day sea level. If sea level were to rise rapidly and significantly, the world would experience an economic and social disaster on a magnitude not yet experienced by the civilized world. Many areas would become permanently (on human timescales) flooded or subject to inundation by storms, beach erosion would be accelerated, and water tables would rise.

The Greenland and Antarctic ice sheets have significant differences that cause them to respond differently to changes in air and water temperatures. The Antarctic ice sheet is about 10 times as large as the Greenland ice sheet, and since it sits on the South Pole, Antarctica dominates its own climate. The surrounding ocean is cold even during summer, and much of Antarctica is a cold desert with low precipitation rates and high evaporation potential. Most meltwater in Antarctica seeps into underlying snow and simply refreezes, with little running off into the sea. Antarctica hosts several large ice shelves fed by glaciers moving at rates of up to a thousand feet (300 m) per year. Most ice loss in Antarctica is accomplished through calving and basal melting of the ice shelves, at rates of about 10–15 inches (25–38 cm) per year.

In contrast, Greenland's climate is influenced by warm North Atlantic currents and its proximity to other land masses. Climate data measured from ice cores taken from the top of the Greenland ice cap show that temperatures have varied significantly in cycles of years to decades. Greenland also experiences significant summer melting and abundant snowfall, and has few ice shelves; its glaciers move quickly at rates of up to miles (several km) per year. These fast-moving glaciers can drain a large amount of ice from Greenland in relatively short periods of time.

The Greenland ice sheet is thinning rapidly along its edges, loosing an average of 15–20 feet (4.5–6 m) in the past decade. In addition tidewater glaciers and the small ice shelves in Greenland are melting on an order of magnitude faster than the Antarctic ice sheets, with rates of melting between 25–65 (7.6–20 m) feet per year. About half of the ice lost from Greenland is through surface melting that runs off into the sea. The other half of ice loss is through calving of outlet glaciers and melting along the tidewater glaciers and ice shelf bases. If just the Greenland ice sheet melts, the water released will contribute another 23 feet (7 m) to sea-level rise, to a level not seen since 125,000 years ago.

These differences between the Greenland and Antarctic ice sheets lead them to play different roles in global sea-level rise. Greenland contributes more to the rapid, short-term fluctuations in sea level, responding to short-term changes in climate. In contrast, most of the world's water available for raising sea level is locked up in the slowly changing Antarctic ice sheet. Antarctica contributes more to the gradual, long-term, sea-level rise.

Data released by the IPCC in 2007 suggest that the current melting of glaciers is largely the result of the recent warming of the planet in the past 100 years through greenhouse warming. Greenhouse gases have been increasing at a rate of more than 0.2 percent per year, and global temperatures are rising accordingly. The most significant contributor to the greenhouse gas buildup is CO_2, produced mainly by burning fossil fuels. Other gases that contribute to greenhouse warming include carbon monoxide, nitrogen oxides, methane (CH_4), ozone (O_3), and chlorofluorocarbons. Methane is produced by gas from grazing animals and termites, whereas nitrogen oxides are increasing because of the increased use of fertilizers and automobiles, and chlorofluorocarbons are increasing as a result of release from aerosols and refrigerants. Together the greenhouse gases have allowed short-wavelength incoming solar radiation to penetrate the gas in the upper atmosphere but trapped the solar radiation after it is reemitted from the Earth in a longer wavelength. The trapped radiation causes the atmosphere to heat up, leading to greenhouse warming. Other factors also influence greenhouse warming and cooling, including the abundance of volcanic ash in the atmosphere and solar luminosity variations, as evidenced by sunspot variations.

Measuring global (also called eustatic) sea-level rise and fall is difficult because many factors influence the relative height of the sea along any coastline. Vertical motions of continents, called epeirogenic movements, may be related to plate tectonics; they rebound from being buried by glaciers or to changes in the amount of heat added to the base of the continent by mantle convection. Continents may rise or sink vertically, causing apparent sea-level change, but these sea-level changes are relatively slow compared with changes induced by global warming and glacial melting. Slow, long-term, sea-level changes can also be induced by changes in the amount of seafloor volcanism associated with seafloor spreading. At some times in Earth history seafloor spreading was particularly vigorous, and the increased volume of volcanoes and the midocean ridge system caused global sea levels to rise.

Steady winds and currents can mass water against a particular coastline, causing a local and temporary sea-level rise. Such a phenomena is associated with the El-Nino-Southern Oscillation (ENSO), causing sea levels to rise by 4–8 (10–20 cm) inches in the Australia-Asia region. When the warm water moves east in an ENSO event, sea levels may rise 4–20 inches (10–50 cm) across much of the North and South American coastlines. Other atmospheric phenomena can also change sea level by centimeters to meters locally, on short timescales. Changes in atmospheric pressure, salinity of sea waters, coastal upwelling, onshore winds, and storm surges all cause short-term fluctuations along segments of coastline. Global or local warming of waters can cause them to expand slightly, causing a local sea-level rise. The extraction and use of groundwater and its subsequent release into the sea is likely causing an additional sea-level rise of about 0.05 inches (0.13 cm) per year. Seasonal changes in river discharge can temporarily change sea levels along some coastlines, especially where winter cooling locks up large amounts of snow that melt in spring.

Attempts to estimate eustatic sea-level changes must be able to average out the numerous local and tectonic effects to arrive at a globally meaningful estimate of sea-level change. Most coastlines seem to be dominated by local fluctuations that are larger in magnitude than any global sea-level rise. Recently satellite radar technology has precisely measured sea surface height and documented annual changes in sea level. Radar altimetry can map sea surface elevations to the sub-inch scale, and to do this globally, providing an unprecedented level of understanding of sea surface topography. Satellite techniques support the concept that global sea levels are rising at about 0.1 inches (.25 cm) per decade.

See also ATMOSPHERE; CLIMATE, CLIMATE CHANGE; DESERTS; GLACIER, GLACIAL SYSTEMS.

FURTHER READING

Abrahams, A. D., and A. J. Parsons. *Geomorphology of Desert Environments.* Norwell, Mass.: Kluwer Academic Publishers for Chapman and Hall, 1994.

Ahrens, C. D. *Meteorology Today: An Introduction to Weather, Climate, and the Environment.* 6th ed. Pacific Grove, Calif.: Brooks/Cole, 2000.

Botkin, D., and E. Keller. *Environmental Science.* Hoboken, N.J.: John Wiley & Sons, 2003.

Bryson, R., and T. Murray. *Climates of Hunger.* Canberra: Australian National University Press, 1977.

Culliton, Thomas J., Maureen A. Warren, Timothy R. Goodspeed, Davida G. Remer, Carol M. Blackwell, and John McDonough III. *Fifty Years of Population Growth along the Nation's Coasts, 1960–2010.* Rockville, Md.: National Oceanic and Atmospheric Administration, 1990.

Dawson, A. G. *Ice Age Earth,* London: Routledge, 1992.

Douglas, B., M. Kearney, and S. Leatherman. *Sea Level Rise: History and Consequence.* San Diego, Calif.: Academic Press, International Geophysics Series, vol. 75, 2000.

Intergovernmental Panel on Climate Change 2007. *Climate Change 2007: The Physical Science Basis. Contributions of Working Group I to the Fourth Assessment Report of the Intergovernmental Panel on Climate Change.* S. Solomon, D. Qin, M. Manning, Z. Chen, M. Marquis, K. B. Averyt, M. Tignor, and H. L. Miller, eds. Cambridge: Cambridge University Press, 2007. Available online. URL: http://www.ipcc.ch/ipccreports/ar4-wg1.htm. Accessed October 10, 2008.

Intergovernmental Panel on Climate Change 2007. *Climate Change 2007: Impacts, Adaptation, and Vulnerability. Contributions of Working Group II to the Fourth Assessment Report of the Intergovernmental Panel on Climate Change.* M. Parry, O. Canziani, J. Palutikof, P. van der Linden, and C. Hanson, eds. Cambridge: Cambridge University Press, 2007. Available online.

URL: http://www.ipcc.ch/ipccreports/ar4-wg2.htm. Accessed October 10, 2008.

Intergovernmental Panel on Climate Change 2007. *Climate Change 2007: Mitigation. Contributions of Working Group III to the Fourth Assessment Report of the Intergovernmental Panel on Climate Change.* B. Metz, O. R. Davidson, P. R. Bosch, R. Dave, L. A. Meyer, eds. Cambridge: Cambridge University Press, 2007. Available online. URL: http://www.ipcc.ch/ipccreports/ar4-wg3.htm. Accessed October 10, 2008.

National Aeronautic and Space Administration (NASA). "Earth Observatory." Available online. URL: http://earthobservatory.nasa.gov/. Accessed October 9, 2008, updated daily.

Reisner, M. *Cadillac Desert: The American West and Its Disappearing Water.* New York: Penguin, 1986.

U.S. Environmental Protection Agency. Climate Change homepage. Available online. URL: http://www.epa.gov/climatechange/. Updated September 9, 2008.

Goldschmidt, Victor M. (1888–1942) Swiss *Chemist, Geochemist, Mineralogist*

Victor Goldschmidt studied chemistry, mineralogy, and geology at the University of Kristiania in Norway and is often considered to be the father of modern geochemistry. His work was greatly influenced by the Norwegian petrologist and mineralogist W. C. Bogger and also by earth scientists Paul von Groth and Friedrich Becke. Goldschmidt received a doctorate in geology in 1911 with his thesis "The Contact Metamorphism in the Kristiania Region." He became a full professor of geochemistry and director of the mineralogical institute of the University of Kristiania (later Oslo) in 1914. His doctoral thesis concerned the factors governing the mineral associations in contact-metamorphic rocks and was based on the samples he had collected in southern Norway. In later years he became a professor on the faculty of natural sciences at Göttingen and head of its mineral institute. While at Kristiania he began geochemical investigations on the noble gases and alkali metals, and the siderophilic and lithophilic elements. He produced a model of the Earth that showed how these different elements and metals accumulated in various geological domains based on their charge and size, and on the polarizability of their ions. Goldschmidt is one of the pioneers in geochemistry who explained the composition of the environment.

EARLY LIFE, CAREER, AND SCIENTIFIC CONTRIBUTIONS

Victor Moritz Goldschmidt was born in Zurich, Switzerland, on January 27, 1888, to Heinrich J. Goldschmidt and Amelie Koehne. In 1901 Victor's father accepted a position as professor of chemistry in Kristiania (Oslo), and the family moved from Switzerland to Norway. Victor studied geology and mineralogy in Norway, completing a doctorate at age 23 in 1911, consisting of two papers, "Die Kontaktmetamorphose im Kristianiagebiet" (Contact metamorphism in the Kristiana region) and "Geologisch-petrographische Studien im Hochgebirge des südlichen Norwegens" (Geologic and petrographic studies in the Hochgebirge area, southern Norway). In 1912 he was awarded Norway's most distinguished scientific award, the Fridtjof Nansen medal, for his Ph.D. research. At the same time he was made an associate professor (docent) of mineralogy and petrography at the University of Oslo. After completing his Ph.D. Goldschmidt authored a series of papers widely considered to represent the beginning of modern geochemistry and was highly influential in the fields of mineralogy, geology, crystallography, and theoretical chemistry. Some of his most important work included descriptions of the role of ionic radii in determining the geochemical behavior of the elements. He stayed in Norway at the University of Kristiania as a professor of mineralogy, then moved back to Oslo in 1935.

In 1942 while Norway was under German occupation, Europe was in a state of chaos from the Nazi occupation, and Goldschmidt was arrested on October 26, 1942, for being a Jew. He was sent to the Berg concentration camp near Tonsberg, Norway, and was almost deported to Auschwitz, but he was held in Norway on the condition that he lend his scientific expertise to help the German war effort. He fled to Sweden as soon as he could, then to England, and returned to Oslo after the war in 1945, but died at the age of 59, on March 20, 1947.

See also GEOCHEMISTRY; MINERAL, MINERALOGY.

FURTHER READING

Goldschmidt, V. M. "The Distribution of the Chemical Elements." *Royal Institution Library of Science, Earth Science* 3 (1971): 219–233.
———. "On the Problems of Mineralogy." *Journal of the Washington Academy of Sciences* 51 (1961): 69–76.

Gondwana, Gondwanaland Gondwana is the Late Proterozoic–Late Paleozoic supercontinent of the Southern Hemisphere, named by British/Austrian geologist Eduard Suess after the Gondwana System of southern India. The name Gondwana means "land of the Gonds" (an ancient tribe in southern India), so the more common rendition of the name Gondwanaland for the southern supercontinent is technically improper, meaning "land of the land of the Gonds." The supercontinent includes the present continents and continental fragments of Africa,

South America, Australia, Arabia, India, Antarctica, and many smaller fragments. Most of these continental masses amalgamated in the latest Precambrian during closure of the Mozambique Ocean and several other oceans, seas, and basins. It persisted as a supercontinent until they joined with the northern continents in the Carboniferous to form the supercontinent Pangaea.

Geologists have matched the different fragments of Gondwana with others using alignment between belts of similar-aged deformation, metamorphism, and mineralization, as well as common faunal, floral, and paleoclimatic belts. The formation and breakup of Gondwana is associated with one of the most remarkable explosions of new life-forms in the history of the planet, the change from simple, single-celled organisms and soft-bodied fauna to complex, multicelled organisms. The formation and dispersal of supercontinents strongly influences global climate and the availability of different environmental niches for biological development, linking plate tectonic and biological processes.

Since the early 1990s a consensus has emerged that Gondwana formed near the end of the Neoproterozoic from the fragmented pieces of an older supercontinent, Rodinia, itself assembled near the end of the Grenville cycle (~1,100 Ma). The now standard model of Gondwana's assembly begins with the separation of East Gondwana (Australia, Antarctica, India, and Madagascar) from the western margin of Laurentia, and the fanlike aggregation of East and West Gondwana. The proposed assembly closed several ocean basins, including the very large

Glossopteris leaf fossil from Permian period found in Coohah, New South Wales, Australia. *Glossopteris* is one of the diagnostic flora used by Alfred Wegener and others to match paleoclimate and paleobiological zones across the southern continents, to re-create the former positions of these continents in the proposed supercontinent of Gondwana. *(Martin Land/Photo Researchers, Inc.)*

Mozambique Ocean, and turned the constituents of Rodinia inside-out, such that the external or "passive" margins in Madagascar and elsewhere became collisional margins in latest Precambrian and earliest Cambrian time.

The notion of a single, short-lived collision between East and West Gondwana is an oversimplification, since geologic relations suggest that at least three major ocean basins closed during the assembly of Gondwana (Pharusian, Mozambique, Adamastor), and published geochronology demonstrates that assembly was a protracted affair. Current research aims to understand these relationships. For example, an alternative two-stage model for closure of the Mozambique Ocean has been recently advanced that ascribes an older "East African" Orogeny (~680 Ma) to collision between Greater India (i.e., India-Tibet-Seychelles-Madagascar-Enderby Land) and the conjoined Congo and Kalahari cratons. A younger "Kuunga" event (~550 Ma) that represents the collision of Australia–East Antarctica with proto-Gondwana followed, thus completing Gondwana's assembly near the end of the Neoproterozoic.

An international research journal *Gondwana Research* was launched in the 1990s in which scientists working on a variety of issues about the landmasses of Gondwana have published their research results. This journal has become a widely used forum in which to present and discuss new data related to the former supercontinent, as well as its geology, lifeforms, mineral deposits, and past climates.

See also PROTEROZOIC; SUPERCONTINENT CYCLES.

FURTHER READING

de Wit, Maarten J., Margaret Jeffry, Hugh Bergh, and Louis Nicolaysen. *Geological Map of Sectors of Gondwana Reconstructed to Their Disposition at ~150 Ma*. Tulsa, Okla.: American Association of Petroleum Geologists, Map Scale 1:10,000,000, 1988.

Hoffman, Paul F. "Did the Breakout of Laurentia Turn Gondwana Inside-out?" *Science* 252 (1991): 1409–1412.

Kusky, Timothy M., Mohamad Abdelsalam, Robert Tucker, and Robert Stern, eds. *Evolution of the East African and Related Orogens, and the Assembly of Gondwana*. Precambrian Research Special Issue. Amsterdam: Elsevier, 2003.

Rogers, J. J. W., and M. Santosh. *Continents and Supercontinents*. Oxford: Oxford University Press, 2004.

Grabau, Amadeus William (1870–1946)
German-American Geologist, Paleontologist
Amadeus William Grabau was born on January 9,

1870, in Cedarburgh, Wisconsin. He was a great contributor to systematic paleontology and stratigraphic geology and also a respected professor and writer. He spent half of his professional life in the United States and the last 25 years in China. Grabau studied at the Massachusetts Institute of Technology (MIT) and received a master of science and a doctorate of science degree at Harvard University, then returned as faculty at MIT from 1892 to 1897. He moved to Rensselaer Polytechnic Institute in Troy, New York, from 1899 to 1901, and became a professor in paleontology at Columbia University in New York City in 1901. In 1912 Grabau married Mary Antin, a Russian immigrant from a shtetl who wrote a best-selling autobiography, *The Promised Land*. In World War I Grabau defended Germany's actions, which led to his divorce from Mary and his being fired from Columbia University. In 1919 Grabau moved to China and became a professor at Peking National University (now called Peking University).

In the first 20 years of his career he was one of the country's leading scientists in paleontology, stratigraphy, and sedimentary petrology. The greatest effect of his scientific work has been his contributions to the principles of paleoecology and to the genetic aspects of sedimentary paleontology. Paleoecology uses fossil data to reconstruct information about past ecosystems. Grabau was interested in relating the ecosystems to differences in the organism that made the fossils he was studying. His stratigraphic work was also influential; not only did it bring about a more developed understanding of the subject, but it was the source of understanding Earth movements. The concepts involved in his polar control theory, pulsation theory, and the separation of Pangaea allowed for the imaginative syntheses of geologic evidence. After Grabau moved to China, he conducted a geologic survey of much of the country, and from this work he became known as the father of Chinese geology. Grabau died on March 20, 1956, in what is now called Beijing.

Grabau published more than 10 books during his career, including *North American Index Fossils* (1909, 1910), *Principles of Stratigraphy* (1913), *Textbook of Geology*, 2 vols. (1920–21), *Silurian Fossils of Yunnan* (1920), *Ordovician Fossils of North China* (1921), *Paleozoic Corals of China* (1921), *Stratigraphy of China* (1924–25), *Migration of Geosynclines* (1924), *Early Permian Fossils of China* (1934), and *Rhythm of the Ages* (1940). This influential geologist and paleontologist received numerous awards and was a member of the following institutes: the Geological Society of America, New York Academy of Science, and Geological Society of China. He was also an honorary member of the Peking Society of

Natural History, the China Institute of Mining and Metallurgy, the Academia Sinica, and the Academia Peipinensis.

See also ASIAN GEOLOGY; HISTORICAL GEOLOGY; PALEONTOLOGY; SEDIMENTARY ROCK, SEDIMENTATION.

FURTHER READING

Grabau, Amadeus W. "The Polar-Control Theory of Earth Development." *Association of Chinese and American Engineering Journal* 18 (1937): 202–223.

———. "Fundamental Concepts in Geology and Their Bearing on Chinese Stratigraphy." *Geological Society of China Bulletin* 16 (1937): 127–176.

———. "Revised Classification of the Palaeozoic Systems in the Light of the Pulsation Theory." *Geological Society of China Bulletin* 15 (1936): 23–51.

granite, granite batholith   A coarse-grained igneous plutonic rock with visible quartz, potassium, and plagioclase feldspar, and dark minerals such as biotite or amphibole is generally known as granite, but the International Union of Geological Scientists (IUGS) define granite more exactly as a plutonic rock with 10–50 percent quartz and the ratio of alkali to total feldspar in the range of 65–90 percent.

Granites and related rocks are abundant in the continental crust and may be generated either by melting preexisting rocks or, in lesser quantities, by differentiation via fractional crystallization of basaltic magma. Many granites are associated with convergent-margin or Andean-style magmatic arcs, and include such large plutons and batholiths as those of the Sierra Nevada batholith, Coast Range batholith, and many others along the American Cordillera. Granites are also a major component of Archean cratons and granite greenstone terranes.

Many building stones are granitic, since they tend to be strong, durable, and nonporous and exhibit many color and textural varieties. Granite often forms rounded hills with large round or oblong boulders scattered over the hillside. Many of these forms are related to weathering along several typically perpendicular joint sets, where water infiltrates and reacts with the rock along the joint planes. Three sets of perpendicular joints define cubes in three dimensions; large blocks get weathered out and eventually rounded as the corners weather faster than the other parts of the joint surface, since they have more surface exposed to weathering agents. Granite also commonly forms exfoliation domes, in which large sheets of rock weather off and slide down mountain sides and inselbergs, isolated steep-sided hills that

Mount Whitney and Alabama Hills outside Lone Pine, California. These mountains form part of the Sierra Nevada granite batholith. *(Mike Norton, Shutterstock, Inc.)*

Plutons in Monteregian Hills, Quebec, Canada, taken by Expedition 14 crew member in the *International Space Station*, April 18, 2007 *(Earth Sciences and Image Analysis Laboratory, NASA Johnson Space Center)*

remain on a more weathered plain. Many granites weather ultimately to flat or gently rolling plains covered by erosional detritus including cobbles, boulders, and granitic gravels.

Pluton is a general name for a large, cooled, igneous intrusive body in the Earth. Some plutons are so large that they have special names—batholiths are plutons with a surface area greater than 60 square miles (100 km²). Several types of igneous intrusions are produced by magmas (generated from melting rocks in the Earth) and intrude the crust, taking one of several forms. The specific type of pluton is based on its geometry, size, and relations to the older rocks surrounding the pluton, known as country rock. Concordant plutons have boundaries parallel to layering in the country rock, whereas discordant plutons have boundaries that cut across layering in the country rock. Dikes are tabular but discordant intrusions, and sills are tabular and concordant intrusives. Volcanic necks are conduits connecting a volcano with its underlying magma chamber. A famous example of a volcanic neck is Devils Tower, Wyoming.

Batholiths and plutons have different characteristics and relationships to surrounding country

rocks based on the depth at which they intruded and crystallized. Epizonal plutons are shallow and typically have crosscutting relationships with surrounding rocks and tectonic foliations. They may have a metamorphic aureole surrounding them, where the country rocks have been heated by the intrusion and grew new metamorphic minerals in response to the heat and fluids escaping from the batholith. Rings of hard-contact metamorphic rocks in the metamorphic aureole surrounding batholiths are known as hornfels rocks. Mesozonal rocks intrude the country rocks at slightly deeper levels than the epizonal plutons but not as deep as catazonal plutons and batholiths. Catazonal plutons and batholiths tend to have contacts parallel with layering and tectonic foliations in the surrounding country rocks and do not show such a large temperature gradient with the country rocks as those from shallower crustal levels. This is because all the rocks are at relatively high temperatures. Catazonal plutons tend to be foliated, especially around their margins and contacts with the country rocks.

Batholiths are derived from deep crustal or deeper melting processes and may be linked to surface volcanic rocks. Batholiths form large parts of the continental crust and are associated with some metallic mineral deposits; they are used for building stones.

PLUTON EMPLACEMENT MECHANISMS
The volume of magma that intruded the Earth's crust in some plutons and batholiths is enormous. All the magma in these plutons had to create space in the crust for it to intrude into, since the plutons typically intrude into preexisting continents.

Geologists have long speculated on how such large volumes of magma intrude the crust, and what relationships these magmas have on the style of volcanic eruption. One mechanism that may operate is assimilation, where the magma melts surrounding rocks as it rises, causing them to become part of the magma. In doing so the magma becomes cooler, and its composition changes to reflect the added melted country rock. Assimilation causes magmas to rise only a limited distance. High pressure can force some magmas into the crust. One variation of this forceful emplacement style is diapirism, whereby the weight of surrounding rocks pushes down on the melt layer, which squeezes up through cracks that can expand and extend, forming volcanic vents at the surface. Stoping is a mechanism whereby big blocks are thermally shattered and drop off the top of the magma chamber, falling into it, much like a glass ceiling breaking and falling into the space below. Many if not most plutons seem to be emplaced into structures such as faults, utilizing the weakness provided by the structure for the emplacement of the magma. Some plutons are emplaced into active faults, intruding into spaces created by gaps that open between misaligned segments of the moving fault zone.

See also CRATON; GRANITE, GRANITE BATHOLITH; IGNEOUS ROCKS; PETROLOGY AND PETROGRAPHY; STRUCTURAL GEOLOGY; WEATHERING.

FURTHER READING
Hargraves, R. B. *Physics of Magmatic Processes*. Princeton, N.J.: Princeton University Press, 1980.
Pitcher, W. S. "Granite Type and Tectonic Environment." In *Mountain Building Processes*, edited by K. Hsu. New York: Academic Press, 1982, pp. 19–40.

gravity, gravity anomaly Gravity is the attraction between any body in the universe and all other bodies described by the inverse square law

$$F = (M_1 M_2)/r^2$$

where F represents the force of gravity, M represents the masses of the two bodies that are attracted, and r represents the distance between the objects. Often the term *gravity* refers specifically to the force exerted on any body on or near the surface of the Earth by the mass of the Earth and any centrifugal force resulting from the planet's rotation. A gravity anomaly is the difference between the observed value of gravity at a point and the theoretically calculated value of gravity at that point, based on a simple gravity model. The value of gravity at a point reflects the distribution of mass and rock units at depth, as well as topography. The average gravitational attraction on the surface is 32 feet per second squared (9.8 m/s^2), with one gravity unit (g.u.) being equivalent to one ten-millionth of this value. Another, older unit of measure, the milligal, is equivalent to 10 g.u. The range in gravity on the Earth's surface at sea level is about 50,000 g.u., or from 32.09–32.15 feet per second squared (9.78–9.83 m/s^2). A person would weigh slightly more at the equator than at the poles because the Earth has a slightly larger radius at the equator than at the poles.

Geologically significant variations in gravity are typically only a few tenths of a gravity unit, so instruments to measure gravity anomalies must be very sensitive. Some gravity surveys are made with a series of gravity meters on the surface, whereas others employ observations of the perturbations of satellite orbits.

The determination of gravity anomalies involves subtracting the effects of the overall gravity field of the Earth, accomplished by removing the gravity field at sea level (geoid), leaving an elevation-dependent gravity measurement. This measurement reflects a lower gravitational attraction with height and distance from the center of the Earth, as well as an increase in gravity caused by the gravitational pull of the material between the point and sea level. The free-air gravity anomaly is a correction of the measured gravity calculated by using only the elevation of the point and the radius and mass of the Earth. A second correction, which depends on the shape and density of rock masses at depth, is the Bouguer gravity anomaly. Sometimes a third correction is applied to gravity measurements, known as the isostatic correction. This applies when a load such as a mountain, sedimentary basin, or other mass is supported by mass deficiencies at depth, much like an iceberg floating lower in the water. There are several different mechanisms of possible isostatic compensation, however, and it is often difficult to know which mechanisms are important on different scales. Therefore this correction is often not applied.

Different geological bodies are typically associated with different magnitudes and types of gravity anomalies. Belts of oceanic crust thrust on continents (ophiolites) represent unusually dense material and are associated with positive gravity anomalies of up to several thousand g.u. Likewise, unusually dense massive sulfide metallic ore bodies are also associated with positive gravity anomalies. Salt domes, oceanic trenches, and mountain ranges all represent an increase in the amount of low-density material in the crustal column and are therefore associated with increasingly negative gravity anomalies, with negative values of up to 6,000 g.u. associated with the highest mountains on Earth, the Himalayan chain.

See also GEOID; GEOPHYSICS.

FURTHER READING

Turcotte, Donald L., and Gerald Schubert. *Geodynamics*. 2nd ed. Cambridge: Cambridge University Press, 2002.

Vanicek, Petr, and Nikolaos T. Christou. *Geoid and Its Geophysical Interpretations*. New York: CRC Press, 1994.

gravity wave The term *gravity wave* can be used in two different contexts. The first is in fluid dynamics in reference to waves generated in a fluid medium or at the interface between two fluids, such as air and water. The second context is in astrophysics and general relativity theory.

In fluid dynamics gravity waves form when a unit of fluid at an interface between two different fluids moves to a region with a different density. When this happens gravity attempts to restore the unit of fluid to equilibrium, forming an oscillating wave. Ocean waves generated by the wind are examples of one type of gravity wave known as surface gravity waves. When gravity waves form at boundaries between fluids in the ocean or atmosphere, they are known as internal gravity waves.

The transfer of energy from wind to the ocean surface causes the formation of surface gravity waves on the sea surface by two different mechanisms. If the surface of the ocean is initially flat and a turbulent wind blows across this surface, the fluctuation in the wind imposes fluctuating stresses on the ocean surface, oriented parallel and perpendicular to the surface. These stresses act as a forcing mechanism that may find a matching frequency and wave number for a mode of vibration of the sea surface, then the surface will begin to vibrate as a gravity wave. As more energy is added by the wind, a resonance grows, and the waves grow in amplitude. A resonance is the tendency of a system to oscillate at a maximum amplitude at specific frequencies, known as the resonance frequencies for that system. Once the sea-air surface has this initial roughness, a second process helps the waves grow. In this phase the waves interact with the turbulent flow of the overlying air, with energy transferred to the new waves in a critical boundary layer that forms at a specific height where the wave speed equals the mean turbulent flow. This mechanism continues until the wind stops, or the distance (fetch) it can act on ends (where land is encountered), and the waves continue to grow until either of those points is met.

Gravity waves in the atmosphere transfer momentum from the troposphere (lower 6–7 miles [9.7–11 km] of the atmosphere) to the mesosphere (31–53 [50–85 km] miles above the surface], and often form in response to the movement of frontal systems and to the passage of air over mountain peaks. Low-altitude gravity waves may resemble undulating clouds and do not significantly change the velocity of the moving air mass. Gravity waves at higher altitudes at the boundary with low-density air become higher in amplitude and break, however, transferring significant energy and momentum to the mean flow of air in the mesosphere. They are thus extremely important in controlling the dynamics of the middle atmosphere.

Gravity waves are also prominent in Albert Einstein's theory of relativity, which forms the foundation of modern ideas about gravity. In this usage gravity waves can be thought of as gravitational radiation that results from a change in the strength of a gravitational field. In relativity theory any mass that accelerates through space will produce a small distortion in the space through which it is traveling, causing a change in the gravitational field, or a gravity wave, that should be emitted at the speed of light. However, the gravitational force is so small compared with the electromagnetic force that the distortions produced by gravity waves are expected to be less that the diameter of an atomic nucleus for masses the size of galaxies, and no one has yet detected a gravity wave in this relativistic sense. But, active research programs are attempting to detect gravity waves from interactions of massive objects such as merging binary star systems, black holes swallowing stars, and other galactic collisions. In one example scientists have observed the collapse of a binary star system to produce a shrinking orbit with energy loss from the system consistent with energy being carried away from the system by gravity waves as predicted by Einstein's relativity theory, but the waves themselves have not been detected.

See also ATMOSPHERE; EINSTEIN, ALBERT; GRAVITY, GRAVITY ANOMALY.

FURTHER READING

Center for Gravitational Wave Astronomy. University of Texas at Brownsville. Available online. URL: http://cgwa.phys.utb.edu/. Accessed October 10, 2008.

Gill, A. E. *Gravity Wave: Atmosphere Ocean Dynamics.* New York: Academic Press, 1982.

greenhouse effect The greenhouse effect is commonly used to explain a phenomenon characterized by abnormal warmth on Earth in response to the atmosphere trapping incoming solar radiation. Global climate is a balance between the amount of solar radiation received and the amount of this energy retained in a given area. The planet receives about 2.4 times as much heat in the equatorial regions as in the polar regions. The atmosphere and oceans respond to this unequal heating by setting up currents and circulation systems that redistribute the heat more evenly. These circulation patterns are, in turn, affected by the ever-changing pattern of the distribution of continents, oceans, and mountain ranges.

The amounts and types of gases in the atmosphere can modify the amount of incoming solar radiation. For instance, cloud cover can cause much of the incoming solar radiation to reflect back to space before it reaches the lower atmosphere. On the other hand, certain types of gases (known as greenhouse gases) allow incoming short-wavelength solar radiation to enter the atmosphere but trap this radiation when it tries to escape in its longer-wavelength reflected form. This causes a buildup of heat in the atmosphere and can lead to a global warming known as the greenhouse effect.

The amount of heat trapped in the atmosphere by greenhouse gases has varied greatly over Earth's history. One of the most important greenhouse gases is carbon dioxide (CO_2) that plants (which release oxygen, O_2, to the atmosphere) now take up by photosynthesis. In the early part of Earth history (the Precambrian), before plants covered the land surface, photosynthesis did not remove CO_2 from the atmosphere, so CO_2 levels were much higher than at present. Marine organisms also take up atmospheric CO_2 by removing it from the ocean surface water (which is in equilibrium with the atmosphere) to use in combination with calcium to form their shells and mineralized tissue. These organisms make $CaCO_3$ (calcite), the main component of limestone, a rock composed largely of the remains of dead marine organisms. Approximately 99 percent of the planet's CO_2 is presently removed from the atmosphere/ocean system, because it has been locked in rock deposits of limestone on the continents and on the seafloor. If this amount of CO_2 were released back into the atmosphere, global temperature would increase dramatically. In the early Precambrian, when this CO_2 was free in the atmosphere, global temperatures averaged about 550°F (290°C).

The atmosphere redistributes heat quickly by forming and redistributing clouds and uncondensed water vapor around the planet along atmospheric circulation cells. Oceans are able to hold and redistribute more heat because of the greater amount of water they contain, but they redistribute this heat more slowly than the atmosphere. Surface currents are formed in response to wind patterns, but deep ocean currents (which move more of the planet's heat) follow courses that are more related to the bathymetry (topography of the seafloor) and the spinning of Earth than they are related to surface winds.

The balance of incoming and outgoing heat from Earth has determined the overall temperature of the planet through time. Examination of the geological record has enabled paleoclimatologists to reconstruct periods when Earth had glacial periods, hot and dry periods, hot and wet periods, or cold and dry periods. In most cases Earth has responded to these changes by expanding and contracting its climate belts. Warm periods see an expansion of the warm subtropical belts to high latitudes, and cold periods see an expansion of the cold climates of the poles to low latitudes.

HISTORICAL DEVELOPMENT OF THE GREENHOUSE EFFECT CONCEPT

The historical development of the greenhouse effect theory stems from a concept first coined by the French physicist Edme Mariotte (1620–84) in 1681, who noted that light and heat from the Sun easily passes through a sheet of glass, but that heat from candles and other sources does not. This concept was then extended by the French mathematician Joseph Fourier (1768–1830) in 1824 to the atmosphere by noting that heat and light from the Sun can pass from space through the atmosphere, but heat radiated back to the atmosphere from Earth may get trapped by some of the atmospheric gases, just as heat from a candle is partly blocked by a glass pane. Then in 1861 the Irish physicist John Tyndall (1820–1893) discovered that the complex molecules of water (H_2O) and carbon dioxide (CO_2) were mainly responsible for the absorption of heat radiated back from Earth, and that other atmospheric gases such as nitrogen and oxygen did not play a role in this effect. Tyndall noted that simple changes in the concentrations of CO_2 and H_2O could alternately cool and heat the atmosphere, producing "all the mutations of climate which the researches of geologists reveal." The next step in understanding the greenhouse effect came from the work of Swedish physicist and chemist Svante Arrhenius (1859–1927) in 1896, who calculated that a 40 percent increase or decrease in the atmospheric concentration of CO_2 could trigger the advance or retreat of continental glaciers, setting off the glacial and interglacial ages. Much later a change in the atmospheric CO_2 of this magnitude was documented in cores of the Greenland ice sheet, as predicted by Arrhenius.

Carbon dioxide can vary naturally in the atmosphere through a variety of driving mechanisms, including changes in volcanism, erosion and plate tectonics, and ocean-atmosphere interactions. The modern concept of linking greenhouse gases with the burning of fossil fuels by humans was formulated by steam engineer and amateur meteorologist Guy Stewart Callendar (1898–1964) in 1938, who calculated that a doubling of atmospheric CO_2 by burning fossil fuels would result in an average global temperature increase of about 3°F (2°C), with more heating at the poles. Callendar made prescient predictions that humans are changing the composition of the atmosphere at a rate that is "exceptional" on geological timescales, and he sought to understand what effects these changes might have on climate. His prediction was that the "principal result of increasing carbon dioxide will be a gradual increase in the mean temperature of the colder regions of the Earth." These predictions were first confirmed in 1947 when the Swedish climatologist Hans Wilhelmsson Ahlmann (1889–1974) reported a 1–2° F (1.3°C) increase in the average temperature of the North Atlantic sector of the Arctic. At this time the nature of the complex interactions of the carbon cycle and exchange of CO_2 in the atmosphere-ocean system was not well understood, however, and many scientists attributed the entire temperature rise to human production of greenhouse gases. Later studies of ocean-atmosphere relationships, and biogeochemistry, showed more complex relationships. Later, in the 1970s, effects of aerosols in the atmosphere, principally to reflect solar radiation to space and cooling Earth, began to be appreciated as another component of the greenhouse effect. The current state of knowledge of the complex physical, chemical, biological, and other processes associated with the greenhouse effect are described in the *Climate Change 2007* report issued by the Intergovernmental Panel on Climate Change.

See also ATMOSPHERE; CLIMATE; CLIMATE CHANGE; GLOBAL WARMING.

FURTHER READING

Ahrens, C. D. *Meteorology Today: An Introduction to Weather, Climate, and the Environment*, 6th ed. Pacific Grove, Calif.: Brooks/Cole, 2000.

Ashworth, William, and Charles E. Little. *Encyclopedia of Environmental Studies, New Edition*. New York: Facts On File, 2001.

Intergovernmental Panel on Climate Change home page. Available online. URL: http://www.ipcc.ch/index.htm. Accessed January 30, 2008.

Intergovernmental Panel on Climate Change 2007. *Climate Change 2007: The Physical Science Basis. Contributions of Working Group I to the Fourth Assessment Report of the Intergovernmental Panel on Climate Change*. Edited by S. Solomon, D. Qin, M. Manning, Z. Chen, M. Marquis, K. B. Avery, M. Tignor, and H. L. Miller Cambridge: Cambridge University Press, 2007.

greenstone belt A greenstone belt is an elongate accumulation of generally mafic volcanic and plutonic rocks, typically associated with assemblages of sedimentary rocks that include sandstones, mudstones, banded iron formations, and, less commonly, carbonates and mature sedimentary rocks. Most greenstone belts are Archean or at least Precambrian in age, although similar sequences are known from orogenic belts of all ages. Nearly all are metamorphosed to greenschist through amphibolite facies and intruded by a variety of granitoid rocks. Older gneissic rocks are associated with some greenstone belts, although most of these are in fault contact with the greenstone belts.

Greenstone belts display a wide variety of shapes and sizes and are distributed asymmetrically across Archean cratons in a manner reminiscent of tectonic zonations in Phanerozoic orogens. For instance, the Yilgarn craton in Australia has mostly granitic gneisses in the southwest, mostly 2.9 billion-year-old greenstones throughout the central craton, and 2.7 billion-year-old greenstones in the east. The Slave Province in Canada contains remnants of a circa 4.2–2.9 billion-year-old gneissic terrain in the western part of the province, dominantly mafic greenstone belts in the center, and 2.68 billion-year-old mixed mafic, and intermediate and felsic calc-alkaline volcanic rocks in the eastern part of the province. Other cratons are also asymmetric in this respect; for example, the Zimbabwe craton, in Africa, has mostly granitic rocks in the east and more greenstones in the west. The Superior Province contains numerous subparallel belts, up to thousands of miles long, that are distinct from each other but similar in scale and rock type to Phanerozoic orogens. These distributions of rock types are analogous to asymmetric tectonic zonations, which are products of plate tectonics in younger orogenic belts, and emphasize that greenstone belts are perhaps only parts of once larger orogenic systems.

There are three significantly different end-member regional outcrop patterns of greenstone belts reflecting the distribution of these belts within cratons. These include broad domal granitoids with interdomal greenstones; broad greenstone terrains with internally branching lithological domains and irregular granitoid contacts; and long, narrow, and straight greenstone belts. The first pattern includes mostly granitoid domes with synformal greenstone belts, which result from either interference folding or dome-shaped granitoids. The second pattern includes many of the terranes with thrust belt patterns, including much of the Yilgarn Province in western Australia and the Slave Province of Canada. Contacts with granitoids are typically intrusive. The third pattern includes composite thrust/strike slip belts dominated by late strike-slip shear zones along one or more sides of the belt. Granite-greenstone contacts are typically a fault or shear zone.

Until recently few complete ophiolite-like sequences were recognized in Archean greenstone belts, leading some workers to the conclusion that no Archean ophiolites or oceanic crustal fragments are preserved. Research documenting partial dismembered ophiolites in several greenstone belts and a complete ophiolite sequence in the North China craton recently challenged these ideas. Archean oceanic crust was possibly thicker than Proterozoic and Phanerozoic counterparts, and this resulted in accretion predominantly of the upper basaltic section of oceanic crust. The crustal thickness of Archean oceanic crust may have resembled modern oceanic plateaus. If this were the case, complete Phanerozoic-like ophiolite sequences would have been very unlikely to be preserved from Archean orogenies. In contrast, only the upper, pillow lava–dominated sections would likely be accreted. Archean greenstone belts have an abundance of accreted ophiolitic fragments compared with Phanerozoic orogens, suggesting that thick, relatively buoyant, young Archean oceanic lithosphere may have had a structure favoring separation of the uppermost parts during subduction and collisional events.

GREENSTONE BELT GEOMETRY

Geophysical surveys have shown that greenstone belts are mostly shallow to intermediate in depth, extending to 3–15 miles (5–20 km). Some have flat or irregular bases, and granitic rocks intrude many of them. They are not steep synclinal keels. Gravity models consistently indicate that greenstone belts rarely extend to greater than 6 miles (10 km) in depth, and seismic reflection studies show that the steeply dipping structures characteristic of most greenstone belts disappear into a horizontally layered mid to lower crustal structure. Seismic reflection surveys have also proven useful in demonstrating that boundaries between different "belts" in granite-greenstone terrains are in some cases marked by large-scale crustal discontinuities most easily interpreted as sutures or major strike-slip faults.

Just as greenstone belts are distributed asymmetrically on cratons, many have asymmetrical distributions of rock types and structural vergence within them, and in this respect they are much like younger orogenic belts. For example, the eastern Norse-

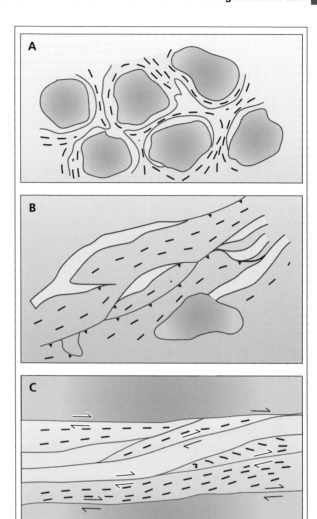

Map showing the main structural elements of greenstone belts: dark green are granites; light green are greenstones; tan are metasediments; red lines are foliations *(modified from T. Kusky, and J. Vearncombe, 1997)*

man-Wiluna belt in the Yilgarn craton contains a structurally disrupted and complex association of oceanic-type mafic and island arc-type volcanic rocks, whereas the western Norseman-Wiluna belt contains disrupted rocks of predominantly oceanic affinity. In other belts it is typical to find juxtaposed rocks from different crustal levels and facies that were originally laterally separated. One of the long-held misconceptions about the structure of greenstone belts is that they simply represent steep synclinal keels of volcanic and sedimentary rocks squeezed between diapiric granitoids. Where studied in detail, there is a complete lack of continuity of strata from either side of the supposed syncline, and the structure is much more complex than the pinched-synform model predicts. The structure and stratigraphy of greenstone belts will be unraveled only when "stratigraphic" methods of mapping are abandoned and techniques commonly applied to gneissic terrains are used for mapping greenstone belts. Greenstone belts should be divided into structural domains, defined by structural style, metamorphic history, distinct lithological associations, and age groupings where these data are available.

One of the most remarkable features of Archean greenstone belts is that structural and stratigraphic dips are in most cases very steep to vertical. These steep dips are evidence of the intense deformation that these belts have experienced, although mechanisms of steepening may be different in different examples. Some belts, including the central Slave Province in Canada and the Norseman-Wiluna belt of Western Australia, appear to have been steepened by a series of thrust faults stacking the rocks end over end. Successive offscraping of the greenstone from oceanic crust in thrust sheets steepens rocks toward the interior of the thrust belt. In other cases intrusions have steepened greenstone belt rocks on the margins of plutons and batholiths. Examples of this mechanism are found in the Pilbara craton of Australia and northern Zimbabwe cratons of Africa. Shortening of the entire crust appears to be an important steepening mechanism in other examples, such as in the Theespruit area of the Barberton Belt of southern Africa's Kaapvaal craton. Tight to isoclinal upright folding, common in most greenstone belts, and fold interference patterns are responsible for other steep dips. In still other cases rotations incurred in strike-slip fault systems (e.g., Norseman-Wiluna belt Superior Province) and on listric normal fault systems (e.g., Quadrilatero Ferrifero, San Francisco craton) may have caused local steepening of greenstone belt rocks. These are the types of structures present throughout Phanerozoic orogenic belts.

STRUCTURAL V. STRATIGRAPHIC THICKNESS OF GREENSTONE BELTS

Many studies of the "stratigraphy" of greenstone belts have assumed that thick successions of metamorphosed sedimentary and volcanic rocks occur without structural repetition, and that they have undergone relatively small amounts of deformation. As fossil control is virtually nonexistent in these rocks, stratigraphic correlations are based on broad similarities of rock types and poorly constrained isotopic dates. In pre-1980 studies it was common to construct single stratigraphic columns that were 6–12 miles (10–20 km) or more thick, but recent advances in the recognition of thin fault zones and precise geochronological ages documenting older rocks thrust over younger rocks in the stratigraphy

of some belts makes reevaluation of these thicknesses necessary. Intact stratigraphic sections more than a couple of miles thick in greenstone belts are rare. Further mapping needs to be structural, based on defining domains of similar structure, lithology, and age, rather than lithological, attempting to correlate multiply-deformed rocks across large distances.

An observation of great importance for interpreting the significance of supposed thick stratigraphic sections in greenstone belts is that there is an apparent lack of correlation between metamorphic grade and inferred thicknesses of the stratigraphic pile. If the purported 6–12-mile (10–20-km) thick sequences were real stratigraphic thicknesses, an increase in metamorphic grade would be detectable with inferred increase in depth. Because this is not observed, the thicknesses must be tectonic and thus reflect stratigraphic repetition in an environment such as a thrust belt or accretionary prism, where stratigraphic units can be stacked end-on-end, with no increase in metamorphic grade in what would be interpreted as stratigraphically downward. Other mechanisms by which apparent stratigraphic thicknesses may be increased are by folding, erosion through listric normal fault blocks, and progressive migration of depositional centers.

GREENSTONE-GNEISS CONTACT RELATIONSHIPS

An important problem in many greenstone belt studies is determining the original structural relationships between greenstone belts and older gneiss terrains. In pre-1990 studies the significance of early thrusting along thin fault zones went unrecognized, leading to a widespread view that many greenstone belts simply rest in depositional contact over older gneisses, or that the older gneisses intruded the greenstone belt. While this may be the case in a few examples, it is difficult to demonstrate, and the classic areas in which such relationships were supposedly clearly demonstrable have recently been shown to contain significant early thrust faults between greenstones and older sedimentary rocks that rest unconformably over older gneisses. Such is the case at Steep Rock Lake in the Superior Province, at Point Lake and Cameron River in the Slave Province, in the Theespruit type section of the Barberton greenstone belt, at Belingwe, Zimbabwe, and in the Norseman-Wiluna belt.

The Norseman-Wiluna, Cameron River, and Point Lake greenstone belts contain up to 1,640-foot (500-m) wide metamorphic aureoles at their bases. The aureoles contain upper-amphibolite facies assemblages, in contrast to greenschist and lower-amphibolite facies assemblages in the rest of the greenstone belts. The aureoles have mylonitic, gneissic, and schistose fabrics parallel to the upper contacts with the greenstone belts, and are locally partially melted, forming granitic anatectites. The aureole at the base of the Norseman-Wiluna belt is an early shear zone structure related to the thrusting of the greenstone belt over the older gneisses. This fault zone is overprinted by greenschist facies metamorphic fabrics and two episodes of regional folding, the second of which is the main regional deformation event and is associated with a strong cleavage. On Cameron River and Point Lake in the Slave Province, the aureoles represent early thrust zones related to the tectonic emplacement of the greenstone belts over the gneisses. Amphibolite-facies mylonites were derived through deformation of mafic and ultramafic rocks at the bases of the greenstone belts, which are largely at greenschist facies. The broad field-scale relationships in these cases are also similar to those found in metamorphic aureoles attached to the bases of many obducted ophiolites.

STRUCTURAL ELEMENTS OF GREENSTONE BELTS

The earliest structures found in greenstone belts formed during deposition of the rocks. In most greenstone belts the mafic volcanic/plutonic section is older than the sedimentary section, so structures that formed during formation of the igneous rocks are older in the magmatic than in the sedimentary rocks. Unequivocal evidence is lacking for large-scale deformation of the igneous rocks of greenstone belts during their deposition. Structures of this generation typically include broken pillow lavas grading into breccias and possible slump folds and faults in interpillow sedimentary horizons.

Two types of early shear zones active before regional contractional deformation have been described from the Jamestown ophiolite in the Barberton greenstone belt of southern Africa. The first are low-angle normal faults located along the lower contacts of the ophiolite-related cherts (so-called Middle Marker); they cause extensive brecciation and alteration of adjacent mafic rocks. The faults and adjacent cherts are cut by subvertical mafic rocks of the Onverwacht Group, showing that these faults were active early, during the formation of the ophiolitic Onverwacht Group. The second type of early faults in the Barberton greenstone belt occur in both the plutonic and the extrusive igneous parts of the Jamestown ophiolite, and may represent steepened continuations (root zones) of the higher-level extensional faults, or they may represent transform faults. Other possible examples of early extensional faults have been described from the Cameron River and Yellowknife greenstone belts in the Slave Province, and from the Proterozoic Purtuniq ophiolite in the Cape Smith Belt of the Ungava Orogen. In the Purtuniq ophiolite early sinuous shear zones locally

separate sheeted dikes from mafic schists, causing rotation of the dike complex. In other places dikes intrude this contact, showing that the shear zones are early features. Although not explicitly interpreted in this way, these shear zones may be related to block faulting in the region of the paleoridge axis.

Detailed mapping in a number of greenstone belts has revealed early thrust faults and associated folds. Most do not have any associated regional metamorphic fabric or axial planar cleavage, making their identification difficult without detailed structural mapping. Examples are known from the Zimbabwe and Kaapvaal cratons, the Yilgarn craton, the Pilbara craton, the Slave Province, and the Superior Province. In these cases it is apparent that early thrust faulting and associated folding is responsible for the overall distribution of rock types in the greenstone belts, and also accounts for what were in some cases previously interpreted as enormously thick stratigraphic sequences. These early thrust faults are responsible for juxtaposing greenstone sequences with older gneissic terrains. Few of these early thrust faults are easy to detect; they occupy thin, poorly exposed structural intervals within the greenstone belts, some are parallel to internal stratigraphy for many miles, and most have been reoriented by later structures. In many greenstone belts there are numerous layer-parallel fault zones, but their origin is unclear because they are not associated with any proven stratigraphic repetition or omission. Although it is possible that these faults are thrust, strike-slip, or normal faults, the evidence so far accumulated in the few well-mapped examples supports the interpretation that they are early thrust faults. In some cases late intrusive rocks have utilized the zone of structural weakness provided by the early thrusts for their intrusion.

Emplacement of the early thrust and fold nappes is typically not associated with any strong fabric or cleavage development or any regional metamorphic recrystallization, making recognition of these early structures even more difficult. Delineation of early thrusts depends critically on very detailed fieldwork with particular attention paid to details of the structural geology. Determining the sense of tectonic transport of early nappes is critical for tectonic interpretation, but is also one of the most elusive goals because of the weak development of critical lineations and reorientation of the earliest fabric elements by younger structures. Kinematic studies of early shear zones have received far less attention than they deserve in Archean greenstone belts.

Folds are in many cases the most obvious outcrop to map-scale structures in greenstone belts. Several different generations of folding are typical, and fold interference patterns are commonplace.

Some greenstone belts show a progression from early recumbent folds (flat-lying folds, associated with thrust/nappe tectonics), through two or more phases of tight to isoclinal upright folds, which are associated with the most obvious mesoscopic and microscopic fabric elements and metamorphic mineral growth. One or more generations of late open folds or broad crustal arches with associated crenulation cleavages also affect many greenstone belts. Fold interference patterns most typically reflect the geometry of second- and third-generation fold (F_2 and F_3) structures because they have similar amplitudes and wavelengths; F_1 recumbent folds are best recognized by reversals in younging directions (the direction toward the original tops of beds), or downward-facing F_2 and F_3 folds.

Relationships of individual fold generations to tectonic events is poorly understood in many greenstone belts, and the orientations of stresses that formed the folds are poorly constrained, largely because of uncertainties associated with correctly unraveling superimposed folding events. The relative importance of "horizontal" versus "vertical" tectonics has been debated, in part because it is difficult to distinguish granite-cored domes produced by the interference of different generations of folds from domes produced by diapirism, or the rising of granite as a crystal-rich magma. Two generations of upright folds with similar amplitudes and wavelengths produce a dome and basin fold interference pattern that closely resembles a pattern formed by intruding granite domes.

Many granite-greenstone terrains are cut by late-stage strike-slip faults, some of which are reactivated structures that may have been active at other times during the history of individual greenstone belts. Strike-slip dominated structural styles are found in the Superior Province, the Yilgarn craton, the Pilbara craton, and southern Africa.

Large-scale lineaments of the Norseman–Wiluna Belt near Kalgoorlie and Kambalda, Australia, show late reverse motion related to regional oblique compression. But their length is often greater than 62 miles (100 km), and consistent indicators of a sinistral component of motion suggest that they are dominantly strike-slip structures. These large-scale structures were present through the deformation history that bound domains within which unique structures are developed. These major faults may represent reactivation of terrane boundaries, since they separate zones of contrasting stratigraphy and structure.

The Superior Province consists of a number of fault-bounded subprovinces containing rocks of different lithological associations, ages, and structural and metamorphic histories. The greenstone terrains consist of several types, including some dominated by oceanic-type igneous rocks, island arc-type volcanic complexes, and continental-style volcanic rocks with associated fluvial deposits. Belts of variably metamorphosed muddy sandstones called turbidites, interpreted as accretionary prisms, are younger than and separate from individual volcanic belts. Most of the greenstone belts are progressively younger toward the south in the central parts of the Superior Province. This observation together with the contemporaneity of deformation events within individual belts suggests that the Superior Province represents an amalgam of oceanic crust and plateaus, island arcs, continental margin arcs, and accretionary prisms, brought together by dextral oblique subduction, which formed the major subprovince-bounding strike-slip faults. Continued late strike-slip motion on some of the faults localized alkalic volcanic and fluvial sedimentary sequences in pull-apart basins on some of these strike-slip faults.

In the Vermillion district of the southern Superior Province the deformation history begins with early nappe-style structures overprinted by the "main" fabric elements related to dextral strike slip along with thrusting. This sequence of structural development is interpreted to reflect dextral-oblique accretion of island arcs and microcontinents of the southern Superior Province. A combination of north-south shortening together with dextral simple shear led to the juxtaposition of zones with constrictional and with flattening strains. Geologists previously interpreted the constrictional strains in this area to be a result of "squeezing" between batholiths, necessitating reevaluation of similar theories for the origin of prolate strains in numerous other greenstone belts.

In some cases strike-slip faults have played an integral role in the localization and generation of greenstone belts in pull-apart basins between the strike-slip faults. Several strike-slip fault systems associated with the formation of greenstone belts in pull-apart basins are known from the Pilbara craton. The Lalla Rookh and circa 2.95 billion-year-old Whim Creek belts are interpreted as "second-cycle greenstones," because they were deposited in strike-slip–related pull-apart basins that formed in already complexly deformed and metamorphosed circa 3.5–3.3 billion-year-old rocks of the Pilbara craton. Thus although these fault systems form an integral part of the structural evolution of the Pilbara craton, they postdate events related to initial formation of the granite-greenstone terrain. One problem that future studies of greenstone belt structure should address is an understanding of the nature of the transition from brittle to ductile strains in pull-apart regions along strike-slip fault systems, such as those of the Pilbara.

Major late-stage strike-slip fault zones cut many cratons, but few have well-constrained kinematic or metamorphic histories. An exception is the 186-mile (300-km) long Koolyanobbing shear zone in the Southern Cross Province of the Yilgarn craton. The Koolyanobbing shear is a 4–9-mile (6–15-km) wide zone with a gradation from foliated granitoid, through protomylonite, mylonite, to ultramylonite, from the edge to the center of the shear zone. Shallowly plunging lineations and a variety of kinematic indicators show that the shear zone is a major sinistral fault, but regional relationships suggest that it does not represent a major crustal boundary or suture. Fault fabrics both overprint and appear coeval with late stages in the development of the regional metamorphic pattern, suggesting that the shear zone was active around 2.7–2.65 billion years ago.

LATE EXTENSIONAL COLLAPSE, DIAPIRISM, DENUDATION

It is widely recognized that in Phanerozoic orogens, late stages of orogenic development are characterized by extensional collapse of structurally overthickened crust. In granite-greenstone terrains late stages of mountain building and orogenesis are characterized by the intrusion of abundant granitic magmas, with chemical signatures indicative of crustal melting. The intrusion of late granitic plutons in greenstone terrains may be related to rapid uplift and crustal melting accompanying extension in the upper crust. Early plutons of the tonalite-trondhjemite-gabbro-granodiorite suite are generated in an island arc setting, and are in turn intruded by continental margin arc magmas after the primitive arcs collide and form larger continental fragments. When plate collision causes further crustal thickening, melting of the deep crust produces thin diapiric plutons that rise partway through the crust but crystallize before rising very far, as they do not contain enough heat to melt their way through the crust. The crustal sections in these collisional orogens gravitationally collapse when the strength of quartz and olivine can no longer support the topography. Decompression in the upper mantle and lower crust related to upper-crustal extension generates significant quantities of basaltic melts. These basaltic melts rise and partially melt the lower or middle crust, becoming more silicic by assimilating crustal material. The hybrid magmas thus formed intrude the middle and upper crust, forming the late to post-kinematic granitoid suite so common in Archean granite-greenstone terrains. If the time interval between crustal thickening and gravitational collapse is short, then magmas related to decompressional melting may quickly rise up the partially solidified crystal/mush pathways provided by the earlier plutons generated during the crustal

thickening phases of orogenesis. Such temporal and spatial relationships easily account for the common occurrence of composite and compositionally zoned plutons in Precambrian and younger orogenic belts.

See also AFRICAN GEOLOGY; ARCHEAN; CONTINENTAL CRUST; CRATON; NORTH AMERICAN GEOLOGY; OPHIOLITES; PRECAMBRIAN.

FURTHER READING

Kious, Jacquelyne, and Robert I. Tilling. "U.S. Geological Survey. This Dynamic Earth: The Story of Plate Tectonics." Available online. URL: http://pubs.usgs.gov/gip/dynamic/dynamic.html. Modified March 27, 2007.

Kusky, T. M., ed. *Precambrian Ophiolites and Related Rocks, Developments in Precambrian Geology 13.* Amsterdam: Elsevier, 2003.

Kusky, Timothy M., Jianghai Li, and Robert T. Tucker. "The Archean Dongwanzi Ophiolite Complex, North China Craton: 2.505 Billion-Year-Old Oceanic Crust and Mantle." *Science* 292 (2001): 1142–1145.

Kusky, Timothy M., and Julian Vearncombe. "Structure of Archean Greenstone Belts." In *Tectonic Evolution of Greenstone Belts*, edited by Maarten J. de Wit and Lewis D. Ashwal, 95–128. Oxford Monograph on Geology and Geophysics. Oxford: Oxford Science Publications, 1997.

Grenville province and Rodinia At several times in the history of the planet, most of the continental landmasses have aggregated or joined together to form large supercontinents. The most recent of these was the fairly familiar supercontinent of Pangaea, which contained most of the planet's continents between 300 and 200 million years ago. Before that, the supercontinent of Gondwana formed at about 570 million years ago and lasted only a short geological time (the exact amount is still under debate and investigation). As we explore further back in geological time, the evidence for older supercontinents becomes harder and harder to interpret. Despite this, in the past decade geologists have been able to reconstruct an older supercontinent, known as Rodinia, that formed about 1 billion years ago and broke up around 700 million years ago.

The Grenville province is the youngest region of the Canadian shield; it is outboard of the Labrador, New Quebec, Superior, Penokean, and Yavapai-Mazatzal provinces. It is the last part of the Canadian shield to experience a major deformational event, this being the Grenville Orogeny, which was responsible for forming many folds and faults throughout the entire region during the amalgamation of numerous continents to form the supercontinent of Rodinia. The other ancient, highly eroded mountain belts around the world that formed during the collision of the other continents to form Rodinia have also become known as Grenvillian belts, named after the excellent type exposures of deeply eroded mountain belts of this age in the Grenville province. The Grenville province has an aerial extent of approximately 600,000 square miles (1,000,000 km²). The subterranean extent of Grenville rocks, however, is much greater in area. Phanerozoic rocks cover their exposure from New York State down the length of the Appalachian Mountains and into Texas.

The Grenville province formed on the margin of the continent of Laurentia (an early or immature stage in the development of North America) in the middle to late Proterozoic. The rocks throughout the province represent a basement and platform sedimentary sequence intruded by igneous rocks. Subsequent to this intrusive event in the late Proterozoic, the entire region underwent high-grade metamorphism and was complexly deformed. But before this high-grade metamorphic event, the rocks of the Grenville province experienced multiple pulses of metamorphism and deformation, including the Elsonian (1,600–1,250-million-year-old) and the Elzevirian (1,250–1,200-million-year-old). Orogenies. The Ottawan Orogeny was the last and most intense in the Grenville province, culminating 1.1 billion years ago, and overprinting much of the earlier tectonic history. This has made it difficult for geologists to describe the earlier orogenies and also to determine the tectonic evolution of the Grenville province. For these reasons, the term *Ottawan Orogeny* is usually used synonymously with the term *Grenville Orogeny*.

The Grenville province is subdivided into numerous subprovinces including the central gneiss belt (CGB), central metasedimentary belt (CMB), and central granulite terrane (CGT), and one major structural feature: the Grenville front (GF).

The central gneiss belt (CGB) is located in the western part of the Grenville province and contains some of the oldest rocks found in the province. The majority of the rocks are 1.8–1.6 billion-year-old gneisses intruded by 1.5–1.4 billion-year-old granitic and monzonitic plutons. Both the metasedimentary and the igneous rocks of the CGB are metamorphosed from upper amphibolite and locally granulite facies. The CGB is bounded by the Grenville front to the northwest and lies in tectonic contact with the central metasedimentary belt to the southeast. The dominant structural trend is northeast, but changes to the northwest near Georgian Bay. The CGB has been divided into smaller terranes including the Nipissing, Algonquin, Tomiko, and Parry Sound, based on lithology, metamorphic grade, and structures, namely, shear zones. These terranes are considered to be mainly parautochthonous (mean-

ing that they have not traveled far from their place of formation) terranes. The shear zones that separate the various terranes contain kinematic indicators that suggest northwest directed tectonic transport, and tectonic transport is thought to have occurred between 1.18 and 1.03 billion years ago.

The Nipissing terrane is located in the western portion of the central gneiss belt. Part of the Nipissing terrane occupies a region known as the Grenville front tectonic zone (GFTZ), an area that lies within 30 miles (50 km) of the Grenville front. The lithologies here are strongly deformed with northeast-striking foliations and zones of cataclasis and moderately plunging southeast lineations. The heterogeneous gneisses of the Nipissing terrane fall into two categories: Archean and Lower Proterozoic migmatitic gneisses that are likely reworked units of the Southern and Superior provinces and Middle Proterozoic metasedimentary gneiss. These rocks were intruded by 1.7 and 1.45 billion-year-old granitic plutonic rocks, both of which are less deformed than the host rocks. Postdating this intrusive event, the region underwent high-grade metamorphism, experiencing temperatures of 1,200°F–1,280°F (650°C–750°C) and pressures of 8.0–8.5 kilobars.

The Tomiko terrane is located in the extreme northwestern portion of the central gneiss belt. The most striking aspect of the Tomiko terrane is the relative abundance of metasedimentary rocks, but it also contains metamorphosed granitic rocks that are Middle Proterozoic in age. The Tomiko terrane is allochthonous (far-traveled) with respect to the Nipissing terrane. Evidence to support this is the distinct detrital zircon population in the Tomiko metaquartzites, dated at 1,687 million years old. This is in sharp contrast to the metaquartzites of the Nipissing terrane, where the detrital zircons are Archean to Lower Proterozoic in age. This suggests that the Nipissing terrane was already adjacent to the Superior province at the time of the Nipissing quartzite formation. Further evidence for the allochthonous nature of the Tomiko terrane is the presence of iron formations in the Tomiko terrane, which are not present elsewhere in the CGB. The metamorphic conditions experienced by the Tomiko terrane are temperatures of fewer than 1,290°F (700°C) and pressures of 6.0–8.0 kilobars.

The Algonquin terrane, the largest terrane in the CGB, consists of numerous domains. The rocks in this terrane are meta-igneous quartzo-feldspathic gneisses and supracrustal gneisses. Generally, the foliations strike northeast and dip to the southeast; down-dip stretching lineations are common. The southern domains have been interpreted as thrust sheets with a clear polarity of southeasterly dips, and the entire Algonquin terrane may be parautochthonous. The metamorphic temperatures and pressures range from 1,240°F–1,520°F (670°C–825°C) and 7.9–9.9 kilobars, respectively.

The Parry Sound terrane, the most studied terrane in the CGB, is located in the south-central portion of the CGB and contains large volumes of mafic rock, marble, and anorthosite. The age of the Parry Sound terrane ranges from 1,425 to 1,350 million years. Both the lithologies and the age of the Parry

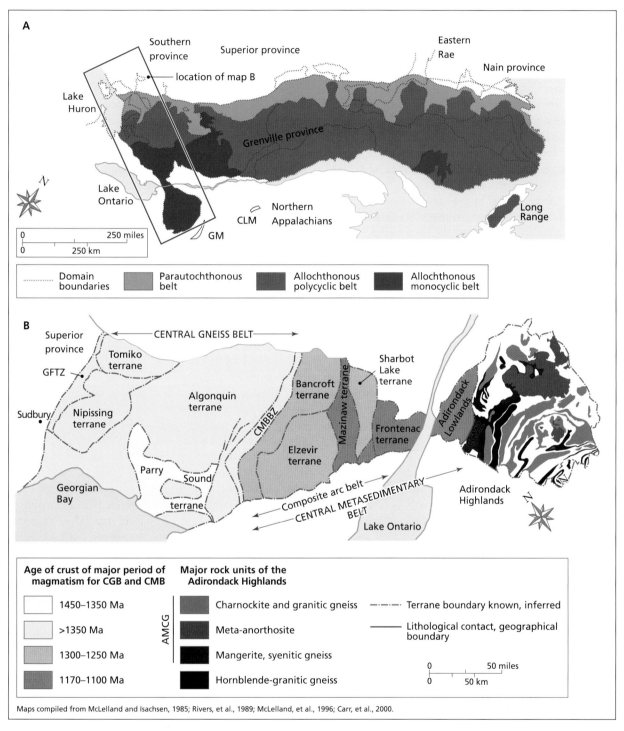

(A) Tectonic subdivisions of the Grenville province according to the classification of Toby Rivers (1989) and Carr, et al. (2000), showing also the older domain boundaries of Wynne-Edwards (1972) and others; (B) Terranes and shear zones of the central gneiss belt (CGB), the central metasedimentary belt (CMB), and major geological features of the Adirondack Highlands. Abbreviations as follows: BCS-Baie Comeau segment; CCMZ-Carthage-Colton mylonite zone; CLM-Chain Lakes massif; CMBBZ-central metasedimentary belt boundary zone; CGB-central gneiss belt; CGT-central granulite terrane; EGP-eastern Grenville province; GFTZ-Grenville Front tectonic zone; GM-Green Mountains.

Sound terrane are different from the rest of the CGB. Not surprisingly, therefore, this terrane is considered as allochthonous and overlying the parautochthonous Algonquin domains. Because the Parry Sound terrane is completely surrounded by the Algonquin terrane, structurally it is considered a klippe. The metamorphic conditions reached by the Parry Sound terrane are in the range of 1,200°F–1,470°F (650°C–800°C) and 8.0–11.0 kilobars.

The central metasedimentary belt (CMB) has a long history of geologic investigation. One of the reasons is the abundance of metasedimentary rocks, which makes it a prime target for locating ore deposits. The CMB was originally named the Grenville series by Sir William Logan in 1863 for an assemblage of rocks near the village of Grenville, Quebec, and is the source of the name for the entire Grenville province. Later, the Grenville series achieved supergroup status, but presently *Grenville Supergroup* is a term limited to a continuous sequence of rocks within the CMB.

The CMB contains Middle Proterozoic metasediments that were subsequently intruded by syn-, late-, and post-tectonic granites. The time of deposition is estimated to have been from 1.3 to 1.1 billion years ago, with the bulk of the material having been deposited before 1.25 billion years ago. After their deposition, the rocks of the CMB underwent deformation and metamorphism from in the Elzevirian Orogeny (1.19–1.06 billion years ago). The effects of the Elzevirian Orogeny were all but wiped out by the later Ottawan Orogeny, which deformed and metamorphosed the rocks to middle-upper amphibolite facies. The CMB contains five distinct terranes: Bancroft, Elzevir, Sharbot Lake, Mazinaw, and Frontenac. The Frontenac is correlative with the Adirondack Lowlands.

The Bancroft terrane is located in the northwestern portion of the CMB. The Bancroft is dominated by marbles but also contains nepheline-bearing gneiss and granodioritic orthogneiss metamorphosed to middle through upper amphibolite facies. The Bancroft terrane contains complex structures, such as marble breccias and high-strain zones. The orthogneiss occurs in thin structural sheets, suggesting that it may occur in thrust-nappe complexes. The thrust sheets generally dip to the southeast with dips increasing toward the dip direction. Rocks of the Bancroft terrane possess a well-developed stretching lineation that also plunges in the southeast direction. Both of these structural orientations suggest northwest directed tectonic transport.

The Elzevir terrane, located in the central portion of the CMB, is known for containing the classic Grenville Supergroup. The Elzevir is composed of 1.30–1.25-billion-year old metavolcanics and metasediments, intruded by 1.27-billion-year-old tonalitic plutons ranging in composition from gabbro to syenite. The largest of these calc-alkaline bodies is the Elzevirian batholith. The calc-alkaline signature of the batholith suggests that it may have been generated in an arc-type setting. The Elzevir terrane also contains metamorphic depressions, areas of lower metamorphic grade, such as greenschist to lower amphibolite facies. These depressions may be related to the region's polyphase deformation history, and in contrast to surrounding high-grade terranes, they contain sedimentary structures enabling the application of stratigraphic principles to determine superposition.

The Mazinaw terrane was once mapped as part of the Elzevir terrane; it also contains some of the classic Grenville Supergroup marbles and the Flinton Group. The rocks encountered here are marbles, calc-alkalic metavolcanic and clastic metasedimentary rocks. The Flinton Group is derived from the weathering of plutonic and metamorphic rocks found in the Frontenac terrane. Furthermore, the complex structural style of the Mazinaw terrane is similar to the Frontenac and the Adirondack Lowlands.

The Sharbot Lake terrane was once mapped as part of the Frontenac terrane but is now considered a separate terrane. The Sharbot Lake principally contains marbles and metavolcanic rocks intruded by intermediate and mafic plutonic rocks and may represent a strongly deformed and metamorphosed carbonate basin. Metamorphic grade ranges from greenschist to lower amphibolite. The lithologies, metamorphic grade, and lack of exposed basement rocks to the Sharbot Lake terrane imply that these rocks may be correlative with the Elzevir terrane.

The Frontenac terrane is located in the southeastern portion of the CMB. This terrane extends into the Northwest Lowlands of the Adirondack Mountains. The Frontenac terrane is composed of marble with pelitic gneisses and quartzites. The relative abundances of the gneisses and quartzites increase toward the southeast, while the relative abundances of metavolcanic rocks and tonalitic plutons decrease in the same direction. A trend also exists in the metamorphic grade from northwest to southeast. In the northwest the metamorphic grade ranges from lower amphibolite to upper amphibolite-granulite facies but then decreases in the southeast to amphibolite facies. Rock attitudes also change, dipping southeast in the northwest, to vertical in the central part, to the northwest in the Northwest Lowlands.

Throughout the CMB, large-scale folds are present. These folds indicate crustal shortening. More important, however, is the recognition of main structural breaks that lie both parallel to and within the CMB. The structural breaks are marked by narrow zones of highly attenuated rocks, such as mylonites. The Robertson Lake mylonite zone (RLMZ), one such structural break, lies between the Sharbot Lake terrane and the Mazinaw terrane. The RLMZ has been interpreted as a low-angle thrust fault and also as a normal fault caused by unroofing.

To the east of the central metasedimentary belt lies the central granulite terrane. These two subprovinces are separated by the Chibougamau-Gatineau Lineament (CGL), a wide mylonite zone. The fact that the CGL is well-defined on aeromagnetic maps suggests that it is a crustal-scale feature. The CGL roughly trends northeast-southwest, where it ranges from about 10 feet (a few meters) to more than four miles (7 km) wide. The CGL may be correlative with the Carthage Colton mylonite zone in the Adirondack Mountains of New York State.

The central granulite terrane (CGT), originally named by Canadian geologist H. R. Wynne-Edwards in 1972, is located in the central and southeastern portion of the Grenville province and is correlative with the Adirondack Highlands. The CGT is often referred to as the core zone of the Grenville orogen, and is the site where the majority of the Grenvillian plutonic activity occurred. This subprovince underwent high-grade metamorphism with paleotemperatures ranging up to 1,470°F (800°C) and paleopressures up to 9.0 kilobars. To explain these high pressures and temperatures, a double thickening of the crust is required. For this reason, geologists have suggested that the Grenville province represents a continent/continent collision zone.

The most abundant rock constituent of the central granulite belt is anorthosite. The larger anorthosite bodies are termed massifs, such as the Morin massif. The anorthosites, along with a whole suite of rocks, known as AMCG (anorthosite, mangerite, charnokite, granite) suite, are thought to have intruded at approximately 1,159–1,126 million years ago according to uranium-lead zircon analysis. These dates are in agreement with uranium-lead zircon ages of the AMCG rocks in the Adirondack Highlands (1,160–1,125 million years old). This places their intrusion as postdepositional with the sediments of the CMB and before the Ottawan Orogeny. The anorthosites were emplaced at shallow levels somewhere between the Grenville supergroup and the underlying basement. A major tectonic event such as continental collision must have occurred to produce the high paleotemperatures and paleopressures recorded in the anorthosites.

For the most part the Grenville front (GF) marks the northwestern limit of Grenville deformation and truncates older provinces and structures. The zone is approximately 1,200 miles (2,000 km) long and is dominated by northwest-directed reverse faulting

that has been recognized since the 1950s. The GF is recognized by faults, shear zones, and metamorphic discontinuities. Faults, foliations, and lineations dip steeply to the southeast. Interpretation of the Grenville front has changed with time. In the 1960s, with the advent of the theory of plate tectonics, the Grenville front was immediately interpreted as a suture. This suggestion was refuted because Archean age rocks of the Superior craton continue south across the Grenville front, implying that the suture should lie to the southeast of the GF. It is possible that the suture is reworked somewhere in the Appalachian orogen. There are still several unresolved questions about the tectonic nature of the Grenville front, considering that the GF marks the limit of Grenvillian deformation: (1) the adjacent foreland to the northwest contains no evidence of supracrustal assemblages associated with the Grenville orogen, (2) the zone lacks Grenville age intrusives that are prevalent to the southeast, and (3) the front divides older rocks from a belt of gneisses that appear to be their reworked equivalents.

TECTONIC EVOLUTION OF THE GRENVILLE PROVINCE

The tectonic framework of the Grenville province is a topic of considerable debate. Many theories and models have been proposed, although there is no one universally accepted model. Nevertheless, researchers do agree upon some aspects of the tectonic framework, in particular, that the Grenville province represents a collisional boundary. Support for this model includes seismic data and the granulite facies metamorphism, both of which suggest that the crust was doubly thickened during peak deformation and metamorphism.

Crustal thickening can occur by a few mechanisms: thrusting, volcanism, plutonism, and homogeneous shortening. One or a combination of these mechanisms must have occurred in the late Proterozoic to produce granulite facies metamorphism in the Grenville province. Two models that account for similar large tectonic crustal thickening are presently occurring on the Earth's surface.

The first model is fashioned after the Andean-type margin. This model suggests that relatively warm, buoyant oceanic crust is subducted under continental crust. Such a model has several implications. The oceanic crust subducted underneath the South American plate is relatively young. Therefore it has not had a sufficient time interval to cool and become dense. The relative low density of the young oceanic crust resists subduction. Consequently the oceanic crust subducts at a relatively shallow angle. A shallow subduction angle creates a compressional stress regime throughout the margin. This has the effect of crustal shortening accommodated by fore-arc frontal thrusts. The subducting oceanic plate also induces plutonism and volcanism that adds to the crustal thickening process.

The second model is based on the Himalayan Orogen. This model thickens the crust by a continent-continent collision. It is somewhat similar to the Andean model, except continental crust replaces the warm, buoyant oceanic crust. The subducting continental crust resists subduction owing to its buoyancy, causing the subducting continental crust to get tucked under the overriding continental crust. The underriding crust never subducts down into the asthenosphere, but rather underplates the overriding continental crust, hence, crustal thickening. The Andean model may predecede the Himalayan model; therefore a combination of the two models may have worked together to produce the Grenville Orogen in the Proterozoic.

A simplistic tectonic model for the Grenville attempts to explain the broad-scale tectonic processes that may be able to account for the large-scale features. An arc-continent collision was followed by a continent-continent collision in the late Proterozoic, probably involving southeastward directed subduction for the continent-continent collision, consistent kinematics in domain boundary shear zones (in the CGB) that preserve an overall northwesterly direction of tectonic transport, consistent with northwestward stacking of crustal slices.

The calc-alkaline trends of the Elzevirian batholith suggest that this is an island arc-type batholith. Thus the Elzevir terrane was probably an island arc before it collided with North America. The Elzevirian age metamorphism resulted from the collision of the CMB and the CGT. Ultimately, the southeastward subduction along the western CMB margin resulted in a continent-continent collision with the CGB.

Plate reconstructions for the Late Proterozoic are currently an area of active investigation. Recent research in geochronology, comparative geology, stratigraphy, and paleomagnetism has provided a wealth of new information that has proven useful in correlating rocks on a global scale. Geologists use these correlations to determine the temporal and spatial plate configurations for the Late Proterozoic. Such plate reconstructions have provided new insight in the study of the Grenville province.

Advances in geochronology have been the greatest contributor in helping to correlate rocks globally. Field mapping in previously unmapped areas and improved techniques in paleomagnetic determination further help to narrow possible plate configurations. With this knowledge geologists take present-day continents and strip away their margins—more precisely, all post–Grenvillian age rocks—and try to piece together the cratons that may once have been conjugate margins.

In 1991 researchers including Canadian Paul Hoffman now at Harvard University, Eldridge Moores from the University of California, and Ian Dalziel from the University of Texas proposed that a supercontinent existed in the Late Proterozoic. This supercontinent, named Rodinia, was formed by the amalgamation of Laurentia (North America and Greenland), Gondwana (Africa, Antarctica, Arabia, Australia, India, and South America), Baltica, and Siberia. The joining of these plates resulted in collisional events along the Laurentian margins. Geologists believe these orogenic events in the Late Proterozoic produced the Grenvillian belts found throughout the world.

Most Late Proterozoic plate reconstructions place the Canadian Grenville province and Amazonian and Congo cratons in close proximity. Therefore Amazonia and Congo were the probable late Proterozoic continental colliders with the eastern margin of Laurentia, resulting in the Ottawan Orogeny. Evidence supporting this correlation includes the similar isotopic ages of 1.4 billion years of the Grenvillian belts found on the Amazonian and Congo cratons, the same as the Laurentian Grenville province.

Plate reconstructions for the Late Proterozoic are not absolute. Unlike the Mesozoic and Cenozoic, the Proterozoic lacks hard evidence, such as hot spot tracks and oceanic magnetic reversal data, to determine plate motions. Furthermore, definitive sutures that would strongly demonstrate a collisional margin, such as ophiolite sequences and blueschist facies terrains, are deformed and few, making it difficult to determine the exact location of the Grenvillian suture. This may be due to the expansive time interval that ensued, later orogenic events, rifting events, and erosion, all of which help to alter and destroy the geologic record.

Most tectonic models for the Grenville province are broadly similar for the late stages of the evolution of this orogenic belt but differ widely in the early stages. The earliest record of arc magmatism in the central metasedimentary belt comes from the Elzevir terrane or composite arc belt, where ca. 1,350–1,225-million-year-old magmatism is interpreted to represent one or more arc/back arc basin complexes. The Adirondack Lowlands terrane may have been continuous with the Frontenac terrane, which together formed the trailing margin of the Elzevirian arc. Isotopic ages for the Frontenac terrane fall in the range of 1,480–1,380 million years, and between 1,450–1,300 million years for the entire central metasedimentary belt, suggesting that the Elzevirian arc is largely a juvenile terrane. The Elzevirian arc is thought to have collided offshore with other components of the composite arc belt by 1,220 million years ago because of widespread northwestward-directed deformation and tectonic repetition in the central metasedimentary belt at that time. Following amalgamation some interpret the subduction to have stepped southeastward to lie outboard of the composite arc, and dipped westward beneath a newly developed active margin. This generated a suite of ca. 1,207-million-year-old calc-alkaline plutons (Antwerp-Rossie suite) and 1,214 ± 21 million-year-old dacitic volcaniclastics, metapelites, and diorite-tonalitic plutons. Other models suggest that the Adirondack Highlands and Frontenac/Adirondack Lowlands terranes remained separated until 1,170–1,150 million years ago, when the Frontenac and Sharbot Lakes domains were metamorphosed and intruded by plutons.

Many geologists regard the Adirondack Highlands–Green Mountains block to be a single arc complex, based on abundant ca. 1,350–1,250-million-year-old calc-alkaline tonalitic to granodioritic plutons in both areas. The Adirondack Highlands–Green Mountains block may have been continuous with the Elzevirian arc as well, forming one large composite arc complex. Neodymium model ages for the Adirondack Mountains–Green Mountain block fall in the range of 1,450–1,350 Ma, suggesting that this arc complex was juvenile, without significant reworking of older material.

Collision of the Adirondack Highlands–Green Mountain block with Laurentia occurred between the intrusion of the ca. 1,207-million-year-old Antwerp-Rossie arc magmas and formation of the 1,172-million-year-old Rockport-Hyde-School-Wellesley-Wells intrusive suite. This inference is based on the observation that peak metamorphic conditions preceded intrusion of the 1,180–1,150-million-year-old intrusive suite in the Frontenac terrane. Also metamorphic zircon and monazite (presumably dating the collision) from the central metasedimentary belt fall in the range of 1,190–1,180 million years. The Carthage-Colton mylonite zone may represent a cryptic suture marking the broad boundary along which the Adirondack Highlands–Green Mountain block is juxtaposed with Laurentia from a collision that emplaced the Lowlands over the Highlands. Early localized delamination beneath the collision zone may have elevated crustal temperatures and generated crustal melts of the ca. 1,172-million-year-old Rockport and Hyde School granites, and the Wells leucocratic gneiss also belongs to this group. But the present geometry with relatively low-grade rocks of the Lowlands juxtaposed with high-grade rocks of the Highlands suggests that the present structure is an extensional fault that may have reactivated an older structure.

The ca. 1,172-million-year-old collisional granites (Rockport, Hyde School gneiss, Wellesley, Wells) are largely syntectonic, and emplacement of these magmas may have slightly preceded formation of large-scale recumbent nappes including F_1 folds. These large nappes may be responsible for complex map patterns and repetition of units in the CMB and CGT. High-temperature deformation of monzonites in the Robertson Lake shear zone took place at ca. 1,162 million years ago and demonstrated that deformation continued for at least 10 million years after intrusion of the 1,172-million-year-old magmatic suite. Deformation had apparently terminated by 1,160 million years ago, however, as shown by the 1,161–1,157 million-year-old Kingston dikes and Frontenac suite plutons, which cross-cut Elzevirian fabrics and cut the Robertson Lake shear zone.

The widespread monzonitic, syenitic, and granitic plutons (AMCG suite) that intruded the Frontenac terrane in the period from 1,180 to 1,150 million years ago swept eastward across the orogen forming the AMCG suite in the Highlands at 1,155–1,125 million years ago. Jim McLelland, Tim Kusky, and others have suggested that separation of the subcontinental lithospheric mantle that started around 1,180–1,160 million years ago may have proceeded to large-scale delamination beneath the orogen. This would have exposed the base of the crust to hot asthenosphere, causing melting and triggering the formation of the AMCG suite. The 1,165-million-year-old metagabbro units are related to this widespread melting and intrusive event in the Adirondacks.

The culminating Ottawan Orogeny from ca. 1,100–1,020 million years ago in the Adirondacks and Grenville orogen is widely thought to result from the collision of Laurentia with another major craton, probably Amazonia. This collision is one of many associated with the global amalgamation of continents to form the supercontinent Rodinia. The event is associated with large-scale thrusting, high-grade metamorphism, recumbent folding, and intrusion of a second generation of crustal melts associated with orogenic collapse. The putative suture (Carthage-Colton mylonite zone) between the accreted Highlands–Green Mountain block and Laurentia was reactivated as an extensional shear zone in this event, partly accommodating the orogenic collapse and exhumation of deep-seated rocks in the Adirondack Highlands. The relative timing of igneous events and folding in the Adirondacks has shown that the F_2 and F_3 folding events in the southern part of the Highlands postdated 1,165 and predated 1,052 million years ago, demonstrating that these folds, and later generations of structures, are related to the Ottawan Orogeny. The Ottawan Orogeny in this area is therefore marked by the formation of early recumbent fold nappes overprinted by upright folds.

The regional chronology and overprinting history of folding related to the Ottawan Orogeny are generally poorly known. In 1939 Buddington noted isoclinal folds dated ca. 1,149 million years old in the Hermon granite gneiss in the Adirondack Highlands, and very large granulite facies fold nappes have been emplaced throughout the Adirondack region. These folds refold an older isoclinal fold generation, thus are F_2 folds and are related to this regional event. The youngest rocks that show widespread development of fabrics attributed to the Ottawan orogeny are the ca. 1,100–1,090-million-year-old Hawkeye suite, that show "peak" conditions of about 1,470°F at 12–15 miles depth (800°C at 20–25 km depth). These conditions existed from about 1,050 through approximately 1,013 million years ago. Older thrust faults along the CMB boundary zone were reactivated at about 1,080–1,050 million years ago. The latter parts of the Ottawan Orogeny (1,045–1,020 million years ago) are marked by extensional collapse of the orogen, with low-angle normal faults accommodating much of this deformation. Crustal melts associated with orogenic collapse are widespread.

GRENVILLE BELTS AND THE RODINIA SUPERCONTINENT

The Proterozoic saw the development of many continental-scale orogenic belts, many of which have been recently recognized to be parts of global-scale systems that reflect the formation, breakup, and reassembly of several supercontinents. Paleoproterozoic orogens include the Wopmay in northern Canada, interpreted to be a continental margin arc that rifted from North America, then collided soon afterward, closing the young back arc basin. There are many 1.9–1.6-Ga orogens in many parts of the world, including the Cheyenne belt in the western United States, interpreted as a suture that marks the accretion of the Proterozoic arc terrains of the southwestern United States with the Archean Wyoming Province.

The supercontinent Rodinia formed in Mesoproterozoic times by the amalgamation of Laurentia, Siberia, Baltica, Australia, India, Antarctica, and the Congo, Kalahari, West Africa, and Amazonia cratons between 1.1 and 1.0 Ga ago. The joining of these cratons resulted in the terminal collisional events at convergent margins on many of these cratons, including the ca. 1.1–1.0-Ga Ottawan and Rigolet Orogenies in the Grenville Province of Laurentia's southern margin. Globally, these events have become known as the Grenville orogenic period, named after the Grenville orogen of eastern North America. Grenville-age orogens are preserved along eastern North America, as the Rodinia-Sunsas belt in Amazonia,

the Irumide and Kibaran belts of the Congo craton, the Namaqua-Natal and Lurian belts of the Kalahari craton, the Eastern Ghats of India, and the Albany-Fraser belt of Australia. Many of these belts now preserve deep-crustal metamorphic rocks (granulites) tectonically buried to 20–25 miles (30–40-km) in depth, then the overlying crust was removed by erosion, forcing the deeply buried rocks to the surface. Since 20–25 miles (28–30 km) of crust still underlies these regions, they may have had double crustal thickness during the peak of metamorphism. Such thick crust is today produced in regions of continent-continent collision, and locally in Andean arc settings. Since the Grenville-aged orogens are so linear and widely distributed, they are generally interpreted to mark the sites of continent-continent collisions where the various cratonic components of Rodinia collided between 1.1 and 1.0 Ga.

See also CONVERGENT PLATE MARGIN PROCESSES; PRECAMBRIAN; STRUCTURAL GEOLOGY; SUPERCONTINENT CYCLES.

FURTHER READING

Dalziel, Ian W. D. "Neoproterozoic-Paleozoic Geography and Tectonics: Review, Hypothesis, Environmental Speculation." *Geological Society of America* 109 (1997): 16–42.

———. "Pacific Margins of Laurentia and East Antarctica-Australia as a Conjugate Rift Pair: Evidence and Implications for an Eocambrian Supercontinent." *Geology* 19 (1991): 598–601.

Davidson, Anthony. "An Overview of Grenville Province Geology, Canadian Shield." In *"Geology of North America, vol. C-1, Geology of the Precambrian Superior and Grenville provinces and Precambrian Fossils in North America,"* edited by S. B. Lucas and M. R. St-Onge, 205–270. Denver, Colo.: Geological Society of America, 1998.

———. "A Review of the Grenville Orogen in its North American Type Area." *Journal of Australian Geology and Geophysics* 16 (1995): 3–24.

Hoffman, Paul F. "Did the Breakout of Laurentia Turn Gondwanaland Inside-Out?" *Science* 252 (1991): 1,409–1,411.

Kusky, Timothy M., and Dave P. Loring. "Structural and U/Pb Chronology of Superimposed Folds, Adirondack Mountains: Implications for the Tectonic Evolution of the Grenville province." *Journal of Geodynamics* 32 (2001): 395–418.

McLelland, Jim M., J. Stephen Daly, and Jonathan M. McLelland. "The Grenville Orogenic Cycle (ca. 1350–1000 Ma): An Adirondack Perspective." In *"Tectonic Setting and Terrane Accretion in Precambrian Orogens,"* edited by Timothy M. Kusky, Ben A. van der Pluijm, Kent Condie, and Peter Coney. *Tectonophysics* 265 (1996): 1–28.

Moores, Eldredge M. "Southwest United States–East Antarctic (SWEAT) Connection: A Hypothesis." *Geology* 19 (1991): 425–428.

Rogers, J. J. W., and M. Santosh. *Continents and Supercontinents.* Oxford: Oxford University Press, 2004.

groundwater Groundwater encompasses all of the water contained within spaces in bedrock, soil, and regolith. The volume of groundwater is 35 times the volume of freshwater in lakes and streams, but overall this water accounts for less than 1 percent of the planet's water. Much of the world's population gets its freshwater from the groundwater system, pumping the water from beneath the surface or pulling buckets up from wells dug into the ground. Any body of rock or unconsolidated sediment that can hold and transmit water is known as an aquifer. Units that restrict the flow of water are known as aquitards.

Groundwater comes from rainfall and surface flow, where it seeps into the ground and slowly makes its way downhill toward the sea. Water exists everywhere beneath the ground surface, and most of this occurs within 2,500 feet (750 m) of the surface. The volume of groundwater is estimated to be equivalent to a layer 180 feet (55 m) thick spread evenly over the Earth's land surface. The distribution of water in the ground can be divided into unsaturated and saturated zones. The top of the water table is defined as the upper surface of the saturated zone; below this surface, all openings are filled with water.

Increasingly, groundwater is being used for more functions than simply drinking or watering plants and animals. Water has a high specific heat capacity, meaning that it takes a long time and a lot of heat energy to heat and cool the water. Additionally the insulating effects of the surrounding soil and bedrock means that the groundwater tends to remain at a similar temperature year round, in the low 50°'s Fahrenheit (10–12°C). Based on these properties, engineers are beginning to use water to help heat and cool buildings. During hot weather water is pumped through radiators, cooling the building. In cool weather water is heated, and since it stores heat energy more efficiently than air, energy savings result.

Freshwater is one of the most important resources in the world. Wars are fought over freshwater, and water rights are political issues in places where it is scarce like the American West and the Middle East. Since we live in a world with a finite amount of freshwater and the global population is growing rapidly, it is likely that freshwater will become an increasingly important topic for generations to come. Much of the groundwater in the world is at increasing risk of being contaminated by industrial and human pollutants. Efforts must be undertaken to protect adequately this scarce resource.

America and other nations have realized that groundwater is a vital resource for national survival and are only recently beginning to appreciate that much of the world's groundwater resources have become contaminated by natural and human-aided processes. Approximately 40 percent of drinking water in the United States comes from groundwater reservoirs. About 80 billion gallons of groundwater are pumped out of these reservoirs every day in the United States.

MOVEMENT OF GROUNDWATER

Most of the water under the ground does not just sit there—it is constantly in motion, although rates are typically only an inch or two (2–5 cm) per day. The rates of movement are controlled by the amount of open space in the bedrock or regolith, and how the spaces are connected. The groundwater system also includes water beneath the ground that is immobile, such as water locked in soil moisture, permafrost, plus geothermal and oil-formation water.

Porosity is the percentage of total volume of a body that consists of open spaces. Sand and gravel typically have about 20 percent open spaces, while clay has about 50 percent. The sizes and shapes of grains determine porosity, which is also influenced by how much they are compacted, cemented together, or deformed.

In contrast, permeability is a body's capacity to transmit fluids or to allow the fluids to move through its open pore spaces. Permeability is not directly related to porosity. For instance, all the pore spaces in a body could be isolated from each other (high porosity), and thus the water may be trapped and unable to move through the body (low permeability). Molecular attraction, the force that makes thin films of water stick to objects instead of being forced to the ground by gravity, also affects permeability. If the pore spaces in a material are small, as in a clay layer, then the force of molecular attraction is strong enough to stop the water from flowing through the body. When the pores are large, the water in the center of the pores is free to move.

After a rainfall much of the water stays near the surface, because clay in the near-surface horizons of the soil retains much water due to molecular attraction. This forms a layer of soil moisture in many regions able to sustain seasonal plant growth.

Some of this near-surface water evaporates and is used by plants. Other water runs directly off into streams. The remaining water seeps into the saturated zone, or into the water table. Once in the saturated zone it moves slowly by percolation, from high areas to low areas, under the influence of gravity. These lowest areas are usually lakes or streams. Many streams form where the water table intersects the surface of the land.

Once in the water table the paths that individual particles follow vary. The transit time from surface to stream may vary from days to thousands of years along a single hillside. Water can flow upward because of high pressure at depth and low pressure in streams.

THE GROUNDWATER SYSTEM

Groundwater is best thought of as a system of many different parts, some of which act as conduits and reservoirs, and others that serve as offramps and onramps into the groundwater system.

Recharge areas are where water enters the groundwater system, and discharge areas are where water leaves the groundwater system. In humid climates recharge areas encompass nearly the land's entire surface (except for streams and floodplains), whereas in desert climates recharge areas consist mostly of the mountains and alluvial fans. Discharge areas consist mainly of streams and lakes.

The level of the water table changes with different amounts of precipitation. In humid regions it reflects the topographic variation, whereas in dry

times or locations it tends to flatten out to the level of the streams and lakes. Water flows faster when the slope is greatest, so groundwater flows faster during wet times. The fastest rate of groundwater flow observed in the United States is 800 feet per year (250 m/yr).

Aquifers include any body of permeable rock or regolith saturated with water through which groundwater moves. Gravel and sandstone make good aquifers, as do fractured rock bodies. Clay is so impermeable that it makes bad aquifers, and typically forms aquicludes that stop the movement of water.

Springs are places where groundwater flows out at the ground surface. They can form where the ground surface intersects the water table or at a vertical or horizontal change in permeability, such as where water in gravels on a hillslope overlays a clay unit and the water flows out on the hill along the gravel/clay boundary.

Water wells fill with water simply because they intersect the water table. The rocks below the surface are not always homogeneous, however, which can result in a complex type of water table known as a perched water table. Perched water tables result from impermeable bodies in the subsurface that create bodies of water at elevations higher than the main water table.

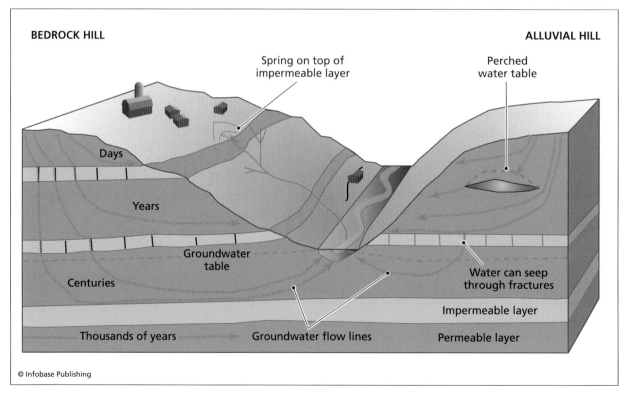

Schematic diagram of the groundwater system. Water enters the system on hillslopes and emanates lower on hills as springs and in effluent streams.

WATER WELLS AND USE OF GROUNDWATER

Most wells fill with water simply because they intersect the water table. But the rocks below the surface are not always homogeneous, which can result in a complex type of water table know as a perched water table. These result from discontinuous bodies in the subsurface, which create bodies of water at elevations higher than the main water table.

Aquifers

Aquifers are any body of permeable rock or regolith saturated with water through which groundwater moves. The term *aquifer* is usually reserved for rock or soil bodies that contain economical quantities of water that are extractable by existing methods. The quality of an aquifer depends on two main qualities, porosity and permeability. Porosity is a measure of the total amount of open void space in the material. *Permeability* refers to the ease at which a fluid can move through the open-pore spaces, and depends in part on the size, shape, and how connected individual pore spaces are in the material. Gravels and sandstone make good aquifers, as do fractured rock bodies. Clay is so impermeable that it makes bad aquifers and typically forms aquicludes, which stop the movement of water.

There are several main types of aquifers. In uniform, permeable rock and soil masses aquifers form as a uniform layer below the water table. In these simple situations wells fill with water simply because they intersect the water table. But the rocks below the surface are not always homogeneous and uniform, which can result in a complex type of water table know as a perched water table. These result from discontinuous impermeable rock or soil bodies in the subsurface, which create domed pockets of water at elevations higher than the main water table, resting on top of the impermeable layer.

When the upper boundary of the groundwater in an aquifer is the water table, the aquifer is said to be unconfined. In many regions, a permeable layer, typically a sandstone, is confined between two impermeable beds, creating a confined aquifer. In these systems water enters the system only in a small recharge area, and if this is in the mountains, then the aquifer may be under considerable pressure. This is known as an artesian system. Water that escapes the system from the fracture or well reflects the pressure difference between the elevation of the source area and the discharge area (hydraulic gradient), and rises above the aquifer as an artesian spring or artesian well. Some of these wells have made fountains that have spewed water 200 feet (60 meters) high. One example of an artesian system is that in Florida, where water enters in the recharge area, and is released near Miami about 19,000 years later.

Big Spring, Missouri *(Jose Azel/Aurora/Getty Images)*

Fracture Zone Aquifers

Hydrologists, geologists, and municipalities in need of water have increasingly appreciated that significant quantities of fresh water is stored in fractures within otherwise impermeable crystalline rocks beneath the ground. Many buried granite and other bedrock bodies are cut by many fractures, faults, and other cracks, some of which may have open spaces along them. Fractures at various scales represent zones of increased porosity and permeability. They may form networks, and therefore, are able to store and carry vast amounts of water. These groundwater systems, called fracture zone aquifers, are similar in some ways to karst systems.

The concept of fracture zone aquifers explains the behavior of groundwater in large fault-controlled watersheds. Fault zones in this case serve as collectors and transmitters of water from one or more recharge zones with surface and subsurface flow strongly controlled by regional fault patterns.

Both the yield and the quality of water in these zones are usually higher than average wells in any type of rock. High-grade water for such a region would be 250 gallons (950 liters) per minute or greater. In addition the total dissolved solids measured in the water from such high-yielding wells will be lower than the average for the region.

The quality and amount of water obtainable from fracture zone aquifers are influenced by the pattern of fractures and their related secondary porosity over an entire watershed area. It is important to understand how the fracture pattern varies across a basin to be able to determine the unique effects of secondary porosity on the processes of groundwater flow, infiltration, transmissivity, and storage, and ultimately find and use the water in the fractures.

Variations in precipitation over the catchment area can determine how a fracture zone aquifer system is recharged. One example is orographic effects where the precipitation over the mountains is substantially greater than at lower elevations. The rainfall is collected over a large catchment area, which contains zones with high permeability because of intense bedrock fracturing associated with major fault zones. The multitude of fractures within these highly permeable zones "funnel" the water into other fracture zones down gradient. These funnels may be in a network hundreds of square miles in area.

The fault and fracture zones serve as conduits for groundwater and often act as channelways for surface flow. Intersections form rectilinear drainage patterns sometimes exposed on the surface but are also represented below the surface and converge down gradient. In some regions these rectilinear patterns are not always visible on the surface due to vegetation and sediment cover. The convergence of these groundwater conduits increases the amount of water available as recharge. The increased permeability, water volume, and ratio of water to minerals within these fault/fracture zones help to maintain the quality of water supply. These channels occur in fractured, nonporous media (crystalline rocks) as well as in fractured, porous media (sandstone, limestone).

At some point in the groundwater course, after convergence, the gradient decreases. The sediment cover over the major fracture zone becomes thicker and acts as a water storage unit with primary porosity. The major fracture zone acts as both a transmitter of water along conduits and a water storage basin along connected zones with secondary (and/or primary) porosity. Groundwater within this layer or lens often flows at accelerated rates. The result can be a pressurization of groundwater both in the fracture zone and in the surrounding material. Rapid flow in the conduit may be replenished almost instantaneously from precipitation. The surrounding materials are replenished more slowly, but also release the water more slowly and serve as a storage unit to replenish the conduit between precipitation events.

Once the zones are saturated, any extra water that flows into them will overflow, if an exit is available. In a large area watershed, it is likely that this water flows along subsurface channelways under pressure until some form of exit is found in the confining environment. Substantial amounts of groundwater may flow along the main fault zone controlling the watershed and may vent at submarine extensions of the fault zone forming coastal or offshore freshwater springs.

Fracture zone aquifers are most common in areas underlain by crystalline rocks and where these rocks have undergone a multiple deformational history that includes several faulting events. It is especially applicable in areas where recharge is possible from seasonal and/or sporadic rainfall on mountainous regions adjacent to flat desert areas.

Fracture zone aquifers are distinguished from horizontal alluvial or sedimentary formation aquifers in that (1) they drain extensive areas and many extend for tens of miles (several tens of km); (2) they constitute conduits to mountainous regions where the recharge potential from rainfall is high; (3) some may connect several horizontal aquifers, and thereby increase the volume of accumulated water; (4) because the source of the water is at higher elevations, the artesian pressure at the groundwater level may be high; and (5) they are usually missed by conventional drilling because the water is often at the depth of hundreds of meters.

The characteristics of fracture zone aquifers make them an excellent source of groundwater in arid and semiarid environments. Fracture zone aquifers are being increasingly used in arid regions. Groundwater resources in arid and semiarid lands are scarce and must be properly used and thoughtfully managed. Most of these resources are "fossil," having accumulated under wet climates during the geological past. The present rates of recharge from the occasional rainfall are not enough to replenish the aquifers. Therefore the resources must be used sparingly without exceeding the optimum pumping rates for each water well field.

GROUNDWATER DISSOLUTION

Groundwater also reacts chemically with the surrounding rocks; it may deposit minerals and cement together grains, causing a reduction in porosity and permeability, or form features like stalagtites and stalagmites in caves. In other cases, particularly when acidic water moves through limestone, it can dissolve the rock, forming caves and underground tunnels. Sinkholes form where these dissolution cavities intersect the surface of the Earth.

Groundwater dissolution leads to the development of a distinctive class of landforms called karst terranes where caves can collapse, leaving sinkholes and valleys on the surfaces, and eventually evolve into spectacular towers and pinnacles of nondissolved rock surrounded by the former cave passageways. South China is famous for highly evolved karst terranes, whereas parts of the U.S. Midwest are known for well-developed underground cave systems.

GROUNDWATER CONTAMINATION

Natural groundwater is typically rich in dissolved elements and compounds derived from the soil, regolith, and bedrock through which the water has migrated. Some of these dissolved elements and compounds are poisonous, whereas others are tolerable in small concentrations but harmful in high concentrations. Human and industrial waste contamination of the groundwater is increasing, and the overuse of groundwater resources has caused groundwater levels to drop and has led to other problems, especially along coastlines. Seawater may move in to replace depleted freshwater, and the ground surface may subside when the water is removed from the pore spaces in aquifers.

The U.S. Public Health Service has established limits on the concentrations of dissolved substances (called total dissolved solids, or t.d.s.) in natural waters that are used for domestic and other purposes. The table of "Drinking Water Standards for the United States" lists these limits for the United States. Many other countries, particularly those with chronic water shortages such as many in the Middle East, have much more lenient standards. Sweet water is preferred for domestic use and has fewer than 500 milligrams (mg) of total dissolved solids per liter (L) of water. Fresh and slightly saline water, with t.d.s. of 1,000–3,000 mg/L, is suitable for use by livestock and irrigation. Water with higher concentrations of t.d.s. is unfit for humans or livestock. Irrigation of fields using waters with high concentrations of t.d.s.

DRINKING WATER STANDARDS FOR THE UNITED STATES

Water Classification	Total Dissolved Solids (T.D.S.)
Sweet	< 500 mg/L
Fresh	500–1,000 mg/L
Slightly saline	1,000–3,000 mg/L
Moderately saline	3,000–10,000 mg/L
Very Saline	10,000–35,000 mg/L
Brine	> 35,000 mg/L

is also not recommended, as the water will evaporate but leave the dissolved salts and minerals behind, degrading and eventually destroying the productivity of the land.

Either a high amount of total dissolved solids or the introduction of a specific toxic element can reduce the quality of groundwater or contaminate it. Most of the total dissolved solids in groundwater are salts derived from dissolution of the local bedrock or soils derived from the bedrock. Salts can also seep into groundwater supplies from the sea along coastlines, particularly if the water is being pumped out for use. In these cases seawater often moves in to replace the depleted freshwater. This process is known as seawater intrusion, or seawater incursion.

Dissolved salts in groundwater commonly include bicarbonate (HCO_3^-) and sulfate (SO_4^{2-}) ions, often associated with other ions. Dissolved calcium (Ca^{2+}) and magnesium (Mg^+) ions can cause the water to become "hard." Hard water is defined as containing more than 120 parts per million dissolved calcium and magnesium. The dissolved ions in hard water make it difficult to lather soap, and they form a crusty mineralization buildup on faucets and pipes. Adding sodium (Na^+) in a water softener can soften hard water, but people with heart problems or those who are on a low-salt diet should not do this. Hard water is common in areas where the groundwater has moved through limestone or dolostone rocks, which contain high concentrations of Ca^{2+} and Mg^{2+}–rich rocks that groundwater easily dissolves.

Groundwater may have many other contaminants, some natural and others the result of human activity. Human pollutants including animal and human waste, pesticides, industrial solvents, road salts, petroleum products, and other chemicals are a serious problem in many areas. Some of the biggest and most dangerous sources of groundwater contamination include chemical and gasoline storage tanks, septic systems, landfills, hazardous waste sites, military bases, and the general widespread use of road salt and chemicals such as fertilizers or pesticides.

The Environmental Protection Agency has led the cleanup from spills from leaking chemical storage tanks in the United States. There are estimated to be more than 10 million buried chemical storage tanks in the United States, containing chemicals such as gasoline, oil, and hazardous chemicals. These tanks can leak over time, and many of the older ones have needed to be replaced in the past two decades, bringing in a new generation of tanks that should last longer and corrode less.

Home and commercial septic systems pose serious threats to some groundwater systems. Most are designed to work effectively and harmlessly, but some were not installed properly or were poorly designed.

In many cases groundwater supplies have been contaminated by chemicals and other contaminants that were poured down drains, entering the septic system and then the groundwater system.

There are more than 20,000 known and abandoned hazardous waste sites in the United States. Some of these contain many barrels of chemicals and hazardous materials that can and do leak, contaminating the water supply. Landfills may also contain many hazardous chemicals—when landfills are designed they are supposed to incorporate a protective impermeable bottom layer to prevent chemicals from entering the groundwater system. But some chemicals that are erroneously placed in the landfill sometimes burn holes in the basal layer, making their way (with a myriad of other chemicals) into the groundwater system.

In parts of the country that freeze, road salts are commonly used to reduce the amount of ice on the roads. These salts dissolve in rainwater and can eventually make their way down into the aquifers as well, turning an aquifer salty. Together with chemicals from lawn and farm field fertilization and application of pesticides, the amount of these chemicals starts to become significant for the safety of the water quality below ground.

Groundwater contamination, whether natural or human-induced, is a serious problem because of the importance of the limited water supply. Pollutants in the groundwater system do not simply wash away with the next rain, as many dissolved toxins in the surface water system do. Groundwater pollutants typically have a residence time, or average length of time that it remains in the system, of hundreds or thousands of years. Many groundwater systems are capable of cleaning themselves from natural biological contaminants using bacteria, but other chemical contaminants have longer residence times.

Arsenic in Groundwater

In parts of the world many people have become sick from arsenic dissolved in the groundwater. Arsenic poisoning leads to a variety of horrific diseases, including hyperpigmentation (abundance of red freckles), hyperkeratosis (scaly lesions on the skin), cancerous lesions on the skin, and squamous cell carcinoma. Arsenic may be introduced into the food chain and body in several ways. In Guizhou Province, China, villagers dry their chili peppers indoors over coal fires. Unfortunately, the coal is rich in arsenic (containing up to 35,000 parts per million arsenic), and much of this arsenic is transferred to the chili peppers during the drying process. Thousands of the local villagers now suffer arsenic poisoning, with cancers and other forms of the disease ruining families and entire villages.

Most naturally occurring arsenic is introduced into the food chain through drinking contaminated groundwater. Arsenic in groundwater is commonly formed by the dissolution of minerals from weathered rocks and soils. In Bangladesh and West Bengal, India, 25–75 million people are at risk for arsenosis, because of high concentrations of natural arsenic on groundwater.

Since 1975 the maximum allowable level of arsenic in drinking water in the United States has been 50 parts per billion. The EPA has been considering adopting new standards on the allowable levels of arsenic in drinking water. Scientists from the National Academy recommend that the allowable levels of arsenic be lowered to 10 parts per billion, but this level was overruled by the Bush administration. The issue is cost: the EPA estimates that it would cost businesses and taxpayers $181 million per year to bring arsenic levels to the proposed 10 parts per billion level, although some private foundations suggest that this estimate is too low by a factor of three. They estimate that the cost would be passed on to the consumer, and residential water bills would quadruple. The EPA estimates that the health benefits from such a lowering of arsenic levels would prevent between 7 and 33 deaths from arsenic-related bladder and lung cancer per year. These issues reflect a delicate and difficult choice for the government. The EPA tries to "maximize health reduction benefits at a cost that is justified by the benefits." How much should be spent to save 7–33 lives per year? Would the money be better spent elsewhere?

Arsenic is not concentrated evenly in the groundwater system of the United States, or anywhere else in the world. The U.S. Geological Survey issued a series of maps in 2000 showing the concentration of arsenic in tens of thousands of groundwater wells in the United States. Arsenic is concentrated mostly in the Southwest, with a few peaks elsewhere such as southern Texas, parts of Montana (due to mining operations), and parts of the upper plains states. Perhaps a remediation plan that attacks the highest concentrations of arsenic would be the most cost-effective and have the highest health benefit.

Contamination by Sewage

A major problem in groundwater contamination is sewage. If chloroform bacteria get into the groundwater, the aquifer is ruined, and care must be taken and samples analyzed before water is used for drinking. In many cases sand filtering can remove bacteria, and aquifers contaminated by chloroform bacteria and other human waste can be cleaned more easily than aquifers contaminated by many other elemental and mineral toxins.

Although serious, detailed discussion of groundwater contamination by human waste is beyond the scope of this encyclopedia, the reader is referred to the sources listed at the end of the chapter for more detailed accounts.

Seawater Intrusion in Coastal Aquifers

Encroachment of seawater into drinking and irrigation wells is an increasing problem for many coastal communities around the world. Porous soils and rocks beneath the groundwater table in terrestrial environments are generally saturated with fresh water, whereas porous sediment and rock beneath the oceans is saturated with salt water. In coastal environments there must be a boundary between the fresh groundwater and the salty groundwater. In some cases this is a vertical boundary, whereas in other cases the boundary is inclined with the denser salt water lying beneath the lighter fresh water. In areas where there is complex or layered stratigraphy, the boundary may be complex, consisting of many lenses.

In normal equilibrium situations the boundary between the fresh and salty water remains rather stationary. In times of drought the boundary may move landward or upward, and in times of excessive precipitation the boundary may move seaward and downward. As sea levels rise the boundary moves inland and wells that formerly tapped fresh water begin to tap salt water. This is called sea water intrusion or encroachment.

Many coastal communities have been highly developed, with many residential neighborhoods, cities, and agricultural users obtaining their water from groundwater wells. When these wells pump more water out of coastal aquifers than is replenished by new rainfall and other inputs to the aquifer, the fresh water lens resting over the salt water lens is depleted. This also causes the salt water to move in to the empty pore spaces to take the place of the fresh water. Eventually as pumping continues the fresh water lens becomes so depleted that the wells begin to draw salt water out of the aquifer, and the well becomes effectively useless. This is another way that salt water intrusion or encroachment can poison groundwater wells. In cases of severe drought the process may be natural, but in most cases seawater intrusion in caused by over-pumping of coastal aquifers, aided by drought conditions.

Many places in the United States have suffered from seawater intrusion. For instance, many East Coast communities have lost use of their wells and had to convert to water piped in from distant reservoirs for domestic use. In a more complicated scenario western Long Island of New York experienced severe seawater intrusion into its coastal aquifers because of intense overpumping of its aquifers in the late 1800s and early 1900s. Used water that was once returned to the aquifer by septic systems began to be dumped directly into the sea when sewers were installed in the 1950s, with the result that the water table dropped more than 20 feet over a period of 20 years. This drop was accompanied by additional seawater intrusion. The water table began to recover in the 1970s when much of the area converted to using water pumped in from reservoirs in the Catskill Mountains to the north of New York City.

SUMMARY

Most of the world's freshwater is locked in glaciers or ice caps, and about 25 percent of the freshwater is stored in the groundwater system. Water in the groundwater system is constantly but slowly moving, being recharged by rain and snow that infiltrates the system, and discharging in streams, lakes, springs, and extracted from wells. Water that moves through a porous network forms aquifers, and underground layers that restrict flow are known as aquicludes. Fracture zone aquifers comprise generally nonpermeable, nonporous crystalline rock units, but faults and fractures that cut the rock create new or secondary porosity along the fractures. If exposed to the surface, these fractures may become filled with water and serve as excellent sources of water in dry regions.

The groundwater system is threatened by pollutants that range from naturally dissolved but deadly elements such as arsenic, to sewerage, to industrial wastes and petroleum products that have leaked from underground storage containers, or were carelessly dumped. Some chemical elements have a short residence time in the groundwater system and are effectively cleaned before long, but other elements may last years or thousands of years before the groundwater is drinkable again.

See also HYDROSPHERE; METEORIC; SOILS.

FURTHER READING

Alley, William M., Thomas E. Reilly, and O. L. Franke. *Sustainability of Ground-Water Resources.* Reston, Va.: United States Geological Survey Circular 1186, 1999.

Ford, D., and P. Williams. *Karst Geomorphology and Hydrology.* London: Unwin-Hyman, 1989.

Keller, Edward A. *Environmental Geology.* 8th ed. Englewood Cliffs, N.J.: Prentice Hall, 2000.

Kusky, T. M. *Floods: Hazards of Surface and Groundwater Systems.* New York: Facts On File, 2008.

Skinner, Brian J., and Stephen C. Porter. *The Dynamic Earth, an Introduction to Physical Geology.* 5th ed. New York: John Wiley & Sons, 2004.

United States Geological Survey. "Water Resources." Available online. URL: http://water.usgs.gov/. Accessed December 10, 2007.

Halley, Edmond (1656–1742) British *Astronomer, Geophysicist, Mathematician, Meteorologist, Physicist*   Edmond Halley was born on November 8, 1656, in Shoreditch, England, and is best known for the comet bearing his name, Halley's Comet. Halley married Mary Tooke in 1682, and the couple had three children. He died on January 14, 1742.

Edmond Halley studied mathematics at an early age while at St. Paul's School in London, before moving to the Queen's College at Oxford in 1673. During his undergraduate years at Oxford, Halley published several scientific papers on sunspots and the solar system.

After graduating from Oxford Halley visited the South Atlantic Ocean island of St. Helena to examine the southern stars, then returned to England in 1678, publishing his observations of the southern sky as his *Catalogus Stellarum Australium* in 1679. This work led to his being awarded a master of arts degree from Oxford and his election as a fellow of the Royal Society.

Some eight years after his voyage to the South Atlantic Halley published the second volume from his field observations, this one on the southern trade winds and monsoons, and his deduction that atmospheric motions were ultimately driven by solar heating of the atmosphere. Halley became interested in gravity and studied 16th century Austrian mathematician Kepler's laws of planetary motion and met with Sir Isaac Newton on the matter in 1684. He found that Newton had derived proof of Kepler's laws. Halley convinced Newton to publish his works and even paid the cost of printing.

In 1691 Halley applied for the position of Savilian Professor of Astronomy at Oxford, but since his views on religion were atheistic, the archbishop of Canter-

bury opposed his appointment and gave it instead to David Gregory, who was supported by Newton.

After working to develop actuarial models for the British government, Halley returned to science and was given command of a vessel to sail to the South Atlantic to study variations in the magnetic compass. After an initial voyage was terminated because of insubordination by the crew in 1698, he sailed from September 1699 to September 1700, then in 1701 published his observations of the magnetic field as his *General Chart of the Variation of the Compass,* the first chart ever to show magnetic isogonic lines, contour lines that show places of constant magnetic declination.

After his detractors at Oxford died, Halley was appointed Savilian Professor of Geometry in 1703 and was given an honorary doctor of law in 1710. While he was in the Savilian Professorship, Halley pursued historical astronomy and published his analysis of past accounts of comet sightings as *Synopsis Astronomia Cometicae* in 1705. He noted comet sightings from the years 1456, 1531, 1607; 1682, and noting the 75–76 year repeat cycle of the comets, he suggested that these sightings were of the same comet, and that it would return in 1758. When the comet did return, it became known as Halley's comet.

See also ASTRONOMY; COMET; SUN.

FURTHER READING
Cook, Alan H. *Edmond Halley: Charting the Heavens and the Seas.* Oxford: Clarendon Press, 1998.

Hess, Harry (1906–1969) American *Geologist*
Harry Hammond Hess is best known for formulating a theory on the origin and evolution of ocean

basins. Drawing on observations from which Alfred Wegener proposed his theory of continental drift in 1912, Hess visualized a process occurring deep below the oceanic crust that caused seafloor spreading. In this model the seafloor is created at ridges and sinks at trenches back into the Earth's mantle. This concept provided a model that catapulted the plate tectonics theory into the earth sciences mainstream.

EARLY YEARS

Harry Hammond Hess was born on May 24, 1906, in New York City, to Julian and Elizabeth Engel Hess. His father worked at the New York Stock Exchange. Harry had one brother, Frank. When he was five Harry's parents photographed him in a sailor suit and fittingly titled the portrait "The Little Admiral," foreshadowing Harry's career. He attended Asbury Park High School in New Jersey, where he specialized in foreign languages.

In 1923 Hess enrolled at Yale University, where he planned to major in electrical engineering. He changed his major to geology and received a bachelor's degree in 1927. The Loangwa Concessions, Ltd. mining company hired Hess to perform exploratory geological mapping and to look for mineral deposits in Zimbabwe (then Rhodesia), in southern Africa. He did not enjoy this work because he had to survey where he was told rather than where he believed he would find something valuable. After two years, he returned to the United States to attend graduate school, but this experience had taught him to appreciate the importance of fieldwork in geological research.

As a doctoral candidate at Princeton University he learned about mineralogy (the study of minerals, their identification, distribution, and properties), petrology (the study of the origin, composition, and structure of rocks), and the structure of the ocean basin. In 1931 he accompanied the Dutch geophysicist Felix A. Vening-Meinesz on a mission to carry out gravity measurements in the West Indies and the Bahamas. Information about gravity at different positions over the Earth's surface gives geologists insight into the composition of the rock under the surface because gravitational fields are stronger over areas with greater mass, which are denser. The measurements had to be performed within a submarine because a pendulum system was used, and ships moved around too much from the surface waves and wind. One interesting observation from this voyage was that the gravity over the Caribbean trench was much weaker than expected. A trench is a long, narrow furrow along the edges of the ocean floor; thus the gravity was expected to be weaker, as it is over all valleys, but the extreme weakness told Hess and Vening-Meinesz that the underlying structure was

unusual. Scientists were aware that volcanoes were often located adjacent to trenches and that earthquakes occurred nearby. Hess wondered about the implications of this finding.

Hess obtained a Ph.D. in geology in 1932. His dissertation was on the serpentinization of a large peridotite intrusive located in the Blue Ridge Mountains of Virginia. A peridotite is a type of igneous rock containing the minerals olivine and pyroxene, and intrusive means that the rock formed as magma without reaching the surface of the Earth. Serpentinization of peridotites is a chemical process by which the minerals olivine and pyroxene change to the mineral serpentine. The peridotite rock is changed into serpentinite, a rock composed of the mineral serpentine. Hess remained interested in mineralogy throughout his career and published two papers, "Pyroxenes of Common Mafic Magmas" (1941) and "Stillwater Igneous Complex, Montana" (1960), that both became classics in the field. NASA later named Hess the principal investigator for pyroxene studies of moon rock samples.

After receiving a doctorate Hess taught at Rutgers University in New Jersey from 1932 to 1933, then worked as a research associate at the Geophysical Laboratory of the Carnegie Institution of Washington, D.C., for one year. He obtained a teaching position in geology at Princeton University in 1934 and remained associated with Princeton until 1966. That same year he married Annette Burns, and they eventually had two sons, George and Frank.

FROM THE ATLANTIC TO THE PACIFIC

Hess had joined the U.S. Navy as a lieutenant to facilitate operations on a navy submarine that he used for gravity studies following his research with Vening-Meinesz. He was in the naval reserves when Japan attacked Pearl Harbor on December 7, 1941. Hess reported for active duty the next morning. Because he had submarine experience, he became an antisubmarine warfare officer with responsibility for detecting enemy submarine operations patterns in the North Atlantic. Hess advised the U.S. Navy that German submarines might be using the cloud cover north of the Gulf Stream (a current that runs from the Gulf of Mexico up the U.S. Atlantic coastline) to escape detection during surfacing. This suggestion resulted in the clearing out of submarines in the North Atlantic within two years. Hess arranged a transfer to the decoy vessel USS *Big Horn* to test the effectiveness of the submarine detection program; he then remained on sea duty for the rest of the war. As commanding officer of the transport vessel USS *Cape Johnson*, Hess carefully chose his travel routes to Pacific Ocean landings on the Marianas, Philippines, and Iwo Jima, continuously performing scientific surveying and profiling of the ocean floor across the North Pacific Ocean.

Hess took advantage of his time in the navy by patterning the travel routes to facilitate his studies on the geology of the ocean floor. He installed a deep-sea echo sounder on his transport ship and used it continuously. This equipment measured the depth of the sea bottom over which the ship traveled by sending a sound signal downward from the ship and measuring the time it took for the signal to bounce back from the ocean floor. Using these data Hess constructed bathymetric maps that showed the contours of the ocean floor across a large area of the Pacific. While collecting bathymetric data, Hess discovered flat-topped underwater volcanoes that he named *guyots* after Swiss geologist Arnold Guyot, who had founded the department of geology at Princeton University in 1854, and Princeton named the geology building after him. Hess remained in the naval reserves until his death, attaining the rank of rear admiral in 1961.

BAFFLING MARINE GEOLOGY DISCOVERIES

In July 1950 a group of scientists studying the seafloor of the Pacific Ocean made surprising discoveries that influenced the formulation of Hess's developing ideas about the origin and evolution of ocean basins. Scientists believed that the oceanic crust was mostly flat and extremely thick owing to the accumulation of billions of years of sediment from continental erosion. Using explosives and seismic waves, the crew determined the thickness of the oceanic crust to be about four miles (7 km). This was much thinner than expected, since the continental crust was known to be about five times thicker. Geologists thought the oceans had existed for 4 billion years, so why was there so little accumulated sediment? The crew took samples from guyots and was surprised to find coral rather than rocky sand, as anticipated. Coral is usually found in shallow areas, and some guyots are two miles (3.2 km) underwater. It also proved to be approximately 130 million years old, hundreds of millions of years younger than expected. Fossil evidence further confirmed the relatively young age of the ocean floor.

A few years later American oceanographer Maurice Ewing observed that oceanic ridges, underground mountain ranges, have rifts, or valleys, running through their centers. The seam of the rift appeared to be due to splitting apart, which was interesting because, in contrast, terrestrial mountain ranges were believed to arise from compression, from chunks of land being forced together. The presence of lava and absence of sediment along the ridges also baffled scientists.

PROPOSAL OF SEAFLOOR SPREADING

Hess contemplated these many unexpected discoveries in relation to the theory of continental drift proposed by the German meteorologist and geophysicist Alfred Wegener in 1912. After noticing that the east coast of South America and the west coast of Africa fit together like pieces of a jigsaw puzzle and collecting additional fossil evidence, Wegener concluded that the continents had once been connected, but split and drifted thousands of miles apart. Wegener offered no explanation for the mechanism driving continental drift, but Hess modified Wegener's theory and provided a plausible driving force. Wegener thought that drifting continents somehow plowed through the ocean floor, but Hess believed they rode along passively as the ocean floor was carried away from its ridges.

In 1960 Hess first proposed his theory of seafloor spreading. In doing so he explained the midoceanic ridge splitting, the presence of lava surrounding ridges, and the thinness of the oceanic crust. The attractive model suggested that the oceanic crust was indeed splitting at a seam, the rift in the center of the midoceanic ridge. As it split, melted magma rose through these weak areas and erupted as lava from the spreading ridges, forming new crust composed mostly of basalt as it cooled. The newly formed crust then was carried away by the mantle spreading out laterally beneath the crust, away from the ridge where it would eventually dive back beneath the surface of the Earth at trenches, located along faults formed by compression. At the trenches slabs of crust would be forced under other slabs of crust, back into the Earth's mantle.

Hess proposed convection, heat transfer by fluid motion, as the driving force behind the process of seafloor spreading. The mantle, located just underneath the crust, is solid rock formed mostly from iron and magnesium minerals, but just below the surface it may melt into magma, which cools into the igneous rock basalt. Though solid, the molten mantle can flow slowly like a fluid, forming convection cells. The hotter rock is less dense and rises, and as the rock cools, it sinks. When it cools and sinks, it pulls the overlying crust down into the mantle with it. The crust that disappeared into the trenches was constantly replenished by newly erupted lava at the ridge. Hess called this paper "an essay in geopoetry" and issued preprints of it in 1960. The official paper, "History of Ocean Basins," was published in 1962 by the Geological Society of America in a symposium volume called *Petrologic Studies: A Volume in Honor of A. F. Buddington.* Buddington had been Hess's petrology professor at Princeton and became a close friend.

Hess's paper was well received, and additional paleomagnetic evidence supporting his theory of seafloor spreading soon came out. Paleomagnetism is the science of reconstruction of the Earth's ancient magnetic field and the positions of the continents from the evidence of magnetization in ancient rocks. Because the Earth acts like a giant spherical magnet, rocks containing iron-rich minerals are magnetized in alignment with the Earth's poles. When magnetic rocks solidify, they become a permanent indicator of the direction of the Earth's magnetic field at the time of their solidification. Every few million years the Earth's poles reverse, thus the age of the rocks and their direction of magnetization provide geophysicists with information regarding the history of the direction of the Earth's magnetic poles.

In 1963 two young British geologists, Fred J. Vine and Drummond H. Matthews, and Lawrence Morley of the Canadian Geological Survey independently described magnetic anomalies that lent further evidence in support of seafloor spreading. Stripes parallel to the mid-oceanic ridge extended laterally from it, with reversals in the magnetic direction every several hundred kilometers. Vine thought that when the magma that erupted from the rift in the ridge cooled, it magnetized in the direction of the current magnetic field, then was carried away laterally from the ridge. Hess wholeheartedly accepted this hypothesis. Further studies performed in 1966 on the magnetic stripes showed that the patterns were indeed parallel to the ridges, and they were bilaterally symmetrical in magnetics and age. This evidence, along with evidence showing the seafloor was older the farther away it was from the ridges, confirmed the lateral movement of the crust and further established Hess's theory of seafloor spreading, as did future evidence of fossils and underwater core samples.

ADMIRED AND HONORED

Beginning in 1962 Hess chaired the Space Science Advisory Board of the National Academy of Sciences. The board's responsibility was to advise NASA. In 1966 he was at Woods Hole, Massachusetts, chairing a meeting to discuss the scientific objectives of lunar exploration when he began having chest pains. He died of a heart attack on August 25, 1969, and was buried at the Arlington National Cemetery.

Hess was elected to membership of several academic societies including the National Academy of Sciences (1952), the American Philosophical Society (1960), and the American Academy of Arts and Sciences (1968). He served as president of the Geodesy Section (1951–53) and the Tectonophysics Section (1956–58) of the American Geophysical Union, the Mineralogical Society of America (1955), and the Geological Society of America (1963). Because he was well respected as a scientist, he also was appointed chairman of the Committee for Disposal

of Radioactive Wastes, chairman of the Earth Sciences Division of the National Research Council, and chairman of the Space Science Advisory Board of the National Academy of Sciences. Along with oceanographer Walter Munk, he was a principal player in the Mohole Project, the goal of which was to drill beneath the oceanic crust into the mantle. The Geological Society of America awarded Hess the Penrose Medal for distinguished achievement in the geological sciences in 1966, and NASA awarded Hess a Distinguished Public Service Award posthumously. Because of his outstanding achievements the American Geophysical Union created the Harry H. Hess Medal for outstanding achievements in research in the constitution and evolution of the Earth and sister planets.

Hess's friends have described his personality as puckish and courageous. Part of Hess's greatness as a scientist was his willingness to entertain new ideas, even if they conflicted with his own previous conclusions. After all, sometimes wrong ideas ushered in new eras of scientific accomplishment. In suggesting that the ocean basins were continuously recycled, Hess explained why seafloor spreading did not cause Earth to grow, why the layer of sediment on the ocean floor was thinner than expected, and why oceanic rocks are younger than continental rocks. Hess's model of seafloor spreading has become part of the foundation knowledge of the geological sciences and has evolved into the theory of plate tectonics. Many questions regarding the forces that occur deep within the Earth are still being actively investigated today.

See also CONVERGENT PLATE MARGIN PROCESSES; DIVERGENT PLATE MARGIN PROCESSES; PLATE TECTONICS; WEGENER, ALFRED.

FURTHER READING

Carruthers, Margaret W., and Susan Clinton. *Pioneers of Geology: Discovering Earth's Secrets*. New York: Franklin Watts, 2001.

Hess, Harry H. "Comments on the Pacific Basin." *Geological Survey of Canada*. Special Paper (1966): 311–316.

———. "The Oceanic Crust." *Journal of Marine Research* 14 (1955): 423–439.

———. "Geological Hypotheses and the Earth's Crust Under the Oceans." *Proceedings of the Royal Society of London, Series A: Mathematical and Physical Sciences* 222 (1954): 341–348.

Hipparchus (160–120 B.C.E.) Greek *Astronomer, Geographer, Mathematician* Hipparchus was one of the greatest astronomical observers of ancient Greece. He is best known for his accurate quantitative models for the motion of the Sun and Moon. He used a trigonometric table and techniques from the Chaldeans (a Semitic people from Mesopotamia) to derive the sizes of Sun and Moon, to determine the latitude and longitude of places on the Earth, and to predict solar eclipses. He is credited also with compilation of the first star catalog and invention of the astrolabe.

Hipparchus was born in Nicaea in the ancient Greek district of Bithynia (now Iznik, Turkey). Historical accounts of his life from Ptolemy and Pliny the Elder suggest that he was born in 190 B.C.E., and it is known from his writings that he visited Alexandria, Egypt, and Babylon, but the dates of these trips are not known. Hipparchus is thought to have spent the later years of his life, and to have died in about 120 B.C.E., on the Greek island of Rhodes in the Aegean Sea.

Much of Hipparchus's scientific work has been lost to antiquity, and most is known from writings by scientists from more recent times, especially Ptolemy. From these writings it is known that Hipparcus wrote at least 14 books and published a star catalog later incorporated into Ptolemy's star catalog.

The world's first known trigonometric table is that of Hipparchus, who used the tables to calculate the eccentricity and orbits of the Sun and Moon. He was also concerned with calculating the distances to the Sun and Moon, and the sizes of these objects. He wrote about his trigonometric methods in the book *Toon en kuklooi eutheioon* (Of lines inside a circle), which has been lost to civilization. Hipparchus worked on stereographic projections (a mapping function that projects a sphere onto a plane), showing that the projections can be made to preserve angles so that they are the same on the projection as in the physical world, and that circles on the sphere that do not pass through the center of the sphere project as circles on the projection (i.e., they are not great circles). Hipparchus used these principles to develop the astrolabe, a historical astronomical instrument used to locate the positions of the Sun, Moon, stars, and planets. Astrolabes also proved useful in calculating location and time for ships at sea, and for navigation.

One of the topics of great interest to Hipparchus was the motion of the Moon. He calculated accurately the Moon's period and predicted eclipses with great accuracy. He was also concerned with the length of the year and the apparent motion of the Sun, and observed the summer solstice in 135 B.C.E., as well as many solar equinoxes. He used these measurements to calculate the length of the year. Toward the end of his career Hipparchus wrote a book on his solar observations and calculations, *Peri eniausíou megéthous* (On the length of the year). In this book he concluded that the year lasted 365 1/4–365 1/300 days.

Hipparchus worked to determine the size of the Sun and the Moon, and the distance to these objects, publishing his work in two books, *Peri megethoon kai 'apostèmátoon* (On sizes and distances). Although his books do not survive, later accounts suggest that he calculated the Sun to be 2,550 Earth radii and the mean distance to the Moon to be 60.5 Earth radii.

Hipparchus may be most famous for his discovery of the precession of the equinoxes through his observations of the Sun and Moon. He published books on precession titled *On the Displacement of the Solsticial and Equinoctial Points* and *On the Length of the Year*, and Ptolemy and others later used these works in their celestial observations and catalogs.

See also ASTRONOMY; SOLAR SYSTEM; SUN.

historical geology   Historical geology is the science that uses the principles of geology to reconstruct and interpret the history of the Earth. It includes study of the changes in the Earth's surface, the record of life, stratigraphy, dating of different geologic units, the history of the motions of plates and past positions of continents, the formation of mountain belts, basins, and past climates. The Earth can be like a jigsaw puzzle, and historical geology attempts to understand the causes and sequence of events that led to the observable features in the geological record, including the origin and destiny of living things and establishing the chronology of events in Earth history by examining the rock record. Geologists who have studied the history of Earth have gradually come to realize that the planet is very old and has had a complex history. The rock record shows that the life-forms on the planet have evolved from preexisting species by means of slow, gradual changes over long periods of time, and this evolution has been punctuated by several major episodes of mass extinctions, where large numbers of species and individual organisms within species have suddenly died off, to be replaced by totally new species in the next layer of younger strata.

STRATIGRAPHIC PRINCIPLES AND THE ROCK RECORD AS INDICATORS OF EARTH HISTORY

Stratigraphy is the study of rock strata or layers and is concerned with aspects of the rock layers such as their succession, age relationships, lithologic composition, geometry, distribution, correlation, fossil content, and environments of deposition. The main aim of stratigraphy is to understand and interpret the rock record in terms of paleoenvironments, mode of origin of the rocks, and the causes of similarities and differences between different stratigraphic units. These units can then be compared across regions, continents, and oceans to obtain an understanding of the conditions on Earth at the time the rocks were deposited.

The most basic unit of stratigraphy is the formation, a distinctive series of strata that originated through the same formative processes. Formations must be distinctive in appearance and easily recognizable. They can be recognized on the basis of lithology, which consists of the composition of the mineral grains, color, texture of grains, thickness and geometry of stratification, character of organic remains (fossils), and outcrop character. According to the code of stratigraphy, there is a hierarchy of naming different layers in rocks. A single layer is called a stratum, and the many layers within a formation are called strata. Groups consist of several related formations, whereas systems of strata consist of several groups of strata.

Early stratigraphers tended to regard Earth history and the stratigraphic record in a simplistic way. Most believed that the stratigraphic divisions being named in Europe were present and of the same age worldwide, developing a model of the Earth that was like a layered cake, with uniform stratigraphic layers around the globe. Gradually it became recognized that there may be some lateral variations between formations. For instance, in 1789 the French polymath, chemist, geologist, farmer, and engineer Antoine Lavoisier (1743–1794) suggested that similarities of fossils in similar sedimentary rocks might reflect environmental factors more than age. Lavoisier showed that shallow, near-shore marine sediments are coarser and contain organisms adapted to rough water, whereas deeper marine, quiet-water sediments are finer and contain delicate bottom-dwelling organisms, as well as floaters and swimmers. Lavoisier showed that different sedimentary products may form in different environments and have different groups of fossils, even though they formed at the same time. Five years later Lavoisier was beheaded during the French revolution, since he was one of 28 tax collectors and branded a traitor.

In the 1830s the British geologists Adam Sedgwick (1785–1873) and Sir Roderick Impey Murchison (1792–1871) also found that in Britain, the nonmarine Old Red Sandstone was laterally equivalent, in part, with marine sandstones. These two famous geologists found that the marine and nonmarine sandstones had an interfingering relationship, meaning one sediment type or formation grades laterally into another.

DEPOSITIONAL ENVIRONMENTS AND FACIES AS A RECORD OF HISTORICAL GEOLOGIC CONDITIONS

The importance of the lateral variations in strata was not appreciated until 1838 when interfinger-ing of two different lithologies with different fossil assemblages was described from a mountainside in Switzerland. The term *sedimentary facies* described the difference in lithology between the two interfingering rock units, and the two were interpreted as products of different environments of deposition, with the gradual change in the character of the rocks reflecting a gradual change in the aspects of the original environments, such as a transition from a sandy beach to a muddy offshore shelf. It is customary to name the facies of a rock unit by the dominant lithology of the unit, for instance, a mud-rich rock might be called a muddy facies.

The analysis of sedimentary facies is aimed primarily at interpreting the environment of deposition of the rocks, and therefore includes descriptions of both rock lithologies and the fossils contained in the rocks. An important aspect of the analysis and interpretation of sedimentary facies is using the principal of uniformitarianism, where features in old rocks are compared with modern environments to understand the origin of the old features.

Facies patterns are best understood by examining changes on regional scales, and can be very useful if not essential for reconstructing past environments. By examining many different facies patterns and determining the paleoenvironments, it is possible to understand the history of environments and life on Earth, one of the goals of historical geology.

Analysis of sedimentary facies shows that sea level is constantly moving up and down, and that the land surface is also moving up and down relative to sea level. Sinking of the land relative to sea level is called subsidence, whereas uplift refers to land rising relative to sea level. Some places in the world are currently subsiding, such as Venice, Italy; New Orleans, Louisiana; and the North Sea coast of Holland. The North Sea coast has been sinking for the past 10,000 years, since the last glacial advance, such that more and more ancient shorelines are located farther and farther offshore. Likewise, the coastal environment along the Mississippi Delta of southern Louisiana is rapidly subsiding, such that many square miles of land sink below sea level every year, and the shoreline is gradually moving inland toward the city of New Orleans. As this process of subsidence continues, the sedimentary environments of the shoreline, the shallow marine and deeper marine all move landward, with the deeper facies migrating on top of the shallow ones. This pattern is produced in what is known as a marine transgression, and can be caused by the land sinking relative to the sea, or by a global sea-level rise.

In contrast to a transgression, a regression of the sea occurs when sea level falls relative to the land, and the shoreline moves away from the present-

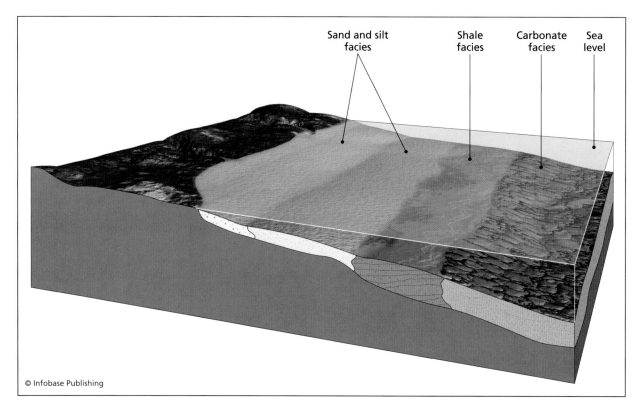

Sand and silt facies   Shale facies   Carbonate facies   Sea level

© Infobase Publishing

Cross section of beach and near-shore environment showing sedimentary facies change from subaerial dunes to beach to shallow marine to transitional to deep marine. Note how the rocks deposited in each facies interfinger with each other.

day coast. Like a transgression, a regression causes a continuous and gradual shift of the sedimentary environments, as well as the sedimentary and biologic products that form a wedge of material that overlies older deposits from the facies that previously existed in that location. In both transgressions and regressions the facies boundaries (and lithology changes in the rock record) form inclined surfaces, whereas time lines, corresponding to the seafloor surface at any time, may be subhorizontal surfaces, and cut across the lithologies as the paleoenvironments changed from beach to nearshore to offshore. Thus even though formations may be defined with attributes such as "near-shore sand" or "offshore mud," time lines cut across formation boundaries.

To interpret the history of an area one must often pick out characteristics of facies patterns to determine whether the sea level was rising in a transgression during deposition, or whether it was falling in a regression. Transgressive patterns are characterized by shrinking land areas, and typically transgressive rock packages are underlain by unconformity surfaces. They show a landward shift of facies through time, and become finer upward at any given location. In contrast regressive patterns that form from the uplift of the land or the fall of the sea show an enlargement of the land area, typically have an erosional unconformity at their top, show a seaward shift of facies with time, and are coarser upward at any given location.

Patterns of facies were studied extensively in the late 1800s by the German geologist Johannes Walther, who noted that facies tend to shift with time in the geological record, and that as the facies shift, adjacent environments succeed each other in the vertical sequence. This understanding led to the formation of "Walther's law," which states that the vertical progression of facies is the same as the corresponding lateral facies changes.

Understanding the local patterns of marine transgressions and regressions was a major accomplishment for geologists, but the next step, correlating the patterns from one shoreline or continent to establish a global pattern, was more difficult. In some cases it seemed as if one continent was rising or falling while others were not, and in other cases the geologic evidence suggested that sea levels were rising or falling at the same time in most locations across the world. These global sea-level rise or fall events are referred to as eustatic changes and may be caused by changes in the amount of water in the oceans from melting or freezing glaciers, or changes in the volume of the deep ocean basins by changing the volume of the midocean ridges through increased or decreased seafloor spreading. Continental collisions can also uplift large portions of the continents, effectively decreasing the amount of continental material in the oceans, expanding the volume of the ocean basins,

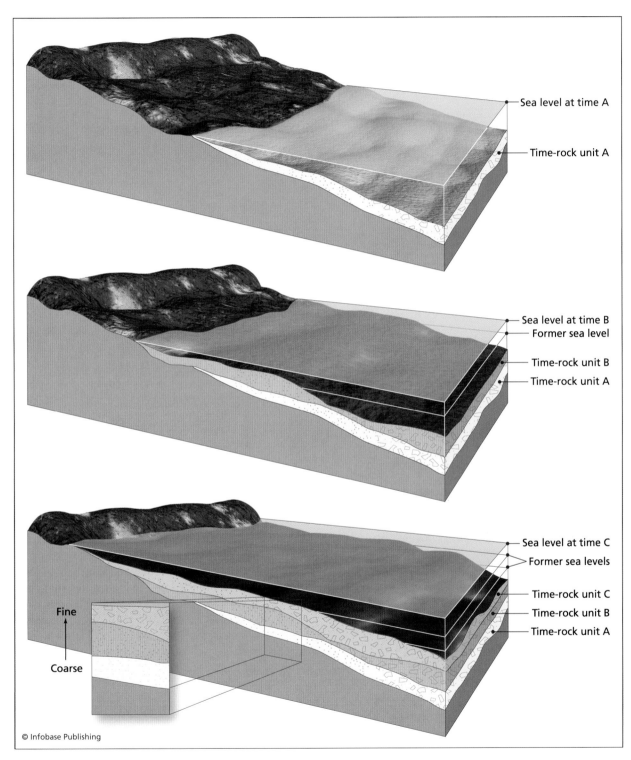

Rising sea level causes the sedimentary facies to migrate shoreward and be deposited one on top of the other and move progressively shoreward. A regression shows the opposite direction of migration of facies. Note how a vertical profile through these sections would yield a sequence of facies that is the same as the horizontal sequence of facies on the surface and that the order of succession can be used to tell the difference between a regressive and a transgressive sequence.

A Disconformity

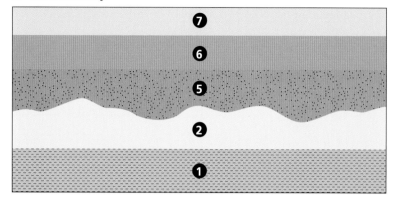

1. Beds 1, 2 deposited
2. Erosion
3. Beds 5–7 deposited

B Angular unconformity

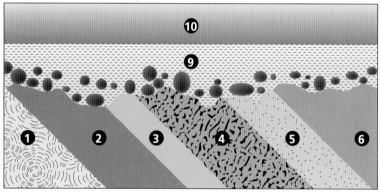

1. Beds 1–6 deposited
2. Beds 1–6 tilled
3. Erosion
4. Beds 9, 10 deposited

C Angular nonconformity

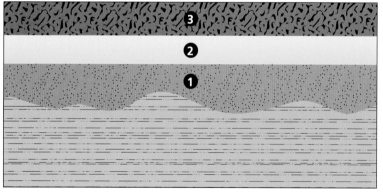

1. Granite formed
2. Granite exposed by erosion
3. Beds 1–3 deposited

Three types of unconformities, including disconformity, angular unconformity, and nonconformity

continent, which, over time, has led to the establishment of global eustatic sea-level curves showing the heights of sea level with time.

BIOSTRATIGRAPHY AND RELATIVE AGES OF STRATA

Biostratigraphy is the branch of stratigraphy that uses fossil assemblages and index fossils to correlate and assign relative ages to strata. Index fossils are used to identify and define geological periods, or faunal stages. Ideal index fossils are short-lived, have a broad distribution, and are easy to identify. Most good index fossils are floating or swimming organisms that live independently of the bottom environment. Many have floating larval stages that are dispersed by currents, seeds or spores blown by the wind, and that evolve rapidly. In contrast, some organisms are restricted to certain sedimentary facies and as such can indicate the paleoenvironment but not necessarily the age of the strata. A fossil zone is an interval of strata characterized by a distinctive index fossil, or fossil assemblage. These index fossils and fossil zones have the same age on a global scale and allow geologists to correlate strata from continent to continent, and to piece together a picture of the globe at any interval in the history of the Earth.

The distribution of and changes in the fossil record is controlled both by the evolution and the extinction of index fossils, and by changing environmental conditions or facies, which may cause individual organisms to migrate to a new habitat. Environmental changes are usually short-term

and causing global sea levels to fall. To determine whether sea-level changes are local or global, a well-correlated geological timescale is needed for each compared with extinctions, and the two effects are easily distinguished. Fossil zones do not necessarily correspond to formation boundaries.

UNCONFORMITIES AND GAPS IN THE HISTORICAL GEOLOGICAL RECORD

Unconformities are regional surfaces that extend for large distances and represent periods of time missing from the geological record at that location. To interpret unconformities and understand what each means for the history of the region and Earth, it is important to determine how much of a time gap is represented, and what caused the stratum that would have been deposited in that interval to not be preserved. In some cases the stratum was once there and has since been eroded, and in other cases the stratum was never deposited.

Unconformities are classified into three major types. Angular unconformities separate rocks below that were deformed and tilted, then eroded, from a new sequence of flat-lying rocks on top (that can be later deformed). Regional angular unconformity surfaces show that there was a major deformation event, typically an orogenic or mountain-building episode, between when the lower and upper rocks sequences were deposited. The age of the unconformity is typically the approximate age of the mountain-building event. Disconformities represent periods of nondeposition or erosion, but have no angle between the lower and upper rock sequences. Disconformities may be harder to recognize than angular unconformities, and typically a prior knowledge of the stratigraphic sequence, biostratigraphic zones, or geochronologic data must be used to establish the existence of the disconformity. Nonconformities are boundaries where sedimentary strata are laid down upon underlying igneous or metamorphic strata, typically in a marine transgression. There may be long time gaps of no history preserved along nonconformity surfaces, typically hundreds of millions or even billions of years.

UNCONFORMITY BOUND SEQUENCES

Understanding of stratigraphy underwent a major change in the 1950s when the American geologist Laurence L. Sloss (1913–96) proposed the concept of sequence stratigraphy. Sloss and many colleagues and subsequent workers recognized numerous large, laterally extensive rock units that they named sequences that are bounded by unconformities of global significance. Some of these unconformities are so significant that they are found in almost all shallow water deposits of that particular age around the world. Studies revealed that these always occur where sea level has dropped from high to low, and the overlying sequence is always transgressive. Sloss and coworkers used index fossils to show that these unconformities have the same age on all continents, and are clearly related to changes in sea level. Sea level has been as much as 1,200 feet (350 m) higher than at present, and as many as 650 feet (200 m) lower than present. By correlating different unconformity-bound sequences on the different continents, Sloss and coworkers produced curves that showed the relative height of the sea for the past 600 million years of Earth history.

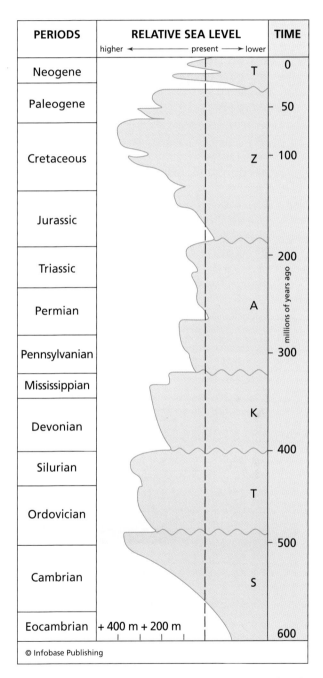

Sea-level curve showing the global average sea-level height through Phanerozoic geologic time, along with the six main unconformity bounded sequences deposited during transgressive and sea level high-stands. These sequences include the Sauk (S), Tippicanoe (T), Kaskasia (K), Absaroka (A), Zuni (Z), and Tejas (T) sequences.

The simple and slow rise and fall of sea level through geologic time produces unconformity-bound sequences. When sea level is high, sediments accumulate on the continental shelf, and when sea level falls, the shelf is exposed and eroded, and the sediments move off the shelf, producing an unconconformity. When sea level rises again the new sequence is transgressive, deposited unconformably over the eroded shelf.

The history of the Earth and its life-forms has been based largely on the study of the stratigraphic record as described above, and by correlating the ages of different events such as changes in sea level from continent to continent using index fossils, fossil zones, geochronology, and a few other timescales such as the magnetic polarity timescale, geochemical timescales, and data from other fields such as plate reconstructions using paleomagnetism, paleoclimate, and paleoenvironmental reconstructions. From this, geologists have been able to piece together a robust history of the planet Earth, although there is still much to be learned and debated by generations of future geologists.

BIOGEOGRAPHY AND PALEOGEOGRAPHY

Biogeography is the study of the geographic distribution of plants and animals, and paleogeography is the study of the past distribution of plants, animals, landmasses, mountains, basins, and climate belts. The distribution of organisms may be explained by one of two general theories. The dispersal theory in biogeography states that a specific group of organisms was created at an initial center spot and radiated outward, during which time specific lineages evolved as they migrated. An alternative model is called vicariance biogeography, where initially primitive groups were widely distributed and were broken up by processes such as rifting and divergent plate tectonics, leading to evolution in individual isolated groups. Both examples have been shown to explain the distribution of species in different cases.

In the following sections of this entry the principles of historical geology and stratigraphy are used to discuss a brief history of the planet Earth and life. The discussion is focused mostly on North America and events that affected North America, but examples from around the planet are brought in to the discourse as appropriate.

PRECAMBRIAN GEOLOGIC HISTORY

The oldest rock record on Earth belongs to the Precambrian, including all rocks formed before 543 million years ago, making up about 80 percent of Earth history. The Precambrian is divided in different ways in different parts of the world. The most common usage is to call the youngest period in the Precambrian the Ediacaran (although it is also known as the Vendian, and in China the Sinian), ranging from 650 to 543 million years ago. This period is named after the Ediacaran Hills in Australia, where fossils from this age are exceptionally well preserved. This was preceeded by the Proterozoic, from 2,500 million to 650 million years ago, divided into the Paleoproterozoic (2.5–1.6 Ga, or billion years ago), the Mesoproterozoic (1.6–1.0 billion years ago), and the Neoproteroic (1.Ga-543 Ma) equivalent, in some usages to the early, middle, and late Proterozoic. The Proterozoic was preceded by the Archean, from 2.5 billion years (Ga) ago to the age of the oldest known rocks (presently 4.2 Ga), and the Archean was preceded by the Hadean, the time from which there is no preserved geologic record.

ARCHEAN

In the Archean the surface of the planet looked very different than it does now. Life was limited to primitive bacteria, so the land had no vegetative cover. The Earth was also producing more heat in the Archean than it is now, so it is likely that heat loss mechanisms, particularly plate tectonics, were operating much more vigorously then than now, with more seafloor volcanism, perhaps greater ridge length, and faster plate motion.

Precambrian rocks form about 50 percent of the continental crust, and most of these are preserved in stable cratons. One of the best-known cratons is the Canadian shield, which extends from the northern United States to northern Canada, although much of it is covered by a thin veneer of younger sedimentary rocks. Maps of North America reveal that the continent is made up of old continental cratons surrounded by Proterozoic and younger orogenic belts representing places where oceans have closed, bringing the cratons together. Most of these cratons were parts of older continents that broke up and then reassembled, by plate tectonic mechanisms, to form the current distribution in the continent.

Most cratons are made of two fundamentally different suites of rocks, including granite-greenstone terranes and granulite-gneiss belts. Granite-greenstone terranes consist of arcuate belts of metamorphosed volcanics and sediments in an overall terrane consisting of 80 percent granitoids and gneisses. Most of the greenstone (volcanic) belts in these terranes are 3–30 miles (5–50 km) wide, and 60–200 miles (100–300 km) long. Most of the greenstone belts in these terranes include gabbro, pillow lavas, banded iron formations, and a type of muddy sandstone called greywacke. Greenstone belts and granite-greenstone terranes show complex structure with many faults and folds, and are metamorphosed to greenschist to amphibolite facies. These resemble younger island arc, accretionary prism, and ophiolitic rocks. Thus Archean granite-greenstone terranes formed by the generation and collision of island arcs, with folds and thrusts formed during the collision stage.

The other main type of Archean terrane is known as the high-grade granulite-gneiss assemblage, and examples include much of Greenland and Labrador, Scotland, southern India, and the Limpopo Province of southern Africa. These provinces contain assemblages of rocks similar to those in granite-greenstone terranes, but the assemblages have been buried to 20–30 miles (35–50 km) depth and are now strongly metamorphosed. The strong deformation and metamorphism in granulite-gneiss belts, plus their common location between older cratonic blocks, suggests that they represent places where continents have collided and are now deeply eroded. As such, granulite-gneiss belts provide natural laboratories to study what happens deep in the crust during continental collision events. High-grade gneiss complexes are also common in Proterozoic terranes (such as the Grenville Province), and are presently forming beneath the Himalayan Mountains, where India and Asia are colliding.

PROTEROZOIC

The Proterozoic rock record shows many features that suggest that large continents were present in this interval, including the first appearance in the geological record of very thick and abundant mature quartz sandstones, although some smaller examples are known from the Archean. These sedimentary rocks include well-sorted quartzites and quartz-rich graywackes, and indicate long periods of abrasion on stable continental shelves, and in some cases are depositionally related to abundant limestones that contain shallow-water fossils called stromatolites. These types of sedimentary assemblages indicate long-term sedimentation on a big stable continent, and are considered to mark the time when many small plates characteristic of the Archean graded into larger plates of the Proterozoic and younger times.

PRECAMBRIAN ATMOSPHERE, OCEANS, AND CLIMATE

The compositions of the Earth's early atmospheres and oceans are not well known, but most models fall into two groups. One is that the early atmosphere was relatively oxygen free (anaerobic), and the other, that it was aerobic, or had oxygen levels approaching modern values.

The early anaerobic atmosphere-ocean model was suggested by biochemists to support their model for the origin of life. This is supported by many dark-colored Archean sedimentary rocks that contain unoxidized carbon, iron sulfides (FeS_2), and iron carbonate ($FeCO_3$) minerals. These minerals should have been oxidized if oxygen were present, although some conditions, such as rapid sedimentation, allow these minerals to exist under present atmospheric conditions. Some Archean terranes also contain copper, manganese, zinc, vanadium, and uranium in their least oxidized states, supporting an early, anaerobic atmosphere. The anaerobic model suggests that any oxygen produced by early photosynthesis or dissociation of water in the upper atmosphere was caught in oxygen sinks and ended up making water (H_2O) and carbon dioxide (CO_2).

The early aerobic (oxygen-rich) atmosphere-ocean model notes that stromatolites were abundant by 3.4 Ga, and these produced a lot of oxygen. This model notes that most of the unoxidized minerals found in Archean sediments can also be found in much younger sediments, and are found in oxygen-poor environments such as swamps. Also oxidized Archean soils and rocks have been found, including some more than 2.6 billion years old in southern Africa. There are other models that suggest that the atmosphere saw a slow increase in oxygen, from low levels to higher from Archean times to present.

Present models for the climate in the Precambrian are changing as new evidence is accumulating. Most models suggest that there was more CO_2 in the atmosphere, so perhaps the climate was warmer, with more acid rains on the surface changing the weathering conditions. At times it seemed the Earth was either all hot, or all cold, with some periods of global glaciations such as a major glaciation in which most of the planet's water was locked up in ice between 700 and 800 million years ago.

ORIGIN OF LIFE

The Earth is believed to be unique in the solar system in that it supports life, yet how life appeared is still unknown. A popular model for the origin of life was formulated in the 1920s. This model suggests that life originated as a consequence of chemical reactions on the early, nonliving Earth. The planet naturally contains a lot of carbon, hydrogen, oxygen, and nitrogen, the major building blocks for life in most organisms. Water is a universal solvent, and early ideas for the origin of life used that property to suggest that the chemical elements for life may have combined into more and more complex molecules, until finally life and organisms emerged.

All life on Earth shares a common chemical system of proteins and nucleic acids. The nucleic acid DNA can copy itself, and it carries a code that directs which proteins the organism assembles. The specific proteins that are assembled determine the structure and function of each cell and, ultimately, what kind of organism it turns into. The DNA code itself is so complicated that it can be rearranged in 4×10^{109} different ways. The origin of life, however, lies in the synthesis of the monomeric components of the proteins—the amino acids. The proteins are what directly determine an organism's characteristics.

The origin and production of amino acids required special conditions on the early Earth. They needed elevated temperature, high-energy source, and appropriate chemical elements. Experiments first done by Stanley L. Miller (American chemist and biologist, 1930–2007) and Harold C. Urey (American chemist, 1893–1981) in 1953 at the University of Chicago show that a primordial soup made of methane, ammonia, hydrogen, and water can, if exposed to electrical discharges similar to ultraviolet light, give rise to amino acids and nearly all the complex organic molecules essential to DNA. These alone do not, however, constitute life. These kinds of "hot soup" environments have been found in hot springs and volcanic lakes. An alternative theory on the origin of complex organic molecules (and life in some models) is that they came to Earth from outer space by meteorite and comets, having formed somewhere else.

Complex organic molecules including amino acids do not constitute life. After the simple amino acids form, it is no easy task to combine them into larger molecules and complex molecules necessary for life. These need additional stimuli, such as hot acidic water, or ultraviolet radiation, or perhaps lightning. A mechanism for initiating the ability for molecules to transmit information so that they can replicate themselves is also necessary. One idea is that this may have first been done on the surfaces of clay minerals, such as those found in some submarine hot spring environments such as those along the mid-ocean ridges. Somehow, in the early Precambrian, life emerged from these complex organic molecules and simple amino acids, but the origin of life remains one of life's biggest mysteries.

EMERGENCE OF PHOTOSYNTHESIS

The first single-celled organisms were heterotrophs; they could not manufacture their own food. These organisms consumed the inorganically formed amino acids or other chemicals existing in their immediate environment. These early organisms received energy from these amino acids by fermentation, the processes of breaking down food molecules (sugars) through a series of chemical steps, into carbon dioxide and alcohol. When photosynthesis began, organisms became able to manufacture their own food from inorganic elements, and they became autotrophs, typically following the reaction

$$CO_2 + H_2O + light \rightarrow CH_2O \text{ (food)} + O_2$$

Photosynthesizing organisms thus produced food for other heterotrophs and also released oxygen into the atmosphere. The exact timing of when organisms became able to produce chlorophyll is unknown, but the oldest record of chlorophyll is found in 3.5 billion-year-old stromatolites from Australia.

LIFE IN THE PRECAMBRIAN

The earliest known fossils are found in 3.5 billion-year-old cherts of the Warrawoona Group at North Pole, Australia. The fossils consist of stromatolites that formed reeflike mounds made of small, single-celled organisms called cyanobacteria, which are still common today. Stromatolites are known from many other Archean locations, and they are essentially the only macro-scale fossil found in Archean rocks. There are a variety of microscopic organisms known, but most of these fall into varieties of single-celled organisms such as cyanobacteria.

The first truly diverse fauna is found in the 1.8–1.6 billion-year-old Gunflint Formation on the north shore of Lake Superior. The Gunflint Formation contains layers of stromatolites, but between the stromatolite layers many new species of cyanobacteria have been found. The Gunflint contains the first possible eukaryotic cells, containing a nucleus, as opposed to all earlier cells, which were prokaryotic and contained no organized nucleus. This represents a very significant evolutionary advance.

The late Precambrian had a more diverse fauna and saw the first definitely eukaryotic cells, found in the 1.3 billion-year-old Beck Springs Dolomite of Death Valley, California. In this period stromatolites became very abundant, and some grew to enormous sizes, some tens of feet (several m) across. Trace fossils also first appeared in the late Precambrian in the form of worm burrows from the late Precambrian rocks at the base of the Grand Canyon Series.

Rocks of the Ediacarian (also known as the Vendian, Sinian, and Eocambrian) range between 700 million years and 543 million years old at the base of the Cambrian and the Paleozoic. This period is named after fossil-rich beds in the Ediacarian Hills of Australia, containing a wide variety of fossils of jellyfish (*Coelenterate medusae*), sea fans (octocorals), worms, and other species without skeletons. All of the Ediacarian fauna became extinct at the end of the period, and none shows any relationship to any younger fauna. There is a 200-million-year break in the fossil record between Precambrian stromatolites and the diverse plant and animal fossils in the Ediacarian Hills, and in this short gap of time for which no record is known, life had to change dramatically from the single-celled organisms that had inhabited the Earth for the past 3 billion years to complex life-forms found in these strata. To accomplish this the eukaryotes had to clump together in colonies; become consumers of food rather than manufacturers; develop specialized cells for reproduction, locomotion, and other functions; and develop a body sack with tissues and organs.

LATE PRECAMBRIAN PALEOGEOGRAPHY AND TECTONICS

In the late Proterozoic North America was part of the large supercontinent Gondwana that included Antarctica, Australia, India, Africa, Baltica, and many other cratons. By the Late Proterozoic this supercontinent began rifting apart, forming narrow seas similar to the present-day Red Sea between Africa and Arabia. These much older narrow seas were between North America and Antarctica and evolved into major oceans. Huge amounts of subsidence along the margins of these rifts formed very thick passive margin sedimentary wedges. By the end of the Precambrian all the margins around North America were passive, as the other continents drifted away. These and other passive margins formed during the breakup of the late Precambrian supercontinent and contributed to a general global rise in sea level as the amount of young ridges was large, and the average elevation of continents was decreased by continental extension, placing a larger volume of continental material below sea level and displacing an equivalent volume of water onto the continents in a global transgression.

EARLY PALEOZOIC HISTORY

One of the greatest changes in Earth history is marked by the Precambrian-Phanerozoic transition. At this time the Earth witnessed the first widespread appearance of organisms with hard shells, and there was a huge adaptive radiation unparalleled in the rest of Earth history. By this time most of the cratons on the planet had formed and large continents existed, and plate tectonics had already been through several supercontinent cycles.

The history of the early Paleozoic can be interpreted from shallow seas that repeatedly flooded the cratons and from orogenic belts on the margins where island arcs and other terranes collided with and were sutured to the cratons. The principles of sequence stratigraphy prove useful for interpreting the history of the Paleozoic, as this era is marked by many transgressions and regressions caused by global rise and fall of sea level, and these formed several globally recognized unconformity-bounded sequences.

SAUK TRANSGRESSION

From Ediacarian through Cambrian times, sea level was constantly rising until it covered almost all of the cratons. On North America all but the central part of the Canadian shield was covered by shallow seas. The Cambrian isopach map shows a prominent feature called the transcontinental arch, thought to have formed by the bending of the North American plate under the great weight of the sedimentary sequences deposited on the margins of the craton. The distribution of Cambrian sedimentary facies around this arch shows that the facies are generally parallel to the arch, and that sand facies are generally next to the arch whereas deeper water shale and limestone facies are further offshore, showing that this is an original feature and not formed by erosion. The name given to the Eocambrian through late Cambrian rise of sea level is the Sauk transgression, which formed a shallow or epeiric sea over more than 75 percent of North America. During this period sea levels rose about one foot (0.3 m) every 20 years. As the seas rose they deposited a layer of quartz-rich sandstone over an unconformity that migrated toward the center of the craton with time, forming one of the major transgressive sand sequences of the past 500 million years of Earth history. Most of these sands were derived from the previous 500 million years of weathering products that accumulated on the cratonic interior of North America, and as the sea level rose, the high energy beach environment reworked these soils, sands, and other products in the regolith to a stable quartz-rich assemblage now preserved as the basal transgressive sand. Many of the sand grains in this basal quartzite are very rounded, and well sorted (meaning they have similar size to each other), suggesting that some of them were derived from windblown sand deposits before they were transported by rivers and reworked in the high-energy beach environment.

As the transgression continued most of the craton soon became covered with water, and less sand was available to be eroded and contribute to the sediments being deposited in the Sauk transgressive sequence. At this time the climate was favorable to the production of carbonate sediments, and the major type of deposition in the Sauk sequence switched to carbonates by the beginning of the Ordovician. Before this time carbonates were already being deposited along the deeper, outer parts of the continental shelves and seas, and worked their way toward the center of the craton. These carbonates consist of limestone and dolostone, largely made of shell fragments, limey muds, algae, and carbonate-secreting organisms. One special type of carbonate is known as oolitic limestone, which consists of many sand-sized carbonate grains with a texture resembling onion skin. These carbonates form by rolling about on the seafloor in shallow agitated waters, continuously being precipitated around a hard nucleus. Oolites form in waters saturated with respect to calcium carbonate, typically

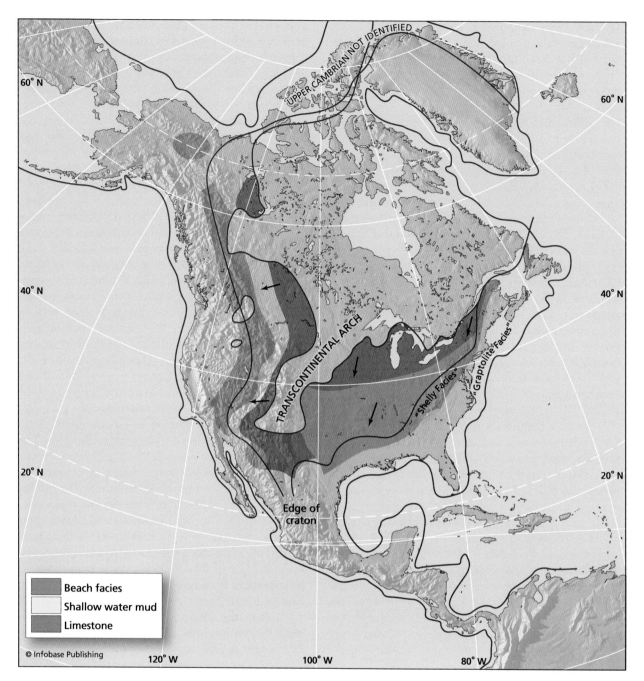

Upper Cambrian map of sedimentary facies of North America showing areas of beach facies (red), shallow-water mud (yellow), limestone (purple), and deepwater facies

in areas of high evaporation and agitation, which releases carbon dioxide. The Sauk Sea was shallow, as indicated by sedimentary structures formed by waves, shallow water fossils called stromatolites, and rare mud-cracks indicating that the sea bottom was occasionally exposed to the air.

Thick limestone sequences of the Sauk sequence indicate that North America was located within about 20 degrees of the equator in the early Paleozoic, and the paleoequator must have run approximately up the center of the continent (with the continent drift-

ing into that position by plate tectonics). The climate was subhumid, warm, and rainy.

The Sauk sequence was terminated abruptly about 490 million years ago when sea level suddenly dropped (on geological timescales, taking a few million years), leading to widespread erosion and the formation of a worldwide unconformity surface on top of the Sauk sequence.

Life in the Cambrian included worms (preserved largely as worm tubes), the first mollusks, echinoderms, sponges, archaeocyathids, and coelenterates, and saw the widespread development of hard skeletons, generally phosphatic. In all probability there were also a large number of soft-bodied organisms in the Cambrian seas, but conditions did not favor their preservation. The evolutionary step of forming skeletons was very important to organisms in the Cambrian. The outer skeleton (exoskeleton) offers protection from ultraviolet rays and predators. It is also possible to make larger organisms with the presence of a skeleton since the skeleton can support the larger structures and act as a base for the development of muscles, which are needed for mobility.

Other organisms common in the Cambrian seas included the trilobites, a bottom-dwelling scavenger creature that resembled the modern horseshoe crab. These were the most abundant organism in the Cambrian, including more than 600 genera, some of which make very good index fossils. The Archaeocyathids were also bottom-dwelling organisms, shaped like a vase, with a double skeletal wall. These became extinct at the end of the Cambrian. Brachiopods, resembling clams, and mollusks, forming caplike shells, were common in shallow-water settings of the Cambrian seas.

TIPPECANOE SEQUENCE

Sea levels rose again and deposited a new transgressive sequence, known as the Tippecanoe sequence, from 490 to 410 million years ago during the Ordovician and Silurian Periods. Life in the Ordovician changed drastically from what it was in the Cambrian, and a great number of organisms flourished in the Ordovician. This may be attributed to the large amount of continental submergence (under the shallow seas), which created a large number of ecological niches, and also from the mild, steady climate of the times. Trilobites were not as numerous in the Ordovician as they were in the Cambrian, and they were less abundant than the Brachiopods and Bryozoans. Other organisms saw major changes in the Ordovician. The phosphatic shells of brachiopods were beginning to be gradually replaced by calcareous shells, and mollusks (snails) experienced rapid evolution, with new groups appearing, including the cephalopod (Nautiloids), which appeared and went through a very rapid evolution, including the appearance of giant forms reaching 30–40 feet (9–12 m). Since the cephalopods were also common, they make excellent index fossils for the Ordovician. Bryozoans appeared for the first time, making hard skeletal structures out of calcium carbonate, and various types of corals appeared including the solitary rugose coral, and more colonial forms that gradually developed prismatic forms that allowed them to grow more closely together. Graptolites are distinctive index fossils since they were widely distributed and evolved quickly. Graptolites were probably floating colonies that resembled seaweed but are generally preserved as single blades about a quarter inch to inch (0.5–3 cm) long. The Ordovician also saw the development of the first vertebrates, with fish containing armored plates found in a few fossil locations.

The shallow Ordovician seas had diverse ecosystems, and many of the modern ecosystem types such as continental shelves and reefs were firmly established in the Ordovician. At this time most life on Earth was based on the seafloor, between the shoreline and the deep abyssal plains. These included a number of different types of organisms, such as infaunal organisms that live within the bottom sediments, and epifaunal organisms that live on the bottom surface. Sessile benthic dwellers are those that are attached to some object on the bottom, whereas burrowers move through the sediment. Vagrant benthic organisms move around on the bottom, whereas deposit feeders eat small organic particles in the bottom sediments. Suspension or filter feeders capture and eat other organisms that float in the water, whereas planktonic organisms are passive floaters that live above the bottom. Nektons are actively swimming organisms, most of whom live in the photic zone, through which light can penetrate in the oceans.

Like the Sauk Sequence, the base of the Tippecanoe Sequence is marked by a pure quartz sandstone or quartzite, known as the St. Peter sandstone, forming a volume of quartz of some 4,800 cubic miles (20,000 km³). The base of the St. Peter sandstone has slightly different ages in different places, being slightly older at the outer edges of the craton than in the interior of the continent, showing that unconformities can be time-transgressive, having different ages in different places.

Life in the Tippecanoe Sea was quite different from that in the Sauk Sea of the early Ordovician. The Tippecanoe ecosystems included diverse types of corals, stromatoporoids, bryozoan cephalopods, brachiopods, and armored fish. Upper Ordovician limestones are typically very shelly. The Upper Ordovician Seas reached all-time highs, representing the most complete flooding of continent ever in the geological past.

Middle to Late Orodovician deposition on the eastern side of the North American craton changed from limestone- to graptolite-bearing black shales, indicating a deepening of the water conditions. This deepening reflects a drastic and important change related to the approach of the Taconic island arc that would soon collide with the then North American craton in the Middle Ordovician. Other evidence also suggests that mountains were being uplifted on the eastern side of the craton during the Middle to Late

Ordovician. These include the presence of volcanic ashes interbedded with the shales, and the fact that the black shales are succeeded eastward and upward by a clastic wedge, containing sandstone, conglomerate shale, and so on eroded from mountains in the east. This clastic wedge represents a foreland basin, where many layers of sandstone, shale, and conglomerate were deposited by rivers that flowed out of the rising mountain range in the east, then redeposited by turbidity currents in deeper water environments during active deformation of the mountain range in a sequence of rocks known as flysch. The flysch, deposited during active deformation of the mountain range in the east is succeeded upward by deposits of molasse, which is nonmarine irregularly stratified conglomerate, sandstone, shale, and coal deposited in the late stages of mountain building.

TACONIC OROGENY IN THE APPALACHIANS

The deepening of the eastern shelf of North America, the presence of volcanic ash, and the clastic wedge of flysch and molasse all indicate middle to late Ordovician mountain building in the Appalachians. The cause of the abrupt deepening of the passive margin was by thrust loading from the weight of Appalachian Mountains being pushed up out of the Iapetus Ocean and onto the passive margin of North America during the collision of an island arc with the North American continent. During this mountain-building event, known as the Taconic Orogeny, the Taconic thrust belt was formed and in it pieces of an island arc and accretionary prism are preserved, representing an oceanic convergent margin that had been active in the Iapetus Ocean since the Cambrian, and collided with eastern North America in the Middle Ordovician, closing the oceanic segment between North America and the Taconic island arc.

POST-TACONIC PALEOGEOGRAPHY AND PALEOCLIMATE (SILURIAN AND DEVONIAN)

The Middle to Late Ordovician Taconic Orogen was eroded during early Silurian times, shedding early Silurian clastic sediments, including molassic sands and gravels. Erosion of the Taconic Orogen was fairly complete by middle Silurian times when the sea was again able to advance over the lands once covered with mountains.

The Silurian and Devonian atmosphere was strongly oxidizing, as indicated by abundant iron oxides in rocks of this age. Sediments deposited on the North American continent at this time include abundant evaporates, so the climate was likely warm for this period. In addition there are numerous reefs from North America, and modern-day reef systems form only at +/- 30° of the equator. Interestingly, the paleobiogeographic record shows that similar plants existed virtually everywhere at this time, suggesting that the Devonian climate was uniform on a near-global basis. Comparison with many past climates reveals that many were much warmer than Earth's present climate, lending support to the idea that the Earth can support very warm climates, and any changes that the planet may be experiencing presently from global warming could lead to major changes in planetary ecosystems.

DEVONIAN STRATA

In the late Silurian and Devonian a major regression affected most of the craton, exposing the underlying rocks to subaerial erosion, except for a few deep basins and narrow seaways. This major unconformity is overlain by a new transgressive sequence known as the Kaskaskia Sequence, which, like the two preceding sequences, is marked by a basal quartz sandstone, overlain in turn by a thick carbonate sequence. Much of the continent was again covered by carbonate and shale deposition, and areas that saw enhanced subsidence during the Tippecanoe Sequence deposition also experienced greater than normal subsidence during the evolution of the Kaskaskia Sequence. These areas became deep basins, including the Michigan and Illinois basins in the midcontinent region.

Maps of the Upper Devonian sedimentary facies show a >1-mile (2-km) thick sequence of rocks on the eastern side of the craton known as the Catskill clastic wedge. This group of rocks provides evidence for another episode of mountain building or orogenesis in the Appalachians. The Catskill delta consists largely of molasse (conglomerate, red beds, mudstones) deposited by river systems in an alluvial fan/braided, meandering river complex, with the rocks derived from uplifted mountains in the east during the Devonian. The Catskill clastic wedge formed in front of the Acadian orogen, which was active (as shown by the ages of igneous rocks) from 360 to 330 million years ago. The Acadian Orogeny is thought to have been caused by closure of an ocean basin between North America and Avalonia (part of northwest Africa), along two subduction zones that dipped beneath both continental masses. The convergent margins collided, forming thick clastic wedges, then the collision continued as mostly sideways or strike slip motions through the late Devonian. The Catskill clastic wedge has an equivalent, but slightly older, sequence of rocks in Europe, known as the Old Red Sandstone, which also represents rapid erosion of a mountain range that formed by the north-south closure of an ocean basin in the Devonian. At this time North America and Europe were still connected in one landmass, and the ocean that closed between the combined North American/European continents and the colliding masses to the south (largely thought

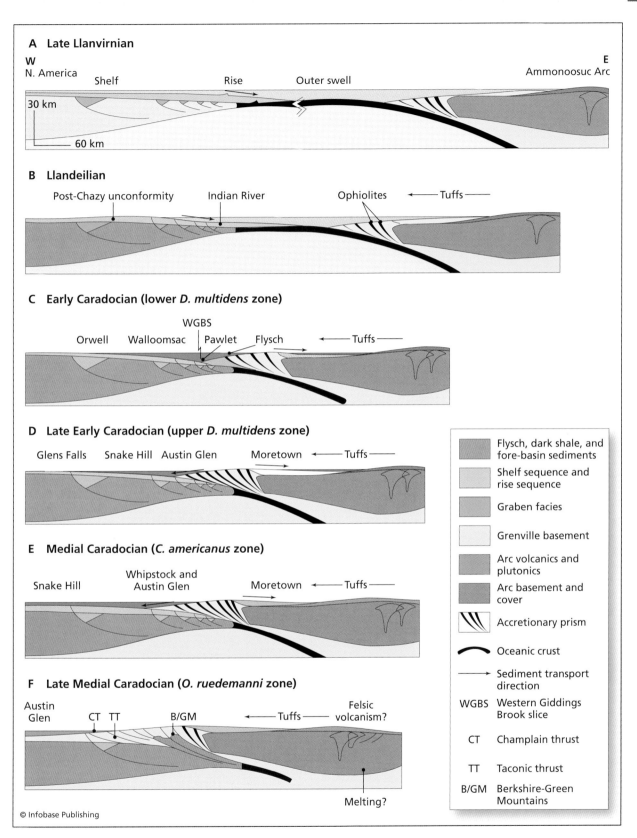

A Late Llanvirnian

W
N. America
Shelf   Rise   Outer swell   E
Ammonoosuc Arc

30 km

60 km

B Llandeilian

Post-Chazy unconformity   Indian River   Ophiolites   ← Tuffs →

C Early Caradocian (lower *D. multidens* zone)

WGBS
Orwell   Walloomsac   Pawlet   Flysch   ← Tuffs →

D Late Early Caradocian (upper *D. multidens* zone)

Glens Falls   Snake Hill   Austin Glen   Moretown   ← Tuffs →

E Medial Caradocian (*C. americanus* zone)

Whipstock and
Austin Glen
Snake Hill   Moretown   ← Tuffs →

F Late Medial Caradocian (*O. ruedemanni* zone)

Austin
Glen   CT   TT   B/GM   ← Tuffs →   Felsic
volcanism?

Melting?

© Infobase Publishing

	Flysch, dark shale, and fore-basin sediments
	Shelf sequence and rise sequence
	Graben facies
	Grenville basement
	Arc volcanics and plutonics
	Arc basement and cover
	Accretionary prism
	Oceanic crust
→	Sediment transport direction
WGBS	Western Giddings Brook slice
CT	Champlain thrust
TT	Taconic thrust
B/GM	Berkshire-Green Mountains

Geological evolution of the Taconic arc-trench collision during the Ordovician Taconic Orogeny in the Appalachian Mountains (modified from D. Rowley and W. S. F. Kidd, 1981) to be African and South America) was the remaining open part of the Iapetus Ocean that did not close during the Middle Ordovician Taconic Orogeny.

SILURIAN-DEVONIAN (436–360 MILLION YEARS AGO) HISTORY OF LIFE

In the Silurian and Devonian, organisms continued to evolve rapidly in the shallow sea that covered much of the continent, and like the late Ordovician, brachiopods and bryozoans were the most common organisms in the shallow seas. However, echinoderms became increasingly more important and abundant in the Silurian. By Silurian times the nautiloids and cephalopods had nearly disappeared, and the grapto-

lites were virtually extinct. One line of descent of the nautiloids survived and evolved into groups of coiled ammonoids in the Devonian. These ammonoids were swimming and floating organisms, and they evolved rapidly, so they formed useful index fossils for this period. Another unusual but widespread group of animals, the small, toothlike conodonts, were abundant in this period, and these seem to represent some disaggregated parts of a larger type of floating or swimming organism.

In Silurian-Devonian times coral reefs became widespread and were populated mainly by rugose and tabulate corals. The Devonian was also important for the evolution of life in that fish experienced

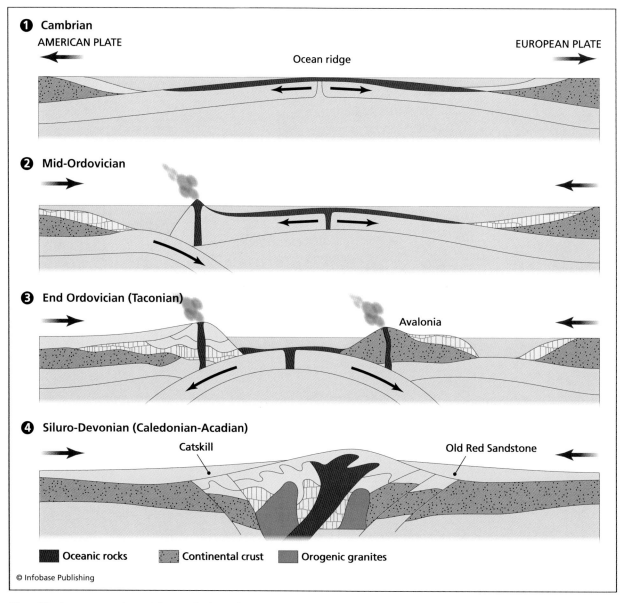

Simplified cross section of the evolution of the Iapetus Ocean, including an Ordovician arc-continent collision and the Devonian collision with Avalonia, in the Acadian Orogeny

widespread development, and the land was extensively invaded by plants. The Devonian has many examples of armored fish fossils, which show the development of internal skeletons, jaws, and evolved bodies better suited for swimming than their older Paleozoic counterparts. Many plants began to invade low-lying areas near the seas and lakes, and by the end of the Devonian, huge forests of thick trees covered much of the land, and the land became inhabited by insects, scorpions, and spiders that evolved from sea creatures. As an example, scorpions evolved from giant, three-foot (1-m) long eurypterids known as sea scorpions. By the late Devonian amphibians were crawling on the land, and these may have evolved from fish that crawled out of the sea.

LATE PALEOZOIC HISTORY

The last major deposition of carbonates on the North American craton was in Mississippian times, after which most of the North American continent has remained emergent, or exposed above sea level. The Mississippian was a time that saw vast proliferation of plants and animals across the continent, with many swamp deposits becoming peat bogs, and later when buried, being converted to coal. For this reason the time periods encompassing the Mississippian and Pennsylvanian are commonly referred to as the Carboniferous Period (355–290 million years ago).

The Late Paleozoic was also a time of continental amalgamation. North America finally collided with Africa and South America, forming Gondwana during the Appalachian or Alleghenian Orogeny (about 300 million years ago), and similar collisions worldwide formed the supercontinent of Pangaea (meaning all-lands, including the southern continents in Gondwana). The evolution of life progressed rapidly on land, with reptiles becoming strong and dominant on the land.

PANGAEA

The Late Paleozoic saw the formation of Pangaea, which included the southern continents amassed in Gondwana and the northern continents grouped in Laurasia. Most of the evidence for the formation of the supercontinent of Pangaea comes from the southern continents, since nearly all of these contain nearly identical fossils and stratigraphy. These were studied extensively by Alex Du Toit from South Africa, and Alfred Wegener from Germany. Separately these two scientists pieced together evidence that eventually formed a compelling case for the existence of a Late Paleozoic supercontinent, encompassing virtually all of the planet's land masses. One of the most compelling arguments was Du Toit's observation of synchronous glaciations in Permian-Triassic times throughout Gondwana. He reasoned that if all of the

continents were together, the patterns of glaciation and the volume of ice needed to explain the synchronous glaciation were reasonable, but if the continents had their present configuration, he calculated that there was not enough water on the planet to make all of the glaciers needed to cover the continents. Furthermore, geological matches of offset features across the oceans are restored when the oceans are closed. Later, paleomagnetic data were found to support the idea that the continents were formerly together and have been drifting apart since the breakup of Pangaea. Pangaea can now be reconstructed by looking at seafloor magnetic anomalies and restoring them to their older, late Paleozoic configuration.

HISTORY OF GONDWANA

Gondwana is the name given to the southern continents that amalgamated before they joined, as a group, to the northern continents of Laurentia to form Pangaea. Most of Gondwana was assembled in the period between 600 million and 400 million years ago, when many Archean cratons were joined together by suturing during closure of several ocean basins that lay between them, forming a series of orogenic belts across Africa, India, South America, Madagascar, Antarctica, and Australia known as the Pan-African belts. The last of these oceans to close was the Mozambique Ocean, which lay between western Gondwana (eastern Africa and South America) on one side and eastern Gondwana (Madagascar, India, Australia, Antarctica) on the other side.

The fauna of Gondwana (southern fauna) were distinct from those in North America (northern fauna) until the Devonian, when they became similar. In the late Paleozoic Gondwana collided with North America during the Appalachian (also called the Alleghenian) Orogeny, forming Pangaea. In addition southern South America, western Antarctica, and New Zealand collided with southern Gondwana to form orogenic belts there and further increasing the size of Pangaea. Like the Taconic and Acadian orogens these orogenic belts also have clastic wedges associated with them, recording the history of uplift and erosion of these mountain ranges.

Gondwana shows a remarkably similar stratigraphy on all of the major southern continents, grading from Devonian tillite (glacial gravels), through Carboniferous and Permian coal, Triassic red beds (sandstone and conglomerate), and Jurassic through Cretaceous volcanic rocks. The Devonian glacial tillites are remarkable rocks, found on all five southern continents. These were deposited unconformably on older rocks, and in many places the unconformity surface preserves scratches made by glaciers hundreds of millions of years ago. Boulders in the tillite also locally preserve glacial striations, plus isolated

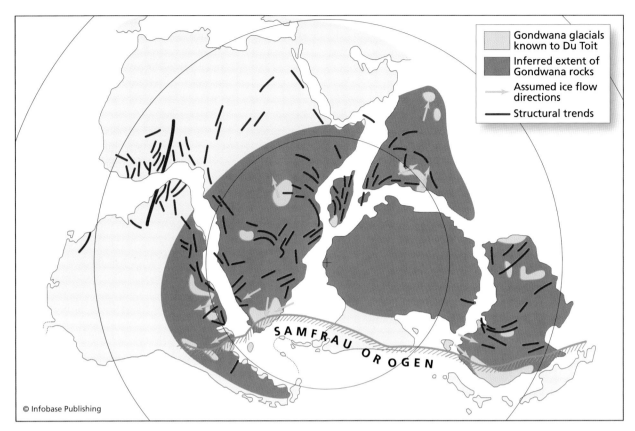

Alex Du Toit's reconstruction of Gondwana showing areas affected by glaciation, striation directions, and other tectonic features

pebbles called dropstones, dropped by glaciers into otherwise muddy sediments, showing beyond reasonable doubt that these rocks were deposited in a glacial environment. Most of these tillites are interbedded with nonmarine rocks that bear a distinctive fossil assemblage including a seed fern called Glossopteris that inhabited low swampy areas next to glaciated terrane. The *Glossopteris* flora is found across much of Gondwana, and was used by Du Toit and Wegener to support their reconstructions of the late Paleozoic supercontinent. The Gondwanan stratigraphy grades up into Carboniferous to Permian coals, and then Triassic red beds, and finally Jurassic-Cretaceous volcanic rocks consisting mostly of basalt flows. These basalts are related to the breakup of Gondwana and Pangaea, and spreading of the modern oceans.

The animals of Gondwana included a great variety of reptiles, amphibians, fish, and invertebrates, some with key correlations across present-day oceans. Since these animals could not swim across such vast expanses of ocean, they provide additional supporting evidence that the continents were once joined together in a supercontinent mass. One of these animals, the Mesosauris, is a reptile that inhabited swampy areas and lived in present-day Brazil and southern Africa. Other animals that show similar matches across the oceans include the dinosaurs and the reptiles Lystrosaurus and Cynognathus.

The climate of Pangaea was diverse. There are glacial deposits in some places and times (e.g., the Permian), indicating a cold climate, and in other places and times there were swamps (*Glossopteris*) forming coals, indicating warm and swampy climate conditions, and in still other places evaporates formed, indicating hot, dry conditions.

The breakup of Pangaea began after the supercontinent had been united for about 200 million years and is marked by the eruption of the voluminous Jurassic and Cretaceous volcanics across the continents. Laurasia saw its breakup in the Late Triassic, as indicated by the formation of many rift basins such as the Triassic rifts along the eastern seaboard of North America. The rifts were filled with coarse sandstones and conglomerates as well as volcanic rocks, and they are generally colored red by the iron oxides formed during the intense hot climate of these times. Many of the rift basins preserve dinosaur footprints and other interesting fossils. As waters from the ocean occasionally spilled into the rifts during short-lived sea-level rises, evaporates formed as these waters evaporated in the hot climate. Rifting continued until the Middle Jurassic, when North

America or Laurentia began drifting apart from Gondwana and the supercontinent began breaking up all over. Some models suggest that the reason Gondwana broke up, and saw so much magmatism during breakup, is that it came to rest over a mantle hot spot that heated the lithosphere and caused the volcanics to erupt.

In the latest Cretaceous (about 66 million years ago, at the Cretaceous-Tertiary boundary) North America began moving away from Europe, opening the North Atlantic Ocean. During the evolution of Pangaea a large ocean, the Tethys Ocean, separated parts of the Northern and Southern continents (which were connected elsewhere). During breakup of Pangaea and opening of the Atlantic Ocean, the Tethys began closing and is now largely closed except for remnants in the Mediterranean, Black, and Caspian Seas.

CRETACEOUS-TERTIARY BOUNDARY AND THE CENOZOIC

The Cenozoic Era marks the emergence of the modern Earth, starting at 66 million years ago and continuing until the present. The Cenozoic includes the Tertiary (Paleogene and Neogene) and Quaternary Periods, and the Paleocene, Eocene, Oligocene, Miocene, Pliocene, Pleistocene, and Holocene Epochs. Many modern ecosystems developed in the Cenozoic, with the appearance of mammals, advanced mollusks, birds, modern snakes, frogs, and angiosperms such as grasses and flowering weeds. Mammals developed rapidly and expanded to inhabit many different environments. Unlike the terrestrial fauna and flora, the marine biota underwent only minor changes, with the exception of the origin and diversification of whales.

The Cenozoic began after a major extinction at the Cretaceous-Tertiary boundary, marking the boundary between the Mesozoic and Cenozoic eras. This extinction event was probably caused by a large asteroid impact that hit the Yucatán Peninsula near Chicxulub, Mexico, at 66 million years ago. Dinosaurs, ammonites, many marine reptile species, and a large number of marine invertebrates suddenly died off, and the planet lost about 26 percent of all biological families and numerous species. Some organisms were dying off slowly before the dramatic events at the close of the Cretaceous, but a clear, sharp event occurred at the end of this time of environmental stress and gradual extinction. Geochemical anomalies including the mineral iridium, generally found only in meteorites, have been found along most of the clay layers that mark this boundary, considered by many to be the "smoking gun," indicating an impact origin for the cause of the extinctions. An estimated one-half million tons of iridium are present in the Cretaceous-Tertiary boundary clay, equivalent to the amount that would be contained in a meteorite with a six-mile (10-km) diameter. Some scientists have argued that volcanic processes within the Earth can produce iridium, and an impact is not necessary to explain the iridium anomaly. However, the presence of other rare elements and geochemical anomalies along the Cretaceous-Tertiary boundary supports the idea that a huge meteorite hit the Earth at this time.

Many features found around and associated with an impact crater on Mexico's Yucatán Peninsula suggest that this site is the crater associated with the death of the dinosaurs. The Chicxulub crater is about 66 million years old and lies half-buried beneath the waters of the Gulf of Mexico and half on land. Tsunami deposits of the same age are found in inland Texas, much of the Gulf of Mexico, and the Caribbean, recording a huge tsunami perhaps several hundred feet (a hundred meters) high generated by the impact. The crater is at the center of a huge field of scattered spherules that extends across Central America and through the southern United States. The large structure is the right age to be the crater that resulted from the impact at the Cretaceous-Tertiary boundary, recording the extinction of the dinosaurs and other families.

The 66-million-year-old Deccan flood basalts, also known as traps, cover a large part of western India and the Seychelles. They are associated with the breakup of India from the Seychelles during the opening of the Indian Ocean. Slightly older flood basalts (90–83 million years old) are associated with the breaking away of Madagascar from India. The volume of the Deccan traps is estimated at 5 million cubic miles (20,841,000 km^3), and the volcanics are thought to have been erupted within about 1 million years, starting slightly before the great Cretaceous-Tertiary extinction. Most workers now agree that the gases released during eruption of the flood basalts of the Deccan traps stressed the global biosphere to such an extent that many marine organisms had gone extinct, and many others were stressed. Then the massive Chicxulub impactor hit the planet, causing the extinction including the end of the dinosaurs. Faunal extinctions have been correlated with the eruption of the Deccan flood basalts at the Cretaceous-Tertiary (K-T) boundary. There is still considerable debate about the relative significance of flood basalt volcanism and impacts of meteorites for the Cretaceous-Tertiary boundary. Most scientists would now agree, however, that the global environment was stressed shortly before the K-T boundary by volcanic-induced climate change, and then a huge meteorite hit the Yucatán Peninsula, forming the Chicxulub impact crater, causing the K-T boundary extinction and the death of the dinosaurs.

CENOZOIC TECTONICS AND CLIMATE

Cenozoic global tectonic patterns are dominated by the opening of the Atlantic Ocean, closure of the Tethys Ocean, and formation of the Alpine-Himalayan Mountain System, and mountain building along the western North American cordillera. Uplift of mountains and plateaus and the movement of continents severely changed oceanic and atmospheric circulation patterns, changing global climate patterns.

As the North and South Atlantic Oceans opened in the Cretaceous, western North America was experiencing contractional orogenesis. In the Paleocene (66–58 Ma) and Eocene (58–37 Ma), shallow dipping subduction beneath western North America caused uplift and basin formation in the Rocky Mountains, with arc-type volcanism resuming from later Eocene through late Oligocene (about 40–25 Ma). In the Miocene (starting at 24 Ma), the Basin and Range Province formed through crustal extension, and the formerly convergent margin in California was converted to a strike-slip or transform margin, causing the initial formation of the San Andreas fault.

The Cenozoic saw the final breakup of Pangaea and closure of the tropical Tethys Ocean between Eurasia and Africa, Asia, and India and a number of smaller fragments that moved northward from the southern continents. Many fragments of Tethyan Ocean floor (ophiolites) were thrust upon the continents during the closure of Tethys, including the Semail ophiolite (Oman), Troodos (Cyprus), and many Alpine bodies. Relative convergence between Europe and Africa, and Asia and Arabia plus India continues to this day, and is responsible for the uplift of the Alpine-Himalayan chain of mountains. The uplift of these mountains and the Tibetan Plateau has had important influences on global climate, including changes in the Indian Ocean monsoon and the cutting off of moisture that previously flowed across southern Asia. Vast deserts such as the Gobi were thus born.

The Tertiary began with generally warm climates, and nearly half of the world's oil deposits formed at this time. By the mid-Tertiary (35 Ma) the Earth began cooling again, culminating in the Ice House climate of the Pleistocene, with many glacial advances and retreats. The Atlantic Ocean continued to open during the Tertiary, which helped lower global temperatures. The Pleistocene experienced many fluctuations between warm and cold climates, called glacial and interglacial stages (the Earth is currently in the midst of an interglacial stage). These fluctuations are rapid—for instance, in the past 1.5 million years the Earth has experienced 10 major and 40 minor periods of glaciation and interglaciation. The most recent glacial period peaked 18,000 years ago when huge ice sheets covered most of Canada and the northern United States, and much of Europe.

The human species developed during the Holocene Epoch (since 10,000 years ago). The Holocene is just part of an extended interglacial period in the planet's current ice house event, raising important questions about how the human species will survive if climate suddenly changes back to a glacial period. Since 18,000 years ago the climate has warmed by several or more degrees, sea level has risen 500 feet (150 m), and atmospheric CO_2 has increased. Some of the global warming is human induced. One scenario of climate evolution is that global temperatures will rise, causing some of the planet's ice caps to melt and raising the global sea level. This higher sea level may increase the Earth's reflectance of solar energy, suddenly plunging the planet into an ice house event and a new glacial advance.

See also CLIMATE; CLIMATE CHANGE; GEOCHRONOLOGY; NORTH AMERICAN GEOLOGY; PALEOMAGNETISM; PALEONTOLOGY; PLATE TECTONICS; SEQUENCE STRATIGRAPHY; STRATIGRAPHY, STRATIFICATION, CYCLOTHEM; SUPERCONTINENT CYCLES.

FURTHER READING

Kious, Jacquelyne, and Robert I. Tilling. U.S. Geological Survey. "This Dynamic Earth: The Story of Plate Tectonics." Available online. URL: http://pubs.usgs.gov/gip/dynamic/dynamic.html. Updated March 27, 2007.

Pomerol, Charles. *The Cenozoic Era: Tertiary and Quaternary.* Chichester, U.K.: Ellis Horwood, 1982.

Proterero, Donald, and Robert Dott. *Evolution of the Earth.* 6th ed. New York: McGraw Hill, 2002.

Stanley, Steven M. *Earth and Life Through Time.* New York: W. H. Freeman, 1986.

Holmes, Arthur (1890–1965) English *Geochronologist and Geologist* Arthur Holmes is considered to be the father of geochronology and study of the age of the Earth. He was the first to complete a uranium-lead radiometric dating experiment to determine the age of a rock. He was one of the early proponents of the model of continental drift at a time when most of the scientific community was strongly against the idea, and in 1956 he received one of geoscience's highest awards, the Wollaston Medal, and also has the European Geoscience Union's "Arthur Holmes Medal" named after him.

DEBATE ON THE AGE OF THE EARTH

Arthur Holmes was born on January 14, 1890, in Gateshead, England, to a cabinetmaker David Holmes and a former schoolteacher Emily Dickinson. Gateshead High School provided Arthur with a strong background in the sciences and an opportunity to develop his musical abilities in the Operatic Society. His teacher introduced him to the age of the Earth debate that had been recently refueled by the discovery of radioactivity. In 1897 Lord Kelvin (1824–1907), a professor of natural history at Glasgow University and an eminent expert of thermodynamics, announced his newest estimation for the age of the Earth. Believing that the Earth had been gradually cooling from its molten genesis, Lord Kelvin calculated that the Earth's crust consolidated 20 million years ago, based on experimentally determined temperatures at which rocks melt and their rate of cooling. Now his long accepted estimation was being challenged, and not by geologists, who seemed to be intimidated by his stature, but by physicists.

Near the end of the 19th century the age of the Earth was a popular topic for research and discussion among geologists, who thought the Earth was an order of magnitude older than Lord Kelvin claimed. Professor John Joly of Trinity College, Dublin, supported the salinity method for estimating the age of the Earth. As the newly formed globe cooled, water condensed and formed the oceans. The water would initially be pure, but as rocks decomposed and washed over the land into the seas, the water would become saltier. Using this assumption, if one measured the salinity of the oceans at two time points separated by a few hundred years, then one could extrapolate back to estimate how much time has passed since the water was pure, that is, when the Earth's crust solidified. From the rate calculated for salt accumulation from erosion, Joly estimated the oceans to be more than 90 million years old. One criticism of this method was that it required the rocks to lose more salt than they ever contained to supply the calculated amounts to the oceans each year. Alternatively, Irish geologist Samuel Haughton employed the simple concept that thicker strata took longer to form to estimate the Earth's age. After figuring that sediments accumulated on the ocean floor at a rate of one foot (30.5 cm) in 8,616 years, he estimated that it would require at least 200 million years, or possibly 10 times longer, to lay down the total thickness of rock covering the planet. Problems with this method included inaccurate estimations of the total thickness of rock on the Earth's surface and sedimentation rates that differed significantly according to time and place. Even without a satisfactory means of measurement, Lord Kelvin's revised approximation of 20–40 million years appeared to be a major underestimate.

Then in 1896 French physicist Henri Becquerel discovered natural radioactivity when he observed that uranium emitted invisible rays of energy. Polish physicist Marie Skłodowska Curie studied the ema-

nations for her doctoral dissertation and found that thorium also emitted such rays, and furthermore, the emanations were a property of atoms and not due to a chemical reaction. Curie named the revolutionary phenomenon radioactivity, and she and her husband, Pierre Curie, proceeded to discover two new radioactive elements, radium and polonium. Ernest Rutherford (1871–1937) and Frederick Soddy (1877–1956) explained radioactivity as the result of the instability of an element that spontaneously emitted particles from its nucleus. For example, uranium released helium atoms as it decayed. In the process an element could transform into another element. Some radioactive elements had very long half-lives; uranium took 4.5 billion years to decay to half of its original amount. In 1905 Rutherford suggested that radioactive decay could be used as a geological timekeeper. Using uranium/helium ratios, he determined the age of a sample of pitchblende to be 90 million years, but he incorrectly assumed that helium did not escape over time.

Pierre Curie and colleague Albert Laborde announced in 1903 that radium emitted enough heat to melt its own weight in ice in less than one hour. This finding that radioactive elements generate heat refueled the debate over the age of the Earth. Lord Kelvin's calculations depended on the Earth's slow cooling in an absence of any external heat source, and these physicists claimed that radioactive elements within the Earth provided enough heat to make his calculations worthless. Holmes was intrigued by these scientists who challenged the authoritative Lord Kelvin and by the potential utility of this phenomenon called radioactivity. As a teenager, witnessing the debate between one of the world's most established scholars and a few lesser known but equally accomplished physicists made a great impression upon Holmes. Now, he was also interested in radioactivity, and these two curiosities would merge to become his lifelong passion, using radioactive decay to determine the age of the Earth.

A YOUNG GEOCHRONOLOGIST

Holmes earned a National Scholarship Award in physics and enrolled at the Royal College of Science in London in 1907. The curriculum required all students to take mathematics, mechanics, chemistry, and physics during their first year, and Holmes took an elective geology course in his second year. The president of the Geological Society, William Watts, taught the course and enticed Holmes to change his course of study during his third year. Fortuitously, Robert J. Strutt (1875–1947) from the Cavendish Laboratory at Cambridge University had joined the college at the same time Holmes enrolled. Strutt was one of the physicists who made public his belief that radioactive elements provided a source of heat sufficient to discredit Lord Kelvin's young estimation for the age of the Earth. Strutt invited Holmes to assist him in examining helium trapped in rocks following radioactive decay. He thought that if they could measure the amount of accumulated helium and establish its rate of production, then they could calculate the age of the rock. The concept seemed simple, but determining the rate of helium production was not a straightforward process. Because helium is a gas, an unknown but significant quantity escapes as it is produced, so only the minimum age could be estimated. (Uranium/helium measurements later were considered unreliable since the helium was not retained consistently.) After graduating from Imperial College (formerly the Royal College of Science) in 1910, Holmes assumed this research project with Strutt as a postgraduate student.

Across the ocean, American chemist Bertram Boltwood (1870–1927) had recently determined that lead was the final product of uranium decay, and he attempted to date several rocks using uranium/lead ratios. From 26 rocks he obtained ages ranging from 92 to 570 million years. Since helium could escape from rocks over time, he thought focusing on the end product would yield more accurate results. Unknown to chemists at the time, Boltwood's analysis was flawed because of the existence of several isotopes of both uranium and lead. From his results Boltwood constructed a rough list of geological ages.

Holmes was anxious to use radioactivity to measure the age of a rock, selecting a Devonian rock from Norway that contained 17 different radioactive minerals so he could check each result against the others. After crushing the rock, extracting the minerals, and chemically separating them for analysis, he determined the ratios of uranium and lead and estimated the rock to be 370 million years old. He analyzed several others, dating the oldest at 1,640 million years, then calculated ages of geological periods from measurements published by Boltwood. Holmes wrote up his results showing that as the ratio of lead to uranium increased, so did the age of the rock (since uranium decays into lead), but he wondered if some lead was already present, which would have rendered his analysis flawed.

Strutt presented Holmes's results in April 1911 at a Royal Society meeting, where fellow geologists seemed interested but were wary of the radiometric dating technique. They questioned whether it was acceptable to assume the uranium decay rate was constant, and had trouble accepting the possibility that the Earth was more than 1 billion years old. Though geologists were looking for evidence indicating the Earth was older than 20 million years and knew the old techniques relied on rates of nonuni-

form processes, they were expecting a value closer to 100 million years as suggested by rates of sedimentation and salt accumulation, the so-called hourglass methods.

MOZAMBIQUE

In 1911 Holmes obtained a position as a geological prospector for Memba Minerals Limited. After giving his research results to Strutt, Holmes left England for Mozambique in March, beginning a physically difficult six-month expedition in search of economically valuable minerals. While there, Holmes contracted malaria, and high fevers occasionally forced him to rest for several days. Lying in bed, he could not stop thinking about radiometric dating and contemplated how he could reconcile data obtained by radiometric methods with data calculated from sedimentation rates. Without access to geology textbooks or journals, he used his memory to approximate the amount of original igneous rocks from which sediments had been derived, then figured out how long it would have taken for the sediments to be deposited. His estimate was 325 million years since the base of the Cambrian period, not too far from the value of 543 million years that he obtained using radiometric methods. He wrote a friend asking him to publish his results.

Though prospecting for precious minerals was unsuccessful in Mozambique, during the trip Holmes developed a lifelong interest in Precambrian time and a new commitment to constructing a geological timescale. He collected zircons, minerals good for age determinations, and several samples of never-examined Precambrian rock types. As the prospectors headed home, Holmes became gravely ill with black water fever. The nuns at the hospital in Mozambique prematurely telegraphed news of his death to London, but Holmes miraculously recovered and arrived in Southampton in November 1911. He continued to suffer from bouts of malaria for years afterward.

THE PROBLEM WITH LEAD

In 1912 Imperial College offered Holmes a position as a demonstrator in geology, and in July 1914 the 23-year-old geologist married Margaret Howe. Holmes kept busy lecturing and researching the petrographical material he brought back from Mozambique. When World War I broke out in August, the military declared Holmes unfit for military service because of his recurring bouts of malaria. His contributions toward the war effort included making scaled topography maps for naval intelligence and researching alternative sources of potash, an ingredient of fertilizer formerly supplied to Great Britain by Germany.

To convince his contemporaries of the usefulness of radiometric dating, he composed *The Age of the Earth* (1913), a review of the historical methods for estimating ages of geological materials that also presented all the current related evidence and contrasted the results obtained by different techniques. In this book he pointed out problems with the other approaches and defended his own estimation of 1,600 million years based on uranium/lead measurements.

The possibility that "ordinary" lead, in existence since the formation of the Earth, was already present in the rock samples before any radioactive decaying took place was troublesome. Another difficulty in using lead measurements was that in addition to uranium, the radioactive element thorium also decayed to lead. To overcome this Robert Lawson, a friend from childhood who worked at the Radium Institute of Vienna, determined the atomic masses of the three lead isotopes, enabling one to adjust age calculations accordingly based on the proportions of each type.

Holmes thought he resolved a means to determine the age of rocks with accuracy, but he did not know yet that uranium also had another isotope. Uranium 238 makes up 99 percent of the total uranium, but uranium 235 decays at a faster rate, and Holmes unknowingly included its end product as part of the ordinary lead. This isotope hitch prevented skeptics from recognizing the promise of this new dating technique.

WORK IN THE PETROLEUM INDUSTRY

By the end of World War I Holmes had written three books but was still only a demonstrator at Imperial College. In 1918 the Holmeses had their first child, Norman, and a demonstrator's income was not sufficient to support the family. The Yomah Oil Company hired Holmes as chief geologist with the promise of a much larger salary. His family moved to Burma in November 1920 and settled in Yenangyaung, where Holmes spent two years frantically searching for new oil finds to save the struggling company. Loyalty to the company kept him working long after the then bankrupt company stopped paying him, and before they finally returned to England in late 1922, Norman died from severe dysentery.

Without an institutional affiliation, Holmes could not secure funding to continue his research. For a while he worked in a fur, brass goods, and knick-knack shop that he opened with Maggie's cousin. His marriage was deteriorating, but soon Maggie was pregnant with their second son, born in February 1924. The year Geoffrey was born, the University of Durham happened to be expanding its science programs and they needed a reader for geology. Holmes gratefully accepted the offered position. The following year he became head of the geology department, of which he was the only faculty member. He was a popular lecturer, and the few students who came through the geology department each year thought he was a fair teacher and a caring mentor.

THE POWERFUL ENGINE OF RADIOACTIVITY

While Holmes's major passion was finding absolutes for geological time, he was also knowledgeable about other subjects. In 1915 German meteorologist Alfred Wegener (1880–1930) proposed the theory of continental drift, suggesting all the continents once were part of an enormous supercontinent that broke into pieces leading to the present distribution of continental masses. This model was exciting because it explained many unusual geological (and biological and climatological) phenomena, but most geologists hesitated to accept it without a plausible mechanism. The English translation of Wegener's book *The Origin of the Continents and Oceans* in 1924 aroused a heated debate, and Holmes was among the few geologists progressive enough to entertain the idea.

Aware of the enormous energy provided by radioactivity, Holmes suggested that the intense heat generated by the radioactive decay of unstable elements within the Earth's interior was a sufficiently powerful engine for moving continents. The substratum, or mantle, was solid, but he thought that over millions of years it behaved like a thick liquid. Holmes proposed thermal convection as a means to dissipate the heat, causing the cooler material close to the surface to sink, leaving space for hotter, less dense material to rise into and fill. In December 1929 Holmes proposed to the Geological Society of Glasgow that convection currents were responsible for continental drift. He explained that as convection currents in the mantle cooled and descended, they could drag continents horizontally across the Earth's surface. "Radioactivity and Earth Movements," his seminal paper, was published in the *Transactions of the Geological Society of Glasgow* in 1931. Though Holmes had a respectable scientific reputation, his ideas were mostly ignored until the 1960s, when American geophysicists Harry Hammond Hess and Robert Sinclair Dietz independently proposed the concept of seafloor spreading, which in combination with continental drift has evolved into the well-supported theory of plate tectonics.

Holmes was invited to lecture around the world, including in the United States in 1932. He took advantage of this opportunity to solicit help in constructing a geological timescale from American scientists. His demanding schedule forced Durham to hire another lecturer in the geology department in 1933; they selected Doris L. Reynolds, a notable petrologist with whom Holmes had been having an affair since

1931. In 1938 Maggie died from stomach cancer, and Holmes married Reynolds in 1939.

A FOURTH ISOTOPE

Since his days in Mozambique Holmes wanted to develop a geological timescale with dates defining the beginning of each period and epoch. The work of previous geologists allowed for the construction of a geological column organized by characteristics of the layers of rock and by the distinctive index fossils contained within the strata. These historical developments permitted relative ordering but not the assignment of specific dates for the time periods. Holmes needed a framework on which to build and asked chemistry professor Fritz Paneth in Berlin for assistance. In 1928 Paneth developed a precise assay for measuring very small amounts of helium and used it to analyze two famous rocks from known geological periods: the Whin Sill from the late Carboniferous period and the Cleveland Dyke from the middle to early Tertiary period. Paneth dated the Whin Sill at 182 million years and the Cleveland Dyke at 26 million years, ages that seemed to agree with the geological evidence. (Today the Whin Sill and Cleveland Dyke are believed to be approximately 295 and 60 million years old respectively.) These two rocks did not have enough lead to determine lead ages as controls, but Holmes was anxious to make progress and confident that a geological timescale was possible.

Geologists were now more accepting of the longer estimates for the age of the Earth, which Holmes reported in his second edition of *The Age of the Earth* (1927) to be between 1,600 and 3,000 million years based on the uranium and lead measurements. The book also contained a geological timescale based on lead ratios and helium ratios, but two decades brought little progress—Holmes summarized all the computed mineral ages in a single short table.

The invention of the first mass spectrograph by English chemist Francis William Aston enabled the identification of isotopes with different atomic weights. Aston used his mass spectrograph to discover no fewer than 212 naturally occurring isotopes and was awarded the Nobel Prize in chemistry for 1922. The mass spectrograph evolved into the more advanced modern mass spectrometer that separates isotopes by passing them through a magnetic field that deflects them to different degrees based on their mass and the charge of the field. In the late 1920s Aston clearly identified three known lead isotopes, a finding with major implications for radioactive dating. He also unexpectedly noted that the isotope believed to be ordinary lead was in fact an end product from the decay of another less abundant uranium isotope that Rutherford helped identify as uranium 235. Rutherford estimated the uranium 235 decay rate, assumed that at the time of the Earth's formation uranium 235 and uranium 238 were present in equal amounts, and calculated the time it would have taken for the equal amounts of the two isotopes to decay to their current ratios. He obtained an astounding value of 3,400 million years, but his results and similar results from a few other geologists were mostly ignored.

If uranium 235 decayed to lead 207, then did so-called ordinary lead exist at all? In 1937 a physicist from Harvard University, Alfred Nier, began exploring the questionable existence of ordinary lead using a new mass spectrometer kept ultraclean to ensure no contaminating lead was present. He easily identified the three known lead isotopes, but he also observed a tiny amount of a fourth lead isotope with an atomic mass of 204 not produced by radioactive decay.

PRIMEVAL LEAD COMPOSITION

By the 1940s geologists had accepted the billion-year range for the age of the Earth, but no one had assigned absolute times to the geological timescale. One problem was that accurate ages could be obtained only from igneous rocks, since they contained high amounts of lead, but it was hard to know their geological age. Nier surmised that since two isotopes of uranium decayed to lead (uranium 238 decayed to lead 206 with a half-life of 700 million years, and uranium 235 decayed to lead 207 with a half-life of 4.5 billion years), comparison of both lead isotope growth rates to the constant value of the fourth isotope, lead 204, would reveal an accurate age. To test the validity of this lead-lead method, he dated 25 ancient lead ores with very low lead ratios and found the calculated ages from the different "clocks" generally agreed.

Holmes employed a novel approach for determining an accurate age of the Earth. His method depended on primeval isotope ratios, the ratios present during the Earth's genesis. Since uranium and thorium were formed, they have been decaying continually, while the amount of ordinary lead has remained stable. Since the primeval lead composition would have been trapped and fossilized inside minerals as the newly formed crust solidified, one should be able to locate minerals with ancient parts of the Earth's crust from which the primeval composition of lead could be identified. Then one could calculate the time elapsed since the Earth's primeval mix of lead isotopes began to be contaminated by radiogenic lead. With the assistance of a newly acquired Marchant calculating machine, Holmes used this method to calculate confidently the age of a rock sample from an ancient galena (lead ore) from Greenland to be 3 billion years. He rightly felt this was a defining moment in his efforts to date the age of the Earth.

Under the assumption that the galena represented primeval lead, Holmes used Nier's measurements to calculate more than 1,400 solutions for the age of the Earth. When he plotted the frequencies of each computed age, he obtained a well-defined peak at 3,350 million years. Though the rock samples that he used actually did not contain the primeval lead values, the mathematical approach he developed formed the basis of the one used today. Because the German Fiesel Houtermans used a similar technique, the method is referred to as the Holmes-Houtermans model for dating the Earth. By this time in 1946 geologists accepted isotope dating but still debated the best means for applying its use.

Next Holmes constructed a geological timescale that reconciled his new radiometric results with the hourglass estimations. He began by collecting information on sediment thicknesses from around the world, then plotted the thicknesses to scale for the whole geological column from the present day to the base of the Cambrian period. He incorporated dates calculated from Nier's data by using primeval lead ratios, plotted the five most probable ages as control points, and extrapolated to estimate dates for the bases of other geological periods. He published his results, "The Construction of a Geological Time Scale," in *Transactions of the Geological Society of Glasgow* in 1947, aware that accumulation rates were not constant. He frequently adjusted his scale to accommodate corrected information and to include additional data obtained by new techniques, and in 1959 he published "A Revised Geological Time Scale" in *Transactions of the Edinburgh Geological Society*. In 1953 Clair Patterson and Harrison Brown obtained accurate primeval lead isotope measurements from an iron meteorite formed at the same time as the Earth and computed the value of 4.55 billion years for the age of the Earth. Though in the last 50 years scientists have obtained new data and made adjustments to improve accuracy, the assessment of 4.55 billion years remains the currently accepted value.

FINAL YEARS

Having accomplished his two career goals of dating the Earth and constructing a geological timescale that could be applied to common rocks, Holmes concentrated on his professorial duties. In 1943 the University of Edinburgh appointed him regius professor of geology, a position subsidized by the king of England. The outbreak of World War II forced Holmes to reduce his geology course from one year to six months. Though it would have saved lecture time to assign students reading material before coming to class, no geology textbook contained information about the recent developments in the field, such as radiometric dating and continental drift. Holmes took it upon himself to compose a book based on his lecture notes. When he published *Principles of Physical Geology* in 1944, it became an immediate best seller and was reprinted more than 18 times during the following 20 years.

Holmes became ill in 1948 and lost all interest in his work. The doctor ordered complete rest, and he and his wife spent the summer in Ireland. After recovering, Holmes focused on Precambrian geology and revised Africa's geological map based on radiometric dates. His heart began to deteriorate, and in 1956 he retired from the University of Edinburgh.

The distinguished geologist belonged to numerous scientific organizations and received many honors and awards during his career. The Geological Society of London gave Holmes their Murchison Medal in 1940 and their highest award, the Wollaston Medal, in 1956. The Geological Society of America awarded Holmes their Penrose Medal in 1956 for his outstanding contributions in the science of geology. In 1964 Holmes received the Vetlesen Prize, the greatest honor for a geologist, for his "uniquely distinguished achievement in the sciences resulting in a clearer understanding of the Earth, its history, and its relation to the universe." His health was too frail to travel to Columbia University for the award ceremony, but he did find the strength to tackle one more major project, revising *Principles of Physical Geology*. He finished just a few months before he died of bronchial pneumonia on September 20, 1965, in London.

Arthur Holmes was a quiet man but did not avoid the controversial topics of geology in his day, in particular, the antiquity of the Earth and continental drift. His background in physics convinced him radiometric dating was the most accurate means for determining the age of rocks and the Earth. This made him the right man for the job of providing actual ages for Earth's geological episodes. The implications of Holmes's estimate for an ancient Earth were widespread; they forced astronomers to reexamine the age of the universe and gave biologists reasonable time to allow for the occurrence of evolutionary processes. Though his contributions toward advancing the idea of drifting continents are often overlooked, Holmes was the first to propose convection currents as a plausible moving force. Today scientists believe the Earth formed 4.55 billion years ago because the father of geological time had a passion for seeking the truth and dedicated himself to laying the groundwork for using the natural geological clocks within the rocks.

See also GEOCHRONOLOGY; HESS, HARRY; PRECAMBRIAN.

FURTHER READING

Dunham, K. C. "Arthur Holmes (1890–1965)." In *Biographical Memoirs of Fellows of the Royal Society.* Vol. 12 (November 1966), 291–310. London: Royal Society, 1966.

Lewis, Cherry. *The Dating Game: One Man's Search for the Age of the Earth.* New York: Cambridge University Press, 2000.

U.S. Geological Survey. *Geologic Time: Online Edition.* Available online. URL: http://pubs.usgs.gov/gip/geotime/. Accessed January 30, 2009.

hot spot A hot spot is a center of volcanic and plutonic activity not associated with an arc and generally not associated with an extensional boundary. Most hot spots are 60–125 miles (100–200 km) across and are located in plate interiors. A few, such as Iceland, are found on oceanic ridges and are identified on the basis of unusually large amounts of volcanism on the ridge. Approximately 200 hot spots are known, and many others have been proposed but their origin is uncertain.

Hot spots are thought to be the surface expression of mantle plumes that rise from deep in the Earth's mantle, perhaps as deep as the core/mantle boundary. As the plumes rise to the base of the lithosphere, they expand into huge, up to 600-mile (1,000-km) wide plume heads, parts of which partially melt the base of the lithosphere and rise as magmas in hot spots of plate interiors.

HAWAIIAN HOT SPOT

The most famous hot spot in the world consists of the chain of the Hawaiian Islands, extending northwest to the Emperor Seamount chain. Hawaii is a group of eight major and about 130 smaller islands in the central Pacific Ocean. The islands are volcanic in origin, having formed over a magmatically active hot spot that has melted magmatic channels through the Pacific plate as it moves over the hot spot, forming a chain of southeastward younging volcanoes over the hot spot. Kilauea volcano on the big island of Hawaii is the world's most active volcano and often has a lava lake with an actively convecting crust developed in its caldera. The volcanoes are made of low-viscosity basalt and form broad shield types of cones that rise from the seafloor. Only the tops are exposed above sea level, but if the entire height of the volcanoes above the seafloor is taken into account, the Hawaiian Islands form the tallest mountain range on Earth.

Halemaumau crater on Kilauea, Hawaii *(G. Brad Lewis/Riser/Getty Images)*

Pillow lava off the coast of Hawaii, formed when lava from the Hawaiian hot spot flowed into the ocean *(OAR/National Undersea Research Program/Photo Researchers, Inc.)*

The Hawaiian Islands are part of the Hawaiian-Emperor seamount chain that extends all the way to the Aleutian-Kamchatka trench in the northwest Pacific, showing both the great distance the Pacific plate has moved and the longevity of the hot spot magmatic source. From east to west (and youngest to oldest) the Hawaiian Islands include Hawaii, Maui and Kahoolawe, Molokai and Lanai, Oahu, Kauai, and Niihau.

The volcanic islands are fringed by coral reefs and have beaches with white coral sands, black basaltic sands, and green olivine sands. The climate on the islands is generally mild, and numerous species of plants lend a paradiselike atmosphere to the islands, with tropical fern forests and many species of birds. The islands have few native mammals (and no snakes), but many have been introduced. Some of the islands such as Niihau and Molokai have drier climates, and Kahoolawe is arid.

ICELAND HOT SPOT

The mid-Atlantic ridge rises above sea level on the North Atlantic island of Iceland, lying 178 miles (287 km) off the coast of Greenland and 495 miles (800 km) from the coast of Scotland. Iceland has an average elevation of more than 1,600 feet (500 m), and owes its elevation to a hot spot interacting with the midocean ridge system beneath the island. The mid-Atlantic ridge crosses the island from southwest to northeast and has a spreading rate 1.2 inches per year (3 cm/yr), with the mean extension oriented toward an azimuth (compass direction) of 103 degrees east of north. The oceanic Reykjanes ridge and sinistral transform south of the island rises to the surface and continues as the Western Rift zone. Active spreading is transferred to the Southern Volcanic zone across a transform fault called the South

Iceland Seismic zone, then continues north through the Eastern Rift zone. Spreading is offset from the oceanic Kolbeinsey ridge by the dextral Tjornes fracture zone off the island's northern coast.

During the past 6 million years the Iceland hot spot has drifted toward the southeast relative to the north Atlantic, and the oceanic ridge system has made a succession of small jumps so that active spreading has remained coincident with the plume of hottest, weakest mantle material. These ridge jumps have caused the active spreading to propagate into regions of older crust that have been remelted, forming alkalic and even silicic volcanic rocks deposited unconformably over older tholeiitic basalts. Active spreading occurs along a series of 5–60-mile (8–100 km) long zones of fissures, graben, and dike swarms, with basaltic and rhyolitic volcanoes rising from central parts of fissures. Hydrothermal activity is intense along the fracture zones, with diffuse faulting and volcanic activity merging into a narrow zone within a few kilometers depth beneath the surface. Detailed geophysical studies have shown that magma episodically rises from depth into magma chambers located a few miles (kilometers) below the surface, then dikes intrude the overlying crust and flow horizontally for tens of miles to accommodate crustal extension of several to several tens of feet over several hundred years.

Many Holocene volcanic events are known from Iceland, including 17 eruptions of Hekla from the Southern Volcanic zone. Iceland has an extensive system of glaciers and has experienced a number of eruptions beneath the glaciers that cause water to infiltrate the fracture zones. The mixture of water and magma induces explosive events including Plinian eruption clouds, phreo-magmatic, tephra-producing eruptions, and sudden floods known as jokulhlaups, induced when the glacier experiences rapid melting from contact with magma. Many Icelanders have learned to use the high geothermal gradients to extract geothermal energy for heating, and to enjoy the many hot springs on the island.

YELLOWSTONE HOT SPOT

The northwest corner of Wyoming and adjacent parts of Idaho and Montana were established as Yellowstone National Park in 1872 by President Ulysses S. Grant, and it remains the largest national park in the conterminous United States. The park serves as a large nature preserve and has large populations of moose, bear, sheep, elk, bison, numerous birds, and a diverse flora. The park sits on a large upland plateau resting at about 8,000 feet (2,400 m) elevation straddling the continental divide. The plateau is surrounded by mountains that range from 10,000 to 14,000 feet (3,000–4,250 m) above sea level. Most of the rocks in the park formed from a massive volcanic eruption that occurred 600,000 years ago, forming a collapse caldera 28 miles (45 km) wide and 46 miles (74 km) long. Yellowstone Lake now largely occupies the deepest part of the caldera. The region is still underlain by molten magma that heats the groundwater system, which boasts more than 10,000 hot springs, 200 geysers, and numerous steaming fumaroles, and hot mud pools. The most famous geyser in the park is Old Faithful, which erupts an average of once every 64.5 minutes blowing 11,000 gallons (41,500 L) of water 150 feet (46 m) into the air. The most famous hot springs include Mammoth hot springs, on the northern side of the park, where giant travertine and mineral terraces have formed from the spring, and where simple heat-loving (thermophilic) organisms live in the hot waters. Other remarkable features of the park include the petrified forests buried and preserved by the volcanic ash, numerous volcanic formations including black obsidian cliff, and waterfalls and canyons including the spectacular Lower Falls in the Grand Canyon of the Yellowstone.

The massive eruption from Yellowstone caldera 600,000 years ago covered huge amounts of the western United States with volcanic ash. If such an eruption were to occur today, the results would be devastating, with perhaps 20 percent of the lower United States covered with thick, hardened ash and burning fumes extending across the whole country. There has been some concern recently about an increase in some of the thermal activity in Yellowstone, although it is probably related to normal changes within the complex system of heated groundwater and seasonal or longer changes in the groundwater system. First, Steamboat geyser, which had been quiet for two decades, began erupting in 2002. New lines of fumaroles formed around Nymph Lake, including one line 250 feet long (75 m) that forced the closure of the visitor trail around the geyser basin. Other geysers, have seen temperature increases from 152°F (67°C) to 190°F (88°C) over a several-month period. Other changes include a greater discharge of steam from some geysers, changes in the frequency of eruptions, and a greater turbidity of thermal pools. Perhaps most worrisome is the discovery of a large bulge beneath Yellowstone Lake, although its age and origin are uncertain. Fears are that the bulge may be related to the emplacement of magma to shallow crustal levels, a process that sometimes precedes eruptions. But the bulge was recently discovered because new techniques are being used to map the lake bottom. The feature has an unknown age and may have been there for decades to hundreds of years.

Yellowstone Park is underlain by a hot spot, the surface expression of a mantle plume. As the North American plate has migrated 280 miles (450 km) southwestward with respect to this hot spot in the past 16 million years, the volcanic effects migrated from the Snake River Plain to the Yellowstone Plateau. There is currently a parabolic-shaped area of seismicity, active faulting, and centers of igneous intrusion centered around the parabolic area, all of which are migrating northeastward. Heat and magma from this mantle plume has emplaced as much as 7.5 miles (12 km) of mafic magma into the continental crust over the plume along this trace, causing the surface eruptions of the massive Snake River Plain flood basalts, and the Yellowstone volcanics. On geological timescales massive volcanism and other effects of this hot spot will likely continue and also slowly move northeast.

See also CONVECTION AND THE EARTH'S MANTLE; ENERGY IN THE EARTH SYSTEM; VOLCANO.

FURTHER READING

Fisher, R. V. *Out of the Crater: Chronicles of a Volcanologist.* Princeton, N.J.: Princeton University Press, 2000.

Francis, Peter. *Volcanoes: A Planetary Perspective.* Oxford: Oxford University Press, 1993.

Hawaiian Volcano Observatory. Available online. URL: http://hvo.wr.usgs.gov/. Accessed November 2, 2008.

MacDougall, J. D., ed. *Continental Flood Basalts.* Dordrecht, Germany: Kluwer Academic Publishers, 1988.

Mahoney, J. J., and M. F. Coffin, eds. *Large Igneous Provinces, Continental, Oceanic, and Planetary Flood Volcanism.* Washington, D.C.: American Geophysical Union, 1997.

Morgan, Lisa A., David J. Doherty, and William P. Leeman. "Ignimbrites of the Eastern Snake River Plain: Evidence for Major Caldera Forming Eruptions." *Journal of Geophysical Research* 89 (1984): 8,665–8,678.

Morgan, W. Jason. "Deep Mantle Convection Plume and Plate Motions." *American Association of Petroleum Geologists Bulletin* 56 (1972): 202–213.

Rogers, David W., R. William Hackett, and H. Thomas Ore. "Extension of the Yellowstone Plateau, Eastern Snake River Plain, and Owyhee Plateau." *Geology* 18 (1990): 1,138–1,141.

Volcanoworld. Available online. URL: http://volcano.und.edu/. Accessed August 27, 2006.

Hubble, Edwin (1889–1953) American *Astronomer* Edwin Powell Hubble was born on November 20, 1889, in Marshfield, Missouri, but his family moved to Wheaton, Illinois, the same year. In his career Hubble made two important discoveries that changed scientists' understanding of the universe. He was the first to prove the existence of galaxies beyond the Milky Way, and he discovered that the redshift of galaxies increased with their distance from the Milky

Way, showing that the universe is expanding in all directions.

Hubble performed well in grade and high school but paid more attention to sports than academics. He completed a bachelor of science degree at the University of Chicago in 1910, with concentrations in mathematics, astronomy, and philosophy. From 1910 to 1913 Hubble was a Rhodes Scholar at Oxford, England, where he studied jurisprudence and Spanish, then returned to the United States. He was inducted into the Kentucky bar association though he never practiced law. Instead he taught high school and became a basketball coach, until he served in World War I. After the war Hubble returned to astronomy studies at the Yerkes Observatory at the University of Chicago, earning a Ph.D. in 1917 for his dissertation, "Photographic Investigations of Faint Nebula." In 1919 Hubble took a position at the Mount Wilson Observatory near Pasadena, California, where he was the first person to use Palomar's 200-inch Hale Telescope. Edwin Hubble died suddenly on September 28, 1953, of a cerebral thrombosis.

When Hubble arrived at the Mount Wilson observatory in 1919, astronomers believed that the universe did not extend beyond the Milky Way, but Hubble soon made discoveries that dramatically expanded the known universe. Using the 100-inch (254-cm) Hooker Telescope (then the largest telescope in the world) Hubble identified a new type of star, a Cepheid variable, that varied in luminosity with a specific period correlated with the luminosity. On January 1, 1925, Hubble announced a correlation between the distance of these objects and their period/luminosity, showing that they were located at very distant places beyond the Milky Way Galaxy. This discovery fundamentally changed the way astronomers viewed the universe.

Hubble next spent time examining the redshift of distant galaxies. Redshifts of the electromagnetic spectrum occur when the emitted or reflected light from an object is shifted toward the less energetic (red) end of the electromagnetic spectrum by the Doppler effect. This happens for objects moving away from the observer, since the radiation needs to travel a greater distance and increases its wavelength as the object moves away from the observer. Conversely, blueshifts occur when an object is moving toward the observer and the wavelengths of radiation from the object are compressed into a smaller area, causing the wavelength to decrease. Redshift was known for some time, with general knowledge that larger redshifts meant that objects were moving away faster from the observer. Hubble and his colleague Milton Humason plotted the redshifts of 46 distant objects against their distance from Earth

and found a rough proportionality with increasing redshifts with distance. They found a proportionality constant to explain this correlation and stated that the farther the object or galaxy is located from Earth, the faster it is moving away—a statement that later became known as Hubble's law. The current estimate of the constant of proportionality for Hubble's Law is 70.1 +/- 1.3 km/sec/Megaparsec, although Hubble initially estimated it to be higher. The redshift means that the more distant the galaxies, the faster they are moving away from Earth and from one another. This was found to agree with Albert Einstein's equations of general relativity and supported his ideas for a homogeneous isotropic expanding universe. Interestingly, when Einstein formulated his laws of general relativity in 1917, he did not know about the redshift and Hubble's law, so he introduced a cosmological constant (a "fudge factor") into his equations to counter the result that his calculations showed the universe must be expanding. When Hubble announced his results, Einstein retracted his cosmological constant, calling it the biggest blunder of his life, then his calculations agreed with Hubble's observations. Hubble's law is now commonly stated as "the greater the distance between any two galaxies, the greater their speed of separation." Hubble's law is one of the major observations that supports the idea that the universe was created in a big bang, and that all matter is moving away from other matter in a homogenous, isotropically expanding universe.

See also ASTRONOMY; GALAXIES; ORIGIN AND EVOLUTION OF THE UNIVERSE; UNIVERSE.

FURTHER READING

Chaisson, Eric, and Steve McMillan. *Astronomy Today.* 6th ed. Upper Saddle River, N.J.: Addison-Wesley, 2007.

Christianson, Gale. *Edwin Hubble: Mariner of the Nebulae.* New York: Farrar, Straus & Giroux, 1995.

Comins, Neil F. *Discovering the Universe.* 8th ed. New York: W.H. Freeman, 2008.

Snow, Theodore P. *Essentials of the Dynamic Universe: An Introduction to Astronomy.* 4th ed. St. Paul, Minn.: West, 1991.

hurricane   Intense tropical storms with sustained winds of more than 74 miles per hour (119 km/hr) are known as hurricanes if they form in the northern Atlantic or eastern Pacific Oceans, cyclones if they form in the Indian Ocean near Australia, or typhoons if they form in the western North Atlantic Ocean. Most large hurricanes have a central eye with calm or light winds and clear skies or broken clouds, surrounded by an eye wall, a ring of very tall and intense thunderstorms that spin around the eye, with some of the most intense winds and rain of the entire storm system. The eye is surrounded by spiral rain bands that spin counterclockwise in the Northern Hemisphere (clockwise in the Southern Hemisphere) in toward the eye wall, moving faster and generating huge waves as they approach the center. Wind speeds increase toward the center of the storm, and the atmospheric pressure decreases to a low in the eye, uplifting the sea surface in the storm center. Surface air flows in toward the eye of the hurricane, then moves upward, often above nine miles (15 km), along the eye wall. From there it moves outward in a large outflow, until it descends outside the spiral rain bands. Air in the rain bands is ascending, whereas between the rain bands belts of descending air counter this flow. Air in the very center of the eye descends to the surface. Hurricanes drop enormous amounts of precipitation, typically spawn numerous tornadoes, and cause intense coastal damage from winds, waves, and storm surges, where the sea surface may be elevated 10 to 30 feet (3–10 m) above its normal level.

Most hurricanes form in summer and early fall over warm tropical waters when winds are light and the humidity is high. In the North Atlantic hurricane season generally runs from June through November, when the tropical surface waters are warmer than 80°F (26.5°C). They typically begin when a trigger acts on a group of unorganized thunderstorms, causing the air to begin converging and spinning. These triggers are found in the intertropical convergence zone that separates the northeast trade winds in the Northern Hemisphere from the southeast trade winds in the Southern Hemisphere. Most hurricanes form within this zone, between 5° and 20° latitude. When a low-pressure system develops in this zone during hurricane season, the isolated thunderstorms can develop into an organized convective system that strengthens to form a hurricane. Many Atlantic hurricanes form in a zone of weak convergence on the eastern side of tropical waves that form over North Africa, then move westward, where they intensify over warm tropical waters.

For hurricanes to develop, high-level winds must be mild, otherwise they might disperse the tops of the growing thunderclouds. In addition high-level winds must not be descending, since this would also inhibit the upward growth of the thunderstorms. Once the

Hurricane Frances over Florida, September 5, 2004 *(Carolina K. Smith, M.D., Shutterstock, Inc.)*

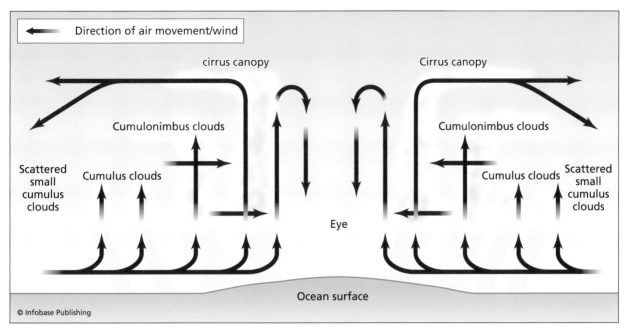

Cross section of typical hurricane showing eye, eye wall, and circulating spiral bands of cumulonimbus clouds

mass of thunderstorms is organized, hurricanes gain energy by evaporating water from the warm tropical oceans. When the water vapor condenses inside the thunderclouds, this heat energy is then converted to wind energy. The upper-level clouds then move outward, causing the storm to grow stronger and decreasing the pressure in the storm's center. The low pressure in the center draws the outlying thunderstorms in toward the surface low, and these rain bands then spiral inward because of the Coriolis force. The clouds spin progressively faster as they move inward, owing to the law of conservation of angular momentum.

The Saffir-Simpson scale classifies the strength of hurricanes by measuring the damage potential of a storm, considering factors such as the central barometric pressure, maximum sustained wind speeds, and potential height of the storm surge.

- Category 1 hurricanes have central pressures greater than 980 millibars, sustained winds between 74 and 95 miles per hour (119–153 km/hr), and a likely 4–5 foot (1–1.5 m) storm surge. Damage potential is minimal, with likely effects including downed power lines, ruined crops, and minor damage to weak buildings.
- Category 2 hurricanes have central barometric pressures between 979 and 965 millibars, maximum sustained winds between 96 and 110 miles per hour (155–177 km/hr), and 6–8 foot (1.8–2.4 m) storm surges. Dam-age is typically moderate, including roof and chimney destruction, beached and splintered boats, destroyed crops, road signs, and traffic lights.
- Category 3 hurricanes have central barometric pressures falling between 964 and 945 millibars, sustained winds between 111 and 130 miles per hour (179–209 km/hr), and storm surges between nine and 12 feet (2.7–3.6 m). Category 3 hurricanes are major storms capable of extensive property damage including uprooting large trees, destroying mobile homes, and demolishing poorly constucted coastal houses. For comparison, Hurricane Katrina was a category 3 storm when it struck New Orleans in 2005.
- Category 4 storms can be devastating, with central barometric pressures falling between 940 and 920 millibars, sustained winds between 131 and 155 miles per hour (211–249 km/hr), and storm surges between 13 and 18 feet (4–5.5 m). These storms typically rip the roofs off homes and businesses, destroy piers, and throw boats well inland. Waves may breach sea walls causing large-scale coastal flooding.
- Category 5 storms are massive, with central barometric pressures dropping below 920 millibars, maximum sustained winds above 155 miles per hour (249 km/hr), and storm surges of more than 18 feet (5.5 m). Storms with this power rarely hit land, but when

they do they can level entire towns, moving large amounts of coastal sediments, and causing large death tolls.

Hurricanes inflict some of the most rapid and severe damage and destruction to coastal regions, and can cause numerous deaths. The number of deaths from hurricanes has been reduced dramatically in recent years owing to an increased ability to forecast the strength and landfall of hurricanes, and the ability to monitor their progress with satellites. The cost of hurricanes in terms of property damage has greatly increased, however, as more and more people build expensive homes along the coast. The greatest number of deaths from hurricanes has been from storm surges. Storm surges typically come ashore as a wall of water that rushes onto land at the forward velocity of the hurricane, as the storm waves on top of the surge are pounding the coastal area with additional energy. For instance, when Hurricane Camille hit Mississippi in 1969 with 200-mile-per-hour winds (322 km/hr), a 24-foot (7.3-m) high storm surge moved into coastal areas, killing most of the 256 who perished in this storm. Winds and tornadoes account for more deaths. Heavy rains from hurricanes also cause considerable damage. Flooding and severe erosion is often accompanied by massive mudflows and debris avalanches, such as those caused by Hurricane Mitch in Central America in 1998. In a period of several days Mitch dropped 25 to 75 inches (63.5–190.5 cm) of rain on Nicaragua and Honduras, initiating many mudslides that were the main cause of the more than 11,000 deaths from this single storm. One of the worst events was the filling and collapse of a caldera on Casitas volcano. When the caldera could hold no more water, it gave way, sending mudflows (lahars) cascading down on several villages and killing 2,000.

Storm surges are water pushed ahead of storms and moving typically on to land as exceptionally high tides in front of severe ocean storms such as hurricanes. Storms and storm surges can cause some of the most dramatic and rapid changes in the coastal zones and are one of the major, most unpredictable hazards to those living along coastlines. Storms that produce surges include hurricanes (which form in late summer and fall) and extratropical lows (which form in late fall through spring). Hurricanes originate in the Tropics and (for North America) migrate westward and northwestward before turning back to the northeast to return to the cold North Atlantic, weakening the storm. North Atlantic hurricanes are driven to the west by the trade winds and bend to the right because the Coriolis force makes objects moving above Earth's surface appear to curve to the right in the Northern Hemisphere. Other weather conditions further modify hurricane paths, such as the location of high- and low-pressure systems and their interaction with weather fronts. Extratropical lows (also known as coastal storms, and north-easters) move eastward across North America and typically intensify when they hit the Atlantic and move up the coast. Both types of storms rotate counterclockwise, and the low pressure at storm center raises the water up to several tens of feet. This extra water moves ahead of the storms as a storm surge that is an additional height of water above the normal tidal range. The wind from the storms adds further height to the storm surge, with the total height of the surge being determined by the length, duration, and direction of wind, plus how low the pressure becomes in the center of the storm. The most destructive storm surges are those that strike low-lying communities at high tide, as the effects of the storm surge and the regular astronomical tides are cumulative. Add high winds and large waves to the storm surge and coastal storms and hurricanes are masters of disaster, as well as powerful agents of erosion. They can remove entire beaches and rows of homes, causing extensive cliff erosion and, significantly, redistributing sands in dunes and the back beach environment. Precise prediction of the height and timing of the approach of the storm surge is necessary to warn coastal residents when they need to evacuate and leave their homes.

Like many natural catastrophic events, the heights of storm surges to strike a coastline are statistically predictable. If the height of the storm surge is plotted on a semilogarithmic plot, with the height plotted in a linear interval and the frequency (in years) plotted on a logarithmic scale, then a linear slope results. This means that communities can plan for storm surges of certain height to occur once every 50, 100, 300, or 500 years, although there is no way to predict when the actual storm surges will occur. One must remember, however, that this is a long-term statistical average, and that one, two, three, or more 500-year events may occur over a relatively short period, but averaged over a long time, the events average out to once every 500 years.

EXTRATROPICAL CYCLONES

Extratropical cyclones, also known as wave cyclones, are hurricane-strength storms that form in middle and high latitudes at all times of the year. Examples of these strong storms include the famous northeasters of New England, storms along the east slopes of the Rockies and in the Gulf of Mexico, and smaller hurricane-strength storms that form in arctic regions. These storms develop along polar fronts that form semicontinuous boundaries between cold polar air and warm subtropical air. Troughs of low pressure

DO BAY OF BENGAL CYCLONES HAVE TO BE SO DEADLY?

Officials estimate that more than 100,000 people perished from tropical cyclone Nargis that hit Myanmar on May 2, 2008. Ninety-five percent of buildings in the coastal region were destroyed by the 12-foot (3.7-m) high storm surge, and more than a million were made homeless and without access to electricity, clean water, or medical care for many months following the disaster. When Cyclone Nargis blew into coastal Myanmar on May 2, it was unfortunately not the first time in recent years that a cyclone has extracted a huge toll on the generally poor, coastal residents of the region.

Bangladesh and the recently devastated coastal Myanmar are densely populated, low-lying regions mostly at or near sea level at the head of the Bay of Bengal. They are delta environments, built where the Ganges, Brahmaputra, and Irrawaddy Rivers drop their sediment eroded from the Himalaya Mountains. These areas sit directly in the path of many Bay of Bengal tropical cyclones (another name for a hurricane), and have been hit by seven of the 10 deadliest hurricane disasters in recorded history.

On November 12 and 13, 1970, a category 3 storm known as the Bhola cyclone hit Bangladesh with 115 mph (184 kph) winds, and a 23-foot (7-m) high storm surge that struck at the astronomically high tides of a full moon. The devastating result caused 500,000 human deaths and half a million farm animals perished. The death toll is hard to estimate in this rural region, with estimates ranging from 300,000 to 1 million people lost in this one storm alone. Most perished from flooding associated with the storm surge that covered most of the deltaic islands on the Ganges River. Again in 1990 another cyclone hit the same area, this time with a 20-foot (6-m) storm surge and 145-mile-per-hour (232 km/hr) winds, killing another 140,000 people and another half-million farm animals. In November 2007 Bangladesh was hit by a powerful category 5 cyclone, with 150 mph (240 km/hr) winds, and was inundated with a 20-foot (6-m) high storm surge. Since the 1990 storm, the area had a better warning system in place, so many more people evacuated low-lying areas before the storm. Still, it is estimated that 5,000–10,000 people perished during Typhoon Sidr, most from the effects of the storm surge.

Why do so many continue to move to areas prone to repeated strikes by tropical cyclones? Bangladesh and coastal Myanmar are overpopulated regions with densities 50 times as great as that of farmlands typical of the midwestern United States. Bangladesh's per capita income is only $200, whereas Myanmar's is $1,900. The delta regions of Bangladesh and Myanmar are the respective country's most fertile. Farmers can expect to yield three rice crops per year, making them attractive place to live despite the risk of storm surges. With the continued population explosion in coastal regions of the Bay of Bengal and the paucity of fertile soils in higher grounds, the delta regions continue to be farmed by millions and continue to be hit by tropical cyclones like the 1970, 1990, 2007, and 2008 disasters. The lower death toll in the 2007 category 5 cyclone in Bangladesh compared with similar earlier storms demonstrates that investment in better warning systems and planned evacuations can save tens to hundreds of thousands of lives. The government of Myanmar has not opened itself to international aid, advice on emergency planning, and better protection of its population.

FURTHER READING

Davis, R., and D. Fitzgerald. *Beaches and Coasts.* Malden, Mass.: Blackwell, 2004.

Kusky, T. M. *The Coast: Hazardous Interactions within the Coastal Environment.* New York: Facts On File, 2008.

---

can develop along these polar fronts, and winds that blow in opposite directions to the north and south of the low set up a cyclonic (counterclockwise in the Northern Hemisphere) wind shear that can cause a wavelike kink to develop in the front. This kink, an incipient cyclone, includes (in the Northern Hemisphere) a cold front that pushes southward and counterclockwise and a warm front that spins counterclockwise and moves to the north. A comma-shaped band of precipitation develops around a central low that develops where the cold and warm fronts meet, and the whole system will migrate east or northeast along the polar front, driven by high-altitude steering winds.

The energy for extratropical cyclones to develop and intensify comes from warm air rising and cold air sinking, transforming potential energy into kinetic energy. Condensation also provides extra energy as latent heat. These storms can intensify rapidly and are especially strong when the cold front overtakes the warm front, occluding the system. The point at which the cold front, warm front, and occluded front meet is known as a triple point. It is often the site of the formation of a new secondary low-pressure system to the east or southeast of the main front. This new secondary low often develops into a new cyclonic system and moves eastward or northeastward, and may become the stronger of the two lows. In the case of New England's northeasters, the secondary lows typically develop off the coast of the Carolinas or Virginia, then rapidly intensify as they move up the coast, bringing cyclonic winds and moisture in from the northeast off the Atlantic Ocean.

POLAR LOW

Polar lows are hurricane- or gale-strength storms that form over water behind (poleward) the main polar front. They can form over either the Northern or Southern Hemisphere oceans but are a larger menace to the more populated regions around the North Atlantic, North Sea, and Pacific Ocean, as well as the Arctic Ocean. Most polar lows are much smaller than tropical and midlatitude cyclones, with diameters typically fewer than 600 miles (1,000 km). Like hurricanes, many polar lows have spiral bands of precipitation (snow in this case) that circle a central warmer low-pressure eye, whereas other polar lows develop a comma-shaped system.

Most polar lows develop during winter months. In the Northern Hemisphere they form along an arctic front, where frigid air blows off landmasses and encounters relatively warm current–fed ocean water, producing a rising column of warm air and sinking columns of cold air. This situation sets up an instability that induces condensation of water vapor in the rising air, along with the associated release of latent heat that then warms the atmosphere. The warming lowers the surface pressure, adding convective updrafts to the system and starting the classical spiral cloud band formation. Polar lows can attain central barometric pressures comparable to hurricanes (28.9 inches or 980 mbars) but tend to dissipate more quickly when they move over the cold polar landmasses.

Storms can open new tidal inlets where none previously existed (without regard to whether any homes were present in the path of the new tidal inlet) and can close inlets previously in existence. Storms also tend to remove large amounts of sand from the beach face and redeposit it in the deeper water offshore (below wave base), but this sand tends gradually to move back onto the beach in the intervals between storms when the waves are smaller. In short, storms are extremely effective modifiers of the beach environment, although they are unpredictable and dangerous.

Many cyclones are spawned in the Indian Ocean. Bangladesh is a densely populated, low-lying country mostly at or near sea level between India and Myanmar. A delta environment, built where the Ganges and Brahmaputra Rivers drop their sediment eroded from the Himalaya Mountains, Bangladesh sits directly in the path of many Bay of Bengal tropical cyclones and has been hit by seven of the nine most deadly hurricane disasters in recorded history. On November 12

and 13, 1970, a category 5 typhoon hit Bangladesh with 155-mile-per-hour (249.5 km/hr) winds and a 23-foot (7-m) high storm surge that struck at the astronomically high tides of a full moon. The result was devastating; 500,000 human deaths and half a million farm animals dead. Again in 1990, another cyclone hit the same area, this time with a 20-foot (6-m) storm surge and 145-mile-per-hour (233 km/hr) winds, killing another 140,000 people and another half-million farm animals.

HURRICANE KATRINA AND THE CONTINUED THREAT TO NEW ORLEANS

On August 28 and early September, 2005, Hurricanes Katrina and Rita devastated the Gulf Coast, inundating New Orleans with up to 20 feet (6 m) of water. While the storm reached category 5 strength offshore, it struck as a category 3, sparing the region the worst potential damage. More than 1,050 were killed in New Orleans alone, however, and large sections of the Gulf Coast in Louisiana, Mississippi, Alabama, and Florida were devastated, with hundreds more dead in these areas. Large sections of the city and coastal regions are now uninhabitable, having been destroyed by floods and subsequent decay by contaminated water and toxic mold. Damage estimates and costs of rebuilding are astronomical, some reaching $300 billion, making the Katrina/Rita disaster the costliest in U.S. history.

One of the initial responses from the residents of the Gulf Coast and much of the rest of the nation was to vow to rebuild the city grander and greater than before. This scientifically unsound response could lead to even greater human catastrophes and financial loss in the future. New Orleans sits on a coastal delta in a basin that is up to 10 feet (3 m) below sea level and is sinking at rates of one-third of an inch to two inches (1–5 cm) per year. Much of the city could be eight feet (2–3 m) farther below sea level by the end of the century. As New Orleans continues to sink, 17.5-foot (5-m) tall levees built to keep the gulf and Lake Pontchartrain out of the city have to be raised repeatedly, and the higher they are built, the greater the likelihood of failure and catastrophe.

Eighteen-foot (5-m) tall flood-protection levees built along the Mississippi keep the river level about 14 feet (4 m) above sea level at New Orleans. If these levees were breached, water from the river would quickly fill in the 10-foot (3-m) deep depression with 25 feet (7–8 m) of water and leave a path of destruction where the torrents of water would rage through the city. These levees also channel the sediments that would naturally get deposited on the floodplain and delta far out into the Gulf of Mexico, with the result that the land surface of the delta south of New Orleans has been sinking below sea level at an alarming rate. A total land area the size of Manhattan disappears every year; this means New Orleans will be directly on the gulf by the end of the century. Current assessments of damage from Katrina are alarming and push that estimate forward by years.

The projected setting of the city in 2100 is in a hole up to 18 feet (5–6 m) below sea level, directly on the hurricane-prone coast, and south of Lake Pontchartrain (by then part of the gulf). The city will need to be surrounded by 50- to 100-foot (15–30-m) tall levees that will make it look like a fish tank submerged off the coast. The levee system will not be able to protect the city from hurricanes any stronger than Katrina. Hurricane storm surges and tsunamis could easily initiate catastrophic collapse of any levee system, causing a major disaster. Advocates of rebuilding are suggesting elevating buildings on stilts or platforms, but they forget that the city will be 10–20 feet (3–6 m) feet below sea level by 2090 and that storm surges may be 30 feet (10 m) above sea level. A levee failure in this situation would be catastrophic, with a debris-laden 50-foot (15-m) tall wall of water sweeping through the city at 30 miles per hour (50 km/hr), hitting these proposed buildings-on-stilts with the force of Niagara Falls and causing devastation equivalent to an Indian Ocean tsunami.

Sea-level rise is rapidly becoming a major global hazard with which we must contend, since most of the world's population lives near the coast in the reach of the rising waters. The current rate of rise of an inch every 10 years will have enormous consequences. Many of the world's large cities, including New York, Houston, New Orleans, and Washington, D.C., are located within a few feet (1 m) of sea level. If sea levels rise a few feet (1 m), many of the city streets will be underwater, not to mention basements, subway lines, and other underground facilities. New Orleans will be the first to go under, lying a remarkable 10–20 feet (3–6 m) below projected sea level, on the coast, at the turn of the next century. These sobering facts suggest that government and private planners should not rebuild major coastal cities in 15-foot (5-m) deep holes along the sinking, hurricane-prone coast. Governments, planners, and scientists must make more sophisticated plans to protect against rising sea levels. The first step would be to use reconstruction money for rebuilding New Orleans as a bigger, better, stronger city in a location where it is above sea level and will last for more than a couple of decades, saving the lives and livelihoods of hundreds of thousands of people.

HURRICANE ANDREW, 1992

Hurricane Andrew was the second-most destructive hurricane in U.S. history, causing more than $30 billion in damage in August 1992. Andrew began to form over North Africa and grew in strength as trade winds drove it across the Atlantic. On August 22 Andrew had grown to hurricane strength and moved across the Bahamas with 150-mile-per-hour (241 km/hr) winds, killing four people. On August 24 Andrew smashed into southern Florida with a nearly 17-foot (5.2-m) high storm surge, steady winds of 145 miles per hour (233 km/hr), and gusts up to 200 miles per hour (322 km/hr). Andrew's path traversed a part of south Florida that had hundreds of thousands of poorly constructed homes and trailer parks, and hurricane winds caused intense and widespread destruction. Andrew destroyed 80,000 buildings, severely damaged another 55,000, and demolished thousands of cars, signs, and trees. In southern Florida 33 people died. Andrew lost much of its strength as it traveled across Florida, but it later moved back into the warm waters of the Gulf of Mexico and regained much of that strength. On August 26 Andrew made landfall again, this time in Louisiana, with 120-mile-per-hour (193-km/hr) winds, where it killed another 15 people. Andrew's winds stirred up the fish-rich marshes of southern Louisiana, where the muddied waters were agitated so much that the decaying organic material overwhelmed the oxygen-rich surface layers, suffocating millions of fish. Andrew then continued to lose strength but dumped flooding rains over much of Mississippi.

THE 1900 GALVESTON HURRICANE

The deadliest natural disaster to affect the United States was a category 4 hurricane that hit Galveston Island, Texas, on September 8, 1900. Galveston is a low-lying barrier island south of Houston, and in 1900 it was a wealthy port city. Residents of coastal Texas received early warning of an approaching hurricane from a Cuban meteorologist, but most people ignored this advice. Later, perhaps too late, U.S. forecasters warned of an approaching hurricane, and many people then evacuated the island to relative safety inland. But many others remained on the island. In the late afternoon the hurricane moved into Galveston, and the storm surge hit at high tide, covering the entire island with water. Even the highest point on the island was covered with one foot (255 cm) of water. Winds of 120 miles per hour (190 km/hr) destroyed wooden buildings as well as many of the stronger brick buildings. Debris from destroyed buildings crashed into other structures, demolishing them and creating a moving mangled mess for residents trapped on the island. The storm continued through the night, battering the island and city with 30-foot (9-m) high waves. In the morning residents who found shelter emerged to see half of the city totally destroyed, and the other half severely damaged. But worst of all, thousands of bodies were strewn everywhere—6,000 on Galveston Island and another 1,500 on the mainland. With no way off the island (all boats and bridges were destroyed), survivors faced the additional danger of disease from the decaying bodies. When help arrived from the mainland, the survivors needed to dispose of the bodies before cholera set in, so they put the decaying corpses on barges and dumped them at sea. But the tides and waves soon brought the bodies back, and they eventually had to be burned in giant funeral pyres built from wood from the destroyed city. Galveston was rebuilt with a seawall made of stones to protect the city; however, in 1915 another hurricane struck Galveston, claiming 275 additional lives.

The Galveston seawall has since been reconstructed and is higher and stronger, although some forecasters believe that even this seawall will not be able to protect the city from a category 5 hurricane. The possibility of a surprise storm hitting Galveston again is not so remote, as demonstrated by the surprise tropical storm of early June 2001. Weather forecasters did not successfully predict the rapid strengthening and movement of this storm, which dumped 23–48 inches (58–122 cm) of rain on different parts of the Galveston-Houston area and attacked the seawall and coastal structures with huge waves and 30-mile-per-hour (48-km/hr) winds. Twenty-two died in the area from the surprise storm, showing that even modern weather forecasting cannot always adequately predict tropical storms. It is best to heed early warnings and prepare for rapidly changing conditions when hurricanes and tropical storms approach vulnerable areas.

See also ATMOSPHERE; BEACHES AND SHORELINES; CLOUDS; GEOLOGICAL HAZARDS.

FURTHER READING

Ahrens, C. Donald. *Meteorology Today: An Introduction to Weather, Climate, and the Environment.* 7th ed. Pacific Grove, Calif.: Thomson/Brookscole, 2003.

Davis, R., and D. Fitzgerald. *Beaches and Coasts.* Malden, Mass.: Blackwell, 2004.

Godschalk, D. R., D. J. Brower, and T. Beatley. *Catastrophic Coastal Storms: Hazard Mitigation and Development Management.* Durham, N.C. and London: Duke University Press, 1989.

Federal Emergency Management Agency. Available online. URL: http://www.fema.gov. Accessed May 23, 2007.

Kusky, T. M. *The Coast: Hazardous Interactions within the Coastal Zone.* New York: Facts On File, 2008.

Longshore, David. *Encyclopedia of Hurricanes, Typhoons, and Cyclones.* New Ed. New York: Facts On File, 2008.

National Research Council. *Drawing Louisiana's New Map: Addressing Land Loss in Coastal Louisiana.* Washington, D.C.: National Academies Press, 2005.

Pilkey, O. H., and W. J. Neal. "Coastal Geologic Hazards." In *The Geology of North America, Volume 1–2, The Atlantic Continental Margin,* edited by R. E. Sheridan, and J. A. Grow. Boulder, Colo.: U.S. Geological Survey and Geological Society of America, 1988.

Pielke, R. A. Jr., and R. A. Pielke, Sr. *Hurricanes: Their Nature and Impacts on Society.* New York: John Wiley & Sons, 1998.

Hutton, James (1726–1797) Scottish *Geologist*
James Hutton was a major 18th-century theorist who studied the Earth and the processes by which it was shaped. He proposed a system for shaping the surface of the Earth. He believed the chief agent for major geological changes was heat generated deep underneath the planet's crust. Proponents of Hutton's theory of the Earth were called vulcanists, after Vulcan, the Roman god of fire, owing to the emphasis of his system on the action of heat and volcanoes, or plutonists, after the ruler of the underworld, Pluto. His proposal of a continuously acting cyclical system of degradation of the land into sediment deposited into strata under the sea, followed by upheaval from volcanic activity, led to decades of debate.

James Hutton's most important contribution to science was his book *Theory of the Earth.* The theory was simple yet contained such fundamental ideas that he was later known as the founder of modern geology. In the 30-year period when he was developing and writing his ideas, other areas of earth sciences had been explored. However, geology had not really been recognized as an important science. Hutton's *Theory of the Earth* contained three main ideas:

- estimating the amount of time the Earth had existed as a "habitable world"
- determining the changes it had undergone in the past
- speculating on whether any end to the present state of affairs could be foreseen

He also talked about how rocks were good indications of the different periods and how they could tell geologists roughly when the Earth was formed. Initially his ideas were not well received by his fellow scientists, because they contradicted the natural conservatism of many geologists, including a reluctance to abandon belief in the biblical account of creation and the widespread catastrophism. By the 1830s, even though geologists were still conservative, they were better equipped to assess the value of Hutton's theory.

CHILDHOOD AND COLLEGE
James Hutton was born in Edinburgh, Scotland, on June 3, 1726, the son of William Hutton and Sarah

Balfour. William died when James was only three, leaving an inheritance that was sufficient for Sarah to raise James and his three sisters, and to send James to the University of Edinburgh when he was 14 years old. While in college James studied humanities but became interested in chemistry when a professor performed an experiment demonstrating that a single acid could dissolve inferior metals, but that two acids were necessary to dissolve gold.

Despite demonstrating academic potential and natural intellectual curiosity, James began an apprenticeship in a lawyer's office when he was 17 years old, but it was obvious that his interests lay elsewhere. His boss released him from their agreement. Because coursework in medicine offered the best opportunity to learn chemistry, Hutton enrolled in medical school at the University of Edinburgh. After three years he moved to Paris to study anatomy. In 1749 he was awarded a doctor of medicine degree after transferring to Leiden, the Netherlands, but he was not interested in practicing medicine.

GEOLOGICAL AND AGRICULTURAL STUDIES

Hutton toured a little, and his interest in chemistry grew into a love of geology and mineralogy. Between 1752 and 1753 he lived with a farmer in Norfolk, England, where he was fascinated by the rows of black flints embedded in the white chalk. He spent time gazing at heaps of seashells on the east coast and noticed chalk and foreign stones embedded in cliffs to the north. In the west he observed red-colored chalk in the strata. With such geological variety above ground, he wondered what was underneath the ground.

Hutton also gained practical knowledge of agriculture during his travels. After spending two years on a farm in Norfolk, he stopped by Flanders to compare animal husbandry techniques with what he learned in England. In 1754 he decided to cultivate farmland in Berwickshire that his father had left him. For 14 years Hutton farmed and applied scientific principles to improve crop yields. While farming he became captivated with the fate of the soil. Over time the soil was washed away, ending up in streams that eventually channeled into the oceans. He must have wondered why this process did not result in a completely flat Earth over long periods of time. He was successful not only in his farming but in a business venture manufacturing ammonium chloride from soot, a process he had helped to devise with a friend. In 1768, with his finances secure, he leased his farmlands and moved to Edinburgh, where he could spend more time studying science.

While in Edinburgh he made many friends among scientific acquaintances including the chemist Joseph Black, who discovered fixed air (carbon dioxide) and the economist Adam Smith. These three founded the Oyster Club, a small society that met weekly to discuss various issues over dinner and to go on field trips. The affable environment afforded frequent scholarly discussions, and Hutton read voraciously on all matters of science. Enjoyable promenades opened his eyes to the idea that the lands on which he walked were not as they had always been. He also joined a society that became the Royal Society of Edinburgh in 1783. A deepening interest in geology sent him all over Scotland, England, and Wales study the rocks, the strata, and the landscapes, searching for clues to the Earth's history. He made observations and collected data, formulating a theory that changed the course of geological science.

"THEORY OF THE EARTH"

On March 7, 1785, Hutton was to read his paper, "Theory of the Earth; or an Investigation of the Laws Observable in the Composition, Dissolution, and Restoration of Land upon the Globe" to the newly chartered Royal Society of Edinburgh. But he became overly nervous from anticipation and the task fell to his friend Joseph Black. He had recovered by the next meeting on April 4, when he read the remainder of his paper.

At the time the prevailing theory for the Earth's formation emphasized the importance of water. The German geologist Abraham Gottlob Werner piloted this group of neptunists, who believed that the Earth's surface was formed by sedimentary deposition in a great turbulent ocean. Of course, many believed this great ocean was the result of a giant flood as described in the biblical book of Genesis. Literal interpretation of the Bible led many to accept the age of the Earth as approximately 6,000 years. Hutton did not think this corresponded with the evidence he had observed.

In a stroke of genius Hutton resolved that the history of the Earth must be explained by events of the present. In other words the natural processes observed today were the same processes that sculpted the Earth's surface into its current form. He thought the Earth was constantly but slowly changing, and that the changes still were occurring. Events such as volcanic eruptions, weathering, and erosion must have had tremendous effects over long periods of time. Hutton was a deist; he believed that nature itself provided evidence of wisdom and design, but after creation, God did not assume control over it. Because Hutton did not believe in a literal interpretation of the Bible, his thinking was not confined to a 6,000-year time frame. In fact, such a short time frame would not allow for the completion of the indeterminate number of cycles, the occurrence of which he had seen evidence. He imagined a much older Earth and postulated that the natural shaping processes of degradation and upheaval were timeless.

This offended many scientists, but Hutton was not just randomly brainstorming or trying to stir up controversy. He based his conclusions on years of careful observation, from which his theory logically germinated. After formulating a hypothesis, Hutton tried to predict what would be observed if the hypothesis were true. Then he roamed the land in search of evidence.

THE EFFECT OF NATURAL PROCESSES ON THE SHAPE OF THE EARTH

Hutton noticed that rocks consisted of strata, parallel orderly layers of consolidated sediment. The layers were composed of different materials that must have been derived from rocks even older than themselves. He thought this was similar to what was currently happening on the ocean floor, where a new layer of sediment was forming. This new layer of sediment contained bits and pieces of material that had been worn away from the land of some preexisting continent and carried out to sea by the natural flowing of waters. Hutton believed the subterranean heat emanating from the interior of the Earth transformed these layers into solid structures. He was aware of the roles that pressure also played, the compaction from upper layers, and the prevention of volatile substances from escaping. Thus the layers of sediment became consolidated after being compacted and cooked for long periods of time.

How could this mechanism account for the formation of mountains? If strata were formed under the sea, what about landmasses that existed thousands of feet above sea level? Hutton again believed that some force underneath the surface of the Earth was responsible. He witnessed powerful volcanic eruptions that he thought were the result of great expansion of the burning igneous matter in the interior of the Earth. He proposed that these great expansions also occurred in the geological past. They caused convulsions that ripped up through the ground and forced rock and crust upward, causing bending and folding, forming mountains and hills. Magma that had not penetrated the Earth's surface during volcanic activity cooled and solidified, forming granite or other crystalline rocks. This in itself was a novel proposition, since at the time the existence of igneous rocks as a type of rock completely separate from sedimentary rock was not recognized. If this were all true, then he predicted that arrangements should exist in the strata such that some upturned strata would be vertical or tilted relative to undisturbed layers. One would expect the slanted strata to have been eroded, and then eventually be overlaid by a new layer of horizontal sedimentary rock. Such

structures, called unconformities, are quite prevalent. One famous locality, called Hutton's Unconformity, was located near his home in Berwickshire, along the west coast at Siccar Point.

Erosion played a key role in Hutton's theory of how the Earth was sculpted. Dry land decayed unremittingly. Flowing water and pounding waves ate away at rock beds. Wind and weathering acted on exposed surfaces of mountains, producing new soils. Glaciers broke loose and transported chunks of rocky matter with them. Loose soil containing mineral components and organic matter was washed away by rain, and silt was carried by rivers. Chemical reactions in water caused particulate matter to precipitate out of solution. Eventually all of the loose particulate matter made its way to the oceans, where the sediment settled and was compacted to form new layers of strata, completing the geological cycle.

Hutton was the first to recognize that many igneous rocks were younger than the rocks in which they were found. Veins of unstratified rock-filling fractures were described from many locations. Hutton had predicted these would occur if the granite in these veins was once melted, and that during the forceful convulsions it was pushed up and outward into cracks that resulted from the violent tremors. The problem with this idea was that no one understood the origin or composition of granite.

OPPOSITION

The neptunists repudiated everything Hutton claimed. They argued that melted rocks cooled into glassy rather than crystalline forms. But, substances that precipitated out of aqueous solutions would form crystals. The action of intense heat on stones such as limestone would cause decomposition before being allowed to cool. Regarding the formation of mountains, neptunists also proposed that sedimentary layers were formed under water, but that the ocean was formerly much deeper. Sometimes particulate matter was deposited in a vertical manner, forming peaks under the water. Then the water subsided, the neptunists said, leaving peaks behind as the water level decreased. Neptunists explained veins of unstratified rocks as cracks into which aqueous material containing mineral deposits trickled down. The material hardened over time and left behind the observed intrusions.

When Hutton first presented these ideas, he also published an awkwardly written, anonymous, 30-page pamphlet, "Abstract of a Dissertation . . . Concerning the System of the Earth, Its Duration, and Stability," summarizing his conclusions. As monumental as Hutton's notions were, they seemed to go largely unnoticed. Those who did notice found them complicated. Opposition not only came from neptunists but from catastrophists, who believed that periodic earth-shattering events, not slow-working incessant forces, were responsible for creating geological structures. Three years after his initial presentation in Edinburgh, "Theory of the Earth" was published in the first volume of *Transactions of the Royal Society of Edinburgh* (1788).

The Irish chemist Richard Kirwan published an attack on Hutton's theory in 1793. Kirwan was the president of the Royal Irish Academy and a staunch Werner devotee. He attacked Hutton's works from religious and scientific standpoints. Scientifically he refuted the significance of erosion, the importance of heat in consolidation of sediment, and the idea that granite could crystallize from a melt. Hutton felt his ideas were misrepresented in Kirwan's account.

This printed assault induced Hutton finally to write a full account of his theory for the formation of the Earth with fuller explanations than he had included in the paper published by the Royal Society. In 1795 Hutton published a 1,204-page, two-volume text, *Theory of the Earth with Proofs and Illustrations*. A 267-page third volume was found and published by the Geological Society in 1899, more than 100 years after Hutton's death.

After 1791 Hutton was frequently ill with a kidney and bladder ailment. When he died on March 26, 1797, he was preparing an agricultural volume. He had amassed a respectable rock collection, which was donated by one of his sisters to the Royal Society of Edinburgh, then later passed to the university museum. The collection has since been lost. Though he devoted himself to geological studies, Hutton was widely read and well versed in a variety of subjects. He had published works on agriculture, meteorology, chemistry, the theory of matter, moral philosophy, and metaphysics.

James Hutton was a sociable man with a winning personality. Though he never married, he had many close friends and staunch supporters. He had one illegitimate son, James, born around 1747, with whom he remained in contact throughout his life. His activities were not limited to promoting his own interests; he volunteered his talents to a variety of causes. He was an active member on a committee involved in managing a project to join the Forth and Clyde Rivers with a canal. In 1788 he was elected a foreign member of the French Royal Society of Agriculture.

SUPPORT FOR HUTTON'S THEORY

After Hutton's death Scottish geologist and chemist James Hall published evidence supporting his friend's theory. Hall initially had rejected Hutton's theory, but over time and after numerous discussions and tours around Britain with Hutton, he became convinced of the truth of the geological cycle. Hall tried to persuade Hutton to perform certain experiments that would support his claims, but Hutton did not deem them necessary. Hutton felt the principles he delineated were clearly evident by observing nature and that nothing in a laboratory could significantly replicate the exquisite power of nature. Out of respect for Hutton, Hall refrained from pushing the issue. But after his death Hall published several experiments that challenged the objections to Hutton's ideas and brought his theory into public prominence.

Hall's first experiment demonstrated that igneous rocks could be converted to crystalline rocks. Neptunists did not believe that igneous rocks were once liquid but thought that if they were, they should turn into glass upon cooling, not crystalline rocks. By slowing down the cooling process, Hall was able to form an opaque crystalline material after melting fused basalt. Some were impressed by this. Others thought the results were sketchy, that perhaps some component was lost during the reaction that changed the actual chemical composition of the material. Next, Hall attacked the idea that marble could not be produced from limestone. It was thought that carbon dioxide would escape as a gas and only quicklime would result. So Hall heated limestone (a powdered chalk) in a gun barrel, sealed in order to allow immense pressure to build up. This prevented the escape of any volatile components during the heating. Then he slowly cooled it and found marble inside! In another experiment Hall produced consolidated sandstone from heating sand with salt water. Eventually, the series of more than 500 experiments Hall performed in his quest to validate Hutton's claims earned him the title of founder of experimental geology and geochemistry.

In 1802 one of Hutton's loyal contemporaries, John Playfair, wrote a biography of Hutton titled *Illustrations of the Huttonian Theory*. In it he not only discussed the life of the now famous geologist but explained Hutton's theory much more clearly than Hutton had managed to do himself. Thirty years later, when society was better prepared to accept the notion that the world was more than 6,000 years old and that it was continually evolving, Playfair's interpretation helped the scientific community adopt Hutton's vision.

Today Hutton's ideas are summarized in the principle of uniformitarianism, which holds that the physical and chemical processes that occur today are the same as those that formed geological structures in the past, though perhaps on an altered timescale. Uniformitarianism may be summarized as "the present is the key to the past," a tenet that forms the basis of modern geology.

See also GRANITE, GRANITE BATHOLITH; HISTORICAL GEOLOGY; UNCONFORMITIES; WERNER, A. G.

FURTHER READING
Carruthers, Margaret W., and Susan Clinton. *Pioneers of Geology: Discovering Earth's Secrets.* New York: Franklin Watts, 2001.
Hutton, James. *Theory of the Earth with Proofs and Illustrations.* 2 Vols. London: Messrs Cadell, Junior, Davies, 1795.
———. *An Investigation of the Principles of Knowledge and of the Progress of Reason, from Sense to Science and Philosophy.* Edinburgh: A. Strahan, and T. Cadell, 1794.
———. "Observations on Granite." *Transactions of the Royal Society of Edinburgh* 3 (1794): 77–81.
Repcheck, Jack. *The Man Who Found Time: James Hutton and the Discovery of the Earth's Antiquity.* Cambridge, Mass.: Perseus, 2003.

Huygens, Christian (1629–1695) Dutch *Mathematician, Astronomer, Physicist* Christian Huygens was born on April 14, 1629, in The Hague, the Netherlands, son of Constantijn Huygens. He attended the University of Leiden and the College of Breda in the southern Netherlands, where he studied law and mathematics before becoming interested in science. Huygens is best known for his contributions showing that light consists of waves, which led to the current model explaining light using the concept of wave-particle duality. This is a quantum mechanical concept whereby all matter and energy exhibit both wave- and particle-like properties. Christian is also credited with developing many concepts of modern calculus.

Huygens was fascinated with mathematics, and in 1657 he wrote and published the first book on probability theory. For some years he experimented with clocks and pendulums, being one of the first to notice the resonance of pendulums, whereby two pendulums mounted on the same beam will adjust to swing in perfectly opposite directions. Huygens made many astronomical observations, including descriptions of Saturn's rings, and one of the first observations of Mercury making a transit over the Sun.

The Royal Society of London elected Huygens a member in 1663, then in 1666 he took a position with the French Academy of Sciences in Paris under the patronage of King Louis XIV. He worked from the Paris Observatory making many astronomical observations and published these in 1684 as his *Astroscopia Compendiaria.* Huygens was a firm believer in life on other planets, and in the liberal political atmosphere of the Netherlands in the 1600s he published his ideas on extraterrestrial life in his *Cosmotheoros* (The celestial worlds discovered: or, conjectures concerning the inhabitants, plants and productions of the worlds in the planets). This is in stark contrast to the political climate elsewhere in Europe at that time; by contrast, Giordano Bruno was burned at the stake in Italy for publicizing similar ideas in 1600.

In 1681 Huygens became seriously ill and returned to The Hague. He was not allowed to return to France later because of the political climate there. Christian Huygens died in The Hague on July 8, 1695.

See also ASTRONOMY; CONSTELLATION; COPERNICUS, NICOLAUS; EINSTEIN, ALBERT; LIFE'S ORIGINS AND EARLY EVOLUTION; MERCURY; PTOLEMY, CLAUDIUS PTOLEMAEUS; SATURN.

hydrocarbons and fossil fuels Hydrocarbons are gaseous, liquid, or solid organic compounds consisting of hydrogen and carbon. Petroleum is a mixture of different types of hydrocarbons (a type of fossil fuel) derived from the decomposed remains of plants and animals trapped in sediment, and can be used as fuel. When plants and animals are alive, they absorb energy from the Sun (directly through photosynthesis in plants, indirectly through consumption in animals) to make complex organic molecules that, after they die, decay to produce hydrocarbons and other fossil fuels. If organic matter is buried before it is completely decomposed, some "solar energy" may become stored in the rocks as fossil fuels (less than 1 percent of total organic matter gets buried). In most industrial nations the chief source of energy is fossil fuels.

The type of organic matter that gets buried in sediment plays an important role in the type of fossil fuel that forms. Shales and muds bury oceanic organisms (such as bacteria and phytoplankton), and the biomolecules (including proteins, carbohydrates, and lipids) produced by them form oil and natural gas when heated. The resins, waxes, lignins, and cellulose common to terrigenous plants (such as trees and bushes) form coals. Incompletely broken down organics in shale form kerogens, or oil shales, which require additional heat to convert to oil. Although this process could be done in the lab, it would take exceedingly long under normal laboratory conditions. In some tectonic settings in the Earth the conditions of temperature and pressure are just right to convert buried organic material to fossil fuels, forming the deposits in use today.

The first people on the planet to use oil were the ancient Iraqis, 6,000 years ago. Oil is fluid and is lighter than water, which strongly influences where it is found. An oil "pool" is an underground accumulation of oil and gas occurring in the pore spaces of rock. An oil field is a group of oil pools of similar type.

Once oil forms from the organic material, it migrates upward until it seeps out at the surface or encounters a trap. The migration of oil is like the movement of groundwater. Migration is slow, and since petroleum is lighter than water, water forces it upward to the tops of the traps. Because oil eventually finds its way to the surface, most oil is found in relatively young rocks.

The formation of oil requires that the source has been through a critical range of pressure and temperature conditions, known as the oil window. If the geothermal gradient is too low or too high, oil will not form. Oil and gas can accumulate only if five basic requirements are met. First, an appropriate source rock is needed to provide the oil. Second and third, a permeable reservoir with an impermeable roof rock is required. Fourth, a trap (stratigraphic or structural) is needed to hold the oil, and, finally, the formation of the trap must have occurred before the oil has escaped from the system. Thus it is statistically unlikely but fortunate if all five criteria are met and a petroleum deposit is formed.

Geologists know the location of approximately 1,000 gigabarrels (one gigabarrel is one thousand billion barrels), but much more remains to be discovered. Many of the unknown reserves include small deposits, but not tars, tar sands, and oil shales, which must be heated and extensively processed to make them useful, and thus are very expensive. The world now consumes 84 million barrels of oil per day; therefore known reserves will last an estimated 33 years. Oil is running out and is becoming an increasingly powerful political weapon. The oil-rich nations can effectively hold the rest of the world hostage, being that the world has become so dependent on oil. Future energy sources may include nuclear fuels, solar energy, hydroelectric power, geothermal energy, biomass, wind, gas hydrates, and tidal energy.

Oil platform off California coast *(Susan Quinland-Stringer, Shutterstock, Inc.)*

COAL

The most abundant fossil fuel, coal is a combustible rock that contains more than 50 percent (by weight) carbonaceous material formed by the compaction and induration of plant remains. Coal is a black sedimentary rock that consists chiefly of decomposed plant matter, with less than 40 percent inorganic material. Most coal formed in ancient swamps, where stagnant oxygen-deficient water prevented rapid decay and allowed burial and trapping of organic matter. In addition anaerobic bacteria in these environments attack the organic matter, releasing more oxygen and forming peat, a porous mass of organic matter that preserves recognizable twigs and other plant parts. Peat contains about 50 percent carbon and burns readily when dried. With increasing temperature and pressure, coal increases in rank, along with an increase of carbon content. In this process peat is transformed into lignite, bituminous coal, and eventually anthracite. Anthracite contains more than 90 percent carbon and is much shinier, brighter, and harder than bituminous coal and lignite. Coal is classified according to its rank and by the amount of impurities present.

NATURAL GAS

Natural gas forms naturally under normal conditions of temperature and pressure in the ground. Composed mostly of methane, natural gas also contains ethane, propane, butane, and pentane, with common impurities of inorganic gases including nitrogen, carbon dioxide, and hydrogen sulfide. Natural gas's origin is similar to that of other hydrocarbons, being derived from the decomposition of buried organic matter, but is simply the lighter end member of the spectrum of compositions of hydrocarbons. Being a gas, it contains only gaseous hydrocarbons having between one and five carbon atoms and no compounds with six or more carbon atoms. All types of organic matter can contribute to the formation of natural gas, when buried and heated to more than 320°F (160°C). Some natural gas is generated during the decomposition of coal and petroleum when they are heated above 320°F, whereas other gas is produced along with the generation of other hydrocarbons. An additional type of natural gas is biogenic methane, produced at shallow levels by the biodegradation of petroleum and when bacteria reduce carbon dioxide to methane in shallow sediments. Natural gas, abundant in shallow crustal reservoirs, is useful as a fossil fuel. Because it generally burns much cleaner than petroleum or coal, it is increasingly sought as an energy source. Reserves of natural gas are huge and may greatly exceed the remaining world reserves of petroleum.

GAS HYDRATES

Gas hydrates, or clathrates, are solid, icelike, water-gas mixtures that form at cold temperatures (40–43°F, or 4–6°C) and pressures above 50 atmospheres. They form on deep marine continental margins and in polar continental regions, often below the sea floor. The gas component is typically methane but may also contain ethane, propane, butane, carbon dioxide, or hydrogen sulfide, with the gas occurring inside rigid cages of water molecules. Anaerobic bacteria produce the methane by the biodegradation of organic material.

Estimates suggest that gas hydrates contain twice the amount of carbon present in all of the planet's known fossil fuel deposits, and as such they represent a huge, virtually untapped potential source of energy. However, the gases expand by more than 150 times the volume of the hydrates, they are located deep in the ocean, and methane is a significant greenhouse gas. These obstacles present significant technical problems to overcome before gas hydrates are widely mined as an energy source.

See also ARABIAN GEOLOGY; ASIAN GEOLOGY; BASIN, SEDIMENTARY BASIN; CARBON CYCLE; CARBONIFEROUS; ECONOMIC GEOLOGY; OCEAN BASIN; PASSIVE MARGIN; SEQUENCE STRATIGRAPHY.

FURTHER READING

North, F. K. *Petroleum Geology*. Dordrecht, Germany: Kluwer Academic Publishers, 1986.
Seeley, Richard. *Elements of Petroleum Geology*. New York: Academic Press, 1998.

hydrosphere The hydrosphere is one of the Earth's external layers, consisting of the oceans, lakes, streams, glaciers, groundwater, and part of the atmosphere. The Earth is a water-rich planet, and the hydrosphere is a dynamic mass of liquid continuously on the move. The term *hydrologic cycle* describes changes, both long- and short-term, in the Earth's hydrosphere. Also known as the water cycle, the hydrologic cycle is powered by heat from the Sun, which causes evaporation, transpiration from plants, and accumulation of water into clouds. This water then moves in the atmosphere and precipitates as rain or snow, which then drains off in streams, evaporates, or moves as groundwater, eventually to begin the cycle over and over again. The time required for individual molecules of water to complete the cycle varies greatly; it may range from a few weeks for some molecules to many thousands of years for others.

Hydrology is the study of water in liquid, solid, and vapor form, on local to global scales. Studies performed by hydrologists include analysis of the properties of water, its circulation, and distribution on and below the surface in reservoirs, streams, lakes, oceans, and the groundwater system. Many hydrologists assess the movement of water through different parts of the hydrologic system and evaluate the influence of human activities on the system in attempts to maximize the benefits to society. Hydrology may also involve environmental and economic aspects of water use.

A view of the Earth from space reveals that water covers most of the surface, in stark contrast to every other planet. The water on the planet is responsible for many things that allow humans to habitate the Earth. Water lubricates the upper layers of the planet, allows plate tectonics to operate, controls climate and weathering, and is part of life itself, found in the bodies of complex fauna and flora to the interior of cells of the simplest single-celled organisms. The surface of the Earth is covered by about 70 percent water, and the human bodies are also composed of about 70 percent water.

Water is the most precious resource on the planet, needed for sustaining all life, yet has also become the most threatened natural resource because of pollution and overuse. It can pose risks for catastrophic floods with increased urbanization of areas that used to store water on floodplains. Wars and political conflicts have always been fought over the ability to obtain freshwater, navigate rivers, irrigate farmland, and develop floodplains. Water rights pose difficult political issues in places where water is scarce, as in the American West and the Middle East. Since this is a finite world with a finite amount of freshwater, and the global population is growing rapidly, the management of freshwater will likely become an increasingly important issue for generations to come, yet public and political understanding of the science behind decisions on water use lags far behind political decision making and land use.

PROPERTIES OF WATER IN THE GROUND AND SURFACE WATER SYSTEMS

Even though most of the surface of the planet is covered with water, more than 97 percent of this water is salty and not readily usable for most human purposes. While the oceans may produce enormous amounts of food, offer transportation, and eventually provide a source of energy, humans require freshwater for drinking, agriculture, and industry. Freshwater is becoming the most valuable resource in the world, and because of its uneven distribution in a world with a rapidly growing population, there will be debate and outright conflict over the rights to and use of freshwater. At present, humans use more than half of all the freshwater that flows in rivers or that is stored in lakes, and this percentage is growing rapidly. Other sources of water are being exploited, including extraction from beneath the ground, to building reservoirs behind large dam projects.

The volume of groundwater is 35 times the volume of freshwater in lakes and streams, but overall freshwater accounts for less than 3 percent of the planet's water. The United States and other nations have come to realize that freshwater is a vital resource for their survival and are only recently beginning to appreciate that much of the world's water resources have become contaminated by natural and human-aided processes. Most drinking water in the United States comes from surface reservoirs or is purified from rivers, yet approximately 40 percent of drinking water in the country comes from groundwater reservoirs; about 80 billion gallons of groundwater are pumped out of these reservoirs every day in the United States. Groundwater is a limited resource since it is being pumped out of the ground faster than it is being replenished by natural processes.

Water is one of the most unusual substances in the entire solar system. Its unusual properties are responsible for controlling climate, life, and many

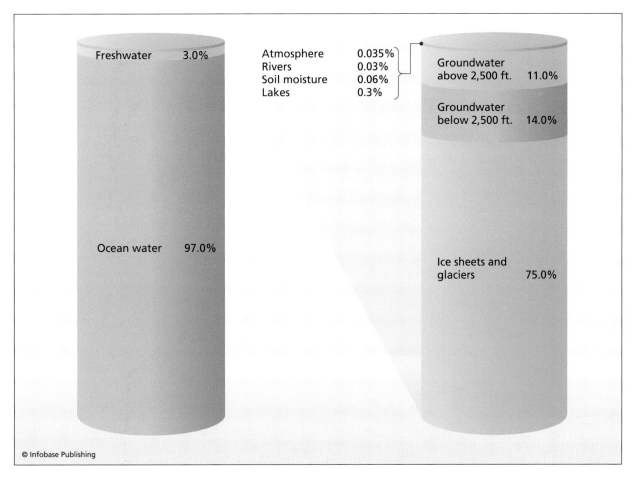

Bar graphs showing the distribution of water in hydrosphere. Ninety-seven percent of the planet's water in the hydrosphere is salt water located in the oceans. Three percent of the planet's water is fresh, but 75 percent of that small fraction is locked up in glacial ice. Most of the remaining freshwater is located underground in the groundwater system, and less than one-tenth of the world's freshwater is readily accessible to people in freshwater streams, rivers, and lakes.

processes on Earth. A water molecule (H_2O) consists of two hydrogen atoms bonded to one oxygen atom and is a polar molecule with a partial positive charge at the end with hydrogen, and a partial negative charge at the end containing the oxygen atom. This allows different water molecules to form weak bonds known as hydrogen bonds with other water molecules, as the positive end of one molecule bonds with the negative end of an adjacent molecule. The nature of this bond determines the properties of water, such as its melting/freezing point of 32°F (0°C) and boiling point of 212°F (100°C). Water may exist in three different states: as a solid (ice), liquid (water), or vapor (water vapor or steam). Since most of the planet has a temperature between 32° and 212°F (0°–100°C), most water exists in the liquid form.

One of the most unusual properties of water is that, like other compounds, it contracts and becomes denser as it cools, until about 39°F (4°C), at which point the cold water begins to become less dense again, becoming less dense than the warmer water. This property of water allows ice to float and causes water bodies to freeze from the top downward. If water was not so unusual, lakes and oceans would freeze from the bottom up and eventually become solid ice. No life would live beneath the seas, and the planet would become a giant cold iceball.

Water can also absorb a lot of heat or solar energy without becoming much warmer. This is because water has a high heat capacity; large water bodies do not change temperature rapidly and have a moderating climatic effect on nearby landmasses. It also takes a large amount of energy to change water from a liquid state to a vapor. To change liquid water to vapor, the water absorbs a lot of this heat energy from the source (say, the ocean) and carries this heat (the heat of vaporization) to the atmosphere, warming it. This is one of the most important heat-transfer processes on the surface of the planet, and it plays a large role in many atmospheric and climate effects. The processes can be appreciated on a personal level by feeling the cooling effect of allowing perspiration to evaporate from the body. Water may also sublimate, or move directly from the solid to the vapor state, or move from the groundwater system to water vapor in the atmosphere through the aid of transpiration in plants.

Water can also dissolve many substances with time, and can carry many substances in solution. Most water contains many mineral salts (such as sodium chloride, sea salt) derived from erosion of the landmasses and many dissolved gases from the atmosphere. The amount of gases dissolved in seawater is partly a function of temperature and plays a large role in climate and global warming.

THE HYDROLOGIC CYCLE

The water cycle describes the sum of processes operative in the hydrosphere, a dynamic mass of liquid continuously on the move between the different reservoirs on land and in the oceans and atmosphere. The hydrosphere includes all the water in oceans, lakes, streams, glaciers, atmosphere, and groundwater, although most water is in the oceans. The hydrologic, or water, cycle encompasses all of the changes, both long- and short-term, in the Earth's hydrosphere. It is powered by heat from the Sun, which causes water to change its state through evaporation and transpiration.

The water cycle can be thought of as beginning in the ocean, where energy from the Sun causes surface waters to evaporate, changing from the liquid to the gaseous states. Evaporation takes heat from the ocean and transfers it into the atmosphere. An estimated 102 cubic miles (425 cubic km^3) of water evaporate from the ocean each year, leaving the salts behind in the ocean. The water vapor then condenses into water droplets in clouds and eventually falls back to the Earth as precipitation. Ninety-two cubic miles (385 cubic km^3) falls directly back into the ocean, but about 26 cubic miles (111 km^3) of precipitation falls as rain or snow on the continents, transforming salty water of the oceans into freshwater on the land. Nearly three-fourths of this water (17 cubic miles, or 71 cubic km/yr) evaporates back to the atmosphere or is aided by the transpiration from plants, returning the water to the atmosphere. The other estimated 10 cubic miles (40 km^3) per year of water runs across the surface, some merging to form streams and rivers that eventually flow back into the ocean, and other parts of this 10 cubic miles per year seeping into the ground to recharge the groundwater system. Humans are now intercepting approximately half of the fresh surface water for drinking, agriculture, and other uses, making a significant impact on the hydrological cycle. Water that seeps into the groundwater system is said to infiltrate, whereas the water that flows across the surface is called runoff.

Understanding the water cycle reveals that freshwater is a renewable resource, replenished and cleaned every year, and available for reuse. It must be used wisely, however, since the quantities are limited, and small amounts of contamination can make entire parts of the system unusable. Freshwater is also supplied unevenly in space and time, with some areas receiving little, and other receiving freshwater in abundance. The water may come in floods or may be withheld, causing drought and suffering. Controlling the flow and usage of freshwater across large regions is one of the major challenges facing humans as the population of the planet grows exponentially and the water supply remains the same.

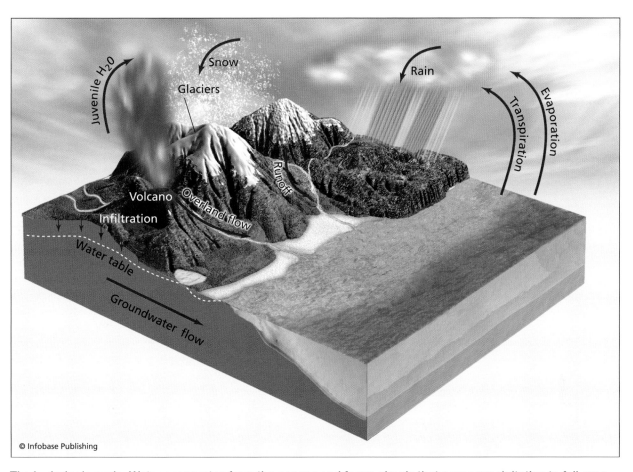

The hydrologic cycle. Water evaporates from the oceans and forms clouds that cause precipitation to fall over the land and oceans. The rain and snow that falls on land can run off in rivers to the oceans, seep into the groundwater system, or be used by plants that then transpire the moisture back to the atmosphere. Water is continuously moving between the different parts of the hydrosphere.

RUNNING WATER AS AN EROSIVE AGENT

Water is an extremely effective erosional agent, including when it falls as rain and runs across the surface in finger-sized tracks called rivulets, and when it runs in organized streams and rivers. Water begins to erode as soon as the raindrops hit the surface—the raindrop impact moves particles of rock, breaking them free from the surface and setting them in motion.

During heavy rains the runoff is divided into overland flow and stream flow. Overland flow is the movement of runoff in broad sheets. Overland flow usually occurs over short distances before it concentrates into discrete channels as stream flow. Erosion performed by overland flow is known as sheet erosion. Stream flow is the flow of surface water in well-defined channels. Vegetative cover strongly influences the erosive power of overland flow by water. Plants that offer thicker ground cover and have extensive root systems prevent erosion much more than thin plants and crops that leave barren soil

exposed between crop rows. Ground cover between that found in a true desert and savanna grasslands tends to be eroded the fastest, while tropical rain forests offer the best land cover to protect from erosion. First, the leaves and branches break the force of the falling raindrops, and the roots form an interlocking network that holds soil in place.

Under normal flow regimes streams attain a kind of equilibrium, eroding material from one bank and depositing it on another. Small floods may add material to overbank and floodplain areas, typically depositing layers of silt and mud over wide areas. During high-volume floods, however, streams may become highly erosive, removing entire floodplains that may have taken centuries to accumulate. The most severely erosive floods are found in confined channels with high flow, as where mountain canyons have formed downstream of many small tributaries that have experienced a large rainfall event, or in rivers that have been artificially channelized by levees. Other severely erosive floods have resulted from dam failures and, in the geological past, from the release of large volumes of water from ice-dammed lakes about 12,000 years ago. The erosive power of these floodwaters dramatically increases when they reach a velocity known as supercritical flow, at which time they can cut through alluvium with ease and even erode bedrock channels. Luckily supercritical flow cannot be sustained for long periods of time, as the effect of increasing the channel size causes the flow to self-regulate and become subcritical.

Cavitation in streams can also cause severe erosion. Cavitation occurs when the stream's velocity is so high that the vapor pressure of water is exceeded and bubbles begin to form on rigid surfaces. These bubbles alternately form then collapse with tremendous pressure, and are thus an extremely effective erosive agent. Cavitation is visible on some dam spillways, where bubbles form during floods and high discharge events, but it is different from the more common and significantly less erosive phenomenon of air entrapment by turbulence, which accounts for most air bubbles observed in white water streams.

WATER AS A RESOURCE

Since freshwater is essential for life, it may be considered an economic resource to manage effectively. Water is needed for drinking, irrigation, household, recreational, and industrial applications. In the United States agriculture uses about 43 percent of all water resources, and industry uses another 38 percent. On a global scale irrigation for agriculture accounts for an even higher percentage of water use, an estimated 69 percent of total water consumption, whereas industry uses only about 15 percent of water on a global scale. Most of the rest of the water is used by households, for drinking, washing, watering lawns, pools, and other benefits of affluent society. Americans use an average of 1,585 gallons (6,000 L) of water a day, compared with a bare one-half gallon (~2 L) a day needed for survival. Americans use about two to four times as much water as inhabitants of Western Europe, and much more than people in drought- and poverty-stricken countries in Africa and the rest of world.

Water resources include any source of water potentially available for human use, including lakes, rivers, rainfall, in reservoirs, and the groundwater system. About two-thirds of the freshwater available on the planet is currently stored in the frozen polar ice caps and in glaciers. Many nations and regions are using the remaining water in streams, lakes, rivers, reservoirs, and the groundwater system faster than the systems are being resupplied by rainfall. Such use is not sustainable, and in time the cost of water will skyrocket, to reflect this supply-and-demand problem. Taxes on water use in Western European countries are already much higher than in the United States. Taxes may be imposed to develop more economical sources of water, such as more energy-efficient desalination plants along coastal communities.

WATER AS A HAZARD

While the supply of clean freshwater is barely able to meet present demands and is expected to become a bigger problem, sometimes there is too much water in one place at one time, creating hazards of another kind. When rains, heavy snowmelts, or combinations of these events bring more water than normal into populated areas, floods result. Many floods cause significant damage and destruction because over the past couple of centuries many cultures have moved large segments of their populations onto floodplains. Floodplains are the flat areas adjacent to rivers that naturally flood. Ancient cultures used these floods and the rich organic mud that covered the floodplains during the floods as natural fertilizers for farmlands. Now that many population centers have been built on floodplains, people regard these natural flood cycles as disasters. In fact about nine out of every 10 disaster proclamations by the U.S. government are for flood disasters, typically to provide funds to those who have built on floodplains. Additionally, many types of natural vegetation have been removed from hillsides, particularly in urban areas. This reduces the amount of infiltration of water into the hillsides and increases the amount and rate of surface runoff thereby increasing the danger of floods.

There are many categories of flood. Flash floods are characterized by huge volumes of water rushing out of mountain canyons, carrying mud, boulders, and every other kind of debris into valleys and lowlands. Floods associated with coastal storms bring high tides into coastal lowlands and back up river systems across deltas and coastal plains. Many regions experience slowly rising, long-lasting regional floods associated with spring snowmelts and unusually heavy rains that can last for weeks or months. Some high-latitude climate zones also experience floods in association with the spring breakup of ice on rivers. As the ice melts blocks move downriver, occasionally jamming and forming ice dams, which can cause rapidly rising ice-cold floodwater to cover the floodplains.

THE WORLD'S DIMINISHING FRESHWATER SUPPLY

On a global scale only about half of the world's population has a connection to a piped-water supply in the home, whereas 30 percent rely on wells or local village pipes, and about 20 percent have no access at all to clean water. World population is expected to grow by another 50 percent (another 3–4

A levee on the Mississippi River being overtopped, with water flooding farmlands in Winfield, Missouri, during Midwest floods of summer 2008 *(T. Kusky)*

billion people) in the next 50 years, so huge investments are needed to maintain the existing water supply infrastructure and develop new supply networks. As population grows and water supplies remain the same or diminish, it is expected that in the next 10 years about half the world's population will not have access to clean drinking water. Most of those without access to clean water will live in Africa, South and Central America, and Southeast Asia. Many of the countries of the Middle East face a different problem—an extreme paucity of water of any kind. Many of these countries have other economic resources such as petroleum, and will have to invest in desalination to meet the needs of their populations.

Presently the highest water consumption per capita is in the United States, followed by the nations of Western Europe. As some developing nations, India and China in particular, grow in affluence, per capita water use is expected to rise dramatically. The huge populations of these countries will further stress the water resources and supply system on a global scale. As populations continue to move into urban areas, countries need to invest in huge water-supply and waste-water treatment facilities to ensure clean water for residents. In many places this has meant pumping

water out of the ground, but the rate of extraction from groundwater aquifers is in many cases faster than the rate at which the water is being replenished, so this resource is being depleted.

Global climate change is starting to impact the hydrologic cycle in patterns of global rainfall and water supply. In some places rainfall is expected to diminish, while in other the amount of rainfall will increase. These changes in climate patterns may dictate massive changes in agricultural patterns and even in population trends within and among nations. It is time for scientists, politicians, and planners to discuss how to handle the coming changes on a dynamic planet.

MODIFICATIONS AND CHANNELIZATION OF RIVER SYSTEMS TO ALLEVIATE WATER SHORTAGES

Some desert and semiarid regions of the world have undergone rapid population explosions, necessitating the alteration of river courses to bring water to thirsty cities and to provide irrigation to farmlands to feed this growing population. In the American desert Southwest, California, and the Middle East, riverways have been extensively modified, regulated, and sometimes diverted hundreds of miles from their

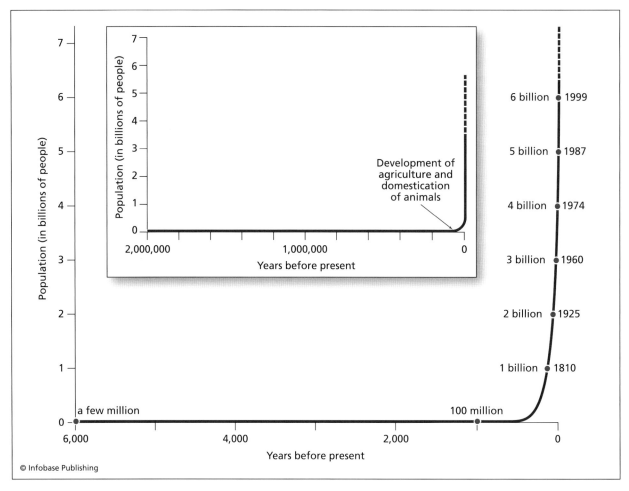

Population curve showing the number of humans on Earth

natural course to provide water to places where people prefer to live.

Many examples of the effects of urbanization on flood intensity have been documented from California and the American desert Southwest. Urban areas like Los Angeles, San Diego, Tucson and Phoenix have documented the speed and severity of floods from similar rainfall amounts along the same drainage basin. These studies have documented that the floodwaters rise much more quickly after urbanization, and they rise up to four times the height of preurbanization, depending on the amount of paving over of the surface. The increased speed at which the floodwaters rise and the increased height to which they rise are directly correlated with the amount of land surface now covered over by roads, houses, and parking lots, blocking infiltration.

In natural systems floods gradually wane after the highest peak passes, and the slow fall of the floodwaters is related to the stream system being recharged by groundwater that seeped into the shallow surface area during the heavy rainfall event. In urbanized areas, however, the floodwaters not only rise quickly but also recede faster than in the natural environment. This is attributed to the lack of groundwater continuing to recharge the stream after the flood peak in urbanized areas.

Many other modifications in stream channels have been made in urbanized areas, with limited success in changing nature's course to suit human needs. Many stream channels have been straightened. This only causes the water to flow faster and have more erosive power. Straightening the stream course also shortens the stream length and thereby steepens the gradient. The stream may respond to this by aggrading and filling the channel with sediment in an attempt to regain the natural gradient.

American Desert Southwest

The history of development the American desert Southwest was crucially dependent on bringing water resources into this semiarid region. Much of California, especially the Los Angeles region, was regarded as worthless desert scrubland until huge water projects designed by the Bureau of Land Reclamation diverted rivers and resources from all over the West. In the years between 1911 and 1923 the California water department under the leadership of William Mulholland quietly purchased most of the water rights to the Owens Valley at the foot of the Sierra Nevada, then constructed a 233-mile (373-km) long aqueduct to bring this water to Los Angeles. When the local Owens Valley ranchers saw their water supplies dry up, they repeatedly dynamited the aqueduct, until Mulholland effectively declared war on the ranchers of the Owens Valley, protecting the aqueduct with a massive show of armed forces. This was the beginning of the present-day California aqueduct system, forming the branch known as the Los Angeles aqueduct.

The California aqueduct is presently 444 miles (715 km) long, and much of it consists of a concrete-lined channel typically 40 feet (12 m) wide and 30 feet (9 m) deep. The aqueduct has several sections, one starting at the San Joaquin–Sacramento River delta, to the San Luis Reservoir, then south to Los Angeles, with a branch heading to the coast in between. The California aqueduct meets the Los Angeles aqueduct north of Los Angeles, and the two systems distribute their water to the valley and thirsty residents of the city.

In the late 1800s geologist and explorer John Wesley Powell explored the West and warned that the water resources in the region were not sufficient for extensive settlement. But Congress went forward with a series of massive dam projects along the Colorado River, including the Hoover Dam, Glen Canyon Dam, and countless others across the region. These dams changed natural canyons and wild rivers into passive reservoirs that now feed large cities including Phoenix, Tucson, Las Vegas, Los Angeles, and San Diego. Use of water from the Colorado became so extensive that by 1969, where the river once flowed to the sea, no more water was flowing in the lower Colorado, the delta environment was destroyed, and water that Mexico used to rely on was no longer available.

Reliance on distant water sources to live in a desert may not seem the wisest of decisions, but much of California and the desert Southwest lives off of water diverted from resources in the Owens Valley, the Trinity River, the Colorado River, and many other western sources. Some conservationists, such as M. Reisner (author of *Cadillac Desert*, 1986) paint an ominous picture of development of the American desert Southwest that has many parallels to ill-fated societies elsewhere in world history. It is becoming increasingly difficult to continue to expand development in the desert and demand increasingly more water resources from a depleting source. Many of the soils are becoming too salty to sustain agriculture. With predictions of global climate change and expanding deserts, the future of the region must be critically examined so the nation can prepare for greater water crises.

Map of California showing the locations of the aqueducts bringing water from the mountains of the Sierra Nevada and from Mono Lake to water-thirsty Los Angeles

Water, Politics, and the Middle East

Water shortage, or drought, coupled with rapid population growth provides for extreme volatility in any region. In the Middle East water shortage issues are coupled with long-standing political and religious differences. The Middle East, stretching from North Africa and the Arabian Peninsula, through Israel and Lebanon to Turkey, and along the Tigris-Euphrates valleys, has only three major river systems and a few smaller rivers. The population stands at about 160 million. The Nile has an annual discharge of about 82 billion cubic yards (62.7 billion m^3), whereas the combined Tigris-Euphrates system has an annual discharge of 93 billion cubic yards (71 billion m^3). Some of the most intense water politics and drought issues in the Middle East arise from the four states that share the relatively small amounts of water of the Jordan River, with an annual discharge of fewer than 2 billion cubic yards (1.5 billion m^3). It has been estimated that with current water usage and population growth, many nations in this region have only 10–15 years before their agriculture and their security will be seriously threatened.

The region is arid, receiving 1–8 inches (2.5–20 cm) of rain per year, and has many drought years with virtually no rain. The Middle East has an annual population growth rate of about 3.5 percent, one of the highest in the world. Many countries in the region have inefficient agricultural practices that contribute to the growing problem of desertification in the region. Some of the problems include planting water-intensive crops, common flooding and furrow methods of irrigation, as well as spraying types of irrigation that lose much water to evaporation, and poor management of water and crop resources. These growing demands on the limited water supply, coupled with political strife resulting from shared usage of waterways that flow through multiple countries, has primed the region for a major confrontation over water rights. Many of the region's leaders have warned that water may be the cause of the next major conflict in the region—in the words of the late King Hussein of Jordan, water issues "could drive nations of the region to war."

Water use by individuals is by necessity much less in the Middle East than in the United States and other Western countries. For instance, every American has about 11,000 cubic yards (8,410 m^3) of freshwater potential to use each year, while citizens of Iraq (pre war) have about 6,000 (4,590 m^3), Turkey 4,400 (3,364 m^3), and Syria about 3,000 (2,294 m^3). Along the Nile Egyptians have about 1,200 cubic yards (917 m^3) available for each citizen. In the Levant Israel's have a freshwater potential of 500 cubic yards (382 m^3) per person per year, and Jordanians have only 280 cubic yards (214 m^3) per year.

The Nile, the second-longest river on Earth, forms the main water supply for nine North African nations, and disputes have grown over how to share this water as demand increases. The Blue Nile flows out of the Ethiopian Highlands and meets the White Nile in the Sudan north of Khartoum, then flows through northern Sudan into Egypt. The Nile is dammed at Aswan, forming Lake Nasser, then flows north through the fertile valley of Egypt to the Mediterranean.

The Nile is the only major river in Egypt, and nearly all of Egypt's population lives in the Nile Valley. About 3 percent of the nation's arable land stretches along the Nile Valley, but 80 percent of Egypt's water use goes to agriculture in the valley. The government has attempted to improve agricultural and irrigation techniques, which in many places have not changed appreciably for 5,000 years. If the Egyptians embraced widespread use of drip irrigation and other modern agricultural practices, then the demand for water could easily be reduced by 50 percent or more.

Egypt has initiated a massive construction and national reconstruction project whose aim to establish a new second branch of the Nile River, extending from Lake Nasser in the south, across the scorching Western Desert, and emerging at the sea at Alexandria. This ambitious project starts in the Tushka Canal area, where water is drained from Lake Nasser and is steered into a topographic depression that winds northward through some of the hottest, driest desert landscape on Earth. The government has been moving thousands of farmers and industrialists from the familiar Nile Valley into this national frontier, hoping to alleviate overcrowding. Cairo's population of 15 million is increasing at a rate of nearly 1 million per year. If successful, this plan could reduce the water demands on the limited resources of the river.

There are many obstacles to this plan. Will people stay in a desert where temperatures regularly exceed 120°F (49°C)? Will the water make it to Alexandria, having to flow through unsaturated sands and through a region where the evaporation rate is 200 times greater than the precipitation rate? How will drifting sands and blowing dust affect plans for agriculture in the Western Desert? Much of the downriver part of the Nile suffers from lower water and silt levels than needed to sustain agriculture and even the current land surface. So much water is used, diverted or dammed upstream that parts of the Nile Delta have actually started to subside (sink) beneath sea level. These regions desperately need to receive the annual silt layer from the flooding Nile to rebuild the land surface and prevent it from disappearing beneath the sea.

There are also political problems with establishing the new river through the Western Desert. Ethiopia contributes about 85 percent of the water to the Nile, yet it is experiencing severe drought and famine in the eastern part of the country. There is no infra- structure to transfer the water from the Nile to the thirsty lands to the east. Sudan and Egypt have long- standing disputes over water allotments, and Sudan

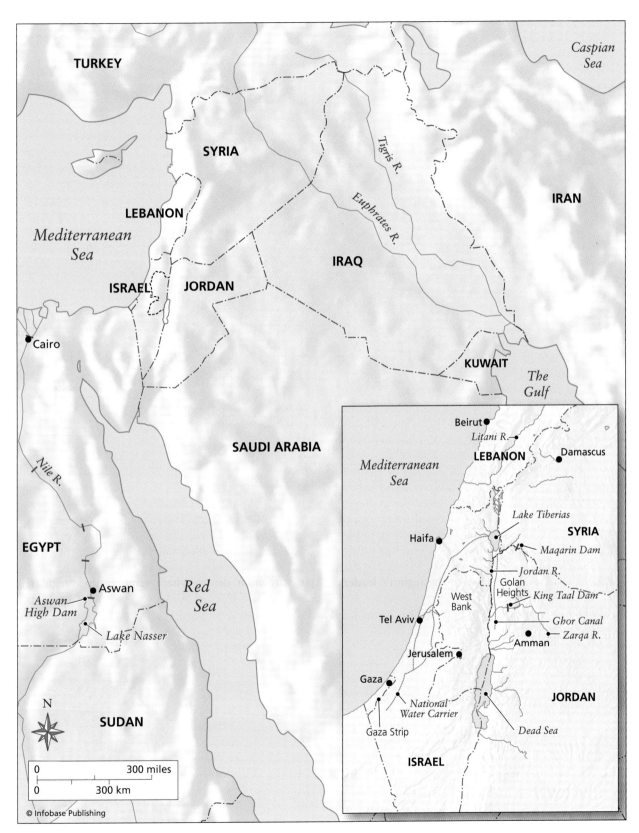

Map of the Middle East showing the main river systems. The area faces a rapidly growing population and a severe lack of water.

is not happy that Egypt is establishing a new river that will further Egyptians' use of the water. Water is currently flowing out of Lake Nasser, filling up several small lake depressions to the west, and sinking into and evaporating between the sands.

The Jordan River basin is host to some of the most severe drought and water-shortage issues in the Middle East. Israel, Jordan, Syria, Lebanon, and the Palestinians share the Jordan River water, and the resource is much more limited than water along the Nile or in the Tigris-Euphrates system. The Jordan River is only 100 miles (160 km) long and is made of three main tributaries, each with different characteristics. The Hasbani River has a source in the mountains of Lebanon and flows south to Lake Tiberias, and the Banias flows from Syria into the lake. The smaller Dan River flows from Israel. The Jordan River then flows out of Lake Tiberias and is joined by water from the Yarmuk flowing out of Syria into the Dead Sea, where any unused water evaporates.

The Jordan River is the source for about 60 percent of the water used in Israel, and 75 percent of the water used in Jordan. The other water used by these countries is largely from groundwater aquifers. Israel has almost exclusive use of the coastal aquifer along the Mediterranean shore, whereas disputes arise over use of aquifers from the West Bank and Golan Heights. These mountainous areas receive more rain and snowfall than other parts of the region and have some of the richest groundwater deposits. Since the 1967 war Israel has tapped the groundwater beneath the West Bank and now gets approximately 30–50 percent of its water supply from groundwater reserves beneath the mountains of the West Bank. The Palestinians get about 80 percent of their water from this mountain aquifer. A similar situation exists for the Golan Heights, though with lower amounts of reserves. These areas therefore have attained a new significance in terms of regional negotiations for peace in the region.

The main problems of water use stem from the shortage of water compared with the population, effectively making drought conditions. The situation is not likely to improve, given the alarming 3.5 percent annual population growth rate. Conservation efforts have only marginally improved the water use problem, and it is unlikely that there will be widespread rapid adoption of many of the drip-irrigation techniques used in Israel throughout the region. This is partly because it takes a larger initial investment in drip irrigation than in conventional furrow and flooding types of irrigation systems. Many farmers cannot afford this investment, even if it would improve their long-term yields and decrease their water use. When the Gaza Strip was turned over to Palestinian control, the authorities ripped out the drip irrigation systems and greenhouses set up by the Israelis and sold the parts for scrap. Now more water is needed to yield the same amount of crops.

Sporadic droughts have worsened this situation in recent years; in 1999 Israel cut in half the amount of water it supplies to Jordan, and Jordan declared drought conditions and mandated water rationing. Jordan currently uses 73 percent of its water for irrigation. If this number could be reduced by adoption of more efficient drip-irrigation, the current situation would be largely in control.

One possible way to alleviate the drought and water shortage would be to explore for water in unconventional aquifer systems such as fractures or faults, which are plentiful in the region. Many faults are porous and permeable structures that are several tens of meters wide, and thousands of meters long and deep. They may be thought of as vertical aquifers, holding as much water as conventional aquifers. If these countries were to be successful in exploring for and exploiting water in these structures, the water shortage and regional tensions might be reduced. This technique has proven effective in many other places in the Middle East, Africa, and elsewhere, and would probably work here as well.

Another set of problems plague the Tigris-Euphrates drainage basin and the countries that share water along their course. There are many political differences among Turkey, Syria, and Iraq, and the Kurdish people have been fighting for an independent homeland in this region. One of the underlying causes of dispute is also the scarce water supply in a drought-plagued area. Turkey is completing a massive dam construction campaign, with the largest dam being the Attatürk on the Euphrates. Overall Turkey is spending an estimated $32 billion on 22 dams and 19 hydroelectric plants. The aim is to increase the irrigated land in Turkey by 40 percent and to supply 25 percent of the nation's electricity through hydroelectric plants. This system of dams also now allows Turkey to control the flow of the Tigris and the Euphrates Rivers. If it pleases, Turkey can virtually shut off the water supply to downstream neighbors. At present Turkey is supplying Syria and Iraq with what it considers to be a reasonable amount of water, but Syria and Iraq claim the amount is inadequate. Political strife and even military action has resulted. Turkey is currently building a pipeline to bring water to drought-stricken Cyprus. Turkey and Israel are forging new partnerships and have been exploring ways to export water from Turkey and import it to Israel, which could help the drought in the Levant.

CAN DESALINATION HELP SOLVE THE WATER CRISIS?

With the hydrologic cycle changing through climate change and increased water use, it is important to find new sources of water. Desalination includes a group of water-treatment processes that remove salt from water; it is becoming increasingly more important as freshwater supplies dwindle and population grows, yet desalination is exorbitantly expensive and cannot be afforded by many countries. A number of different processes can accomplish desalination of salty water, whether it comes from the oceans or the ground. These are divided broadly into thermal processes, membrane processes, and minor techniques such as freezing, membrane distillation, and solar humidification. All existing desalination technologies require energy input to work and end up separating a clear fraction or stream of water from a stream enriched in concentrated salt that must be disposed of, typically by returning it to the sea.

Thermal distillation processes produce about half of the desalted water in the world. In this process salt water is heated or boiled to produce vapor that is then condensed to collect freshwater. There are many varieties of this technique, including processes that reduce the pressure and boiling temperature of water to cause flash vaporization effectively, using less energy than simply boiling the water. The multistage flash-distillation process is the most widely used around the world. In this technique steam is condensed on banks of tubes that carry chemically treated seawater through a series of vessels known as brine heaters with progressively lower pressures, and this freshwater is gathered for use. Multieffect distillation has been used for industrial purposes for many years. Multieffect distillation uses a series of vessels with reduced ambient pressure for condensation and evaporation, and operates at lower temperatures than multistage flash distillation. Salt water is generally preheated then sprayed on hot evaporator tubes to promote rapid boiling and evaporation. The vapor and steam is then collected and condensed on cold surfaces, where the concentrated brines run off. Vapor compression condensation is often used in combination with other processes or by itself for small-scale operations. Water is boiled, and the steam is ejected and mechanically compressed to collect freshwater.

Membrane processes operate on the principle of membranes being able to separate salts selectively from water. Reverse osmosis, commonly used in the United States, is a pressure-driven process in which water is pressed through a membrane, leaving the salts behind. Electrodialysis uses electrical potential, driven by voltage, to move salts selectively through a membrane, leaving freshwater behind. Electrodialysis operates on the principle that most salts are ionic and carry an electrical charge, so they can be driven to migrate toward electrodes with the opposite charge. Membranes are built that allow passage of only certain types of ions, typically either positively (cation) or negatively (anion) charged. Direct-current sources with positive and negative charge are placed on either side of the vessel, with a series of alternate cation and anion selective membranes placed in the vessel. Salty water is pumped through the vessel, the salt ions migrate through the membranes to the pole with the opposite charge, and freshwater is gathered from the other end of the vessel. Reverse osmosis appeared technologically feasible only in the 1970s. The main energy required for this process is for applying the pressure to force the water through the membrane. The salty feed water is preprocessed to remove suspended solids and chemically treated to prevent microbial growth and precipitation. As the water is forced through the membrane, a portion of the salty feed water must be discharged from the process to prevent the precipitation of supersaturated salts. Presently membranes are made of hollow fibers or spiral wound. Improvements in energy recovery and membrane technology has decreased the cost of reverse osmosis, and this trend may continue, particularly with the use of new nanofiltration membranes that can soften water in the filtration process by selectively removing calcium (Ca^{2+}) and magnesium (Mg^{2+}) ions.

Several other processes have been less successful in desalination. These include freezing, which naturally excludes salts from the ice crystals. Membrane distillation uses a combination of membrane and distillation processes, which can operate at low temperature differentials but require large fluxes of salt water. Solar humidification was used in World War II for desalination stills in life rafts, but these are not particularly efficient because they require large solar collection areas, have a high capital cost, and are vulnerable to weather-related damage.

See also ATMOSPHERE; CLOUDS; FLOOD; GLACIER, GLACIAL SYSTEMS; OCEAN BASIN; RIVER SYSTEM.

FURTHER READING

Botkin, D., and E. Keller. *Environmental Science.* Hoboken, N.J.: John Wiley & Sons, 2003.

Buros, O. K. *The ABCs of Desalting.* Topsfield, Mass.: International Desalination Association, 2000.

Gordon, N. D., T. A. McMahon, and B. L. Finlayson. *Stream Hydrology: An Introduction for Ecologists.* New York: John Wiley & Sons. 1992.

Intergovernmental Panel on Climate Change home page. Available online. URL: http://www.ipcc.ch/index.htm. Accessed January 30, 2008.

Intergovernmental Panel on Climate Change 2007. *Climate Change 2007: The Physical Science Basis. Contributions of Working Group I to the Fourth Assessment*

Report of the Intergovernmental Panel on Climate Change, edited by S. Solomon, D. Qin, M. Manning, Z. Chen, M. Marquis, K. B. Averyt, M. Tignor, and H. L. Miller. Cambridge: Cambridge University Press, 2007. Also available online. URL: http://www.ipcc.ch/index.htm. Accessed October 10, 2008.

Leopold, L. B. *A View of the River.* Cambridge, Mass.: Harvard University Press, 1994.

Ritter, D. F., R. C. Kochel, and J. R. Miller. *Process Geomorphology.* 3rd ed. Boston: WCB–McGraw Hill, 1995.

Schumm, S. A. *The Fluvial System.* New York: Wiley-Interscience, 1977.

U.S. Geological Survey. Water Resources of the United States home page. Available online. URL: http://water.usgs.gov/. Accessed October 8, 2008. Updated daily.

INDEX